THE PHYSICAL UNIVERSE

EIGHTH EDITION

KONRAD B. KRAUSKOPF

Professor Emeritus of Geochemistry, Stanford University

ARTHUR BEISER

Boston, Massachusetts   Burr Ridge, Illinois   Dubuque, Iowa
Madison, Wisconsin   New York, New York   San Francisco, California   St. Louis, Missouri

WCB/McGraw-Hill

A Division of The **McGraw·Hill** *Companies*

Photo credits appear on pages A-28 to A-30, and on this page by reference.

This book is printed on acid-free paper.

2 3 4 5 6 7 8 9 0   DOW   DOW   9 0 9 8 7

P/N 036170-3
Part of
ISBN 0-07-913011-9

This book was set in New Aster by The Clarinda Company.
The editors were Karen J. Allanson, James W. Bradley, and David A. Damstra;
the designer was Joseph A. Piliero;
the production supervisor was Kathryn Porzio.
The photo editor was Kathy Bendo;
the photo researcher was Elyse Rieder.
R. R. Donnelley & Sons Company was printer and binder.

Cover painting: *North*, by Rockwell Kent, courtesy of the Rockwell Kent Legacies.

Library of Congress Cataloging-in-Publication Data

Krauskopf, Konrad Bates, (date)
    The physical universe / Konrad B. Krauskopf, Arthur Beiser. — 8th ed.
       p.   cm.
    Includes index.
    ISBN 0-07-036170-3   ISBN 0-07-913011-9 (set)
    I. Beiser, Arthur. II. Title.
Q161.2.K7 1997
500.2—dc20   96-20187

BRIEF

CONTENTS

CONTENTS

PREFACE

THE TEXT

The aim of *The Physical Universe* is to present, as simply and clearly as possible, the essentials of physics, chemistry, earth science, and astronomy.

Because of the scope of these sciences and because we assume little preparation on the part of the reader, our choice of topics and how far to develop them had to be limited. The emphasis throughout is on the basic concepts of each discipline. We also wanted to show how science works, how scientists approach problems, and why science is a never-ending quest rather than a fixed set of facts. We hope to equip readers of the book to appreciate major developments in science in their later years and to understand questions of public policy related to science.

There are many possible ways to organize a book of this kind. We chose the one that provides the most logical progression of ideas, so that each new subject builds on the ones that came before.

Our first concern in *The Physical Universe* is the scientific method, using as illustration the steps that led to today's picture of the universe and the earth's place in it. Next we consider motion and the influences that affect moving bodies. Gravity, energy, and momentum are examined and the theory of relativity is introduced. Matter in its three states next draws our attention, and we pursue this theme from the kinetic-molecular model to the laws of thermodynamics and the significance of entropy. A grounding in electricity and magnetism follows, and then an exploration of wave phenomena that includes the electromagnetic theory of light. We go on from there to the atomic nucleus and elementary particles, followed by a discussion of the quantum theories of light and of matter that lead to the modern view of atomic structure.

The transition from physics to chemistry is made via the periodic table. A look at chemical bonds and how they act to hold together molecules, solids, and liquids is followed by a survey of chemical reactions, organic chemistry, and the chemistry of life.

Our concern now shifts to the planet on which we live, and we begin by inquiring into the oceans of air and water that cover it. From there we proceed to the materials of the earth, to its ever-evolving crust, and to its no-longer-mysterious interior. After a brief narrative of the earth's geological history we go on to what we know about our nearest neighbors in space—planets and satellites, asteroids, meteoroids, and comets.

Now the sun, the monarch of the solar system and the provider of nearly all our energy, claims our notice. We go on to broaden our astronomical sights to include the other stars, both individually and as members of the immense assemblies called galaxies. The evolution of the universe starting from the big bang is the last major subject, and we end with the origin of the earth and the likelihood that other inhabited planets exist in the universe and how we might communicate with them.

Because physical scientists rely on experimental data as both the source and the test of their findings, we think it important to include a few quantitative discussions. However, they are kept simple and supplement rather than dominate our arguments.

STUDENT AIDS

A variety of aids are provided in *The Physical Universe* to help the reader master the text.

Important Terms and Ideas. The meanings of important terms are given at the end of each chapter; this also serves as a chapter summary. A list of the formulas needed to solve problems based on the chapter material is also given where appropriate.

Multiple-Choice Exercises. An average chapter has 36 multiple-choice exercises (with answers) that act as a quick, painless check on understanding. Correct answers provide reinforcement and encouragement; incorrect ones identify areas of weakness.

Questions. Some of the questions are meant to find out how well the reader has understood the chapter material. Others ask the reader to apply what he or she has learned to new situations. Answers to the odd-numbered questions are given at the back of the book.

Problems. The physics and chemistry chapters include problems that range from quite easy to moderately challenging. Although not essential, being able to work out such problems signifies a real grasp of these subjects. Outline solutions (not just answers) for the odd-numbered problems are given at the back of the book.

Math Refresher. Although the mathematical level of the book has been kept low, a little algebra is needed and is reviewed here. Powers-of-ten notation for small and large numbers is carefully explained. This section is self-contained and can provide all the math background needed.

Study Guide. Additional review material of various kinds can be found in a separate *Study Guide* prepared by Steven D. Carey of the University of Mobile.

Ancillaries. Available to instructors using the text are an *Instructor's Manual/Test Bank,* a computerized Test Bank in both Windows and Macintosh formats, and Overhead Transparencies of important illustrations found in the main text.

THIS EDITION

In revising *The Physical Universe* for its eighth edition we have updated its content to reflect recent advances and tried to make the book as a whole more interesting and digestible than before. The illustrations, which are full partners of the text, have been increased in number. Three new features have been added:

Sidebars. These are brief accounts of topics related to the main text. Examples are Magnetic Navigation by Animals, Naming the Elements, Buckyballs, Weather Forecasting, and Ripples in Space-Time.

Biographies. Thirty-four biographical sketches of important contributors to the physical sciences are sprinkled through the book to furnish historical and human perspectives.

Essays. Four distinguished young scientists—a physicist (Timothy C. Miller), a chemist (Cynthia M. Friend), a geologist (Andrea Donellan), and an astronomer (Wendy Freedman)—describe what a typical day at work involves for each of them. Though the intellectual (and sometimes physical) effort is considerable, their joy at being at the frontier of knowledge is evident.

ACKNOWLEDGMENTS

In preparing this edition of *The Physical Universe* we had the benefit of suggestions by:

Karl Fritz Bruder, Grand Rapids Community College
Diane Bruscato, Northeast Louisiana University
Marcellas Coltharp, Kentucky State College
David Dorado, Evergreen Valley College
Ron Harris, Lee College
Ruth Major, Hudson Valley Community College
Ben de Mayo, West Georgia College
Robert McRae, Eastern Montana College
Jack Messer, Pratt Community College/Area Vocational School
Michael Rogers, Daytona Community College
Jan Siemieniewski, Riverside City College
Gregory Smith, Edison Community College
Lynn Thompson, Ricks College
Jai-Ching Wang, Alabama A & M University

We are grateful to Nancy Woods of Des Moines Area Community College for compiling the Videolists in the *Instructor's Manual/Test Bank* for *The Physical Universe.*

Finally, we want to thank our friends at McGraw-Hill for their skilled and dedicated help in producing this edition.

Konrad B. Krauskopf
Arthur Beiser

THE PHYSICAL UNIVERSE

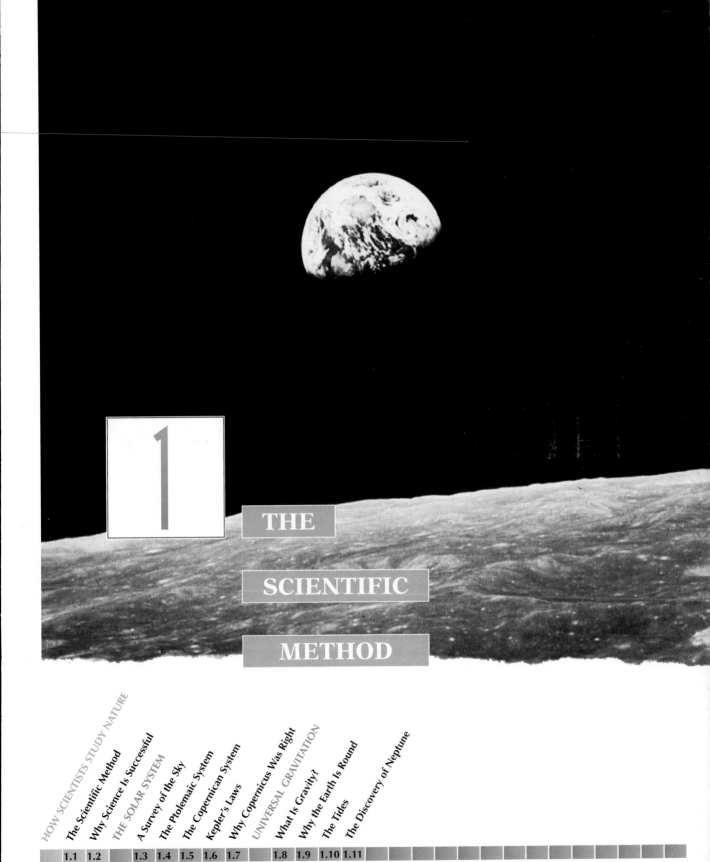

1

THE
SCIENTIFIC
METHOD

*A*ll of us belong to two worlds, the world of people and the world of nature. As members of the world of people, we take an interest in human events of the past and present and find such matters as politics and economics worth knowing about. As members of the world of nature, we also owe ourselves some knowledge of the sciences that seek to understand this world. It is not idle curiosity to ask why the sun shines, why the sky is blue, how old the earth is, why things fall down. These are serious questions, and to know their answers adds an important dimension to our personal lives.

We are made of atoms linked together into molecules, and we live on a planet circling a star—the sun—that is a member of one of the many galaxies of stars in the universe. It is the purpose of this book to survey what physics, chemistry, geology, and astronomy have to tell us about atoms and molecules, stars and galaxies, and everything in between. No single volume can cover all that is significant in this vast span, but the basic ideas of each science can be summarized along with the raw material of observation and reasoning that led to them.

Like any other voyage into the unknown, the exploration of nature is an adventure. This book records that adventure and contains many tales of wonder and discovery. The search for knowledge is far from over, with no end of exciting things still to be found. What some of these things might be and where they are being looked for are part of the story in the chapters to come.

HOW SCIENTISTS STUDY NATURE

Every scientist dreams of lighting up some dark corner of the natural world—or, almost as good, of finding a dark corner where none had been suspected. The most careful observations, the most elaborate calculations will not be fruitful unless the right questions are asked. Here is where creative imagination enters science, which is why most of the greatest scientific advances have been made by young, nimble minds.

Scientists study nature in a variety of ways. Some approaches are quite direct: a geologist takes a rock sample to a laboratory and, by inspection and analysis, finds out what it is made of and how and when it was probably formed. Other approaches are indirect: nobody has ever visited the center of the earth or ever will, but by combining a lot of thought with clues from different sources, a geologist can say with near certainty that the earth has a core of molten iron. No matter what the approaches to particular problems may be, however, the work scientists do always fits into a certain pattern of steps. This pattern, a general scheme for looking at the universe, has become known as the **scientific method.**

1-1 THE SCIENTIFIC METHOD

Four Steps We can think of the scientific method in terms of four steps: (1) formulating a problem, (2) observation and experiment, (3) interpreting the data, and (4) testing the interpretation by further observation and experiment. These steps are often carried out by different scientists, sometimes many years apart and not always in this order. Whatever way it is carried out, though, the scientific method

Hermann von Helmholtz, a German physicist and biologist of a century ago, summed up his experience of scientific research in these words: "I would compare myself to a mountain climber who, not knowing the way, ascends slowly and toilsomely and is often compelled to retrace his steps because his progress is blocked; who, sometimes by reasoning and sometimes by accident, hits upon signs of a fresh path, which leads him a little farther; and who, finally, when he has reached his goal, discovers to his annoyance a royal road on which he might have ridden up if he had been clever enough to find the right starting point at the beginning."

is not a mechanical process but a human activity that needs creative thinking in all its steps. Looking at the natural world is at the heart of the scientific method, because the results of observation and experiment serve not only as the foundations on which scientists build their ideas but also as the means by which these ideas are checked (Fig. 1-1).

1. **Formulating a problem** may mean no more than choosing a certain field to work in, but more often a scientist has in mind some specific idea he or she wishes to investigate. In many cases formulating a problem and interpreting the data overlap. The scientist has a speculation, perhaps only a hunch, perhaps a fully developed concept, about some aspect of nature but cannot come to a definite conclusion without further study.

2. **Observation and experiment** are carried out with great care. Facts about nature are the building blocks of science and the ultimate test of its results. This insistence on accurate, objective data is what sets science apart from other modes of intellectual endeavor.

3. **Interpretation** may lead to a general rule to which the data seem to conform. Or it may be a more ambitious attempt to account for what has been found in terms of how nature works. In any case, the interpretation must be able to cover new data obtained under different circumstances. As put forward orginally, a scientific interpretation is usually called a **hypothesis**.

4. **Testing the interpretation** involves making new observations or performing new experiments to see whether the interpretation correctly predicts the results. If the results agree with the predictions, the scientist is clearly on the right track. The new data may well lead to refinements of the original idea, which in turn must be checked, and so on indefinitely.

FIG. 1-1 The scientific method. No hypothesis is ever final because future data may show that it is incorrect or incomplete. Unless it turns out to be wrong, a hypothesis never leaves the loop of experiment, interpretation, testing. Of course, the more times the hypothesis goes around the loop successfully, the more likely it is to be a valid interpretation of nature. A hypothesis that has survived testing is called a law or theory.

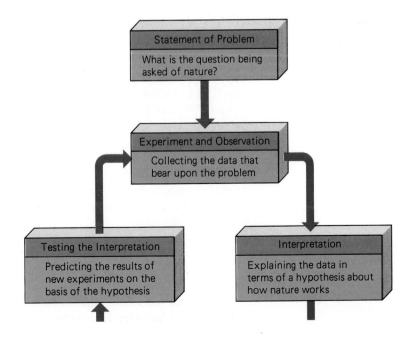

The Laws of Nature

The laws of a country tell its citizens how they are supposed to behave. Different countries have different laws, and even in one country laws are changed from time to time. Furthermore, though he or she may be caught and punished for doing so, anybody can break any law at any time.

The laws of nature are different. Everything in the universe, from atoms to galaxies of stars, behaves in certain regular ways, and these regularities are the laws of nature. To be considered a law of nature, a given regularity must hold everywhere at all times within its range of applicability.

The laws of nature are worth knowing for two reasons apart from satisfying our curiosity about how the universe works. First, we can use them to predict phenomena not yet discovered. Thus Newton's law of gravity was applied over a century ago to apparent irregularities in the motion of the planet Uranus, then the farthest known planet from the sun. Calculations not only showed that another, more distant planet should exist but also indicated where in the sky to look for it. Astronomers who looked there found a new planet, which was named Neptune.

Second, the laws of nature can give us an idea of what goes on in places we cannot examine directly. We will never visit the sun's interior (much too hot) or the interior of an atom (much too small), but we know a lot about both regions. The evidence is indirect but persuasive.

Theories and Models

A **law** tells us *what*; a **theory** tells us *why*. A theory explains why certain events take place and, if they obey a particular law, how that law originates in terms of broader considerations. For example, Albert Einstein's general theory of relativity interprets gravity as a distortion in the properties of space and time around a body of matter. This theory not only accounts for Newton's law of gravity but goes further, including the prediction—later confirmed—that light should be affected by gravity.

As the French mathematician Henri Poincaré once remarked, "Science is built with facts just as a house is built with bricks, but a collection of facts is not a science any more than a pile of bricks is a house."

It may not be easy to get a firm intellectual grip on some aspect of nature. Therefore a **model**—a simplified version of reality—is often part of a hypothesis or theory. In developing the law of gravity, Newton considered the earth to be perfectly round, even though it is actually more like a grapefruit than like a billiard ball. Newton regarded the path of the earth around the sun as an oval called an **ellipse,** but the actual orbit has wiggles no ellipse ever had. By choosing a sphere as a model for the earth and an ellipse as a model for its orbit, Newton isolated the most important features of the earth and its path and used them to arrive at the law of gravity. If he had started with a more realistic model—a somewhat squashed earth moving somewhat irregularly around the sun—he probably would have made little progress. Once he had formulated the law of gravity, Newton was then able to explain how the spinning of the earth causes it to become distorted into the shape of a grapefruit and how the attractions of the other planets cause the earth's orbit to differ from a perfect ellipse.

1-2 WHY SCIENCE IS SUCCESSFUL

Science Is a Living Body of Knowledge, Not a Set of Frozen Ideas What has made science such a powerful tool for investigating nature is the constant testing and retesting of its findings. As a result, science is a living body of information and not a collection of dogmas. The laws and theories of science are not necessarily the final word on a subject: they are valid only as long as no contrary evidence comes to light. If such contrary evidence does turn up, the law or theory must be modified or even discarded. To rock the boat is part of the game; to overturn it is one way to win.

Scientists are open about the details of their work, so that others can follow their thinking and repeat their experiments and observations. Nothing is accepted on anybody's word alone, or because it is part of a religious or political doctrine. "Common sense" is not a valid argument, either. What counts are definite measurements and clear reasoning, not vague notions that vary from person to person.

The power of the scientific approach is shown not only by its success in understanding the natural world but also by the success of the technology based on science. It is hard to think of any aspect of life today untouched in some way by science. The synthetic clothing we wear, the medicines that lengthen our lives, the cars and airplanes we travel in, the telephone, radio, and television by which we communicate—all are ultimately the products of a certain way of thinking. Curiosity and imagination are part of that way of thinking, but the most important part is that nothing is ever taken for granted but is always subject to test and change.

Religion and Science In the past, scientists were burned at the stake for daring to make their own interpretations of what they saw. Galileo, the first modern scientist, was forced by the Roman Catholic Church in 1633 under threat of torture to deny that the earth moves about the sun. Even today, attempts are being made to compel the teaching of religious beliefs—for instance, the story of the Creation as given in the Bible—under the name of science.

Many people find religious beliefs important in their lives, but such beliefs are not part of science because they are matters of faith with principles that are meant to be accepted without question. Skepticism, on the other hand, is at the heart of science. To mix the religious and the scientific ways of looking at the world is good for neither, particularly if compulsion is involved. A free marketplace of ideas is one of the glories of our society, earned with great struggle, and all of us will be the losers if this marketplace is allowed to disappear.

THE SOLAR SYSTEM

Each day the sun rises in the east, sweeps across the sky, and sets in the west. The moon, planets, and most stars do the same. These heavenly bodies also move relative to one another, though more slowly.

There are two ways to explain the general east-to-west motion. The most obvious is that the earth is stationary and all that we see in the sky revolves around it. The other possibility is that the earth itself turns once a day, so that the heavenly bodies only appear to circle it. How the second alternative came to be seen as correct and how this finding led to the discovery of the law of gravity are important chapters in the history of the scientific method.

1-3   A SURVEY OF THE SKY

Everything Seems to Circle the North Star

One star in the northern sky seems barely to move at all. This is the North Star, or **Polaris,** long used as a guide by travelers because of its nearly unchanging position. Stars near Polaris do not rise or set but instead move around it in circles. These circles carry the stars under Polaris from west to east and over it from east to west. Farther from Polaris the circles get larger and larger, until eventually they dip below the horizon. Sun, moon, and stars rise and set because their circles lie partly below the horizon. Thus, to an observer north of the equator, the whole sky appears to revolve once a day about this otherwise ordinary star.

Why does Polaris occupy such a central position? The earth rotates once a day on its axis, and Polaris happens by chance to lie almost directly over the north pole. As the earth turns, everything else around it seems to be moving. Except for their circular motion around Polaris, the stars appear fixed in their positions with respect to one another. Stars of the Big Dipper move halfway around Polaris between every sunset and sunrise, but the shape of the Dipper itself remains unaltered. (Actually, as discussed later, the stars *do* change their relative positions, but the stars are so far away that these changes are not easy to detect.)

Easily recognized groups of stars, like those that form the Big Dipper, are called **constellations** (Fig. 1-2). Near the Big Dipper is the less conspicuous Little Dipper with Polaris at the end of its handle. On the other side of Polaris from the Big Dipper are Cepheus and the W-shaped Cassiopeia, named for an ancient king and queen of Ethiopia. Next to Cepheus is Draco, which means dragon.

Elsewhere in the sky are dozens of other constellations that represent animals, heroes, and beautiful women. An especially easy one to recognize on winter evenings is Orion, the mighty hunter of legend. Orion has four stars, three of them quite bright, at the corners of a warped rectangle with a belt of three stars in line across its middle (Fig. 1-3). Except for the Dippers, a lot of imagination is needed to connect a given star pattern with its corresponding figure, but the constellations nevertheless are useful as convenient labels for regions of the sky.

Sun, Moon, and Planets

In their daily east-west crossing of the sky, the sun and moon move more slowly than the stars and so appear to drift eastward relative to the constellations. In the same way, a person on a train traveling west who walks toward the rear car is moving east relative to the train although still moving west relative to the ground. In the sky, the apparent eastward motion is most easily observed for the moon. If the moon is seen near a bright star on one evening, by the next evening it will be some distance east of that

Time exposure of stars in the northern sky. The trail of Polaris is the bright arc slightly above and to the right of the center of the circles.

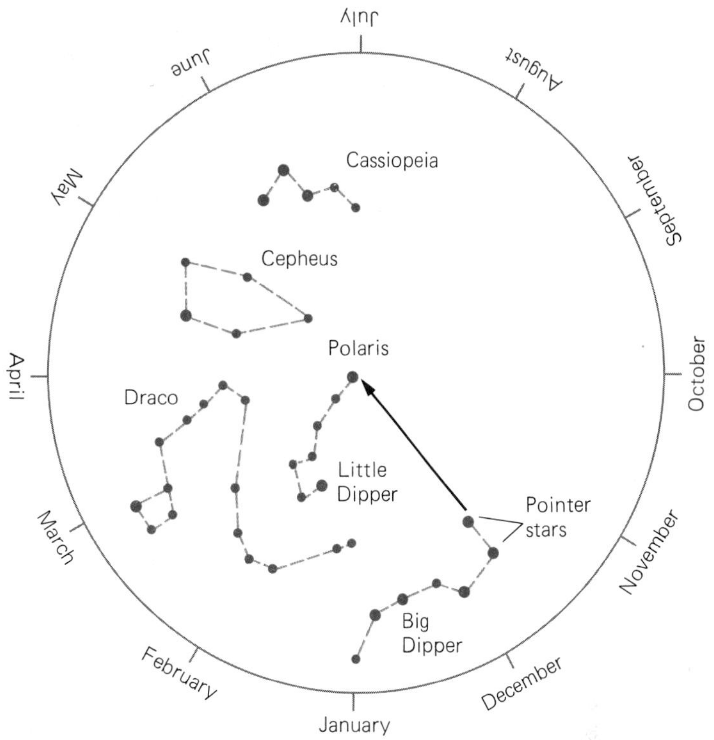

FIG. 1-2 Constellations near Polaris as they appear in the early evening to an observer who faces north with the figure turned so that the current month is at the bottom. Polaris is located on an imaginary line drawn through the two "pointer" stars at the end of the bowl of the Big Dipper. The brighter stars are shown larger in size.

FIG. 1-3 Orion, the mighty hunter. Betelgeuse is a bright red star, and Bellatrix and Rigel are bright blue stars.

star, and on later nights it will be farther and farther to the east. In about 4 weeks the moon drifts eastward completely around the sky and returns to its starting point.

The sun's relative motion is less easy to follow because we cannot observe directly which stars it is near. But if we note which constellations appear where the sun has just set, we can estimate the sun's location among the stars and follow it from day to day. We find that the sun drifts eastward more slowly than the moon, so slowly that the day-to-day change is scarcely noticeable. Because of the sun's motion each constellation appears to rise about 4 min earlier each night, and so, after a few weeks or months, the appearance of the night sky becomes quite different from what it was when we started our observations.

By the time the sun has migrated eastward completely around the sky, a year has gone by. In fact, the year is defined as the time needed for the sun to make such an apparent circuit of the stars.

Five other celestial objects visible to the naked eye also shift their positions with respect to the stars. These objects, which themselves resemble stars, are the **planets** (Greek for "wanderer") and are named for the Roman gods Mercury, Venus, Mars, Jupiter, and Saturn. Like the sun, the planets shift their positions so slowly that their day-to-day motion is hard to detect. Unlike the sun, they move in complex paths. In general, each planet drifts eastward among the stars, but its relative speed varies and at times the planet even reverses direction to head westward briefly. Thus the path of a planet appears to consist of loops that recur regularly, as in Fig. 1-4.

West   East

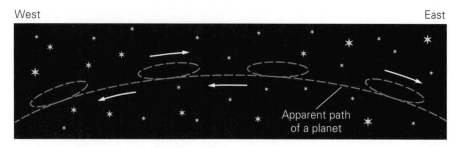

FIG. 1-4 Apparent path of a planet in the sky looking south from the northern hemisphere of the earth. The planets seem to move eastward relative to the stars most of the time, but at intervals they reverse their motion and briefly move westward.

Apparent path of a planet

THE PTOLEMAIC SYSTEM

The Earth as the Center of the Universe
Although the philosophers of ancient Greece knew that the apparent daily rotation of the sky could be explained by a rotation of the earth, most of them preferred to regard the earth as stationary. The scheme most widely accepted was originally the work of Hipparchus. Ptolemy of Alexandria later included Hipparchus's ideas into his *Almagest*, a survey of astronomy that was to be the standard reference on the subject for over a thousand years. This model of the universe became known as the **ptolemaic system.**

The model was intricate and ingenious (Fig. 1-5). Our earth stands at the center, motionless, with everything else in the universe moving about it either in circles or in combinations of circles. (To the Greeks, the circle was the only "perfect" curve, hence the only possible path for a celestial object.) The fixed stars are embedded in a huge crystal sphere that makes a little more than a complete turn around the earth each day. Inside the crystal sphere is the sun, which moves around the earth exactly once a day. The difference in speed between sun and stars is just enough so that the sun appears to move eastward past the stars, returning to a given point among them once a year. Near the earth in a small orbit is the moon, revolving more slowly than the sun. The planets Venus and Mercury come between moon and sun, the other planets between sun and stars.

To account for irregularities in the motions of the planets, Ptolemy imagined that each planet moves in a small circle about a point that in turn follows a large circle about the earth. By a combination of these circular motions a planet travels in a series of loops. Since we observe these loops edgewise, it appears to us as if the planets move with variable speeds and sometimes even reverse their directions of motion in the sky.

From observations made by himself and by others, Ptolemy calculated the speed of each celestial object in its assumed orbit. Using these speeds he could then figure out the location in the sky of any object at any time, past or future. These calculated positions checked fairly well, though not perfectly, with positions that had been recorded centuries earlier, and the predictions also agreed at first with observations made in later years. So Ptolemy's system fulfilled all the requirements of a scientific theory: it was based on observation, it accounted for the celestial motions known in his time, and it made predictions that could be tested in the future.

Ptolemy (A.D. 100–170)

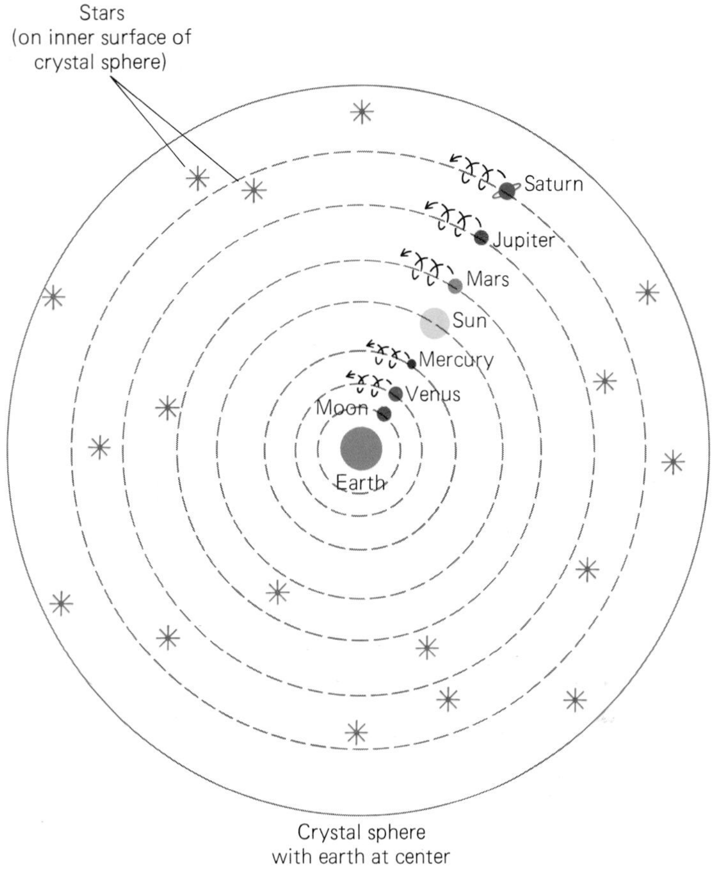

Stars
(on inner surface of
crystal sphere)

Saturn
Jupiter
Mars
Sun
Mercury
Venus
Moon
Earth

Crystal sphere
with earth at center

FIG. 1-5 The ptolemaic system, showing the assumed arrangement of the members of the solar system within the celestial sphere. Each planet is supposed to travel around the earth in a series of loops, while the orbits of the sun and moon are circular. Only the planets known in Ptolemy's time are shown. The stars are all supposed to be at the same distance from the earth.

1-5 THE COPERNICAN SYSTEM

A Spinning Earth That Circles the Sun

By the sixteenth century it had become clear that something was seriously wrong with the ptolemaic model. The planets were simply not in the positions in the sky predicted for them. The errors could be removed in two ways: either the ptolemaic system could be made still more complicated, or it could be replaced by a different model of the universe.

Nicolaus Copernicus, a versatile and energetic Pole of the early sixteenth century, chose the second approach. Let us consider the earth, said Copernicus, as one of the planets, a sphere rotating once a day on its axis. Let us imagine that all the planets, including the earth, circle the sun (Fig. 1-6), that the moon circles the earth, and that the stars are all far away. In this model, it is the earth's rotation that explains the daily rising and setting of celestial objects, not the motions of these objects. The apparent shifting of the sun among the stars is due to the earth's motion in its orbit. As the earth swings around the sun, we see the sun changing its position against the background of the stars. The moon's gradual eastward drift is mainly due to its orbital motion. Apparently irregular movements of the planets are really just combinations of their motions with our own shifts of position as the earth moves.

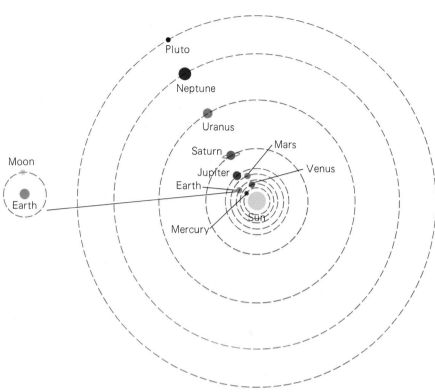

FIG. 1-6 The copernican system. The planets, including the earth, are supposed to travel around the sun in circular orbits. The earth rotates daily on its axis, the moon revolves around the earth, and the stars are far away. All planets known today are shown here. The actual orbits are ellipses and are not spaced as shown here, though they do lie in approximately the same plane.

THE TEMPLE OF THE SUN   Here is how Copernicus summed up his picture of the solar system: "Of the moving bodies first comes Saturn, who completes his circuit in thirty years. After him Jupiter, moving in a twelve-year revolution. Then Mars, who revolves biennially. Fourth in order an annual cycle takes place, in which we have said is contained the Earth, with the lunar orbit as an epicycle, that is, with the moon moving in a circle around the earth. In the fifth place Venus is carried around in nine months. Then Mercury holds the sixth place, circulating in the space of eighty days. In the middle of all dwells the Sun. Who indeed in this most beautiful temple would place the torch in any other or better place than one whence it can illuminate the whole at the same time?"

The **copernican model** offended both Protestant and Catholic religious leaders, who did not want to see the earth taken from its place at the hub of the universe. The publication of Copernicus's manuscript began a long and bitter argument. To us, growing up with the knowledge that the earth moves, it seems odd that this straightforward idea was so long and so violently opposed. But in the sixteenth century good arguments were available to both sides.

Consider, said supporters of Ptolemy, how fast the earth's surface must move to complete a full turn every 24 h. Would not everything loose be flung into space by this whirling ball, just as mud is thrown from the rim of a carriage wheel? And would not such dizzying speeds produce a great wind to blow down buildings, trees, plants? The earth does spin rapidly, replied the followers of Copernicus, but the effects are counterbalanced by whatever force it is that holds our feet to the ground. Besides, if the speed of the earth's rotation is a problem, how much more of a problem would be the tremendous speeds of the sun, stars, and planets if they revolve, as Ptolemy thought, once a day around a fixed earth.

BIOGRAPHY *Nicolaus Copernicus* (1473–1543) was a student in his native Poland when Columbus made his first voyage to the New World. In the years that followed intellectual as well as geographical horizons receded before eager explorers. In 1496 Copernicus went to Italy to learn medicine, theology, and astronomy. Italy was then an exciting place to be, a place of business expansion and conflicts between rival cities, great fortunes and corrupt governments, brilliant thinkers and inspired artists such as Leonardo da Vinci and Michelangelo. After 10 years in Italy Copernicus returned to Poland where he practiced medicine, served as a canon in the cathedral of which his uncle was the bishop, and

became involved in currency reform, but much of his time was devoted to developing the idea that the planets move around the sun rather than around the earth. The idea was not new—the ancient Greeks were aware of it—but Copernicus went further and worked out the planetary orbits and speeds in detail. Although a summary of his results had been circulated in manuscript form earlier, not until a few weeks before his death was Copernicus's *De Revolutionibus*

Orbium Coelestium published in book form. Today *De Revolutionibus* is recognized as one of the foundation stones of modern science, but soon after its appearance it was condemned by the Catholic Church (which did not lift its ban until 1835) and had little impact on astronomy until Kepler further developed its concepts a half century later.

1-6 KEPLER'S LAWS

How the Planets Actually Move

Fortunately, improvements in astronomical measurements—the first since the time of the Greeks—were not long in coming. Tycho Brahe (1546–1601), an astrologer to the Danish king, built an observatory on the island of Hven near Copenhagen in which the instruments were remarkably precise. With the help of these instruments, Tycho, blessed with exceptional eyesight and patience, made thousands of measurements, a labor that occupied much of his life. Even without the telescope, which had not yet been invented, Tycho's observatory was able to determine celestial angles to better than $\frac{1}{100}$ of a degree.

At his death in 1601, Tycho left behind his own somewhat peculiar model of the solar system, a body of superb data extending over many years, and an assistant named Johannes Kepler. Kepler regarded the copernican scheme "with incredible and ravishing delight," in his words, and fully expected that Tycho's improved figures would prove Copernicus correct once and for all. But this was not the case; after 4 years of work on the orbit of Mars alone, Kepler could not get Tycho's data to fit any of the models of the solar system that had by then been proposed. If the facts do not agree with the theory, then the scientific method requires that the theory, no matter how attractive, must be discarded. Kepler then began to look for a new cosmic design that would fit Tycho's observations better.

A 1598 portrait of Tycho Brahe in his laboratory. The man at the right is determining the position of a celestial body by shifting a sighting vane along a giant protractor until the body is visible through the aperture at upper left.

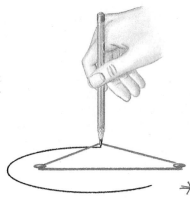

FIG. 1-7 To draw an ellipse, place a loop of string over two tacks a short distance apart. Then move the pencil as shown, keeping the string taut. By varying the length of the string, ellipses of different shapes can be drawn. The points in an ellipse corresponding to the positions of the tacks are called **focuses**; the orbits of the planets are ellipses with the sun at one focus, which is Kepler's first law.

The First Law

After considering every possibility, which meant years of drudgery in making calculations by hand, Kepler found that circular orbits for the planets were out of the question even when modified in various ways. He abandoned circular orbits reluctantly, for he was something of a mystic and believed, like Copernicus and the Greeks, that circles were the only fitting type of path for celestial bodies. Kepler then examined other geometrical figures, and here he found the key to the puzzle (Fig. 1-7). According to **Kepler's first law:**

The paths of the planets around the sun are ellipses.

ellipse-

The Second Law

Even this crucial discovery was not enough, as Kepler realized, to establish the courses of the planets through the sky. What was needed next was a way to relate the speeds of the planets to their positions in their elliptical orbits. Kepler could not be sure a general relationship of this kind even existed, and he was overjoyed when he had figured out the answer, known today as **Kepler's second law:**

Johannes Kepler (1571–1630) was born in Germany and as a child was much impressed by seeing a comet and a total eclipse of the moon. In college, where astronomy was his worst subject, Kepler concentrated on theology, but his first job was as a teacher of mathematics and science in Graz, Austria. There he pondered the copernican system and concluded that the sun must exert a force (which he later thought was magnetic) on the planets to keep them in their orbits. Kepler also devised a geometrical scheme to account for the spacing of the planetary orbits and put all his ideas into a book called *The Cosmic Mystery*. Tycho Brahe, the Danish astronomer, read the book and took Kepler on as an assistant in his new observatory in Prague in what was then Bohemia. Upon Brahe's death (the result of drinking too much at a party given by the Emperor of Bohemia), Kepler replaced him at the observatory and gained access to all of Brahe's data, the most complete and accurate set then in existence.

Kepler felt that the copernican model of the solar system was not only capable of better agreement with the data than had yet been achieved but also contained within it yet-undiscovered regularities. Many years of labor resulted in three laws of planetary motion that fulfilled Kepler's vision and were to bear their ultimate fruit in Newton's law of gravity. Kepler also found time to prepare new tables of planetary positions, to explain how telescopes produce magnified images, to father 13 children, and to prepare horoscopes for the Emperor of Bohemia, the main reason for his employment (as it had been for Brahe).

> A planet moves so that its radius vector sweeps out equal areas in equal times.

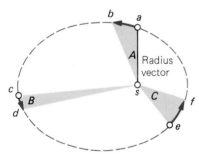

Area *A* = area *B* = area *C*

FIG. 1-8 Kepler's second law. As a planet goes from *a* to *b* in its orbit, its radius vector (an imaginary line joining it with the sun) sweeps out the area *A*. In the same amount of time the planet can go from *c* to *d*, with its radius vector sweeping out the area *B*, or from *e* to *f*, with its radius vector sweeping out the area *C*. The three areas *A*, *B*, and *C* are equal.

The radius vector of a planet is an imaginary line between it and the sun. Thus in Fig. 1-8 each of the shaded areas is covered in the same period of time. This means that each planet travels faster when it is near the sun than when it is far away. The earth, for instance, has a speed of 30 km/s when it is nearest the sun and 29 km/s when it is farthest away, a difference of over 3 percent.

The Third Law

A great achievement, but Kepler was not satisfied. He was obsessed with the idea of order and regularity in the universe, and spent 10 more years making calculations. It was already known that, the farther a planet is from the sun, the longer it takes to orbit the sun. **Kepler's third law** of planetary motion gives the exact relationship:

> The ratio between the square of the time needed by a planet to make a revolution around the sun and the cube of its average distance from the sun is the same for all the planets.

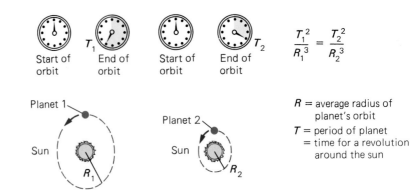

FIG. 1-9   Kepler's third law states that the ratio T^2/R^3 is the same for all the planets.

In equation form, this law states that

$$\frac{(\text{Period of planet})^2}{(\text{Average orbit radius})^3} = \text{same value for all the planets}$$

The period of a planet is the time needed for it to go once around the sun; in the case of the earth, the period is 1 year. Figure 1-9 illustrates Kepler's third law. Table 16-1 gives the values of the periods and average orbit radii for the planets.

At last the solar system could be interpreted in terms of simple motions. Planetary positions computed from Kepler's ellipses agreed not only with Tycho's data but also with observations made thousands of years earlier. Predictions could be made of positions of the planets in the future—accurate predictions this time, no longer approximations. Furthermore, Kepler's laws showed that the speed of a planet in different parts of its orbit was governed by a simple rule and that the speed was related to the size of the orbit.

ASTROLOGY   To our ancestors of thousands of years ago, things happened in the world because gods caused them to happen. Famine and war, earthquake and eclipse —any conceivable catastrophe— all occurred under divine control. In time the chief gods were identified with the sun, the moon, and the five planets visible to the naked eye: Mercury, Venus, Mars, Jupiter, and Saturn. Early observers of the sky were primarily interested in finding links between celestial events and earthly ones, a study that became known as **astrology.**

Until only a few hundred years ago, astronomy was almost entirely in the service of astrology. The wealth of precise astronomical measurements that ancient civilizations compiled had as their purpose interpreting the ways of the gods.

Almost nobody today takes seriously the mythology of old. Although the basis of the connection has disappeared, however, some people still believe that the position in the sky of various celestial bodies at certain times controls the world we live in and our individual destinies as well. It does not seem very gracious for contemporary science to dismiss astrology in view of the great debt astronomy owes its practitioners of long ago. However, it is hard to have confidence in a doctrine that, for all its internal consistency and often delightful notions, nevertheless lacks any basis in scientific theory or observation and has proved no more useful in predicting the future than a crystal ball.

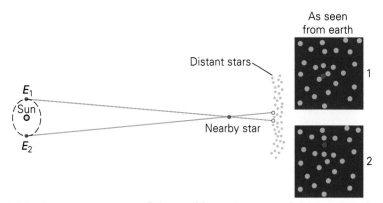

FIG. 1-10 As a consequence of the earth's motion around the sun, nearby stars shift in apparent position relative to distant stars.

1-7 WHY COPERNICUS WAS RIGHT

Evidence Was Needed That Supported His Model While Contradicting Ptolemy's Model

It is often said that Kepler proved that Copernicus was "right" and that Ptolemy was "wrong." True enough, the copernican system, by having the planets move around the sun rather than around the earth, was simpler than the ptolemaic system. As modified by Kepler, the copernican system was also more accurate. However, the ptolemaic system could also be modified to be just as accurate, though in a very much more complicated way. Astronomers of the time squared themselves both with the practical needs of their profession and with the Church by using the copernican system for calculations while asserting the truth of the ptolemaic system.

The copernican system is attractive because it accounts in a straightforward way for many aspects of what we see in the sky. However, only observations that contradict the ptolemaic system can prove it wrong. The copernican system is today considered correct because there is direct evidence of various kinds for the motions of the planets around the sun and for the rotation of the earth. An example of such evidence is the change in apparent position of nearby stars relative to the background of distant ones as the earth revolves around the sun (Fig. 1-10). Shifts of this kind are small because all stars are far away, but they have been found.

UNIVERSAL GRAVITATION

As we know from everyday experience, and as we shall learn in a more precise way in Chap. 2, a force is needed to cause something to move in a curved path (Fig. 1-11). The planets are no exception to this rule: a force of some kind must be acting to hold them in their orbits around the sun. Three centuries ago Isaac Newton had the inspired idea that this force must have the same character as the familiar force of gravity that pulls things to the earth's surface.

FIG. 1-11 An inward force is needed to keep an object moving in a curved path. The force here is provided by the string.

A Fundamental Force

Perhaps, thought Newton, the moon revolves around the earth much as the ball in Fig. 1-11 revolves around the hand holding the string, with gravity taking the place of the pull of the string. In other words, perhaps the moon is a falling object, pulled to the earth just as we are, but moving so fast in its orbit that the earth's pull is just enough to keep the moon from flying off (Fig. 1-12). The earth and its sister planets might well be held in their orbits by a stronger gravitational pull from the sun. These notions turned out to be true, and Newton was able to show that his detailed theory of gravity accounts for Kepler's laws.

It is worth noting that Newton's discovery of the law of gravity depended on the copernican model of the solar system. "Common sense" tells us that the earth is the stationary center of the universe, and people were once severely punished for believing otherwise. Clearly the progress of our knowledge about the world we live in depends upon people, like Copernicus, who are able to look behind the screen of appearances that make up everyday life and who are willing to think for themselves.

Gravity is a **fundamental force** in the sense that it cannot be explained in terms of any other force. Only four fundamental forces are known: gravitational, electromagnetic, weak, and strong. These forces are responsible for everything that happens in the universe. Gravitational forces act between all bodies everywhere and hold together planets, stars, and the giant groups of stars called galaxies. Electromagnetic forces, which (like gravity) are unlimited in range, act between electrically charged particles and govern the structures and behavior of atoms, molecules, solids, and liquids. When a bat hits a ball, the interaction between them can be traced to electromagnetic forces. The weak and strong forces have very short ranges and act inside atomic nuclei.

FIG. 1-12 The gravitational pull of the earth on the moon causes the moon to move in an orbit around the earth. If the earth exerted no force on the moon, the moon would fly off into space. If the moon had no orbital motion, it would fall directly to the earth.

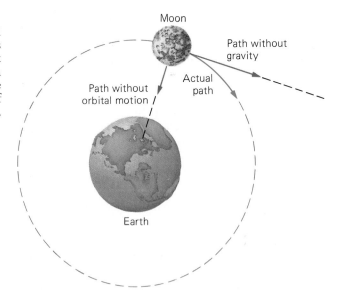

BIOGRAPHY

Isaac Newton (1642–1727) was born to an English farming family. Although his mother wanted him to stay on the farm, the young Newton showed a talent for science and went to Cambridge University for further study. An outbreak of plague led the university to close in 1665, the year Newton graduated, and he returned home for 18 months. In that period Newton came up with the binomial theorem of algebra; invented calculus, which gave science and engineering a new and powerful mathematical tool; discovered the law of gravity, thereby not only showing why the planets move as they do but also providing the key to understanding much else about the universe; and demonstrated that white light is a composite of light of all colors—an amazing list.

When Cambridge University reopened, Newton went back and 2 years later became professor of mathematics there. He lived quietly and never married, carrying out experimental as well as theoretical research in many areas of physics; a reflecting telescope he made with his own hands was widely admired. Especially significant was Newton's development of the laws of motion (see Chap. 2), which showed exactly how force and motion are related, and his application of them to a variety of problems. Newton collected the results of his work on mechanics in the *Principia*, a scientific classic that was published in 1687. A later book, *Opticks*, summarized his efforts in this field. Newton also spent much time on chemistry, though here with little success.

After writing the *Principia*, Newton began to turn away from science. He became a member of Parliament in 1689 and later an official, eventually the Master, of the British Mint. At the Mint Newton helped reform the currency (one of Kepler's interests, too) and fought counterfeiters. Newton's spare time in his last 30 years was mainly spent in trying to date events in the Bible. He died at 85, a figure of honor whose stature remains great to this day.

Gravity Is the Same Everywhere

How can we be sure that Newton's law of gravity, which fits data on the solar system, also holds throughout the rest of the universe? The evidence for this generalization is indirect but persuasive. For instance, many double stars are known in which each member of the pair revolves around the other, which means some force holds them together. Throughout the universe stars occur in galaxies, and only gravity could keep them assembled in this way.

But is the gravity that acts between stars the same as the gravity that acts in the solar system? Analyzing the light and radio waves that reach us from space shows that the matter in the rest of the universe is the same as the matter found on the earth. If we are to believe that the universe contains objects that do not obey Newton's law of gravity, we must have evidence for such a belief—and there is none. This line of thought may not seem as positive as we might prefer, but taken together with various theoretical arguments, it has convinced nearly all scientists that gravity is the same everywhere.

**LATITUDE AND
LONGITUDE** We can imagine any flat surface as a sheet of graph paper and specify a location on it in terms of x (horizontal) and y (vertical) coordinates. Because the earth is round, to specify a location on its surface is a little more complicated.

We need a few definitions to begin with. A **great circle** is any circle on the earth's surface whose center is the earth's center. The **equator** is a great circle midway between the north and south poles (Fig. 1-13). A **meridian** is a great circle that passes through both poles, and it forms a right angle with the equator. The meridian that passes through Greenwich, England, is called the **prime meridian** (Fig. 1-14).

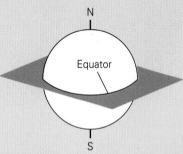

FIG. 1-13   The equator is a great circle around the earth halfway between the poles.

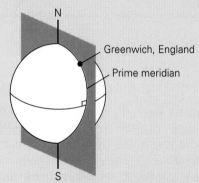

FIG. 1-14   The prime meridian is a great circle perpendicular to the equator that passes through Greenwich, England.

The **longitude** of a point on the earth's surface is the angular distance between a meridian through this point and the prime meridian. The prime meridian itself has the longitude 0°, and longitudes are given in degrees east or west of the prime meridian (Fig. 1-15). For instance, the longitude of New York is 74°W, which means that its meridian is 74° westward from the prime

meridian. The longitude of Moscow is 38°E, which means that its meridian is 38° eastward from the prime meridian.

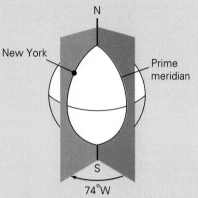

FIG. 1-15   The longitude of a place on the earth's surface is the angle between the meridian it lies on and the prime meridian. The longitude of New York is 74°W because its meridian is 74° west of the prime meridian. The longitude of the prime meridian is 0°.

The **latitude** of a point on the earth's surface is the angle between a line drawn from the earth's center to that point and another line drawn from the center to a point on the equator on the same meridian (Fig. 1-16). Thus the latitude of the north pole is 90°N, that of the south pole is 90°S, and that of the equator is 0°. The latitude of New York is 41°N, which means it lies on a circle (smaller than a great circle) called a **parallel of latitude** whose angular distance north of the equator is 41°.

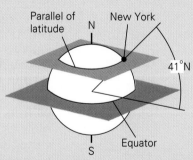

FIG. 1-16   New York lies on a parallel of latitude 41° north of the equator.

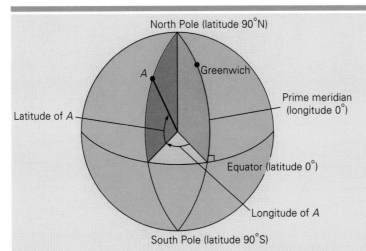

North Pole (latitude 90°N)

Greenwich

A

Latitude of A

Prime meridian
(longitude 0°)

Equator (latitude 0°)

Longitude of A

South Pole (latitude 90°S)

FIG. 1-17 The latitude and longitude of a point A on the earth's surface.

Figure 1-17 summarizes the definitions of latitude and longitude. Degrees of latitude and longitude are further divided into **minutes** (1° = 60 minutes = 60′). A **nautical mile,** equal to 1853 m or 6076 ft, is very nearly the length of 1° of latitude and is universally used for air and sea navigation. A statute mile, used on land in the United States, is 5280 ft in length, so a nautical mile equals 1.151 statute miles. The **knot** is a unit of speed equal to 1 nautical mile per hour.

A map is a representation of all or part of the earth's surface on a sheet of paper. Because the earth is round and the paper is flat, all maps are distorted to some extent. Figure 1-18 shows how the earth appears on a Mercator projection, in which lines of constant latitude and longitude are straight.

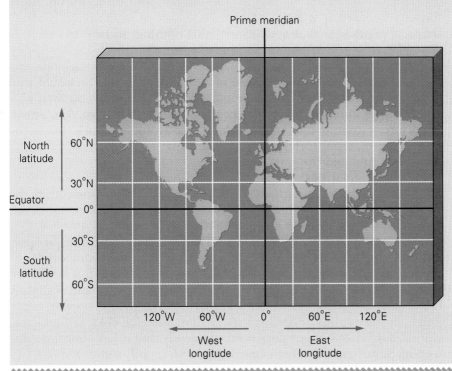

Prime meridian

North latitude

60°N

30°N

Equator

0°

30°S

South latitude

60°S

120°W   60°W   0°   60°E   120°E

West longitude

East longitude

FIG. 1-18 Latitude and longitude on a Mercator projection of the globe. In this projection, the spacing of parallels of latitude increases from equator to poles in order to avoid distorting the shapes of geographic features. As a result, the scale of the map varies with latitude; Greenland is actually smaller than South America, for instance. Other projections vary less in scale but do not preserve shapes.

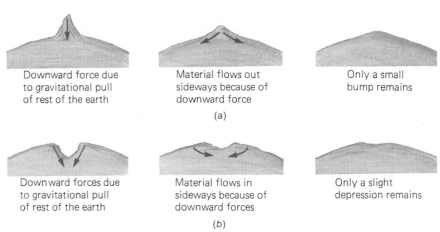

Downward force due
to gravitational pull
of rest of the earth

Material flows out
sideways because of
downward force

Only a small
bump remains

(a)

Downward forces due
to gravitational pull
of rest of the earth

Material flows in
sideways because of
downward forces

Only a slight
depression remains

(b)

FIG. 1-19   Gravity forces the earth to be round. (a) How a large bump would be pulled down. (b) How a large hole would be filled in.

1-9   WHY THE EARTH IS ROUND

The Big Squeeze   A sign of success of any scientific theory is its ability to account for previously mysterious findings. One such finding is the roundness of the earth. Early thinkers believed it was round because a sphere is the only "perfect" shape, a vague idea that actually explains nothing. In fact, the earth is round because gravity squeezes it into this shape.

As shown in Fig. 1-19, if any part of the earth were to stick out very much, the gravitational attraction of the rest of the earth would pull downward on the projection. The material underneath would then flow out sideways until the projection became level or nearly so. The downward forces around the rim of a deep hole would similarly cause the underlying material to flow into it. The same argument applies to the moon, the sun, and the stars.

Such irregularities as mountains and ocean basins are on a very small scale compared with the earth's size. The total range from the Pacific depths to the summit of Everest is less than 20 km, not much compared with the earth's radius of 6400 km.

The earth is not a perfect sphere. The reason was apparent to Newton: since the earth is spinning rapidly, inertia causes the equatorial portion to swing outward, just as a ball on a string does when it is whirled around. As a result the earth bulges slightly at the equator and is slightly flattened at the poles, much like a grapefruit. The total distortion is not great, for the earth is only 43 km wider than it is high (Fig. 1-20). Venus, whose "day" is 243 of our days, turns so slowly that it has almost no distortion. Saturn, at the other extreme, spins so rapidly that it is almost 10 percent out of round.

FIG. 1-20   The influence of its rotation distorts the earth. The effect is greatly exaggerated in the figure; the equatorial diameter of the earth is actually only 43 km (27 mi) more than its polar diameter.

Perfect sphere

Actual shape of earth

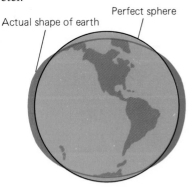

Astronauts in the Apollo 11 spacecraft saw this view of the earth as they orbited the moon, part of whose bleak landscape appears in the foreground.

1-10 **THE TIDES**

Up and Down Twice a Day

Those of us who live near an ocean know well the rhythm of the tides, the twice-daily rise and fall of water level. Usually the change in height is no more than a few meters, but in some regions—the Bay of Fundy in eastern Canada is one—the total range can be over 20 m. What causes the endless advance and retreat of the oceans on such a grand scale?

Tides occur because the moon gravitationally attracts different parts of the earth unequally. In Fig. 1-21 the moon's attraction is strongest at *A*, which is closest, so water there is pulled toward the moon. At *B*, which is farthest from the moon, the attraction is weakest, and the earth itself is pulled away from *B* to leave water heaped up there. As a result, there are bulges of water at *both A* and *B*. The bulges stay in place as the earth rotates under them to produce two high tides and two low tides at a given place every day.

There is more to the story. The sun also affects the waters of the earth, but to a smaller extent than the moon even though the gravitational pull of the sun is greater than that of the moon. The reason is that what causes the tides is the *difference* between the attractions on the near and far sides of the earth, and this difference is greater for the

FIG. 1-21 The origin of the tides. The moon's attraction is greatest at *A*, and so water there is pulled toward the moon. The moon's attraction is weakest at *B*; hence the solid earth is pulled away from *B* to leave water heaped up there. As the earth rotates, the water bulges stay in place to produce two high and two low tides every day.

Earth

Moon

High and low water at Campobello Island, Canada. Two tidal cycles occur every day. The moon causes tides by its unequal attraction for parts of the earth at different distances from it.

moon because it is closer to the earth than the sun. About twice a month—when the sun, moon, and earth are in a straight line (new moon and full moon in Fig. 16-8)—solar tides add to lunar tides to give the especially high (and low) **spring tides;** see Fig. 1-22. When the line between moon and earth is perpendicular to that between sun and earth (first and last quarters in Fig. 16-8), the tide-raising forces partly cancel to give **neap tides,** whose range is smaller than average.

FIG. 1-22 Variation of the tides. Spring tides are produced when the moon is at M_1 or M_2, neap tides when the moon is at M_3 or M_4. The range between high and low water is greatest for spring tides.

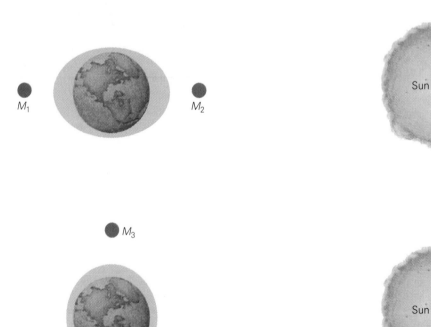

1-11 THE DISCOVERY OF NEPTUNE

Another Triumph for the Law of Gravity

In Newton's time, as in Ptolemy's, only six planets were known: Mercury, Venus, Earth, Mars, Jupiter, and Saturn. In 1781 a seventh, Uranus, was identified. Measurements during the next few years enabled astronomers to work out details of the new planet's orbit and to predict its future positions in the sky. To make these predictions, not only the sun's attraction but also the smaller attractions of the nearby planets Jupiter and Saturn had to be considered. For 40 years, about half the time needed for Uranus to make one complete revolution around the sun, calculated positions of the planet agreed well with observed positions.

Then a discrepancy crept in. Little by little Uranus moved away from its predicted path among the stars. The calculations were checked and rechecked, but no mistake could be found. There were two possibilities: either the law of gravity, on which the calculations were based, was wrong, or else some unknown body was pulling Uranus away from its predicted path.

So firmly established was the law of gravitation that two young men, Urbain Leverrier in France and John Couch Adams in England, set themselves the task of calculating the position of an unknown body that might be responsible for the discrepancies in Uranus's position. Adams, completing his computations first, sent them to England's Astronomer Royal. Busy with other matters, the Astronomer Royal put the calculations away to examine later.

Meanwhile, Leverrier sent his paper to a young German astronomer, Johann Gottfried Galle, who lost no time in turning his telescope to the part of the sky where the new planet should appear. Very close to the position predicted by Leverrier, Galle found a faint object, which proved to be the eighth member of the sun's family and was called Neptune. A little later the Astronomer Royal showed that Neptune's position had been correctly given in Adams's work also. The theory of gravity had again successfully gone around the loop of the scientific method shown in Fig. 1-1.

IMPORTANT TERMS AND IDEAS

The **scientific method** of studying nature has four steps: (1) formulating a problem; (2) observation and experiment; (3) interpreting the results; (4) testing the interpretation by further observation and experiment. When first proposed, a scientific interpretation is called a **hypothesis.** After thorough checking, it becomes a **law** if it states a regularity or relationship, or a **theory** if it uses general considerations to account for specific phenomena.

Polaris, the North Star, lies almost directly above the north pole. A **constellation** is a group of stars that form a pattern in the sky. The **planets** are heavenly bodies that shift their positions regularly with respect to the stars.

In the **ptolemaic system,** the earth is stationary at the center of the universe. In the **copernican system,** the earth rotates on its axis and, with the other planets, revolves around the sun. Observational evidence supports the copernican system.

Kepler's laws are three regularities that the planets obey as they move around the sun.

Newton's **law of gravity** describes the attraction all bodies in the universe have for one another. The gravitational forces the sun exerts on the planets are what hold them in their orbits. Kepler's laws are explained by the law of gravity.

The **tides** are periodic rises and falls of sea level caused by differences in the gravitational pulls of the moon and sun. Water facing the moon is attracted to it more than the earth itself is, and the earth is attracted to the moon more than water on the far side of the earth. The corresponding effect of the sun is smaller than that of the moon and acts to increase or decrease tidal ranges, depending on the relative positions of the moon and sun.

EXERCISES: MULTIPLE CHOICE

1. The "scientific method" is
   a. a continuing process
   b. a way to arrive at ultimate truth
   c. a laboratory technique
   d. based on accepted laws and theories

2. A scientific law or theory is valid
   a. forever
   b. for a certain number of years, after which it is retested
   c. as long as a committee of scientists says so
   d. as long as it is not contradicted by new experimental findings

3. A hypothesis is
   a. a new scientific idea
   b. a scientific idea that has been confirmed by further experiment and observation
   c. a scientific idea that has been discarded because it disagrees with further experiment and observation
   d. a group of linked scientific ideas

4. The object in the sky that apparently moves least in the course of time is
   a. Polaris   c. the sun
   b. Venus   d. the moon

5. The stars in a constellation are
   a. about the same age
   b. about the same distance from the earth
   c. members of the solar system
   d. unrelated except for proximity in the sky as seen from the earth

6. Which of the following is no longer considered valid?
   a. the ptolemaic system
   b. the copernican system
   c. Kepler's laws of planetary motion
   d. Newton's law of gravity

7. A planet not visible to the naked eye is
   a. Mercury   c. Neptune
   b. Saturn   d. Jupiter

8. Arrange the following planets in the order of their distance from the sun:
   a. Mars   c. Uranus
   b. Jupiter   d. Saturn

9. The planet closest to the sun is
   a. Earth   c. Mars
   b. Venus   d. Mercury

10. Kepler modified the copernican system by showing that the planetary orbits are
    a. ellipses
    b. circles
    c. combinations of circles forming looped orbits
    d. the same distance apart from one another

11. The speed of a planet in its elliptical orbit around the sun
    a. is constant
    b. is highest when the planet is closest to the sun
    c. is lowest when the planet is closest to the sun
    d. varies, but not with respect to the planet's distance from the sun

12. According to Kepler's third law, the time needed for a planet to complete an orbit around the sun
    a. is the same for all the planets
    b. depends on the planet's size
    c. depends on the planet's distance from the sun
    d. depends on how fast the planet spins on its axis

13. The law of gravity
    a. applies only to large bodies such as planets and stars
    b. accounts for all known forces
    c. holds only in the solar system
    d. holds everywhere in the universe

14. The earth bulges slightly at the equator and is flattened at the poles because
    a. it spins on its axis
    b. it revolves around the sun
    c. of the sun's gravitational pull
    d. of the moon's gravitational pull

15. The longitude of Greenwich, England, is
    a. 0°   c. 90°W
    b. 90°E   d. 180°

16. A latitude of 0° corresponds to
    a. the north pole
    b. the south pole
    c. the equator
    d. the prime meridian

17. The usual tidal pattern in most parts of the world consists of
    a. a high tide one day and a low tide on the next

b. one high tide and one low tide daily
c. two high tides and two low tides daily
d. three high tides and three low tides daily

18. Tides are caused
    a. only by the sun
    b. only by the moon
    c. by both the sun and the moon
    d. sometimes by the sun and sometimes by the moon

19. High tide occurs at a given place
    a. only when the moon faces the place
    b. only when the moon is on the opposite side of the earth from the place
    c. both when the moon faces the place and when the moon is on the opposite side of the earth from the place
    d. when the place is halfway between facing the moon and being on the opposite side of the earth from the moon

QUESTIONS

1. What role does "common sense" play in the scientific method?

2. What is the basic distinction between the scientific method and other ways of looking at the natural world?

3. What is the difference between a hypothesis and a law? Between a law and a theory?

4. Scientific models do not correspond exactly to reality. Why are they nevertheless so useful?

5. What must be your location if the stars move across the sky in circles centered directly overhead?

6. In terms of what you would actually observe, what does it mean to say that the moon apparently moves eastward among the stars?

7. From observations of the moon, why would you conclude that it is a relatively small body revolving around the earth rather than another planet revolving around the sun?

8. The sun, moon, and planets all follow approximately the same path from east to west across the sky; none of them is ever seen in the far northern or far southern sky. What does this suggest about the arrangement of these members of the solar system in space?

9. What is the basic difference between the ptolemaic and copernican models? Why is the ptolemaic model considered incorrect?

10. Ancient astronomers were troubled by variations in the brightnesses of the various planets with time. Does the ptolemaic or the copernican model account better for these variations?

11. Compare the ptolemaic and copernican explanations for (a) the rising and setting of the sun; (b) the eastward drift of the sun relative to the stars that takes a year for a complete circuit; (c) the eastward drift of the moon relative to the stars that takes about 4 weeks for a complete circuit.

12. What do you think is the reason scientists use an ellipse rather than a circle as the model for a planetary orbit?

13. The average distance from the earth to the sun is called the astronomical unit (AU). If an asteroid is 4 AU from the sun and its period of revolution around the sun is 8 years, does it obey Kepler's third law?

14. How many fundamental forces are there? Which one is responsible for holding the moon to the earth? The earth to the sun? When you press against a wall, which force transmits your push to the wall?

15. What, if anything, would happen to the shape of the earth if it were to rotate on its axis faster than it does today?

16. What is the difference between spring and neap tides? Under what circumstances does each occur?

17. The earth takes almost exactly 24 h to make a complete turn on its axis, so we might expect each high tide to occur 12 h after the one before. However, the actual time between high tides is 12 h 25 min. Can you account for the difference?

ANSWERS TO MULTIPLE CHOICE

1. a	**6.** a	**11.** b	**16.** c
2. d	**7.** c	**12.** c	**17.** c
3. a	**8.** a, b, d, c	**13.** d	**18.** c
4. a	**9.** d	**14.** a	**19.** c
5. d	**10.** a	**15.** a	

2

MOTION

*E*verything in the universe is in nonstop movement. Whatever the scale of size, from the tiny particles inside atoms to the huge galaxies of stars far away in space, motion is the rule, not the exception. In order to understand the universe, we must begin by understanding motion and the laws it obeys.

The laws of motion that govern the behavior of atoms and stars apply just as well to the objects of our daily lives. Engineers need these laws to design cars and airplanes, machines of all kinds, even roads—how steeply to bank a highway curve is calculated from the same basic formula that Newton combined with Kepler's findings to arrive at the law of gravity. Terms such as speed and acceleration, force and weight, are used by everyone. Let us now see exactly what these terms mean and how the quantities they refer to are related.

DESCRIBING MOTION

When an object goes from one place to another, we say it moves. If the object gets there quickly, we say it moves fast; if the object takes a long time, we say it moves slowly. The first step in analyzing motion is to be able to say just how fast is fast and how slow is slow.

2-1 SPEED

How Fast Is Fast

The **speed** of something is the rate at which it covers distance. The higher the speed, the faster it travels and the more distance it covers in a given period of time.

If a car goes through a distance of 40 miles in a time of 1 hour, its speed is 40 miles per hour, usually written 40 miles/hour.

What if the time interval is not exactly 1 hour? For instance, the car might travel 60 miles in 2 hours on another trip. The general formula for speed is distance divided by time:

$$\text{Speed} = \frac{\text{distance}}{\text{time}}$$

Hence the car's speed in the second case is

$$\text{Speed} = \frac{\text{distance}}{\text{time}} = \frac{60 \text{ miles}}{2 \text{ hours}} = 30 \text{ miles/hour}$$

The same formula works for times of less than a full hour. The speed of a car that covers 24 miles in half an hour is, since $\frac{1}{2}$ hour = 0.5 hour,

$$\text{Speed} = \frac{\text{distance}}{\text{time}} = \frac{24 \text{ miles}}{0.5 \text{ hour}} = 48 \text{ miles/hour}$$

These speeds are all **average speeds,** because we do not know the details of how the cars moved during their trips. They probably went slower than the average during some periods, faster at others, and even came to a stop now and then at traffic lights. What the speedometer of a car shows is the car's **instantaneous speed** at any moment, that is, how fast it is going at that moment.

For the sake of convenience, quantities such as distance, time, and speed are often abbreviated and printed in italics:

The speedometer of a car shows its instantaneous speed. This speedometer is calibrated in both mi/h and km/h.

$$d = \text{distance} \qquad t = \text{time} \qquad v = \text{speed}$$

In terms of these symbols the formula for speed becomes

$$v = \frac{d}{t} \qquad speed$$

Distance   The above formula can be rewritten in two ways. Suppose we want to know how far a car whose average speed is v goes in a time t. To find out, we must solve $v = d/t$ for a distance d. According to one of the rules of algebra (see the Math Refresher at the back of this book), a quantity that divides one side of an equation can be shifted to multiply the other side. Thus

$$v = \frac{d}{t}$$

becomes

$$v = \frac{d}{t}$$

$$vt = d$$

which is the same as

$$d = vt$$

$$\text{Distance} = (\text{speed})(\text{time})$$

How far does a car travel in 6 hours when its average speed is 40 miles/hour? We put the given quantities into the above formula to find that (Fig. 2-1)

$$d = vt = \left(40 \, \frac{\text{miles}}{\text{hour}}\right)(6 \text{ hours}) = 240 \left(\frac{\text{miles}}{\text{hour}}\right)(\text{hours}) = 240 \text{ miles}$$

We see that, since hours/hour = 1, the hours cancel out to give just miles in the answer.

Time   In another situation we might want to know how long it takes something moving at a certain speed to cover a certain distance. In other words, we know v and d and want to find the time t. What we do here is solve $d = vt$ for the time t. From basic alge-

FIG. 2-1 A car whose average speed is 40 miles/hour travels 240 miles in 6 hours.

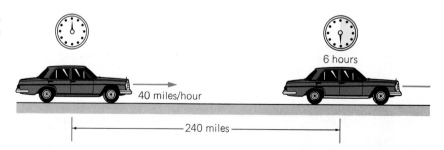

bra we know that something that multiplies one side of an equation can be shifted to divide the other side. What we do, then, is shift the v in the formula $d = vt$ to divide the d:

$$d = \overset{\curvearrowleft}{v}t$$

$$\frac{d}{v} = t$$

which is the same as

$$t = \frac{d}{v}$$

$$\text{Time} = \frac{\text{distance}}{\text{speed}}$$

How much time is needed for a car whose average speed is 40 miles/hour to travel 100 miles? The answer is

$$t = \frac{d}{v} = \frac{100 \text{ miles}}{40 \text{ miles/hour}} = 2.5 \frac{\text{miles}}{\text{miles/hour}}$$

$$= 2.5 \frac{(\cancel{\text{miles}})(\text{hour})}{\cancel{\text{miles}}} = 2.5 \text{ hours}$$

We notice that

$$\frac{1}{1/(\text{hour})} = (1)\left(\frac{\text{hour}}{1}\right) = \text{hour}$$

Quantities such as miles and hours are always carried along in calculations and are treated like ordinary algebraic quantities.

2-2 UNITS

How Many of What   When we say that the distance between Chicago and Minneapolis is 405 miles, what we are really doing is comparing this distance with a certain standard length called the mile. Standard quantities such as the mile are known as **units.** The result of every measurement thus has two parts. One is a number (405 for the Chicago-Minneapolis distance) to answer the question "How many?" The other is a unit (the mile in this case) to answer the question "Of what?"

The most widely used units today are those of the International System, abbreviated SI after its French name Système International d'Unités. Examples of SI units are the **meter** (m) for length, the **second** (s) for time, the **kilogram** (kg) for mass, the **joule** (J) for energy, and the **watt** (W) for power. SI units are used universally by scientists and in most of the world in everyday life as well. Although the British system of units, with its familiar foot and pound, remains in common use in English-speaking countries, it is on the way out and eventually will be replaced by the SI. Since this is a book about science, only SI units will be used from here on.

METER,
KILOGRAM,
SECOND

SI units are derived from the units of the older **metric system.** This system was introduced in France two centuries ago to replace the hodgepodge of traditional units, often different in different countries and even in different parts of the same country, that was making commerce and industry difficult.

The **meter,** the standard of length, was originally defined as one ten-millionth of the distance from the equator to the north pole. The **gram,** the standard of mass, was defined as the mass of 1 cubic centimeter (cm^3) of water; 1 cm^3 is the volume of a cube 1 cm (0.01 m) on each edge, and 1 kilogram = 1000 grams. The meter and gram were new units. The ancient division of a day into 24 hours, an hour into 60 minutes, and a minute into 60 seconds was kept for the definition of the second as 1/(24)(60)(60) = 1/86,400 of a day.

As more and more precision became needed, these definitions were modified several times. Today the second is specified in terms of the microwave radiation given off under certain circumstances by one type of cesium atom, ^{133}Cs: 1 s equals the time needed for 9,192,631,770 cycles of this radiation to be emitted. The meter, which for convenience had become the distance between two scratches on a platinum-iridium bar kept at Sèvres, France, is now the distance traveled in 1/299,792,458 s by light in a vacuum. There are approximately 3.28 feet in a meter. The kilogram is the mass of a platinum-iridium cylinder 39 mm in diameter and 39 mm high at Sèvres. As discussed in Sec. 2-10, mass and weight are not the same. The weight of a given mass is the force with which gravity attracts it to the earth; the weight of 1 kg is 2.2 pounds.

The great advantage of SI units is that their subdivisions and multiples are in steps of 10, 100, 1000, and so on, in contrast to the irregularity of British units. In the case of lengths, for instance (Fig. 2-2),

$$1 \text{ meter (m)} = 100 \text{ centimeters (cm)}$$

$$1 \text{ kilometer (km)} = 1000 \text{ meters}$$

whereas

$$1 \text{ foot (ft)} = 12 \text{ inches (in.)}$$

$$1 \text{ mile (mi)} = 5280 \text{ feet}$$

Table 2-1 lists the most common subdivisions and multiples of SI units. Each is designated by a prefix according to the corresponding power of 10. (Powers of 10 are reviewed in the Math Refresher at the back of the book.)

Table 2-2 contains conversion factors for changing a length expressed in one system to its equivalent in the other. (More conversion factors are given inside the back cover of this book.) We note from the table that there are about 2½ centimeters in an inch, so a centimeter is

FIG. 2-2   There are 1000 meters in a kilometer and 100 centimeters in a meter.

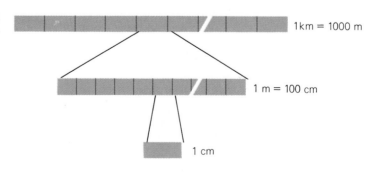

1 km = 1000 m

1 m = 100 cm

1 cm

TABLE 2-1

Subdivisions and Multiples of SI Units. (The Symbol μ Is the Greek Letter "mu")

Prefix	Power of 10	Abbreviation	Pronunciation
Pico-	10^{-12}	p	pee′ koe
Nano-	10^{-9}	n	nan′ oe
Micro-	10^{-6}	μ	my′ kroe
Milli-	10^{-3}	m	mil′ i
Centi-	10^{-2}	c	sen′ ti
Hecto-	10^{2}	h	hec′ toe
Kilo-	10^{3}	k	kil′ oe
Mega-	10^{6}	M	meg′ a
Giga-	10^{9}	G	ji′ ga
Tera-	10^{12}	T	ter′ a

roughly the width of a shirt button; a meter is a few inches longer than 3 feet; and a kilometer is nearly $\frac{2}{3}$ mile.

For practice, let us convert the 405-mi distance between Chicago and Minneapolis to km. From Table 2-2, the conversion factor we need is 1.61 km/mi, so

$$d = (405 \text{ mi}) \left(1.61 \ \frac{\text{km}}{\text{mi}} \right) = 652 \text{ km}$$

2-3 VECTORS

Which Way as Well as How Much

Some quantities need only a number and a unit to be completely specified. It is enough to say that the area of a farm is 600 acres, that the frequency of a sound wave is 440 cycles per second, that a lightbulb uses electric energy at the rate of 75 watts. These are examples of **scalar**

TABLE 2-2

Conversion Factors for Length

Multiply a Length Expressed in	By	To Get the Same Length Expressed in
Centimeters	$0.394 \ \frac{\text{in.}}{\text{cm}}$	Inches
Meters	$39.4 \ \frac{\text{in.}}{\text{m}}$	Inches
Meters	$3.28 \ \frac{\text{ft}}{\text{m}}$	Feet
Kilometers	$0.621 \ \frac{\text{mi}}{\text{km}}$	Miles
Inches	$2.54 \ \frac{\text{cm}}{\text{in.}}$	Centimeters
Feet	$30.5 \ \frac{\text{cm}}{\text{ft}}$	Centimeters
Feet	$0.305 \ \frac{\text{m}}{\text{ft}}$	Meters
Miles	$1.61 \ \frac{\text{km}}{\text{mi}}$	Kilometers

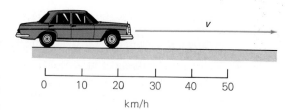

FIG. 2-3   The vector **v** represents a velocity of 40 km/s to the right. The scale is 1 cm = 10 km/h.

quantities. The **magnitude** of a quantity refers to how large it is. Thus the magnitudes of the scalar quantities given above are respectively 600 acres, 440 cycles per second, and 75 watts.

A **vector quantity,** on the other hand, has a direction as well as a magnitude associated with it, and this direction can be important. Displacement (change in position) is an example of a vector quantity. If we drive 1000 km north from Denver, we will end up in Canada; if we drive 1000 km south, we will end up in Mexico. Force is another example of a vector quantity. Applying enough upward force to this book will lift it from the table; applying a force of the same magnitude downward on the book will press it harder against the table, but the book will not move.

The **speed** of a moving object tells us only how fast the object is going, regardless of its direction. Speed is therefore a scalar quantity. If we are told that a car has a speed of 40 km/h, we do not know where it is headed, or even if it is moving in a straight line—it might well be going in a circle. The vector quantity that includes both speed and direction is called **velocity.** If we are told that a car has a constant velocity of 40 km/h toward the west, we know all there is to know about its motion and can easily figure out where it will be in an hour, or 2 hours, or at any other time.

A handy way to represent a vector quantity on a drawing is to use a straight line called a **vector** that has an arrowhead at one end to show the direction of the quantity. The length of the line is scaled according to the magnitude of the quantity. Figure 2-3 shows how a velocity of 40 km/h to the right is represented by a vector on a scale of 1 cm = 10 km/h. All other vector quantities can be pictured in a similar way.

Vector quantities are usually printed in boldface type (**F** for force, **v** for velocity). Italic type is used for scalar quantities (f for frequency, V for volume). Italic type is also used for the magnitudes of vector quantities: F is the magnitude of the force **F;** v is the magnitude of the velocity **v.** For instance, the magnitude of a velocity **v** of 40 km/h to the west is the speed $v = 40$ km/h. A vector quantity is usually indicated in handwriting by an arrow over its symbol, so that $\vec{F}$ means the same thing as **F.**

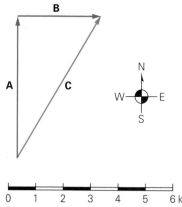

FIG. 2-4   Adding vector **B** (3 km east) to vector **A** (5 km north) gives vector **C** whose length corresponds to 5.8 km.

Adding Vectors

To add scalar quantities of the same kind, we just use ordinary arithmetic. For example, 5 kg of onions plus 3 kg of onions equals 8 kg of onions. The same method holds for vector quantities of the same kind whose directions are the same. If we drive north for 5 km and then continue north for another 3 km, we will go a total of 8 km to the north.

What if the directions are different? If we drive north for 5 km and then east for 3 km, we will not end up 8 km from our starting point. The vector diagram of Fig. 2-4 provides the answer. To add the vectors **A** and **B,** we draw **B** with its tail at the head of **A.** Connecting the tail of **A** with the head of **B** gives us the vector **C,** which corresponds to our net displacement from the start of our trip to its finish. The length of **C** tells us that our displacement was slightly less than 6 km. Any number of vectors of the same kind can be added in this way by stringing them together tail to head and then joining the tail of the first with the head of the last one.

2-4   ACCELERATION

Vroom!   An **accelerated** object is one whose velocity is changing. As in Fig. 2-5, the change can be an increase or a decrease in speed—the object can be going faster and faster, or slower and slower. A change in direction, too, is an acceleration, as discussed later. Acceleration in general is a vector quantity. For the moment, though, we will stick to straight-line motion, where acceleration is the rate of change of speed. That is,

$$a = \frac{v_f - v_i}{t} \quad \textit{straight-line motion}$$

$$\text{Acceleration} = \frac{\text{change in speed}}{\text{time interval}}$$

where the symbols mean the following:

a = acceleration
t = time interval
v_i = speed at start of time interval = initial speed
v_f = speed at end of time interval = final speed

FIG. 2-5 Three cases of accelerated motion, showing successive positions of a body after equal periods of time. At the top the intervals between the positions of the body increase in length because the body is traveling faster and faster. Below it the intervals decrease in length because the body is slowing down. At the bottom the intervals are the same in length because the speed is constant, but the direction of motion is constantly changing.

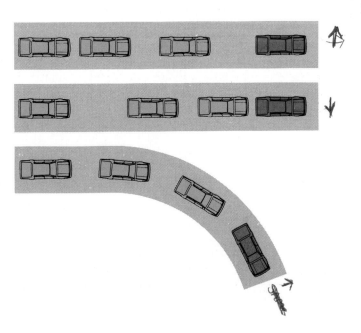

Express elevators in tall buildings have accelerations of no more than 1 m/s^2 to prevent passenger discomfort. The world's fastest elevator, in the Yokohama Land- mark Tower (Japan's highest building), climbs 69 floors in 40 s, but only 5 s is spent at its top speed of 12.5 m/s (28 mi/h). The elevator reaches this speed at the 27th floor and then begins to slow down at the 42nd floor.

Suppose the speed of a car changes from 15 m/s (about 34 mi/h) to 25 m/s (about 56 mi/h) in 20 s when its gas pedal is pressed hard (Fig. 2-6). Here

$$v_i = 15 \text{ m/s} \qquad v_f = 25 \text{ m/s} \qquad t = 20 \text{ s}$$

and so the car's acceleration is

$$a = \frac{v_f - v_i}{t} = \frac{25 \text{ m/s} - 15 \text{ m/s}}{20 \text{ s}} = \frac{10 \text{ m/s}}{20 \text{ s}} = 0.5 \frac{\text{m/s}}{\text{s}} = 0.5 \text{ m/s}^2$$

This result means that the speed of the car increases by 0.5 m/s during each second the acceleration continues. It is customary to write (m/s)/s

FIG. 2-6 A car whose speed increases from 15 m/s to 25 m/s in 20 s has an acceleration of 0.5 m/s^2.

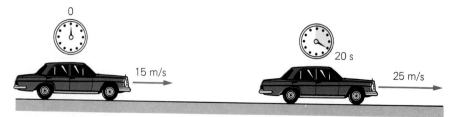

(meters per second per second) as just m/s^2 (meters per second squared) since

$$\frac{m/s}{s} = \frac{m}{(s)(s)} = \frac{m}{s^2}$$

In Sec. 2-9 we will see why acceleration is such an important quantity in physics.

ACCELERATION OF GRAVITY

Drop a stone, and it falls. Does the stone fall at a constant speed, or does it go faster and faster? Does the stone's motion depend on its weight, or its size, or its shape?

Before Galileo, philosophers tried to answer such questions in terms of supposedly self-evident principles, concepts so seemingly obvious that there was no need to test them. This was the way in which Aristotle (384–322 B.C.), the famous thinker of ancient Greece, approached the subject of falling bodies. To Aristotle, every kind of material had a "natural" place where it belonged and toward which it tried to move. Thus fire rose "naturally" toward the sun and stars, whereas stones were "earthy" and so fell downward toward their home in the earth. A big stone was more earthy than a small one and so ought to fall faster.

BIOGRAPHY

Galileo Galilei (1564–1642) fathered modern science by clearly stating the central idea of the scientific method: the study of nature must be based on observation and experiment. He also pioneered the use of mathematical reasoning to interpret and generalize his findings. Galileo was born in Pisa, Italy; following a local custom, his given name was a variation of his family name. Although his father thought that medicine would be a more sensible career choice, Galileo studied physics and mathematics and soon became a professor at Pisa and afterward at Padua. His early work was on accelerated motion, falling bodies, and the paths taken by projectiles. Later, with a telescope he had built, Galileo was the first person to see sunspots, the phases of Venus, the four largest satellites of Jupiter, and the mountains of the moon. Turning his telescope to the Milky Way, he found that it consisted of individual stars. To Galileo, as to his contemporaries, these discoveries were "infinitely stupendous."

By then famous, in 1610 Galileo went to Florence as court mathematician to Cosimo, Duke of Tuscany. Here Galileo expounded the copernican model of the universe and pointed out that his astronomical observations supported this model. Because the ptolemaic model with the earth as the stationary center of the universe was part of the doctrine of the Catholic Church, the Holy Office told Galileo to stop advocating the contrary, and he obeyed. Then, when a friend of his became pope in 1631, Galileo resumed teaching the merits of the copernican system. But the pope turned against him, and in 1633, when he was 70, Galileo was convicted of heresy by the Inquisition. Although he escaped being burnt at the stake, the fate of other heretics, Galileo was sentenced to house arrest for the remainder of his life and was forced to publicly deny that the earth moves. (According to legend, Galileo then muttered, "Yet it does move.") Not until 1992 did Pope John Paul II officially concede that Galileo had been right after all.

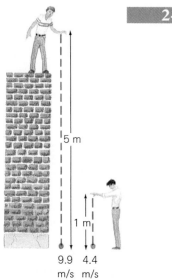

9.9   4.4
m/s   m/s

FIG. 2-7 Falling bodies are accelerated downward. A stone dropped from a height of 5 m strikes the ground with a speed more than double that of a stone dropped from a height of 1 m.

2-5   FREE FALL

What Goes Up Must Come Down   Two thousand years later the Italian physicist Galileo found that the higher a stone is when it is dropped, the greater its speed when it reaches the ground (Fig. 2-7). This means the stone is accelerated. Furthermore, the acceleration is the same for *all* stones, big and small. For more accuracy with the primitive instruments of his time, Galileo measured the accelerations of balls rolling down an inclined plane rather than their accelerations in free fall, but his conclusions were perfectly general.

Galileo's experiments showed that, if there were no air for them to push their way through, all falling objects near the earth's surface would have the same acceleration of 9.8 m/s^2. This acceleration is usually abbreviated g:

Acceleration of gravity = g = 9.8 m/s^2

Ignoring for the moment the effect of air resistance, something that drops from rest has a speed of 9.8 m/s at the end of the first second, a speed of (9.8 m/s^2)(2 s) = 19.6 m/s at the end of the next second, and so on (Fig. 2-8).

The downward acceleration g is the same whether an object is just

FIG. 2-8 All falling objects have a downward acceleration of 9.8 m/s^2. (The distance an object will have fallen in each time interval is not shown to scale here.)

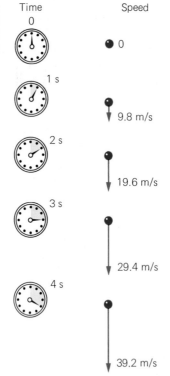

Time
0

1 s

2 s

3 s

4 s

Speed

0

9.8 m/s

19.6 m/s

29.4 m/s

39.2 m/s

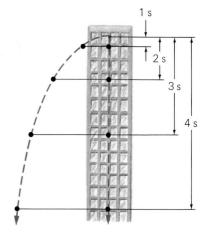

FIG. 2-9 The acceleration of gravity does not depend upon horizontal motion. When one ball is thrown horizontally from a building at the same time that a second ball is dropped vertically, the two reach the ground at the same time because both have the same downward acceleration.

FIG. 2-10 When a ball is thrown upward, its downward acceleration reduces its original speed until it comes to a momentary stop. At this time the ball is at the top of its path, and it then begins to fall as if it had been dropped from there. The ball is shown after equal time intervals.

The Leaning Tower of Pisa, which is 58 m high, was begun in 1174 and took over two centuries to complete. During its construction the tower started to sink into the clay soil under its south side, and corrections were made to the upper floors to try to make them level. Today the tilt is 5.5° and continues to increase, leaving the tower in danger of collapse if efforts to stabilize the ground under it fail. According to legend, Galileo dropped a bullet and a cannonball from the tower to show that all objects fall with the same acceleration.

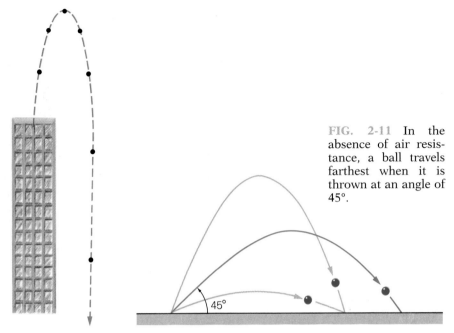

FIG. 2-11 In the absence of air resistance, a ball travels farthest when it is thrown at an angle of 45°.

dropped or is thrown upward, downward, or sideways. If a ball is held in the air and dropped, it goes faster and faster until it hits the ground. If the ball is thrown horizontally, we can imagine its velocity as having two parts, a horizontal one that stays constant and a vertical one that is affected by gravity. The result, as in Fig. 2-9, is a curved path that becomes steeper as the downward speed increases.

When a ball is thrown upward, as in Fig. 2-10, the effect of the downward acceleration of gravity is at first to reduce the ball's upward speed. The upward speed decreases steadily until finally it is zero. The ball is then at the top of its path, when the ball is momentarily at rest. The ball then begins to fall at ever-increasing speed, exactly as though it had been dropped from the highest point. Interestingly enough, something thrown upward at a certain speed will return to its starting point with the same speed, although the object is now moving in the opposite direction.

What happens when a ball is thrown downward? Now the ball's original speed is steadily increased by the downward acceleration of gravity. When the ball reaches the ground, its final speed will be the sum of its original speed and the speed increase due to the acceleration.

When a ball is thrown upward at an angle to the ground, the result is a curved path called a **parabola** (Fig. 2-11). The maximum range (horizontal distance) for a given initial speed occurs when the ball is thrown at an angle of 45° above the ground. At higher and lower angles, the range will be shorter. As the figure shows, for every range up to the maximum there are two angles at which the ball can be thrown and land in the same place.

The terminal speeds of sky divers are greatly reduced when their parachutes open, which permits them to land safely.

2-6   AIR RESISTANCE

Why Rain-drops Don't Kill   Air resistance keeps falling things from developing the full acceleration of gravity. Without this resistance raindrops would reach the ground with bulletlike speeds and even a light shower would be dangerous.

In air, a stone falls faster than a feather because air resistance affects the stone less. In a vacuum, however, there is no air, and the stone and feather fall with the same acceleration of 9.8 m/s^2 (Fig. 2-12).

FIG. 2-12 In a vacuum all bodies fall with the same acceleration.

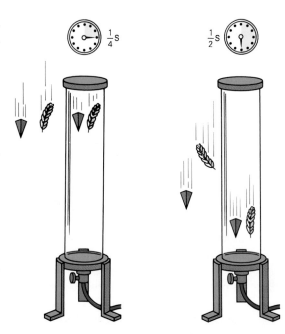

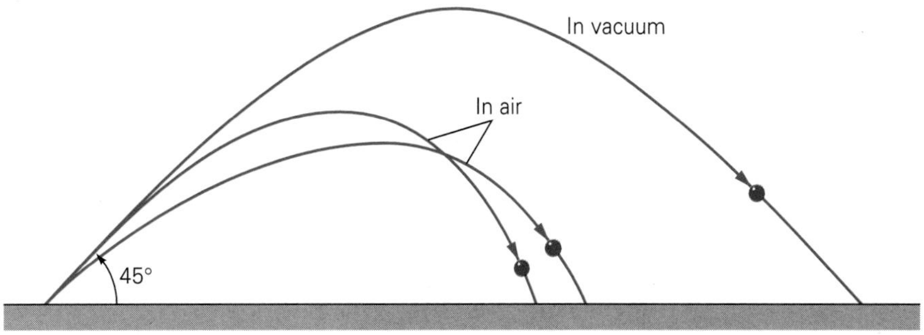

In vacuum

In air

45°

FIG. 2-13 Effect of air resistance on the path of a thrown ball. An angle of less than 45° now gives the greatest range.

TABLE 2-3

Some Terminal Speeds
(1 m/s = 2.2 mi/h)

Object	Terminal Speed
16-lb shot	145 m/s
Baseball	42
Golf ball	40
Tennis ball	30
Basketball	20
Ping-Pong ball	9

The faster something moves, the more the air in its path resists its motion. At 100 km/h (62 mi/h), the drag on a car due to air resistance is about 5 times as great as the drag at 50 km/h (31 mi/h). In the case of a falling object, the air resistance increases with speed until the object cannot go any faster. It then continues to drop at a constant **terminal speed** that depends on its size and shape and on how heavy it is (Table 2-3). A person in free fall has a terminal speed of about 54 m/s (120 mi/h), but with an open parachute the terminal speed of only about 6.3 m/s (14 mi/h) permits a safe landing.

Air resistance reduces the range of a projectile. Figure 2-13 shows how the path of a ball is affected. In a vacuum, as we saw in Fig. 2-11, the ball goes farthest when it is thrown at an angle of 45°, but in air (that is, in real life), the maximum range occurs for an angle of less than 45°. For a baseball struck hard by a bat, an angle of 40° will take it the greatest distance.

FORCE AND MOTION

What can make something originally at rest begin to move? Why do some things move faster than others? Why are some accelerated and others not? Questions like these led Isaac Newton to formulate three principles that summarize so much of the behavior of moving bodies that they have become known as the **laws of motion.** Based on observations made by Newton and others, these laws are as valid today as they were when they were set down about three centuries ago.

2-7   FIRST LAW OF MOTION

Constant Velocity Is as Natural as Being at Rest

Imagine a ball lying on a level floor. Left alone, the ball stays where it is. If you give it a push, the ball rolls a short way and then comes to a stop. The smoother the floor, the farther the ball rolls before stopping. With a perfectly round ball and a perfectly smooth and level floor, and no air to slow down its motion, would the ball ever stop rolling?

There will never be a perfect ball and a perfect surface for it to roll on, of course. But we can come close. The result is that, as the resistance to its motion becomes less and less, the ball goes farther and farther for the same push. We can reasonably expect that, under ideal conditions, the ball would keep rolling forever.

This conclusion was first reached by Galileo. Later it was stated by Newton as his **first law of motion:**

> **If no net force acts on it, an object at rest remains at rest and an object in motion remains in motion at constant velocity (that is, at constant speed in a straight line).**

According to this law, an object at rest never begins to move all by itself—a force is needed to start it off. If it is moving, the object will continue going at constant velocity unless a force acts to slow it down, to speed it up, or to change its direction. Motion at constant velocity is just as "natural" as staying at rest.

Force

In thinking about **force,** most of us think of a car pulling a trailer or a person pushing a lawn mower or lifting a crate. Also familiar are the force of gravity, which pulls us and things about us downward, the pull of a magnet on a piece of iron, and the force of air pushing against the sails of a boat. In these examples the central idea is one of pushing, pulling, or lifting. Newton's first law gives us a more precise definition:

> **A force is any influence that can change the speed or direction of motion of an object.**

When we see something accelerated, we know that a force must be acting upon it.

It is worth emphasizing that just applying a force to an object at rest will not necessarily set the object in motion. If we push against a stone wall, the wall is not accelerated as a result. Only if a force is applied to an object that is able to respond will the state of rest or motion of the object change. On the other hand, *every* acceleration can be traced to the action of a force.

An object continues to be accelerated only as long as a **net force**— a force not balanced out by one or more other forces—acts upon it (Fig. 2-14). An ideal car on a level road would therefore need its engine only to be accelerated to a particular speed, after which it would keep moving at this speed forever with the engine turned off. Actual cars are not so cooperative because of the retarding forces of friction and air resistance, which require counteracting by a force applied by the engine to the wheels.

FIG. 2-14 When several forces act on an object, they may cancel one another out to leave no net force. Only a net (or *unbalanced*) force can accelerate an object.

2-8 MASS

A Measure of Inertia

The reluctance of an object to change its state of rest or of uniform motion in a straight line is called **inertia.**

When you are in a car that starts to move, you feel yourself pushed back in your seat (Fig. 2-15). What is actually happening is that your

The more mass an object has, the greater its resistance to a change in its state of motion, as this shot-putter knows.

inertia tends to keep your body where it was before the car started moving. When the car stops, on the other hand, you feel yourself pushed forward. What is actually happening now is that your inertia tends to keep your body moving while the car comes to a halt.

The name **mass** is given to the property of matter that shows itself as inertia. The inertia of a bowling ball exceeds that of a basketball, as you can tell by kicking them in turn, so the mass of the bowling ball exceeds that of the basketball. Mass may be thought of as quantity of matter: the more mass something has, the greater its inertia and the more matter it contains.

The SI unit of mass is the kilogram (kg). A liter of water, which is a little more than a quart, has a mass of 1 kg (Fig. 2-16).

FIG. 2-15 *(a)* When a car suddenly starts to move, the inertia of the passengers tends to keep them at rest relative to the earth, and so their heads move backward relative to the car. *(b)* When the car comes to a sudden stop, inertia tends to keep the passengers moving, and so their heads move forward relative to the car.

(*a*) Sudden start   (*b*) Sudden stop

FIG. 2-16 A liter, which is equal to 1.057 quarts, represents a volume of 1000 cubic centimeters (cm^3). One liter of water has a mass of 1 kg.

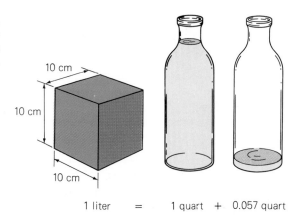

10 cm

10 cm

10 cm

1 liter   =   1 quart + 0.057 quart

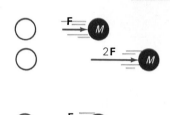

2-9 SECOND LAW OF MOTION

FIG. 2-17 Newton's second law of motion. When different forces act upon the same mass, the greater force produces the greater acceleration. When the same force acts upon different masses, the greater mass receives the smaller acceleration.

Force and Acceleration Throw a baseball hard, and it leaves your hand going faster than if you toss it gently. This suggests that the greater the force, the greater the acceleration while the force acts. Experiments show that doubling the net force doubles the acceleration, tripling the net force triples the acceleration, and so on (Fig. 2-17).

Do all balls you throw with the same force leave your hand with the same speed? Heave an iron shot instead, and it is clear that the more mass something has, the less its acceleration for a given force. Experiments make the relationship precise: for the same net force, doubling the mass cuts the acceleration in half, tripling the mass cuts the acceleration to one-third its original value, and so on.

Newton's **second law of motion** is a statement of these findings. If we let F = force and m = mass, this law states that

$$a = \frac{F}{m} \quad \textit{second law of motion}$$

$$\text{Acceleration} = \frac{\text{force}}{\text{mass}}$$

Another way to express the second law of motion is in the form of a definition of force:

$$F = ma$$

$$\text{Force} = (\text{mass})(\text{acceleration})$$

An important aspect of the second law concerns direction. The direction of the acceleration is always the same as the direction of the net force. A car is going faster and faster—therefore the net force on it is in the same direction as that in which the car is headed. The car then slows down—therefore the net force on it is now in the direction *opposite* that in which it is headed (Fig. 2-18).

Thus we can say that

> **The net force on an object equals the product of the mass and the acceleration of the object. The direction of the force is the same as that of the acceleration.**

FIG. 2-18 The direction of a force is significant. A force applied in the direction in which a body is moving produces a positive acceleration (increase in speed). A force applied opposite to the direction of motion produces a negative acceleration (decrease in speed).

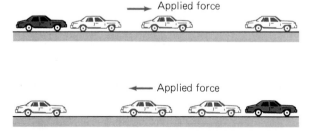

The relationship between force and acceleration means that the less acceleration something has, the smaller the net force on it. If you drop to the ground from a height, as this jumper has, you can reduce the force of the impact by bending your knees as you hit the ground so you come to a stop gradually instead of suddenly. The same reasoning can be applied to make cars safer. If a car's body is built to crumple progressively in a crash, the forces acting on the passengers will be smaller than if the car were rigid.

The second law of motion is the key to understanding the behavior of moving objects because it links cause (force) and effect (acceleration) in a definite way. When we speak of force from now on, we know exactly what we mean, and we know exactly how an object free to move will respond when a given force acts on it.

The Newton The second law of motion shows us how to define a unit for force. If we express mass m in kilograms and acceleration a in m/s², force F is given in terms of (kg)(m/s²). This unit is given a special name, the **newton** (N). Thus

$$1 \text{ newton} = 1 \text{ N} = 1 \text{ (kg)(m/s}^2)$$

When a force of 1 N is applied to a 1-kg mass, the mass is given an acceleration of 1 m/s² (Fig. 2-19).

In the British system, the unit of force is the **pound** (lb). The pound and the newton are related as follows:

$$1 \text{ N} = 0.225 \text{ lb}$$

$$1 \text{ lb} = 4.45 \text{ N}$$

FIG. 2-19 A force of 1 newton gives a mass of 1 kilogram an acceleration of 1 m/s².

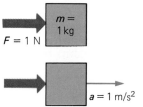

As an example, let us find the force needed to serve a 60-g tennis ball at 30 m/s. The ball will be in contact with the racket for a time that is typically 5 thousandths of a second, 0.005 s. Since the ball starts from rest, $v_i = 0$, and the acceleration required is

$$a = \frac{v_f - v_i}{t} = \frac{30 \text{ m/s} - 0}{0.005 \text{ s}} = 6000 \text{ m/s}^2$$

Because the ball's mass is 60 g = 0.06 kg, the force the racket must exert on it is

MUSCULAR FORCES The forces an animal exerts result from contractions of its skeletal muscles, which occur when the muscles are electrically stimulated by nerves. The maximum force a muscle can exert is proportional to its cross-sectional area and can be as much as 70 N /cm² (100 lb/in.²). An athlete might have a biceps muscle in his arm 8 cm across, so it could produce up to 3500 N (790 lb) of force. This is a lot, but the geometry of an animal's skeleton and muscles favors range of motion over force. As a result the actual force a person's arm can exert is much smaller than the forces exerted by the arm muscles themselves, but the person's arm can move through a much greater distance than the amount the muscles contract.

An animal whose length is L has muscles that have cross-sectional areas and hence strengths roughly proportional to L^2. But the mass of the animal depends on its volume, which is roughly proportional to L^3. Therefore the larger an animal is, in general, the weaker it is relative to its mass. This is obvious in nature. For instance, even though insect muscles are intrinsically weaker than human muscles, many insects can carry loads several times their weights, whereas animals the size of humans are limited to loads comparable with their weights.

60 g

30 m/s

FIG. 2-20 A person serving a tennis ball must exert a force of 360 N on it for the ball to have a speed of 30 m/s if the racket is in contact with the ball for 0.005 s.

$$F = ma = (0.06 \text{ kg})(6000 \text{ m/s}^2) = 360 \text{ N}$$

In more familiar units, this force is 81 lb. Of course, it does not seem so great to the person serving because the duration of the impact is so brief (Fig. 2-20).

2-10 MASS AND WEIGHT

Weight Is a Force The **weight** of an object is the force with which it is attracted by the earth's gravitational pull. If you weigh 160 lb (711 N), the earth is pulling you down with a force of 160 lb. Weight is different from mass, which refers to how much matter something contains. There is a very close relationship between weight and mass, however.

Let us look at the situation in the following way. Whenever a net force F is applied to a mass m, Newton's second law of motion tells us

The more mass a planet has and the smaller it is, the greater the acceleration of gravity *g* at its surface. The values of *g* for the various planets are listed in Table 16-1. From these values, if you know your mass you can figure out your weight on any planet using the formula $w = mg$. In familiar units, if you weigh 150 lb on the earth, here is what you would weigh on the planets and the moon:

Mercury	57 lb	Saturn	180 lb
Venus	135	Uranus	165
Earth	150	Neptune	180
Mars	57	Pluto	65
Jupiter	390	Moon	25

that the acceleration *a* of the mass will be in accord with the formula

$$F = ma$$

Force = (mass)(acceleration)

In the case of an object at the earth's surface, the force gravity exerts on it is its weight *w*. This is the force that causes the object to fall with the constant acceleration $g = 9.8$ m/s^2 when no other force acts. We may therefore substitute *w* for *F* and *g* for *a* in the formula $F = ma$ to give

$$w = mg \qquad \textit{weight and mass}$$

Weight = (mass)(acceleration of gravity)

The weight *w* of an object and its mass *m* are always proportional to each other: twice the mass means twice the weight, and half the mass means half the weight.

In the SI system, mass rather than weight is normally specified. A customer in a French grocery might ask for a kilogram of bread or 5 kg of potatoes. To find the weight in newtons of something whose mass in kilograms is known, we simply turn to $w = mg$ and set $g = 9.8$ m/s^2. Thus the weight of 5 kg of potatoes is

$$w = mg = (5 \text{ kg})(9.8 \text{ m/s}^2) = 49 \text{ N}$$

This is the force with which the earth attracts a mass of 5 kg.

At the earth's surface, the weight of a 1-kg mass in British units is 2.2 lb. The weight in pounds of 5 kg of potatoes is therefore (5 kg) (2.2 lb/kg) = 11 lb. The mass that corresponds to a weight of 1 lb is 454 g.

The mass of something is a more basic property than its weight because the pull of gravity on it is not the same everywhere. This pull is less on a mountaintop than at sea level and less at the equator than near the poles because the earth bulges slightly at the equator. A person who weighs 200 lb in Lima, Peru, would weigh nearly 201 lb in Oslo, Norway. On the surface of Mars the same person would weigh only 76 lb, and he or she would be able to jump much higher than on the earth. However, the person would not be able to throw a ball any faster: because the force *F* the person exerts on the ball and the ball's mass *m* are the same on both planets, the acceleration *a* would be the same, too.

2-11 THIRD LAW OF MOTION

Action and Reaction Suppose you push against a heavy table and it does not move. This must mean that the table is pushing back just as hard as you push against it. The table stays in place because your force on it is matched by the opposing force of friction between the table legs and the floor. You don't move because the force of the table on you is matched by a similar opposing force between your shoes and the floor.

Now imagine that you and the table are on a frozen lake whose surface is so slippery on a warm day that there is no friction. Again you

Force exerted   Force exerted
by table on person   by person on table

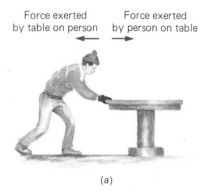

(a)

(b)

FIG. 2-21 Action and reaction forces act on different bodies. Pushing a table on a frozen lake results in person and table moving apart in opposite directions.

push on the table, which this time moves away as a result (Fig. 2-21). But you can stick to the ice no better than the table can, and you find yourself sliding backward. No matter what you do, pushing on the table always means that the table pushes back on you.

Considerations of this kind led Newton to his **third law of motion:**

> **When one object exerts a force on a second object, the second object exerts an equal force in the opposite direction on the first object.**

No force ever occurs singly. A chair pushes downward on the floor; the floor presses upward on the chair (Fig. 2-22). The firing of a rifle exerts a force on the bullet; at the same time the firing exerts a backward push (recoil) on the rifle. A pear falls from a tree because of the earth's pull on the pear; there is an equal upward pull on the earth by the pear that is not apparent because the earth has so much more mass than the pear, but this upward force is nevertheless present.

Newton's third law always applies to two different forces on two different objects—the **action force** that the first object exerts on the second, and the opposite **reaction force** the second exerts on the first.

The third law of motion permits us to walk. When you walk, what is actually pushing you forward is not your own push on the ground but instead the reaction force of the ground on you. As you move forward, the earth itself moves backward, though by too small an amount (by virtue of its enormous mass) to be detected.

Sometimes the origin of the reaction force is not obvious. A book lying on a table exerts the downward force of its weight; but how can an apparently rigid object like the table exert an upward force on the book? If the tabletop were made of rubber, we would see the book push it down, and the upward force would result from the elasticity of the rubber. A similar explanation actually holds for tabletops of wood or metal, which are never perfectly rigid, although the depressions made in them may be extremely small.

FIG. 2-22 Some examples of action-reaction pairs of forces.

Reaction force   Action force
on rifle   on bullet

Reaction forces
of floor
on chair

Action forces of
chair on floor

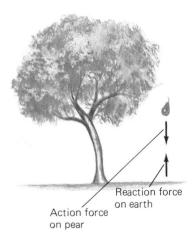

Reaction force
on earth

Action force
on pear

When a cannon is fired, the forward force on the cannonball is matched by an equal backward force on the cannon itself. This is an example of Newton's third law of motion.

GRAVITATION

Left to itself, a moving object travels in a straight line at constant speed. Because the moon circles the earth and the planets circle the sun, forces must be acting on the moon and planets. As we learned in Chap. 1, Newton discovered that these forces are the same in nature as the gravitational force that holds us to the earth. Before we consider how gravity works, we must look into exactly how curved paths come about.

2-12   CIRCULAR MOTION

A Curved Path Requires an Inward Pull

Tie a ball to the end of a string and whirl the ball around your head, as in Fig. 2-23. What you will find is that your hand must pull on the string to keep the ball moving in a circle. If you let go of the string, there is no longer an inward force on the ball, and it flies off to the side.

The force that has to be applied to make something move in a curved path is called **centripetal** ("toward the center") **force:**

FIG. 2-23 A centripetal force is necessary for circular motion. An inward centripetal force F_c acts upon every object that moves in a curved path. If the force is removed, the object continues moving in a straight line tangent to its original path.

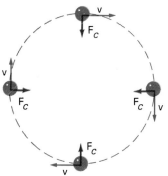

String provides centripetal force

When these people push backward on the ground with their
feet, the ground pushes forward
on them. The latter reaction force
is what leads to their forward
motion.

Centripetal force = inward force on an object moving in a curved path

The centripetal force always points toward the center of curvature of the object's path, which means the force is at right angles to the object's direction of motion at each moment. In Fig. 2-23 the ball is moving in a circle, so its velocity vector **v** is always tangent to the circle and the centripetal force vector $\mathbf{F}_c$ is always directed toward the center of the circle.

A detailed calculation shows that the centripetal force $\mathbf{F}_c$ needed for something of mass m and speed v to travel in a circle of radius r has the magnitude

$$\text{Centripetal force} = F_c = \frac{mv^2}{r}$$

This formula tells us three things about the force needed to cause an object to move in a circular path: (1) the greater the object's mass, the greater the force; (2) the faster the object, the greater the force; and (3) the smaller the circle, the greater the force (Fig. 2-24).

From the formula for F_c we can see why cars rounding a curve are so difficult to steer when the curve is sharp (small r) or the speed is high (a large value for v means a very large value for v^2). On a level road, the centripetal force is supplied by friction between the car's tires and the road. If the force needed to make a particular turn at a certain speed is more than friction can supply, the car skids outward.

Suppose a 1000-kg car traveling at 5 m/s goes around a turn 30 m in radius, as in Fig. 2-25. The centripetal force needed to make the turn is

FIG. 2-24 The centripetal force needed to keep an object moving in a circle depends upon the mass and speed of the object and upon the radius of the circle. The direction of the force is always toward the center of the circle.

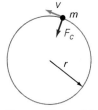

The magnitude of the
centripetal force is
$F_c = mv^2/r$

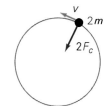

Doubling the mass *m* doubles
the needed centripetal force

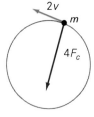

Doubling the speed *v*
quadruples the needed
centripetal force
because of the *v²* factor

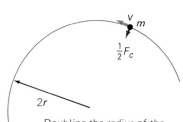

Doubling the radius of the
circle halves the needed
centripetal force

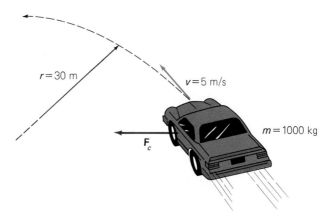

FIG. 2-25 A centripetal force of 833 N is needed by this car to make the turn shown.

$$F_c = \frac{mv^2}{r} = \frac{(1000 \text{ kg})(5 \text{ m/s})^2}{30 \text{ m}} = 833 \text{ N}$$

This force is easily transferred from the road to the car's tires if the road is dry and in good condition. However, if the car's speed were 20 m/s, the force needed would be 16 times as great, and the car would probably skid outward.

To reduce the chance of skids, particularly when the road is wet and therefore slippery, highway curves are often **banked** so that the roadbed tilts inward. A car going around a banked curve has an inward reaction force on it provided by the road itself, apart from friction.

2-13 NEWTON'S LAW OF GRAVITY

What Holds the Solar System Together In outline, here are the steps by which Newton arrived at the law of gravity that bears his name.

If we imagine that a planet's orbit is a circle, the centripetal force on the planet must originate in the sun, since the sun is at the center of the circle. This suggests that the force of gravity **F** between two objects acts along the line between them. Newton used Kepler's second law (Sec. 1-6) to show

A wall of snow provides this bobsled with the centripetal force it needs to round the turn.

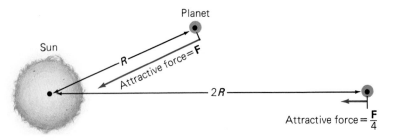

FIG. 2-26 The gravitational force between two bodies depends upon the square of the distance between them. The gravitational force on a planet would drop to one-fourth its usual amount if the distance of the planet from the sun were to be doubled. If the distance is halved, the force would increase to 4 times its usual amount.

that this conclusion holds even though the planetary orbits are actually ellipses, not circles.

Newton then combined Kepler's third law with the formula for centripetal force to find that F, the magnitude of the gravitational force **F**, varies inversely with the square of the distance R between planet and sun. That is, F varies as $1/R^2$. If a given planet were twice as far from the sun as the planet's normal distance, it would feel only $1/2^2 = \frac{1}{4}$ as much attractive force. If the planet were half as far away, it would feel a force $1/(\frac{1}{2})^2 = 1/\frac{1}{4} = 4$ times greater (Fig. 2-26). Kepler's first law supported this conclusion, since only a force of this kind can lead to a circular or elliptical orbit.

Galileo's work on falling bodies supplied the final clue. An object in free fall at the earth's surface has the acceleration g. Therefore the weight mg of the object, which is the force of gravity upon it, is always proportional to its mass m. Newton's third law of motion (action-reaction) requires that, if the earth attracts an object, the object also attracts the earth. If the earth's attraction for a stone depends upon the stone's mass, then the reaction force exerted by the stone on the earth depends upon the earth's mass. Hence the gravitational force between two bodies is proportional to *both* of their masses.

We can summarize the above conclusions in a single statement:

> **Every object in the universe attracts every other object with a force proportional to both of their masses and inversely proportional to the square of the distance between them.**

FIG. 2-27 For computing gravitational effects, spherical bodies (such as the earth and moon) may be regarded as though their masses are located at their geometrical centers, provided that they are uniform spheres or consist of concentric uniform spherical shells.

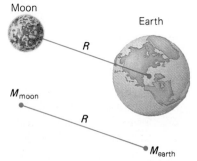

In equation form, **Newton's law of gravity** states that the force F that acts between two objects whose masses are m_1 and m_2 is

$$\text{Gravitational force} = F = \frac{Gm_1m_2}{R^2} \qquad \textit{law of gravity}$$

Here R is the distance between the objects and G is a constant of nature, the same number everywhere in the universe. The value of G is 6.670×10^{-11} N $\cdot$ m^2/kg^2.

The point in an object from which R is to be measured depends on the object's shape and on the way in which its mass is distributed. The **center of mass** of a uniform sphere is its geometric center (Fig. 2-27).

The inverse square—$1/R^2$—variation of gravitational force with distance R means that this force drops off rapidly with increasing R. Fig-

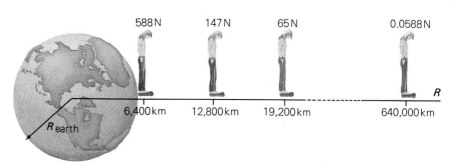

FIG. 2-28 The weight of a person near the earth is the gravitational force the earth exerts upon her. As she goes farther and farther away from the earth's surface, her weight decreases inversely as the square of her distance from the earth's center. The mass of the person here is 60 kg.

ure 2-28 shows how this variation affects the weight of a 60-kg astronaut who leaves the earth on a spacecraft. At the earth's surface she weighs 588 N (132 lb). When she is 100 times farther from the center of the earth, her weight is $1/100^2$ or $\frac{1}{10,000}$ as great, only 0.0588 N—the weight of a cigar on the earth's surface.

<table>
<tr><td>2-14</td><td></td></tr>
</table>

2-14 MASS OF THE EARTH

Weighing Our Planet On the basis of what we know already, we can find the mass of the earth. This sounds, perhaps, like a formidable job, but it is really fairly easy to do. It is worth following as an example of the indirect way in which scientists go about performing such seemingly impossible feats as "weighing" the earth, the sun, other planets, and even distant stars.

Let us focus our attention on an apple of mass m on the earth's surface. The downward force of gravity on the apple is its weight of mg:

$$\text{Weight of apple} = F = mg$$

We can also use Newton's law of gravity to find F, with the result

$$\text{Gravitational force on apple} = F = \frac{GmM}{R^2}$$

FIG. 2-29 The gravitational force of the earth on an apple at the earth's surface is the same as the force between masses M and m the distance R apart. This force equals the weight of the apple.

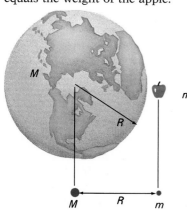

Here M is the earth's mass and R is the distance between the apple and the center of the earth, which is the earth's radius of 6400 km = 6.4×10^6 m (Fig. 2-29). The two ways to find F must give the same result, so

$$\text{Gravitational force on apple} = \text{weight of apple}$$

$$\frac{GmM}{R^2} = mg$$

We note that the apple's mass m appears on both sides of this equation, hence it cancels out. Solving for the earth's mass M gives

$$M = \frac{gR^2}{G} = \frac{(9.8 \text{ m/s}^2)(6.4 \times 10^6 \text{ m})^2}{6.67 \times 10^{-11} \text{ N·m}^2/\text{kg}^2} = 6 \times 10^{24} \text{ kg}$$

The number 6×10^{24} is 6 followed by 24 zeros! Enormous as it is, the earth is one of the least massive planets: Saturn has 95 times as much mass, and Jupiter 318 times as much. The sun's mass is more than 300,000 times that of the earth.

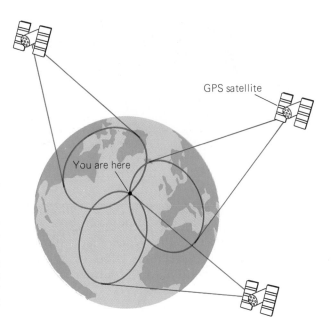

FIG. 2-30 In the Global Positioning System (GPS), each of a fleet of orbiting satellites sends out coded radio signals that enable a receiver on the earth to determine both the exact position of the satellite in space and its exact distance from the receiver. Given this information, a computer in the receiver then calculates the circle on the earth's surface on which the receiver must lie. Data from three satellites give three circles, and the receiver must be located at the one point where all three intersect.

2-15 ARTIFICIAL SATELLITES

Thousands Circle the Earth

The first artificial satellite, Sputnik I, was launched by the Soviet Union in 1957. Since then thousands of others have been put into orbits around the earth, most of them by the United States and the former Soviet Union. Men and women have been in orbit regularly since 1961, when a Soviet cosmonaut circled the globe at an average height of 240 km.

About 3000 satellites are now in orbits that range from 130 to 36,000 km above the earth. The closer satellites are "eyes in the sky" that survey the earth's surface, both for military purposes and to provide information on weather and earth resources such as mineral deposits, crops, and water. Twenty-four satellites at an altitude of 17,600 km are used in the new Global Positioning System (GPS) developed by the United States (Fig. 2-30). GPS receivers smaller than this book enable users to find their positions, including altitude, with uncertainties of as little as 15 m anywhere in the world at any time.

The most distant satellites circle the equator exactly once a day, so they remain in place indefinitely over a particular location on the earth. A satellite in such a geostationary orbit can "see" about a third of the earth's surface. About 200 of the satellites now in geostationary orbits are used to relay radio and television communications from one place to another, which is often cheaper than using cables between them.

The exact location of this GPS receiver is shown on its screen. The receiver processes radio signals sent by satellites that orbit the earth.

Why Satellites Don't Fall Down

What keeps all these satellites up there? The answer is that a satellite *is* actually falling down but, like the moon (which is a natural satellite), at exactly such a rate as to circle the earth in a stable orbit. "Stable" is a relative term, to be sure, since friction due to the extremely thin atmosphere present at the altitudes of actual satellites

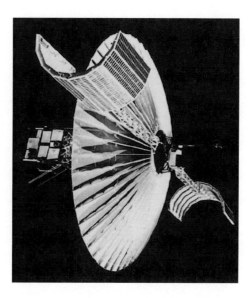

This Landsat satellite circles the earth at an altitude of 915 km. The satellite carries a television camera and a scanner system that provides images of the earth's surface in four color bands. The data radioed back provide information valuable in geology, water supply, agriculture, and land-use planning.

will eventually bring them down. Satellite lifetimes in orbit range from a matter of days to hundreds of years.

Let us think about a satellite in a circular orbit. The gravitational force on the satellite is its weight mg, where g is the acceleration of gravity at the satellite's altitude (the value of g decreases with altitude). The centripetal force a satellite of speed v needs to circle the earth at the distance r from the earth's center is mv^2/r. Since the earth's gravity is providing this centripetal force,

$$\text{Centripetal force} = \text{gravitational force}$$

$$\frac{mv^2}{r} = mg$$

$$v^2 = rg$$

$$\text{Satellite speed} = v = \sqrt{rg}$$

The mass of the satellite does not matter.

For an orbit a few kilometers above the earth's surface, the satellite speed turns out to be about 28,400 km/h. Anything sent off around the earth at this speed will become a satellite of the earth. (Of course, at such a low altitude air resistance will soon bring it down.) At a lower speed than this an object sent into space would simply fall to the earth, while at a higher speed it would have an elliptical rather than a circular orbit (Fig. 2-31). A satellite initially in an elliptical orbit can be given a circular orbit if it has a small rocket motor to give it a further push at the required distance from the earth.

If its original speed is high enough, at least 40,000 km/h, a space-craft can escape entirely from the earth. The speed required for something to leave the gravitational influence of an astronomical body permanently is called the **escape speed.** Readers of *Through the Looking Glass* may recall the Red Queen's remark, "Now, here, you see, it takes all the running you can do to stay in the same place. If you want to get

FIG. 2-31 The minimum speed an earth satellite can have is 28,400 km/h. The escape velocity from the earth is 40,000 km/h.

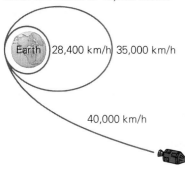

Earth 28,400 km/h 35,000 km/h

40,000 km/h

An earth satellite is always falling toward the earth. As a result, an astronaut inside feels "weightless," just as a person who jumps off a diving board feels "weightless." But a gravitational force does act on both people—what is missing is the upward reaction force of the ground, the diving board, the floor of a room, the seat of a chair, or whatever each person would otherwise be pressing on. In the case of an astronaut, the floor of the satellite falls just as fast as he or she does instead of pushing back.

somewhere else, you must run at least twice as fast as that!" The ratio between escape speed and minimum orbital speed is actually $\sqrt{2}$, about 1.41. Escape speeds for the planets are listed in Table 16-1.

IMPORTANT TERMS

To measure something means to compare it with a standard quantity of the same kind called a **unit.** The **SI** system of units is used everywhere by scientists and in most of the world in everyday life as well. The SI unit of length is the **meter** (m).

The **speed** of an object is the rate at which it covers distance. The object's **velocity** specifies both its speed and the direction in which it is moving. The **acceleration** of an object is the rate at which its speed changes. Changes in direction are also accelerations.

A **scalar quantity** has magnitude only; mass and speed are examples. A **vector quantity** has both magnitude and direction; force and velocity are examples. An arrowed line that represents the magnitude and direction of a quantity is called a **vector.**

The **acceleration of gravity** is the downward acceleration of a freely falling object near the earth's surface. Its value is $g = 9.8$ m/s^2.

The **inertia** of an object is the resistance the object offers to any change in its state of rest or motion. The property of matter that shows itself as inertia is called **mass;** mass may be thought of

as quantity of matter. The unit of mass is the **kilogram** (kg).

A **force** is any influence that can cause an object to be accelerated. The unit of force is the **newton** (N). The **weight** of an object is the gravitational force with which the earth attracts it.

Newton's **first law of motion** states that, if no net force acts on it, every object continues in its state of rest or uniform motion in a straight line. Newton's **second law of motion** states that when a net force F acts on an object of mass m, the object is given an acceleration of F/m in the same direction as that of the force. Newton's **third law of motion** states that when one object exerts a force on a second object, the second object exerts an equal but opposite force on the first. Thus for every **action force** there is an equal but opposite **reaction force.**

The **centripetal force** on an object moving along a curved path is the inward force needed to cause this motion. Centripetal force acts toward the center of curvature of the path.

Newton's **law of gravity** states that every object in the universe attracts every other object with a force directly proportional to both their masses and inversely proportional to the square of the distance separating them.

IMPORTANT FORMULAS

Speed: $v = \dfrac{d}{t}$

Acceleration: $a = \dfrac{v_f - v_i}{t}$

Second law of motion: $F = ma$

Weight: $w = mg$

Centripetal force: $F_c = \dfrac{mv^2}{r}$

Law of gravity: $F = \dfrac{Gm_1m_2}{R^2}$

EXERCISES: MULTIPLE CHOICE

1. Which of the following quantities is not a vector quantity?
   a. velocity
   b. acceleration
   c. mass
   d. force

2. Which of the following units could be associated with a vector quantity?
   a. kilogram
   b. hour
   c. liter
   d. meter/second

3. In which of the following examples is the motion of the car not accelerated?
   a. A car turns a corner at the constant speed of 20 km/h
   b. A car climbs a steep hill with its speed dropping from 60 km/h at the bottom to 15 km/h at the top
   c. A car climbs a steep hill at the constant speed of 40 km/h
   d. A car climbs a steep hill and goes over the crest and down on the other side, all at the same speed of 40 km/h

4. Two objects have the same size and shape but one of them is twice as heavy as the other. They are dropped simultaneously from a tower. If air resistance is negligible,
   a. the heavy object strikes the ground before the light one
   b. they strike the ground at the same time, but the heavy object has the higher speed
   c. they strike the ground at the same time and have the same speed
   d. they strike the ground at the same time, but the heavy object has the lower acceleration because it has more mass

5. The acceleration of a stone thrown upward is

   a. greater than that of a stone thrown downward
   b. the same as that of a stone thrown downward
   c. less than that of a stone thrown downward
   d. zero until it reaches the highest point in its path

6. When an object is accelerated,
   a. its direction never changes
   b. its speed always increases
   c. it always falls toward the earth
   d. a net force always acts on it

7. If we know the magnitude and direction of the net force on an object of known mass, Newton's second law of motion lets us find its
   a. position
   b. speed
   c. acceleration
   d. weight

8. The weight of an object
   a. is the quantity of matter it contains
   b. is the force with which it is attracted to the earth
   c. is basically the same quantity as its mass but is expressed in different units
   d. refers to its inertia

9. Compared with her mass and weight on the earth, an astronaut on Venus, where the acceleration of gravity is 8.8 m/s^2, has
   a. less mass and less weight
   b. less mass and the same weight
   c. less mass and more weight
   d. the same mass and less weight

10. In Newton's third law of motion, the action and reaction forces
    a. act on the same object
    b. act on different objects
    c. need not be equal in strength but must act in opposite directions
    d. must be equal in strength but need not act in opposite directions

11. A car that is towing a trailer is accelerating on a level road. The magnitude of the force the car exerts on the trailer is
    a. equal to the force the trailer exerts on the car
    b. greater than the force the trailer exerts on the car
    c. equal to the force the trailer exerts on the road
    d. equal to the force the road exerts on the trailer

12. When a boy pulls a cart, the force that causes him to move forward is
    a. the force the cart exerts on him
    b. the force he exerts on the cart

c. the force he exerts on the ground with his feet
d. the force the ground exerts on his feet

13. In order to cause something to move in a circular path, it is necessary to provide
    a. a reaction force
    b. an inertial force
    c. a centripetal force
    d. a gravitational force

14. A body moving in a circle at constant speed is accelerated
    a. in the direction of motion
    b. toward the center of the circle
    c. away from the center of the circle
    d. any of the above, depending upon the circumstances

15. A car rounds a curve on a level road. The centripetal force on the car is provided by
    a. inertia
    b. gravity
    c. friction between the tires and the road
    d. the force applied to the steering wheel

16. The centripetal force that keeps the earth in its orbit around the sun is provided
    a. by inertia
    b. by the earth's rotation on its axis
    c. partly by the gravitational pull of the sun
    d. entirely by the gravitational pull of the sun

17. The gravitational force with which the earth attracts the moon
    a. is less than the force with which the moon attracts the earth
    b. is the same as the force with which the moon attracts the earth
    c. is more than the force with which the moon attracts the earth
    d. varies with the phase of the moon

18. An astronaut inside an orbiting satellite feels weightless because
    a. he or she is wearing a space suit
    b. the satellite is falling toward the earth just as fast as the astronaut is, so there is no upward reaction force on him or her
    c. there is no gravitational pull from the earth so far away
    d. the sun's gravitational pull balances out the earth's gravitational pull

19. Of the following, the shortest is
    a. 1 mm   c. 0.001 m
    b. 0.01 in.   d. 0.001 ft

20. Of the following, the longest is

a. 1000 ft   c. 1 km
b. 500 m   d. 1 mi

21. A bicycle travels 12 km in 40 min. Its average speed is
    a. 0.3 km/h   c. 18 km/h
    b. 8 km/h   d. 48 km/h

22. Which one or more of the following sets of displacements might be able to return a car to its starting point?
    a. 5, 5, and 5 km   c. 5, 10, and 10 km
    b. 5, 5, and 10 km   d. 5, 5, and 20 km

23. An airplane whose airspeed is 200 km/h is flying in a wind of 80 km/h. The airplane's speed relative to the ground is between
    a. 80 and 200 km/h   c. 120 and 200 km/h
    b. 80 and 280 km/h   d. 120 and 280 km/h

24. How long does a car whose acceleration is 2 m/s^2 need to go from 10 m/s to 30 m/s?
    a. 10 s   c. 40 s
    b. 20 s   d. 400 s

25. A ball is thrown upward at a speed of 12 m/s. It will reach the top of its path in about
    a. 0.6 s   c. 1.8 s
    b. 1.2 s   d. 2.4 s

26. When a net force of 1 N acts on a 1-kg body, the body receives
    a. a speed of 1 m/s
    b. an acceleration of 0.1 m/s^2
    c. an acceleration of 1 m/s^2
    d. an acceleration of 9.8 m/s^2

27. When a net force of 1 N acts on a 1-N body, the body receives
    a. a speed of 1 m/s
    b. an acceleration of 0.1 m/s^2
    c. an acceleration of 1 m/s^2
    d. an acceleration of 9.8 m/s^2

28. A car whose mass is 1600 kg (including the driver) has a maximum acceleration of 1.2 m/s^2. If three 80-kg passengers are also in the car, its maximum acceleration will be
    a. 0.5 m/s^2 c. 1.04 m/s^2
    b. 0.72 m/s^2 d. 1.2 m/s^2

29. A 300-g ball is struck with a bat with a force of 150 N. If the bat was in contact with the ball for 0.020 s, the ball's velocity is
    a. 0.01 m/s   c. 2.5 m/s
    b. 0.1 m/s   d. 10 m/s

30. A bicycle and its rider together have a mass of 80

kg. If the bicycle's speed is 6 m/s, the force needed to bring it to a stop in 4 s is
a. 12 N   c. 120 N
b. 53 N   d. 1176 N

31. The weight of 400 g of onions is
    a. 0.041 N   c. 3.9 N
    b. 0.4 N   d. 3920 N

32. A salami weighs 3 lb. Its mass is
    a. 0.31 kg   c. 6.6 kg
    b. 1.36 kg   d. 29.4 kg

33. The radius of the circle in which an object is moving at constant speed is doubled. The required centripetal force is
    a. one-quarter as great as before
    b. one-half as great as before
    c. twice as great as before
    d. 4 times as great as before

34. A car rounds a curve at 20 km/h. If it rounds the curve at 40 km/h, its tendency to overturn is
    a. halved   c. tripled
    b. doubled   d. quadrupled

35. A 1200-kg car whose speed is 6 m/s rounds a turn whose radius is 30 m. The centripetal force on the car is
    a. 48 N   c. 240 N
    b. 147 N   d. 1440 N

36. If the earth were 3 times as far from the sun as it is now, the gravitational force exerted on it by the sun would be
    a. 3 times as large as it is now
    b. 9 times as large as it is now
    c. one-third as large as it is now
    d. one-ninth as large as it is now

37. A woman whose mass is 60 kg on the earth's surface is in a spacecraft at an altitude of one earth's radius above the surface. Her mass there is
    a. 15 kg   c. 60 kg
    b. 30 kg   d. 120 kg

38. A man whose weight is 800 N on the earth's surface is also in the spacecraft of Question 37. His weight there is
    a. 200 N   c. 800 N
    b. 400 N   d. 1600 N

QUESTIONS

1. In the following pairs of length units, which is the shortest: Inch and centimeter? Yard and meter? Mile and kilometer?

2. What kind of quantity is the magnitude of a vector quantity?

3. Is it correct to say that scalar quantities are abstract, idealized quantities with no precise counterparts in the physical world, whereas vector quantities properly represent reality because they take directions into account?

4. Can a rapidly moving object have the same acceleration as a slowly moving one?

5. What happens to the speed of an object as it falls freely? To its acceleration?

6. A rifle is aimed directly at a squirrel in a tree. Should the squirrel drop from the tree at the instant the rifle is fired or should it remain where it is? Why?

7. Suppose you are in a barrel going over Niagara Falls and during the fall you drop an apple inside the barrel. Would the apple appear to move toward the top of the barrel or toward the bottom, or would it remain stationary within the barrel?

8. You throw a ball upward from a roof at the same time as a friend drops another ball from there. How do the speeds of the balls compare when they reach the ground? How do their accelerations compare?

9. A person in a stationary elevator drops a coin and the coin reaches the floor of the elevator 0.6 s later. Would the coin reach the floor in less time, the same time, or more time if it were dropped when the elevator was (a) falling at a constant speed? (b) falling at a constant acceleration? (c) rising at a constant speed? (d) rising at a constant acceleration?

10. When a football is thrown, it follows a curved path through the air like the ones shown in Fig. 2-11. If there were no air resistance, where in its path would the ball's speed be greatest? Where would it be least?

11. Every acceleration is the result of a force. Does every applied force produce an acceleration?

12. What, if anything, is wrong with saying that a certain person weighs 70 kg?

13. Distinguish between mass and weight.

14. Are Newton's first and second laws of motion related in any way? His second and third laws?

15. Since the opposite forces of the third law of

motion are equal in magnitude, how can anything ever be accelerated?

16. When you whirl a ball at the end of a string, the ball seems to be pulling outward away from your hand. When you let the string go, however, the ball moves along a straight path perpendicular to the direction of the string at the moment you let go. Explain each of these effects.

17. A book rests on a table. What is the reaction force to the force the book exerts on the table? To the force gravity exerts on the book?

18. Two children wish to break a string. Are they more likely to succeed if each takes one end of the string and they pull against each other, or if they tie one end of the string to a tree and both pull on the free end? Why?

19. An engineer designs a propeller-driven spacecraft. Because there is no air in space, the engineer includes a supply of oxygen as well as a supply of fuel for the motor. What do you think of the idea?

20. If you were set down in the center of a frozen lake whose surface was so smooth and slippery that it offered no frictional resistance, how could you get off the lake?

21. Under what circumstances, if any, can something move in a circular path without a centripetal force acting on it?

22. A car makes a clockwise turn on a level road at too high a speed and overturns. Do its left or right wheels leave the ground first?

23. A track team on the moon could set new records for the high jump or pole vault (if they did not need space suits, of course) because of the smaller gravitational force. Could sprinters also improve their times for the 100-m dash?

24. If the moon were half as far from the earth as it is today, how would the gravitational force it exerts on the earth compare with the force it exerts today?

25. Compare the weight and mass of an object at the earth's surface with what they would be at an altitude of two earth's radii.

26. A hole is bored to the center of the earth and a stone is dropped into it. How do the mass and weight of the stone at the earth's center compare with their values at the earth's surface?

27. Is the sun's gravitational pull on the earth the same at all seasons of the year? Explain.

28. According to the theory of gravitation, the earth must be continually "falling" toward the sun. If this is true, why does the average distance between earth and sun not grow smaller?

29. According to Kepler's second law, the earth travels fastest when it is closest to the sun. Is this consistent with the law of gravitation? Explain.

30. A "geostationary" satellite does not appear to move across the sky but remains over the same place on the earth's equator. Why doesn't it fall to the ground?

31. An artificial satellite is placed in orbit half as far from the earth as the moon is. Would the satellite's time of revolution around the earth be longer or shorter than the moon's if its orbit is to be stable?

32. Is an astronaut in an orbiting spacecraft actually "weightless"?

33. An airplane makes a vertical circle in which it is upside down at the top of the loop. Will the passengers fall out of their seats if there is no belt to hold them in place?

PROBLEMS

1. The distance from Paris to Brussels is 291 km. How many meters is this? How many miles?

2. The Empire State Building in New York City is 450 m high. How high is this in feet? In kilometers?

3. The speedometer of a European car gives its speed in kilometers per hour. What is the car's speed in miles per hour when the speedometer reads 80?

4. You are watching a batter hit a pitched baseball. If $\frac{1}{3}$ s passes between the time you see him strike the ball and the time you hear the sound, how far from the batter are you sitting? The speed of sound in air at ordinary temperatures is 343 m/s. (The speed of light is nearly a million times greater than this, so you see the impact of the bat on the ball almost at the instant it occurs.)

5. If a lightning flash occurs 1.5 km from you, how long will it take before you hear the thunder?

6. A pitcher throws a baseball at 40 m/s. How much time does the ball take to reach the batter 18 m away?

7. In 1977 Steve Weldon ate 91 m of spaghetti in 29

s. At the same speed, how long would it take Mr. Weldon to eat 5 m of spaghetti?

8. A motorist drives for 11 h per day and averages 80 km/h. How many days will she need to cover the 3700-km distance by road between Chicago and San Francisco?

9. A woman walks 70 m to an elevator and then rises upward 40 m. What is her displacement from her starting point?

10. In going from one city to another, a car whose driver tends to get lost goes 30 km north, 50 km west, and 20 km southeast. Approximately how far apart are the cities?

11. A DC-8 airplane reaches its takeoff speed of 78 m/s in a time of 35 s after starting from rest. What is its average acceleration?

12. (a) Find the acceleration of a car whose speed increases from 8 m/s to 20 m/s in 10 s. (b) Find the acceleration of the car when its speed decreases from 20 m/s to 10 m/s in 5 s.

13. A car whose acceleration is constant reaches a speed of 80 km/h in 20 s starting from rest. How much more time is required for it to reach a speed of 130 km/h?

14. The tires of a car begin to lose their grip on the road at an acceleration of 5 m/s^2. At this acceleration, how long does the car need to reach a speed of 25 m/s starting from 10 m/s?

15. The brakes of a car moving at 14 m/s are applied, and the car comes to a stop in 4 s. (a) What was the car's acceleration? (b) How long would the car take to come to a stop starting from 20 m/s with the same acceleration? (c) How long would the car take to slow down from 20 m/s to 10 m/s with the same acceleration?

16. A ball is thrown upward at 12 m/s. What is its speed 1.0 s later?

17. A ball is thrown downward at 12 m/s. What is its speed 1.0 s later?

18. The air resistance experienced by a falling object becomes important when its speed becomes about half its terminal speed. The terminal speed of a golf ball is 40 m/s. How much time is needed for a dropped golf ball to reach a speed of half this?

19. A ball is thrown vertically upward with an initial speed of 30 m/s. How long will it take the ball to reach the highest point in its path? How long

will it take the ball to return to its starting place? What will the ball's speed be there?

20. A 430-g soccer ball lying on the ground is kicked and flies off at 25 m/s. If the impact on the ball lasted 0.01 s, what was the average force that acted on it?

21. A bicycle and its rider together have a mass of 80 kg. If the bicycle's speed is 6 m/s, how much force is needed to bring it to a stop in 4 s?

22. An empty truck whose mass is 2000 kg has a maximum acceleration of 1 m/s^2. What is its maximum acceleration when it is carrying a load of 1000 kg?

23. A force of 20 N gives a brick an acceleration of 5 m/s^2. (a) What force would be needed to give the brick an acceleration of 1 m/s^2? (b) An acceleration of 10 m/s^2?

24. A 1200-kg car goes from 10 m/s to 24 m/s in 16 s. What is the average force acting upon it?

25. The brakes of the above car exert a force of 4000 N. How long will it take for them to slow the car to a stop from an initial speed of 24 m/s?

26. What is the weight in newtons of a 75-kg person? How many pounds is this?

27. What is the mass of a 700-N man? With how much force is he attracted to the earth? If he falls from a cliff, what will his downward acceleration be?

28. An 80-kg man slides down a rope at constant speed. (a) What is the minimum breaking strength the rope must have? (b) If the rope has precisely this strength, will it support the man if he tries to climb back up?

29. A 60-g tennis ball approaches a racket at 15 m/s, is in contact with the racket for 0.005 s, and then rebounds at 20 m/s. What was the average force the racket exerted on the ball?

30. A 12,000-kg airplane launched by a catapult from an aircraft carrier is accelerated from 0 to 200 km/h in 3 s. (a) How many times the acceleration of gravity is the airplane's acceleration? (b) What is the average force the catapult exerts on the airplane?

31. A woman whose mass is 60 kg is riding in an elevator whose upward acceleration is 2 m/s^2. What force does she exert on the floor of the elevator?

32. The cable supporting a 1500-kg elevator can safely provide a force of 18,000 N. What is the maximum upward acceleration the elevator can have if the force the cable provides is not to exceed this figure?

33. What centripetal force is needed to keep a 1-kg ball moving in a circle of radius 2 m at a speed of 5 m/s?

34. The 200-g head of a golf club moves at 40 m/s in a circular arc of 1.2 m radius. How much force must the player exert on the handle of the club to prevent it from flying out of her hands at the bottom of the swing? Ignore the mass of the club's shaft.

35. The string of a certain yo-yo is 80 cm long and will break when the force on it is 10 N. What is the highest speed the 200-g yo-yo can have when it is being whirled in a circle? Ignore the gravitational pull of the earth on the yo-yo.

36. The maximum force a road can exert on the tires of a certain 1200-kg car is 8000 N. What is the greatest speed with which the car can round a turn of radius 200 m?

37. A 40-kg crate is lying on the flat floor of the rear of a station wagon moving at 15 m/s. A force of 150 N is needed to slide the crate against the friction between the bottom of the crate and the floor. What is the minimum radius of a turn the station wagon can make if the box is not to slip?

38. Find the minimum radius at which an airplane flying at 300 m/s can make a U-turn if the centripetal force on it is not to exceed 4 times the airplane's weight.

39. A road has a hump 12 m in radius. What is the minimum speed at which a car will leave the road at the top of the hump?

40. A 2-kg mass is 20 cm away from a 5-kg mass. What is the gravitational force (a) that the 5-kg mass exerts on the 2-kg mass, and (b) that the 2-kg mass exerts on the 5-kg mass? (c) If both masses are free to move, what are their respective accelerations if no other forces are acting?

41. A dishonest grocer installs a 100-kg lead block under the pan of his scale. How much gravitational force does the lead exert on 2 kg of cheese placed on the pan if the centers of mass of the lead and cheese are 0.3 m apart? Compare this force with the weight of 1 g of cheese to see if putting the lead under the scale was worth doing.

42. A bull and a cow elephant, each of mass 2000 kg, attract each other gravitationally with a force of 1×10^{-5} N. How far apart are they?

43. With the help of the data in Table 16-1, find the minimum speed artificial satellites must have to pursue stable orbits about Jupiter.

ANSWERS TO MULTIPLE CHOICE

1. c	11. a	21. c	31. c
2. d	12. d	22. a, b, c	32. b
3. c	13. c	23. d	33. b
4. c	14. b	24. a	34. d
5. b	15. c	25. b	35. d
6. d	16. d	26. c	36. d
7. c	17. b	27. d	37. c
8. b	18. b	28. c	38. a
9. d	19. b	29. d	
10. b	20. d	30. c	

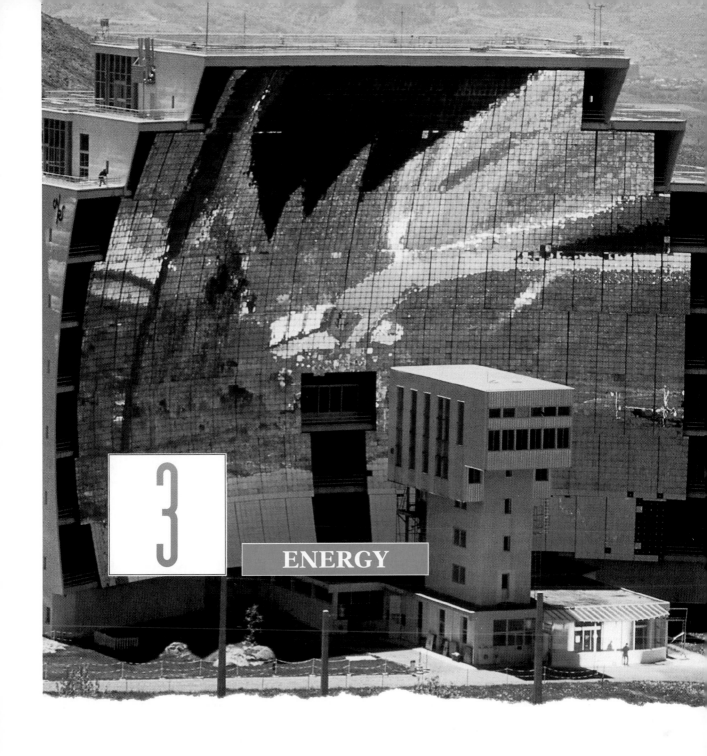

3

ENERGY

The word energy has become part of everyday life. We say that an active person is energetic. We hear a candy bar described as being full of energy. We complain about the cost of the electric energy that lights our lamps and turns our motors. We worry about some day running out of the energy stored in coal and oil. We argue about whether nuclear energy is a blessing or a curse. Exactly what is meant by energy?

In general, energy refers to an ability to accomplish change. When almost anything happens in the physical world, energy is somehow involved. But "change" is not a very precise notion, and we must be sure of exactly what we are talking about in order to go further. Our procedure will be to begin with the simpler idea of work and then use it to relate change and energy in the orderly way of science.

Changes that take place in the physical world are the result of forces. Forces are needed to pick things up, to move things from one place to another, to squeeze things, to stretch things, and so on. However, not all forces produce changes, and it is the distinction between forces that accomplish change and forces that do not that is central to the idea of work.

3-1 THE MEANING OF WORK

FIG. 3-1 Work is done by a force when the object it acts on moves while the force is applied. No work is done by pushing against a stationary wall. Work is done when throwing a ball because the ball moves while being pushed during the throw.

No work done

Work done

A Measure of the Change a Force Produces

Suppose we push against a wall. When we stop, nothing has happened even though we exerted a force on the wall. But if we apply the same force to a stone, the stone flies through the air when we let it go (Fig. 3-1). The difference is that the wall did not move during our push but the stone did. A physicist would say that we have done work on the stone, and as a result it was accelerated and moved away from our hand.

Or we might try to lift a heavy barbell. If we fail, the world is exactly the same afterward. If we succeed, though, the barbell is now up in the air, which represents a change. As before, the difference is that in the second case an object moved while we exerted a force on it, which means that work was done on the object.

To make our ideas definite, **work** is defined in this way:

> The work done by a force acting on an object is equal to the magnitude of the force multiplied by the distance through which the force acts.

If nothing moves, no work is done, no matter how great the force. And even if something moves, work is not done on it unless a force is acting on it.

What we usually think of as work agrees with this definition. However, we must be careful not to confuse becoming tired with the amount of work done. Pushing against a wall for an afternoon in the hot sun is certainly tiring, but we have done no work because the wall didn't move.

In equation form,

$W = Fd = \text{work}$
$= NM = \text{joule}$

Work is done when a barbell is lifted, but no work is done while it is being held in the air even though this can be very tiring.

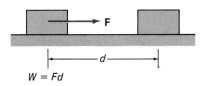

$W = Fd$

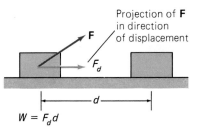

Projection of **F** in direction of displacement

$W = F_d d$

FIG. 3-2 When a force and the distance through which it acts are parallel, the work done is equal to the product of F and d. When they are not in the same direction, the work done is equal to the product of d and the projection of **F** in the direction of **d**.

$$W = Fd \qquad work$$

Work done = (applied force)(distance through which force acts)

The direction of the force **F** is assumed to be the same as the direction of the distance **d.** If not, for example in the case of a child pulling a wagon with a rope not parallel to the ground, we must use for F the projection of the applied force **F** that acts in the direction of motion (Fig. 3-2).

A force that is perpendicular to the direction of motion of an object can do no work on the object. Thus gravity, which results in a downward force on everything near the earth, does no work on objects moving horizontally along the earth's surface. However, if we drop an object, work is definitely done on it as it falls to the ground.

The Joule   The SI unit of work is the **joule** (J), where one joule is the amount of work done by a force of one newton when it acts through a distance of one meter. That is,

1 joule (J) = 1 newton-meter (N · m)

The joule is named after the English scientist James Joule and is pronounced "jool." To raise an apple from your waist to your mouth takes about 1 J of work.

If we push a box for 8 m across a floor with a force of 100 N (22.5 lb), the work we perform is $W = Fd = (100\text{ N})(8\text{ m}) = 800\text{ J}$. The mass of the box does not matter in this case; what counts is the applied force and the distance through which the box moves. However, if we raise the box, the force we must use is the box's weight of $w = mg$, and so the work we do in the lifting does depend on the box's mass.

3-2   **POWER**

The Rate of   The time needed to carry out a job is often as impor-
Doing Work   tant as the amount of work needed. If we have enough
                   time, even the tiny motor of a toy train can lift an ele-
vator as high as we like. However, if we want the elevator to take us up

THE
HORSEPOWER

THE
HORSEPOWER

The **horsepower** (hp) is the traditional unit of power in engineering.
The origin of this unit is interesting.
In order to sell the steam engines he had perfected two
centuries ago, James Watt had to compare their power
outputs with that of a horse, a source of work his cus-
tomers were familiar with. After various tests he found
that a typical horse could perform work at a rate of 497
W for as much as 10 hours per day. To avoid any disputes, Watt increased this figure by one-half to establish

the unit he called the horsepower. Watt's horsepower
therefore represents a rate of doing work of 746 W:

1 horsepower (hp) = 746 W = 0.746 kW

1 kilowatt (kW) = 1.34 hp

Few horses can develop this much power for very long.
The early steam engines ranged from 4 to 100 hp, with
the 20-hp model being the most popular.

fairly quickly, we must use a motor whose output of work is rapid in
terms of the total work needed. Thus the rate at which work is being
done is significant. This rate is called **power:** The more powerful something is, the faster it can do work.

If the amount of work W is done in a period of time t, the power
involved is

$$P = \frac{W}{t} \quad power$$

$$\text{Power} = \frac{\text{work done}}{\text{time interval}}$$

The SI unit of power is the **watt** (W), where

1 watt (W) = 1 joule/second (J/s)

Thus a motor with a power output of 500 W is capable of doing 500 J of
work per second. The same motor can do 250 J of work in 0.5 s, 1000 J
of work in 2 s, 5000 J of work in 10 s, and so on. The watt is quite a
small unit, and often the **kilowatt** (kW) is used instead, where 1 kW =
1000 W.

A person in good physical condition is usually capable of a continuous power output of about 75 W, which is 0.1 horsepower. A runner or
swimmer during a distance event may have a power output 2 or 3 times
greater. What limits the power output of a trained athlete is not muscular development but the supply of oxygen from the lungs through the
bloodstream to the muscles, where oxygen is used in the metabolic
processes that enable the muscles to do work. However, for a period of
less than a second, an athlete's power output may exceed 5 kW, which
accounts for the feats of weightlifters and jumpers.

ENERGY

We now go from the straightforward idea of work to the complex and
many-sided idea of **energy:**

Energy is that property something has that enables it to do work.

When we say that something has energy, we mean it is able, directly or
indirectly, to exert a force on something else and perform work. When

work is done on something, energy is added to it. Energy is measured in the same unit as work, the joule.

3-3   KINETIC ENERGY

The Energy of Position   Energy occurs in several forms. One of them is the energy a moving object has because of its motion. Every moving object has the capacity to do work. By striking something else, the moving object can exert a force and cause the second object to shift its position, to break apart, or to otherwise show the effects of having work done on it. It is this property that defines energy, so we conclude that all moving things have energy by virtue of their motion. The energy of a moving object is called **kinetic energy** (KE). ("Kinetic" is a word of Greek origin that suggests motion is involved.)

The kinetic energy of a moving thing depends upon its mass and its speed. The greater the mass and the greater the speed, the more the KE. A train going at 30 km/h has more energy than a horse galloping at the same speed and more energy than a similar train going at 10 km/h. The exact way KE varies with mass m and speed v is given by the formula

$$KE = \tfrac{1}{2}mv^2 \qquad \textit{kinetic energy}$$

The v^2 factor means the kinetic energy increases very rapidly with increasing speed. At 30 m/s a car has 9 times as much KE as at 10 m/s—and requires 9 times as much force to bring to a stop in the same distance (Fig. 3-3). The fact that KE, and hence the ability to do work (in this case, damage), depends upon the square of the speed is what is responsible for the severity of automobile accidents at high speeds. The variation of KE with mass is less marked: a 2000-kg car going at 10 m/s has just twice the KE of a 1000-kg car with the same speed.

The kinetic energy of a 1000-kg car whose speed is 10 m/s is

$$KE = \tfrac{1}{2}mv^2 = (\tfrac{1}{2})(1000 \text{ kg})(10 \text{ m/s})^2$$

$$= (\tfrac{1}{2})(1000 \text{ kg})(10 \text{ m/s})(10 \text{ m/s}) = 50,000 \text{ J} = 50 \text{ kJ}$$

In order to bring the car to this speed from rest, 50 kJ of work had to be done by its engine. To stop the car from this speed, the same amount of work must be done by its brakes.

FIG. 3-3   Kinetic energy is proportional to the square of the speed. A car traveling at 30 m/s has 9 times the KE of the same car traveling at 10 m/s.

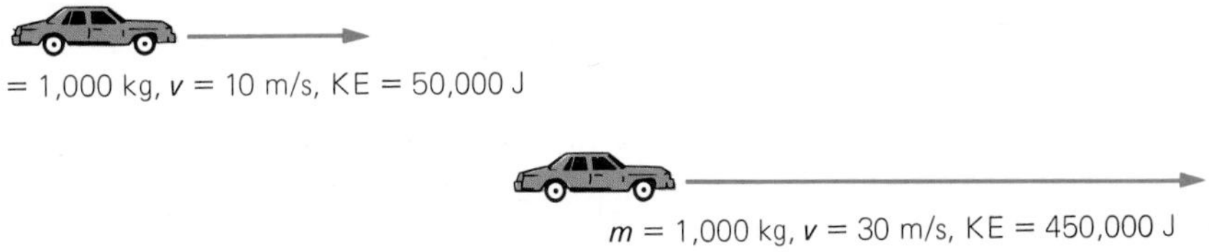

$m = 1,000$ kg, $v = 10$ m/s, KE = 50,000 J

$m = 1,000$ kg, $v = 30$ m/s, KE = 450,000 J

FORCE ON A NAIL

Have you ever wondered how much force a hammer exerts on a nail? Suppose you hit a nail with a hammer and drive the nail 5 mm into a wooden board (Fig. 3-4). If the hammer's head has a mass of 0.6 kg and it is moving at 4 m/s when it strikes the nail, what is the average force on the nail?

The KE of the hammer head is $\frac{1}{2}mv^2$, and this amount of energy becomes the work Fd done in driving the nail the distance $d = 5$ mm = 0.005 m into the board. Hence

$$\text{KE of hammer head} = \text{work done on nail}$$

$$\frac{1}{2}mv^2 = Fd$$

and

$$F = \frac{mv^2}{2d} = \frac{(0.6 \text{ kg})(4 \text{ m/s})^2}{2(0.005 \text{ m})} = 960 \text{ N}$$

This is 216 lb—watch your fingers!

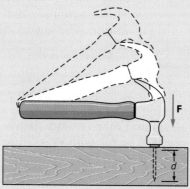

FIG. 3-4 When a hammer strikes this nail, the hammer's kinetic energy is converted into the work done to push the nail into the wooden board.

3-4 POTENTIAL ENERGY

The Energy of Motion

When we drop a stone, it falls faster and faster and finally strikes the ground. If we lift the stone afterward, we see that it has done work by making a shallow hole in the ground (Fig. 3-5). In its original raised position, the stone must have had the capacity to do work even though it was not moving at the time and therefore had no KE.

The amount of work the stone could do by falling to the ground is called its **potential energy** (PE). Just as kinetic energy may be thought

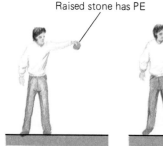
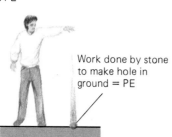

FIG. 3-5 A raised stone has potential energy because it can do work on the ground when dropped.

Raised stone has PE

Work done by stone to make hole in ground = PE

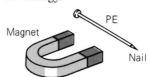

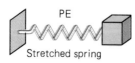

FIG. 3-6 Two examples of potential energy.

Magnet   PE   Nail   No PE

PE   Stretched spring   No PE   Normal spring

of as energy of motion, potential energy may be thought of as energy of position.

Examples of potential energy are everywhere. A book on a table has PE since it can fall to the floor. A skier at the top of a slope, water at the top of a waterfall, a car at the top of the hill, anything able to move toward the earth under the influence of gravity has PE because of its position. Nor is the earth's gravity necessary: a stretched spring has PE since it can do work when it is let go, and a nail near a magnet has PE since it can do work in moving to the magnet (Fig. 3-6).

Gravitational Potential Energy

It is easy to find a formula for the gravitational PE an object has near the earth's surface. The work W needed to raise an object of mass m to a height h above its original position, as in Fig. 3-7, is

$$W = Fd = mgh$$

Work = (force)(distance) = (weight)(height)

since the force needed is the object's weight of mg. This work is equal to the PE of the raised object. Hence the PE at the height h is

$$\text{Potential energy} = \text{PE} = mgh$$

This result for PE agrees with our experience. Consider a pile driver, a simple machine that lifts a heavy weight (the "hammer") and allows it to fall on the head of a pile, which is a wooden or steel post, to drive the pile into the ground. From the formula PE = mgh we would expect the effectiveness of a pile driver to depend on the mass m of its hammer and the height h from which it is dropped, which is exactly what experience shows.

The PE of a 1000-kg car when it is at the top of a 45-m cliff is

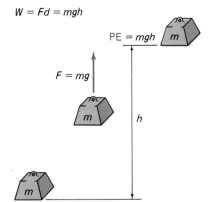

$W = Fd = mgh$

PE = mgh m

$F = mg$

m

h

m

FIG. 3-7 The increase in the potential energy of a raised object is equal to the work used to lift it.

In the operation of a pile driver, the gravitational potential energy of the raised hammer becomes kinetic energy as it falls. The kinetic energy in turn becomes work as the pile is pushed into the ground.

$$PE = mgh = (1000 \text{ kg})(9.8 \text{ m/s}^2)(45 \text{ m}) = 441{,}000 \text{ J} = 441 \text{ kJ}$$

This is less than the KE of the same car when it moves at 30 m/s (Fig. 3-3). Thus a crash at 30 m/s into a wall or tree will yield more work— that is, do more damage—than dropping the car from a cliff 45 m high.

PE Is Relative   It is worth noting that the gravitational PE of an object depends on the level from which it is reckoned. Often the earth's surface is convenient, but sometimes other references are more appropriate.

Suppose you lift this book as high as you can above the table while remaining seated. It will then have a PE *relative to the table* of about 12 J. But the book will have a PE *relative to the floor* of about twice that, or 24 J. And if the floor of your room is, say, 50 m above the ground, the book's PE *relative to the ground* will be about 760 J.

What is the book's true PE? The answer is that there is no such thing as "true" PE. Gravitational PE is a relative quantity. However, the *difference* between the PEs of an object at two points *is* significant, since it is this difference that can be changed into work or KE.

3-5   ENERGY TRANSFORMATIONS

Easy Come, Easy Go   Nearly all familiar mechanical processes involve interchanges among KE, PE, and work. Thus when the car of Fig. 3-8 is driven to the top of a hill, its engine must do work in order to raise the car. At the top, the car has an amount of PE equal to the work done in getting it up there (neglecting friction). If the engine is turned off, the car can still coast down the hill, and its KE at the bottom of the hill will be the same as its PE at the top.

Changes of a similar nature, from kinetic energy to potential and back, occur in the motion of a planet in its orbit around the sun (Fig. 3-9) and in the motion of a pendulum (Fig. 3-10). The orbits of the planets are ellipses with the sun at one focus (Fig. 1-7), and each planet is therefore at a constantly varying distance from the sun. At all times the total of its potential and kinetic energies remains the same. When close to the sun, the PE of a planet is low and its KE is high. The additional speed due to increased KE keeps the planet from being pulled into the sun by the greater gravitational force on it at this point in its path. When the planet is far from the sun, its PE is higher and its KE lower, with the reduced speed exactly keeping pace with the reduced gravitational force.

A pendulum (Fig. 3-10) consists of a ball suspended by a string. When the ball is pulled to one side with its string taut and then released, it swings back and forth. When it is released, the ball has a PE relative to the bottom of its path of *mgh*. At its lowest point all this PE has become kinetic energy $\frac{1}{2}mv^2$. After reaching the bottom, the ball continues in its motion until it rises to the same height h on the opposite side from its initial position. Then, momentarily at rest since all its KE is now PE, the ball begins to retrace its path back through the bottom to its initial position.

FIG. 3-8 In the absence of friction, a car can coast from the top of one hill into a valley and then up to the top of another hill of the same height as the first. During the trip the initial potential energy of the car is converted into kinetic energy as the car goes downhill, and this kinetic energy then turns into potential energy as the car climbs the next hill. The total amount of energy (KE + PE) remains unchanged.

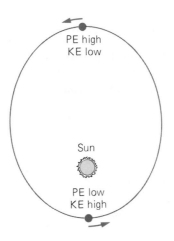

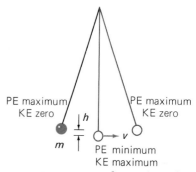

FIG. 3-9 Energy transformations in planetary motion. The total energy (KE + PE) of the planet is the same at all points in its orbit. (Planetary orbits are much more nearly circular than shown here.)

FIG. 3-10 Energy transformations in pendulum motion. The total energy of the ball stays the same but is continuously exchanged between kinetic and potential forms.

Other Forms of Energy Energy can exist in a variety of forms besides kinetic and potential. The *chemical energy* of gasoline is used to propel our cars and the chemical energy of food enables our bodies to perform work. *Heat energy* from burning coal or oil is used to form the steam that drives the turbines of power stations. *Electric energy* turns motors in home and factory. *Radiant energy* from the sun performs work in lifting water from the earth's surface into clouds, in producing differences in air temperature that cause winds, and in promoting chemical reactions in plants that produce foods.

Just as kinetic energy can be converted to potential energy and potential to kinetic, so other forms of energy can readily be transformed. In the cylinders of a car engine, for example, chemical energy stored in gasoline and air is changed first to heat energy when the mixture is ignited by the spark plugs, then to kinetic energy as the expanding gases push down on the pistons. This kinetic energy is in large part transmitted to the wheels, but some is used to turn the generator and thus produce electric energy for charging the battery, and some is changed to heat by friction in bearings. Energy transformations go on constantly, all about us.

The elastic potential energy of the bent bow becomes kinetic energy of the arrow when the bowstring is released.

The potential energy of these skiers at the top of the slope turns into kinetic energy and eventually into heat as they slide downhill.

3-6 CONSERVATION OF ENERGY

A Funda-mental Law of Nature

A skier slides down a hill and comes to rest at the bottom. What became of the potential energy he or she had at the top? The engine of a car is shut off while the car is allowed to coast along a level road. Eventually the car slows down and comes to a stop. What became of its original kinetic energy?

All of us can give similar examples of the apparent disappearance of kinetic or potential energy. What these examples have in common is that heat is always produced in an amount just equivalent to the "lost" energy. One kind of energy is simply being converted to another; no energy is lost, nor is any new energy created. Exactly the same is true when electric, magnetic, radiant, and chemical energies are changed into one another or into heat. Thus we have a law from which no deviations have ever been found:

Energy cannot be created or destroyed, although it can be changed from one form to another.

This generalization is the **law of conservation of energy.** It is the principle with the widest application in science, applying equally to distant stars and to biological processes in living cells.

We shall learn later in this chapter that matter can be transformed into energy and energy into matter. The law of conservation of energy still applies, however, with matter considered as a form of energy.

3-7   THE NATURE OF HEAT

Count Rumford (1753–1814)

James Prescott Joule (1818–1889)

The Downfall of Caloric

Although it comes as little surprise to us today to learn that heat is a form of energy, in earlier times this was not so clear. Less than two centuries ago most scientists regarded heat as an actual substance called **caloric.** Absorbing caloric caused an object to become warmer; the escape of caloric caused it to become cooler. Because the weight of an object does not change when the object is heated or cooled, caloric was considered to be weightless. It was also supposed to be invisible, odorless, and tasteless, properties that, of course, were why it could not be observed directly.

Actually, the idea of heat as a substance was fairly satisfactory for materials heated over a flame, but it could not account for the unlimited heat that could be generated by friction. One of the first to appreciate this difficulty was the American Benjamin Thompson, who had supported the British during the Revolutionary War and thought it wise to move to Europe afterward, where he became Count Rumford.

One of Rumford's many occupations was supervising the making of cannon for a German prince, and he was impressed by the large amounts of heat given off by friction in the boring process. He showed that the heat could be used to boil water and that heat could be produced again and again from the same piece of metal. If heat was a fluid, it was not unreasonable that boring a hole in a piece of metal should allow it to escape. However, even a dull drill that cut no metal produced a great deal of heat. Also, it was hard to imagine a piece of metal as containing an infinite amount of caloric, and Rumford accordingly regarded heat as a form of energy.

James Prescott Joule was an English brewer who performed a classic experiment that settled the nature of heat once and for all. Joule's experiment used a small paddle wheel inside a container of water (Fig. 3-11). Work was done to turn the paddle wheel against the resistance of the water, and Joule measured exactly how much heat was supplied to the water by friction in this process. He found that a given amount of work always produced exactly the same amount of heat. This was a clear demonstration that heat is energy and not something else.

Joule also carried out chemical and electrical experiments that agreed with his mechanical ones, and the result was his announcement of the law of conservation of energy in 1847, when he was 29. Although Joule was a modest man ("I have done two or three little things, but nothing to make a fuss about," he later wrote), many honors came his way, including naming the SI unit of energy after him.

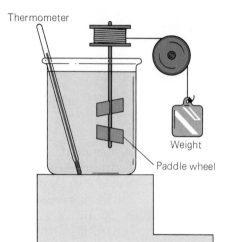
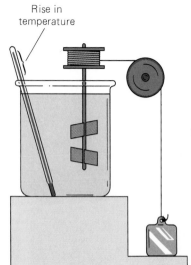

FIG. 3-11 Joule's experimental demonstration that heat is a form of energy. As the weight falls, it turns the paddle wheel, which heats the water by friction. The potential energy of the weight is converted first into the kinetic energy of the paddle wheel and then into heat.

MOMENTUM

Because the universe is so complex, a variety of different quantities besides the basic ones of length, time, and mass are useful to help us understand its many aspects. We have already found velocity, acceleration, force, work, and energy to be valuable, and more are to come. The idea behind defining each of these quantities is to single out something that is involved in a wide range of observations. Then we can boil down a great many separate findings about nature into a brief, clear statement, for example, the law of conservation of energy. Now we shall learn how the concepts of linear and angular momenta can give us further insights into the behavior of moving things.

3-8   LINEAR MOMENTUM

Another Conservation Law

As we know, a moving object tends to continue moving at constant speed along a straight path. The **linear momentum** of such an object is a measure of this tendency. The more linear momentum something has, the more effort is needed to slow it down or to change its direction. Another kind of momentum is **angular momentum,** which reflects the tendency of a spinning body to continue to spin. When there is no question as to which is meant, linear momentum is usually referred to simply as momentum.

The linear momentum **p** of an object of mass m and velocity **v** (we recall that velocity includes both speed and direction) is defined as

$$\mathbf{p} = m\mathbf{v} \qquad \textit{linear momentum}$$

Linear momentum = (mass)(velocity)

The greater m and **v** are, the harder it is to change the object's speed or direction.

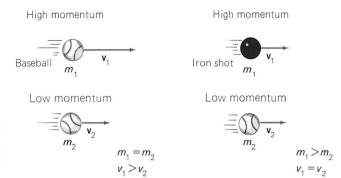

FIG. 3-12   The linear momentum $m\mathbf{v}$ of a moving object is a measure of its tendency to continue in motion at constant velocity. The symbol > means "greater than."

This definition of momentum is in accord with our experience. A baseball hit squarely by a bat (large $\mathbf{v}$) is more difficult to stop than a baseball thrown gently (small $\mathbf{v}$). The heavy iron ball used for the shot-put (large m) is more difficult to stop than a baseball (small m) when their speeds are the same (Fig. 3-12).

Conservation of Momentum

Momentum considerations are most useful in situations that involve explosions and collisions. When outside forces do not act on the objects involved, their combined momentum (taking directions into account) is conserved, that is, does not change:

> **In the absence of outside forces, the total momentum of a set of objects remains the same no matter how the objects interact with one another.**

This statement is called the **law of conservation of momentum.** What it means is that, if the objects interact only with one another, each object can have its momentum changed in the interaction, provided that the total momentum after it occurs is the same as it was before.

Momentum is conserved when a running girl jumps on a stationary sled, as in Fig. 3-13. Even if there is no friction between the sled and the snow, the combination of girl and sled moves off more slowly than the girl's running speed. The original momentum, which is that of the girl alone, had to be shared between her and the sled when she jumped on it. Now that the sled is also moving, the new speed must be less than before in order that the total momentum stay the same.

Let us now see what happens when an object breaks up into two parts. Suppose that an astronaut outside a space station throws away a

$$\text{Total momentum} = m_1\,\mathbf{v}_1 = (m_1 + m_2)\,\mathbf{v}_2$$

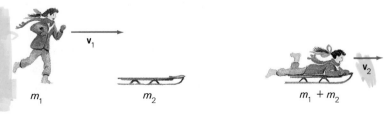

FIG. 3-13   When a running girl jumps on a stationary sled, the combination moves off more slowly than the girl's original speed. The total momentum of girl + sled is the same before and after she jumps on it.

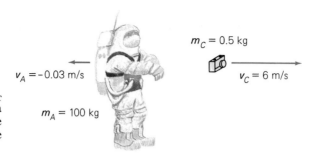

FIG. 3-14 The momentum $m_C v_C$ to the right of the thrown camera is equal in magnitude to the momentum $m_A v_A$ to the left of the astronaut who threw it away.

0.5-kg camera in disgust when it jams (Fig. 3-14). The mass of the space-suited astronaut is 100 kg, and the camera moves off at 6 m/s. What happens to the astronaut?

The total momentum of the astronaut and camera was zero originally. According to the law of conservation of momentum, their total

COLLISIONS   Applying the law of conservation of momentum to collisions gives some interesting results. These are shown in Fig. 3-15 for an object of mass m and speed v that strikes a stationary object of mass M and does not stick to it. Three situations are possible:

1. The target object has more mass, so that $M > m$. What happens here is that the incoming object bounces off the heavier target one and they move apart in opposite directions.
2. The two objects have the same mass, so that $M = m$. Now the incoming object stops and the target object moves off with the same speed v the incoming one had.
3. The target object has less mass, so that $m > M$. In this case the incoming object continues in its original direction after the impact but with reduced speed while the target object moves ahead of it at a faster pace. The greater m is compared with M, the closer the target object's final speed is to $2v$.

The third case corresponds to a golf club striking a golf ball. This suggests that the more mass the clubhead has for a given speed, the faster the ball will fly off when struck. However, a heavy golf club is harder to swing fast than a light one, so a compromise is neces-

The speed of a golf ball is greater than the speed of the clubhead that struck it because the mass of the ball is smaller than that of the clubhead.

sary. Experience has led golfers to use clubheads with masses about 4 times the 46-g mass of a golf ball when they want maximum distance. A good golfer can swing a clubhead at 50 m/s or even more.

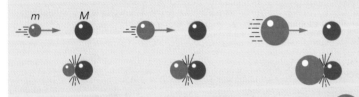

FIG. 3-15 How the effects of a head-on collision with a stationary target object depend on the relative masses of the two objects.

momentum must therefore be zero afterward as well. If we call the astronaut A and the camera C, then

$$\text{Momentum before} = \text{momentum afterward}$$

$$0 = m_A v_A + m_C v_C$$

Hence

$$m_A v_A = - m_C v_C$$

where the minus sign signifies that $\mathbf{v}_A$ is opposite in direction to $\mathbf{v}_C$. Throwing the camera away therefore sets the astronaut in motion as well, with camera and astronaut moving in opposite directions. Newton's third law of motion (action-reaction) tells us the same thing, but conservation of momentum enables us to find the astronaut's speed at once:

$$v_A = - \frac{m_C v_C}{m_A} = - \frac{(0.5 \text{ kg})(6 \text{ m/s})}{100 \text{ kg}} = -0.03 \text{ m/s}$$

After an hour, which is 3600 s, the camera will have traveled $v_C t =$ 21,600 m = 21.6 km, and the astronaut will have traveled $v_A t =$ 108 m in the opposite direction if not tethered to the space station.

3-9 ROCKETS

Momentum Conservation Is the Basis of Space Travel   The operation of a rocket is based on conservation of linear momentum. When the rocket stands on its launching pad, its momentum is zero. When it is fired, the momentum of the exhaust gases that rush downward is balanced by the momentum in the other direction of the rocket moving upward. The total momentum of the entire system, gases and rocket, remains zero, because momentum is a vector quantity and the upward and downward momenta cancel (Fig. 3-16).

Thus a rocket does not work by "pushing" against its launching pad, the air, or anything else. In fact, rockets function best in space where no atmosphere is present to interfere with their motion.

The ultimate speed a rocket can reach is governed by the amount of fuel it can carry and by the speed of its exhaust gases. Because both these quantities are limited, **multistage rockets** are used in the exploration of space. The first stage is a large rocket that has a smaller one mounted in front of it. When the fuel of the first stage has burnt up, its motor and empty fuel tanks are cast off. Then the second stage is fired. Since the second stage is already moving rapidly and does not have to carry the motor and empty fuel tanks of the first stage, it can reach a much higher final speed than would otherwise be possible.

Depending upon the final speed needed for a given mission, three or even four stages may be required. The Saturn V launch vehicle that carried the Apollo 11 spacecraft to the moon in July 1969 had three stages. Just before takeoff the entire assembly was 111 m long and had a mass of nearly 3 million kg.

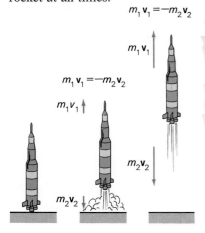

FIG. 3-16 Rocket propulsion is based upon conservation of momentum. If gravity is absent, the downward momentum of the exhaust gases is equal in magnitude and opposite in direction to the upward momentum of the rocket at all times.

$$m_1 \mathbf{v}_1 = -m_2 \mathbf{v}_2$$

$$m_1 \mathbf{v}_1$$

$$m_1 \mathbf{v}_1 = -m_2 \mathbf{v}_2$$

$$m_1 \mathbf{v}_1$$

$$m_2 \mathbf{v}_2$$

$$m_2 \mathbf{v}_2$$

Apollo 11 lifts off its pad to begin the first human visit to the moon. The spacecraft's final speed was 10.8 km/s, which is equivalent to 6.7 mi/s. Conservation of linear momentum underlies rocket propulsion.

3-10   ANGULAR MOMENTUM

The More a Spinning Object Has, the Greater Is Its Tendency to Continue to Spin

We have all noticed the tendency of rotating objects to continue to spin unless they are slowed down by an outside agency. A top would spin indefinitely but for friction between its tip and the ground. Another example is the earth, which has been turning for billions of years and is likely to continue doing so for many more to come.

The rotational quantity that corresponds to linear momentum is called **angular momentum,** and **conservation of angular momentum** is the formal way to describe the tendency of spinning objects to keep spinning.

The precise definition of angular momentum is complicated because it depends not only upon the mass of the object and upon how fast it is turning, but also upon how the mass is arranged in the body. As we might expect, the greater the mass of a body and the more rapidly it rotates, the more angular momentum it has and the more pronounced its tendency to continue to spin. Less obvious is the fact that, the farther away from the axis of rotation the mass is distributed, the more the angular momentum.

An illustration of both the latter fact and the conservation of angular momentum is a skater doing a spin (Fig. 3-17). When the skater starts the spin, she pushes against the ice with one skate to start turning. Initially both arms and one leg are extended, so that her mass is spread as far as possible from the axis of rotation. Then she brings her arms and the outstretched leg in tightly against her body, so that now all her mass is as close as possible to the axis of rotation. As a result, she spins faster. To make up for the change in the mass distribution, the speed must change as well to conserve angular momentum.

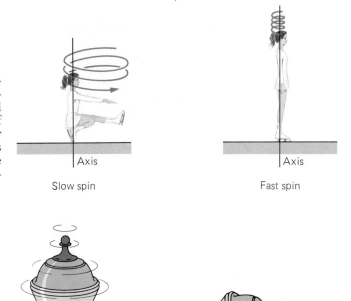

FIG. 3-17 Conservation of angular momentum. Angular momentum depends upon both the speed of turning and the distribution of mass. When the skater pulls in her arms and extended leg, she spins faster to compensate for the change in the way her mass is distributed.

Slow spin

Fast spin

FIG. 3-18 The faster a top spins, the more stable it is. When all its angular momentum has been lost through friction, the top falls over.

Spin Stabilization

Like linear momentum, angular momentum is a vector quantity with direction as well as magnitude. Conservation of angular momentum therefore means that a spinning body tends to maintain the *direction* of its spin axis in addition to the amount of angular momentum it has. A stationary top falls over at once, but a rapidly spinning top stays upright because its tendency to keep its axis in the same orientation by virtue of its angular momentum is greater than its tendency to fall over (Fig. 3-18). Footballs and rifle bullets are sent off spinning to prevent them from tumbling during flight, which would increase air resistance and hence shorten their range.

RELATIVITY

In 1905 a young physicist of 26 named Albert Einstein published an analysis of how measurements of time and space are affected by motion between an observer and what he or she is studying. To say that Einstein's **theory of relativity** revolutionized science is no exaggeration.

Relativity links not only time and space but also energy and matter. From it have come a host of remarkable predictions, all of which have been confirmed by experiment. Eleven years later Einstein took relativity a step further by interpreting gravity as a distortion in the structure of space and time, again predicting extraordinary effects that were verified in detail.

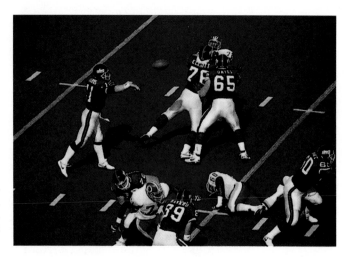

Conservation of angular momentum keeps a spinning football from tumbling end-over-end, which would slow it down and reduce its range.

3-11   SPECIAL RELATIVITY

Things Are Seldom What They Seem

Thus far in this book no special point has been made about how such quantities as length, time, and mass are measured. In particular, who makes a certain measurement would not seem to matter—everybody ought to get the same result. Suppose we want to find the length of an airplane when we are on board. All we have to do is put one end of a tape measure at the airplane's nose and look at the number on the tape at the airplane's tail.

But what if we are standing on the ground and the airplane is in flight? Now things become more complicated because the light that carries information to our instruments travels at a definite speed. According to Einstein, our measurements from the ground of length, time, and mass in the airplane would differ from those made by somebody moving with the airplane.

Einstein began with two postulates. The first concerns **frames of reference.** When we say something is moving, we mean that its position relative to something else—the frame of reference—is changing. A passenger walking down the aisle moves relative to an airplane, the airplane moves relative to the earth, the earth moves relative to the sun, and so on.

If we are in the windowless cabin of a cargo airplane, we cannot tell whether the airplane is in flight at constant velocity or is at rest on the ground, since without an external frame of reference the question has no meaning. To say that something is moving always requires a frame of reference. From this follows Einstein's first postulate:

> **The laws of physics are the same in all frames of reference moving at constant velocity with respect to one another.**

If the laws of the physics were different for different observers in relative motion, the observers could find from these differences which of them were "stationary" in space and which were "moving." But such a distinction does not exist, hence the above postulate.

The second postulate, which follows from the results of a great many experiments, states that

The speed of light in free space has the same value for all observers.

The speed of light in free space is $c = 3 \times 10^8$ m/s, about 186,000 mi/s.

Length, Time, and Mass

Let us suppose I am in an airplane moving at the constant velocity **v** relative to you on the ground. I find that the airplane is L_0 long, that it has a mass of m_0, and that a certain time interval (say an hour on my watch) is t_0. Einstein showed from the above postulates that you, on the ground, would find that

1. The length L you measure is shorter than L_0.

2. The time interval t you measure is longer than t_0.

3. The mass m you measure is greater than m_0.

That is, to you on the ground, the airplane appears shorter and more massive than to me, and to you, my watch appears to tick more slowly.

The differences between L and L_0, t and t_0, and m and m_0 depend on the ratio v/c between the relative speed v of the frames of reference (here the speed of the airplane relative to the ground) and the speed of light c. Because c is so great, these differences are too small to detect at speeds like those of airplanes. However, they must be taken into account in spacecraft flight. And, at speeds near c, which often occur in the sub-

All motion is relative to a chosen frame of reference. Here the photographer has turned the camera to keep pace with one of the cyclists. Relative to him, both the road and the other cyclists are moving. There is no fixed frame of reference in nature, and therefore no such thing as "absolute motion"; all motion is relative.

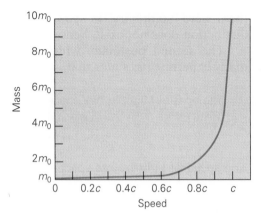

FIG. 3-19   The relativity of mass. The greater the speed of an object relative to an observer, the greater the object's mass appears to the observer. This effect is conspicuous only at speeds near the speed of light c, which is 3×10^8 m/s, about 186,000 mi/s.

atomic world of such tiny particles as electrons and protons, relativistic effects are conspicuous (Fig. 3-19). At a speed of 99.99 percent of c, which an electron can easily be given, its mass is 71 times its mass measured at rest.

The closer v gets to c, the closer m gets to infinity. Since an infinite mass is impossible, this conclusion means that nothing can travel as fast as light or faster: c is the absolute speed limit in the universe. The implications of this limit for space travel are discussed in Chap. 18.

Einstein's 1905 theory, which led to the above results among others, is called **special relativity** because it is restricted to constant velocities. His later theory of **general relativity,** which deals with gravity, includes accelerations.

3-12   REST ENERGY

Matter Is a Form of Energy   The most far-reaching conclusion of special relativity is that mass and energy are related to each other so closely that matter can be converted into energy and energy into matter. The **rest energy** of a body is the energy equivalent of its mass. If a body has the mass m_0 when it is at rest, its rest energy is

$$E_0 = m_0 c^2 \qquad \textit{rest energy}$$

Rest energy = (rest mass)(speed of light)^2

The rest energy of a 1.5-kg object, such as this book, is

$$E_0 = m_0 c^2 = (1.5 \text{ kg})(3 \times 10^8 \text{ m/s})^2 = 1.35 \times 10^{17} \text{ J}$$

quite apart from any kinetic or potential energy it might have. If liberated, this energy would be more than enough to send a million tons to the moon. By contrast, the PE of this book on top of Mt. Everest, which is 8850 m high, relative to its sea-level PE is less than 10^4 J.

How is it possible that so much energy can be bottled up in even a little bit of matter without anybody having known about it until Einstein's work? In fact, we do see matter being converted into energy

BIOGRAPHY

Albert Einstein (1879–1955), bitterly unhappy with the rigid discipline of the schools of his native Germany, went to Switzerland at 16 to complete his education and later got a job examining patent applications at the Swiss Patent Office in Berne. Then, in 1905, ideas that had been in his mind for years when he should have been paying attention to other matters blossomed into three short papers that were to change decisively the course of not only physics but modern civilization as well.

The first paper proposed that light has a dual character with particle as well as wave properties.

This work is described in Chap. 8 together with the quantum theory of the atom that flowed from it. The subject of the second paper was brownian motion, the irregular zigzag motion of tiny bits of suspended matter such as pollen grains in water (Fig. 3-20). Einstein arrived at a formula that related brownian motion to the bombardment of the particles by randomly moving molecules of the fluid in which they were suspended. Although the molecular theory of matter had been proposed many years before, this formula was the long-awaited definite link with experiment that convinced the remaining doubters. The third paper introduced the theory of relativity.

Although much of the world of physics was originally either indifferent or skeptical, even the most unexpected of Einstein's conclusions were soon confirmed and the development of what is now called modern physics began in earnest. After university posts in Switzerland and Czechoslovakia, in 1913 Einstein took up an appointment at the Kaiser Wilhelm Institute in Berlin that left him able to do research free of financial worries and routine duties. His interest was now mainly in gravity, and he began where Newton had left off more than 200 years earlier.

The general theory of relativity that resulted from Einstein's work provided a deep understanding of gravity, but his name remained unknown to the general public. This

FIG. 3-20 The irregular path of a microscopic particle bombarded by molecules. The line joins the positions of a single particle observed at 10-s intervals. This phenomenon is called brownian movement and is direct evidence of the reality of molecules and their random motions. It was discovered in 1827 by the British botanist Robert Brown.

changed in 1919 with the dramatic discovery that gravity affects light exactly as Einstein had predicted. He immediately became a world celebrity, but his well-earned fame did not provide security when Hitler and the Nazis came to power in Germany in the early 1930s. Einstein left in 1933 and spent the rest of his life at the Institute for Advanced Study in Princeton, New Jersey, thereby escaping the fate of millions of other European Jews at the hands of the Germans. Einstein's last years were spent in a fruitless search for a "unified field theory" that would bring together gravitation and electromagnetism in a single picture. The problem was worthy of his gifts, but it remains unsolved to this day although progress is being made.

around us all the time. We just do not normally think about what we find in these terms. All the energy-producing reactions of chemistry and physics, from the lighting of a match to the nuclear fusion that powers the sun and stars, involve the disappearance of a small amount of matter and its reappearance as energy. The simple formula $E_0 = m_0 c^2$ has led not only to a better understanding of how nature works but also to the nuclear power plants—and nuclear weapons—that are so important in today's world.

The discovery that matter and energy can be converted into each other does not affect the law of conservation of energy provided we include mass as a form of energy.

3-13 GENERAL RELATIVITY

Gravity Is a Warping of Spacetime Einstein's general theory of relativity, published in 1916, related gravitation to the structure of space and time. What is meant by "the structure of space and time" can be given a quite precise meaning mathematically, but unfortunately no such precision is possible using ordinary language. All the same, we can legitimately think of the force of gravity as arising from a warping of spacetime around a body of matter so that a nearby mass tends to move toward the body, much as a marble rolls toward the bottom of a saucer-shaped hole (Fig. 3-21). It may seem as though one abstract concept is merely replacing another, but in fact the new point of view led Einstein and other scientists to a variety of remarkable discoveries that could not have come from the older way of thinking.

Perhaps the most spectacular of Einstein's results was that light ought to be subject to gravity. The effect is very small, so a large mass, such as that of the sun, is needed to detect the influence of its gravity on light. If Einstein was right, light rays that pass near the sun should be bent toward it by 0.0005°—the diameter of a dime seen from a mile away. To check this prediction, photographs were taken of stars that appeared in the sky near the sun during an eclipse in 1919, when they could be seen because the moon obscured the sun's disk (see Chap. 16). These photographs were then compared with photographs of the same region of the sky taken when the sun was far away (Fig. 3-22), and the observed changes in the apparent positions of the stars matched Einstein's calculations. Other predictions based on general relativity have also been verified, and the theory remains today without serious rival.

FIG. 3-21 General relativity pictures gravity as a warping of the structure of space and time due to the presence of a body of matter. An object nearby experiences an attractive force as a result of this distortion in space-time, much as a marble rolls toward the bottom of a saucer-shaped hole in the ground.

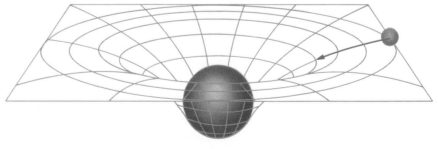

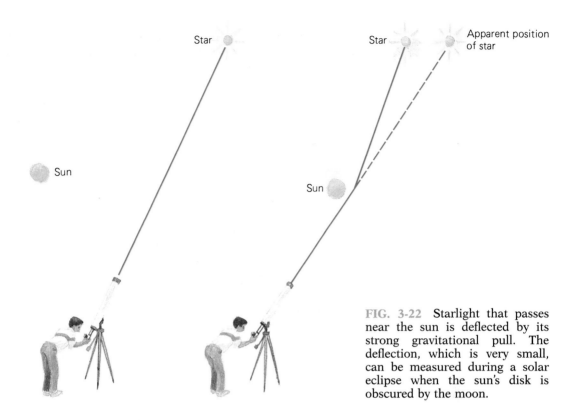

FIG. 3-22 Starlight that passes near the sun is deflected by its strong gravitational pull. The deflection, which is very small, can be measured during a solar eclipse when the sun's disk is obscured by the moon.

ENERGY AND CIVILIZATION

The rise of modern civilization would have been impossible without the discovery of vast resources of energy and the development of ways to transform it into useful forms. All that we do requires energy. The more energy we have at our command, the better we can satisfy our desires for food, clothing, shelter, warmth, light, transport, communication, and manufactured goods.

Unfortunately oil and gas, the most convenient fuels, although currently abundant and not too expensive, have limited reserves. Other energy sources all have serious handicaps of one kind or another and nuclear fusion, the ultimate energy source, remains a technology of the future. At the same time, world population is increasing and with this increase comes a need for more and more energy. The choice of an appropriate energy strategy for the future is therefore one of the most critical of today's world problems.

3-14   THE ENERGY PROBLEM

Limited Supply, Unlimited Demand

Almost all the energy available to us today has a single source—the sun. Light and heat reach us directly from the sun; food and wood owe their energy content to sunlight falling on plants; water power exists because the sun's heat evaporates water from the oceans to fall later as rain and snow on high ground; wind power comes from motions in the atmosphere due to unequal heating of the earth's

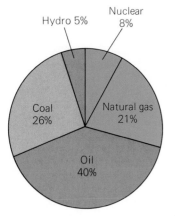

FIG. 3-23 A total of about 10^{20} J of energy was produced in the United States in 1995 from the sources shown.

surface by the sun. The fossil fuels coal, oil, and natural gas were formed from plants and animals that lived and stored energy derived from sunlight millions of years ago. Only nuclear energy and heat from sources inside the earth cannot be traced to the sun's rays (Fig. 3-23).

In the advanced countries, the standard of living is already high and populations are stable, so their need for energy is not likely to grow very much. Indeed, this need may even decline as energy use becomes more efficient. Elsewhere energy consumption is still low, less than 1 kW per person for more than half the people of the world compared with 11 kW per person in the United States. These people seek better lives, which means more energy, and their numbers are increasing rapidly, which means still more energy. Where is the energy to come from?

The fossil fuels coal, oil, and natural gas, which today furnish by far the greatest part of the world's energy, cannot last forever. Oil and natural gas will be the first to be exhausted. At the current rate of consumption, known oil reserves will last only about another century. More oil will certainly be found, and better technology will increase the yield from existing wells, but even so oil will inevitably become scarce sooner or later. The same is true for natural gas. This situation will be a real pity because oil and gas burn efficiently and are easy to extract, process, and transport. Half the oil used today goes into fuels that power ships, trains, aircraft, cars, and trucks, and oil and gas are superb feedstocks for synthetic materials of all kinds. Although liquid fuels can be made from coal and coal can serve as the raw material for synthetics, these technologies involve greater expense and greater risk to health and to the environment.

Even though the coal we consume every year took about 2 million years to accumulate, enough remains to last several hundred more years at the present rate of consumption. Coal reserves are equivalent in energy content to 5 times oil reserves. Before 1941, coal was the world's chief fuel, and it is likely to return to first place as oil and gas run out.

Bright sunlight can deliver over 1 kW of power to each square meter it falls on. At this rate, an area the size of a tennis court receives solar energy equivalent to a gallon of gasoline every 10 min or so. The sunlight concentrated by this mirror system in France produces steam that powers a turbine connected to an electric generator. Another approach under development uses photoelectric cells to covert solar radiation directly to electric energy. Although the efficiency of such cells is being steadily improved, they are still too expensive for widespread use.

The kinetic energy of falling water is converted into electric energy as the water turns turbine blades connected to generators in this dam on the Niagara River in New York State. Hydroelectric plants in many parts of the world have flooded large areas and turned once fertile river valleys into wastelands unfit for agriculture. Few sites remain that would not lead to such ecological damage.

But coal is far from being a desirable fuel. Not only is mining it dangerous and usually leaves large areas of land unfit for further use, but also the air pollution due to burning coal adversely affects the health of millions of people. Acid rain (see Chap. 11) from the same source harms plant and animal life on a large scale. Most estimates put the number of deaths in the United States from cancer and respiratory diseases caused by burning coal at over 10,000 per year. Coal-burning power plants expose the people living around them to more radioactivity—from traces of uranium, thorium, and radon in their smoke—than do normally operating nuclear plants.

Nuclear fuel reserves much exceed those of fossil fuels. Besides their having an abundant fuel supply, properly built and properly operating nuclear plants are in many respects excellent energy sources. Nuclear energy is already responsible for nearly a fifth of the electricity generated in the United States, and in a number of other countries the proportion is even higher; in France it is nearly three-quarters.

To be sure, nuclear energy has serious drawbacks. Nuclear plants are expensive and the potential for large-scale disaster is always present. Two major reactor accidents, at Three Mile Island in Pennsylvania and at Chernobyl in Ukraine, have already occurred. On a smaller scale, above-average cancer and leukemia rates have been reported near badly run nuclear installations here and abroad. Although the overall public-health record of nuclear plants is still far better than that of coal-burning ones, there remains much to worry about. Furthermore, a reactor produces many tons of radioactive wastes each year whose safe disposal is still an unsettled issue.

Windmill "farm" near Palm Springs, California. Windmills are noisy, cost more than conventional power stations, and are practical only where winds are steady and strong. On the other hand, they are nonpolluting and use up no resources. Helped by favorable tax treatment, windmills produce 1.5 percent of California's electricity.

3-15

Geothermal power station in Sonoma County, California, uses the heat of the earth's interior as its energy source. Other such plants are located in Indonesia, the Philippines, New Zealand (where they provide 11 percent of the total energy supply), Italy, and on Caribbean islands. Geothermal energy is practical today in only a few places, but it has considerable potential for the future.

THE FUTURE

What to Do Until Fusion Comes
In the long run, practical ways to utilize the energy of nuclear fusion may well be developed. As described in Chap. 7, a fusion reactor will get its fuel from the sea, will be safe and nonpolluting, and cannot be adapted for military purposes. But nobody can predict when, or even if, this ultimate source of energy will become an everyday reality.

Assuming that fusion energy is really on the way, the big question today is how to manage until it arrives. Coal can be used more widely, but only at the cost of more human suffering and more damage to the environment. And the continued burning of fossil fuels will increase the carbon dioxide content of the atmosphere which, as we will see in Chap. 13, is likely to produce global warming with possibly catastrophic consequences. Nuclear energy can help bridge the gap, but memories of Three Mile Island and Chernobyl will have to fade before more reactors can be built on a large enough scale.

What about the energy of sunlight, of winds and tides, of falling water, of trees and plants, of the earth's own internal heat? After all, the technologies needed to make use of these renewable resources already exist. But a close look shows that such alternative energy is not likely to provide more than a small fraction (though a welcome one) of future needs. In every case the required installation either is very expensive for the energy obtained, or is practical only in a few favorable locations in the world, or both. And all of them need a lot of space.

A city of medium size might use 1000 MW of power. Less than 150 acres is enough for a 1000-MW nuclear plant, whereas solar collectors of the same capacity might need 5000 acres, windmills over 10,000 acres, and to grow crops for conversion to fuel might require 200 square miles of farmland to give 1000 MW averaged over a year. All this is not to say that such energy sources are without value, particularly where local conditions are suitable, only that they are unlikely to satisfy by themselves the world's future energy appetite.

Clearly there is no simple solution possible in the near future to the problem of safe, cheap, and abundant energy. The sensible course is to practice conservation and try to get the best from the various available technologies while pursuing fusion energy as rapidly as possible. Although each of these technologies has limitations, it may be a reasonable choice in a given situation. If their full potential is realized and population growth slows down, social disaster (starvation, war) and environmental catastrophe (a planet unfit for life) may well be avoided during the wait for fusion.

IMPORTANT TERMS AND IDEAS

Work is a measure of the change, in a general sense, a force causes when it acts upon something. The work done by a force acting on an object is the product of the magnitude of the force and the distance through which the object moves while the force acts on it. If the direction of the force is not the same as the direction of motion, the projection of the force in the direction of motion must be used. The unit of work is the **joule** (J).

Power is the rate at which work is being done. Its unit is the **watt** (W).

Energy is the property that something has that enables it to do work. The unit of energy is the joule. The three broad categories of energy are **kinetic energy,** which is the energy something has by virtue of its motion, **potential energy,** which is the energy something has by virtue of its position, and **rest energy,** which is the energy something has by virtue of its mass. According to the law of **conservation of energy,** energy cannot be created or destroyed, although it can be changed from one form to another (including mass).

Linear momentum is a measure of the tendency of a moving object to continue in motion along a straight line. **Angular momentum** is a measure of the tendency of a rotating object to continue spinning about a fixed axis. Both are vector quantities. If no outside forces act on a set of objects, then their linear and angular momenta are **conserved,** that is, remain the same regardless of how the objects interact with one another.

According to the **special theory of relativity,** when there is relative motion between an observer and what is being observed, lengths are shorter than when at rest, time intervals are longer, and masses are greater. Nothing can travel faster than the speed of light.

The **general theory of relativity,** which relates gravitation to the structure of space and time, correctly predicts that light should be subject to gravity.

IMPORTANT FORMULAS

Work: $W = Fd$

Power: $P = \dfrac{W}{t}$

Kinetic energy: $\text{KE} = \frac{1}{2}mv^2$

Gravitational potential energy: $\text{PE} = mgh$

Linear momentum: $\mathbf{p} = m\mathbf{v}$

Rest energy: $E_0 = m_0 c^2$

EXERCISES: MULTIPLE CHOICE

1. Which of the following is not a unit of power?
   a. joule-second
   b. watt
   c. newton-meter/second
   d. horsepower

2. An object at rest may have
   a. velocity
   b. momentum
   c. kinetic energy
   d. potential energy

3. A moving object does not necessarily have
   a. velocity
   b. momentum
   c. kinetic energy
   d. potential energy

4. An object that has linear momentum must also have
   a. acceleration

b. angular momentum
c. kinetic energy
d. potential energy

5. The total amount of energy (including the rest energy of matter) in the universe
   a. cannot change
   b. can decrease but not increase
   c. can increase but not decrease
   d. can either increase or decrease

6. When the speed of a body is doubled,
   a. its kinetic energy is doubled
   b. its potential energy is doubled
   c. its rest energy is doubled
   d. its momentum is doubled

7. Two balls, one of mass 5 kg and the other of mass 10 kg, are dropped simultaneously from a window. When they are 1 m above the ground, the balls have the same
   a. kinetic energy
   b. potential energy
   c. momentum
   d. acceleration

8. A bomb dropped from an airplane explodes in midair.
   a. Its total kinetic energy increases
   b. Its total kinetic energy decreases
   c. Its total momentum increases
   d. Its total momentum decreases

9. The operation of a rocket is based upon
   a. pushing against its launching pad
   b. pushing against the air
   c. conservation of linear momentum
   d. conservation of angular momentum

10. When a spinning skater pulls in her arms to turn faster,
    a. her angular momentum increases
    b. her angular momentum decreases
    c. her angular momentum remains the same
    d. any of the above, depending on the circumstances

11. According to the principle of relativity, the laws of physics are the same in all frames of reference
    a. at rest with respect to one another
    b. moving toward or away from one another at constant velocity
    c. moving parallel to one another at constant velocity
    d. all of the above

12. When an object whose rest mass is 1 kg approaches the speed of light, its mass approaches

a. 0	**c.** 2 kg
b. 0.5 kg	**d.** infinite mass

13. A spacecraft has left the earth and is moving toward Mars. An observer on the earth finds that, relative to measurements made when the spacecraft was at rest, its
    a. length is greater
    b. mass is smaller
    c. clocks tick faster
    d. total energy is greater

14. In the formula $E_0 = m_0c^2$, the symbol c represents
    a. the speed of the body
    b. the speed of the observer
    c. the speed of sound
    d. the speed of light

15. It is not true that
    a. light is affected by gravity
    b. the mass of a moving object depends upon its speed
    c. the maximum speed anything can have is the speed of light
    d. momentum is a form of energy

16. Albert Einstein did not discover that
    a. the mass of a moving object is greater than its mass at rest
    b. the acceleration of gravity g is a universal constant
    c. light is affected by gravity
    d. gravity is a warping of space-time

17. The rate at which sunlight delivers energy to an area of 1 m^2 is roughly

a. 1 W	**c.** 1000 W
b. 10 W	**d.** 1,000,000 W

18. The chief source of energy in the world today is

a. coal	**c.** natural gas
b. oil	**d.** uranium

19. The source of energy whose reserves are greatest is

a. coal	**c.** natural gas
b. oil	**d.** uranium

20. The work done in holding a 50-kg object at a height of 2 m above the floor for 10 s is

a. 0	**c.** 1000 J
b. 250 J	**d.** 98,000 J

21. The work done in lifting 30 kg of bricks to a height of 20 m is

a. 61 J	**c.** 2940 J
b. 600 J	**d.** 5880 J

22. A total of 4900 J is used to lift a 50-kg mass. The mass is raised to a height of

a. 10 m   **c.** 960 m
b. 98 m   **d.** 245 km

23. A 40-kg boy runs up a flight of stairs 4 m high in 4 s. His power output is
    a. 160 W   **c.** 40 W
    b. 392 W   **d.** 1568 W

24. Car A has a mass of 1000 kg and is moving at 60 km/h. Car B has a mass of 2000 kg and is moving at 30 km/h. The kinetic energy of car A is
    a. half that of car B
    b. equal to that of car B
    c. twice that of car B
    d. 4 times that of car B

25. A 1-kg object has a potential energy of 1 J relative to the ground when it is at a height of
    a. 0.102 m   **c.** 9.8 m
    b. 1 m   **d.** 98 m

26. A 1-kg object has kinetic energy of 1 J when its speed is
    a. 0.45 m/s   **c.** 1.4 m/s
    b. 1 m/s   **d.** 4.4 m/s

27. Car A has a mass of 1000 kg and is moving at 60 km/h. Car B has a mass of 2000 kg and is moving at 30 km/h. The momentum of car A is
    a. half that of car B
    b. equal to that of car B
    c. twice that of car B
    d. 4 times that of car B

28. A 10,000-kg freight car moving at 2 m/s collides with a stationary 15,000-kg freight car. The two cars couple together and move off at
    a. 0.8 m/s   **c.** 1.3 m/s
    b. 1 m/s   **d.** 2 m/s

29. A 30-kg girl and a 25-kg boy are standing on frictionless roller skates. The girl pushes the boy, who moves off at 1.0 m/s. The girl's speed is
    a. 0.45 m/s   **c.** 0.83 m/s
    b. 0.55 m/s   **d.** 1.2 m/s

30. An object has a rest energy of 1 J when its mass is
    a. 1.1×10^{-17} kg   **c.** 1 kg
    b. 3.3×10^{-9} kg   **d.** 9×10^{16} kg

31. A 1-kg ball is thrown in the air. When it is 10 m above the ground, its speed is 3 m/s. At this time most of the ball's total energy is in the form of
    a. kinetic energy
    b. potential energy relative to the ground
    c. rest energy
    d. momentum

32. The smallest part of the total energy of the ball of Question 27 is
    a. kinetic energy
    b. potential energy relative to the ground
    c. rest energy
    d. momentum

33. The 2-kg blade of an ax is moving at 60 m/s when it strikes a log. If the blade penetrates 2 cm into the log as its KE is turned into work, the average force it exerts is
    a. 3 kN   **c.** 72 kN
    b. 90 kN   **d.** 180 kN

34. The lightest particle in an atom is an electron, whose rest mass is 9.1×10^{-31} kg. The energy equivalent of this mass is approximately
    a. 10^{-13} J   **c.** 3×10^{-23} J
    b. 10^{-15} J   **d.** 10^{-47} J

QUESTIONS

1. Is it correct to say that all changes in the physical world involve energy transformations of some sort? Why?

2. Under what circumstances (if any) is no work done on a moving object even though a net force acts upon it?

3. In what part of its orbit is the earth's potential energy greatest with respect to the sun? In what part of its orbit is the earth's kinetic energy greatest? Explain your answers.

4. Does every moving body possess kinetic energy? Does every stationary body possess potential energy?

5. A golf ball and a Ping-Pong ball are dropped in a vacuum chamber. When they have fallen halfway to the bottom, how do their speeds compare? Their kinetic energies? Their potential energies? Their momenta?

6. The potential energy of a golf ball in a hole is negative with respect to the ground. Under what circumstances (if any) is the ball's kinetic energy negative? Its rest energy?

7. Two identical balls move down a tilted board. Ball A slides down without friction and ball B rolls down. Which ball reaches the bottom first? Why?

8. The kilowatt-hour is a unit of what physical quantity or quantities?

9. Why does a nail become hot when it is hammered into a piece of wood?

10. Two identical spring-driven watches, one wound and the other unwound, are dissolved in acid. What do you think becomes of the potential energy of the spring in the wound watch?

11. Is it possible for an object to have more kinetic energy but less momentum than another object? Less kinetic energy but more momentum?

12. When the momentum of an object is doubled, what happens to its kinetic energy?

13. When the kinetic energy of an object is doubled, what happens to its momentum?

14. What, if anything, happens to the speed of a fighter plane when it fires a cannon at an enemy plane in front of it?

15. An empty dump truck coasts freely with its engine off along a level road. (a) What happens to the truck's speed if it starts to rain and water collects in it? (b) The rain stops and the accumulated water leaks out. What happens to the truck's speed now?

16. A railway car is at rest on a frictionless track. A man at one end of the car walks to the other end. (a) Does the car move while he is walking? (b) If so, in which direction? (c) What happens when the man comes to a stop?

17. If the polar ice caps melt, the length of the day will increase. Why?

18. All helicopters have two propellers. Some have both propellers on vertical axes but rotating in opposite directions, and others have one on a vertical axis and one on a horizontal axis perpendicular to the helicopter body at the tail. Why is a single propeller never used?

19. What are the two postulates from which Einstein developed the special theory of relativity?

20. What quantity will all observers always find the same value for?

21. The length of a rod is measured by several observers, one of whom is stationary with respect to the rod. What must be true of the value obtained by the stationary observer?

22. Is the mass of an object the same whether it is moving toward you or away from you at the same speed? Is the mass the same whether the object moves toward you or you move toward it at the same speed? What is the connection between the answers to these questions and the first of Einstein's postulates?

23. In the formula $E_0 = m_0 c^2$ for rest energy, what does the symbol c represent? What is the difference between m_0 and m?

24. Why is it impossible for an object to move faster than the speed of light?

25. What is the effect on the law of conservation of energy of the discovery that matter and energy can be converted into each other?

26. Which three fuels provide most of the world's energy today?

27. What energy sources, if any, cannot be traced to sunlight falling on the earth?

28. What are some of the disadvantages shared by all renewable-energy sources such as solar cells and windmills?

PROBLEMS

1. A person holds 5 kg of onions 1.5 m above the ground for 2 min. How much work is done?

2. A horizontal force of 80 N is used to move a 20-kg crate across a level floor. How much work is done when the crate is moved 5 m? How much work would have been done if the crate's mass were 30 kg?

3. How much work is needed to raise a 110-kg load of bricks 12 m above the ground to a building under construction?

4. A centripetal force of 5 N is used to keep a 200-g ball moving in a circle with uniform speed at the end of a string 80 cm long. How much work is done by the force in each revolution of the ball?

5. A total of 490 J of work is needed to lift a body of unknown mass through a height of 10 m. What is its mass?

6. A weightlifter raises a 150-kg barbell from the floor to a height of 2.2 m in 0.8 s. What is the average power output during the lift?

7. An 80-kg mountaineer climbs a 3000-m mountain in 10 h. What is the average power output during the climb?

8. A 15-kW electric motor provides power for the elevator of a six-story building. If the total mass of the loaded elevator is 1000 kg, what is the minimum time needed for it to rise the 30 m from the ground floor to the top floor?

9. A moving object whose initial KE is 10 J is subject to a frictional force of 2 N that acts in the

opposite direction. How far will the object move before coming to a stop?

10. Find the KE of a 60-kg runner who covers 400 m in 45 s at constant speed.

11. Is the work needed to bring a car's speed from 0 to 10 km/h less than, equal to, or more than the work needed to bring its speed from 10 to 20 km/h?

12. A hammer with a 500-g head exerts an average force of 810 N on a nail as it drives the nail 4 mm into a wall. What was the initial speed of the hammer head?

13. A 1-kg salmon is hooked by a fisherman and it swims off at 2 m/s. The fisherman stops the salmon in 50 cm by braking his reel. How much force does the fishing line exert on the fish?

14. What is the power output of a 1200-kg car if it can go from rest to 20 m/s in 10 s?

15. A 70-kg athlete runs up the stairs from the ground floor of the Empire State Building to its one hundred second floor, a height of 370 m, in 25 min. How much power did the athlete develop?

16. A 500-kg pile-driver hammer is dropped from a height of 4 m above the head of a pile. If the pile is driven 10 cm into the ground with each impact of the hammer, what is the average force on the pile when it is struck?

17. A 3-kg stone is dropped from a height of 100 m. Find its kinetic and potential energies when it is 50 m from the ground.

18. An 800-kg car coasts down a hill 40 m high with its engine off and the driver's foot pressing on the brake pedal. At the top of the hill the car's speed is 6 m/s and at the bottom it is 20 m/s. How much energy was converted into heat on the way down?

19. A girl on a swing is 2 m above the ground at her highest point and 1 m above the ground at her lowest point. What is the girl's maximum speed?

20. A force of 500 N is used to lift a 20-kg object to a height of 10 m. There is no friction present. (a) How much work is done by the force? (b) What is the change in the potential energy of the object? (c) What is the change in the kinetic energy of the object?

21. An 80-kg crate is raised 2 m from the ground by a man who uses a rope and a system of pulleys.

He exerts a force of 220 N on the rope and pulls a total of 8 m of rope through the pulleys while lifting the crate, which is at rest afterward. How much work does the man do? What is the change in the potential energy of the crate? If the answers to these questions are different, explain why.

22. A man drinks a bottle of beer and proposes to work off its 460 kJ by exercising with a 20-kg barbell. If each lift of the barbell from chest height to over his head is through 60 cm and the efficiency of his body is 10 percent under these circumstances, how many times must he lift the barbell?

23. In an effort to lose weight, a person runs 5 km per day at a speed of 4 m/s. While running, the person's body processes consume energy at a rate of 1.4 kW. Fat has an energy content of about 40 kJ/g. How many grams of fat are metabolized during each run?

24. A girl throws a 5-kg pumpkin at 10 m/s to a 50-kg boy, who catches it. If the boy is on a frictionless frozen lake, how fast does he move backward?

25. A 70-kg person dives horizontally from a 200-kg boat with a speed of 2 m/s. What is the recoil speed of the boat?

26. A 2000-kg car traveling at 8 m/s strikes a stationary car whose mass is the same. The two cars stick together after the crash. What is their speed just afterward?

27. The 176-g head of a golf club is moving at 45 m/s when it strikes a 46-g golf ball and sends it off at 65 m/s. Find the final speed of the clubhead after the impact, assuming that the mass of the club's shaft can be neglected.

28. A 30-kg girl who is running at 3 m/s jumps on a stationary 10-kg sled on a frozen lake. How fast does the sled with the girl on it then move?

29. One kilogram of water at 0°C contains 335 kJ of energy more than 1 kg of ice at 0°C. What is the mass equivalent of this amount of energy?

30. When 1 kg of gasoline is burned in the engine of a car, 47 MJ of energy is liberated. How much mass is lost in the process? Would you expect to be able to measure this mass change directly?

31. Approximately 5.4×10^6 J of chemical energy is released when 1 kg of dynamite explodes. What fraction of the total energy of the dynamite is this?

32. Approximately 4×10^9 kg of matter is converted into energy in the sun per second. Express the power output of the sun in watts.

ANSWERS TO MULTIPLE CHOICE

1. a	**4.** c	**7.** d	**10.** c
2. d	**5.** a	**8.** a	**11.** d
3. d	**6.** d	**9.** c	**12.** d

13. d	**19.** d	**25.** a	**31.** c
14. d	**20.** a	**26.** c	**32.** a
15. d	**21.** d	**27.** b	**33.** d
16. b	**22.** a	**28.** a	**34.** a
17. c	**23.** b	**29.** c	
18. b	**24.** c	**30.** a	

4

MATTER

AND

ENERGY

Suppose our microscopes had no limit to their power, so that we could examine a drop of water at any magnification we like. What would we find if the drop were enlarged a million or more times? Would we still see a clear, structureless liquid? If not, what else?

The answer is that, on a very small scale of size, our drop of water consists of billions of tiny separate particles. Indeed, *all* matter does, whether in the form of a solid, a liquid, or a gas. This much was suspected over 2000 years ago in ancient Greece. Modern science has not only confirmed this suspicion but extended it: the particles that make up all matter are in constant random motion, and the kinetic energy of this motion is what constitutes heat. In everyday life, matter shows no direct sign of either the particles or their motion. However, plenty of indirect signs support this picture, and we shall consider some of them in this chapter.

TEMPERATURE AND HEAT

Temperature and heat are easy to confuse. Certainly the higher its temperature, the more heat something contains. But we cannot say that an object at one temperature contains more heat than another object at a lower temperature just because of the temperature difference. A cup of boiling water is at a higher temperature than a pailful of cool water, but the pailful of cool water would melt more ice (Fig. 4-1). And the same masses of different substances at the same temperature contain different amounts of heat. A kilogram of boiling water can melt 33 times as much ice as 1 kg of gold at the temperature of the water, for instance.

4-1 TEMPERATURE

Putting Numbers to Hot and Cold

Temperature, like force, is a physical quantity that means something to us in terms of our sense impressions. And, as with force, a certain amount of discussion is needed before a statement of exactly what temperature signifies can be given. Such a discussion appears later in this chapter. For the time being, we can simply regard temperature as that which gives rise to sensations of hot and cold.

A **thermometer** is a device that measures temperature. Most substances expand when heated and shrink when cooled, and the ther-

FIG. 4-1 The heat content of a given substance depends upon both its mass and its temperature. A pail of cool water contains more heat than a cup of boiling water.

Allowance must be made in the design of a bridge for its expansion and contraction as the temperature changes. The Golden Gate Bridge in San Francisco varies by over a meter in length between summer and winter.

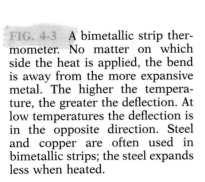

FIG. 4-2 A liquid-in-glass thermometer. Mercury or a colored alcohol solution responds to temperature changes to a greater extent than glass does, and so the length of the liquid column is a measure of the temperature of the thermometer bulb.

mometers we use in everyday life are designed around this property of matter. More precisely, they are based upon the fact that different materials react to a given temperature change to different extents. The familiar mercury-in-glass thermometer (Fig. 4-2) works because mercury expands more than glass when heated and contracts more than glass when cooled. Thus the length of the mercury column in the glass tube provides a measure of the temperature around the bulb.

Another common thermometer used for high temperatures, such as in ovens and furnaces, makes use of the different rates of expansion of different kinds of metals. Two straight strips of dissimilar metals are joined together at a particular temperature (Fig. 4-3). At higher temperatures the bimetallic strip bends so that the metal with the greater expansion is on the outside of the curve, and at lower temperatures it

FIG. 4-3 A bimetallic strip thermometer. No matter on which side the heat is applied, the bend is away from the more expansive metal. The higher the temperature, the greater the deflection. At low temperatures the deflection is in the opposite direction. Steel and copper are often used in bimetallic strips; the steel expands less when heated.

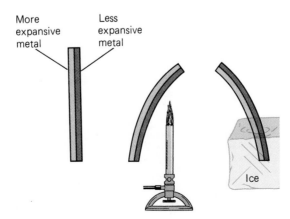

More expansive metal

Less expansive metal

Ice

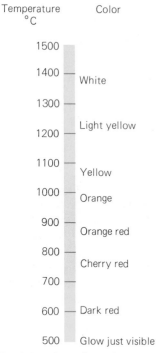

Temperature °C	Color
1500	
1400	White
1300	
1200	Light yellow
1100	
1000	Yellow
900	Orange
800	Orange red
700	Cherry red
600	Dark red
500	Glow just visible

FIG. 4-4 The color of an object hot enough to glow varies with its temperature roughly as shown here.

bends in the opposite direction. In each case the exact amount of bending depends upon the temperature. Bimetallic strips of this kind are used in the **thermostats** that switch on and off heating systems, refrigerators, and freezers at preset temperatures.

Thermal expansion is not the only property of matter that can be used to make a thermometer. As another example, the color and amount of light emitted by an object vary with its temperature. A poker thrust into a fire first glows dull red, then successively bright red, orange, and yellow. Finally, if the poker achieves a high enough temperature, it becomes "white hot." The precise color of the light given off by a glowing object is thus a measure of its temperature (Fig. 4-4).

Temperature Scales

Two temperature scales are used in the United States. On the **fahrenheit scale** the freezing point of water is 32° and the boiling point of water is 212°. On the **celsius scale** these points are 0° and 100° (Fig. 4-5). The fahrenheit scale is used only in a few English-speaking countries that cling to it with the same obstinacy that preserves the equally awkward British system of units. The rest of the world, and all scientists, use the more convenient celsius (or centigrade) scale.

To go from a fahrenheit temperature T_F to a celsius temperature T_C, and vice versa, we note that 180°F separates the freezing and boiling points of water on the fahrenheit scale. On the celsius scale, however, the difference is 100°C. Therefore fahrenheit degrees are $\frac{100}{180}$, or $\frac{5}{9}$, as large as celsius degrees:

$$5 \text{ celsius degrees} = 9 \text{ fahrenheit degrees}$$

Taking into account that the freezing point of water is 0°C = 32°F, we see that

The color and brightness of objects heated until they glow, such as these steel billets, depend on their temperatures. An object that glows white is hotter than one that glows red and gives off more light as well.

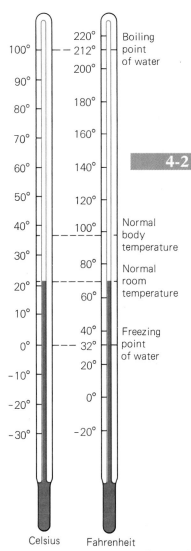

FIG. 4-5 Comparison of the celsius and fahrenheit temperature scales.

$$\text{Fahrenheit temperature} = T_F = \tfrac{9}{5}\,T_C + 32°$$
$$\text{Celsius temperature} = T_C = \tfrac{5}{9}(T_F - 32°)$$

Thus the celsius equivalent of the normal body temperature of 98.6°F is

$$T_C = \tfrac{5}{9}(98.6° - 32.0°) = \tfrac{5}{9}(66.6°) = 37.0°C$$

4-2   HEAT

Different Substances Need Different Amounts of Heat for the Same Temperature Change

The **heat** in a body of matter is the sum of the kinetic energies of all the separate particles that make up the body. The more energy these particles have, the more heat the body contains, and the higher its temperature.

Since heat is a form of energy, the joule is the proper unit for it. The amount of heat needed to raise or lower the temperature of 1 kg of a substance by 1°C depends on the nature of the substance. In the case of water, 4.2 kJ of heat is required per kilogram per °C (Fig. 4-6). To heat 1 kg of water from, say, 20°C to 60°C means raising its temperature by 40°C. The amount of heat that must be added is therefore (1 kg)(4.2 kJ/kg · °C)(40°C) = 168 kJ.

For a given temperature change, liquid water must have more heat added to or taken away from it per kilogram than nearly all other materials. For instance, to change the temperature of 1 kg of ice by 1°C, we must transfer to or from it 2.09 kJ, about half as much as for water. To do the same for 1 kg of gold takes only 0.13 kJ. To put it another way, the 4.2 kJ of heat that will change the temperature of 1 kg of water by 1°C will change the temperature of 1 kg of ice by 2°C and that of 1 kg of gold by 33°C (Fig. 4-7).

FIG. 4-6 To raise the temperature of 1 kg of water by 1°C, 4.2 kJ of heat must be added to it. The same amount of heat must be removed to cool the water by 1°C.

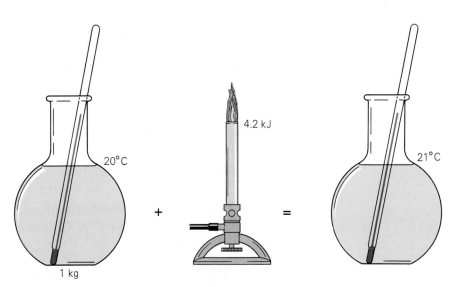

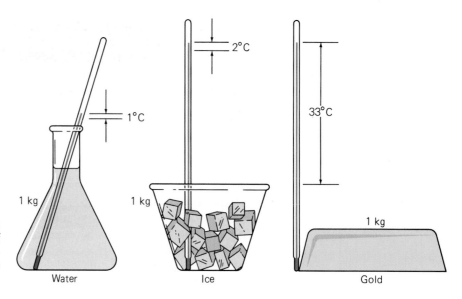

FIG. 4-7 If 4.2 kJ of heat is added to or removed from 1 kg of other substances, their temperatures change by more than the 1°C change produced in water.

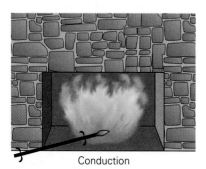

Conduction

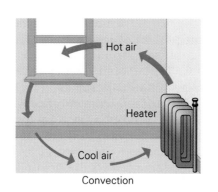

Convection

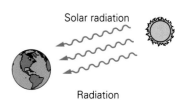

Radiation

FIG. 4-8 The three mechanisms of heat transfer.

Heat Transfer

Heat can be transferred from one place to another in three ways (Fig. 4-8). If we put one end of a poker in a fire, the other end becomes warm as heat flows through the poker. Such **conduction** is inefficient in air, and a heater warms a room mainly through the actual movement of hot air. This process is called **convection** (see Sec. 13-7). In space, which is virtually empty, neither conduction nor convection can occur to any real extent. Instead, the earth receives heat from the sun in the form of **radiation,** which consists of electromagnetic waves (light and radio waves are examples of such radiation; see Secs. 6-13 and 13-5).

4-3   METABOLIC ENERGY

The Energy of People and Animals

Metabolism refers to the biochemical processes by which the energy content of the food an animal eats is liberated. Table 4-1 lists the energy contents of some common foods. The unit is the **kilocalorie** (kcal), which is the amount of heat needed to change the temperature of 1 kg

TABLE 4-1

Energy Contents of Some Common Foods (1 kcal = 4.2 kJ)

Food	kcal
1 raw onion	5
1 dill pickle	15
6 asparagus	20
1 gum drop	35
1 poached egg	75
8 raw oysters	100
1 banana	120
1 cupcake	130
1 broiled hamburger patty	150
1 glass milk	165
1 cup bean soup	190
$\frac{1}{2}$ cup tuna salad	220
1 ice cream soda	325
$\frac{1}{2}$ broiled chicken	350
1 lamb chop	420

BANANA POWER A 50-kg girl eats a banana and she then proposes to work off its 500 kJ energy content by climbing a hill. If her body is 10 percent efficient, how high must she climb?

From Sec. 3-4 we know that the amount of work needed to raise something whose mass is m through the height h is $W = mgh$. Here the available energy is 10 percent of 500 kJ, so $W = (0.10)(500\ kJ) = 50\ kJ = 50,000\ J$, and the height is

$$h = \frac{W}{mg} = \frac{50,000\ J}{(50\ kg)(9.8\ m/s^2)} = 102\ m$$

of water by 1°C. Thus 1 kcal = 4.2 kJ. The "calorie" used by dieticians is actually the kilocalorie.

The efficiency with which metabolic energy is converted to mechanical work by muscular activity is not very high, only 10 or 20 percent. (An electric motor with the same power output as a person is typically 50 percent efficient, and larger electric motors are more efficient still.) The rest of the energy goes into heat, most of which escapes through the animal's skin. The maximum power output an animal is capable of depends upon its maximum metabolic rate, which in turn depends upon its ability to dissipate the resulting heat, and therefore upon its surface area.

A large animal has more surface area than a small one and so is capable of a higher power output. However, a large animal also has more mass than a small one, and because an animal's mass goes up faster with its size than its skin area does, its metabolic rate per kilogram decreases. Typical basal metabolic rates, which correspond to an animal resting, are 5.2 W/kg for a pigeon, 1.2 W/kg for a person, and 0.67 W/kg for a cow. African elephants partly overcome the limitation of the small surface/mass ratio of their huge bodies by their enormous ears, which help them get rid of metabolic heat. Birds are small because, with increasing size, a bird's metabolic rate (and hence power output) per kilogram decreases while the work it must perform per kilogram to fly stays the same. Large birds such as ostriches and emus are not notable for their flying ability.

When an animal is active, its metabolic rate may be much greater than its basal rate. A 70-kg person, to give an example, has a basal metabolic rate of around 80 W. When the person is reading or doing light work while sitting, the rate will go up to perhaps 125 W. Walking means a metabolic rate of 300 W or so, and running hard increases it to as much as 1200 W.

An animal's brain as well as its muscles uses energy to function. A person's brain typically uses 20 percent of his or her metabolic energy—

Egret coming in for a landing in the Florida Everglades. Birds are limited in size because the larger an animal is, the less is its ability to dissipate waste metabolic energy per unit of body mass.

thinking is hard work! The proportion is only about 9 percent for monkeys and 5 or 6 percent for cats and dogs.

When the intake of energy from food exceeds a person's metabolic needs, the excess goes into additional tissue: muscle if there is enough physical activity, otherwise fat. The energy stored in fat is available should metabolic needs not be provided by food at a later time.

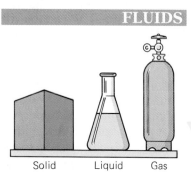

FLUIDS

Solid   Liquid   Gas

FIG. 4-9 Solids, liquids, and gases. A solid maintains its shape and volume no matter where it is placed; a liquid assumes the shape of its container while maintaining its volume; a gas expands indefinitely unless stopped by the walls of a container.

The particles of matter in a solid vibrate around fixed positions, so the solid has a definite size and shape (Fig. 4-9). In a liquid, the particles are about as far apart as those in a solid but are able to move about. Hence a liquid sample has a definite volume but flows to fit its container. In a gas, the particles can move freely, so a gas has neither a definite volume nor shape but fills whatever container it is in.

Liquids and gases are together called **fluids** because they flow readily. This ability has some interesting consequences. One is buoyancy, which permits objects to float in a fluid under certain conditions—a balloon in air, a ship in water. In order to understand how buoyancy comes about, we must first look into the ideas of density and pressure.

4-4   DENSITY

A Characteristic Property of Every Material

The **density** of a material is its mass per unit volume:

$$d = \frac{m}{V} \quad density$$

$$\text{Density} = \frac{\text{mass}}{\text{volume}}$$

When we say that lead is a "heavy" metal and aluminum a "light" one, what we really mean is that lead has a higher density than aluminum: the density of lead is 11,300 kg per cubic meter (kg/m^3) whereas that of aluminum is only 2700 kg/m^3, a quarter as much.

Although the proper SI unit of density is the kg/m^3, densities are often given instead in g/cm^3 (grams per cubic centimeter), where 1 g/cm^3 = 1000 kg/m^3 = 10^3 kg/m^3. Thus the density of lead can also be expressed as 11.3 g/cm^3. Table 4-2 lists the densities of some common substances.

Let us find the mass of the water in a bathtub whose interior is 1.300 m long and 0.600 m wide and that is filled to a height of 0.300 m (Fig. 4-10). The water's volume is

FIG. 4-10 The volume of water in this bathtub is equal to the product (length) (width) (height).

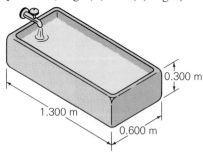

$$V = (\text{length})(\text{width})(\text{height}) = (1.300 \text{ m})(0.600 \text{ m})(0.300 \text{ m}) = 0.234 \text{ m}^3$$

According to Table 4-2 the density of water is 1000 kg/m^3. Rewriting $d = m/V$ as $m = dV$ gives

$$\text{Mass} = (\text{density})(\text{volume})$$

$$m = dV = \left(1000 \ \frac{\text{kg}}{\text{m}^3}\right)(0.234 \text{ m}^3) = 234 \text{ kg}$$

The weight of this amount of water is a little over 500 lb.

TABLE 4-2

Densities of Various Substances at Room Temperature and Atmospheric Pressure

Substance	Density, kg/m³	Substance	Density, kg/m³
Air	1.3	Hydrogen	0.09
Alcohol (ethyl)	7.9×10^2	Ice	9.2×10^2
Aluminum	2.7×10^3	Iron	7.8×10^3
Balsa wood	1.3×10^2	Lead	1.1×10^4
Concrete	2.3×10^3	Mercury	1.4×10^4
Gasoline	6.8×10^2	Oak	7.2×10^2
Gold	1.9×10^4	Water, pure	1.00×10^3
Helium	0.18	Water, sea	1.03×10^3

4-5 PRESSURE

How Much of a Squeeze We next look into what is meant by **pressure.** When a force F acts perpendicular to a surface whose area is A, the pressure acting on the surface is the ratio between the force and the area:

$$p = \frac{F}{A} \qquad \textit{pressure}$$

$$\text{Pressure} = \frac{\text{force}}{\text{area}}$$

The SI unit of pressure is the **pascal** (Pa), where

$$1 \text{ pascal} = 1 \text{ Pa} = 1 \text{ newton/meter}^2$$

This unit honors the French scientist and philosopher Blaise Pascal (1623–1662). The pascal is a very small unit: the pressure exerted by pushing hard on a table with your thumb is about a million pascals. For this reason the **kilopascal** (kPa) is often used, where 1 kPa = 1000 Pa = 10^3 Pa.

A hydraulic ram converts pressure in a liquid into an applied force. The pressure is provided by an engine-driven pump.

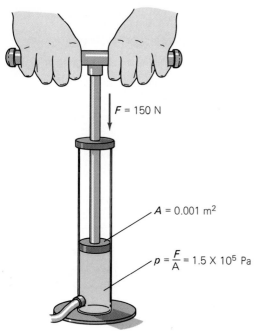

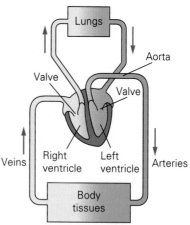

FIG. 4-11 Pressure is force per unit area. A force of 150 N applied to a piston of area 0.001 m² in a tire pump results in a pressure of 1.5×10^5 N/m² = 1.5×10^5 Pa.

FIG. 4-12 The human circulatory system. The heart consists of two pumps, called **ventricles.** The contraction and relaxation of the muscular walls of the ventricles take the place of the piston strokes of an ordinary pump. The right ventricle pumps blood from the veins through the lungs, where it absorbs oxygen from the air that has been breathed in and gives up carbon dioxide. The oxygenated blood then goes to the more powerful left ventricle, which pumps it via the aorta and the arteries to the rest of the body. A typical rate of flow of blood in a resting person is 6 L/min.

Suppose we have a tire pump with a piston whose cross-sectional area is 0.001 m² (Fig. 4-11). If we apply a force of 150 N to the piston, the pressure on the air in the pump is

$$p = \frac{F}{A} = \frac{150 \text{ N}}{0.001 \text{ m}^2} = 1.5 \times 10^5 \text{ N/m}^2 = 1.5 \times 10^5 \text{ Pa} = 150 \text{ kPa}$$

The pressures that force blood through the lungs and body are produced by the heart, as we see in Fig. 4-12.

FIG. 4-13 An aneroid barometer. The flexible ends of a sealed metal chamber are pushed in by a high atmospheric pressure. Under low atmospheric pressure, the air inside the chamber pushes the ends out.

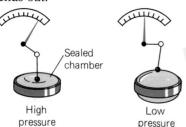

Pressure and Depth In the pump of Fig. 4-11, the air inside is under greater pressure at the bottom of the cylinder than at the top because of the weight of the air in the cylinder. In the case of a tire pump, the pressure difference is very small, of course, but elsewhere it can be significant. At sea level on the earth, for instance, the pressure due to the weight of the air above us averages 101 kPa, nearly 15 lb/in². This corresponds to a force of 10.1 N on every square centimeter of our bodies. We are not aware of this pressure because the pressures inside our bodies are the same. Atmospheric pressures are measured with instruments called **barometers,** one type of which is shown in Fig. 4-13.

Because pressure increases with depth, most submarines cannot go down more than a few hundred meters without the danger of collapsing. At a depth of 10 km in the ocean, the pressure is about 1000 times sea-level atmospheric pressure, enough to compress water by 3 percent of

Arterial blood pressures are measured by using an inflatable cuff that is pumped up until the flow of blood stops, as monitored by a stethoscope. Air is then let out of the cuff until the flow just begins again, which is recognized by a gurgling sound in the stethoscope. The pressure at this point is called *systolic* and corresponds to the maximum pressure the heart produces in the arteries. Next, more air is let out until the gurgling stops, which corresponds to normal blood flow. The pressure now, called *diastolic*, corresponds to the arterial pressure between strokes of the heart. Blood pressures are usually expressed in *torr*, which is the pressure exerted by a column of mercury 1 mm high; 1 torr = 133 Pa. In a healthy person the systolic and diastolic blood pressures are about 120 and 80 torr, respectively.

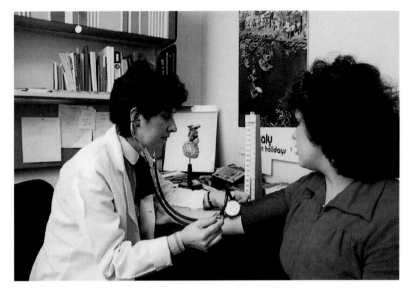

its volume. Fish that live at such depths are not crushed for the same reason we can survive the pressure at the bottom of our ocean of air: pressures inside their bodies are kept equal to pressures outside.

Scuba divers carry tanks of compressed air with regulator valves that provide air at the same pressure as the water around them. ("Scuba" stands for *s*elf-*c*ontained *u*nderwater *b*reathing *a*pparatus.) When a diver returns to the surface, he or she must breathe out continuously to allow the air pressure in the lungs to decrease at the same rate as the external pressure decreases. If this is not done, the pressure difference may burst the lungs. When brought to the surface quickly, deep-sea fish sometimes explode because of their high internal pressures.

Air at high pressure in the tank of a scuba diver is reduced by a regulator valve to the pressure at the depth of the water. The diver must wear lead weights to overcome his or her buoyancy. The deeper the diver goes, the greater the water pressure, and the faster the air in the tank is used up.

4-6   BUOYANCY

**Sink
or Swim**

An object immersed in a fluid is acted upon by an upward force that arises because pressures in a fluid increase with depth. Hence the upward force on the bottom of the object is greater than the downward force on its top. The difference between the two forces is the **buoyant force.**

Buoyancy enables balloons to float in the air and ships to float in the sea. If the buoyant force on an immersed object is greater than its weight, the object floats; if the force is less than its weight, the object sinks.

Imagine a solid object of any kind whose volume is V that is in a tank of water (Fig. 4-14). A body of water of the same size and shape in the tank is supported by a buoyant force F_b equal to its weight of $W_{water} = dVg$. The buoyant force is the result of all of the forces that the rest of the water in the tank exert on this particular body of water. This force is always upward because the pressure underneath the body of water is greater than the pressure above it. The pressures on the sides cancel one another out, as shown in the figure.

If we now replace the body of water by the solid object, the forces on the object are the same as before. The buoyant force therefore remains dVg. In this formula, d is the density of the fluid (which need not be water, of course), V is the volume of fluid displaced by the solid object, and g is the acceleration of gravity, 9.8 m/s^2.

Thus we have **Archimedes' principle:**

> **Buoyant force on an object in a fluid = weight of fluid displaced by the object**

Archimedes' principle holds whether the object floats or sinks. If the object's weight W_{object} is greater than the buoyant force F_b, it sinks. If its

FIG. 4-14 Archimedes' principle. The buoyant force F_b on an object immersed in water (or other fluid) is equal to the weight of the body of water displaced by the object.

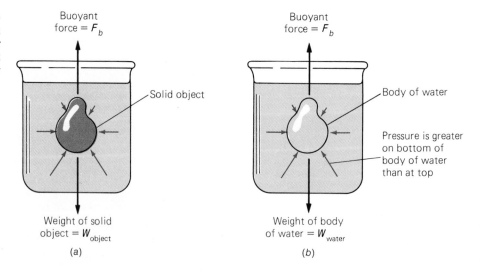

Buoyant force = F_b

Solid object

Weight of solid object = W_{object}

(a)

Buoyant force = F_b

Body of water

Pressure is greater on bottom of body of water than at top

Weight of body of water = W_{water}

(b)

Archimedes (287?–212 B.C.), the preeminent scientist/mathematician of the ancient world, was born in Syracuse, Sicily, at that time a Greek colony. He went to Alexandria, Egypt, to study under a former pupil of Euclid and then returned to Syracuse where he spent the rest of his life. Many stories have come down through the ages about Archimedes, some of them possibly true. The principle named after him, that the buoyant force on a submerged object equals the weight of the fluid it displaces, is supposed to have come to him while in his bath, whereupon he rushed naked into the street crying "Eureka!" ("I've got it!") He had been trying to think of a way to determine whether a new crown made for King Hieron of Sicily was pure gold without damaging it and used his discovery to show that it was not; the goldsmith was then executed.

Archimedes worked out the theory of the lever and remarked (so it is said), "Give me a place to stand and I can move the world." King Hieron then challenged him to move something really large, even if not the world,

and Archimedes responded by using a system of pulleys, which are developments of the lever, to haul a laden ship overland all by himself. Archimedes was an able mathematician, and among his other accomplishments he calculated that the value of π lay between 223/71 and 220/70, an extremely good approximation. Archimedes is given credit for keeping the Roman fleet that attacked Syracuse in 215 B.C. at bay for 3 years by a variety of clever devices, including giant lenses that focused sunlight on the ships and set them afire. Finally the Romans did conquer Syracuse, and a Roman soldier found Archimedes bent over a mathematical problem scratched in the sand. When Archimedes would not stop work immediately, the soldier killed him.

weight is less than the buoyant force, it floats, in which case the volume V refers only to the part of the object that is in the fluid.

The condition for an object to float in a given fluid, then, is that its average density be lower than the density of the fluid. Why does a steel ship float when the density of steel is nearly 8 times that of water? The answer is that the ship is a hollow shell, so its average density is less than that of water even when loaded with cargo. If the ship springs a leak and fills with water, its average density goes up, and the ship sinks. The purpose of a life jacket is to reduce the average density of a person in the water so that she floats higher and is less likely to get water in her lungs and drown as a result.

4-7 THE GAS LAWS

All Gases Obey Them In many ways the gaseous state is the one whose behavior is the easiest to describe and account for. As an important example, the pressures, volumes, and temperatures of gas samples are related by simple formulas that have no counterpart in the cases of liquids and solids. The discovery of these formulas led to a search for their explanation in terms of the basic nature of gases, a search that resulted in the kinetic theory of matter.

106

Because water expands when it freezes, ice floats. Nearly 90 percent of the volume of this iceberg near Greenland lies below the surface.

Boyle's Law   Suppose that a sample of some gas is placed in the cylinder of Fig. 4-15, and a pressure of 100 kPa is applied. The final volume of the sample is 1 m³. If we double the pressure to 200 kPa, the piston will move down until the gas volume is 0.5 m³, half its original amount, provided the gas temperature is kept unchanged. If the pressure is made 10 times greater, the piston will move down farther, until the gas occupies a volume of 0.1 m³, again if the gas temperature is kept unchanged.

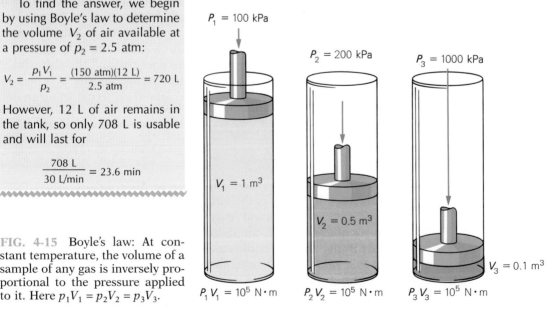

FIG. 4-15 Boyle's law: At constant temperature, the volume of a sample of any gas is inversely proportional to the pressure applied to it. Here $p_1V_1 = p_2V_2 = p_3V_3$.

At constant pressure, heating a gas causes it to expand. The density of hot air therefore is less than the density of cool air at the same pressure, which is why a hot-air balloon is buoyant. These balloons have propane burners in their gondolas to supply the needed heat.

These findings can be summarized by saying that the volume of a given quantity of a gas at constant temperature is inversely proportional to the pressure applied to it. (By "inversely proportional" is meant that as the pressure increases, the volume decreases by the same proportion.) If the volume of the gas is V_1 when the pressure is p_1 and the volume changes to V_2 when the pressure is changed to p_2, the relationship among the various quantities is

$$\frac{p_1}{p_2} = \frac{V_2}{V_1} \qquad \text{(at constant temperature)} \qquad \textit{Boyle's law}$$

This relationship is called **Boyle's law,** in honor of the English physicist who discovered it.

Charles's Law

Changes in the volume of a gas sample are also related to temperature changes in a simple way. If a gas is cooled steadily, starting at 0°C, while its pressure is maintained constant, its volume decreases by $\frac{1}{273}$ of its volume at 0°C for every degree the temperature falls. If the gas is heated, its volume increases by the same fraction (Fig. 4-16). If volume rather than pressure is kept fixed, the pressure increases with rising temperature and decreases with falling temperature, again by the fraction $\frac{1}{273}$ of its 0°C value for every degree change.

These figures suggest an obvious question: What would happen to a gas if we could lower its temperature to −273°C? If we kept the gas at constant volume, the pressure at this temperature ought to fall to zero. If the pressure stayed constant, the volume ought to fall to zero.

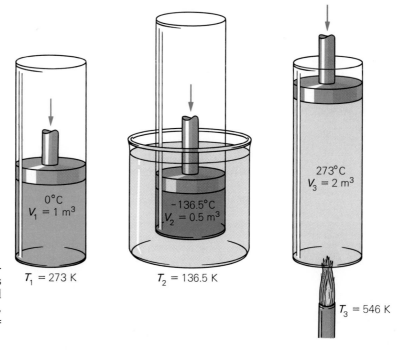

FIG. 4-16 Charles's law: At constant pressure, the volume of a gas sample is directly proportional to its absolute temperature T_K, where $T_K = T_C + 273°$. Here $V_1/T_1 = V_2/T_2 = V_3/T_3$.

Lord Kelvin (1824–1907)

It is hardly likely, however, that our experiments would have such results. In the first place, we should find it impossible to reach quite so low a temperature. In the second place, all known gases turn into liquids before that temperature is reached. Nevertheless, a temperature of −273°C has a special significance, a significance that will become clearer shortly. This temperature is called **absolute zero.**

For many scientific purposes it is convenient to begin the temperature scale at absolute zero. Temperatures on such a scale, given as degrees celsius above absolute zero, are called **absolute temperatures.** Thus the freezing point of water is 273° absolute, written as 273 K in honor of the English physicist Lord Kelvin, and the boiling point of water is 373 K. Any celsius temperature T_C can be changed to its equivalent absolute temperature T_K by adding 273 (Fig. 4-17):

$$T_K = T_C + 273$$

Absolute temperature = celsius temperature + 273

Using the absolute scale, we can express the relationship between gas volumes and temperatures quite simply: the volume of a gas is directly proportional to its absolute temperature (Fig. 4-18). This relation may be expressed in the form

$$\frac{V_1}{V_2} = \frac{T_1}{T_2} \qquad \text{(at constant pressure)} \qquad \textit{Charles's law}$$

FIG. 4-17   The absolute temperature scale.

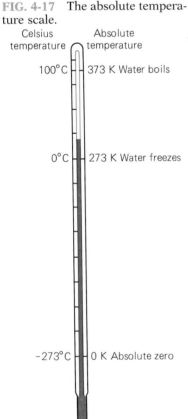

where the T's are absolute temperatures. Discovered by two eighteenth-century French physicists, Jacques Alexandre Charles and Joseph Gay-Lussac, this relation is commonly known as **Charles's law.**

Ideal Gas Law   Boyle's and Charles's laws can be combined in a single formula known as the **ideal gas law:**

$$\frac{p_1 V_1}{T_1} = \frac{p_2 V_2}{T_2} \qquad \textit{ideal gas law}$$

At constant temperature, $T_1 = T_2$ and we have Boyle's law. At constant pressure, $p_1 = p_2$ and we have Charles's law. Another way to write the ideal gas law is

FIG. 4-18   Graphic representation of Charles's law, showing the proportionality between volume and absolute temperature for a gas at constant pressure. If the temperature of the gas could be reduced to absolute zero, its volume would fall to zero. Actual gases liquefy at temperatures above absolute zero.

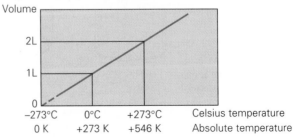

$$\frac{pV}{T} = \text{constant} \qquad \textit{ideal gas law}$$

since this particular combination of quantities does not change in value for a gas sample even though the individual quantities p, V, and T may vary.

KINETIC THEORY OF MATTER

The **kinetic theory of matter** accounts for a wide variety of physical and chemical properties of matter in terms of a simple model. According to this model, all matter is composed of tiny particles called **molecules** that are in constant motion. Later on we shall discuss molecules in some detail, but for the time being we can simply regard them as the basic particles characteristic of a substance.

4-8 KINETIC THEORY OF GASES

Why Gases Behave as They Do

Today we know a great deal about the sizes, speeds, even shapes of the molecules in various kinds of matter.

For example, a molecule of nitrogen, the chief constituent of air, is about 0.18 billionth of a meter (1.8×10^{-10} m) across and has a mass of 4.7×10^{-26} kg. It travels (at 0°C) at an average speed of 500 m/s, about the speed of a rifle bullet, and in each second collides with more than a billion other molecules. Of similar dimensions and moving with similar speeds in each cubic centimeter of air are 2.7×10^{19} other molecules. If all the molecules in such a thimbleful of air were divided equally among the 5.6 billion people on the earth, each person would receive almost 5 billion molecules.

The three basic assumptions of the kinetic theory for gas molecules, which have been verified by experiment, are these:

1. Gas molecules are small compared with the average distance between them.

2. Gas molecules collide without loss of kinetic energy.

3. Gas molecules exert almost no forces on one another, except when they collide.

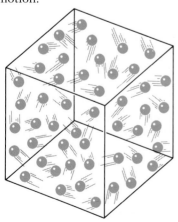

FIG. 4-19 The molecules of a gas are in constant random motion.

A gas, then, is mostly empty space, with its isolated molecules moving helter-skelter like a swarm of angry bees in a closed room (Fig. 4-19). Each molecule collides with others several billion times a second, changing its speed and direction at each collision but unaffected by its neighbors between collisions. If a series of collisions brings it momentarily to a stop, new collisions will set it in motion. If its speed becomes greater than the average, successive collisions will slow it down. There is no order in the motion, no uniformity of speed or direction. All we can say is that the molecules have a certain average speed and that at any instant as many molecules are moving in one direction as in another.

This animated picture explains the more obvious properties of gases. The ability of a gas to expand and to leak through small openings

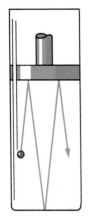

FIG. 4-20 Gas pressure is the result of molecular bombardment. For simplicity, only vertical molecular motions are shown.

follows from the rapid motion of its molecules and their lack of attraction for one another. Gases are easily compressed because the molecules are, on the average, widely separated. One gas mixes with another because the spaces between molecules leave plenty of room for others. Because a given volume of a gas consists mainly of empty space, the mass of that volume is much less than that of the same volume of a liquid or a solid.

Origin of Boyle's Law The pressure of a gas on the walls of its container is the result of bombardment by billions and billions of molecules, the same bombardment that causes brownian movement (Fig. 4-20). The many tiny, separate blows affect our senses and our measuring instruments as a continuous force.

The kinetic theory accounts nicely for Boyle's law, $p_1V_1 = p_2V_2$ (at constant temperature). As in Fig. 4-21, we can think of the molecules of a gas in a cylinder as moving in a regular manner, some of them vertically between the piston and the base of the cylinder and the others horizontally between the cylinder walls. If the piston is raised so that the gas volume is doubled, the vertically moving molecules have twice as far to go between collisions with top and bottom and hence will strike only half as often. The horizontally moving molecules must spread their blows over twice as great an area, and hence the number of impacts per unit area will be cut in half. Thus the pressure in all parts of the cylinder is exactly halved, as Boyle's law predicts. It is not hard to extend this reasoning to a real gas whose molecules move at random.

4-9 MOLECULAR MOTION AND TEMPERATURE

FIG. 4-21 Origin of Boyle's law. Expanding a gas sample means that its molecules must travel farther between successive impacts on the container wall and that their blows are spread over a larger area, so the gas pressure drops.

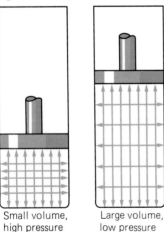

Small volume, high pressure

Large volume, low pressure

The Faster the Molecules, the Higher the Temperature To account for the effect of a temperature change on a gas, the kinetic theory requires a further concept:

4. The absolute temperature of a gas is proportional to the average kinetic energy of its molecules.

That temperature should be related to molecular energies and thus to molecular speeds follows from the increase in the pressure of a confined gas as its temperature rises. Increases in pressure must mean that the molecules are striking the walls of their container more forcefully and so must be moving faster.

Earlier in this chapter we learned that the pressure of a gas approaches zero as its temperature falls toward 0 K, which is –273°C. For the pressure to become zero, molecular bombardment must stop. Thus absolute zero is interpreted as the temperature at which gas molecules would lose their kinetic energies completely, as shown in Fig. 4-22. (This is a simplification of the actual situation: in reality, even at 0 K a molecule will have a very small amount of KE that cannot be reduced.) There can be no lower temperature, simply because there can be no smaller amount of energy. The regular increase of gas pressure with absolute temperature if the volume is constant and the similar increase of volume if the pressure is constant (Charles's law) follow from this definition of absolute zero.

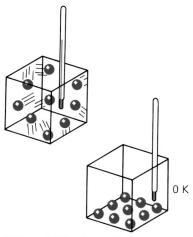

FIG. 4-22 According to the kinetic theory of gases, at absolute zero the molecules of a gas would not move. More advanced theories show that even at 0 K a very slight movement will persist.

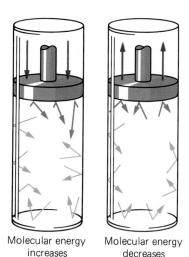

Molecular energy increases   Molecular energy decreases

FIG. 4-23 Compressing a gas causes its temperature to rise because molecules rebound from the piston with more energy. Expanding a gas causes its temperature to drop because molecules rebound from the piston with less energy.

Origin of Charles's Law

If temperature is a measure of average molecular energy, then compressing a gas in a cylinder ought to cause its temperature to rise. While the piston of Fig. 4-23 is moving down, molecules rebound from it with increased energy just as a baseball rebounds with increased energy from a moving bat. To verify this prediction, all we have to do is pump up a bicycle tire and notice how hot the pump becomes after the air in it has been compressed a few times. On the other hand, if a gas is expanded by pulling a piston outward, its temperature falls, since each molecule that strikes the retreating piston gives up some of its kinetic energy.

The cooling effect of gas expansion explains the formation of clouds from rising moist air, as discussed in Chap. 13. Atmospheric pressure decreases with altitude, and the water vapor in the moist air cools as it moves upward until it condenses into the water droplets that constitute clouds.

Why Molecules Keep Moving

The picture of a gas as a collection of tiny molecules that fly about in all directions brings up the question, What keeps them moving? We might think of molecules as being like billiard balls that move rapidly and collide with one another many times a second. But real billiard balls would soon lose all their kinetic energy and come to rest. All other motions of our experience, except those of the moon and planets, similarly stop sooner or later unless some outside force keeps them going. No outside force acts on molecular motions, yet they keep on indefinitely. Why is it that friction of some kind does not affect molecules?

The answer is that friction changes kinetic energy into heat, which is molecular energy. Friction between molecules would mean molecular energy turning into molecular energy—which is not a change at all. The question of what keeps molecules moving is really meaningless. The

In the operation of a snowmaking machine, a mixture of compressed air and water is blown through a set of nozzles. The expansion of the air cools the mixture sufficiently to freeze the water into the ice crystals of snow.

energy of motion in the molecular world cannot become heat, as it does in the larger world of everyday life, because this energy *is* heat.

The kinetic theory of matter is clearly a success in explaining the behavior of gases. Let us now see what this theory has to say about liquids and solids. In particular, changes of state between gas and liquid and between liquid and solid are extremely interesting in molecular terms.

4-10 LIQUIDS AND SOLIDS

Intermolecular Forces Hold Them Together   If a gas is like a swarm of angry bees, the molecules in a liquid are more like bees in a hive, crawling over one another constantly. Liquids flow because their molecules slide past one another easily, but they flow less readily than gases because of intermolecular attractions that act only over short distances.

The forces between the molecules of a solid are stronger than those in a liquid, so strong that the molecules are not free to move about (Fig. 4-24). They are hardly at rest, however. Held in position as if by springs attached to its neighbors, each molecule vibrates back and forth rapidly (Fig. 4-25). Each spring represents a bond between two adjacent molecules. Such bonds are electrical in nature. A solid is elastic because its molecules return to their normal separations after being pulled apart or pushed together when a moderate force is applied. If the force is great, the solid may be permanently deformed. In this process, the molecules shift to new positions and find new partners for their attractive forces. Too much applied force, of course, may break the solid apart.

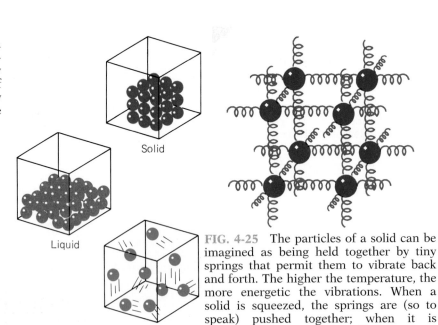

FIG. 4-24 Molecular models of a solid, a liquid, and a gas. The molecules of a solid are firmly attached to one another; those of a liquid can move about but stay close together; those of a gas have no restrictions on their motion.

Solid

Liquid

Gas

FIG. 4-25 The particles of a solid can be imagined as being held together by tiny springs that permit them to vibrate back and forth. The higher the temperature, the more energetic the vibrations. When a solid is squeezed, the springs are (so to speak) pushed together; when it is stretched, the springs are pulled apart.

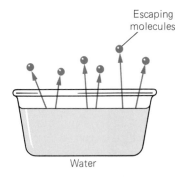

Escaping molecules

Water

Alcohol

FIG. 4-26 Evaporation. Alcohol evaporates more rapidly than water because the attractive forces between its molecules are smaller. In each case, the faster molecules escape. Hence the average kinetic energy of the remaining molecules is lower and the liquid temperature drops.

4-11 EVAPORATION AND BOILING

Liquid into Gas Suppose we have two liquids, water and alcohol, in open dishes (Fig. 4-26). Molecules in each dish are moving in all directions, with a variety of speeds. At any instant some molecules are moving fast enough upward to escape into the air in spite of the attractions of their slower neighbors. By this loss of its faster molecules each liquid gradually evaporates. Since the remaining molecules are the slower ones, evaporation leaves cool liquids behind. The alcohol evaporates more quickly than the water and cools itself more noticeably because the attraction of its particles for one another is smaller and a greater number can escape.

When we add heat to a liquid, eventually a temperature is reached at which even molecules of average speed can overcome the forces binding them together. Now bubbles of gas form throughout the liquid, and it begins to boil. This temperature is accordingly called the **boiling point** of the liquid. As we would expect, the boiling point of alcohol (78°C) is lower than that of water (100°C).

Thus evaporation differs from boiling in two ways:

1. Evaporation occurs only at a liquid surface; boiling occurs in the entire volume of liquid.

2. Evaporation occurs at all temperatures; boiling occurs only at the boiling point.

Heat of Vaporization Whether evaporation takes place by itself from an open dish or is aided by heating, forming a gas from a liquid requires energy. In the first case energy is supplied from the heat content of the liquid itself (since the liquid grows cooler), in the other case from the outside source of heat. For water at its boiling point of 100°C, 2260 kJ (the **heat of vaporization**) is needed to change each kilogram of liquid into gas (Fig. 4-27). With no difference in temperature between liquid and gas, there is no difference in their average molecular kinetic energies. If not into kinetic energy, into what form of molecular energy does the 2260 kJ of heat go?

FIG. 4-27 The heat of vaporization of water is 2260 kJ/kg.

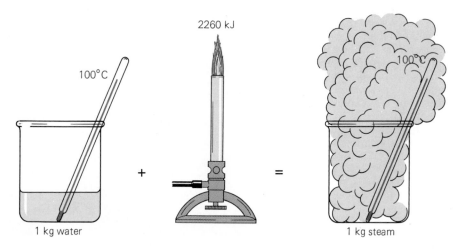

2260 kJ

100°C

100°C

+   =

1 kg water   1 kg steam

Boiling points vary with pressure: the higher the pressure, the higher the boiling point. At twice sea-level atmospheric pressure, for instance, water boils at 120°C. The pressure cooker is based on this observation. When water is heated in a closed container, the pressure in the container increases, and the temperature at which the water inside boils will be correspondingly higher than 100°C. In this way food can be cooked faster than in an open pan. Similarly, lowering the pressure reduces the boiling point. Atmospheric pressure decreases with altitude (see Fig. 13-1), so the boiling point of water in Denver, which is well above sea level, is only 96°C.

Intermolecular forces provide the answer. In a liquid these forces are strong because the molecules are close together. To tear the molecules apart, to separate them by the wide distances that exist in the gas, requires that these strong forces be overcome. Each molecule must be moved, against the pull of its neighbors, to a new position in which their attraction for it is very small. Just as a stone thrown upward against the earth's gravity gains potential energy, so molecules moved apart in this way gain potential energy—potential energy with respect to intermolecular forces. When a gas becomes a liquid, the process is reversed. The molecules "fall" toward one another under the influence of their mutual attractions, and their potential energy is taken up as heat by the surroundings.

The high heat of vaporization of water is what makes steam dangerous. If a gram of water (a third of a teaspoonful) at 100°C falls on a person's skin, the heat given to the skin as the water cools to body temperature will be about 0.26 kJ. But if a gram of steam at 100°C strikes the person's skin, 2.26 kJ of heat of vaporization will also be given off, for a total of 2.52 kJ—almost 10 times as much heat. Hence great care must always be taken near steam.

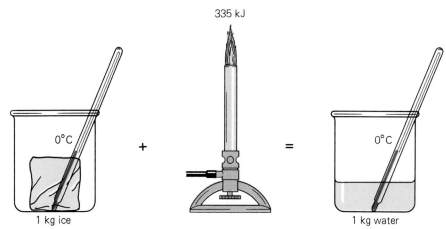

4-12 MELTING

Solid into Liquid   Just as heat must be added at its boiling point to turn a liquid into a gas, so heat at its melting point is needed to turn a solid into a liquid. The heat required to change 1 kg of a solid at its melting point into a liquid is called the **heat of fusion** of the substance. The same amount of heat must be given off by 1 kg of the substance when it is a liquid at its melting point for it to harden into a solid. The heat of fusion of water is 335 kJ/kg (Fig. 4-28). Most other substances have lower heats of fusion.

The heat of fusion of a substance is always much smaller than its heat of vaporization. The molecules of a solid are arranged in a fixed pattern such that the forces holding each one to its neighbors are as large as possible. To overcome these forces and give the molecules the random, constantly shifting arrangement of a liquid, additional energy must be given to them (Fig. 4-29). However, the molecules still stick together sufficiently to give a liquid sample a definite volume. To turn a liquid into a gas, enough energy must be added to pull the molecules permanently apart, a much harder job. The molecules of the gas can then move about freely, and the gas expands. At atmospheric pressure 1 L of water becomes 1680 L of steam; in a vacuum, it would expand indefinitely.

Figure 4-30 shows what happens as we supply heat to 1 kg of ice originally at −50°C. The ice warms up until it reaches 0°C, when it begins to melt. Then the temperature remains steady at 0°C until all the ice has melted. When all the ice has become water, the temperature rises

FIG. 4-29 The orderly arrangement of particles in a crystalline solid changes to the random arrangement of particles in a liquid when enough energy is supplied to the solid to overcome the bonding forces within it.

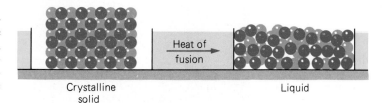

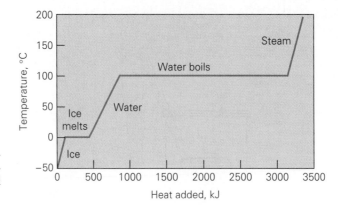

FIG. 4-30 A graph of the temperature of 1 kg of water, originally ice at −50°C, as heat is added to it.

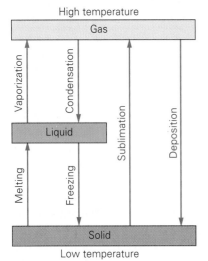

FIG. 4-31 The various changes of state.

again. When the water reaches 100°C, it starts to turn into steam, which takes a lot of heat, much more than the ice needed to melt. Finally all the water has become steam at 100°C and the temperature of the steam now increases as more heat is added.

Sublimation Most substances change directly from the solid to the vapor state, a process called **sublimation,** under the right conditions of temperature and pressure. Usually pressures well under atmospheric pressure are needed for sublimation. A familiar exception is solid carbon dioxide ("dry ice"), which turns into a gas without first becoming a liquid even at atmospheric pressure. Figure 4-31 summarizes all the changes of state we have been discussing.

The best instant coffee is prepared with the help of sublimation. The brewed coffee is first frozen and then put in a vacuum chamber. The ice in the frozen coffee sublimes to water vapor, which is pumped away. Freeze drying affects the flavor of coffee much less than drying it by heating. The same process is also used to preserve other materials of biological origin, such as blood plasma.

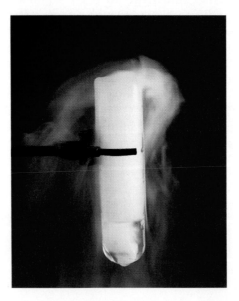

Solid carbon dioxide ("dry ice") vaporizes into a gas at atmospheric pressure, a process called sublimation.

Any form of energy, including heat, can be converted to any other form. But heat is unusual in that it cannot be converted *efficiently*. We routinely obtain mechanical energy from the heat given off by burning coal and oil in engines of various types, but a large part of the heat is always wasted—about two-thirds in the case of electric power stations, for example. The losses are serious because nearly all the "raw" energy available to modern civilization is liberated from its sources as heat.

The basic inefficiency of all engines whose energy input is heat was discovered in the nineteenth century, at the start of the Industrial Revolution. It is not a question of poor design or construction; the transformation of heat simply will not take place without such losses. Research both by engineers trying to get as much mechanical energy as possible from each ton of fuel and by scientists whose interest was in the properties of heat eventually brought to light why such transformations are so wasteful. As we shall see, the ultimate reason is that what we call heat is actually the kinetic energy of random molecular motion.

4-13 HEAT ENGINES

Turning Heat into Work

Heat is the easiest and cheapest form of energy to obtain, since all we have to do to obtain it is to burn an appropriate fuel. A device that turns heat into mechanical energy is called a **heat engine.** Examples are the gasoline and diesel engines of cars, the jet engines of aircraft, and the steam turbines of ships and power stations. All these engines operate in the same basic way: a gas is heated and allowed to expand against a piston or the blades of a turbine.

Figure 4-32 shows a gas being heated in a cylinder. As the temperature of the gas increases, its pressure increases as well, and the piston is pushed upward. The energy of the upward-moving piston can be used to propel a car or turn a generator or for any other purpose we wish. When the piston reaches the top of the cylinder, however, the conversion of heat into mechanical energy stops. In order to keep the engine working, we must now push the piston back down again in order to begin another energy-producing expansion.

If we push the piston down while the gas in the cylinder is still hot, we will find that we have to do exactly as much work as the energy provided by the expansion. Thus there will be no net work done at all. To make the engine perform a net amount of work in each cycle, we must first cool the gas so that less work is needed to compress it. It is in this cooling process that heat is lost. There is no way to avoid throwing away some of the heat added to the gas in the expansion if the engine is to continue to work. The wasted heat usually ends up in the atmosphere around the engine, in the water of a nearby river, or in the ocean.

What happens in a complete cycle, then, is that heat flows in and out of the engine, and during the flow we manage to change some of the heat into mechanical energy. We need *both* a hot reservoir and a cold one in order to produce a flow of heat from which we can extract some of it, as in Fig. 4-33. In a gasoline or diesel engine the hot reservoir is

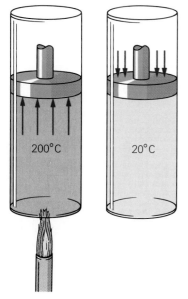

FIG. 4-32 An idealized heat engine. A gas at 200°C gives out more energy in expanding than is required to compress the gas at 20°C. This excess energy is available for doing work.

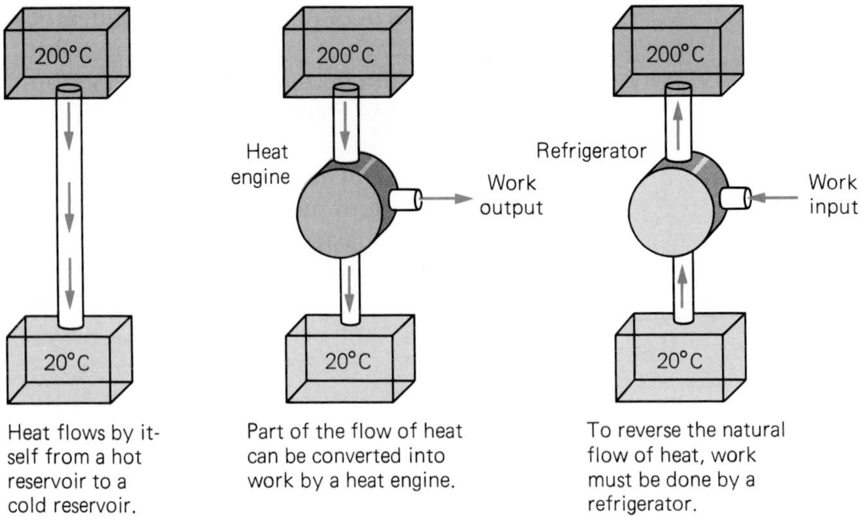

FIG. 4-33 A heat engine converts part of the heat flowing from
a hot reservoir to a cold one into
work. A refrigerator extracts heat
from a cold reservoir and delivers
it to a hot one by doing work that
is converted into heat.

Heat flows by it-
self from a hot
reservoir to a
cold reservoir.

Part of the flow of heat
can be converted into
work by a heat engine.

To reverse the natural
flow of heat, work
must be done by a
refrigerator.

GASOLINE AND The operating
DIESEL ENGINES cycle of a
four-stroke
gasoline engine is shown in Fig.
4-34. In the intake stroke, a mixture of gasoline and air from the
carburetor is sucked into the
cylinder as the piston moves
downward. In the compression
stroke, the fuel-air mixture is
compressed to one-seventh or
one-eighth of its original volume.
At the end of the compression
stroke the spark plug is fired,
which ignites the fuel-air mixture.
The expanding gases force the
piston downward in the power
stroke. Finally the piston moves
upward again to force the spent
gases out through the exhaust
valve.

In a diesel engine, only air is
drawn into the cylinder in the
intake stroke. At the end of the
compression stroke, diesel fuel is
injected into the cylinder and is
ignited by the high temperature
of the compressed air. No spark
plug is needed. Diesel engines
are more efficient but also heavier and more expensive than
gasoline engines.

the burning gases of the power stroke and the cold reservoir is the
atmosphere.

A vast amount of heat is contained in the molecular motions of the
atmosphere, the oceans, and the earth itself, but only rarely can we use
it because we need a colder reservoir nearby to which the heat can flow.
What about using a refrigerator as the cold reservoir? A **refrigerator** is
the reverse of a heat engine, as we see in Fig. 4-33. It uses mechanical
energy to push heat "uphill" from a cold reservoir (the inside of the
refrigerator) to a warm reservoir (the air of the kitchen), a path opposite
to the normal direction of heat flow. Because of the energy needed to
drive a refrigerator, using one as the cold reservoir for a heat engine
would be a losing proposition.

FIG. 4-34 A four-stroke gasoline engine.

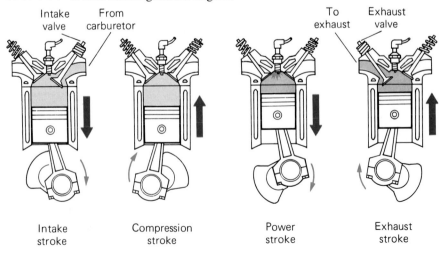

Intake   Compression   Power   Exhaust
stroke   stroke   stroke   stroke

HOW A REFRIGERATOR WORKS
A refrigerator takes in heat at a low temperature and exhausts it at a higher one. Figure 4-35 shows a refrigerator whose working substance is an easily liquefied gas called a **refrigerant.** The most common refrigerants are CFCs, which are compounds of chlorine, fluorine, and carbon such as CCl_2F_2, whose trade name is Freon 12. (Because they damage the ozone layer of the atmosphere, CFCs are being replaced by other gases; see Sec. 13-1.) The operation of the refrigeration system of Fig. 4-35 proceeds as follows:

1. The **compressor,** usually driven by an electric motor, brings the refrigerant to a high pressure, which raises its temperature as well.
2. The hot refrigerant passes through the **condenser,** an array of thin tubes that give off heat from the refrigerant to the atmosphere. The condenser is on the back of most household refrigerators. As it cools, the refrigerant becomes a liquid under high pressure.
3. The liquid refrigerant now goes into the **expansion valve,** from which it emerges at a lower pressure and temperature.
4. In the **evaporator** the cool liquid refrigerant absorbs heat from the storage chamber and vaporizes. Farther along in the evaporator the refrigerant vapor absorbs more heat and becomes warmer. The warm vapor then goes back to the compressor to begin another cycle.

In step 4 of this cycle, heat is extracted from the storage chamber by the refrigerant. In step 1, work is done on the refrigerant by the compressor. In step 2, heat from the refrigerant leaves the system. A refrigerator might remove two or more times as much heat from its storage chamber as the amount of work done.

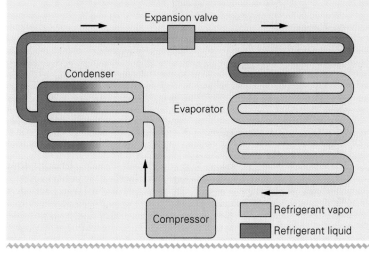

FIG. 4-35 A typical refrigeration system. Heat is absorbed by the refrigerant from the storage chamber in the evaporator and is given up by the refrigerant in the condenser.

4-14   THERMODYNAMICS

You Can't Win

Thermodynamics is the science of heat transformation, and it has two fundamental laws:

1. Energy cannot be created or destroyed, but it can be converted from one form to another.
2. It is impossible to take heat from a source and change all of it to mechanical energy or work; some heat must be wasted.

The first law of thermodynamics is the same as the law of conservation of energy discussed in Chap. 3. What it means is that we can't get something for nothing. The second law singles out heat from other kinds of energy and recognizes that all conversions of heat into any of the others must be inefficient.

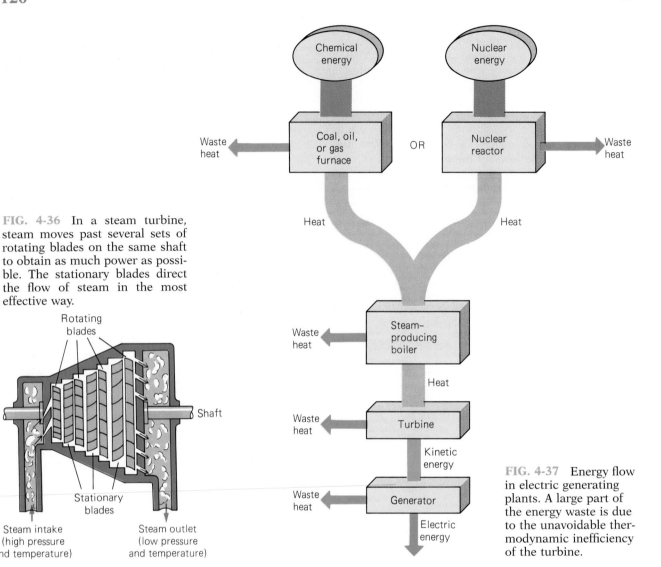

FIG. 4-36 In a steam turbine, steam moves past several sets of rotating blades on the same shaft to obtain as much power as possible. The stationary blades direct the flow of steam in the most effective way.

FIG. 4-37 Energy flow in electric generating plants. A large part of the energy waste is due to the unavoidable thermodynamic inefficiency of the turbine.

Thermodynamics is able to specify the maximum efficiency of a heat engine, ignoring losses to friction and other practical difficulties. The maximum efficiency turns out to depend only on the absolute temperatures T_{hot} and T_{cold} of the hot and cold reservoirs between which the engine operates:

$$\text{Maximum efficiency} = 1 - \frac{T_{cold}}{T_{hot}} \qquad \textit{engine efficiency}$$

The greater the ratio between the two temperatures, the less heat is wasted and the more efficient the engine.

Figure 4-36 shows the basic design of a steam turbine. In a power station, the steam comes from a boiler heated by a coal, oil, or gas furnace or by a nuclear reactor, and the turbine shaft is connected to an electric generator. In a typical power station, steam enters a turbine at about 570°C and leaves at about 95°C into a partial vacuum. The corresponding absolute temperatures are 843 K and 368 K, so the maximum efficiency of such a turbine is

$$\text{Maximum efficiency} = 1 - \frac{T_{\text{cold}}}{T_{\text{hot}}} = 1 - \frac{368 \text{ K}}{843 \text{ K}} = 0.56$$

which is 56 percent. The actual efficiency is less than 40 percent because of friction and other sources of energy loss (Fig. 4-37).

Why a Heat Engine Must Be Inefficient   On a molecular level, it is not hard to see why heat resists being changed into other forms of energy. When heat is added to the gas of a heat engine, its molecules increase their average speeds. But the molecules are moving in random directions, whereas the engine can draw upon the increased energies of only those molecules that are moving in more or less the same direction as the piston or turbine blades. If we could line up the molecules and aim them all, like miniature bullets, right at the piston or turbine blades, all the added energy could be turned into mechanical energy. Because this is impossible, only a fraction of any heat given to a gas can be extracted as energy of orderly motion. The nature of heat is responsible for the inefficiency of heat engines, and there is no way around it.

4-15   FATE OF THE UNIVERSE

Order into Disorder   Other kinds of energy can be entirely converted into heat, whereas only part of a given amount of heat can be converted the other way. As a result, there is an overall tendency toward an increase in the heat energy of the universe at the expense of the other kinds it contains. We see this tendency all around us in everyday life. When coal or oil burns in an engine, much of its chemical energy becomes heat; when any kind of machine is operated, friction turns some of its energy into heat; an electric lightbulb emits heat as well as light; and so on. Most of the lost energy is dissipated in the atmosphere, the oceans, and the earth itself where it is largely unavailable for recovery.

In the world of nature a similar steady degradation of energy into unusable heat occurs. In the universe as a whole, the stars (for instance, the sun) constitute the hot reservoir and everything else (for instance, the earth) constitutes the cold reservoir from a thermodynamic point of view. As time goes on, the stars will grow cooler and the rest of the universe will grow warmer, so that less and less energy will be available to power the further evolution of the universe. On a molecular level, order will become disorder. If this process continues indefinitely, the entire universe will be at the same temperature and all its particles will have the same average energy. This condition is sometimes called the "heat death" of the universe. However, as we shall see in Sec. 18-10, this is not the only possible fate of the universe.

4-16 ENTROPY

The Arrow of Time   Although "disorder" may not seem a very specific concept, a quantity called **entropy** can be defined that is a measure of the disorder of the molecules that make up any body of matter. For instance, the entropy of liquid water is nearly 3 times that of ice, a fact that reflects the greater disorder of the molecules in the liquid state than in the solid state. The entropy of steam is still greater.

In terms of entropy, the second law of thermodynamics becomes

> **The entropy of a system of some kind isolated from the rest of the universe cannot decrease.**

If a puddle of water were to rise from the ground by itself while turning into ice, the process could conserve energy (and so obey the first law of thermodynamics) as the heat lost by the water becomes kinetic energy of the ice (Fig. 4-38). However, such an event would decrease the entropy of the water, and so cannot occur. The advantage of expressing the second law in terms of entropy is that, because entropy can be determined for a variety of systems, this law can be applied in a precise way to such systems.

Biological systems might seem to violate the second law. Certainly entropy decreases when a plant turns carbon dioxide and water into leaves and flowers. But this transformation of disorder into order needs the energy of sunlight to take place. If we take into account the increase in the entropy of the sun as it produces the required sunlight, the net

FIG. 4-38 The second law of thermodynamics provides a way to distinguish between processes that conserve energy and (*a*) increase entropy, hence are possible, and those that (*b*) decrease entropy, hence are impossible.

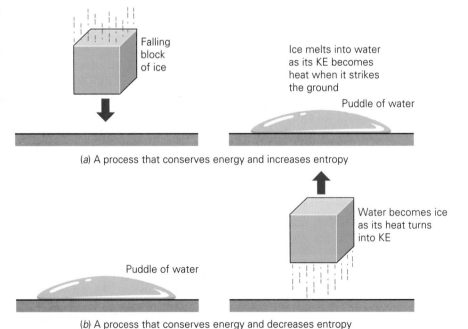

Falling block of ice

Ice melts into water as its KE becomes heat when it strikes the ground

Puddle of water

(*a*) A process that conserves energy and increases entropy

Puddle of water

Water becomes ice as its heat turns into KE

(*b*) A process that conserves energy and decreases entropy

BIOGRAPHY

Ludwig Boltzmann (1844–1906)

was born in Vienna and attended the university there. He then taught and carried out research at a number of institutions in Austria and Germany, moving from one to another every few years. Boltzmann was interested in poetry, music, and travel, and visited the United States three times, something unusual in those days. Of Boltzmann's many contributions to physics, the most important were to the kinetic theory of gases and to the foundations of thermodynamics. The constant k in the formula $KE_{av} = \frac{3}{2}kT$ for the average molecular energy in a gas at the absolute temperature T is called Boltzmann's constant in honor of his work. The mathematical relationship between molecular disorder and entropy was developed by Boltzmann; a monument to him in Vienna is inscribed with this relationship.

Boltzmann was a champion of the atomic theory of matter, still controversial in the late nineteenth century because there was then only indirect evidence for the existence of atoms and molecules. Battles with nonbelieving scientists deeply upset Boltzmann, and in his later years asthma, headaches, and poor eyesight further depressed his spirits. He committed suicide in 1906, not long after Albert Einstein published a paper on brownian motion that was to convince the remaining doubters of the correctness of the atomic theory.

result is an increase in the entropy of the universe. There is no way to avoid the second law.

The second law is an unusual physical principle in several respects. It does not apply to individual particles, only to assemblies of many particles. It does not say what can occur, only what cannot occur. And it is unique in that it is closely tied to the direction of time.

Events that involve only a few individual particles are always reversible. Nobody can tell whether a film of a billiard ball bouncing around is being run forward or backward. But events that involve many-particle systems are not always reversible. The film of an egg breaking when dropped makes no sense when run backward. The arrow of time always points in the direction of entropy increase. Time never runs backward; a broken egg never reassembles itself; in the universe as a whole, entropy—and the disorder it mirrors—marches on.

IMPORTANT TERMS AND IDEAS

Temperature is that property of a body of matter that gives rise to sensations of hot and cold; it is a measure of average molecular kinetic energy. **Heat** is molecular kinetic energy.

A **thermometer** is a device for measuring temperature. In the **celsius scale**, the freezing point of water is given the value 0°C and the boiling point of water the value 100°C. In the **fahrenheit scale**, these temperatures are given the values 32°F and 212°F.

In **conduction**, heat is carried from one place to another by molecular collisions. In **convection**, the transport is by the motion of a volume of hot fluid. **Radiation** transfers heat by means of **electromagnetic waves**, which require no material medium for their passage.

The **density** of a substance is its mass per unit volume.

The **pressure** on a surface is the perpendicular force per unit area acting on the surface. The unit of pressure is the **pascal**, which is equal to the newton/meter^2. The pressure at any point in a fluid depends on the weight of the fluid above the point as well as on any applied pressure.

According to **Archimedes' principle**, the buoyant force on an object immersed in a fluid is equal to the weight of fluid displaced by the object.

Boyle's law states that, at constant temperature, the volume of a gas sample is inversely proportional to its pressure.

The **absolute temperature scale** has its zero point at –273°C; temperatures in this scale are designated K. **Absolute zero** is 0 K = –273°C. **Charles's**

law states that, at constant pressure, the volume of a gas sample is directly proportional to its absolute temperature.

The **ideal gas law**—which states that $pV/T =$ constant for a gas sample regardless of changes in p, V, and T—is a combination of Boyle's and Charles's laws and is approximately obeyed by all gases.

According to the **kinetic theory of matter,** all matter consists of tiny individual **molecules** that are in constant random motion. The ideal gas law can be explained by the kinetic theory on the basis that the absolute temperature of a gas is proportional to the average kinetic energy of its molecules. At absolute zero, gas molecules would have no kinetic energy.

The **heat of vaporization** of a substance is the amount of heat needed to change 1 kg of it at its boiling point from the liquid to the gaseous state. The **heat of fusion** of a substance is the amount of heat needed to change 1 kg of it at its melting point from the solid to the liquid state.

Sublimation is the direct conversion of a substance from the solid to the vapor state without it first becoming a liquid.

A **heat engine** is a device that converts heat into mechanical energy or work. The **first law of thermodynamics** is the law of conservation of energy. The **second law of thermodynamics** states that some of the heat input to a heat engine must be wasted in order for the engine to operate.

Entropy is a measure of the disorder of the particles that make up a body of matter. In a system of any kind isolated from the rest of the universe, entropy cannot decrease.

IMPORTANT FORMULAS

Temperature scales: $T_F = \frac{9}{5}T_C + 32°$

$$T_C = \frac{5}{9}(T_F - 32°)$$

Density: $d = \frac{m}{V}$

Pressure: $p = \frac{F}{A}$

Boyle's law: $\dfrac{p_1}{p_2} = \dfrac{V_2}{V_1}$ (at constant temperature)

Absolute temperature scale: $T_K = T_C + 273$

Charles's law: $\dfrac{V_1}{V_2} = \dfrac{T_1}{T_2}$ (at constant pressure; temperatures on absolute scale)

Ideal gas law: $\dfrac{p_1 V_1}{T_1} = \dfrac{p_2 V_2}{T_2}$ (temperatures on absolute scale)

Maximum efficiency of heat engine:

$$\text{Eff}_{max} = 1 - \frac{T_{cold}}{T_{hot}}$$ (temperatures on absolute scale)

EXERCISES: MULTIPLE CHOICE

1. Two thermometers, one calibrated in °F and the other in °C, are used to measure the same temperature. The numerical reading on the fahrenheit thermometer
   a. is less than that on the celsius thermometer
   b. is equal to that on the celsius thermometer
   c. is greater than that on the celsius thermometer
   d. may be any of the above, depending on the temperature

2. The fluid at the bottom of a container is
   a. under less pressure than the fluid at the top
   b. under the same pressure as the fluid at the top
   c. under more pressure than the fluid at the top
   d. any of the above, depending upon the circumstances

3. The pressure of the earth's atmosphere at sea level is due to
   a. the gravitational attraction of the earth for the atmosphere
   b. the heating of the atmosphere by the sun
   c. the fact that most living things constantly breathe air
   d. evaporation of water from the seas and oceans

4. A cake of soap placed in a bathtub of water sinks. The buoyant force on the soap is
   a. 0
   b. less than its weight
   c. equal to its weight
   d. more than its weight

5. The density of freshwater is 1.00 g/cm³ and that of seawater is 1.03 g/cm³. A ship will float
   a. higher in freshwater than in seawater
   b. lower in freshwater than in seawater
   c. at the same level in freshwater and seawater
   d. any of the above, depending on the shape of its hull

6. At constant pressure, the volume of a gas sample is directly proportional to
   a. the size of its molecules
   b. its fahrenheit temperature
   c. its celsius temperature
   d. its absolute temperature

7. Which of the following statements is not correct?
   a. Matter is composed of tiny particles called molecules
   b. These molecules are in constant motion, even in solids
   c. All molecules have the same size and mass
   d. The differences between the solid, liquid, and gaseous states of matter lie in the relative freedom of motion of their respective molecules

8. Molecular motion is not responsible for
   a. the pressure exerted by a gas
   b. Boyle's law
   c. evaporation
   d. buoyancy

9. Absolute zero may be regarded as that temperature at which
   a. water freezes
   b. all gases become liquids
   c. all substances become solid
   d. molecular motion in a gas would be the minimum possible

10. On the molecular level, heat is
   a. kinetic energy
   b. potential energy
   c. rest energy
   d. all of the above, in proportions that depend on the circumstances

11. At a given temperature
   a. the molecules in a gas all have the same average velocity
   b. the molecules in a gas all have the same average energy
   c. light gas molecules have lower average energies than heavy gas molecules
   d. heavy gas molecules have lower average energies than light gas molecules

12. The temperature of a gas sample in a container of fixed volume is raised. The gas exerts a higher pressure on the walls of its container because its molecules
   a. lose more PE when they strike the walls
   b. lose more KE when they strike the walls
   c. are in contact with the walls for a shorter time
   d. have higher average velocities and strike the walls more often

13. The volume of a gas sample is increased while its temperature is held constant. The gas exerts a lower pressure on the walls of its container because its molecules strike the walls

   a. less often
   b. with lower velocities
   c. with less energy
   d. with less force

14. When evaporation occurs, the liquid that remains is cooler because
   a. the pressure on the liquid decreases
   b. the volume of the liquid decreases
   c. the slowest molecules remain behind
   d. the fastest molecules remain behind

15. When a vapor condenses into a liquid,
   a. its temperature rises
   b. its temperature falls
   c. it absorbs heat
   d. it gives off heat

16. A heat engine takes in heat at one temperature and turns
   a. all of it into work
   b. some of it into work and rejects the rest at a lower temperature
   c. some of it into work and rejects the rest at the same temperature
   d. some of it into work and rejects the rest at a higher temperature

17. In any process, the maximum amount of heat that can be converted to mechanical energy
   a. depends on the amount of friction present
   b. depends on the intake and exhaust temperatures
   c. depends on whether kinetic or potential energy is involved
   d. is 100 percent

18. In any process, the maximum amount of mechanical energy that can be converted to heat
   a. depends on the amount of friction present
   b. depends on the intake and exhaust temperatures
   c. depends on whether kinetic or potential energy is involved
   d. is 100 percent

19. A frictionless heat engine can be 100 percent efficient only if its exhaust temperature is
   a. equal to its input temperature
   b. less than its input temperature
   c. 0°C
   d. 0 K

20. The heat a refrigerator absorbs from its contents is
   a. less than it gives off
   b. the same amount it gives off
   c. more than it gives off

d. any of the above, depending on its design

21. The second law of thermodynamics does not lead to the conclusion that
    a. on a molecular level, order will eventually become disorder in the universe
    b. all the matter in the universe will eventually end up at the same temperature
    c. no heat engine can convert heat into work with 100 percent efficiency
    d. the total amount of energy in the universe, including rest energy, is constant

22. The greater the entropy of a system of particles,
    a. the less the energy of the system
    b. the more the energy of the system
    c. the less the order of the system
    d. the more the order of the system

23. Ethyl alcohol boils at 172°F. The celsius equivalent of this temperature is
    a. 64°C   **c.** 140°C
    b. 78°C   **d.** 278°C

24. A temperature of 20°C is the same as
    a. −20.9°F   **c.** 68°F
    b. −6.4°F   **d.** 94°F

25. A 400-kg concrete block has the dimensions 1 m × 0.6 m × 0.3 m. Its density is
    a. 72 kg/m^3 **c.** 667 kg/m^3
    b. 222 kg/m^3 **d.** 2222 kg/m^3

26. A 20-kg piston rests on a sample of a gas in a cylinder whose cross-sectional area is 150 cm^2. The pressure the piston exerts on the gas is
    a. 0.13 Pa   **c.** 1.3 kPa
    b. 1.3 Pa   **d.** 13 kPa

27. The block of Question 25 can exert three different pressures on a horizontal surface, depending upon which face it rests on. The highest pressure is
    a. 0.7 kPa   **c.** 13.1 kPa
    b. 2.2 kPa   **d.** 21.8 kPa

28. The pressure on 100 liters of helium is increased from 100 kPa to 400 kPa. The new volume of the helium is
    a. 25 liters   **c.** 400 liters
    b. 50 liters   **d.** 1600 liters

29. Lead melts at 330°C. On the absolute scale this temperature corresponds to
    a. 57 K   **c.** 571 K
    b. 362 K   **d.** 603 K

30. At which of the following temperatures would the molecules of a gas have twice the average kinetic energy they have at room temperature, 20°C?
    a. 40°C   **c.** 313°C
    b. 80°C   **d.** 586°C

31. A heat engine absorbs heat at a temperature of 127°C and exhausts heat at a temperature of 77°C. Its maximum efficiency is
    a. 13 percent   **c.** 61 percent
    b. 39 percent   **d.** 88 percent

QUESTIONS

1. Running hot water over the metal lid of a glass jar makes it easier to open the jar. Why?

2. When a mercury-in-glass thermometer is heated, its mercury column goes down briefly before rising. Why?

3. Is it meaningful to say that an object at a temperature of 200°C is twice as hot as one at 100°C?

4. Why do tables of densities always include the temperature for which the listed values hold? What would be true of the densities of most solids and liquids at a temperature higher than the quoted one?

5. When a person drinks a soda through a straw, where does the force come from that causes the soda to move upward?

6. A jar is filled to the top with water, and a piece of cardboard is slid over the opening so that there is only water in the jar. If the jar is turned over, will the cardboard fall off? What will happen if there is any air in the jar?

7. Some water is boiled briefly in an open metal can. The can is then sealed while still hot. Why does the can collapse when it cools?

8. A U-shaped tube contains water and an unknown liquid separated by mercury, as in Fig. 4-39. How does the density of the liquid compare with the density of water? How do the pressures at *A* and *B* compare?

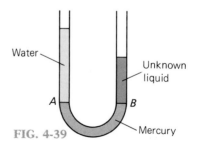

FIG. 4-39

9. The three containers shown in Fig. 4-40 are filled with water to the same height. Compare the pressures at the bottoms of the containers.

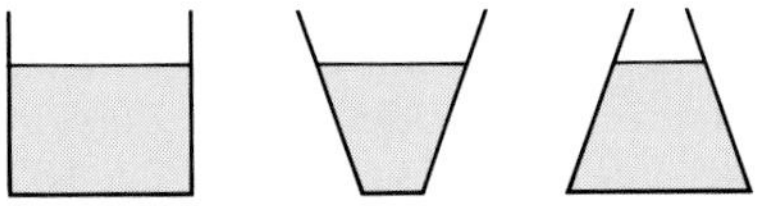

FIG. 4-40

10. A highway bridge in Sweden carries the Göta Canal over a highway. What, if anything, happens to the load on the bridge when a boat passes across it in the canal?

11. As Table 4-2 shows, ice has a lower density than water, which is why ice floats. What will happen when an ice cube floating in a glass of water filled to the brim begins to melt?

12. A wooden block is submerged in a tank of water and pressed down against the bottom of the tank so that there is no water underneath it. The block is released. Will it rise to the surface or stay where it is?

13. An aluminum canoe is floating in a swimming pool. After a while it begins to leak and sinks to the bottom of the pool. What, if anything, happens to the water level in the pool?

14. Why does a hot-air balloon rise?

15. When water is boiled in a pan, the bubbles of steam increase in size as they rise through the water. Why?

16. Gas molecules have speeds comparable with those of rifle bullets, yet it is observed that a gas with a strong odor (ammonia, for instance) takes a few minutes to diffuse through a room. Why?

17. At absolute zero, a sample of an ideal gas would have zero volume. Why would this not be true of an actual gas at absolute zero?

18. A sample of hydrogen is expanded to twice its original volume, while its temperature is held constant. What happens to the average speed of the hydrogen molecules?

19. When they are close together, molecules attract one another slightly. As a result of this attraction, are gas pressures higher or lower than expected from the ideal gas law?

20. Temperatures in both the celsius and fahrenheit scales can be negative. Why is a negative temperature impossible on the absolute scale?

21. How can the conclusion of kinetic theory that molecular motion occurs in solids be reconciled with the observation that solids have definite shapes and volumes?

22. In what state of matter are molecules, in general, farthest apart from one another?

23. Why is a piece of ice at 0°C more effective in cooling a drink than the same mass of cold water at 0°C?

24. When heat is added to a liquid, does its temperature always rise? If not, explain.

25. Canteens often have cloth coverings. On a hot day, what is the advantage to keeping the cloth around a canteen wet?

26. Why does evaporation cool a liquid?

27. Give as many methods as you can think of that will increase the rate of evaporation of a liquid sample. Explain why each method will have this effect.

28. If an egg is dropped into boiling water over a gas flame, the length of time necessary to cook the egg is not changed by turning the gas higher, although more heat is being supplied to the water. Explain. What becomes of the extra heat?

29. Why are both a hot and a cold reservoir needed for a heat engine to operate?

30. The oceans contain an immense amount of heat energy. Why can a submarine not make use of this energy for propulsion?

31. Is it correct to say that a refrigerator "produces cold"? If not, why not?

32. A person tries to cool a kitchen by switching on an electric fan and closing the kitchen door and windows. What will happen?

33. In another attempt to cool the kitchen, the person leaves the refrigerator door open, again with the kitchen door and windows closed. Now what will happen?

34. What is entropy? How is it related to time? How is it related to the second law of thermodynamics?

PROBLEMS

1. The normal temperature of the human body is 37°C. What is this temperature on the fahrenheit scale?

2. The temperature of a room is 70°F. What is this temperature on the celsius scale?

3. Dry ice (solid carbon dioxide) starts to vaporize into a gas at −112°F. What is this temperature on the celsius scale?

4. How many kJ of heat are needed to raise the temperature of 20 kg of water from the freezing point to the boiling point?

5. How many kJ of heat are needed to raise the temperature of 200 g of water from 20°C to 100°C in preparing a cup of coffee?

6. The diet of a 60-kg person provides 12,000 kJ daily. If this amount of energy were added to 60 kg of water, by how much would its temperature be increased?

7. A 10-kg stone is dropped into a pool of water from a height of 100 m. How much energy in joules does the stone have when it strikes the water? If all this energy goes into heat and if the pool contains 10 m³ of water, by how much is its temperature raised? (The mass of 1 m³ of water is 10³ kg.)

8. How high is a waterfall if the water at its base is 1°C higher in temperature than the water at the top? (*Hint:* Start by considering 1 kg of water.)

9. A room is 5 m long, 4 m wide, and 3 m high. What is the mass of the air it contains?

10. A piece of balsa wood 10 cm wide, 5 cm thick, and 50 cm long has a mass of 325 g. Find the density of balsa wood.

11. A 50-g bracelet is suspected of being gold-plated lead instead of pure gold. It is dropped into a full glass of water and 4 cm³ of water overflows. Is the bracelet pure gold?

12. A 1200-kg concrete slab that measures 2 m × 1 m × 20 cm is delivered to a building under construction. Does the slab contain steel reinforcing rods or is it plain concrete?

13. Mammals have approximately the same density as freshwater. Find the volume in liters of a 55-kg woman and the volume in cubic meters of a 140,000-kg blue whale. (*Note:* 1 liter = 10⁻³ m³ = 0.001 m³.)

14. A 60-kg woman balances on the heel of her right shoe, whose area is 0.5 cm². How much pressure does she exert on the ground? How many times atmospheric pressure is this?

15. The two upper front teeth of a person exert a total force of 30 N on a piece of steak. If the biting edge of each tooth measures 10 mm × 1 mm, find the pressure on the steak.

16. The smallest bone in the index finger of a 75-kg circus acrobat has a cross-sectional area of 0.5 cm² and breaks under a pressure of 1.7×10^8 Pa. Is it safe for the acrobat to balance his entire weight on this finger?

17. A certain quantity of hydrogen occupies a volume of 1000 cm³ at 0°C (273 K) and ordinary atmospheric pressure. If the pressure is tripled but the temperature is held constant, what will the volume of the hydrogen be? If the temperature is increased to 273°C but the pressure is held constant, what will the volume of the hydrogen be?

18. A sample of nitrogen occupies 2 m³ at 300 K and a pressure of 200 kPa. (a) The sample is compressed to a volume of 1 m³. What is its pressure at the same temperature? (b) What volume does the sample occupy at the same temperature if its pressure is decreased to 150 kPa? (c) What volume does the sample occupy at a temperature of 400 K and a pressure of 200 kPa?

19. To what celsius temperature must a gas sample initially at 20°C be heated if its volume is to double while its pressure remains the same?

20. The propellant gas that remains in an empty can of spray paint is at atmospheric pressure. If such a can at 20°C is thrown into a fire and is heated to 600°C, how many times atmospheric pressure is the new pressure inside the can? (The higher pressure may burst the can, which is why such cans should not be thrown into a fire.)

21. An air tank used for scuba diving has a safety valve set to open at a pressure of 28 MPa. The normal pressure of the full tank at 20°C is 20 MPa. If the tank is heated after being filled to 20 MPa, at what temperature will the safety valve open?

22. A tank contains 5 kg of carbon dioxide gas at 20°C and a pressure of 100 kPa. What happens to the pressure when another 5 kg of carbon dioxide is added to the tank at the same temperature? Explain this result in terms of the kinetic theory of gases.

23. To what temperature must a gas sample initially at 27°C be raised in order for the average energy of its molecules to double?

24. The average speed of a hydrogen molecule at room temperature is about 1.6 km/s. What is the average speed of an oxygen molecule, whose mass is 16 times greater, at the same temperature?

25. How much heat is given off when 1 kg of steam at 100°C condenses and cools to water at 20°C?

26. How much steam is produced when 1 MJ of heat is added to 1 kg of water at 50°C?

27. How much heat is required to change 50 g of ice at 0°C into water at 20°C?

28. A 1-kg block of ice at 0°C falls into a lake whose water is also at 0°C, and 0.01 kg of the ice melts. What was the minimum altitude from which the ice fell?

29. If all the heat lost by 1 kg of water at 0°C when it turns into ice at 0°C could be turned into kinetic energy, what would the speed of the ice be?

30. A man is resting in the shade on a hot day in which the air temperature is the same as his body temperature of 37°C. In this situation the chief way his body dissipates the 120 W his metabolic processes liberate is by the evaporation of sweat. How much sweat per hour is required? The heat of vaporization of water at 37°C is 2430 kJ/kg.

31. An engine is proposed that is to operate between 200°C and 50°C with an efficiency of 35 percent. Will the engine perform as predicted? If not, what would its maximum efficiency be?

32. A typical temperature for surface water in a tropical ocean is 27°C, whereas at a depth of a kilometer or more it is only about 5°C. It has been proposed to operate heat engines using surface water as the hot reservoir and deep water (pumped to the surface) as the cold reservoir. What would the maximum efficiency of such an engine be? Why might such an engine eventually be a practical proposition even with so low an efficiency?

33. An engine that operates between 2000 K and 700 K has an efficiency of 40 percent. What percentage of its maximum possible efficiency is this?

34. Three designs for an engine to operate between 450 K and 300 K are proposed. Design *A* is claimed to require a heat input of 800 J for each 1000 J of work output, design *B* a heat input of 2500 J, and design *C* a heat input of 3500 J. Which design would you choose and why?

ANSWERS TO MULTIPLE CHOICE			
1. d	**9.** d	**17.** b	**25.** d
2. c	**10.** a	**18.** d	**26.** d
3. a	**11.** b	**19.** d	**27.** d
4. b	**12.** d	**20.** a	**28.** a
5. b	**13.** a	**21.** d	**29.** d
6. d	**14.** c	**22.** c	**30.** c
7. c	**15.** d	**23.** b	**31.** a
8. d	**16.** d	**24.** c	

5

ELECTRICITY

AND

MAGNETISM

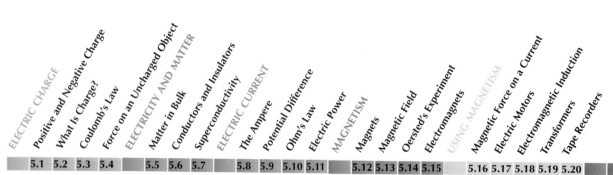

We have now learned about force and motion, mass and energy, the law of gravity, and the concept of matter as being made up of tiny moving molecules. With the help of these ideas we have been able to make sense of a wide variety of observations, from the paths of the planets across the sky to the melting of ice and the boiling of water. Is this enough for us to understand how the entire physical universe works?

For an answer all we need do is run a hard rubber comb through our hair on a dry day. Little sparks occur, and the comb then can pick up small bits of dust and paper. What is revealed in this way is an **electrical** phenomenon, something that neither gravity nor the kinetic theory of matter can account for.

In everyday life electricity is familiar as that which causes our lightbulbs to glow, many of our motors to turn, our telephones and radios to bring us sounds, our television screens to bring us sights. But there is more to electricity than its ability to transport energy and information. All matter turns out to be electrical in nature, and electric forces are what bind electrons to nuclei to form atoms and what hold atoms together to form molecules, liquids, and solids. Most of the properties of the ordinary matter around us—an exception is mass—can be traced to electrical forces.

ELECTRIC CHARGE

The first recorded studies of electricity were made in Greece by Thales of Miletus about 2500 years ago. Thales experimented with amber, called *electron* in Greek, and fur. The name **electric charge** is today given to whatever it is that a piece of amber (or hard rubber) possesses as a result of being rubbed with fur. It is this charge that causes sparks to occur and that attracts light objects such as bits of paper.

5-1 POSITIVE AND NEGATIVE CHARGE

Opposites Attract

Let us begin by hanging a small plastic ball from a thread, as in Fig. 5-1. We touch the ball with a hard rubber rod and find that nothing happens. Next we stroke the rod with a piece of fur and again touch the plastic ball with the rod. This time the ball flies away from the rod. What must have happened is that some of the electric charge on the rod has flowed to the ball, and the fact that the ball then flies away from the rod means that charges of the same kind repel each other.

FIG. 5-1 A rubber rod stroked with fur becomes negatively charged. When it is touched against a plastic ball, some of the negative charge flows to the ball. The plastic ball then flies away because like charges repel each other.

1. A plastic ball held by a string is touched by a hard rubber rod. Nothing happens.

2. The rubber rod is stroked against a piece of fur.

3. The plastic ball is again touched by the rubber rod.

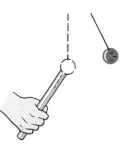
4. After the touch, the plastic ball flies away from the rod.

Is there only one kind of electric charge? To find out, we try other combinations of materials and see what happens when the various charged plastic balls are near each other. Figure 5-2 shows the result when one ball has been charged by a rubber rod stroked with fur and the other ball has been charged by a glass rod stroked with silk: the two balls fly together. We conclude that the charges on the rods are somehow different and that different charges attract each other.

Comprehensive experiments show that *all* electric charges fall into one of these two types. Regardless of origin, charges always behave as though they came either from a rubber rod rubbed with fur or from a glass rod rubbed with silk. Benjamin Franklin suggested names for these two basic kinds of electricity. He called the charge produced on the rubber rod **negative charge** and the charge produced on the glass rod **positive charge.** These definitions are still used today.

The above experiments can be summarized very simply:

> **All electric charges are either positive or negative. Like charges repel one another; unlike charges attract one another.**

FIG. 5-2 A glass rod stroked with silk becomes positively charged. When one plastic ball is touched with a negatively charged rubber rod and another plastic ball is touched with a positively charged glass rod, the two balls fly together because unlike charges attract each other.

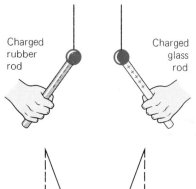

Charged rubber rod

Charged glass rod

Charge Separation

We have thus far been paying attention to the positive charge of the glass and the negative charge of the rubber. However, we do not produce only positive charge by rubbing glass with silk or only negative charge by stroking rubber with fur. If the fur used with the rubber is brought near a negatively

FIG. 5-3 When a rubber rod is stroked against a piece of fur, charges that were originally mixed together evenly become separated so that the rod becomes negatively charged and the fur becomes positively charged.

charged plastic ball, the ball is attracted. Thus the fur must have a positive charge (Fig. 5-3). Similarly the silk used with the glass turns out to have a negative charge. Whenever electric charge is produced by contact between two objects of different materials, one of them ends up with a positive charge and the other a negative charge. Which is which depends on the particular materials used.

The rubbing process does not *create* the electric charges that appear as a result. All "uncharged" objects actually contain equal amounts of positive and negative charge. For some pairs of materials, as we have seen, mere rubbing is enough to separate some of the charges from each other. In most cases, however, the charges are firmly held in place and more elaborate treatment is needed to pull them apart.

An object whose positive and negative charges exactly balance out is said to be electrically **neutral.**

This girl has been given an electric charge by touching the terminal of a static electric generator. Because all her hairs have charge of the same sign, they repel one another.

5-2 WHAT IS CHARGE?

Protons, Electrons, and Neutrons

In our own experience, matter and electric charge seem continuous, so that we can imagine dividing them into smaller and smaller portions without limit. But there is another level beyond reach of our senses, though not beyond reach of our instruments, on which every substance is revealed as being composed of tiny bits of matter called **atoms.** (Atoms are discussed in detail in later chapters along with how they join together to form molecules, liquids, and solids.) Over 100 varieties of atom are known, but each of them is made up of just three kinds of **elementary particles;** nature is very economical. Two of the particles carry electric charges, so that charge, like mass, comes in small parcels of definite size. The third particle has no charge.

The three elementary particles found in atoms are

1. The **proton,** which has a mass of 1.673×10^{-27} kg and is positively charged.
2. The **electron,** which has a mass of 9.11×10^{-31} kg and is negatively charged
3. The **neutron,** which has a mass of 1.675×10^{-27} kg and is uncharged

The proton and electron have exactly the same amounts of charge, although of opposite sign. Protons and neutrons have almost equal masses, which are nearly 2000 times greater than the electron mass.

Every atom has a small, central **nucleus** of protons and neutrons with its electrons moving about the nucleus some distance away (Fig. 5-4). Different types of atoms have different combinations of protons and neutrons in their nuclei. For instance, the most common carbon atom has a nucleus that contains six protons and six neutrons; the most common uranium atom has a nucleus that contains 92 protons and 146 neutrons. The electrons in an atom are normally equal in number to the protons, so the atom is electrically neutral unless disturbed in some way.

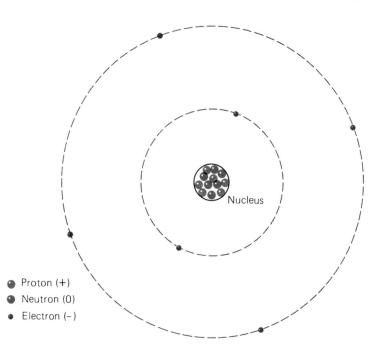

FIG. 5-4 An atom consists of a central nucleus of protons and neutrons with electrons moving around it some distance away. Shown is a simplified model (not to scale) of the most common type of carbon atom, which has six protons, six neutrons, and six electrons. Two of the electrons are relatively near the nucleus, the others are farther away.

● Proton (+)
● Neutron (0)
● Electron (−)

The Coulomb

The unit of electric charge is the **coulomb** (C). The proton has a charge of $+1.6 \times 10^{-19}$ C and the electron has a charge of -1.6×10^{-19} C. All charges, both positive and negative, are therefore found only in multiples of 1.6×10^{-19} C. This basic quantity of charge is abbreviated e:

$$e = 1.6 \times 10^{-19} \text{ C} \qquad \textit{basic unit of charge in nature}$$

Electric charge appears continuous outside the laboratory because e is such a small quantity. A charge of −1 C, for example, corresponds to more than 6 billion billion electrons (Fig. 5-5). Atoms are small, too: coal is almost pure carbon, and 6 billion billion carbon atoms would make a piece of coal only about the size of a pea.

FIG. 5-5 Electric charge is not continuous but occurs in multiples of $\pm e = \pm 1.6 \times 10^{-19}$ coulomb.

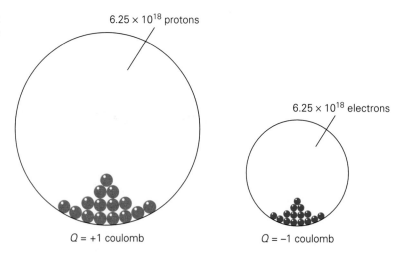

6.25×10^{18} protons

6.25×10^{18} electrons

$Q = +1$ coulomb $Q = -1$ coulomb

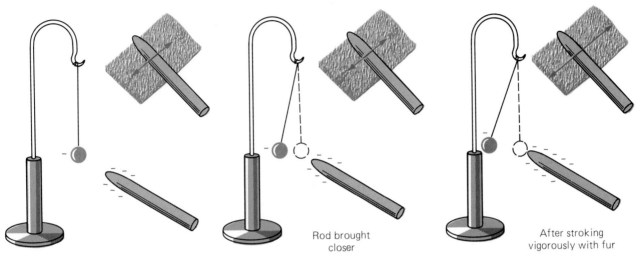

FIG. 5-6   The forces between electric charges. When a rubber rod that has been stroked with fur is brought near a negatively charged plastic ball, the force on the ball is greater when the rod is held close to it and also greater when the rod has been vigorously stroked.

5-3 COULOMB'S LAW

Charles Coulomb   (1736–1806)

The Law of Force for Electric Charges

The forces between electric charges can be studied in rather simple experiments, such as that shown in Fig. 5-6. What we find is that the force between a charged rod and a charged plastic ball depends on two things: how close the rod is to the ball, and how much charge each one has.

Precise measurements show that the force between charges follows the same inverse-square variation with distance that the gravitational force between two masses does (Sec. 2-13). For instance, when the charges are 2 cm apart, the force between them is $\frac{1}{4}$ as great as the force when they are 1 cm apart; it is 4 times greater when they are $\frac{1}{2}$ cm apart (Fig. 5-7a). If R is the distance between the charges, we can say that the force between them is proportional to $1/R^2$.

The force also depends on the magnitude of each charge: if either charge is doubled, the force doubles too, and if both charges are doubled, the force increases fourfold (Fig. 5-7b). If the charges have the respective magnitudes Q_1 and Q_2, then the force varies as their product Q_1Q_2.

The above results are summarized in **Coulomb's law:**

FIG. 5-7   (a) The force between two charges varies inversely as the square of their separation; increasing the distance reduces the force. (b) The force is proportional to the product of the charges.

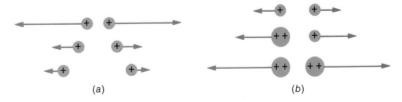

(a)   (b)

$$F = K\frac{Q_1 Q_2}{R^2} \qquad \textit{electric force}$$

which is named in honor of Charles Coulomb, who helped develop it. The quantity K is a constant whose value is almost exactly 9×10^9 N · m²/C². Thus we may rewrite Coulomb's law as

$$F = 9 \times 10^9 \frac{Q_1 Q_2}{R^2} \qquad \textit{electric force}$$

Just by looking at this formula we can see that the force between two charges of 1 C each that are separated by 1 m is 9×10^9 N, 9 billion newtons. This is an enormous force, equal to about 2 billion lb! We conclude that the coulomb is a very large unit indeed, and that even the most highly charged objects that can be produced cannot contain more than a small fraction of a coulomb of net charge of either sign.

5-4   FORCE ON AN UNCHARGED OBJECT

Why a Comb Can Attract Bits of Paper   One sign that a body has an electric charge is that it causes small, uncharged objects such as dust particles, bits of paper, and suspended plastic balls to move toward it. Where does the force come from?

The explanation comes from the fact that the electrons in a solid have some freedom of movement. In a metal this freedom is considerable, but even in other substances the electrons can shift around a little without leaving their parent atoms or molecules. When a comb is given a negative charge by being run through our hair, electrons in a nearby bit of paper are repelled by the negative charge and move away as far as they can (Fig. 5-8). The side of the paper near the comb is left with a positive charge, and the paper is accordingly attracted to the comb. If the comb is removed without actually touching the paper, the disturbed electrons resume their normal positions. Only a small amount of charge separation actually occurs, and so, with little force available, only very light things can be picked up this way.

FIG. 5-8 A charged object attracts an uncharged one by first causing a separation of charge in the latter.

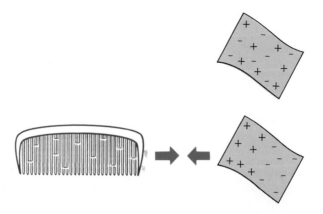

Let us now look into some aspects of the electrical behavior of matter.

5-5 MATTER IN BULK

Gravity versus Electricity

Coulomb's law for the force between charges is one of the fundamental laws of physics, in the same category as Newton's law of gravity. The latter, as we know, is written in equation form as

$$F = G\,\frac{m_1 m_2}{R^2} \qquad gravitational\ force$$

Coulomb's law resembles the law of gravity with the difference that gravitational forces are always attractive but electric forces may be either attractive or repulsive.

This last fact has an important consequence. Because one lump of matter always attracts another lump gravitationally, matter in the universe tends to come together into large masses. Even though dispersive influences of various kinds exist, they must fight against this steady attraction. Galaxies, stars, and planets, which condensed from matter that was originally spread out in space, bear witness to this cosmic herd instinct.

To collect much electric charge of either sign, however, is far more of a feat. Charges of opposite sign attract each other strongly, so it is hard to separate neutral matter into differently charged portions. And charges of the same sign repel each other, so putting together a large amount of charge of one sign is difficult.

To sum up, we can say that a system of electrically neutral particles is most stable (that is, has a minimum potential energy) when the particles make up a single body, while a system of electric charges is most stable when charges of opposite signs pair off to cancel each other out. Hence on a cosmic scale gravitational forces are significant and electric ones are not. On an atomic scale, however, the reverse is true. The masses of subatomic particles are too small for them to interact gravitationally to any appreciable extent, whereas their electric charges are enough for electric forces to exert marked effects.

The hydrogen atom illustrates the above statement. Its nucleus is a single proton, and the electron that circles the proton does so at an average distance of 5.3×10^{-11} m (Fig. 5-9). The electrical force of attraction between the proton and electron is more than 10^{39} times greater than the gravitational force! Clearly, gravitational effects are negligible within atoms as compared with electrical effects.

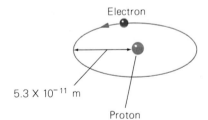

Electron

5.3×10^{-11} m

Proton

FIG. 5-9 A model of the hydrogen atom. The electric force between the electron and proton is more than 10^{39} times as great as the gravitational force between them.

5-6 CONDUCTORS AND INSULATORS

How Charge Flows from One Place to Another

A substance through which electric charge can flow readily is called a **conductor.** Metals are the only solid conductors at room temperature, copper being an especially good one. In a metal, each atom gives up

The earth as a whole, at least that part of it beneath the outer dry soil, is a fairly good electrical conductor. Hence if a charged object is connected with the earth by a piece of metal, the charge is conducted away from the object to the earth. This convenient method of removing the charge from an object is called **grounding** the object. As a safety measure, the metal shells of electrical appliances are grounded through special wires that give electric charges in the shells paths to the earth. The round post in the familiar three-prong electric plug is the ground connection.

one or more electrons to a "gas" of electrons that can move relatively freely inside the metal. The atoms themselves stay in place and are not involved in the movement of charge.

In an **insulator,** charge can flow only with great difficulty. Nonmetallic solids are insulators because all their electrons are tightly bound to particular atoms or groups of atoms. Glass, rubber, and plastics are good insulators.

A few substances, called **semiconductors,** are between conductors and insulators in their ability to let charge move through them. Semiconductors have made possible devices called **transistors,** whose ability to transmit charge can be changed at will. Transistors are widely used in modern electronics, notably in radio and television receivers. A computer contains thousands of transistors that act as miniature switches to perform arithmetic and carry out logical operations. Semiconductor memories are also used in computers, with huge numbers of memory elements built into a "chip" smaller than a fingernail.

Ions

The conduction of electricity through gases and liquids—in a neon sign, for instance, or in the acid of a storage battery—involves the movement of charged atoms and molecules called **ions.** An atom or molecule gains a positive charge (becomes a positive ion) when it loses one or more electrons, and it gains a negative charge (becomes a negative ion) when electrons in excess of its normal number become attached to it.

The process of forming ions, or **ionization,** can take place in a number of ways. A gas like ordinary air, which is normally a poor conductor, becomes ionized when x-rays, ultraviolet light or radiation from a radioactive material passes through it, when an electric spark is produced, or even when a flame burns in it. Air molecules are sufficiently disturbed by these processes that electrons are torn loose from some of

Enlargement of part of a semiconductor "chip" used in a computer. A chip 5 mm square can contain over 2 million circuits that store data and carry out logic operations.

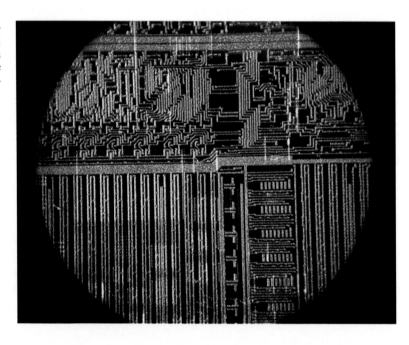

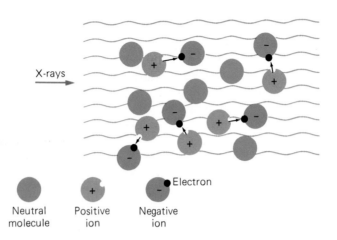

FIG. 5-10 A gas such as air becomes ionized when x-rays disrupt its molecules. A molecule losing an electron becomes a positive ion; a molecule gaining an electron becomes a negative ion. Ultraviolet light, radiation from radioactive substances, sparks, and flames also cause ionization to occur.

them. The electrons thus set free may attach themselves to adjacent molecules, so both positive and negative ions are formed (Fig. 5-10). Eventually oppositely charged ions come together, whereupon the extra electrons on negative ions shift to positive ions to give neutral molecules again. At normal atmospheric pressure and temperature the ions last no more than a few seconds.

In the upper part of the earth's atmosphere, air molecules are so far apart on the average that the ionization produced by x-rays and ultraviolet light from the sun tends to persist. The ability of these ions to reflect radio waves makes possible long-range radio communication (see Chap. 6).

In contrast with gases, certain liquids may be permanently ionized to a greater or lesser extent (see Chap. 10). The conductivity of pure water itself is extremely small, but even traces of some impurities increase its conductivity enormously. Since most of the water we use in daily life is somewhat impure, it is usually considered a fair conductor of electricity.

5-7 SUPERCONDUCTIVITY

A Revolution in Technology May Be Near   Even the best conductors resist to some extent the flow of charge through them at ordinary temperatures. However, when extremely cold, some substances lose all electrical resistance. This phenomenon, called **superconductivity,** was discovered by Kamerlingh Onnes in the Netherlands in 1911. For example, aluminum is a superconductor below 1.2 K, which is –272°C. Such temperatures are difficult and expensive to reach, and as a result superconductivity has not been exploited commercially to any great extent.

If electrons are set in motion in a closed wire loop at room temperature, they will come to a stop in less than a second even in a good conductor such as copper. In a superconducting wire loop, on the other hand, electrons have circulated for years with no outside help.

Superconductivity is important because electric currents—flows of charge—are the means by which electric energy is carried from one place to another. Electric currents are also the means by which magnetic fields are produced in such devices as electric motors, as we shall learn later in this chapter. In an ordinary conductor, some of the energy of a current is lost as heat. Where long distances or large currents are involved, quite a bit of energy can be wasted in this way. About 10 percent of the electric energy generated in the United States is lost as heat in transmission lines.

High-Temperature Superconductors

Despite much effort, until 1986 no substance was known that was superconducting above 23 K. In that year Alex Müller and Georg Bednorz, working in Switzerland, discovered a ceramic that was superconducting up to 35 K. Soon afterward others extended their approach to produce superconductivity at temperatures higher than 150 K. Although still extremely cold by everyday standards, such temperatures are above the 77 K boiling point of liquid nitrogen. Because liquid nitrogen is cheap (cheaper than milk) and readily available, the new superconductors, when perfected (the early ones were mechanically too weak for practical use), are sure to be widely employed with liquid nitrogen to keep them cold.

A room-temperature superconductor would truly revolutionize the world's technology. As just one example, all trains would run suspended above the ground by magnetic forces, resulting in better fuel efficiency and high speeds. In general, less waste of electric energy would mean a lower rate of depletion of fuel resources and reduced pollution. A room-temperature superconductor, though still a dream, does not seem quite so hopeless a dream as it did only a few years ago.

ELECTRIC CURRENT

A flow of charge from one place to another constitutes an **electric current.** Currents and not stationary charges are involved in nearly all the practical applications of electricity.

5-8 THE AMPERE

The Unit of Electric Current

A battery turns chemical energy into electric energy. If we connect a wire between the terminals of a battery to make a complete conducting path, or **circuit,** electrons will flow into the wire from the negative terminal and out of the wire into the positive terminal. Chemical reactions in the battery keep the electrons moving. We do not say, "The electrons carry the current," or "The motion of the electrons produces a current"; the moving electrons *are* the current.

The flow of electricity along a wire is a lot like the flow of water in a pipe (Fig. 5-11). When we describe the rate at which water moves through a pipe, we give the flow in terms of, say, liters per second. If 5 liters of water passes through a given pipe each second, the flow is 5 liters/s.

The description of electric current follows the same pattern. As we know, quantity of electric charge is measured in coulombs, as quantity

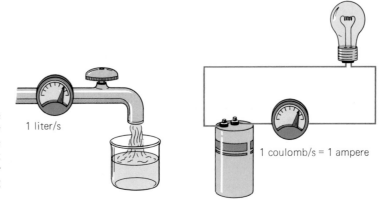

FIG. 5-11 The ampere is the unit of electric current. The flow of charge in a circuit is like the flow of water in a pipe except that a return wire is necessary in order to have a complete conducting path.

1 liter/s

1 coulomb/s = 1 ampere

of water is measured in liters. The natural way to refer to a flow of charge in a wire, then, is in terms of the number of coulombs per second that go past any point in the wire. This unit of electric current is called the **ampere** (A), after the French physicist André Marie Ampère. That is,

$$1 \text{ ampere} = 1 \, \frac{\text{coulomb}}{\text{second}}$$

$$1 \text{ A} = 1 \text{ C/s}$$

The current in the lightbulb of most desk lamps is a little less than 1 A.

A battery is rated according to the total amount of charge it can transfer from one terminal to the other, expressed in ampere-hours (A · h). A typical car battery has a capacity of 60 A · h, which means it can supply a current of 60 A for 1 h, a current of 30 A for 2 h, a current of 1 A for 60 h, and so on. The less the current, the longer the battery can supply it.

5-9   POTENTIAL DIFFERENCE

The Push behind a Current   Consider a liter of water at the top of a waterfall. The water has potential energy there, since it can move downward under the pull of gravity. When the water drops, its PE decreases. As we learned in Chap. 3, the work that can be obtained from the liter of water during its fall is equal to its decrease in PE.

Now consider a coulomb of negative charge on the – terminal of a battery. It is repelled by the – terminal and attracted by the + terminal, and so it has a certain amount of PE. When the coulomb of charge has moved along a wire to the + terminal, its PE is gone. The work the coulomb of charge can perform while flowing from the – to the + terminal of the battery is equal to this decrease in PE.

The decrease in its PE brought about by the motion of 1 C of charge from the – to the + terminal is called the **potential difference** between

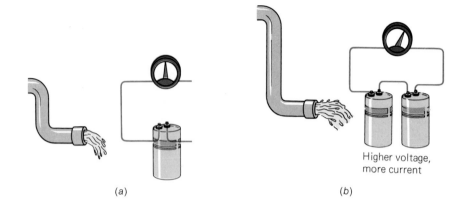

FIG. 5-12 The flow of electric charge in a wire is analogous to the flow of water in a pipe. Thus having the water fall through a greater height at (b) than at (a) yields a greater flow of water, which corresponds to using two batteries to obtain a higher potential difference and thereby a greater current.

(a)   (b)

Higher voltage, more current

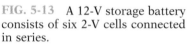
Alessandro Volta (1745–1827)

FIG. 5-13 A 12-V storage battery consists of six 2-V cells connected in series.

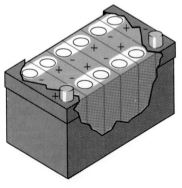

the two terminals. It is analogous to the difference in height in the case of water (Fig. 5-12). The potential difference between two points is equal to the corresponding energy difference per coulomb. We measure difference of height in meters; we measure difference of potential in **volts** (named for the Italian physicist Alessandro Volta).

When 1 coulomb of charge travels through 1 volt of potential difference, the work that it does is equal to 1 joule. By definition:

$$1 \text{ volt} = 1 \; \frac{\text{joule}}{\text{coulomb}}$$

$$1 \text{ V} = 1 \text{ J/C}$$

Potential difference is often called simply **voltage.**

Batteries   The normal potential difference between the terminals of a car's storage battery is about 12 V, in the case of a dry cell about 1.5 V. Every coulomb of electricity at the negative terminal of the storage battery can do 8 times as much work as a coulomb at the negative terminal of a dry cell—just as a liter of water at the top of a waterfall 12 m high can do 8 times as much work as a liter at the top of a 1.5-m fall. If a storage battery and a dry cell are connected in identical circuits, the battery will push 8 times as many electrons around its circuit in a given time as the dry cell, giving a current 8 times as great. We may think of the potential difference between two points as the amount of "push" available to move charge between the points.

If we connect two or more batteries together as shown in Fig. 5-12b, the available voltage is increased. The method of connection is − terminal to + terminal, so that each battery in turn supplies its "push" to electrons flowing through the set. The voltage of a particular cell depends on the chemical reactions that take place in it. In the case of the lead-acid storage battery of a car, each cell has a voltage of 2 V, and six of them are connected together to give the 12 V needed to run the car's electrical equipment (Fig. 5-13). As its name suggests, a storage battery can be **recharged** when the energy it contains is used up. Ordinary dry-cell batteries, such as those used in flashlights and portable radios, have voltages of 1.5 V and cannot be recharged.

Charge separation occurs in a thunderstorm when droplets of water and ice crystals collide, and the separation continues on a larger scale mainly through convection. The result is a region of negative charge sandwiched between regions of positive charge. The charge separation produces potential differences of up to several million volts, and electrical discharges (which are like giant sparks) then occur that appear as lightning strokes. Although cloud-to-ground strokes are the most familiar, lightning is actually much more frequent inside thunderclouds, where it is hidden from our view by the clouds although we can hear the thunder it causes. Thunder is the result of the heating of the air along the path of a lightning stroke, whose sudden expansion gives rise to intense sound waves.

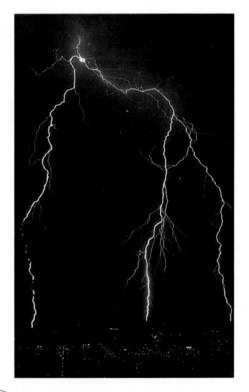

Georg Ohm (1787–1854)

5-10   OHM'S LAW

Current, Voltage, and Resistance

When different voltages are applied to the ends of the same piece of wire, we find that the current in the wire is proportional to the potential difference. Doubling the voltage doubles the current. This generalization is called **Ohm's law** after its discoverer, the German physicist Georg Ohm.

The property of a conductor that opposes the flow of charge in it is called **resistance.** We can think of resistance as a kind of friction. The more the resistance in a circuit, the less the current for a given applied voltage (Fig. 5-14). If we write I for current, V for voltage, and R for resistance, Ohm's law says that

FIG. 5-14 (a) A short, wide pipe yields a large flow of water, which corresponds to using a short, thick wire that offers less resistance to the flow of charge. (b) A long, narrow pipe yields a small flow of water, which corresponds to using a long, thin wire that offers more resistance to the flow of charge.

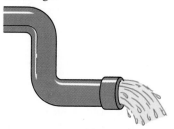

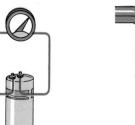

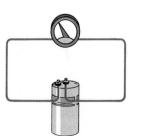

Lower resistance,
more current
(a)

Higher resistance,
less current
(b)

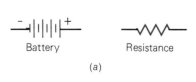

Battery   Resistance

(a)

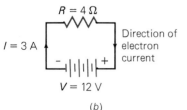

$R = 4 \, \Omega$

$I = 3 \, \text{A}$

Direction of electron current

$V = 12 \, \text{V}$

(b)

FIG. 5-15 (*a*) Symbols for a battery and a resistance. (*b*) A current of 3 A flows in a circuit whose resistance is 4 Ω when a potential difference of 12 V is applied. The current consists of a flow of electrons, so its direction is from the − terminal of the battery to the + terminal.

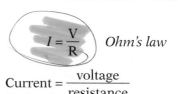

$$I = \frac{V}{R} \qquad \textit{Ohm's law}$$

$$\text{Current} = \frac{\text{voltage}}{\text{resistance}}$$

The unit of resistance is the **ohm,** whose abbreviation is Ω, the Greek capital letter "omega." Hence 1 A = 1 V/Ω and

$$1 \text{ ohm} = 1 \, \frac{\text{volt}}{\text{ampere}}$$

$$1 \, \Omega = 1 \text{ V/A}$$

The resistance of a wire or other metallic conductor depends on the material it is made of (an iron wire has 7 times the resistance of a copper wire of the same size); its length (the longer the wire, the more its resistance); its cross-sectional area (the greater this area, the less the resistance); and the temperature (the higher the temperature, the more the resistance).

Using Ohm's Law Let us see how Ohm's law can be applied in a practical situation. How long can a car with a 12-V battery of 60 A · h capacity have its headlights and taillights, of total resistance 4 Ω, left on before the battery runs down? We assume the engine is not running, so the car's generator is not recharging the battery. To solve the problem, we first calculate the current (Fig. 5-15), which is

$$I = \frac{V}{R} = \frac{12 \text{ V}}{4 \, \Omega} = 3 \text{ A}$$

Because the battery's capacity is 60 A · h, the lights can be left on for 20 h before the battery runs down.

ELECTRICAL SAFETY   Body tissue is a fairly good electrical conductor because it contains ions in solution. Dry skin has the most resistance and can protect the rest of the body in case of accidental exposure to a high voltage. This protection disappears when the skin is wet. An electric current in body tissue stimulates nerves and muscles and produces heat. Most people can feel a current as small as 0.0005 A, one of 0.005 A is painful, and one of 0.01 A or more leads to muscle contractions that may prevent a person from letting go of the source of the current. Breathing becomes impossible when the current is greater than about 0.018 A.

Touching a single "live" conductor has no effect if the body is isolated since a complete conducting path is necessary for a current to occur. However, if a person is at the same time grounded by being in contact with a water pipe, by standing on wet soil, or in some other way, a current will pass through his or her body. The human body's resistance is in the neighborhood of 1000 Ω, so if a potential difference of 120 V is applied via wet skin, the resulting current will be somewhere near $I = V/R = 120 \text{ V}/1000 \, \Omega = 0.12$ A. Such a current is exceedingly dangerous because it causes the heart muscles to contract rapidly and irregularly, which is fatal if allowed to continue.

Electrical devices in bathrooms and kitchens are potential sources of danger because the moisture on a wet finger may be enough to provide a conducting path to the interior of the devices. If a person is in a bathtub and thus is grounded through the tub's water to its drainpipe, or the person has one hand on a faucet, even touching a switch with a wet finger is risky.

Ohm's law can also be used to find the resistance of an electrical appliance when its voltage and current ratings are known. An electric toaster draws a current of 4 A when it is plugged into a 120-V supply line. To find its resistance, we rewrite Ohm's law in the form $R = V/I$ and substitute the values given:

$$R = \frac{V}{I} = \frac{120\ \text{V}}{4\ \text{A}} = 30\ \Omega$$

The resistance of the toaster is 30 Ω.

Despite its name, Ohm's law is not a basic physical principle such as the law of conservation of energy. Ohm's law is obeyed only by metallic conductors, not by gaseous or liquid conductors and not by such electronic devices as transistors.

5-11 ELECTRIC POWER

Current Times Voltage

Electric energy is so useful both because it is conveniently carried by wires and because it is easily converted into other kinds of energy. Electric energy in the form of electric current becomes radiant energy in a lightbulb, chemical energy when a storage battery is charged, kinetic energy in an electric motor, heat in an electric oven. In each case the current performs work on the device it passes through, and the device then turns this work into another kind of energy.

As in the case of the energy lost due to friction, the energy lost due to the resistance of a conductor becomes heat. This is the basis of electric heaters and stoves. In a lightbulb, the filament is so hot that it glows white. In electric circuits it is obviously important to use wires large enough in diameter, and hence small enough in resistance, to prevent the wires becoming so hot that they melt their insulation and start fires. A thin extension cord suitable for a lamp or radio might well be dangerous used for a heater or power tool.

An important quantity in any discussion of electric current is the rate at which a current is doing work—in other words, the **power** of the current. Electric power is given by the product of the current and the voltage of the circuit:

$$P = IV \qquad \textit{electric power}$$

Power = (current)(voltage)

Now we can see why electrical appliances are rated in watts; as we learned in Sec. 3-2, the watt is the unit of power. A 60-W lightbulb uses twice the power of a 30-W bulb, and one-tenth the power of a 600-W electric drill (Table 5-1).

A fuse or circuit breaker interrupts a power line whenever an unsafe amount of current passes through it. Many of the fuses normally used in homes are rated at 15 A. Since the power-line voltage is 120 V, the greatest power a 15-A line can provide without blowing the fuse is

$$P = IV = (15\ \text{A})(120\ \text{V}) = 1800\ \text{W} = 1.8\ \text{kW}$$

TABLE 5-1

Typical Power Ratings of Various Appliances

Appliance	Power, W
Charger for electric toothbrush	1
Clothes dryer	5,000
Coffeemaker	700
Dishwasher	1,600
Fan	150
Fax transmitter/receiver	65
Heater	2,000
Iron	1,000
Personal computer	150
Photocopier	1,400
Portable sander	200
Refrigerator	400
Stove	12,000
TV receiver	120
Vacuum cleaner	750

A circuit breaker will open, cutting off current to its circuit, when the current is too high for the wiring to carry it safely or when there is a defect in the circuit. Each of these breakers is labeled with the current at which it will open.

Because $P = IV$, it is easy to find how much current is needed by an appliance rated in watts when connected to a power line of given voltage. For instance, a 60-W bulb connected to a 120-V line needs a current of

$$I = \frac{P}{V} = \frac{60 \text{ W}}{120 \text{ V}} = 0.5 \text{ A}$$

A kilowatthour meter registers the electric energy that has been supplied to a house or other user of electricity.

The Kilowatthour

Users of electricity pay for the amount of energy they consume. The usual commercial unit of electric energy is the **kilowatthour** (kWh), which is the energy supplied per hour when the power level is 1 kilowatt. If electricity is sold at $0.12 per kilowatthour, the cost of operating a 1.5-kW electric heater for 7 h would be

Cost = (price per unit of energy)(energy used)

= (price per unit of energy)(power)(time)

= ($0.12/kWh)(1.5 kW)(7 h) = $1.26

Table 5-2 summarizes the various electrical quantities we have been discussing.

TABLE 5-2

Electrical Quantities

Quantity	Symbol	Unit	Meaning	Formula
Charge	Q	Coulomb (C)	A basic property of most elementary particles. The electron has a charge of -1.6×10^{-19}C.	
Current	I	Ampere (A) (1 A = 1 C/s)	Rate of flow of charge.	$I = \frac{Q}{t} = \frac{P}{V}$
Potential difference (voltage)	V	Volt (V) (1 V = 1 J/C)	Potential energy difference per coulomb of charge between two points; corresponds to pressure in water flow.	$V = IR = \frac{P}{I}$
Resistance	R	Ohm (Ω) (1 Ω = 1 V/A)	A measure of the opposition to the flow of charge in a particular circuit. For a given voltage, the higher the resistance, the lower the current.	$R = \frac{V}{I}$
Power	P	Watt (W) (1 W = 1 V · A)	Rate of energy flow.	$P = IV$

Electricity and magnetism were once considered as completely separate phenomena. One of the great achievements of nineteenth-century science was the realization that they are really very closely related, a realization that led to the discovery of the electromagnetic nature of light. And one of the great achievements of nineteenth-century technology was the invention of electric motors and generators, whose operation depends upon the connection between electricity and magnetism.

5-12 MAGNETS

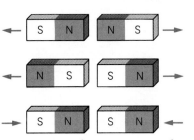

FIG. 5-16 Like magnetic poles repel each other; unlike magnetic poles attract.

Attraction and Repulsion
Ordinary magnets are familiar to everybody. The simplest is a bar of iron that has been magnetized in one way or another, say, by having been stroked by another magnet. A magnetized iron bar is recognized, of course, by its ability to attract and hold other pieces of iron to itself. Most of the force a magnet exerts comes from its ends, as we can see by testing the attraction of different parts of a bar magnet for iron nails.

If we pivot a magnet at its center so that it can swing freely, we will find that it turns so that one end points north and the other south. The north-pointing end is called the **north pole** of the magnet, and the south-pointing end is called its **south pole.** The tendency of a magnet to line up with the earth's axis is the basis of the compass, whose needle is a small magnet. (In Chap. 14 we shall find that the reason for this behavior is that the earth itself is a giant magnet.)

If the north poles of two magnets are brought near each other, the magnets repel. If the north pole of one magnet is brought near the south pole of the other, the magnets attract (Fig. 5-16). This gives us a simple rule like that for electric charges:

Like magnetic poles repel one another; unlike poles attract one another.

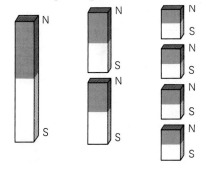

FIG. 5-17 Cutting a magnet in half produces two other magnets. There is no such thing as a single free magnetic pole.

Poles Always Come in Pairs
Positive and negative charges in neutral matter can be separated from each other. Can the north and south poles of a magnet also be separated? It would seem that all we have to do is to saw the magnet in half. But if we do this, as in Fig. 5-17, we find that the resulting pieces each have an N pole and an S pole. We may cut the resulting magnets in two again and continue as long as we like, but each piece, however small, will still have both an N pole and an S pole. There is no such thing as a single free magnetic pole.

Since a magnet can be cut into smaller and smaller pieces indefinitely with each piece a small magnet in itself, we conclude that magnetism is a property of the iron atoms themselves. Each atom of iron behaves as if it has an N pole and an S pole. In ordinary iron the atoms have their poles randomly arranged, and nearby N and S poles cancel out each other's effect. When a bar of iron is magnetized, many or all of the atoms are aligned with the N poles in the same direction, so that the

A magnetic compass uses the earth's magnetic field to establish direction. The magnetic axis of the earth is not quite aligned with its axis of rotation, so a magnetic compass does not point exactly to true north.

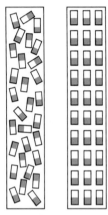

FIG. 5-18   The iron atoms in an unmagnetized iron bar are randomly oriented, whereas in a magnetized bar they are aligned with their north poles pointing in the same direction. The ability of iron atoms to remain aligned in this way is responsible for the magnetic properties of iron.

strengths of all the tiny magnets are added together (Fig. 5-18). A "permanent" magnet can be demagnetized by heating it strongly or by hammering it. Both of these processes agitate the atoms and restore them to their normal random orientations.

Iron is not the only material from which permanent magnets can be made. Nickel, cobalt, and certain combinations of other elements can also be magnetized. Nor is iron the only material affected by magnetism—*all* substances are, though generally only to a very slight extent. Some are attracted to a magnet, but most are repelled. In the case of mercury the repulsion, though weak, is still enough to be easily observed.

5-13   MAGNETIC FIELD

How Magnetic Forces Act

We are so familiar with gravitational, electric, and magnetic forces that we take them for granted. However, if we think about them, it is clear that something remarkable is going on: these forces act without the objects involved touching each other. We cannot move a book from a table by just waving our hand at it, and a golf ball will not fly off until a golf club actually strikes it. An iron nail, however, does not wait until a magnet touches it, but is pulled to the magnet when the two are some distance apart. The region near the magnet is somehow altered by the magnet's presence, just as a mass or an electric charge alters the region around itself, in each case in a different way.

The region of altered space around a mass, an electric charge, or a magnet is called a **force field.** A physicist describes a force field in terms of what it does, which is to exert a force on appropriate objects. Although we cannot see a force field, we can detect its presence by its

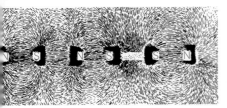

FIG. 5-19 Patterns formed by iron filings sprinkled on a card held over three bar magnets. The filings align themselves in the direction of the magnetic field. It is convenient to think of the pattern in terms of "field lines," but such lines do not actually exist since the field is a continuous property of the region of space it occupies.

effects. In fact, even the forces we think of as being exerted by direct contact turn out to involve force fields. For instance, when a golf club strikes a ball, it is the action of electric forces on the molecular level that leads to the observed transfer of energy and momentum to the ball. There is actually no such thing as "direct contact" since the atoms involved never touch each other.

When iron filings are scattered on a card held over a magnet, they form a pattern that suggests the form of the magnet's field. At each point on the card, the filings line up in the direction in which a piece of iron would move if put there, and the filings gather most thickly where the force on the iron would be greatest. Figure 5-19 shows the patterns of iron filings near three bar magnets.

Field Lines   It is traditional, and convenient, to think of a magnetic field in terms of imaginary **field lines** that correspond to the patterns formed by iron filings. A magnetic field line traces the path that would be taken by a small iron object if placed in the field, with the lines close together where the field is strong and far apart where the field is weak. Although the notion of field lines is helpful in illustrating a number of magnetic effects, we must keep in mind that they are imaginary—a force field is a continuous property of the region of space where it is present, not a collection of strings.

5-14   OERSTED'S EXPERIMENT

Magnetic Fields Originate in Moving Electric Charges   Electric currents may not be familiar to us as sources of magnetic fields, yet every current has such a field around it. To repeat a famous experiment first performed in 1820 by the Danish physicist Hans Christian Oersted, we can connect a horizontal wire to a battery and hold under the wire a small compass needle (Fig. 5-20). The needle at once swings into a position at right angles to the wire. When the compass is placed just above the wire, the needle swings around until it is again perpendicular to the wire but pointing in the opposite direction.

We can use iron filings to study the magnetic field pattern around a wire carrying a current. When we do this, we find that the field lines near the wire consist of circles, as in Fig. 5-21. The direction of the field

FIG. 5-20 Oersted's experiment showed that a magnetic field surrounds every electric current. The field direction above the wire is opposite to that below the wire.

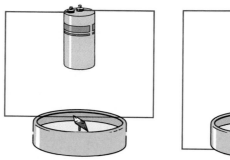
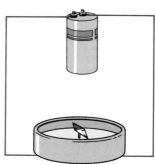

FIG. 5-21 Magnetic field lines around a wire carrying an electric current. The direction of the lines may be found by placing the thumb of the left hand in the direction of electron flow; the curled fingers then point in the direction of the field lines. In the right-hand diagram the electron current flows out of the paper toward you.

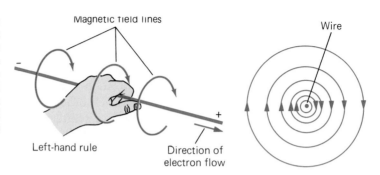

Magnetic field lines

Wire

Left-hand rule

Direction of electron flow

Hans Christian Oersted (1777–1851)

lines (that is, the direction in which the N pole of the compass points) depends on the direction of flow of electrons through the wire. When one is reversed, the other reverses also.

In general, the direction of the magnetic field around a wire can be found by encircling the wire with the fingers of the left hand, so that the extended thumb points along the wire in the direction of the field. According to this **left-hand rule,** the current and the field are perpendicular to each other.

Oersted's discovery showed for the first time that a connection exists between electricity and magnetism. It was also the first demonstration of the principle on which the electric motor is based. Magnetism and electricity are related, but only through moving charges. An electric charge *at rest* has no magnetic properties. A magnet is not influenced by a stationary electric charge near it, and vice versa.

When a current passes through a wire bent into a circle, the resulting magnetic field, shown in Fig. 5-22, is the same as that produced by a bar magnet. One side of the loop acts as a north pole, the other as a south pole. If free to turn, the loop swings to a north-south position. A current loop attracts pieces of iron just as a bar magnet does. Indeed, the magnetic properties of iron and other substances can be traced to tiny currents within their atoms. Thus it is correct to say that

All magnetic fields originate from moving electric charges.

FIG. 5-22 The magnetic field of a loop of electric current is the same as that of a bar magnet.

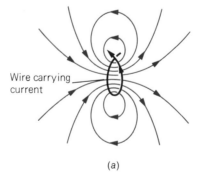

Wire carrying current

(a)

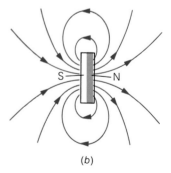

S   N

(b)

André Marie Ampère (1775–1836) was largely self-taught and had mastered advanced mathematics by his early teens. Starting as a teacher in local schools near Lyon, he went on to a series of professorships in Paris and was appointed by Napoleon as inspector-general of the French university system. Ampère's personal life was one misfortune after another: the execution of his father during the French Revolution, the early death of his much-loved first wife, a disastrous second marriage, financial problems.

In contrast were the successes of his scientific career. Upon learning of Oersted's discovery that a magnetic field surrounds every electric current, Ampère carried out experiments of his own that resulted in "a great theory of these phenomena and of all others known for magnets," as he wrote to his son only two weeks later. Ampère's results included the law that describes the magnetic force between two currents and the observation that the magnetic field around a current loop is the same as the field around a bar magnet. He went on to speculate that the magnetism of a material such as iron is due to loops of electric current in its atoms, a concept very much ahead of his time. The unit of electric current is called the ampere because he was the first to distinguish clearly between current and potential difference.

The Electromagnetic Field

An electric charge at rest is surrounded only by an electric field, and when the charge is moving it is surrounded by a magnetic field as well. Suppose we travel alongside a moving charge, in the same direction and at the same speed. All we find now is an electric field—the magnetic field has disappeared. But if we move past a stationary charge with our instruments, we find both an electric and a magnetic field! Clearly the *relative motion* between charge and observer is needed to produce a magnetic field: no relative motion, no magnetic field.

According to the theory of relativity, whatever it is in nature that shows itself as an electric force between charges at rest *must* also show itself as a magnetic force between moving charges. One effect is not possible without the other. Thus the proper way to regard what we think of as separate electric and magnetic fields is that they are both aspects of a single electromagnetic field that surrounds every electric charge. The electric field is always there, but the magnetic field appears only when relative motion is present.

In the case of a wire that carries an electric current, there is only a magnetic field because the wire itself is electrically neutral. The electric field of the electrons is canceled out by the opposite electric field of the positive ions in the wire. However, the positive ions are stationary and therefore have no magnetic field to cancel the magnetic field of the moving electrons. If we simply move a wire that has no current in it, the electric and magnetic fields of the electrons are canceled by the electric and magnetic fields of the positive ions.

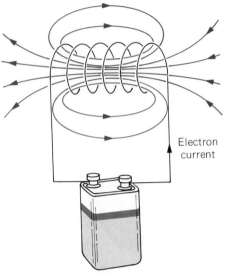

FIG. 5-23   The magnetic field of a coil is like that of a single loop but is stronger.

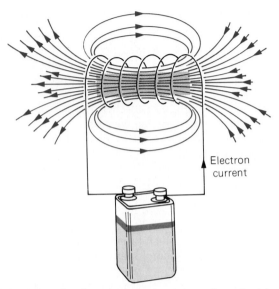

FIG. 5-24   An electromagnet consists of a coil with an iron core, which considerably enhances the magnetic field produced.

5-15   ELECTROMAGNETS

Electromagnet loading scrap iron and steel.

How to Create a Strong Magnetic Field

When several wires that carry currents in the same direction are side by side, their magnetic fields add together to give a stronger total magnetic field. This effect is often used to increase the magnetic field of a current loop. Instead of one loop, many loops of wire are wound into a coil, as in Fig. 5-23, and the resulting magnetic field is as many times stronger than the field of one turn as there are turns in the coil. A coil with 50 turns produces a field 50 times greater than a coil with just one turn.

The magnetic field of the coil is enormously increased if a rod of iron is placed inside it (Fig. 5-24). This combination of coil and iron core is called an **electromagnet.** An electromagnet exerts magnetic force only when current flows through its turns, and so its action can be turned on and off. Also, by using many turns and enough current, an electromagnet can be made far more powerful than a permanent magnet. Electromagnets are widely used and range in size from the tiny coils in telephone receivers to the huge ones that load and unload scrap iron.

USING MAGNETISM

An electric motor uses magnetic fields to turn electric energy into mechanical energy, and a generator uses magnetic fields to turn mechanical energy into electric energy. As we shall find, magnetic fields also play essential roles in television picture tubes, in sound and video recording, and in the transformers used to distribute electric power over large areas.

MAGLEV TRAINS   Electromagnets can provide enough force to support—"levitate"—trains and so eliminate friction (although air resistance remains). Magnetic forces also propel such **maglev** trains. Experimental maglev trains that are based on pioneering work done in the United States 30 years ago are currently operating in Germany and Japan. In Germany, attractive forces are developed by conventional electromagnets that curl under a T-shaped guideway. A $5 billion maglev system of this kind is now being built to link Hamburg and Berlin, which are 260 km apart, by the year 2004. The trains will run at 425 km/h.

In Japan the upward force is provided by magnetic repulsion using superconducting coils for the magnetic fields. Because the coils do not need iron cores, the cars are lighter, and the cars float higher above their guideways than the German ones, so small misalignments are less important. Although the Japanese system needs further development, it holds the speed record of 517 km/h (321 mi/h). In both systems the magnetic field of an alternating current passed through electromagnets along the guideway creates attractive forces that pull the train's own magnets forward and repulsive forces that push them from behind. The higher the frequency of the alternating current, the higher the train's speed. When perfected, maglev trains are projected to use less than half the energy per person per kilometer than jet aircraft do and ought to be competitive in total trip times for distances of up to perhaps 1000 km.

This experimental Japanese train uses magnetic forces for both support and propulsion.

5-16 MAGNETIC FORCE ON A CURRENT

FIG. 5-25 A magnetic field exerts a sidewise push on an electric current. In this arrangement, the wire moves to the side in a direction perpendicular to both the magnetic field and the current. A handy way to figure out the direction of the force is to open your left hand so that the fingers are together and the thumb sticks out. When your thumb is in the direction of the electron current and your fingers are in the direction of the magnetic field, your palm faces in the direction of the force. To remember this rule, think of your thumb in terms of hitchhiking and so as the motion of the electrons, of your parallel fingers as magnetic field lines, and of your palm as pushing on something. The same rule holds for the force on a moving electron or other negative charge. The force on a moving positive charge is in the opposite direction.

A Sidewise Push

Suppose that a horizontal wire connected to a battery is suspended as in Fig. 5-25, so that it is free to move from side to side, and the N pole of a bar magnet is

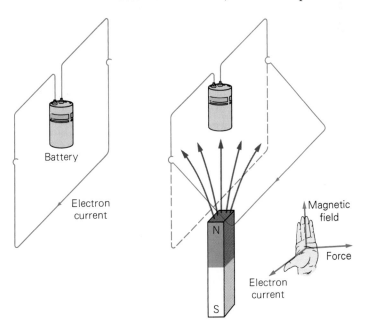

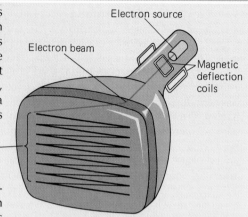

TELEVISION PICTURE TUBE In a television picture tube, a beam of electrons is directed at a fluorescent screen that glows where electrons strike it. A beam of electrons is, of course, an electric current, and magnetic fields produced by coils outside the tube are used to deflect the beam. By changing the currents in the coils, the magnetic field strengths can be changed, which permits the beam to be aimed anywhere on the screen. As in Fig. 5-26, the beam is moved across the screen in a pattern of horizontal lines starting at the upper left. The returns from right to left take about a tenth as long as the sweeps from left to right used to form the picture. In the United States, a complete image is divided into 525 lines, but in many other countries more lines are used to give better picture quality. A complete image is built up by two scans of the screen, each covering alternate lines, a process that takes place 30 times per second. During the scans the electron beam intensity changes to give the variations in brightness that make up the picture. In color television separate signals are used for the red, green, and blue contents of the image. The receiver screen consists of red, green, and blue phosphor dots that flash the appropriate color when the electron beam strikes them.

FIG. 5-26 The electron beam of a television picture tube is directed by magnetic fields to cover the screen in a pattern of horizontal lines.

then placed directly under it. This arrangement is the reverse of Oersted's experiment. Oersted placed a movable magnet near a wire fixed in position, whereas here we have a movable wire near a fixed magnet. We might predict, from Oersted's results and Newton's third law of motion, that in this case the wire will move. It does indeed, swinging out to one side as soon as the current is on. The direction of the wire's motion is perpendicular to the bar magnet's field. Whether the wire swings to one side or the other depends on the direction of flow of electrons in the wire and on which pole of the magnet is used.

Thus the force a magnetic field exerts on an electric current is not a simple attraction or repulsion but a *sidewise push*. The maximum side-

FIG. 5-27 Equal and opposite forces are exerted by parallel currents on each other. The forces are attractive when the currents are in the same direction, repulsive when they are in opposite directions.

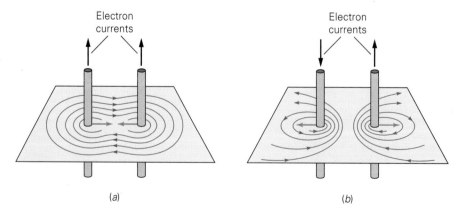

wise push occurs when the current is perpendicular to the magnetic field, as in Fig. 5-25. At other angles the push is less, and it disappears when the current is parallel to the magnetic field.

Every current has a magnetic field around it, and as a result nearby currents exert magnetic forces on each other. When parallel currents are in the same direction, as in Fig. 5-27a, the forces are attractive; when the currents are in opposite directions, as in Fig. 5-27b, the forces are repulsive.

5-17 ELECTRIC MOTORS

Mechanical Energy from Electric Energy The sidewise push of a magnetic field on a current-carrying wire can be used to produce continuous motion in an arrangement like that shown in Fig. 5-28. A magnet gives rise to a magnetic field inside which a wire loop is free to turn. When the plane of the loop is parallel to the magnetic field, there is no force on the two sides of the loop that lie along the magnetic field. The side of the loop at left in the diagram, however, receives an upward push, and the side at right receives a downward one. Thus the loop is turned clockwise.

To produce a continuous movement, the direction of the current in the loop must be reversed when the loop is vertical. The reversed current then interacts with the magnetic field to continue to rotate the loop through 180°. Now the current must have its direction reversed once more, whereupon the loop will again swing around through a half-turn. The device used to automatically change the current direction is called a **commutator;** it is visible on the shaft of a direct-current motor as a cooper sleeve divided into segments. Normally more than two loops and commutator segments are used in order to yield the maximum turning force.

FIG. 5-28 A simple direct-current electric motor. The commutator reverses the current in the loop periodically so that the loop always rotates in the same direction.

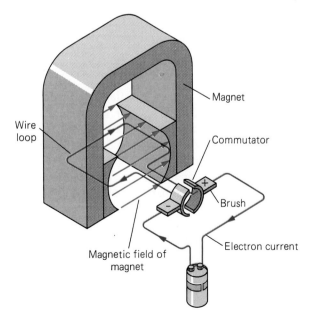

Magnet

Wire loop

Commutator

Brush

Electron current

Magnetic field of magnet

The stationary windings of a large electric motor. Magnetic forces underlie the operation of such motors.

Actual direct-current motors, such as the starter motor of a car, are more complicated than the one shown in Fig. 5-28, but their basic operating principle is the same. Usually electromagnets are employed rather than permanent ones to create the field, and in some motors the coil is fixed in place and the magnet or magnets rotate inside it. Motors built for alternating rather than direct current do not need commutators because the current direction changes back and forth many times per second.

5-18   ELECTROMAGNETIC INDUCTION

Electric Energy from Mechanical Energy

The electric energy that our homes and industries use in such quantity comes from generators driven by turbines powered by running water or, more often, by steam. In the latter case, as we saw in Fig. 4-37, the boilers that supply the steam obtain heat from coal, oil, or natural gas, or from nuclear reactors. Ships and isolated farms have smaller generators operated by gasoline or diesel engines. In all cases the energy that is turned into electricity is the kinetic energy of moving machinery.

The principle of the generator was discovered by the nineteenth-century English physicist Michael Faraday. Faraday's curiosity was aroused by the researches of Ampère and Oersted on the magnetic fields around electric currents. He reasoned that, if a current can produce a magnetic field, then somehow a magnet should be able to generate an electric current.

A wire placed in a magnetic field and connected to a meter shows no sign of a current. What Faraday found instead is that

BIOGRAPHY

Michael Faraday (1791–1867), the son of a blacksmith, was apprenticed to a bookbinder at 13 and taught himself chemistry and physics from the books he was learning to bind. At 21 he became bottle washer for Humphrey Davy at that noted chemist's laboratory in the Royal Institution in London. Within 20 years Faraday succeeded Davy as head of the Institution. During that period Faraday had, among other things, liquefied a number of gases for the first time and formulated what are today called Faraday's laws of electrolysis. In the later years of his life came the remarkable work in electricity and magnetism whose results include building the first electric motor and the discovery of electromagnetic induction.

Faraday realized at once the implications of electromagnetic induction and soon had generators and transformers working in his laboratory. Asked by a politician what use these devices were, Faraday replied, "At present I do not know, but one day you will be able to tax

them." To make sense of the electric and magnetic fields he could not see or feel or represent mathematically (he was poor at math), Faraday invented field lines—lines that do not exist but help us to picture what is going on. At his death he left behind notebooks with over 16,000 entries that testify to his originality, intuition, skill, and diligence.

A current is produced in a wire when there is relative motion between the wire and a magnetic field.

FIG. 5-29 Electromagnetic induction. The direction of the induced current is perpendicular both to the magnetic lines of force and to the direction in which the wire is moving. No current is induced when the wire is at rest.

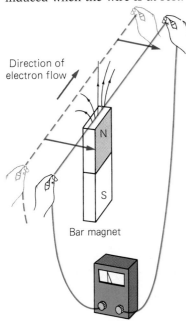

Direction of electron flow

Bar magnet

As long as the wire continues to move across magnetic field lines, the current continues. When the motion stops, the current stops. Because it is produced by motion through a magnetic field, this sort of current is called an **induced current**. The entire effect is known as **electromagnetic induction**.

Let us repeat Faraday's experiment. Suppose that the wire of Fig. 5-29 is moved back and forth across the field lines of force of the bar magnet. The meter will indicate a current first in one direction, then in the other. The direction of the induced current through the wire depends on the relative directions of the wire's motion and of the field lines. Reverse the motion, or use the opposite magnetic pole, and the current is reversed. The strength of the current depends on the strength of the magnetic field and on how rapid the wire's motion is.

Electromagnetic induction is related to the sidewise force a magnetic field exerts on electrons flowing along a wire. In Faraday's experiment electrons are again moved through a field, but now by moving the wire as a whole. The electrons are pushed sidewise as before and, in response to the push, move along the wire as an electric current.

Alternating and Direct Currents

In order to obtain a large induced current, an actual generator uses several coils rather than a single wire and several electromagnets instead of a bar magnet. Turned rapidly between the electromagnets, wires of the coil cut lines of force first one way, then the other. How a generator

MAGNETIC NAVIGATION BY ANIMALS   A number of animals—notably certain birds, turtles, and fish—use the earth's magnetic field to help them find their way on journeys that may cover thousands of kilometers. Sharks detect the field by means of sense organs on their snouts that respond to tiny electric currents induced by their motion through the field. The amount of current depends on the angle between the field and the direction in which the shark is swimming. Other animals have bits of magnetite, a mineral that contains iron and is affected by magnetic fields, in their brains.

Exactly how such animals employ magnetite to sense direction is not known, but experiments leave no doubt that they do. Some bacteria that contain magnetite even use the earth's field to distinguish between up and down when floating in a body of water, which enables them to migrate to their preferred habitat in the ooze at the bottom. The human brain also contains tiny amounts of magnetite: does that mean we have built-in compasses? Nobody knows for sure. A third possible sensing mechanism is based on the presence of certain pigments in the eyes of some animals that become weakly magnetic when light falls on them. The signals such eyes send to the brain seem to be affected by any magnetic field present.

Sharks navigate with the help of the earth's magnetic field. They detect the field with the help of electromagnetic induction.

---

works is shown in Fig. 5-30, where a coil is shown turning between two magnets. During one part of each turn, each side of the coil cuts the field in one direction. Then, during the other part of the turn, each side of the coil cuts the field in the opposite direction. Hence the induced current flows first one way and then the other. Such a back-and-forth current is an **alternating current.**

FIG. 5-30 An alternating-current generator. As the loop rotates, current is induced in it first in one direction (*ABCD*) and then in the other (*DCBA*). No current flows at those times when the loop is moving parallel to the magnetic field.

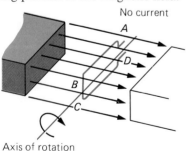

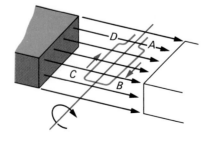

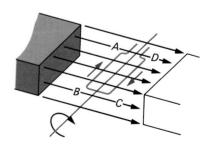

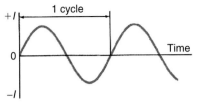

FIG. 5-31 How a 60-Hz (60 cycles/s) alternating current varies with time. The frequency of the current is the number of cycles that occur per second. Here each complete cycle takes $\frac{1}{60}$ s. If a current in one direction in a circuit is considered +, a current in the opposite direction is considered –.

The electric currents that come from such sources as batteries and photoelectric cells are always one-way **direct currents** that can be reversed only by changing the connections. In the 60-Hz (1 Hz = 1 hertz = 1 cycle/second) alternating current that we ordinarily use in our homes, electrons change their direction 120 times each second (Fig. 5-31). The usual abbreviation for alternating current is ac and that for direct current is dc.

By using commutators like those used on dc motors, generators can be built that produce direct current. Another way to obtain direct current from an ac generator (or **alternator**) is to use a **rectifier,** a device that permits current to pass through it in only one direction. Because alternators are simpler to make and more reliable than dc generators, they are often used together with rectifiers to give the direct current needed to charge the batteries of cars and trucks.

5-19 TRANSFORMERS

Stepping Voltage Up or Down

To induce a current requires that magnetic field lines move across a conductor. As in Fig. 5-29, one way to do this is to move a wire past a magnet. Another way is to hold the wire stationary while the magnet is moved. We come now to a third, less obvious, method, which involves no visible motion at all.

Let us connect coil A in Fig. 5-32 to a switch and a battery and connect the separate coil B to a meter. When the switch is closed, a current flows through A, building up a magnetic field around it. The current and field do not reach their full strengths at once. A fraction of a second is needed for the current to increase from zero to its final value, and the magnetic field increases along with the current. As this happens, the field lines from coil A spread outward across the wires of coil B. This motion of the lines across coil B produces in it a momentary current.

FIG. 5-32 A simple transformer. Momentary currents are detected by the meter when the current in coil A is started or stopped.

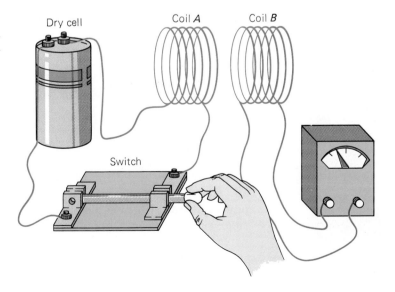

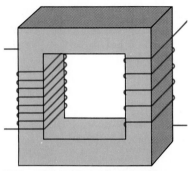

FIG. 5-33 Actual transformers usually have iron cores. The winding with the greater number of turns has the higher voltage across it and carries the lower current. The power in both windings is the same.

Transformers at a power station step up the voltage of the electric power generated there for transmission over long distances. The higher the voltage V, the lower the current I for the same power P, since $P = IV$. The advantage of a low current is that less energy is lost as heat in the transmission lines. Other transformers step down the voltage for the consumer to the usual 220–240 V or 110–120 V.

Once the current in A reaches its normal, steady value, the magnetic field becomes stationary and the induced current in B stops.

Next let us open the switch to break the circuit. In a fraction of a second the current in A drops to zero, and its magnetic field collapses. Once more field lines cut across B to induce a current, this time in the opposite direction since the field lines are now moving the other way past B. Thus starting and stopping the current in A has the same effect as moving a magnet in and out of B. An induced current is generated whenever the switch is opened or closed.

Suppose A is connected not to a battery but to a 60-Hz alternating current. Now we need no switch. Automatically, 120 times each second, the current comes to a complete stop and starts off again in the other direction. Its magnetic field expands and contracts at the same rate, and the field lines cutting B, first in one direction and then in the other, induce an alternating current similar to that in A. An ordinary meter will not respond to these rapid alterations, but an instrument meant for ac will show the induced current.

Thus an alternating current in one coil produces an alternating current in a nearby (but unconnected) coil. To generate an induced current most efficiently, the two coils should be close together and wound on a core of soft iron. Such a combination of two coils and an iron core is a **transformer** (Fig. 5-33). The coil into which electricity is fed from an outside source is the primary coil, and the coil in which an induced current is generated is the secondary coil.

Why Transformers Are Useful

Transformers are useful because the voltage of the induced current can be raised or lowered by suitable winding of the coils. If the secondary coil has the same number of turns as the primary, the induced voltage will be the same as the primary voltage. If the secondary has twice as many turns, its voltage is twice that of the primary; if it has one-third as many turns, its voltage is one-third that of the primary; and so on. By using a suitable transformer, we can obtain any voltage we like, high or low, from a given alternating current.

When the secondary coil of a transformer has a higher voltage than the primary coil, its current is lower than that in the primary (and vice versa), so that the power $P = IV$ is the same in both coils. Thus

$$\frac{\text{Primary turns}}{\text{Secondary turns}} = \frac{\text{primary voltage}}{\text{secondary voltage}} = \frac{\text{secondary current}}{\text{primary current}}$$

$$\frac{N_1}{N_2} = \frac{V_1}{V_2} = \frac{I_2}{I_1} \qquad \textit{transformer}$$

For many purposes in homes, factories, and laboratories it is desirable to change the voltage of alternating currents. But most valuable of all, transformers permit the efficient long-distance transmission of power. Currents in long-distance transmission must be as small as possible, since large currents mean energy lost in heating the transmission wires. Hence at a power plant electricity from the generator is led into a "step-up" transformer, which increases the voltage and decreases the current, each by several hundred times. On high-voltage lines (sometimes carrying currents at voltages exceeding 1 million V) this current is carried to local substations, where other transformers "step down" its voltage to make it safe for local transmission and use.

5-20 TAPE RECORDERS

Storing and Reproducing Sights and Sounds

What we have thus far learned about electricity and magnetism allows us to understand how sounds can be changed into electrical signals by a microphone, recorded on a magnetic tape, and then reproduced by a loudspeaker.

The first step is to change the pressure variations of a sound wave (described in detail in Chap. 6) into an alternating current. This is done by a **microphone,** of which there are several kinds. One of them is shown in Fig. 5-34. The diaphragm is a paper cone that vibrates as sound waves strike it. As the diaphragm vibrates, a coil of thin wire at its neck moves back and forth in the field of a permanent magnet, which induces an alternating current in the coil whose variations match those of the sound waves.

The electrical signal from the microphone is then amplified and, in the tape recorder itself, converted into a magnetic pattern on the magnetizable coating of a plastic tape. As in Fig. 5-35, this tape is magnetized by being passed over an electromagnet connected to the amplifier. The air gap in the electromagnet of the recording head permits its mag-

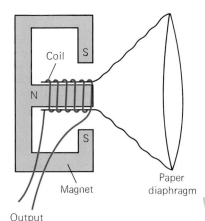

FIG. 5-34 A moving-coil microphone. When sound waves reach the diaphragm, it vibrates accordingly. The motion of the coil through the magnetic field of the magnet induces an alternating current in the coil that corresponds to the original sound. A loudspeaker is similar in construction except that an alternating current in its coil causes the diaphragm to vibrate and thereby produce sound waves.

FIG. 5-35 A tape recorder. The polarity and degree of magnetization of the magnetizable coating on the tape correspond to the pressure variations of the original sound wave.

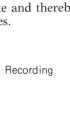
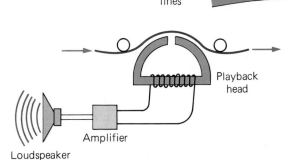

netic field to reach the tape. The polarity and degree of magnetization of the coating on the tape correspond to the pressure variations of the original sound wave.

The playing head of the recorder is similar in design to the recording head but makes use of electromagnetic induction. The changing magnetic fields of the moving tape induce an alternating current in the coil of the playing head, which is amplified and fed into a loudspeaker to reconstruct the original sound. A loudspeaker resembles the moving-coil microphone of Fig. 5-34 but is larger. An alternating current in its coil causes its conical diaphragm to vibrate and thereby produce a sound wave in the air around it.

A videotape recorder uses recording and playback heads like those in Fig. 5-35, but the electrical signals it records come from a television camera and are played back through a television set.

IMPORTANT TERMS AND IDEAS

Electric charge is a fundamental property of certain of the elementary particles of which all matter is composed. The two kinds of charge are called **positive** and **negative.** Charges of the same sign repel each other; charges of opposite sign attract each other. The unit of charge is the **coulomb** (C). All charges, of either sign, occur in multiples of $e = 1.6 \times 10^{-19}$ C.

Atoms are composed of **electrons,** whose charge is $-e$; **protons,** whose charge is $+e$; and **neutrons,** which have no charge. Protons and neutrons have almost equal masses, which are nearly 2000 times greater than the electron mass. Every atom has a small, central **nucleus** of protons and neutrons with its electrons moving about the nucleus some distance away. The number of protons and electrons is equal in a normal atom, which is therefore electrically neutral. An atom that has lost one or more electrons is a **positive ion,** and an atom that has picked up one or more electrons in excess of its usual number is a **negative ion.**

A flow of charge from one place to another is an **electric current.** The unit of electric current is the **ampere** (A), which is equal to a flow of 1 coulomb/second. Charge flows easily through a **conductor,** only with difficulty through an **insulator.** A **superconductor** offers no resistance at all to the flow of charge.

The **potential difference** (or **voltage**) between two points is the work needed to take a charge of 1 C from one of the points to the other. The unit of potential difference is the **volt** (V), which is equal to 1 joule/coulomb.

According to **Ohm's law,** the current in a metal conductor is proportional to the potential difference between its ends and inversely proportional to its **resistance.** The unit of resistance is the **ohm** (Ω), which is equal to 1 volt/ampere.

The **power** of an electric current is the rate at which it does work.

Every electric current (and moving charge) has a **magnetic field** around it that exerts a sidewise force on any other electric current (or moving charge) in its presence. All atoms contain moving electrons, and **permanent magnets** are made from substances, notably iron, whose atomic magnetic fields can be lined up instead of being randomly oriented.

Electromagnetic induction refers to the production of a current in a wire when there is relative motion between the wire and a magnetic field.

The direction of an **alternating current** reverses itself at regular intervals. In a **transformer** an alternating current in one coil of wire induces an alternating current in another nearby coil. Depending on the ratio of turns of the coils, the induced current can have a voltage that is larger, smaller, or the same as that of the primary current.

IMPORTANT FORMULAS

Coulomb's law: $F = K \dfrac{Q_1 Q_2}{R_2}$

Ohm's law: $I = \dfrac{V}{R}$

Electric power: $P = IV$

Transformer: $\dfrac{N_1}{N_2} = \dfrac{V_1}{V_2} = \dfrac{I_2}{I_1}$ (N = number of turns)

EXERCISES: MULTIPLE CHOICE

1. Electric charge
   a. can be subdivided indefinitely
   b. occurs only in separate parcels of $\pm 1.6 \times 10^{-19}$ C
   c. occurs only in separate parcels of ± 1 C
   d. occurs only in separate parcels whose value depends on the particle carrying the charge

2. A negative electric charge
   a. interacts only with positive charges
   b. interacts only with negative charges
   c. interacts with both positive and negative charges
   d. may interact with either positive or negative charges, depending on circumstances

3. A positively charged rod is brought near an isolated metal ball. Which of the sketches in Fig. 5-36 best illustrates the arrangement of charges on the ball?
   a. *a*   c. *c*
   b. *b*   d. *d*

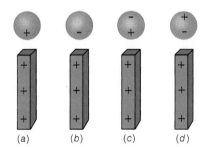

(a)   (b)   (c)   (d)

FIG. 5-36

4. Which of the following statements is not true?
   a. The positive charge in an atomic nucleus is due to the protons it contains
   b. All protons have the same charge
   c. Protons and electrons have charges equal in magnitude although opposite in sign
   d. Protons and electrons have equal masses

5. Coulomb's law for the force between electric charges belongs in the same general category as
   a. the law of conservation of energy
   b. Newton's second law of motion
   c. Newton's law of gravitation
   d. the second law of thermodynamics

6. The electric force between a proton and an electron
   a. is weaker than the gravitational force between them
   b. is equal in strength to the gravitational force between them
   c. is stronger than the gravitational force between them
   d. is any of the above, depending on the distance between the proton and the electron

7. The electrons in an atom
   a. are bound to it permanently
   b. are some distance away from the nucleus
   c. have more mass than the nucleus
   d. may be positively or negatively charged

8. Atoms and molecules are normally
   a. electrically neutral
   b. negatively charged
   c. positively charged
   d. ionized

9. An object has a positive electric charge whenever
   a. it has an excess of electrons
   b. it has a deficiency of electrons
   c. the nuclei of its atoms are positively charged
   d. the electrons of its atoms are positively charged

10. Which of the following statements is correct?
    a. Electrons carry electric current
    b. The motion of electrons produces an electric current
    c. Moving electrons constitute an electric current
    d. Electric currents are carried by conductors and insulators only

11. Match each of the electrical qualities listed below with the appropriate unit from the list on the right:
    a. resistance   volt
    b. current   ampere
    c. potential difference ohm
    d. power   watt

12. Electric power is equal to
    a. (current)(voltage)
    b. current/voltage
    c. voltage/current
    d. (resistance)(voltage)

13. The electric energy lost when a current passes through a resistance
    a. becomes magnetic energy
    b. becomes potential energy
    c. becomes heat
    d. disappears completely

14. All magnetic fields originate in
    a. iron atoms
    b. permanent magnets
    c. stationary electric charges
    d. moving electric charges

15. The force on an electron that moves in a curved path must be
    a. gravitational
    b. electrical
    c. magnetic
    d. one or more of the above

16. A drawing of the field lines of a magnetic field provides information on
    a. the direction of the field only
    b. the strength of the field only
    c. both the direction and the strength of the field
    d. the source of the field

17. Magnetic field lines provide a convenient way to visualize a magnetic field. Which of the following statements is not true?
    a. The path followed by an iron particle released near a magnet corresponds to a field line
    b. The path followed by an electric charge released near a magnet corresponds to a field line
    c. A compass needle in a magnetic field turns until it is parallel to the field lines around it
    d. Magnetic field lines do not actually exist

18. A moving electric charge produces
    a. only an electric field
    b. only a magnetic field
    c. both an electric and a magnetic field
    d. any of the above, depending on its speed

19. The magnetic field of a bar magnet resembles most closely the magnetic field of
    a. a straight wire carrying a direct current
    b. a straight wire carrying an alternating current
    c. a wire loop carrying a direct current
    d. a wire loop carrying an alternating current

20. The magnetic field shown in Fig. 5-37 is produced by
    a. two north poles
    b. two south poles
    c. a north pole and a south pole
    d. a south pole and an unmagnetized iron bar

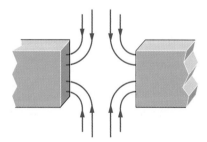

FIG. 5-37

21. The magnetic field lines around a long, straight current are
    a. straight lines parallel to the current
    b. straight lines that radiate from the current like spokes of a wheel
    c. concentric circles around the current
    d. concentric helixes around the current

22. A magnet does not exert a force on
    a. an unmagnetized iron bar
    b. a magnetized iron bar
    c. a stationary electric charge
    d. a moving electric charge

23. A current-carrying wire is in a magnetic field with the direction of the current the same as that of the field.
    a. The wire tends to move parallel to the field
    b. The wire tends to move perpendicular to the field
    c. The wire tends to turn until it is perpendicular to the field
    d. The wire has no tendency to move or to turn

24. An electromagnet
    a. uses an electric current to produce a magnetic field
    b. uses a magnetic field to produce an electric current
    c. is a magnet that has an electric charge
    d. operates only on alternating current

25. The nature of the force that is responsible for the operation of an electric motor is
    a. electric
    b. magnetic
    c. a combination of electric and magnetic
    d. either electric or magnetic, depending on the design of the motor

26. A generator is said to "generate electricity." What it actually does is act as a source of
    a. electric charge   c. magnetism
    b. electrons   d. electric energy

27. The alternating current in the secondary coil of a transformer is induced by
    a. the varying electric field of the primary coil
    b. the varying magnetic field of the primary coil
    c. the varying magnetic field of the secondary coil
    d. the iron core of the transformer

28. A transformer can change
    a. the voltage of an alternating current
    b. the power of an alternating current
    c. alternating current to direct current
    d. direct current to alternating current

29. A positive and a negative charge are initially 4 cm apart. When they are moved closer together so that they are now only 1 cm apart, the force between them is
a. 4 times smaller than before
b. 4 times larger than before
c. 8 times larger than before
d. 16 times larger than before

30. The force between two charges of -3×10^{-9} C that are 5 cm apart is
a. 1.8×10^{-16} N   **c.** 1.6×10^{-6} N
b. 3.6×10^{-15} N   **d.** 3.2×10^{-5} N

31. The potential difference between a certain thundercloud and the ground is 4×10^6 V. A lightning stroke occurs during which 80 C of charge is transferred between the cloud and the ground. The energy dissipated during the lightning stroke is
a. 5×10^{-6} J   **c.** 3.2×10^7 J
b. 2×10^5 J   **d.** 3.2×10^8 J

32. The current in a 12-Ω toaster operated at 120 V is
a. 0.1 A   **c.** 12 A
b. 10 A   **d.** 1440 A

33. The voltage needed to produce a current of 5 A in a resistance of 40 Ω is
a. 0.125 V   **c.** 8 V
b. 5 V   **d.** 200 V

34. The resistance of a lightbulb that draws a current of 2 A when connected to a 12-V battery is
a. 1.67 Ω **c.** 6 Ω
b. 2 Ω **d.** 24 Ω

35. The current in a 40-W 120-V electric lightbulb is
a. $\frac{1}{3}$ A   **c.** 80 A
b. 3 A   **d.** 4800 A

36. A transformer whose primary winding has twice as many turns as its secondary winding is used to convert 240-V ac to 120-V ac. If the current in the secondary circuit is 4 A, the primary current is
a. 1 A   **c.** 4 A
b. 2 A   **d.** 8 A

QUESTIONS

1. What reasons might there be for the universal belief among scientists that there are only two kinds of electric charge?

2. Electricity was once thought to be a weightless fluid, an excess of which was "positive" and a deficiency of which was "negative." What phenomena can this hypothesis still explain? What phenomena can it not explain?

3. Are electrons or protons easier to remove from an atom?

4. Why does the production of electricity by friction always yield equal amounts of positive and negative charge?

5. Is there any distance at which the gravitational force between two electrons is greater than the electric force between them?

6. Nearly all the mass of an atom is concentrated in its nucleus. Where is its charge located?

7. (a) List the similarities and differences between electric and gravitational fields. (b) How could you distinguish experimentally between an electric field and a gravitational field?

8. How do we know that the force holding the earth in its orbit about the sun is not an electric force, since both gravitational and electric forces vary inversely with the square of the distance between centers of force?

9. When two objects attract each other electrically, must both of them be charged? When two objects repel each other electrically, must both of them be charged?

10. A person can be electrocuted while taking a bath if he or she touches a poorly insulated light switch. Why is the electric shock received under these conditions so much more dangerous than usual?

11. Name several conductors and insulators of electricity. How well do these substances conduct heat? What general relationship between the ability to conduct heat and the ability to conduct electricity could you infer from this information?

12. How is the movement of electricity through air different from its movement through a copper wire?

13. Why are two wires used to carry electric current instead of a single one?

14. Why do you think bending a wire does not affect its electrical resistance, even though a bent pipe offers more resistance to the flow of water than a straight one?

15. An electrical appliance is sometimes said to "use

up" electricity. What does it actually use in its operation?

16. What aspect of superconductivity has prevented its large-scale application thus far?

17. A fuse prevents more than a certain amount of current from flowing in a particular circuit. What might happen if too much current were to flow? What determines how much is too much?

18. Heavy users of electric power, such as large electric stoves and clothes driers, are sometimes designed to operate on 240 V rather than 120 V. What advantage do you think the higher voltage has in these applications?

19. Why is a piece of iron attracted to either pole of a magnet?

20. From a drawing of magnetic field lines, how can you tell where the magnetic field described by the drawing is strong and where it is weak?

21. Explain why lines of force can never cross one another.

22. What kind of observations would you have to make in order to prepare a map showing the lines of force of the earth's magnetic field?

23. Figure 5-38 shows a current-carrying wire and a compass. In which direction will the compass needle point?

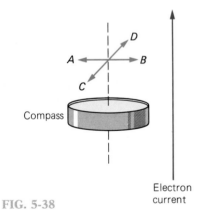

FIG. 5-38

24. An electron current flows east along a power line. Find the directions of the magnetic field above and below the power line. (Neglect the earth's magnetic field.)

25. A current-carrying wire is in a magnetic field. What angle should the wire make with the direction of the field for the force on it to be zero? What should the angle be for the force to be a maximum?

26. A length of copper wire AB rests across a pair of parallel copper wires that are connected to a battery through a switch, as in Fig. 5-39. The arrangement is placed between the poles of a magnet and the switch is closed. In what direction does the wire AB move?

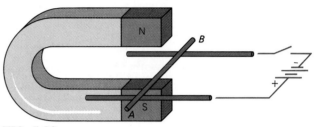

FIG. 5-39

27. When a wire loop is rotated in a magnetic field, the direction of the current induced in the loop reverses itself twice per rotation. Why?

28. The shaft of a generator is much easier to turn when the generator is not connected to an outside circuit than when such a connection is made. Why?

29. The following are three basic electromagnetic phenomena: (a) a magnetic field is produced by an electric current; (b) a current-carrying wire experiences a force in a magnetic field (unless aligned with the field); (c) a current is induced in a wire that moves across a magnetic field (or when the field moves across the wire). Which of these phenomena is involved in the operation of the following devices: electric motor, generator, transformer, tape recording head, tape playback head?

30. Would you expect to find direct or alternating current in (a) the filament of a lightbulb in your home? (b) the filament of a lightbulb in a car? (c) the secondary coil of a transformer? (d) the output of a battery charger? (e) an electromagnet?

31. What acts on the secondary winding of a transformer to cause an alternating voltage to occur across its ends even though the primary and secondary windings are not connected?

32. Given a coil of wire and a small lightbulb, how can you tell whether the current in another coil is direct or alternating without touching the second coil or its connecting wires?

33. What would happen if the primary winding of a transformer were connected to a battery?

PROBLEMS

1. Find the total charge of 1 g of protons.

2. A charge of -5×10^{-7} C is 10 cm from a charge of $+6 \times 10^{-6}$ C. Find the magnitude and direction of the force on each charge.

3. A charge of $+3 \times 10^{-9}$ C is 50 cm from a charge of -5×10^{-9} C. Find the magnitude and direction of the force on each charge.

4. Two charges attract each other with a force of 4×10^{-6} N when they are 4 mm apart. Find the force on each charge when they are 5 mm apart.

5. Two charges originally 80 mm apart are brought together until the force between them is 16 times greater. How far apart are they now?

6. Two small spheres are given identical positive charges. When they are 1 cm apart, the repulsive force on each of them is 0.002 N. What would the force be if (a) the distance is increased to 3 cm? (b) one charge is doubled? (c) both charges are tripled? (d) one charge is doubled and the distance is increased to 2 cm?

7. (a) A metal sphere with a charge of $+1 \times 10^{-5}$ C is 10 cm from another metal sphere with a charge of -2×10^{-5} C. Find the magnitude of the attractive force on each sphere. (b) The two spheres are brought in contact and again separated by 10 cm. Find the magnitude of the new force on each sphere.

8. Two electrons exert forces on each other equal in magnitude to the weight of an electron. How far apart are the electrons?

9. How much positive charge must be added to the earth and the moon so that the resulting electrical repulsion balances the gravitational attraction between them? Assume equal amounts of charge are added to each body. (The mass of the earth is 6.0×10^{24} kg and that of the moon is 7.3×10^{22} kg.)

10. Sensitive instruments can detect the passage of as few as 60 electrons/s. To what current does this correspond?

11. How many electrons per second flow past a point in a wire carrying a current of 2 A?

12. Find the resistance of a 120-V electric toaster that draws a current of 0.8 A.

13. What potential difference must be applied across a 1500-Ω resistance in order that the resulting

current be 50 mA? (1 mA = 1 milliampere = 0.001 A)

14. A 240-V water heater has a resistance of 24 Ω. What must be the minimum rating of the fuse in the circuit to which the heater is connected?

15. An electric drill rated at 400 W is connected to a 240-V power line. How much current does it draw?

16. A power rating of 1 horsepower (hp) is equivalent to 746 W. How much current does a $\frac{1}{4}$-hp electric motor require when it is operated at 120 V? Assume 100 percent efficiency.

17. If your home has a 120-V power line, how much power in watts can you draw from the line before a 30-A fuse will burn out? How many 100-W lightbulbs can you put in the circuit before the fuse will burn out?

18. Wire A has a potential difference of 50 V across it and carries a current of 2 A. Wire B has a potential difference of 100 V across it and also carries a current of 2 A. Compare the resistances, rates of flow of charge, and rates of flow of energy in the two wires.

19. If a 75-W lightbulb is connected to a 120-V power line, how much current flows through it? What is the resistance of the bulb? How much power does the bulb consume?

20. How many coulombs of electric current pass through an electrical appliance in 20 min if the current through the appliance is 0.4 A? If the potential difference across the appliance is 120 V, how much power does it consume? How much energy in joules does it draw from the circuit in 20 min?

21. A 240-V clothes drier draws a current of 15 A. How much energy, in kilowatthours and in joules, does it use in 45 min of operation?

22. How many coulombs of charge can a 12-V battery of 100 A · h capacity supply? How much energy?

23. A 1.35-V mercury cell with a capacity of 1.5 A · h is used to power a cardiac pacemaker. If the power required is 0.1 mW (1 mW = 1 milliwatt = 0.001 W), what is the average current? How long will the cell last?

24. When a certain 1.5-V battery is used to power a 3-W flashlight bulb, it is dead after an hour's use. If the battery costs $0.50, what is the cost of a kilowatthour of electric energy obtained in this way? How does this compare with the cost of electric energy supplied to your home?

25. A hot-water heater employs a 2000-W resistance element. If all the heat from the resistance element is absorbed by the water in the heater, how much water per hour can be warmed from 10 to 70°C?

26. A transformer has a 120-turn primary winding and an 1800-turn secondary winding. A current of 10 A flows in the primary winding when a potential difference of 550 V is placed across it. Find the current in the secondary winding and the potential difference across it.

27. A transformer connected to a 120-V ac power line has 200 turns in its primary winding and 50 turns in its secondary winding. The secondary is connected to a 100-Ω lightbulb. How much current is drawn from the 120-V power line?

28. A transformer has a primary coil with 1000 turns and a secondary coil with 200 turns. If the primary voltage is 660, what is the secondary voltage? What primary current is required if 1000 W is to be drawn from the secondary?

29. A transformer rated at a maximum power of 10 kW is used to couple a 5000-V transmission line to a 240-V circuit. What is the ratio of turns in the

transformer? What is the maximum current in the 240-V circuit?

30. An electric welding machine employs a current of 400 A. The device uses a transformer whose primary coil has 400 turns and that draws 4 A from a 220-V power line. How many turns are there in the secondary coil of the transformer? What is the potential difference across the secondary coil?

ANSWERS TO MULTIPLE CHOICE

1. b	**11. a.** ohm	**18.** c	**28.** a
2. c	**b.** ampere	**19.** c	**29.** d
3. d	**c.** volt	**20.** b	**30.** d
4. d	**d.** watt	**21.** c	**31.** d
5. c	**12.** a	**22.** c	**32.** b
6. c	**13.** c	**23.** d	**33.** d
7. b	**14.** d	**24.** a	**34.** c
8. a	**15.** d	**25.** b	**35.** a
9. b	**16.** c	**26.** d	**36.** b
10. c	**17.** b	**27.** b	

6

WAVES

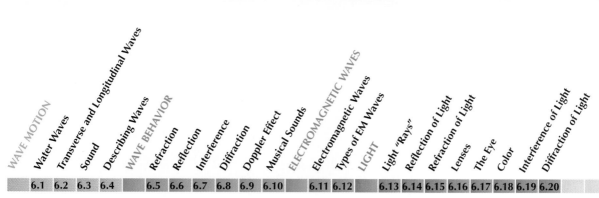

A **wave** is a periodic disturbance—a back-and-forth change of some kind that is repeated regularly as time goes on—that spreads out from a source and carries energy with it. Throw a stone into a lake: water waves move out from the splash. Clap your hands: sound waves carry the noise around you. Switch on a lamp: light waves flood the room. Water waves, sound waves, and light waves are very different from one another in various ways, but all have in common some basic properties that are explored in this chapter.

Two important categories of waves are **mechanical waves** and **electromagnetic waves.** Mechanical waves, such as water waves and sound waves, travel only through matter and involve the motion of particles of the matter they pass through. Electromagnetic waves, such as light waves and radio waves, consist of varying electric and magnetic fields and can travel through a vacuum as well as through matter. Because mechanical waves are the easier of the two kinds to understand, we shall begin by looking at what they are, how they are described, and how they behave.

6-1 WATER WAVES

Crests and Troughs

If we stand on an ocean beach and watch the waves roll in and break one after the other, we might guess that water is moving bodily toward the shore. After a few minutes, though, we see that this cannot be true. Between the breakers, water rushes back out to sea, and there is no piling up of water on the beach. The overall motion is really an endless movement of water to and fro.

We can see what is happening better by moving out beyond the breakers, say to the end of a pier. If we study a piece of seaweed floating

Water molecules in deep water move in circular orbits as waves pass by (see Fig. 6-1). When the waves reach shallow water, the molecules in the lower parts of their orbits touch the bottom, which slows them down. The wave crests, however, continue to move forward as before, which causes the fronts of the waves to become steeper and steeper as the water gets shallower. Finally the wave crests topple over, or "break," in a shower of foam that spills down the wave front. At this stage the water depth is about 1.3 times the wave height, and the wave crests may be moving toward the shore twice as fast as the waves themselves.

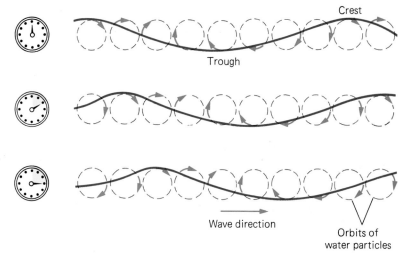

Crest

Trough

Wave direction

Orbits of
water particles

FIG. 6-1 Nature of a water wave in deep water. Each water molecule performs a periodic motion in a small circle. Because successive molecules reach the tops of their circles at slightly later times, their combination appears as a series of crests and troughs moving along the surface of the water. There is no net transfer of water by the wave.

on the water, we find little change in its position. As the crest of a wave passes, the seaweed rises and moves shoreward. In the trough that follows the crest, the seaweed falls and moves the same distance seaward. On the whole the seaweed moves in a roughly circular path perpendicular to the water surface.

The illusion of overall movement toward the shore comes about because each molecule of water undergoes its circular motion a moment later than the molecule behind it (Fig. 6-1). At the crest of a wave the molecules move in the direction of the wave, while in a trough the molecules move in the backward direction.

What *does* move shoreward is not water but energy. Ocean waves are produced by wind, and it is the energy from the wind out at sea that is carried by means of wave motion to the shore. All mechanical waves behave the same way: they transfer energy from place to place by a series of periodic motions of individual particles but cause no permanent shift in the position of matter.

6-2 TRANSVERSE AND LONGITUDINAL WAVES

Across and Back; Toward and Away

Water waves are familiar but complicated. Simpler are the waves set up when we shake one end of a rope whose other end is fixed in place (Fig. 6-2a). Here the rope particles move perpendicular to the direction in which the wave moves. Such waves are said to be **transverse.**

Another type of wave can occur in a long coil spring (Fig. 6-2b). If the left-hand end of the spring is moved back and forth, a series of **compressions** and **rarefactions** move along the spring. The compressions are places where the loops of the spring are pressed together; the rarefactions are places where the loops are stretched apart. Any one loop simply moves back and forth, transmitting its motion to the next in line, and the regular series of back-and-forth movements gives rise to the

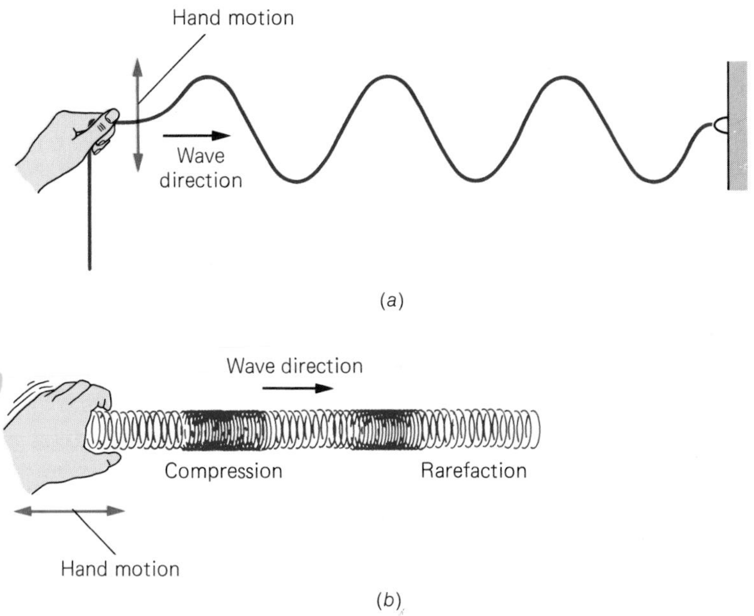

FIG. 6-2 Transverse and longitudinal waves. (*a*) Transverse waves travel along the rope in the direction of the black arrow. The individual particles of the rope move back and forth (color arrows) perpendicular to the direction of the waves. (*b*) In longitudinal waves, successive regions of compression and rarefaction move along the spring. The particles of the spring move back and forth parallel to the spring.

compressions and rarefactions. Waves of this kind, in which the motion of individual units is along the same line that the wave travels, are called **longitudinal waves.**

Water waves are a combination of transverse and longitudinal waves, as Fig. 6-1 shows. Transverse mechanical waves can occur only in solids, whereas longitudinal waves can travel in any medium, solid or fluid. Transverse motion requires that each particle, as it moves, drag with it adjacent particles to which it is tightly bound. This is impossible in a fluid, where molecules easily slide past their neighbors. Longitudinal motion, on the other hand, merely requires that each particle push on its neighbors, which can happen as easily in a gas or liquid as in a solid. (Surface waves on water—in fact any waves at the boundary between two fluids—are an exception to this rule, for in part they involve transverse motion.) The fact that longitudinal waves that originate in earthquakes pass through the center of the earth while transverse earthquake waves cannot is one of the reasons the earth is thought to have a liquid core (Chap. 14).

6-3 SOUND

Pressure Waves in a Solid, Liquid, or Gas

Most sounds are produced by a vibrating object, such as the cone of a loudspeaker (Fig. 6-3). When it moves outward, the cone pushes the air molecules in front of it together to form a region of high pressure that spreads outward. The cone then moves backward, which expands the space available to nearby air mol-

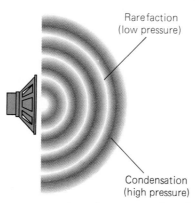

FIG. 6-3 Sound waves produced by a loudspeaker. Alternate regions of compression and rarefaction move outward from the vibrating cone of the loudspeaker.

Sound waves can be generated in various ways, as by the vibrating strings of a violin, the vibrating air column of a clarinet, and the vibrating membrane of a drum.

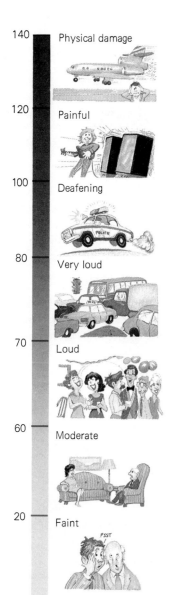

FIG. 6-4 Decibel scale for sounds.

ecules. Some of these molecules now flow toward the cone, leaving a region of low pressure that spreads outward behind the high-pressure region. The repeated vibrations of the loudspeaker cone thus send out a series of compressions and rarefactions that constitute sound waves.

Sound waves are longitudinal, because the molecules in their paths move back and forth in the same direction as that of the waves. The air (or other material) in the path of a sound wave becomes alternately denser and rarer, and the resulting pressure changes cause our eardrums to vibrate, which produces the sensation of sound.

The great majority of sounds consist of waves of this type, but a few—the crack of a rifle, the first sharp sound of a thunderclap—are single, sudden compressions of the air rather than periodic phenomena.

The speed of sound is about 343 m/s (767 mi/h) in sea-level air at ordinary temperatures. Since the particles in liquids and solids are closer together than those in gases and therefore respond more quickly to one another's motions, sound travels faster in liquids and solids than in gases: about 1500 m/s in water and 5000 m/s in iron.

The Decibel   The more energy a sound wave carries, the louder it sounds. However, our ears respond to sound waves in a peculiar way. Doubling the rate of energy flow of a particular sound gives the sensation of an only slightly louder sound, not nearly twice as loud. For this reason a special scale, whose unit is the **decibel** (dB), is used to describe how powerful a sound is. A sound that can barely be heard by a normal person is given the value 0 dB; ordinary conversation is usually about 60 dB (Fig. 6-4).

Sounds of 85 dB or more are unpleasantly loud, and long exposure to them leads to permanent hearing damage. Rock music is often played at levels as high as 125 dB, and many fans of such music have suffered partial hearing loss.

FIG. 6-5 A transverse wave moving in the x direction whose displacements are in the y direction. The wavelength is λ and the amplitude is A.

DESCRIBING WAVES

Wavelength, Frequency, and Speed Are Related

All waves can be represented by a curve like that in Fig. 6-5. The resemblance to transverse wave motion is easiest to see; in fact the curve is an idealized picture of continuous waves in a rope like that of Fig. 6-2a. As the wave moves to the right, each point on the curve can be thought of as moving up or down just as any point on the rope would move. In the case of a longitudinal wave, the high points of the curve represent the maximum shifts of particles in one direction and the low points represent their maximum shifts in the other direction.

With the help of Fig. 6-5 we can assign numbers to certain key properties of a wave, so that different waves can be compared. The distance from crest to crest (or trough to trough) is called the **wavelength,** usually symbolized by the Greek letter λ (lambda). The **speed** v of the waves is the rate at which each crest moves, and the **frequency** f is the number of crests that pass a given point each second. The **period** T is the time needed for a complete wave (crest + trough) to pass a given point (Table 6-1).

The **amplitude** A of a wave is the height of the crests above the undisturbed level (or the depth of the troughs below this level). Not surprisingly, the energy carried by waves depends on amplitude and frequency, that is, on the violence of the waves and the number of them per second. It turns out that the energy is proportional to the square of each of these quantities.

The unit of frequency is the cycle per second (c/s). As mentioned in Chap. 5, this unit is usually called the **hertz** (Hz), after Heinrich Hertz, a pioneer in the study of electromagnetic waves.

Basic Wave Formula

The number of waves that pass a point per second multiplied by the length of each wave gives the speed with which the waves travel (Fig. 6-6). Thus frequency f times wavelength λ gives speed v:

$$v = f\lambda \qquad \textit{wave speed}$$

Wave speed = (frequency)(wavelength)

Table 6-1			
Wave Quantities			
Quantity	**Symbol**	**Formula**	**Meaning**
Speed	v	$v = f\lambda$	Distance through which each wave moves per second
Wavelength	λ	$\lambda = v/f$	Distance between adjacent crests or troughs
Frequency	f	$f = v/\lambda$	Number of waves that pass a given point per second
Period	T	$T = 1/f$	Time needed for a wave to pass a given point
Amplitude	A		Maximum displacement of any particle from its normal position

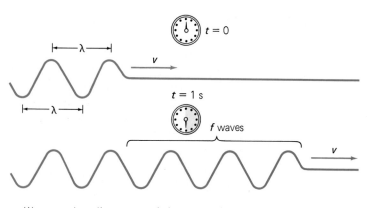

Wave speed = distance traveled per second
= (number of waves per second) (length per wave)
= (frequency) (wavelength)
$v = f\lambda$

FIG. 6-6 Wave speed equals frequency times wavelength.

If 10 waves, each 2 m long, pass in a second, then each wave must travel 20 m during that second to give a speed of 20 m/s.

The above formula applies to waves of all kinds. For example, waves on the open sea whose wavelength is 50 m travel at about 8.5 m/s (Fig. 6-7). The frequency of such waves is

$$f = \frac{v}{\lambda} = \frac{8.5 \text{ m/s}}{50 \text{ m}} = 0.17 \text{ Hz}$$

The period of the waves is

$$T = \frac{1}{f} = \frac{1}{0.17 \text{ Hz}} = 5.9 \text{ s}$$

so a wave passes a given point every 5.9 s.

As another example, sound in air travels at 343 m/s and the musical note A has a frequency of 440 Hz. The corresponding wavelength is

$$\lambda = \frac{v}{f} = \frac{343 \text{ m/s}}{440 \text{ Hz}} = 0.78 \text{ m}$$

FIG. 6-7 Waves whose speed is 8.5 m/s and whose wavelength is 50 m have a frequency of 0.17 Hz. This means that such waves pass an anchored boat once every 5.9 s.

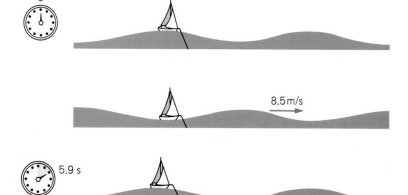

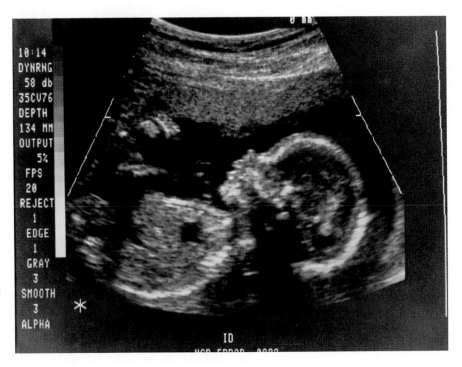

Ultrasound is used in medicine to produce images of internal parts of the body, including unborn babies. Ultrasound is better able than x-rays to distinguish between soft tissues and liquids, and is far less harmful. What is done is to send a pulse of ultrasound waves in a narrow beam into a patient's body through the skin; the reflections of the pulse from interfaces between different materials return to a detector at different times. A picture of the internal structure of the body can be built up by moving the ultrasound beam in a scanning pattern. The fetus shown here is 20 weeks old.

Our ears are most sensitive to sounds whose frequencies are between 3000 and 4000 Hz. Almost nobody can hear sounds with frequencies below about 20 Hz **(infrasound)** and above about 20,000 Hz **(ultrasound),** although for many animals the upper limit is higher. Hearing deteriorates with age, most noticeably at the higher frequencies.

WAVE BEHAVIOR

Throw a ball. It has a certain speed, it carries energy, and it bounces off a wall in its path. A wave has these properties, too, but in other respects waves are unique. For instance, two balls that meet in midair bounce off each other, but if two waves of the same kind meet, they combine to form a new wave different from the original ones. We shall look into the behavior of mechanical waves first, since they are easier to visualize, and then see how what we find helps us to understand the behavior of electromagnetic waves.

6-5   REFRACTION

A Change in Direction Produced by a Change in Speed

No matter from where the wind is blowing, waves always approach a sloping beach very nearly at right angles to the shore (Fig. 6-8). Farther out in open water, the wave direction may be oblique (that is, at a slanting angle) to the shore, but the waves swing around as they move in so that their crests become roughly parallel to the shoreline. This is an example of **refraction.**

Something similar to refraction occurs when a tracked vehicle such as a bulldozer makes a turn. The right-hand track of this bulldozer was slowed down, and the greater speed of its left-hand track then swung the bulldozer around to the right.

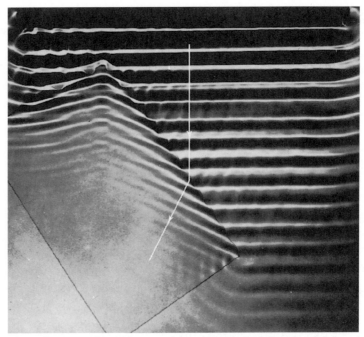

The refraction of water waves in a tank. In the left side of the tank is a glass plate over which the water is shallower than elsewhere. Waves move more slowly in shallow water than in deeper water, and hence refraction occurs at the edge of the plate. The arrows show the direction of movement of the waves.

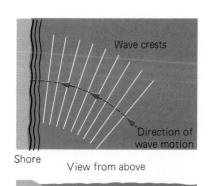

FIG. 6-8 Refraction of water waves. Waves approaching shore obliquely are turned because they move more slowly in shallow water near shore.

The explanation is straightforward. As a wave moves obliquely shoreward, its near-shore end encounters shallow water before its outer end does, and friction between the near-shore end and the sea bottom slows down that part of the wave down. More and more of the wave is slowed as it continues to move toward the shore, and the slowing becomes more pronounced as the water gets shallower. As a result the whole wave turns until it is moving almost directly shoreward. The wave has turned because part of it was forced to move more slowly than the rest. Thus refraction is caused by differences in speed across the wave.

The bending will be at a sharp angle if the waves cross a definite boundary between regions in which they move at different speeds. The photograph on this page shows this effect in ripples that move obliquely from deep water to shallow water in a tank. If the waves approach the boundary at right angles, no refraction occurs because the change in wave speed takes place across each wave at the same time.

Refraction occurs when waves of any kind change speed as they cross a boundary at a slanting angle. The refraction of light waves is responsible for the foreshortened appearance of people standing in shallow water, and the refraction of these waves when moving from air into glass and out again accounts for the focusing of light by lenses.

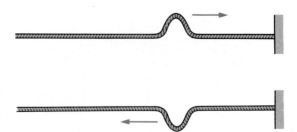

FIG. 6-9 A wave in a stretched rope is reflected when it reaches a fixed end. The reflected wave is inverted.

REFLECTION

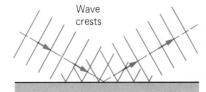

FIG. 6-10 Reflection of water waves. Waves approaching an obstacle obliquely appear to re-form and move away in a different direction.

Rebounding from an Obstacle   Another familiar property of waves is their ability to be **reflected** when they meet an obstacle. When a single wave sent down a stretched rope by a shake of the free end meets the attached end, it re-forms itself and travels back along the rope (Fig. 6-9). Water waves are similarly reflected, as we can see by watching waves strike a breakwater obliquely and re-form themselves into waves that move outward at a similar angle in the other direction (Fig. 6-10).

When a series of waves is sent along the rope, the reflected waves will meet the forward-moving waves head on. Each point on the rope must then respond to two different impulses at the same time. The two impulses add together. If the point on the rope is being pushed in the same direction by both waves, it will move in that direction with an amplitude equal to the sum of the amplitudes of the two waves. If the wave impulses at a point on the rope are in opposite directions, that point will have an amplitude equal to the difference of the two wave amplitudes.

Standing Waves   With the timing just right, the two motions may cancel out completely for some points of the rope while other points move with twice the normal amplitude. In this situation the waves appear not to travel at all. Some parts of the rope simply move up and down, and other parts remain at rest (Fig. 6-11). Waves of this sort are called **standing waves.**

Vibrating strings in musical instruments are the most familiar examples of standing waves (Fig. 6-12). Longitudinal waves traveling in opposite directions over the same path may also set up standing waves, as in the vibrating air columns of whistles, organ pipes, flutes, and clarinets.

FIG. 6-11 Standing waves in a stretched rope.

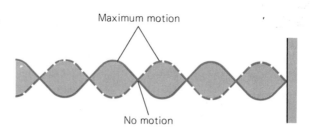

Maximum motion

No motion

Strong winds set up standing waves in the Tacoma Narrows Bridge in Washington State soon after its completion in 1940. The bridge collapsed as a result. Today bridges are stiffened to prevent such disasters.

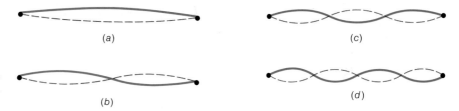

(a)

(c)

FIG. 6-12 A few of the possible standing waves in a stretched string, such as a violin string.

(b)

(d)

6-7 INTERFERENCE

Waves in Step and out of Step

Interference refers to the adding together of two or more waves of the same kind that pass by the same point at the same time. The formation of standing waves is an example of interference.

How interference works is shown in Fig. 6-13. Let us shake the stretched strings *AC* and *BC* at the ends *A* and *B*. The single string *CD* is then affected by *both* sets of waves. Each portion of *CD* must respond to two impulses at the same time, and its motion is therefore the total of

FIG. 6-13 Interference. (a) Waves started along stretched strings *AC* and *BC* will interfere at *C*. (b) Constructive interference. (c) Destructive interference. (d) A mixture of constructive and destructive interference.

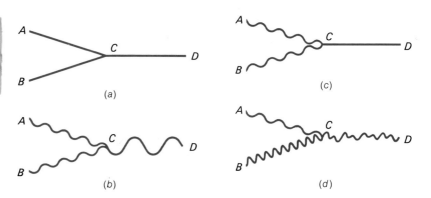

(a)

(c)

(b)

(d)

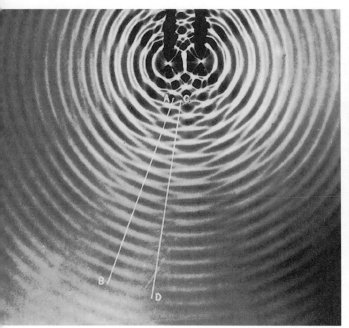

left: The interference of water waves. Ripples spread out across the surface of a shallow tank of water from the two sources at the top. In some directions (for instance *AB*) the ripples reinforce each other and the waves are more prominent. In other directions (for instance *CD*) the ripples are out of step and cancel each other, so that the waves are small or absent.

right: Destructive interference provides a way to reduce or eliminate noise. The procedure is to use an electronic device to analyze a particular noise and then produce mirror-image sound waves that cancel it out. If the device is appropriately programmed, wanted sounds such as conversation or music come through clearly. Powerful antinoise systems are already in use to counter the noise of suction grain unloaders, which are as loud as jet engines, and other systems are being developed to make the cabins of turboprop aircraft noisefree. On a smaller scale, the Noise Buster shown here feeds a pair of headphones with antinoise for personal use.

the effects of the two original waves. Suppose we shake *A* and *B* in step with each other so that, at *C*, crest meets crest and trough meets trough. Then the crests in *CD* are twice as high and the troughs twice as deep as those in *AB* and *BC*. This situation is called **constructive interference.**

On the other hand, if we shake *A* and *B* exactly out of step with each other, wave crests in *AC* will arrive at *C* just when troughs get there from *BC*. As a result, crest matches trough, the wave impulses cancel each other out, and *CD* remains at rest. This situation is called **destructive interference.**

As another possibility, if *A* is shaken through a smaller range than and half as rapidly as *B*, the two waves add together to give the complex waveform of Fig. 6-13*d*. The variations are endless, and the resulting waveforms depend upon the amplitudes, wavelengths, and timing of the incoming waves.

The interference of water waves is shown by ripples in the photograph on this page. Ripples that spread out from the two vibrating rods affect the same water molecules. In some directions, crests from one source arrive at the same time as crests from the other source and the ripples are reinforced. Between these regions of vigorous motion are narrow lanes where the water is quiet. These lanes represent directions in which crests from one source arrive together with troughs from the other so that the wave motions cancel.

6-8 DIFFRACTION

Why Shadows Are Never Completely Dark

An important property of all waves is their ability to bend around the edge of an obstacle in their path. This property is called **diffraction**.

A simple example of diffraction occurs when we hear the noise of a car horn around the corner of a building. The noise could not have reached us through the building, and refraction is not involved since the speed of sound does not change between the source of noise and our ears. What happens is that the sound waves spread out from the corner of the building into the "shadow" as though they come from the corner (Fig. 6-14). The diffracted waves are not as loud as those that proceed directly to a listener, but they go around the corner in a way that a stream of particles, for example, cannot.

FIG. 6-14 Diffraction causes waves to bend around the corner of an obstacle into the "shadow" region. The diffracted waves spread out as though they originated at the corner of the obstacle and are weaker than the direct waves. The waves shown here could be of any kind, for instance, water waves, sound waves, or light waves.

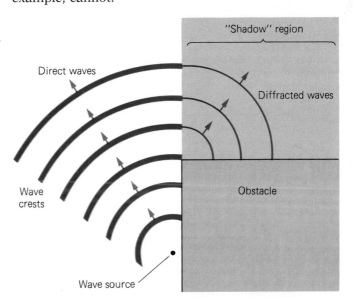

6-9 DOPPLER EFFECT

Higher Pitch When Approaching; Lower Pitch When Receding

We all know that sounds produced by vehicles moving toward us seem higher pitched than usual, whereas sounds produced by vehicles moving away from us seem lower pitched than usual. Anybody who has listened to the siren of a police car as it passes by at high speed is aware of these changes in frequency, called the **doppler effect**.

The doppler effect arises from the relative motion of the listener and the source of the sound. Either or both may be moving. When the motion reduces the distance between source and listener, as in Fig. 6-15b, the wavelength decreases to make the frequency higher. When the motion takes source and listener farther away from each other, as in Fig. 6-15c, the wavelength increases to make the frequency lower.

A simple way to visualize the doppler effect is to imagine traveling in a boat on a windy day. If we head into the wind, waves strike the boat more often than when the boat is at rest, and the ride may be very choppy. On the other hand, if we head away from the wind, waves catch up with us more slowly than when the boat is at rest, so their apparent frequency is less.

An interesting use of the doppler effect is to measure the speed of blood in an artery. When a beam of high-frequency sound waves is directed at an artery, the waves reflected from the moving blood cells show a doppler shift in frequency because the cells then act as moving

FIG. 6-15 The doppler effect. At (a) the police car is standing still, and sound waves from its siren reach you at their normal frequency. At (b) the car approaches you, moving a distance x between two successive waves. To you, the wavelength is shorter by x than before and the frequency higher. At (c) the car moves away from you, again moving a distance x between successive sound waves. Here you find that the wavelength is longer by x and the frequency lower.

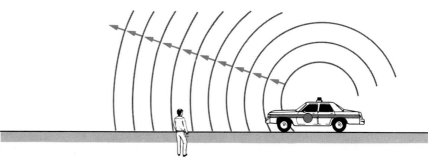

(a) Normal frequency is heard

(b) Higher frequency (shorter wavelength) is heard

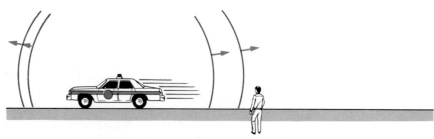

(c) Lower frequency (longer wavelength) is heard

Doppler shifts in radar waves are widely used by police to determine vehicle speeds.

wave sources. From this shift the speed of the blood can be calculated. It is a few centimeters per second in the main arteries, less in the smaller ones.

The doppler effect occurs in light waves and is one of the ways by which astronomers detect and measure motions of the stars. Stars emit light that has only certain characteristic wavelengths. When a star moves either toward or away from the earth, these wavelengths appear, respectively, shorter or longer than usual. From the amount of the shift, it is possible to calculate the speed with which the star is approaching or receding. As we shall learn in Chap. 18, this is how the expansion of the universe was discovered.

6-10 MUSICAL SOUNDS

Fundamentals and Overtones

Musical sounds are produced by vibrating objects—stretched gut or wire in stringed instruments, vocal cords in the throat, membranes in drums, air columns in wind instruments. The simplest vibration of a stretched string is one in which a single standing wave takes up the entire length of the string, as in Fig. 6-12a. The frequency may be varied by changing the tension on the string, as a violinist does when tuning the instrument. The tighter the string, the higher the frequency. For a given tension, the frequency can be varied by changing the length of the string, as the violinist does with the pressure of a finger on the string.

Depending on where the string is plucked, bowed, or struck, more complex vibrations may be set up: standing waves may form with two, three, or even more crests (Fig. 6-12b, c, d). Sound waves set up by these shorter standing waves have higher frequencies, and the frequencies are related to the frequency of the longest wave by simple ratios—2:1, 3:1,

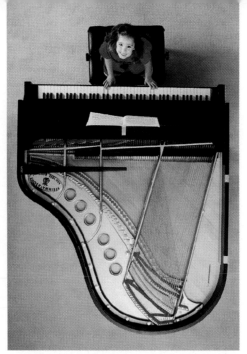

The 88 "strings" of a piano are metal wires whose fundamental frequencies range from 27 Hz for the lowest bass note to 4186 Hz for the highest treble note. The total tension of all the strings is about 20 tons and is borne by a heavy cast iron frame.

and so on. The tone produced by the string vibrating as in Fig. 6-12*a* is called the **fundamental,** and the higher frequencies produced when it vibrates in segments are called **overtones.**

Resonance In practice, the strings of a musical instrument give not just the fundamental or a single overtone but a combination of the fundamental plus several overtones. The motion of the string, and so the form of the sound wave, may be very complex (Fig. 6-16). To the ear a fundamental tone by itself seems flat and uninteresting. As overtones are added the tone becomes richer, with the quality, or timbre, of the tone depending on which particular overtones are emphasized. The emphasis largely depends on the shape of the instrument, which enables it to **resonate** at particular frequencies. The sounding part of the instrument—the belly of the violin or the soundboard of the piano—has certain natural frequencies of vibration, and it is more readily set vibrating at these frequencies than at others. The resulting sounds may include a large number of overtones, but the greater emphasis on certain overtones provides the musical quality characteristic of the instrument.

FIG. 6-16 The waveforms of sounds can be analyzed electronically with the help of an oscilloscope, a device that displays electric signals on the screen of a tube like the picture tube of a television set. A microphone is used to convert sound waves into electric signals, and these in turn can be displayed on the oscilloscope screen. "Pure" tones, like those produced by a tuning fork, have simple waveforms like that of Fig. 6-5, while musical instruments and the human voice produce complex waveforms. Ordinary nonmusical noises consist of waves with complex and rapidly changing forms.

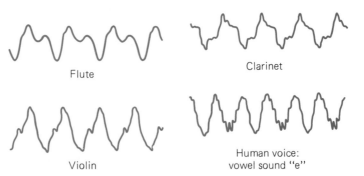

Flute

Clarinet

Violin

Human voice:
vowel sound "e"

The human voice is produced when the vocal cords in a person's throat set in vibration an air column that extends from the throat to the mouth and the nasal cavity above it. The shape of this column, which we adjust while speaking or singing by manipulating the mouth and tongue, determines the different vowel sounds by emphasizing some overtones and suppressing others. This shape also gives rise to the subtle differences in sound quality that enable us to tell one person's voice from another's. Vocal power is typically less than 1 W.

HARMONY AND DISCORD Certain mixtures of frequencies are pleasing to the ear. For instance, the combination of a tone and its first overtone, whose frequency is twice as great, appears harmonious to a listener. Such an interval is called an **octave** in music because it includes eight notes. Also agreeable are tones with a frequency ratio of 2:3, such as C (262 Hz) and G (392 Hz). This interval is called a **fifth** because it includes five notes, here C, D, E, F, G. Somewhat less agreeable are tones whose frequencies are in the ratio 4:5, such as C and E; the interval here spans three notes and is called a **third.** The larger the numbers that express their frequency ratio, the less attractive a combination of tones appears. Thus C and D (ratio 8:9) seems discordant, and E and F (ratio 15:16) more discordant still. Ordinary sounds are mixtures of frequencies that have no special relationship with one another. If the mixture seems particularly harsh, we consider it noise.

Wind instruments produce sounds by means of vibrating air columns. In an organ, there is a separate pipe for each note. The shorter the pipe, the higher the pitch. Woodwinds, such as flutes and clarinets, use a single tube with holes whose opening and closing controls the effective length of the air column. Most brass instruments have valves connected to loops of tubing. Opening a valve adds to the length of the air column and thus produces a note of lower pitch. In a slide trombone, the length of the air column is varied by sliding in or out a telescoping U tube. A bugle has neither holes nor valves, and a bugler obtains different notes with the lips alone.

The fundamental frequencies of the speaking voice average about 145 Hz in men and about 230 Hz in women. Even considering the overtones present, the frequencies in ordinary speech are mostly below 1000 Hz. In singing, the first and second overtones may be louder than their fundamentals, and even higher overtones add to the beauty of the sound.

$V_{em} = C = 3.0 \times 10^{10} \, M/s = (86,000 \, \frac{miles}{sec})$

ELECTROMAGNETIC WAVES

In 1864 the British physicist James Clerk Maxwell suggested that an accelerated electric charge generates combined electrical and magnetic disturbances able to travel indefinitely through empty space. These disturbances are called **electromagnetic waves.** Such waves are hard to visualize because they represent fluctuations in fields that are themselves difficult to form mental images of. But they certainly exist—light, radio waves, and x-rays are examples of electromagnetic waves.

6-11   ELECTROMAGNETIC WAVES

Waves without Matter

We learned in Chap. 5 that a changing magnetic field gives rise by electromagnetic induction to an electric current in a nearby wire. We can reasonably conclude that a changing magnetic field has an electric field associated with it. Maxwell proposed that the opposite effect also exists, so that a changing electric field has an associated magnetic field. The electric fields produced by electromagnetic induction are easy to measure because metals offer little resistance to the flow of electrons. There is no such thing as a magnetic current, however, and it was impossible in Maxwell's time to detect the weak magnetic fields he had predicted. But there is another way to check Maxwell's idea.

If Maxwell was right, then electromagnetic (em) waves must occur in which changing electric and magnetic fields are coupled together by both electromagnetic induction and the mechanism he proposed. The linked fields spread out in space much as ripples spread out when a stone is dropped into a body of water. The energy carried by an em wave is constantly being exchanged between its fluctuating electric and magnetic fields. Calculations show that the wave speed should have the same value as the speed of light, which is 3×10^8 m/s (186,000 mi/s).

BIOGRAPHY

James Clerk Maxwell (1831–1879) was born in Scotland shortly before Michael Faraday discovered electromagnetic induction. While still a student, Maxwell used his ideas on color vision to make the first color photograph. At 24 he showed that the rings of Saturn would not be solid or liquid but must consist of separate small bodies. At about this time Maxwell became interested in electricity and magnetism and soon was convinced that these were not separate phenomena but had an underlying unity of some kind. Starting from the results of Faraday and others, he created a single comprehensive theory of electricity and magnetism that remains the foundation of the subject today. From his equations Maxwell predicted that electromagnetic waves should exist that travel with the speed of light and surmised that light consisted of such waves. Sadly, he did not live to see his work confirmed in the experiments of Heinrich Hertz. Maxwell died of cancer at 48 in 1879, the year in which Albert Einstein was born. Maxwell had been the greatest theoretical physicist of the nineteenth century; Einstein was to be the greatest theoretical physicist of the twentieth century. (By a similar coincidence, Newton was born in the year of Galileo's death.)

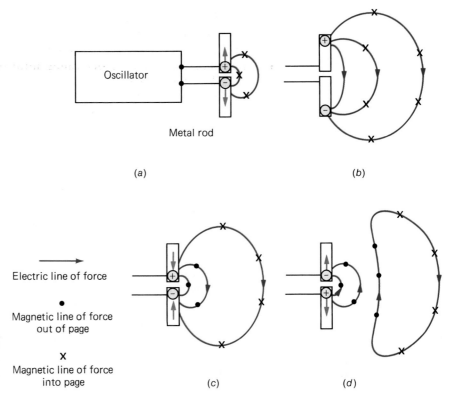

FIG. 6-17 A pair of metal rods connected to an electrical oscillator (source of alternating current) give rise to coupled electric and magnetic fields that constitute electromagnetic waves. The waves spread out from their source with the speed of light.

Creating an Em Wave

To see how an em wave can be created, let us consider what happens when we connect an alternating-current generator to a pair of metal rods, as in Fig. 6-17. For simplicity we will imagine that only a single charge is in each rod at any time. This is not the only way to create em waves, but it is a particularly easy way to visualize how this can occur.

In Fig. 6-17a the charges are moving apart. Some of the electric field lines between them are shown as color lines; the magnetic field lines are in the form of circles perpendicular to the page. In b the charges have stopped, so there is no magnetic field being produced at this moment. The existing magnetic field continues to spread out along with the electric field. Both fields travel with the speed of light. In c the charges are moving toward each other, so the new magnetic field is opposite in direction to the old one although the electric field is in the same direction as before. In d the polarity of the rods has changed and now a negative charge is in the upper rod and a positive charge is in the lower one. These charges are moving apart. The magnetic field of the moving charges is still in the same direction as in c but the direction of the electric field is reversed.

The result of the above sequence is that the outermost electric and magnetic field lines form closed loops that are no longer joined to the oscillating charges. The loops move freely through space and make up an em wave. As the charges in the two rods continue to move back and forth, closed loops of electric and magnetic field lines continue to form and expand outward.

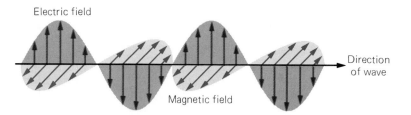

Electric field

Direction of wave

Magnetic field

FIG. 6-18 The electric and magnetic fields in an electromagnetic wave vary together. The fields are perpendicular to each other and to the direction of the wave.

Figure 6-18 shows the relationship between the electric and magnetic fields in an electromagnetic wave. Here the fields are represented by a series of vectors (not field lines) that indicate the magnitude and direction of the fields in the path of the wave. The fields are perpendicular to each other and to the direction of the wave, and they remain in step as they periodically reverse their directions.

6-12   TYPES OF EM WAVES

They Carry Information as well as Energy

During Maxwell's lifetime em waves remained an unproven idea. Finally, in 1887, the German physicist Heinrich Hertz showed experimentally that em waves indeed exist and behave exactly as Maxwell expected them to.

Hertz was not concerned with the commercial possibilities of em waves, and other scientists and engineers developed what we now call radio. A radio signal is sent by means of em waves produced by electrons that move back and forth hundreds of thousands to millions of times per second in the antenna of the sending station. The signal resides either in variations in the strength of the waves (**amplitude modulation,** or AM) or in variations in the frequency of the waves (**frequency modulation,** or FM), as in Fig. 6-19. Frequency modulation is less subject to random disturbances ("static").

When the waves reach the antenna of a receiver, the electrons there vibrate in step with the waves. The receiver can be tuned to respond only to a narrow frequency band. Since transmitters operate on different frequencies, a receiver can pick up the signals sent out by whatever station we wish. The currents set up in the receiving antenna are very weak, but they are strong enough for electronic circuits in the receiver

Heinrich Hertz (1857–1894)

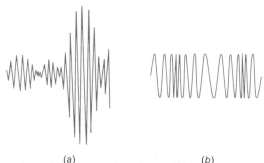

FIG. 6-19   (a) In amplitude modulation (AM), variations in the amplitude of a constant-frequency radio wave constitute the signal being sent out. (b) In frequency modulation (FM), variations in the frequency of a constant-amplitude wave constitute the signal.

(a)   (b)

THE IONOSPHERE Not long after Hertz's experiments with em waves, the Italian engineer Guglielmo Marconi thought of using them for communication. In Hertz's work the transmitter and receiver were only a few meters apart. Marconi was able to extend the range to many kilometers, and in 1899 he sent a radio message across the English Channel. Two years later he sent signals across the Atlantic Ocean from England to Newfoundland, using kites to raise his antennas.

Radio waves, like light waves, tend to travel in straight lines, and the curvature of the earth should therefore prevent radio communication over long distances. For this reason Marconi's achievements came as a surprise. The mystery was solved with the discovery of a region of ionized gas called the **ionosphere** that extends from about 70 km to several hundred km above the earth's surface. The ions are produced by the action of high-energy ultraviolet rays and x-rays from the sun. The ionosphere behaves like a mirror to high-frequency radio waves, which can bounce one or more times between the ionosphere and the earth's surface (Fig. 6-20). Low-frequency radio waves are absorbed by the ionosphere, and very-high frequency (VHF) and ultrahigh frequency (UHF) waves pass through it and so can be used to communicate with satellites and spacecraft.

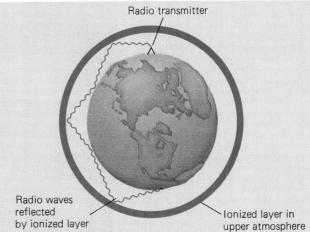

Radio transmitter

Radio waves reflected by ionized layer

Ionized layer in upper atmosphere

FIG. 6-20 The ionosphere is a region in the upper atmosphere whose ionized layers make possible long-range radio communication by their ability to reflect short-wavelength radio waves.

Guglielmo Marconi at his laboratory in Newfoundland with the instruments that detected the first radio transmission across the Atlantic Ocean. The radio waves were reflected by the ionosphere.

to extract the signal from them and turn it into sounds from a loudspeaker.

The frequencies of ordinary radio waves extend up to about 2 MHz (1 MHz = 1 megahertz = 10^6 Hz) and those of waves used in long-range short-wave communication extend up to about 30 MHz. Still higher frequencies have found widespread use in television and radar. Such extremely short waves are not reflected by the ionosphere, so direct reception of television is limited by the horizon unless rebroadcast by a satellite station.

Waves whose frequencies are around 10 GHz (1 GHz = 1 gigahertz = 10^9 Hz), corresponding to wavelengths of a few centimeters, can readily be formed into narrow beams. Such beams are reflected by solid objects such as ships and airplanes, which is the basis of **radar** (from *radio detection and ranging*). A rotating antenna is used to send out a pulsed beam, and the distance of a particular target is found from the time needed for the echo to return to the antenna. The direction of the target is the direction in which the antenna is then pointing.

The rotating scanner of a radar set sends out pulses of high-frequency radio waves in a narrow beam and then detects their reflections from objects in their paths. These reflections are then displayed on a screen. Shown here is the image of a harbor entrance; the center of the screen corresponds to the position of the ship carrying the radar.

Figure 6-21 shows the range (or **spectrum**) of em waves. The human eye can detect only light waves in a very short frequency band, from about 4.3×10^{14} Hz for red light to about 7.5×10^{14} Hz for violet light. Infrared radiation has lower frequencies than those in visible light, and ultraviolet radiation has higher frequencies. Still higher are the frequencies of x-rays and of the gamma radiation from atomic nuclei.

FIG. 6-21 The electromagnetic spectrum. All em waves have the same fundamental character and the same speed in a vacuum, but how they interact with matter depends on their frequency. The range of frequencies is truly enormous, from less than 10^4 (ten thousand) Hz to more than 10^{21} (1000 billion billion) Hz. The corresponding wavelengths range from more than 3×10^6 m (186 miles) to less than 3×10^{-11} m (smaller than an atom).

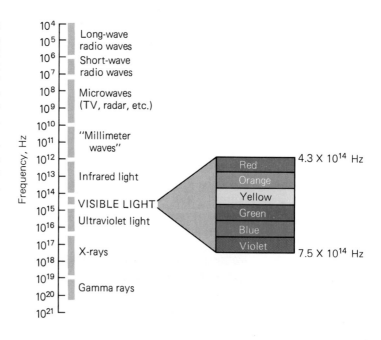

LIGHT   Many aspects of the behavior of light waves can be understood without reference to their electromagnetic nature. In fact, we can even use a yet simpler model in many (but far from all) situations by thinking of light in terms of "rays" rather than waves.

6-13   LIGHT "RAYS"

The Paths Light Takes   Early in life we become aware that light travels in straight lines. A simple piece of evidence is the beam of a flashlight on a foggy night. Actually, our entire orientation to the world about us, our sense of the location of things in space, depends on *assuming* that light follows straight-line paths.

Just as familiar, however, is the fact that light does not always follow straight lines. We see most objects by reflected light, light that has been turned sharply on striking a surface. The distorted appearance of things seen in water or through the heated air rising above a flame further testifies to the ability of light to be bent from a straight path. In these latter cases the light is refracted, and we note that this occurs when light moves from one transparent material to another.

Although the conscious part of our minds recognizes that light can be reflected and refracted, it is easy to be deceived about the true positions of things. When we look in a mirror, for example, we are seeing light that travels to the mirror and then from the mirror to our eyes, but our eyes seem to tell us that the light comes from an image behind the mirror. When we look at the legs of someone standing in shallow water, they appear shorter than they do in air because light going from water into air is bent. Our eyes and brains have no way to take this into account and so we register the illusion rather than the reality.

Much can be learned about the behavior of light by studying the paths that light follows under various circumstances. Since light appears to travel in a straight path in a uniform medium, we can represent its motion by straight lines called **rays.** Rays are a convenient abstraction, and we can visualize what we mean by thinking of a narrow pencil of light in a darkened room.

6-14   REFLECTION OF LIGHT

Mirror, Mirror on the Wall   When we look at ourselves in a mirror, light from all parts of the body (this is reflected light, of course, but we may treat it as if it originated in the body) is reflected from the mirror back to our eyes, as in Fig. 6-22. Light from the foot, for example, follows the path *CC'E*. Our eyes, which see the ray *C'E* and automatically project it in a straight line, register the foot at the proper distance but apparently behind the mirror at *C"*. A ray from the top of the head is reflected at *A'*, and our eyes see the point *A* as if it were behind the mirror at *A"*. Rays from other points of the body are similarly reflected, and in this manner a complete **image** is formed that appears to be behind the mirror.

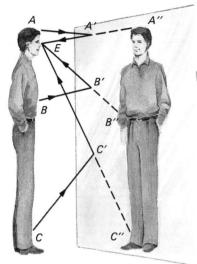

FIG. 6-22 Formation of an image in a mirror. The image appears to be behind the mirror because we instinctively respond to light as though it travels in straight lines.

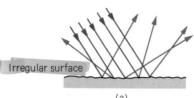
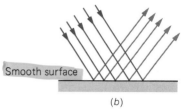

FIG. 6-23 (a) Light that strikes an irregular surface is scattered randomly and cannot form an image. (b) Light that strikes a smooth, flat surface is reflected at an angle equal to the angle of incidence. Such a surface acts as a mirror.

Left and right are interchanged in a mirror image, because front and back have been reversed by the reflection. Thus a printed page appears backward in a mirror, and what seems to be one's left hand is really one's right hand.

Why do we not see images of ourselves in walls and furniture as well as in mirrors? This is simply a question of the relative roughness of surfaces. Rays of light are reflected from walls just as they are from mirrors, but the reflected rays are scattered in all directions by the many surface irregularities (Fig. 6-23). We see the wall by the scattered light reflected from it.

6-15   REFRACTION OF LIGHT

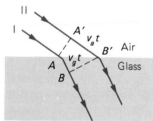

FIG. 6-24 Refraction occurs whenever light passes from one medium to another in which its speed is different. Here two rays of light, I and II, pass from air, in which their speed is v_a, to glass, in which their speed is v_g. Because v_g is less than v_a, $A'B'$ is longer than AB, and the beam of which I and II are part changes direction when it enters the glass.

Bending Light

We all know that the water in a bathtub, a swimming pool, and even a puddle is always deeper than it seems to be. The reason is that light is refracted—changes direction—when it goes from one medium into another medium in which the speed of light is different (Fig. 6-24). The effect is similar to the refraction of water waves described in Sec. 6-5. A ray of light from the stone in Fig. 6-25 follows the bent path ABE to our eyes, but our

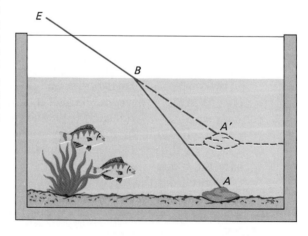

FIG. 6-25 Light is refracted when it travels obliquely from one medium to another. Here the effect of refraction is to make the water appear shallower than it actually is.

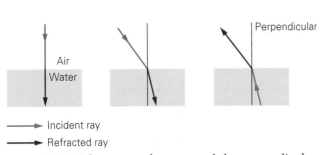

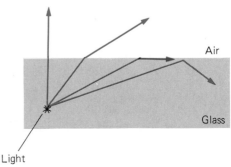

———▶ Incident ray

———▶ Refracted ray

FIG. 6-26 Light rays are bent toward the perpendicular when they enter an optically denser medium, away from the perpendicular when they enter an optically less dense medium. A ray moving along the perpendicular is not bent. The paths taken by light rays are always reversible.

FIG. 6-27 Total internal reflection occurs when the angle through which a light ray going from one medium to a less optically dense medium is refracted by more than 90°.

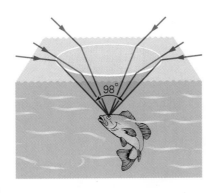

FIG. 6-28 All the light reaching an underwater observer from above the surface is concentrated in a cone 98° wide, so that the observer sees a circle of light at the surface when looking upward.

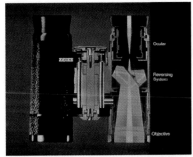

The sharpness and brightness of a light beam are better preserved by internal reflection than by reflection from an ordinary mirror. Because of this, optical devices such as binoculars use prisms instead of mirrors when light is to have its direction changed. The prisms used in binoculars have two purposes: they invert the magnified image so that it is right-side-up, and by reversing the optical path twice, they shorten the length of the instrument.

brain registers that the segment *BE* is part of the straight-line path starting at *A'*.

In general, light rays that go obliquely from one medium to another are bent *away* from a perpendicular to the surface between them if light in the second medium travels more slowly than in the first (Fig. 6-26). If light in the second medium travels faster there, the rays are bent *toward* the perpendicular. Light that enters another medium perpendicular to the surface between them does not change direction.

Internal Reflection

Since light travels more slowly in glass than in air, light that goes from glass to air at a slanting angle is refracted away from the perpendicular to the glass surface. If the angle is shallow enough, the light will be bent back into the glass (Fig. 6-27). This phenomenon is called **internal reflection.**

What does an underwater fish (or diver) see when looking upward? The paths taken by light rays are always reversible. By reversing the rays of light in Fig. 6-27 we can see that light from above the water's surface can reach the fish's eyes only through a circle on the surface. The rays from above are all brought together in a cone whose angular width turns out to be 98° (Fig. 6-28). Outside the cone is darkness.

FIBER OPTICS Internal reflection makes it possible to "pipe" light by means of a series of reflections from the wall of a glass rod, as in Fig. 6-29. If a cluster of thin glass fibers is used instead of a single rod, an image can be transferred from one end to the other with each fiber carrying a part of the image. Because a fiber cluster is flexible, it can be used for such purposes as examining a person's stomach by being passed in through the mouth. Some of the fibers provide light for illumination, and the rest carry the reflected light back outside for viewing.

Glass fibers have been used in telephone systems since 1977. The electric signals that would otherwise be sent along copper wires are converted to a series of pulses according to a standard code and then sent as flashes of infrared light down a hair-thin glass fiber. At the other end the flashes are converted back to electric signals. Modern electronic methods allow at most 32 telephone conversations to be carried at the same time by a pair of copper wires, but tens of thousands can be carried by a single fiber with no problems of electrical interference. Telephone fiberoptic systems today link many cities and exchanges within cities all over the world, and fiberoptic cables span the Atlantic and Pacific oceans.

Each of the thin glass fibers in this cable can carry thousands of telephone conversations in the form of coded flashes of light. The record for transmission capacity is held by a fiber in which 10 different frequencies of light each carried information at a rate of 2 billion flashes per second for a total of 20 billion flashes per second—the equivalent of 300,000 simultaneous conversations.

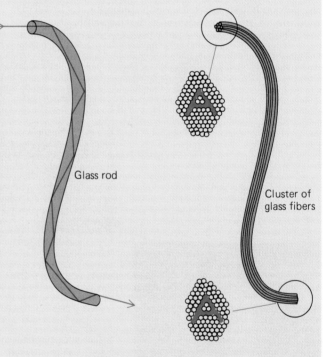

Glass rod

Cluster of glass fibers

FIG. 6-29 Light can be "piped" from one place to another by means of internal reflections in a glass rod. Using a cluster of glass fibers permits an image to be carried in this way.

6-16 LENSES

Bending Light to Form an Image

A **lens** is a piece of glass or other transparent material shaped so that it can produce an image by refracting light that comes from an object. Lenses are used for many purposes: in eyeglasses to improve vision, in cameras to record scenes, in projectors to show

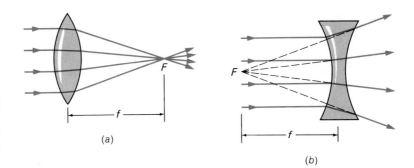

FIGURE 6-30 (a) A converging lens brings parallel rays of light together to focal point F. (b) A diverging lens spreads out parallel rays of light so that they seem to originate at a focal point F. In both cases the distance between F and the lens is the focal length f of the lens.

(a)

(b)

images on a screen, in microscopes to enable small things to be seen, in telescopes to enable distant things to be seen, and so forth.

Lenses are of two kinds, **converging** and **diverging.** A converging lens is thicker in the middle than at its rim; a diverging lens is thinner in the middle. As in Fig. 6-30a, a converging lens brings a parallel beam of light to a single focal point F. If sunlight is used, the concentration of radiant energy may be enough to burn a hole in a piece of paper. The distance from the lens to F is called the **focal length** of the lens. A diverging lens spreads out a parallel beam of light so that the rays seem to have come from a focal point behind the lens, as in Fig. 6-30b.

The Camera   Figure 6-31 shows how the converging lens of a camera brings light rays from a point on an object to a point on the film. All the rays meet at the same point provided the distance from the lens to the film is just right. The focal length of the lens is this distance when the object is very far away. Thus a "50-mm lens" must be exactly 50 mm from the film when taking a picture of a distant scene. The closer the object, the farther the lens must be from the film. To photograph a flower 50 cm away, for instance, a 50-mm lens should be 55.6 mm from the film.

The film in a camera is exposed by opening the shutter for a fraction of a second. A short exposure permits "freezing" a moving object; a long

FIG. 6-31 A camera. The light-sensitive film is exposed by opening the shutter for a fraction of a second. The adjustable diaphragm permits the amount of light entering the camera to be varied to suit the shutter speed and film used. The lens brings light from each point on the object being photographed to a single point on the film.

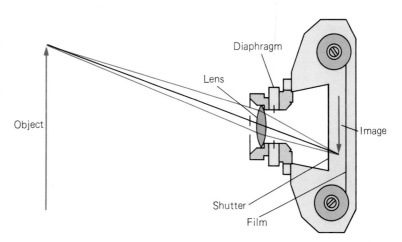

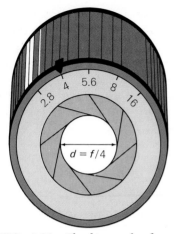

exposure may be needed when the light is weak. For a given shutter speed, the larger the diaphragm opening, the more light reaches the film. The size of the opening is specified in terms of the focal length f of the lens. Thus $f/4$ means that the diameter of the opening is $\frac{1}{4}$ of f (Fig. 6-32). The larger the f-number, the smaller the opening.

The sequence of f-numbers marked on a camera lens is chosen so that each step changes the amount of light reaching the film by a factor of 2. A typical sequence is 2.8, 4, 5.6, 8, 11, 16. Going from $f/2.8$ to $f/4$ doubles the light, going from $f/2.8$ to $f/5.6$ quadruples it, and so on. The same amount of light gets to the film during exposures of $\frac{1}{50}$ s at $f/5.6$, $\frac{1}{100}$ s at $f/4$, and $\frac{1}{200}$ s at $f/2.8$.

The appropriate focal length of a camera lens depends upon the film size and the desired angle of view. A "normal" lens gives an angle of view of about 45°, which means a focal length of 50 mm for a camera using 35-mm film. A lens of shorter focal length is a wide-angle lens because it captures more of a given scene, though at the expense of reducing the sizes of details in the scene. A telephoto lens has a long focal length to give larger images of distant objects, though less of the scene is included. Typical wide-angle lenses for a 35-mm camera have focal lengths of 35 and 28 mm; typical telephoto lenses have focal lengths of 90 and 135 mm. "Zoom" lenses can cover a wide range of focal lengths.

FIG. 6-32 The larger the f-number of a camera diaphragm, the smaller the opening. The sequence of values is such that each change of one f-number changes the amount of light reaching the film by a factor of 2.

6-17   THE EYE

A Remarkable Optical Instrument   The structure of the human eye is shown in Fig. 6-33. The **cornea,** the transparent outer membrane, and the jellylike **lens** together focus incoming light on the sensitive **retina,** which converts what is seen into nerve impulses that are carried to the brain by the **optic nerve.** Focusing on objects different distances away is done when the **ciliary muscle** changes the shape and hence the focal length of the lens. The colored

FIG. 6-33 The human eye, shown larger than life size. In dim light the iris opens wide to let enough light enter through the pupil for good vision.

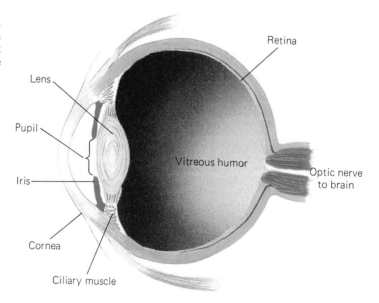

No photoreceptors are near the optic nerve, so that region of the retina is called the **blind spot.** To experience the blind spot, close your left eye and look directly at the cross in Fig. 6-34 with your right eye. When the cross is about 20 cm from your eye, the dot should disappear. We are not usually aware of the blind spot for two reasons: the blind spots of the two eyes obscure different fields of vision, and the eyes are never completely stationary but are in constant scanning motion.

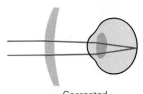

FIG. 6-34 Locating the blind spot.

To verify that cones, which are responsible for color vision, are concentrated near the center of the retina, hold something with a bright color off to one side and slowly move it around until it is in front of you. As you do this, you will find the sensation of its color going from weak to strong.

iris acts like the diaphragm of a camera to control the amount of light entering the **pupil,** which is the opening of the iris. In bright light the pupil is small, in dim light it is large. A fully opened pupil lets in about 16 times as much light as a fully contracted one. The retina itself can also cope with a wide range of brightnesses.

The retina has about a million tiny structures called **cones and rods** that are sensitive to light. The cones are specialized for color vision and occur in three types that respond respectively to red, green, and blue light. Rods need much less light to be activated than cones do, but rods do not distinguish colors. In poor illumination, then, what we see is in shades of gray, like a black-and-white photograph. Because the central region of the retina contains only cones, it is easier to see something in a dim light by looking a bit to one side instead of directly at it.

Defects of Vision

Two common defects of vision are **farsightedness** and **nearsightedness** (Fig. 6-35). In farsightedness the eyeball is too short, and light from nearby objects comes to a focus behind the retina. (Distant objects can be seen clearly, however.) A converging eyeglass lens corrects farsightedness. In nearsightedness the eyeball is too long, and light from distant objects comes to a focus in front of the retina. (Nearby objects can be seen clearly, however.) Here the correction is a diverging eyeglass lens.

Sometimes the cornea or lens of an eye has different curvatures in different planes. When light rays that lie in one plane are in focus on the retina of such an eye, rays in other planes are in focus either in front of or behind the retina. This means that only one of the bars of a cross can be in focus at any time (Fig. 6-36), a condition called **astigmatism.** Astigmatism causes eyestrain because the eye continually varies the focus of the lens as it tries to produce a completely sharp image of what it sees. A cylindrical corrective lens (Fig. 6-37) is the remedy for astigmatism.

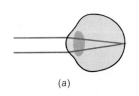
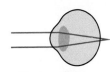

FIG. 6-35 (a) A normal eye. (b) Farsightedness can be corrected with a converging lens. (c) Nearsightedness can be corrected with a diverging lens.

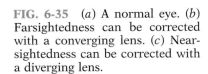
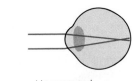

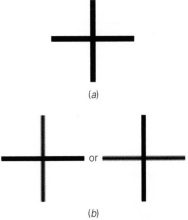

FIG. 6-36 How a cross is seen (a) by a normal eye and (b) by an astigmatic eye.

Refraction at the cornea gives rise to most of the focusing power of the human eye. When the eye is immersed in water rather than air, the amount of refraction is much less. As a result, underwater objects cannot be brought to a sharp focus unless goggles or a face mask are used to keep water away from the cornea. Beavers cope with this problem by having transparent eyelids that adjust the optical systems of their eyes for underwater vision. Fish avoid the problem by having eyes in which most of the focusing is done by very thick lenses, with the cornea playing only a minor part.

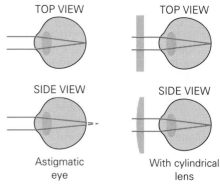

FIG. 6-37 A cylindrical lens can improve the image formed by an astigmatic eye.

| 6-18 | COLOR |

Each Frequency of Light Produces the Sensation of a Different Color

White light is a mixture of light waves of different frequencies, each of which produces the visual sensation of a particular color. To show this, we can direct a narrow beam of white light at a glass prism. Because the speed of light in glass varies slightly with frequency, light of each color is refracted to a different extent. The effect is called **dispersion.** The result is that the original beam is separated by the prism into beams of various colors, with red light bent the least and violet light bent the most.

A beam of white light is separated into its component wavelengths, each of which appears to the eye as a different color, by dispersion in a glass prism.

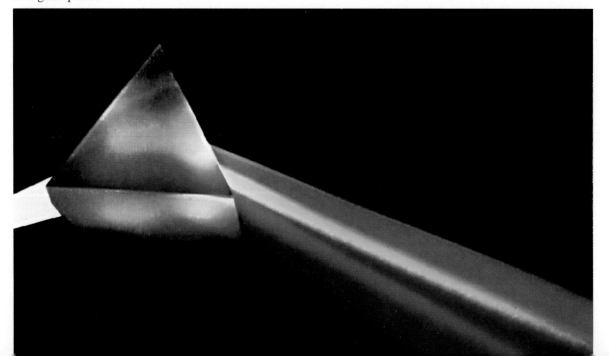

more effectively than red light. When we look at the sky, what we see is light from the sun that has been scattered out of the direct beam and hence appears blue (Fig. 6-41). The sun itself is therefore a little more yellowish or reddish than it would appear if there were no atmosphere. At sunrise or sunset, when the sun's light must make a long passage through the atmosphere, much of its blue light is scattered out. The sun may be a brilliant red as a result. Above the atmosphere the sky is black, and the moon, stars, and planets are visible to astronauts in the daytime.

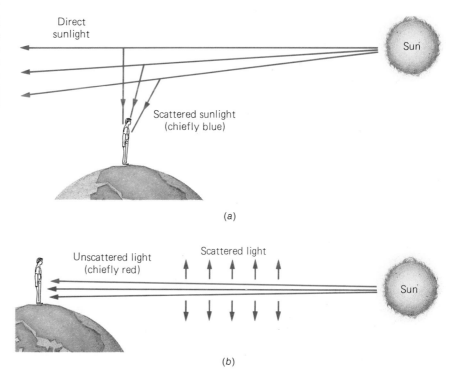

FIG. 6-41 (*a*) The preferential scattering of blue light in the atmosphere is responsible for the blue color of the sky. (*b*) The remaining direct sunlight is reddish, which is the reason for the red color of the sun at sunrise and sunset.

6-19 INTERFERENCE OF LIGHT

Why Thin Films Are Brightly Colored

All of us have seen the brilliant colors that appear in soap bubbles and thin oil films. This effect can be traced to a combination of reflection and interference.

Let us consider what happens when light of only one color, and hence only one wavelength, strikes an oil film. As in Fig. 6-42, part of the light passes right through the film, but some is reflected from the upper surface of the film and some is reflected from the lower surface. The two reflected waves interfere with each other. At some places in the film its thickness is just right for the reflected waves to be exactly out of step (crest-to-trough), as in Fig. 6-42a. This is the same effect as the destructive interference shown in Fig. 6-13c. Little or no reflection can take place in this part of the oil film, and nearly all the incoming light simply passes right through it. The oil film accordingly seems black in this region.

Where the film is slightly thicker or thinner than in Fig. 6-42a, the reflected waves may be exactly in step and therefore reinforce each other, as in b. This corresponds to the constructive interference shown in Fig. 6-13b. Here the film is a good reflector and appears bright. Shining light of one color on a thin oil film gives rise in this way to areas of light and dark whose pattern depends on the varying thickness of the film.

When white light is used, the reflected waves of only one color will be in step at a particular place while waves of other colors will not. The result is a series of brilliant colors. This is the reason for the rainbow effects we see in soap bubbles and in oil films.

COATED LENSES A portion of the light that strikes an air-glass or glass-air interface is reflected, just as in the case of the oil film of Fig. 6-42. Because there are many such interfaces in most optical instruments (10 in the case of each half of a pair of binoculars), the total amount of light lost may be considerable, and the reflections also blur the image.

To reduce reflection at an interface, the glass can be coated with a thin layer of a transparent substance whose thickness is just right for the reflected rays from the top and bottom of the layer to interfere destructively, as in Fig. 6-42a. Of course, the cancellation is exact only for a particular wavelength. What is done is to choose a wavelength in the middle of the visible spectrum, which corresponds to green light, so that partial cancellation occurs over a wide range of colors. The red and violet ends of the spectrum are accordingly least affected, and the light reflected from a coated lens is a mixture of these colors, a purplish hue. Good-quality optical instruments always use coated lenses.

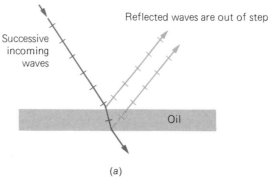

(a)

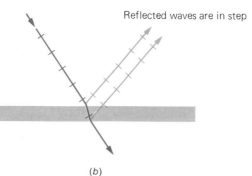

(b)

FIG. 6-42 (a) Destructive and (b) constructive interference in a thin film for light of a particular wavelength. When the film has the thickness in (a), it appears dark; when it has the thickness in (b), it appears bright. Light of other wavelengths undergoes destructive and constructive interference at different film thicknesses.

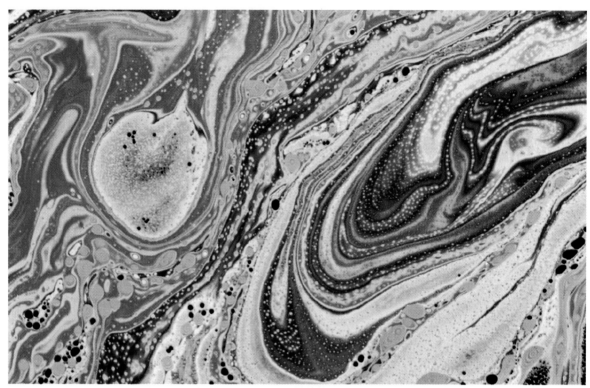

A colored pattern occurs when an uneven thin film of oil floats on water, because light of each wavelength undergoes constructive interference at different oil thicknesses.

6-20 DIFFRACTION OF LIGHT

The Larger the Diameter of a Telescope, the Sharper the Image

Diffraction, too, occurs in light waves. As we recall, diffraction refers to the "bending" of waves around the edges of an obstacle in their path. Because of diffraction, a shadow is never completely dark, although the wavelengths of light waves are so short that the effects of diffraction are largely limited to the border of the shadow region.

Diffraction limits the useful magnifications of microscopes and telescopes. The larger the diameter of a lens (or of a curved mirror that acts like a lens), the less significant is diffraction. For this reason a small telescope cannot be used at a high magnification, since the result would be a blurred image instead of a sharp one. The huge Hale telescope at Mt. Palomar in California has a mirror 5 m in diameter, but even so it can resolve objects only 50 m or more across on the moon's surface.

The resolving power of a telescope depends upon the wavelength of the light that enters it divided by the diameter of the lens or mirror; the smaller the resolving power, the sharper the image (Fig. 6-43). This relationship poses a severe problem for radio telescopes, which are antennas designed to receive radio waves from space (Chap. 18). Because such waves might have wavelengths as much as a million times greater

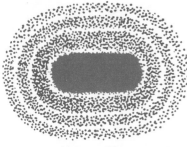

Very small lens

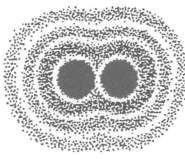

Small lens

Large lens

FIG. 6-43 A large lens or mirror is better able to resolve nearby objects than a small one.

than those in visible light, an antenna would have to be tens or hundreds of kilometers across to be able to separate sources as close together in the sky as optical telescopes can. In fact, modern electronics permits such resolution to be achieved by combining the signals received by a series of widely spaced antennas. Radio astronomers have established such an antenna array extending from the Virgin Islands across the United States to Hawaii, a span of about 8000 km—nearly the diameter of the earth. This array provides better resolution than even the largest optical telescope.

IMPORTANT TERMS AND IDEAS

Waves carry energy from one place to another by a series of periodic motions of the individual particles of the medium in which the waves occur. (Electromagnetic waves are an exception.) There is no net transfer of matter in wave motion.

In a **longitudinal wave** the particles of the medium vibrate back and forth in the direction in which the waves travel. In a **transverse wave** the particles vibrate from side to side perpendicular to the wave direction. Sound waves are longitudinal; waves in a stretched string are transverse; water waves are a combination of both since water molecules move in circular orbits when a wave passes.

The **frequency** of waves is the number of wave crests that pass a particular point per second. The **period** is the time needed for a complete wave to pass a given point. **Wavelength** is the distance between adjacent crests or troughs. The **amplitude** of a wave is the maximum displacement of a particle of the medium on either side of its normal position when the wave passes.

The change in direction of waves when they enter a region in which their speed changes is called **refraction.** In **reflection,** waves strike an obstacle and rebound from it.

Interference refers to the adding together of two or more waves of the same kind that pass by the same point at the same time. In **constructive interference** the new wave has a greater amplitude than any of the original ones; in **destructive interference** the new wave has a smaller amplitude.

The ability of waves to bend around the edge of an obstacle in their path is called **diffraction.**

A **lens** is a piece of glass or other transparent material shaped to produce an image by refracting light that comes from an object. A **converging lens** brings parallel light to a single point at a distance called the **focal length** of the lens. A **diverging lens** spreads out parallel light so that it seems to come from a point behind the lens.

White light is a mixture of different frequencies, each of which produces the visual sensation of a particular color. Because the speed of light in a medium is slightly different for different frequencies, white light is **dispersed** into its separate colors when refracted in a glass prism or a water droplet.

The **doppler effect** refers to the change in frequency of a wave when there is relative motion between its source and an observer.

Resonance occurs when an object (such as a musical instrument) vibrates at a frequency equal to one of its natural frequencies of vibration.

Electromagnetic waves consist of coupled electric and magnetic field oscillations. Radio waves, light waves, and x-rays are all electromagnetic waves that differ only in their frequency.

In **amplitude modulation** (AM), information is contained in variations in the amplitude of a constant-frequency wave. In **frequency modulation** (FM), information is contained in variations in the frequency of the wave.

IMPORTANT FORMULAS

Wave speed: $v = f\lambda$

Wave period: $T = \dfrac{1}{f}$

EXERCISES: MULTIPLE CHOICE

1. The distance from crest to crest of any wave is called its
   a. frequency   c. speed
   b. wavelength   d. amplitude

2. Of the following properties of a wave, the one that is independent of the others is its
   a. frequency   c. speed
   b. wavelength   d. amplitude

3. When waves go from one place to another, they transport
   a. amplitude   c. wavelength
   b. frequency   d. energy

4. Water waves are
   a. longitudinal
   b. transverse
   c. a mixture of longitudinal and transverse
   d. sometimes longitudinal and sometimes transverse

5. Sound waves are
   a. longitudinal
   b. transverse
   c. a mixture of longitudinal and transverse
   d. sometimes longitudinal and sometimes transverse

6. Sound cannot travel through
   a. a solid   c. a gas
   b. a liquid   d. a vacuum

7. Sound travels fastest in
   a. air   c. iron
   b. water   d. a vacuum

8. The amplitude of a sound wave determines its
   a. loudness   c. wavelength
   b. pitch   d. overtones

9. The higher the frequency of a wave,
   a. the lower its speed
   b. the shorter its wavelength
   c. the smaller its amplitude
   d. the lower its pitch

10. An automobile sounding its horn is moving away from an observer. The pitch of the horn's sound relative to its normal pitch is
    a. higher
    b. lower
    c. the same
    d. higher or lower depending upon the exact frequency

11. The doppler effect occurs
    a. only in sound waves
    b. only in longitudinal waves
    c. only in transverse waves
    d. in all types of waves

12. A pure musical note causes a thin wooden panel to vibrate with the same frequency. This is an example of
    a. an overtone   c. resonance
    b. diffraction   d. interference

13. Maxwell based his theory of electromagnetic waves on the hypothesis that a changing electric field gives rise to
    a. an electric current   c. a magnetic field
    b. a stream of electrons d. longitudinal waves

14. The direction of the magnetic field in an electromagnetic wave is
    a. parallel to the electric field
    b. perpendicular to the electric field
    c. parallel to the direction of the wave
    d. random

15. In a vacuum, the speed of an electromagnetic wave
    a. depends upon its frequency
    b. depends upon its wavelength
    c. depends upon the strength of its electric and magnetic fields
    d. is a universal constant

16. Which of the following does not consist of electromagnetic waves?
    a. x-rays   c. sound waves
    b. radar waves   d. infrared waves

17. The energy of an electromagnetic wave resides in its
    a. frequency
    b. wavelength
    c. speed
    d. electric and magnetic fields

18. Light waves
    a. require air or another gas to travel through

b. require some kind of matter to travel through
c. require electric and magnetic fields to travel through
d. can travel through a perfect vacuum

19. The ionosphere is a region of ionized gas in the upper atmosphere. The ionosphere is responsible for
    a. the blue color of the sky
    b. rainbows
    c. long-distance radio communication
    d. the ability of satellites to orbit the earth

20. A pencil in a glass of water appears bent. This is an example of
    a. reflection   **c.** diffraction
    b. refraction   **d.** interference

21. A fish that you are looking at from a boat seems to be 60 cm below the water surface. The actual depth of the fish
    a. is less than 60 cm
    b. is 60 cm
    c. is more than 60 cm
    d. may be any of the above, depending on the angle of view

22. Which of the following combinations of shutter speed and lens opening will let the most light reach the film of a camera?
    a. $\frac{1}{125}$ s at $f/8$ **c.** $\frac{1}{250}$ s at $f/4$
    b. $\frac{1}{125}$ s at $f/16$ **d.** $\frac{1}{250}$ s at $f/5.6$

23. The quality in sound that corresponds to color in light is
    a. amplitude   **c.** waveform
    b. resonance   **d.** pitch

24. Light of which color has the lowest frequency?
    a. red   **c.** yellow
    b. blue   **d.** green

25. Light of which color has the shortest wavelength?
    a. red   **c.** yellow
    b. blue   **d.** green

26. The Spanish flag is yellow and red. When viewed with yellow light it appears
    a. all yellow   **c.** yellow and black
    b. yellow and red   **d.** white and red

27. The Danish flag is red and white. When viewed with red light it appears
    a. all red   **c.** red and white
    b. all white   **d.** red and black

28. Thin films of oil and soapy water owe their brilliant colors to a combination of reflection and

    a. refraction   **c.** diffraction
    b. interference   **d.** doppler effect

29. The sky is blue because
    a. air molecules are blue
    b. the lens of the eye is blue
    c. the scattering of light is more efficient the shorter its wavelength
    d. the scattering of light is more efficient the longer its wavelength

30. Diffraction refers to
    a. the splitting of a beam of white light into its component colors
    b. the interference of light that produces bright colors in thin oil films
    c. the bending of waves around the edge of an obstacle in their path
    d. the increase in frequency due to motion of a wave source toward an observer

31. The useful magnification of a telescope is limited by
    a. the speed of light   **c.** interference
    b. the doppler effect   **d.** diffraction

32. Which one or more of the following will improve the resolving power of a lens?
    a. Increase the wavelength of the light used
    b. Increase the frequency of the light used
    c. Decrease the diameter of the lens used
    d. Increase the diameter of the lens used

33. The speed of sound waves having a frequency of 256 Hz compared with the speed of sound waves having a frequency of 512 Hz is
    a. half as great   **c.** twice as great
    b. the same   **d.** 4 times as great

34. The wavelength of sound waves having a frequency of 256 Hz compared with the wavelength of sound waves having a frequency of 512 Hz is
    a. half as great   **c.** twice as great
    b. the same   **d.** 4 times as great

35. Waves in a lake are observed to be 5 m in length and to pass an anchored boat 1.25 s apart. The speed of the waves is
    a. 0.25 m/s
    b. 4 m/s
    c. 6.25 m/s
    d. impossible to find from the information given

36. A boat at anchor is rocked by waves whose crests are 20 m apart and whose speed is 5 m/s. These waves reach the boat with a frequency of
    a. 0.25 Hz   **c.** 20 Hz
    b. 4 Hz   **d.** 100 Hz

37. One kHz (kilohertz) is equal to 10^3 Hz. What is the wavelength of the electromagnetic waves sent out by a radio station whose frequency is 660 kHz? The speed of light is 3×10^8 m/s.
a. 2.2×10^{-3} m   **c.** 4.55×10^3 m
b. 4.55×10^2 m   **d.** 1.98×10^{14} m

QUESTIONS

1. Distinguish between longitudinal and transverse waves. Do all waves fall into one or the other of these categories? If not, give an example of one that does not.

2. Does the wavelength of a wave depend on its frequency? If so, what is the relationship between these quantities?

3. Why does sound travel fastest in solids and slowest in gases?

4. The speed of sound in a gas depends upon the average speed of the gas molecules. Why is such a relationship reasonable?

5. What types of waves can be refracted? Under what circumstances does refraction occur?

6. What eventually becomes of the energy of sound waves?

7. Even if astronauts on the moon's surface did not need to be enclosed in space suits, they could not speak directly to each other but would have to communicate by radio. Can you think of the reason?

8. If you want to grab a fish underwater, should your hand be above or below where the fish appears to be?

9. In what unit is the loudness of a sound expressed?

10. How can constructive and destructive interference be reconciled with the principle of conservation of energy?

11. In what kinds of waves can the doppler effect occur?

12. The characteristic wavelengths of light emitted by a distant star are observed to be shifted toward the red end of the spectrum. What does this suggest about the motion of the star relative to the earth?

13. If you walk past a bell while it is ringing, you notice no change in pitch, but if you ride past the same bell in a rapidly moving train or car the pitch seems to change markedly. Explain.

14. Why are light waves able to travel through a vacuum whereas sound waves cannot?

15. How could you show that light carries energy?

16. Under what circumstances does an electric charge radiate electromagnetic waves?

17. Light is said to be a transverse wave. What is it that varies at right angles to the direction in which a light wave travels?

18. How are the directions of the electric and magnetic fields of an em wave related to each other and to the direction in which the wave is moving?

19. Light waves carry both energy and momentum. Why doesn't the momentum of the sun diminish with time as its energy content does?

20. Give as many similarities and differences as you can between sound and light waves.

21. The period of daylight is increased by a small amount because of the refraction of sunlight by the earth's atmosphere. Show with the help of a diagram how this effect comes about.

22. When a fish looks up through the water surface at an object in the air, will the object appear to be its normal size and distance above the water? Use a diagram to explain your answer, and assume that the fish's eye and brain, like the human eye and brain, are accustomed to interpreting light rays as straight lines.

23. What is the height of the smallest mirror in which you could see yourself at full length? Use a diagram to explain your answer. Does it matter how far away you are?

24. Use a diagram to show why slides must be inserted upside down in a projector in order to be seen correctly on a screen.

25. Does a lens opening of $f/8$ let more or less light reach the film of a camera than a lens opening of $f/16$?

26. When white light is dispersed by a glass prism, red light is bent least and violet light is bent most. What does this tell you about the relative speeds of red and violet light in glass?

27. When a beam of white light passes perpendicularly through a flat pane of glass, it is not dispersed into a spectrum. Why not?

28. What color would red cloth appear if it were illuminated by (a) white light? (b) red light? (c) green light?

29. What would the American flag look like when viewed with red light? With blue light?

30. Light of what color is scattered most in the atmosphere? Least?

31. If the earth had no atmosphere, what would the color of the sky be during the day?

32. Give two advantages that a telescope lens or mirror of large diameter has over one of small diameter.

33. Radio waves are able to diffract readily around buildings, as anybody with a portable radio receiver can verify. However, light waves, which are also electromagnetic waves, undergo no discernible diffraction around buildings. Why not?

PROBLEMS

1. An opera performance is being broadcast by radio. Who will hear a certain sound first, a member of the audience 30 m from the stage or a listener to a radio receiver in a city 5000 km away?

2. The light-year is an astronomical unit of length equal to the distance light travels in a year. What is the length of a light-year in meters?

3. A person is watching as spikes are being driven to hold a steel rail in place. The sound of each sledgehammer blow arrives 0.14 s through the rail and 2 s through the air after the person sees the hammer strike the spike. Find the speed of sound in the rail.

4. The engineer of a train moving at 30 m/s blows its whistle. How much later does someone in the caboose 300 m behind the engine hear the sound?

5. Find the frequency of sound waves in air whose wavelength is 25 cm.

6. A certain groove in a phonograph record travels past the needle at 40 cm/s and the sound that is produced has a frequency of 3000 Hz. Find the wavelength of the wiggles in the groove.

7. The radio frequency used internationally for distress calls from ships is 2182 kHz. What is the corresponding wavelength?

8. Radio amateurs are allowed to communicate on the "10-meter band." What is the frequency of radio waves whose wavelength is 10 m? Of sound waves whose wavelength is 10 m?

9. Water waves whose crests are 6 m apart reach the shore every 1.2 s. Find the frequency and speed of the waves.

10. A nanosecond is 10^{-9} s. (a) What is the frequency of an em wave whose period is 1 ns? (b) What is its wavelength? (c) To what class of em waves does it belong?

11. A violin string vibrates 1044 times per second. How many vibrations does it make while its sound travels 10 m?

12. At one end of a ripple tank 90 cm across, a 6-Hz vibrator produces waves whose wavelength is 50 mm. Find the time the waves need to cross the tank.

13. A radar sends out 0.05-μs pulses of microwaves whose wavelength is 25 mm. What is the frequency of these microwaves? How many waves does each pulse contain? (1 μs = 1 microsecond = 10^{-6} s)

14. An ultrasonic beam used to scan body tissue has a frequency of 1.2 MHz. If the speed of sound in a particular tissue is 1540 m/s and the limit of resolution is equal to one wavelength, what is the size of the smallest detail that can be resolved?

ANSWERS TO MULTIPLE CHOICE

1. b	**11.** d	**21.** c	**31.** d
2. d	**12.** c	**22.** c	**32.** b, d
3. d	**13.** c	**23.** d	**33.** b
4. c	**14.** b	**24.** a	**34.** c
5. a	**15.** d	**25.** b	**35.** b
6. d	**16.** c	**26.** c	**36.** a
7. c	**17.** d	**27.** a	**37.** b
8. a	**18.** d	**28.** b	
9. b	**19.** c	**29.** c	
10. b	**20.** b	**30.** c	

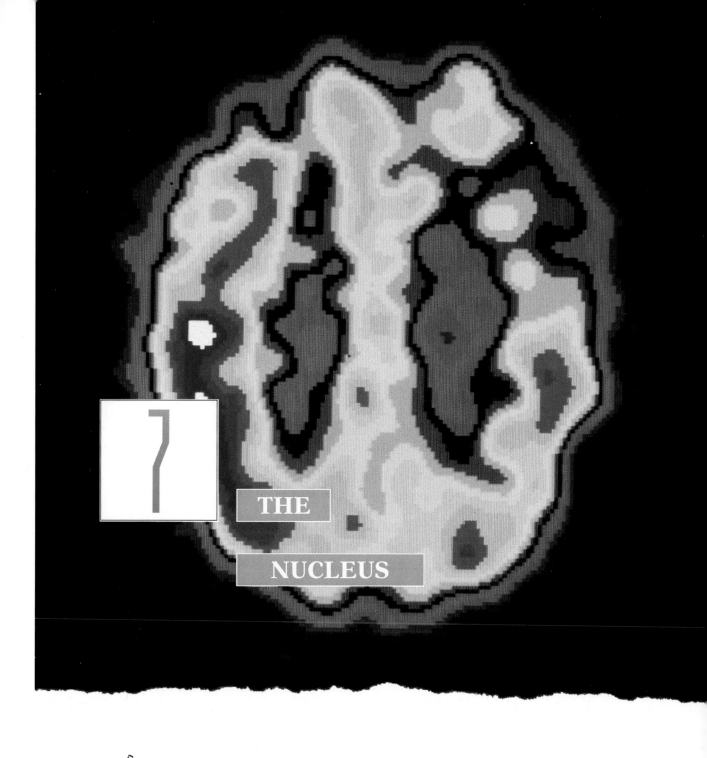

7

THE

NUCLEUS

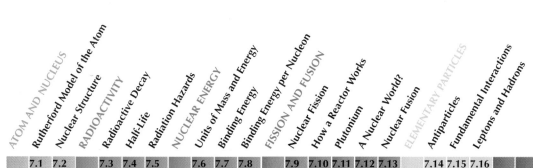

Atoms are the smallest particles of ordinary matter. Every atom has a central core, or **nucleus,** of protons and neutrons that provide nearly all the atom's mass. Moving around the nucleus are the much lighter electrons, the same in number as the number of protons in the nucleus so that the atom as a whole is electrically neutral.

The chief properties (except mass) of atoms, molecules, solids, and liquids can be traced to the behavior of atomic electrons. But the atomic nucleus is also significant in the grand scheme of things. The continuing evolution of the universe is powered by energy that comes from nuclear reactions and transformations. Like other stars, the sun obtains its energy in this way. In turn, the coal, oil, and natural gas of the earth, as well as its wind and falling water, owe their energy contents to the sun's rays. Nuclear processes are responsible for the heat of the earth's interior and for the energy produced by nuclear reactors. Thus *all* the energy at our command has a nuclear origin, except for the energy of the tides, which are the result of the gravitational pull of the moon and sun on the waters of the world.

ATOM AND NUCLEUS

Until 1911 nothing was known about atoms except that they exist and contain electrons. Since electrons carry negative charges but atoms are neutral, scientists agreed that positively charged matter of some kind must be present in atoms. But what kind? And arranged in what way?

One suggestion, made by the British physicist J. J. Thomson in 1898, was that atoms are simply positively charged lumps of matter with electrons embedded in them, like raisins in a fruitcake (Fig. 7-1). Because Thomson had played an important part in discovering the electron, his idea was taken seriously. But the real atom turned out to be very different.

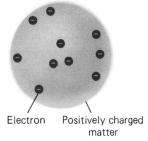

Electron   Positively charged matter

FIG. 7-1 The Thomson model of the atom. Experiment shows it to be incorrect.

7-1 RUTHERFORD MODEL OF THE ATOM

An Atom Is Mostly Empty Space

The most direct way to find out what is inside a fruitcake is to poke a finger into it. A similar method was used in 1911 in an experiment suggested by the British physicist Ernest Rutherford to find out what is inside an atom. Alpha particles were used as probes. (As discussed later in this chapter, alpha particles are emitted by certain substances. For now, all we need to know about these particles is that they are almost 8000 times heavier than electrons and each one has a charge of $+2e$.) A sample of an alpha-emitting substance was placed behind a lead screen with a small hole in it, as in Fig. 7-2, so that a narrow beam of alpha particles was produced. This beam was aimed at a thin gold foil. A zinc sulfide screen, which gives off a visible flash of light when struck by an alpha particle, was set on the other side of the foil.

Rutherford expected the alpha particles to go right through the foil with hardly any deflection. This follows from the Thomson model, in which the electric charge inside an atom is assumed to be uniformly spread through its volume. With only weak electric forces exerted on them, alpha particles that pass through a thin foil ought to be deflected only slightly, 1° or less.

BIOGRAPHY

Ernest Rutherford (1871–1937)

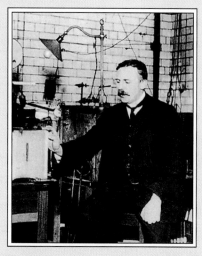

was digging potatoes on his family's farm in New Zealand when he learned that he had won a scholarship for graduate study at Cambridge University in England. "This is the last potato I will ever dig," he said, throwing down his spade. Thirteen years later he received the Nobel Prize for his work on radioactivity, which included discovering that alpha particles are the nuclei of helium atoms and that the radioactive decay of an element can give rise to a different element.

In 1911 Rutherford showed that the nuclear model of the atom was the only one that could explain the observed scattering of alpha particles by thin metal foils. Using alpha particles to bombard nitrogen nuclei, Rutherford was the first to artificially transmute one element into another. Notable achievements made in his laboratory by his associates include the discovery of the neutron in 1932 and the construction of the first high-energy particle accelerator for nu-

clear research. But Rutherford was not infallible: only a few years before the first nuclear reactor was built, he dismissed the idea of practical uses for nuclear energy as "moonshine." Rutherford is buried near Newton in Westminster Abbey.

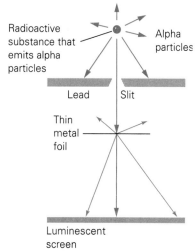

Radioactive substance that emits alpha particles

Alpha particles

Lead   Slit

Thin metal foil

Luminescent screen

FIG. 7-2 Principle of the Rutherford experiment. Nearly all the alpha particles pass through the foil with little or no deflection, but a few of the particles are scattered through large angles, even in the backward direction. This result means that strong electric fields must act on the particles, and such fields can arise only if atoms have very small nuclei in which their positive charge is concentrated.

What was found instead was that, although most of the alpha particles indeed were not deviated by much, a few were scattered through very large angles. Some were even scattered in the backward direction. As Rutherford remarked, "It was as incredible as if you fired a 15-inch shell at a piece of tissue paper and it came back and hit you."

Why the Nucleus Must Be Small

Since alpha particles are relatively heavy and since those used in this experiment had high speeds, it was clear that strong forces had to be exerted upon them to cause such marked deflections. The only way to explain the results, Rutherford found, was to picture an atom as having a tiny nucleus in which the positive charge and nearly all the mass of the atom are concentrated. The electrons are some distance away, as in Fig. 7-3.

With an atom that is largely empty space, it is easy to see why most alpha particles go right through a thin foil. However, when an alpha particle happens to come near a nucleus, the strong electric field there causes the particle to be scattered through a large angle. The atomic electrons, being so light, have little effect on the alpha particles.

Suppose, as an analogy, that a star approaches the solar system from space at great speed. The chances are good that the star will not be deflected. Even a collision with a planet would not change the star's path to any great extent. Only if the star came near the great mass of the sun would the star's direction change by much. Similarly, said Rutherford, an alpha particle plows straight through an atom, unaffected by striking an electron now and then. Only a close approach to the heavy central nucleus of an atom can turn the alpha particle aside.

Ordinary matter, then, is mostly empty space. The solid wood of a table, the steel that supports a bridge, the hard rock underfoot, all are

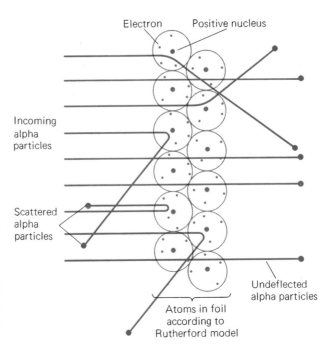

FIG. 7-3 In the Rutherford model of the atom, the positive charge is concentrated in a central nucleus with the electrons some distance away. This model correctly predicts that some alpha particles striking a thin metal foil will be scattered through large angles by the strong electric fields of the nuclei.

just collections of electric charges, comparatively farther away from one another than the planets are from the sun. If all the electrons and nuclei in our bodies could somehow be packed closely together, we would be no larger than specks just visible with a microscope.

7-2 NUCLEAR STRUCTURE

Protons and Neutrons   What gives a nucleus its mass and charge? The simplest nucleus, that of the hydrogen atom, usually consists of a single **proton.** As mentioned in Chap. 5, the proton is a particle whose charge is $+e$ and whose mass is 1836 times the electron mass. Nuclei more complex than that of hydrogen contain **neutrons** as well as protons. The neutron has no charge, and its mass, 1839 times that of the electron, is slightly more than that of the proton. The compositions of several atoms are illustrated in Fig. 7-4.

Elements are the simplest substances in the bulk matter around us. Over 100 elements are known, of which 10 are gases, 2 are liquids, and the rest are solids at room temperature and atmospheric pressure. Hydrogen, helium, oxygen, chlorine, and neon are gaseous elements, and bromine and mercury are the two liquids. Most of the solid elements are metals. Elements are discussed in more detail in Chap. 9.

In a neutral atom of any element, the number of protons equals the number of electrons. This number is called the **atomic number** of the element. Thus the atomic number of hydrogen is 1, of helium 2, of lithium 3, and of beryllium 4. The atomic number of an element is its most basic property since this determines how many electrons its atoms have and how they are arranged, which in turn govern the physical and

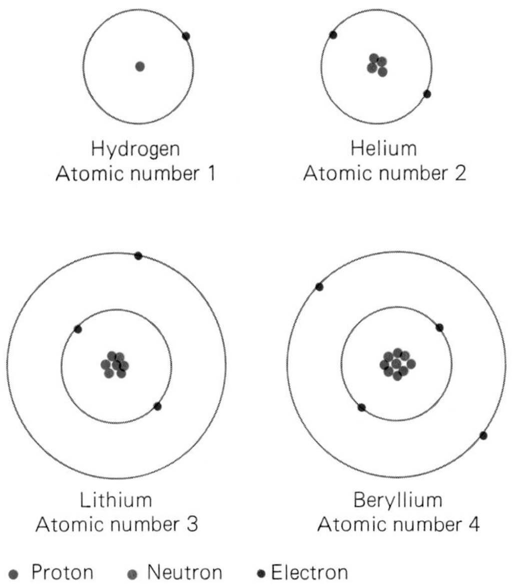

Hydrogen
Atomic number 1

Helium
Atomic number 2

Lithium
Atomic number 3

Beryllium
Atomic number 4

● Proton   ● Neutron   • Electron

FIG. 7-4 The elements that correspond to the atomic numbers 1, 2, 3, and 4 are hydrogen, helium, lithium, and beryllium. The various particles are actually far too small to be seen even on this scale.

chemical behavior of the element. Atomic numbers and symbols of all the elements are given in Table 9-6.

Isotopes

All atoms of a given element have nuclei with the same number of protons but not necessarily the same number of neutrons. For instance, even though more than 99.9 percent of hydrogen nuclei are just single protons, a few also contain a neutron as well, and a very few contain two neutrons along with the single proton (Fig. 7-5). The different kinds of hydrogen atom are called **isotopes.** All elements have isotopes.

A nucleus with a particular composition is called a **nuclide.** Symbols for nuclides follow the pattern

$$^{A}_{Z}X \qquad \textit{nuclide symbol}$$

where X = chemical symbol of element
Z = atomic number of element
  = number of protons in nucleus
A = **mass number** of nucleus
  = number of protons and neutrons in nucleus

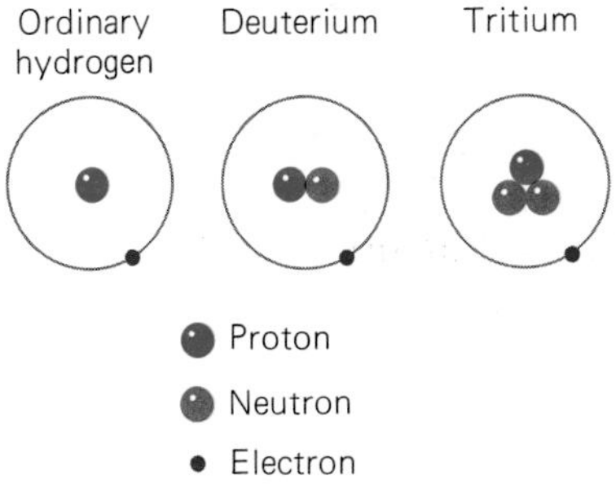

Thus the nucleus of the chlorine isotope that contains 17 protons and 18 neutrons has the atomic number $Z = 17$ and the mass number $A = 17 + 18 = 35$. Its symbol is accordingly

$$^{35}_{17}Cl$$

The symbol of this nuclide is sometimes shortened to ^{35}Cl or Cl-35. The symbols of the nuclides of Fig. 7-4 are $^{1}_{1}H$, $^{4}_{2}He$, $^{7}_{3}Li$, and $^{9}_{4}Be$.

The term **nucleon** refers to both protons and neutrons, so that the mass number A is the number of nucleons in a particular nucleus.

FIG. 7-5 The isotopes of hydrogen.

Ordinary hydrogen   Deuterium   Tritium

● Proton
● Neutron
• Electron

RADIOACTIVITY

In 1896 Henri Becquerel accidentally discovered in his Paris laboratory that the element uranium can expose covered photographic film, can ionize gases, and can cause certain materials (such as the zinc sulfide used in the Rutherford experiment) to glow in the dark. Becquerel concluded that uranium gives off some kind of invisible but penetrating radiation, a property soon called **radioactivity.**

Not long afterward, Pierre and Marie Curie, in the course of extracting uranium from the ore pitchblende at the same laboratory, found two other elements that are also radioactive. They named one polonium, after Marie Curie's native Poland. The other, which turned out to be thousands of times more radioactive than uranium, was called radium.

Chemical reactions do not change the ability of a radioactive material to emit radiation, nor does heating it in an electric arc or cooling it in liquid air. Radioactivity must therefore be associated with atomic nuclei because these are the only parts of atoms not affected by such treatment.

The radioactivity of an element is due to the radioactivity of one or more of its isotopes. Most elements in nature have no radioactive isotopes, though such isotopes can be prepared artificially and are useful in biological and medical research as "tracers." The procedure is to incorporate a radionuclide in a chemical compound and follow what happens to the compound in a living organism by monitoring the radiation from the isotope. Other elements, such as potassium, have some stable isotopes and some radioactive ones. A few, such as uranium, have only radioactive isotopes.

7-3 RADIOACTIVE DECAY

How Unstable Nuclei Change into Stable Ones

Early experimenters found that a magnetic field splits the radiation from a radioactive material such as radium into three parts (Fig. 7-6). One part is deflected as though it consists of positively charged particles. Called **alpha particles,** these turned out to be the nuclei of helium atoms. Such nuclei contain two protons and two neutrons, so their symbol is $_2^4$He. (These were the probes used in Rutherford's discovery of the nucleus.)

Another part of the radiation is deflected as though it consists of negatively charged particles. Called **beta particles,** these are electrons.

The rest of the radiation, which is not affected by a magnetic field, consists of **gamma rays.** Today these are known to be electromagnetic waves whose frequencies are higher than those of x-rays. A gamma ray

FIG. 7-6 The radiations from a radium sample may be analyzed with the help of a magnetic field. Alpha particles are deflected to the left, hence they are positively charged; beta particles are deflected to the right, hence they are negatively charged; and gamma rays are not affected, hence they are uncharged.

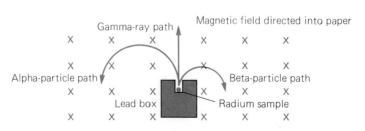

BIOGRAPHY

Marie Sklodowska Curie (1867–1934) was born in Poland. After high school, she worked as a governess until she was 24 so that she could study science in Paris, where she had barely enough money to survive. In 1894 Marie married Pierre Curie, who was 8 years older than she and already a noted physicist. In 1897, just after the birth of her daughter Irene (who, like her mother, was to win a Nobel Prize in physics), Marie began to investigate the newly discovered phenomenon of radioactivity—her word—for her doctoral thesis.

After a search of all the known elements, Marie learned that thorium as well as uranium was radioactive. She then examined various minerals for radioactivity and found that the uranium ore pitchblende was far more radioactive than its uranium content would sug-

gest. Marie and Pierre together went on to identify first polonium, named for her native Poland, and then radium as the sources of the additional activity. With the primitive facilities that were all they could afford (they had to use their own money), they had succeeded by 1902 in purifying a tenth of a gram of radium from several tons of ore, a task that involved immense physical as well as intellectual labor.

Together with Becquerel, the Curies shared the 1903 Nobel Prize in physics. Pierre ended his acceptance speech with these words: "One may also imagine that in criminal hands radium might become very dangerous, and here one may ask if humanity has anything to gain by learning the secrets of nature, if it is ready to profit from them, or if this knowledge is not harmful. . . . I am among those who think . . . that humanity will obtain more good than evil from the new discoveries."

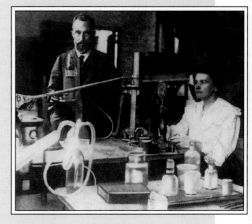

In 1906 Pierre was struck and killed by a horse-drawn carriage in a Paris street. Marie continued work on radioactivity and became world-famous. Even before Pierre's death both Curies had suffered from ill health because of their exposure to radiation, and much of Marie's later life was marred by radiation-induced ailments, including the leukemia from which she died.

is emitted by a nucleus that, for one reason or another, has more than its normal amount of energy. The composition of the nucleus does not change in gamma decay, unlike the cases of alpha and beta decay. Gamma rays are the most penetrating of the three kinds of radiation, alpha particles the least (Fig. 7-7).

Why Decays Occur

A nucleus is said to **decay** when it emits an alpha or beta particle or a gamma ray. Alpha decay occurs in nuclei too large to be stable (Fig. 7-8). The forces that hold protons and neutrons together in a nucleus act only over short distances. As a result these particles interact strongly only with their near-

FIG. 7-7 Alpha particles from radioactive materials are stopped by a piece of cardboard. Beta particles penetrate the cardboard but are stopped by a sheet of aluminum. Even a thick slab of lead may not stop all the gamma rays.

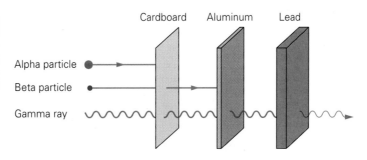

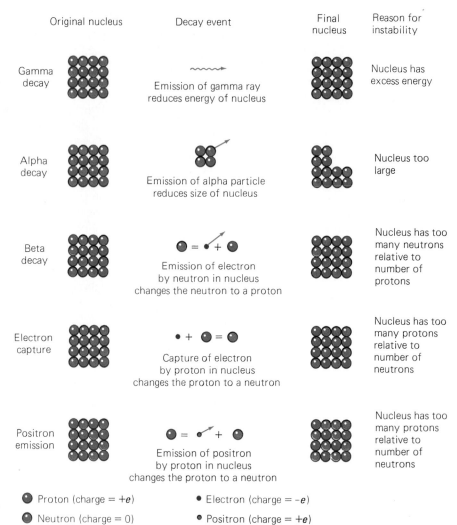

FIG. 7-8 Five kinds of radioactive decay.

est neighbors in a nucleus. Because the electrical repulsion of the protons is strong throughout the entire nucleus, there is a limit to the ability of neutrons to hold together a large nucleus. This limit is represented by the bismuth isotope $^{209}_{83}$Bi, which is the heaviest stable (that is, nonradioactive) nucleus. All larger nuclei become smaller ones by alpha decay.

Another cause of radioactive decay is a ratio of neutrons to protons that is too large or too small. A small nucleus is stable with equal numbers of neutrons and protons. However, larger nuclei need more neutrons than protons in order to overcome the electrical repulsion of the protons. In beta decay, one of the neutrons in a nucleus with too many of them spontaneously turns into a proton with the emission of an electron, as in Fig. 7-8.

In a nucleus with too few neutrons for stability, one of the protons may become a neutron with the emission of a **positron,** which is an electron that has a positive charge rather than a negative one. Alterna-

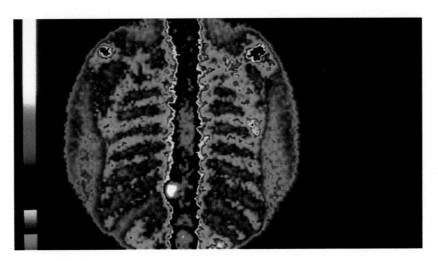

Substances that contain a radionuclide can be traced in living tissue by the radiation they emit. In this image, different levels of radiation intensity are shown in different colors. The gamma-emitting radionuclide used here is absorbed more readily by cancerous bone than by normal bone. The white area in the spine corresponds to a high rate of gamma emission and indicates a tumor there.

tively one of the electrons in the atom may be absorbed by one of the protons to form a neutron. This process is called **electron capture.**

Sometimes a certain nuclide requires a number of radioactive decays before it reaches a stable form. The uranium isotope $^{238}_{92}U$, for instance, undergoes eight alpha decays and six beta decays before it eventually becomes the lead isotope $^{206}_{82}Pb$, which is not radioactive.

Free neutrons outside of nuclei are unstable and undergo radioactive decay into a proton and an electron. Nevertheless it is not correct to think of a neutron as a combination of a proton and an electron: a neutron is a separate particle with unique properties. If we were to try to create a neutron by bringing together a proton and an electron, we would merely get a hydrogen atom as the result, not a neutron. Protons outside a nucleus are apparently stable.

7.4 HALF-LIFE

Nuclei Decay at Random

The **half-life** of a radionuclide is the period of time needed for half of any initial amount of the nuclide to decay. As time goes on, the undecayed amount becomes smaller, but there is some left for many half-lives.

Suppose we start with 1 milligram (mg) of the radium isotope $^{226}_{88}Ra$, which alpha decays to the radon isotope $^{222}_{86}Rn$ with a half-life of about 1600 years. After 1600 years, 0.5 mg of radium will remain, with the rest having turned into radon (which, by the way, is a gas; radium is a metal). During the next 1600 years, half the 0.5 mg of radium that is left will decay, to leave 0.25 mg of radium (Fig. 7-9). After a further 1600 years, which means a total of 4800 years or 3 half-lives, 0.125 mg of radium will be left—still a fair amount. Even after 6 half-lives, more than 1 percent of an original sample will remain undecayed.

Every radionuclide has a characteristic and unchanging half-life. Some half-lives are only a millionth of a second; others are billions of years. Radon, for example, is an alpha emitter like its parent radium,

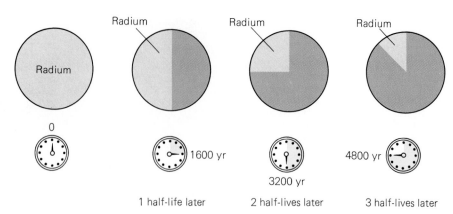

FIG. 7-9 The decay of the radium isotope $^{226}_{88}$Ra. The number of undecayed radium atoms in a sample decreases by one-half in each 1600-year period. This time span is accordingly known as the "half-life" of radium. The radium alpha decays into the radon isotope $^{222}_{86}$Rn, whose own half-life is 3.8 days.

but the half-life of radon is only 3.8 days instead of 1600 years. One of the biggest problems faced by nuclear power plants is the safe disposal of radioactive wastes since some of the isotopes present have long half-lives.

The dating of archaeological specimens and rock samples (including those brought back from the moon) by methods based on radioactive decay is described in Chap. 15.

7-5 RADIATION HAZARDS

Invisible but Dangerous The various radiations from radionuclides ionize matter through which they pass. X-rays ionize matter, too. All ionizing radiation is harmful to living tissue, although if the damage is slight, the tissue can often repair itself with no permanent effect. Radiation hazards are easy to underestimate because there is usually a delay, sometimes of many years, between an exposure and some of its possible consequences. These consequences include cancer, leukemia, and changes in reproductive cells that lead to children with physical deformities and mental handicaps.

Radiation dosage is measured in **sieverts** (Sv), where 1 Sv is the amount of any radiation that has the same biological effects as that produced when 1 kg of body tissue absorbs 1 joule of x-rays or gamma rays. Although radiobiologists disagree about the exact relationship between radiation exposure and the likelihood of developing cancer, there is no question that such a link exists. Natural sources of radiation lead to a dosage rate per person of about 3 mSv/y averaged over the U.S. population (1 mSv = 0.001 Sv). Other sources of radiation add 0.6 mSv/y, with medical x-rays contributing the largest amount. The total per person is about 3.6 mSv/y, about the dose received from 25 chest x-rays.

Natural Sources Figure 7-10 shows the relative contributions to the radiation dosage received by an average person in the United States. The most important single source is the radioactive gas radon, a decay product of radium whose own origin traces back to the decay of uranium. Uranium is found in many com-

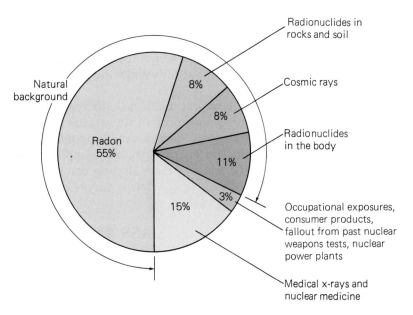

FIG. 7-10 Sources of radiation dosage for an average person in the United States. The total is about equivalent to 25 chest x-rays. Actual dosages vary widely. For instance, radon concentrations are not the same everywhere, some people receive more medical x-rays than others; cosmic rays are more intense at high altitudes; and so on. Nuclear power stations are responsible for 0.08 percent of the total, although accidents can raise the amount in affected areas to dangerous levels.

mon rocks, notably granite. Hence radon, colorless and odorless, is present nearly everywhere, though usually in amounts too small to endanger health. Problems arise when houses are built in uranium-rich regions, since it is impossible to prevent radon from entering such houses from the ground under them. Surveys show that millions of American homes have radon concentrations high enough to pose a small but definite cancer risk. As a cause of lung cancer, radon is second only to cigarette smoking. The most effective method of reducing radon levels in an existing house in a hazardous region seems to be to extract air from underneath the ground floor and disperse it into the atmosphere before it can enter the house.

Other natural sources of radiation dosage include cosmic rays (see Chap. 18) and radionuclides present in rocks and soil. The human body itself contains tiny amounts of radionuclides of such elements as potassium and carbon.

Many useful processes involve ionizing radiation. Some employ such radiation directly, as in the x-rays and gamma rays used in medicine and industry. In other cases the radiation is an unwanted but inescapable by-product, notably in the operation of nuclear reactors and in the disposal of their wastes.

X-Rays

It is not always easy to find an appropriate balance between risk and benefit where radiation is concerned. This seems particularly true for medical x-ray exposures, many of which are made for no strong reason and do more harm than good. In this category are "routine" chest x-rays upon hospital admission, "routine" x-rays as part of regular physical examinations, and "routine" dental x-rays. The once "routine" x-raying of symptomless young women to search for breast cancer is now generally believed to have increased, not decreased, the overall death rate due to cancer. Particularly dangerous is the x-raying of pregnant women, until not long ago another "routine" procedure, which dramatically increases the chance of cancer in their children. Of course, x-rays have many valuable applications in medicine. The point is that every exposure should have a definite justification that outweighs the risk involved.

NUCLEAR ENERGY

The atomic nucleus is the energy source of the reactors that produce more and more of the world's electricity. It is also the energy source of the most destructive weapons ever invented. But there is more to nuclear energy than these applications: nearly all the energy that keeps the sun and stars shining comes from the nucleus as well. Before considering what nuclear energy does, let us look into exactly what it is.

7-6 UNITS OF MASS AND ENERGY

The Atomic Mass Unit and the Electronvolt

Until now we have been using the kilogram as the unit of mass and the joule as the unit of energy. These units are far too large in the atomic world, and physicists find it more convenient to use smaller units for mass and energy in this world.

The **atomic mass unit** (u) has the value

$$1 \text{ atomic mass unit} = 1 \text{ u} = 1.66 \times 10^{-27} \text{ kg}$$

This mass is approximately equal to the mass of the hydrogen atom, whose actual mass is 1.008 u.

The energy unit used in atomic physics is the **electronvolt** (eV), which is the energy gained by an electron accelerated by a potential difference of 1 volt. The joule equivalent of the electronvolt is

$$1 \text{ electronvolt} = 1 \text{ eV} = 1.60 \times 10^{-19} \text{ J}$$

A typical quantity expressed in electronvolts is the energy needed to remove an electron from an atom. In the case of a nitrogen atom this energy is 14.5 eV, for example.

In nuclear physics the electronvolt is too small, and its multiple the **megaelectronvolt** (MeV) is more suitable:

$$1 \text{ megaelectronvolt} = 1 \text{ MeV} = 10^6 \text{ eV} = 1.60 \times 10^{-13} \text{ J}$$

(*Mega* is the prefix for million.) A typical quantity expressed in megaelectronvolts is the energy of the radiation emitted by a radionuclide. The alpha particle emitted by a nucleus of the radium isotope $^{226}_{88}\text{Ra}$ has an energy of 4.9 MeV, for example.

The energy equivalent of the atomic mass unit is 931 MeV.

7-7 BINDING ENERGY

The Missing Energy That Keeps a Nucleus Together

An ordinary hydrogen atom has a nucleus that consists of a single proton, as its symbol ^1_1H indicates. The isotope of hydrogen called deuterium, ^2_1H, has a neutron as well as a proton in its nucleus. Thus we expect the mass of the deuterium atom to equal the mass of

FIG. 7-11 The mass of a deuterium atom ($^2_1$H) is less than the sum of the masses of a hydrogen atom ($^1_1$H) and a neutron. The energy equivalent of the missing mass is called the binding energy of the nucleus.

a $^1_1$H hydrogen atom plus the mass of a neutron:

Mass of $^1_1$H atom	1.0078 u
+ Mass of neutron	+1.0087 u
Expected mass of $^2_1$H atom	2.0165 u

However, the measured mass of the $^2_1$H atom is only 2.0141 u, which is 0.0024 u *less* than the combined masses of a $^1_1$H atom and a neutron (Fig. 7-11).

Deuterium atoms are not the only ones that have less mass than the combined masses of the particles they are composed of—*all* atoms (except $^1_1$H) are like that. We conclude that nuclei are stable because they lack enough mass to break up into separate nucleons.

What happens when a nucleus is formed is that a certain amount of energy is given off due to the action of the forces that hold the neutrons and protons together. Energy is similarly given off due to the action of gravity when a stone strikes the ground or due to the action of intermolecular forces when water freezes into ice. In the case of a nucleus, the energy comes from the mass of the particles that join together. The resulting nucleus therefore has less mass than the total mass of the particles before they interact.

Since the energy equivalent of 1 u of mass is 931 MeV, the energy that corresponds to the missing deuterium mass of 0.0024 u is

$$\text{Missing energy} = (0.0024 \text{ u})(931 \text{ MeV/u}) = 2.2 \text{ MeV}$$

FIG. 7-12 The binding energy of the deuterium nucleus is 2.2 MeV. A gamma ray whose energy is 2.2 MeV or more can split a deuterium nucleus into a proton and neutron. A gamma ray whose energy is less than 2.2 MeV cannot do this.

To test the above interpretation of the missing mass, we can perform experiments to see how much energy is needed to break apart a deuterium nucleus into a separate neutron and proton. The required energy turns out to be 2.2 MeV, as we expect (Fig. 7-12). When less energy than 2.2 MeV is given to a $^2_1$H nucleus, the nucleus stays together. When the added energy is more than 2.2 MeV, the extra energy goes into kinetic energy of the neutron and proton as they fly apart.

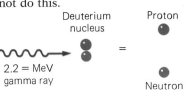

> **The energy equivalent of the missing mass of a nucleus is called the *binding energy* of the nucleus. The greater its binding energy, the more the energy that must be supplied to break up the nucleus.**

Nuclear binding energies are strikingly high. The range for stable nuclei is from 2.2 MeV for $^{2}_{1}$H (deuterium) to 1640 MeV for $^{209}_{83}$Bi (an isotope of the metal bismuth). Larger nuclei are all unstable and decay radioactively. To appreciate how high binding energies are, we can compare them with more familiar energies in terms of kilojoules of energy per kilogram of mass. In these units, a typical binding energy is 8×10^{11} kJ/kg—800 billion kJ/kg. By contrast, to boil water involves a heat of vaporization of a mere 2260 kJ/kg, and even the heat given off by burning gasoline is only 4.7×10^{4} kJ/kg, 17 million times smaller.

7-8 BINDING ENERGY PER NUCLEON

The Most Important Graph in Science

For a given nucleus, the **binding energy per nucleon** is found by dividing the total binding energy of the nucleus by the number of nucleons (protons and neutrons) it contains. Thus the binding energy per nucleon for $^{2}_{1}$H is 2.2 MeV/2 = 1.1 MeV/nucleon, and for $^{209}_{83}$Bi it is 1640 MeV/209 = 7.8 MeV/nucleon.

Figure 7-13 shows binding energy per nucleon plotted against mass number (number of nucleons). The greater the binding energy per nucleon, the more stable the nucleus. The graph has its maximum of 8.8 MeV/nucleon when the number of nucleons is 56. The nucleus that has 56 protons and neutrons is $^{56}_{26}$Fe, an iron isotope. This is the most stable nucleus of them all, since the most energy is needed to pull a nucleon away from it. Larger and smaller nuclei are less stable.

FIG. 7-13 The binding energy per nucleon is a maximum for nuclei of mass number $A = 56$. Such nuclei are the most stable. When two light nuclei join to form a heavier one, a process called *fusion*, the greater binding energy of the product nucleus causes energy to be given off. When a heavy nucleus is split into two lighter ones, a process called *fission*, the greater binding energy of the product nuclei also causes energy to be given off.

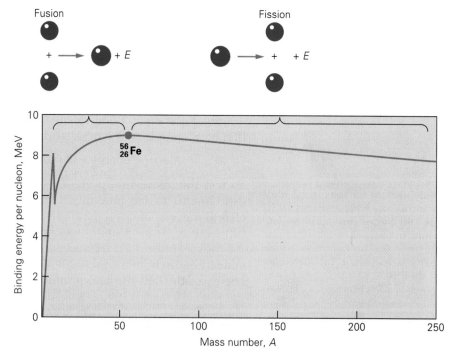

Fission and Fusion

Two remarkable conclusions can be drawn from the curve of Fig. 7-13. The first is that, if we somehow split a heavy nucleus into two medium-size ones, each of the new nuclei will have *more* binding energy per nucleon (and hence less mass per nucleon) than the original nucleus did. The extra energy will be given off, and it can be a lot.

As an example, if the uranium nucleus $^{235}_{92}U$ is broken into two smaller nuclei, the difference in binding energy per nucleon is about 0.8 MeV. Since $^{235}_{92}U$ contains 235 nucleons, the total energy given off is

$$\left(0.8 \; \frac{MeV}{nucleon}\right) (235 \; nucleons) = 188 \; MeV$$

This is a truly enormous amount of energy to come from a single atomic event. For comparison, ordinary chemical reactions involve only a few eV per atom. Splitting a large nucleus, which is called **nuclear fission,** thus involves a hundred million times more energy per atom than, say, burning coal or oil.

The other notable conclusion from Fig. 7-13 is that joining two light nuclei together to give a single nucleus of medium size also means more binding energy per nucleon in the new nucleus. For instance, if two $^{2}_{1}H$ deuterium nuclei combine to form a $^{4}_{2}He$ helium nucleus, over 23 MeV is released. Such a process, called **nuclear fusion,** is also a very effective way to obtain energy. In fact, nuclear fusion is the main energy source of the sun and other stars, as described in Chap. 17.

The graph of Fig. 7-13 has a good claim to being the most significant in all of science because it is the key to energy production in the universe.

Also, the fact that binding energy exists at all means that nuclei more complex than the single proton of hydrogen can be stable. This stability in turn accounts for the existence of the various elements and consequently for the existence of the many and diverse forms of matter we see around us. Because the curve peaks in the middle, we have the explanation for the energy that powers, directly or indirectly, the evolution of much of the universe: this energy comes from the fusion of protons and light nuclei to form heavier nuclei. And the harnessing of nuclear fission in reactors and weapons has irreversibly changed modern civilization.

FISSION AND FUSION

The words "nuclear energy" bring to mind two images. One is of a huge building in which a mysterious thing called a nuclear reactor turns an absurdly small amount of uranium into an absurdly large amount of energy. The other image is of a mushroom-shaped cloud rising from the explosion of a nuclear bomb, an explosion that can level the largest city and kill millions of people.

The first image is a picture of hope, hope for a future of plentiful, cheap, pollution-free energy—a hope only partly fulfilled. The second image is a picture of horror—but such bombs have not been used in war for over 50 years. So nuclear energy has not turned out as yet to be either the overwhelming blessing or the overwhelming curse it might have been.

7-9   NUCLEAR FISSION

Divide and Conquer

As we have seen, a lot of energy will be released if we can break a large nucleus into smaller ones. But nuclei are ordinarily not at all easy to break up. What we need is a way to split a heavy nucleus without using more energy than we get back from the process.

The answer came in 1939 with the discovery that a nucleus of the uranium isotope $^{235}_{92}\text{U}$ undergoes fission when struck by a neutron. It is not the impact of the neutron that has this effect. Instead, the $^{235}_{92}\text{U}$ nucleus absorbs the neutron to become $^{236}_{92}\text{U}$, and the new nucleus is so unstable that almost at once it splits into two pieces (Fig. 7-14). Isotopes of several elements besides uranium were later found to be fissionable by neutrons in similar processes.

Most of the energy set free in fission goes into kinetic energy of the new nuclei. These nuclei are usually radioactive, some with long half-lives. Hence the products of fission, which are found in reactor fuel rods and in the fallout from a nuclear weapon explosion, are extremely dangerous and remain so for many generations.

Chain Reaction

When a nucleus breaks apart, two or three neutrons are set free at the same time. This suggests a remarkable possibility. Perhaps, under the right conditions, the neutrons emitted by one uranium nucleus as it undergoes fission can cause other uranium nuclei to split; the neutrons from these other fissions might then go on to split still more uranium nuclei; and so on, with a series of fission reactions spreading through a mass of uranium. A **chain reaction** of this kind was first demonstrated in Chicago in 1942 under the direction of Enrico Fermi, an Italian physicist who had not long before taken refuge in the United States. Figure 7-15 is a sketch of the events that occur in a chain reaction.

For a chain reaction to occur, at least one neutron produced by each fission must, on the average, lead to another fission and not either

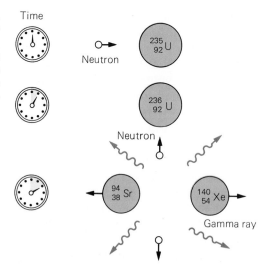

FIG. 7-14 In nuclear fission an absorbed neutron causes a heavy nucleus to split into two parts. Several neutrons and gamma rays are emitted in the process. The smaller nuclei shown here are typical of those produced in the fission of $^{235}_{92}\text{U}$.

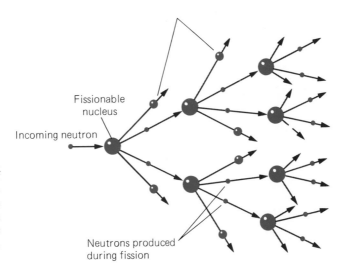

Fissionable nucleus

Incoming neutron

Neutrons produced during fission

FIG. 7-15 Sketch of a chain reaction. The reaction continues if at least one neutron from each fission event on the average induces another fission event. If more than one neutron per fission on the average induces another fission, the reaction is explosive.

Enrico Fermi
(1901–1954)

escape or be absorbed without producing fission. If too few neutrons cause fissions, the reaction slows down and stops. If precisely one neutron per fission causes another fission, energy is released at a steady rate. This is the case in a nuclear reactor, which is an arrangement for producing controlled power from nuclear fission.

Nuclear Weapons   What happens if more than one neutron from each fission causes other fissions? Then the chain reaction speeds up and the energy release is so fast that an explosion results. An "atomic" bomb makes use of this effect. The destructive power of nuclear weapons does not stop with their detonation but continues long afterward through the radioactive debris that is produced and widely dispersed.

In 1945 atomic bombs were exploded over Hiroshima and Nagasaki to end the war with Japan. Since then tens of thousands of nuclear weapons have been built, many times more than enough to destroy all human life. Today, although the United States and several of the nations that used to be part of the Soviet Union, the major nuclear powers, are starting to reduce their arsenals, other nuclear powers are not doing so, and still other countries are continuing efforts to develop such weapons. Eliminating the nuclear threat to the world remains an unfinished task.

7-10   HOW A REACTOR WORKS

From Uranium to Heat to Electricity   For every gram of uranium that undergoes fission in a reactor, 2.6 tons of coal must be burned in an ordinary power plant of the same rating. The energy given off in a nuclear reactor becomes heat, which is removed by a liquid or gas coolant. The hot coolant is then used to boil water, and the resulting steam is fed to a turbine that can power an electric generator, a ship, or a submarine.

The fuel rods of a nuclear reactor are metal tubes filled with pellets of uranium oxide.

In order for a chain reaction to occur at a steady rate, one neutron from each fission must cause another fission to take place. Since each fission in ^{235}U liberates an average of 2.5 neutrons, no more than 1.5 neutrons per fission can be lost on the average. However, natural uranium contains only 0.7 percent of the fissionable isotope ^{235}U. The rest is ^{238}U, an isotope that captures the rapidly moving neutrons emitted during the fission of ^{235}U but usually does not undergo fission afterward. The neutrons absorbed by ^{238}U are therefore wasted, and since 99.3 percent of natural uranium is ^{238}U, too many disappear for a chain reaction to occur in a solid lump of natural uranium.

Fast and Slow Neutrons

There is an ingenious way around this problem. As it happens, ^{238}U tends to pick up only fast neutrons, not slow ones. In addition, slow neutrons are more apt to induce fission in ^{235}U than fast ones. If the fast neutrons from fission are slowed down, then many more will produce further fissions despite the small proportion of ^{235}U present.

To slow down fission neutrons, the uranium fuel in a reactor is mixed with a **moderator,** a substance whose nuclei absorb energy from fast neutrons that collide with them. In general, the more nearly equal in mass colliding particles are, the more energy is transferred. A ball bounces off a wall with little loss of energy, but it can lose all its energy

NUCLEAR POWER PLANTS The fuel for a nuclear reactor consists of uranium oxide pellets sealed in long, thin tubes. Control rods of cadmium or boron, which are good absorbers of slow neutrons, can be slid in and out of the reactor core to adjust the rate of the chain reaction. In the most common type of reactor, water under pressure (to prevent boiling) circulates around the fuel in the core where it acts as both moderator and coolant. As in Fig. 7-16, the pressurized water transfers heat from the chain reaction in the fuel rods to a steam generator. The resulting steam then passes out of the containment shell, which serves as a barrier to protect the outside world from accidents to the reactor, and is piped to a turbine that drives an electric generator.

In a typical plant, the steel reactor vessel is 13.5 m high and

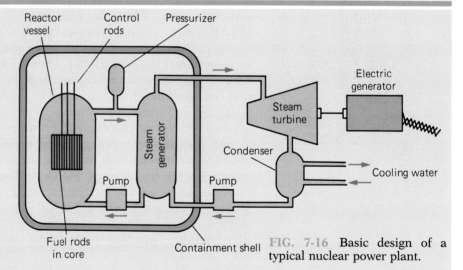

FIG. 7-16 Basic design of a typical nuclear power plant.

4.4 m in diameter and weighs 385 tons. It contains 90 tons of uranium oxide in the form of 50,952 fuel rods, each 3.85 m long and 9.5 mm in diameter. Four steam generators are used, instead of the single one shown in Fig. 7-16, as well as a number

of turbine-generators. The reactor operates at 3400 MW and yields 1100 MW of electric power, enough for the needs of over a million people. The fuel must be replaced every few years as its ^{235}U is used up.

when it strikes another ball. Since hydrogen nuclei are protons with nearly the same mass as neutrons, hydrogen is widely used as a moderator in the form of water, H_2O, each of whose molecules contains two hydrogen atoms along with an oxygen atom.

Unfortunately a neutron striking a proton has a certain tendency to stick to it to form a deuterium nucleus, ^2_1H. As a result, a reactor whose moderator is water cannot use ordinary uranium as fuel but must instead use **enriched** uranium whose ^{235}U content has been increased to about 3 percent. Uranium enriched to about 94 percent of ^{235}U is used in one type of nuclear weapon.

7-11   PLUTONIUM

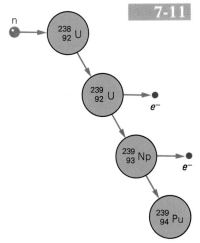

Another Fissionable Material Some nonfissionable nuclides can be changed into fissionable ones by absorbing neutrons. A notable example is ^{238}U, which becomes ^{239}U when it captures a fast neutron. The latter uranium isotope beta-decays soon after its creation into the neptunium isotope $^{239}_{93}Np$. In turn $^{239}_{93}Np$ beta-decays into the plutonium isotope $^{239}_{94}Pu$ (Fig. 7-17). Like ^{235}U, ^{239}Pu undergoes fission when it absorbs a neutron and can support a chain reaction.

Both neptunium and plutonium are called **transuranium elements** because their atomic numbers are greater than the 92 of uranium. Other transuranium elements have been created in the laboratory up to atomic number 110 in high-energy collisions of lighter nuclei. Element 110, the latest to be discovered, was first produced in 1994 by bombarding lead nuclei with nickel nuclei. No transuranium elements of natural origin are found on the earth because all of them decay too fast to have survived even if they had been present when the earth came into being 4.5 billion years ago.

FIG. 7-17 The nonfissionable uranium isotope ^{238}U, which makes up 99.3% of natural uranium, becomes the fissionable plutonium isotope ^{239}Pu by absorbing a neutron and beta-decaying twice. This transformation is the basis of the breeder reactor, which produces many times more nuclear fuel in the form of plutonium than it uses up in the form of ^{235}U.

A certain amount of plutonium is produced in the normal operation of a uranium-fueled reactor, and its fission adds to the energy produced by the reactor. Plutonium separated from the uranium that remains in a used fuel rod can serve as a reactor fuel itself and also, like highly enriched uranium, as the active ingredient in nuclear weapons.

7-12   A NUCLEAR WORLD?

Not Yet, but Perhaps on the Way In 1951 the first electricity from a nuclear plant was generated in Idaho. Today over 500 reactors in 26 countries produce about 200,000 MW of electric power—the equivalent of nearly 10 million barrels of oil per day. France, Belgium, and Taiwan obtain more than half their electricity from reactors, with Finland, Sweden, Switzerland, Bulgaria, and Japan close behind. In the United States, nuclear energy is responsible for about 20 percent of generated electricity, slightly more than the world average. Yet for all the success of nuclear technology, no new nuclear power stations have been planned in this country since 1979. Why not?

In March 1979, failures in its cooling system disabled one of the reactors at Three Mile Island in Pennsylvania, and a certain amount of radioactive material escaped. Although a nuclear reactor cannot explode in the way an atomic bomb does, breakdowns can occur that put large populations at risk. Although a true catastrophe was narrowly avoided, the Three Mile Island incident made it clear that the hazards associated with nuclear energy are real.

After 1979 it was inevitable that greater safety would have to be built into new reactors, adding to their already high cost. In addition, demand for electricity in the United States was not increasing as fast as expected, partly because of efforts toward greater efficiency and partly because of a decline in some of the industries (such as steel, cars, and chemicals) that are heavy users of electricity. As a result of these factors, new reactors made less economic sense than before, which together with widespread public unease led to a halt in the expansion of nuclear energy in the United States.

Chernobyl

Elsewhere the situation was different. Nuclear reactors still seemed the best way to meet the energy needs of many countries without abundant fossil fuel resources. Then, in April 1986, a severe accident destroyed a 1000-MW reactor at Chernobyl in what was then the Soviet Union. Much radioactive material entered the atmosphere and was carried around the world by winds. Tens of thousands of people were evacuated from the reactor vicinity, and radiation levels in many parts of Europe rose well above normal. Hundreds of plant and rescue workers died as a result of exposure to radiation. The widespread contamination with radionuclides, particularly of food supplies, ensures that cancer will raise the death toll to many thousands in the years to come.

As in the United States after Three Mile Island, public anxiety over the safety of nuclear programs grew abroad after Chernobyl. Some countries, for instance Italy, abandoned plans for new reactors and closed down some existing ones. In other countries, for instance France, the logic behind their nuclear programs remained strong enough for them to continue despite Chernobyl. Forty-eight nuclear power plants are currently under construction in various parts of the world.

Nuclear Wastes

Quite apart from the safety of reactors themselves is the issue of what to do with the wastes they produce. Even if old fuel rods are processed to separate out the uranium and plutonium they contain, what is left is still highly radioactive. Although a lot of the activity will be gone in a few months and much of the rest in a few hundred years, some of the radionuclides have half-lives in the millions of years. At present over 15,000 tons of spent nuclear fuel is being stored on a temporary basis in the United States.

Burying nuclear wastes deep underground currently seems to be the best long-term way to dispose of them. The right location is easy to specify but not easy to find: stable geologically with no earthquakes likely, no nearby population centers, a type of rock that does not disintegrate in the presence of heat and radiation but is easy to drill into, with little groundwater that might become contaminated. Studies continue on suitable sites with a view to beginning waste burial early in the next century.

NUCLEAR FUSION

The Energy Source of the Future?
Enormous as the energy produced by fission is, the fusion of small nuclei to form larger ones can give out even more energy per kilogram of starting materials. Nuclear fusion is the energy source of the sun and stars, as discussed in Chap. 17. On the earth, it is possible that fusion will become the ultimate source of energy: safe, almost nonpolluting, and with the oceans supplying limitless fuel.

Three conditions must be met by a successful fusion reactor. The first is a high temperature—100 million °C or more—so that the nuclei are moving fast enough to collide despite the repulsion of their positive electric charges. The second condition is a high concentration of the nuclei to ensure that such collisions are frequent. Third, the reacting nuclei must remain together for a long enough time to give off more energy than the reactor's operation uses. The last two conditions are related, since the more nuclei there are in a given volume, the shorter the minimum confinement time for a net energy output.

The fusion reaction that is the basis of current research involves the combination of a deuterium nucleus and a tritium nucleus to form a helium nucleus:

$$\underset{\text{deuterium}}{{}^{2}_{1}\text{H}} \quad + \quad \underset{\text{tritium}}{{}^{3}_{1}\text{H}} \quad \rightarrow \quad \underset{\text{helium}}{{}^{4}_{2}\text{He}} \quad + \quad \underset{\text{neutron}}{{}^{1}_{0}n} \quad + \quad \underset{\text{energy}}{17.6 \text{ MeV}} \qquad \textit{fusion reaction}$$

Most of the energy given off is carried by the neutron that is emitted. To recover this energy, one proposal is to surround the reactor chamber with lithium to absorb the neutrons. The resulting hot lithium would then act as the heat source for a conventional electric generating system.

About 0.015 percent of the waters of the world is deuterium, which adds up to a total of over 10^{15} tons of ${}^{2}_{1}\text{H}$—no scarcity there. A gallon of seawater has the potential for fusion energy equivalent to the chemical energy in 600 gallons of gasoline. Seawater contains too little tritium for economic recovery, but as it happens, neutrons react with lithium nuclei to yield tritium and helium. Thus, once a fusion reactor is given an initial charge of tritium, it will make enough additional tritium from the surrounding lithium for its further operation.

Two Approaches
The big problem in making fusion energy practical is to achieve the necessary combination of temperature, density, and confinement time. Two approaches are being explored. In one, strong magnetic fields are used to keep the reacting nuclei close together. Four decades of research have led to larger and

ITER The planned International Thermonuclear Experimental Reactor (ITER) represents the final step before practical fusion power is expected to become a reality. Sponsored by the United States and a number of other countries and intended to start operating in 2005, ITER will cost perhaps $7.5 billion and should generate 1 GW from deuterium-tritium reactions. Superconducting magnets will keep the reacting ions in a doughnut-shaped region 12 m in diameter. About 80 percent of the energy released is carried off by the neutrons that are produced, and these neutrons will be absorbed by lithium pellets in tubes that surround the reaction chamber. Circulating water will carry away the resulting heat; this is the heat that could be used in a working reactor to power turbines connected to electric generators.

An experimental fusion reactor at Princeton University. Based on a Soviet design called a tokamak, the reactor uses powerful magnetic fields to confine a hot ionized gas of the hydrogen isotopes deuterium and tritium. When nuclei of these isotopes react, they form a helium nucleus and a neutron in a process that gives off a great deal of energy. If practical fusion reactors can be developed, they might well supply much of the world's future energy needs.

larger experimental magnetic fusion reactors that have brought to light no reasons why eventual success should not be possible. In one encouraging experiment, 6.2 MW of power was produced for 4s.

The other approach to practical fusion energy involves using energetic beams to both heat and compress tiny deuterium-tritium pellets. Laser beams (see Chap. 8) are being tried for this purpose as well as beams of charged particles such as electrons and protons. A series of beams would strike each pellet momentarily from all sides to keep it in place as it is squeezed together. If 10 pellets the size of a grain of sand are ignited every second, the energy output would be enough to provide electric power to a city of 175,000 people.

ELEMENTARY PARTICLES

The electrons, protons, and neutrons of which atoms are composed are **elementary particles** in the sense that they cannot be broken down into anything else. Electrons are simply bits of electrically charged matter, but experiments show that nucleons (protons and neutrons) consist of still smaller particles called **quarks** (Fig. 7-18). The quarks in a nucleon stick together too tightly to permit the nucleon to be split apart, so nucleons are regarded as elementary particles despite their inner structures.

A great many other elementary particles besides electrons and nucleons are known, some composed of quarks and some not. Few of these other particles seem to have anything to do with ordinary matter, although their discovery has helped physicists in their study of how nature works.

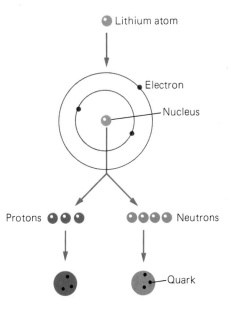

FIG. 7-18 The search for truly elementary particles has led to the discovery of particles within particles. Today all ordinary matter seems to be made up of electrons and quarks. Shown are the various levels of organization of a lithium $_3^7$Li atom.

7-14 ANTIPARTICLES

The Same but Different

Nearly all elementary particles have **antiparticles.** The antiparticle of a given particle has the same mass as the particle and behaves similarly in most respects, but its electric charge is opposite in sign. Thus the positron $e+$ is the antiparticle of the electron $e-$, and the negatively charged **antiproton** $p-$ is the antiparticle of the proton $p+$. Certain uncharged elementary particles, such as the neutron, have antiparticles because they have properties other than charge that are different in the particle and its antiparticle.

BIOGRAPHY

Paul A. M. Dirac (1902–1984) was born in Bristol, England. After studying electrical engineering he switched to physics and obtained his Ph.D. from Cambridge University in 1926. A new and revolutionary theory of the atom called quantum mechanics (see Sec. 8-12) was just then coming into being, and Dirac made a number of major contributions to it. Soon Dirac had joined special relativity to quantum mechanics to give a theory of the electron that predicted the existence of positively charged electrons, or positrons, which were then unknown. At first he thought that protons were the positive antiparticles of electrons despite their much greater mass and the fact that they are not annihilated by electrons. Then, in 1932, the American physicist Carl Anderson found that positrons do exist and have the same mass as electrons. In the same year Dirac became Lucasian Profes-sor of Mathematics at Cambridge, the post Newton had held two and a half centuries earlier. Dirac remained active in physics for the rest of his life, after 1969 in the warmer climate of Florida, but as is often the case in science he will be remembered for the brilliant achievements of his youth.

A positron emission tomography (PET) scan of the brain of a patient with Alzheimer's disease. Different colors correspond to different rates of metabolic activity. In PET, a positron-emitting isotope of an element appropriate to the condition being studied (here an oxygen isotope) is injected and allowed to circulate in a patient's body. When an emitted positron encounters an electron, which it does almost immediately, both are annihilated and a pair of gamma rays is created. Tracing back the directions of the gamma rays gives the location of the annihilation, which is very close to that of the emitting nucleus. In this way, a map of the concentration of the radionuclide can be built up. In a normal brain, metabolic activity produces a similar PET pattern in each side. Here the irregular appearance of the scan indicates that brain tissue has degenerated.

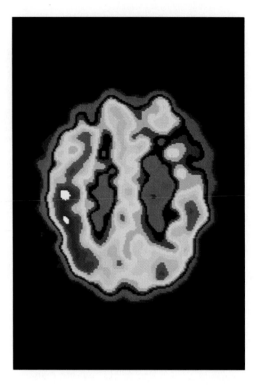

Antiparticles are not easy to find for a very basic reason. When a particle and its antiparticle happen to come together, they destroy each other in a process called **annihilation.** The lost mass reappears as energy in the form of gamma rays when electrons and positrons are annihilated (Fig. 7-19). Unstable particles of various kinds may be produced instead of gamma rays when protons and antiprotons (or neutrons and antineutrons) are annihilated.

The reverse of annihilation can also take place, with energy becoming matter and electric charge being created where none existed before.

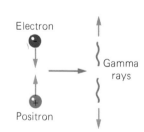

FIG. 7-19   The mutual annihilation of an electron and a positron results in a pair of gamma rays whose total energy is equal to mc^2, where m is the total mass of the electron and positron.

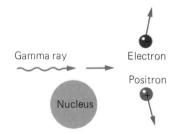

FIG. 7-20   Pair production. (The presence of a nucleus is required in order that both momentum and energy be conserved.) Proton-antiproton and neutron-antineutron pairs can also be produced if the gamma ray has enough energy.

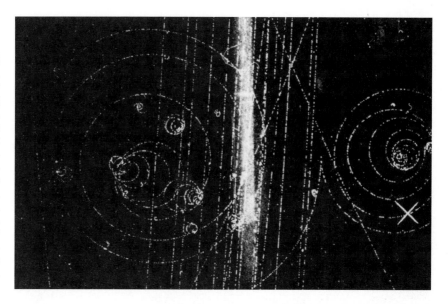

A bubble chamber contains liquid hydrogen under high pressure. When the pressure is suddenly released, tiny bubbles form along tracks of fast electrically charged particles. This photograph shows an electron-positron pair that formed in a bubble chamber in which there was a magnetic field perpendicular to the page. The field deflected the electron and positron into oppositely curved paths. These paths are spirals because the particles lost energy as they moved through the hydrogen in the chamber and so were more and more strongly affected by the field.

In the remarkable process of **pair production,** a particle and its antiparticle materialize when a high-energy gamma ray passes near an atomic nucleus (Fig. 7-20). According to Einstein's formula $E = mc^2$, the energy equivalent of the electron mass is 0.51 MeV. To produce an electron-positron pair therefore requires a gamma ray whose energy is at least 1.02 MeV. If the gamma ray has more energy than 1.02 MeV, the excess goes into the kinetic energies of the electron and positron. The minimum energy needed for a proton-antiproton or neutron-antineutron pair is nearly 2 GeV. The antiparticles formed in pair production exist for only a short time before they meet up with their particle counterparts in ordinary matter and are annihilated.

ANTIMATTER   There seems to be no reason why atoms could not be composed of antiprotons, antineutrons, and positrons. Indeed, hydrogenlike atoms that consist of antiprotons and positrons have already been created in the laboratory. Such **antimatter** ought to behave exactly like ordinary matter. Of course, if antimatter comes in contact with ordinary matter, the same amount of both will disappear in a burst of energy. A postage stamp of antimatter reacting with a similar stamp of matter would release enough energy to send the space shuttle into orbit. But we might imagine that, when the universe was formed, equal quantities of matter and antimatter came into being that became separate galaxies of stars. If this were true, elsewhere in the universe would be stars, planets, and living things made entirely of antimatter.

The idea that the universe consists of both matter and antimatter is an attractive one, but unfortunately it does not seem to be the case. Although galaxies are far apart on the average, now and then two of them collide. A collision between a matter galaxy and an antimatter galaxy would be a violent explosion giving rise to a flood of gamma rays with characteristic energies. No such gamma rays are observed, from which astronomers conclude that there cannot be much, if any, antimatter in the universe. Current theories of elementary particles suggest that matter and antimatter may not be exactly mirror images of each other, with matter having been favored for survival when the universe came into being in the Big Bang described in Chap. 18.

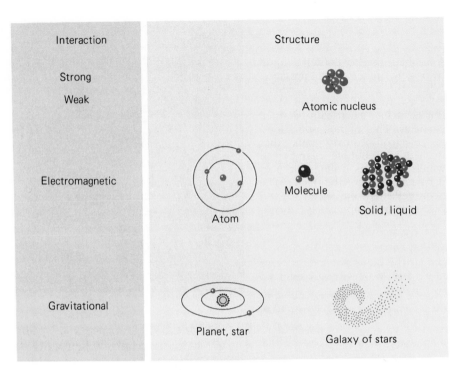

FIG. 7-21 The four fundamental interactions determine how matter comes together to form the characteristic structures of the universe.

7-15   FUNDAMENTAL INTERACTIONS

Only Four Give Rise to All Physical Processes in the Universe

Elementary particles interact with each other in only four ways. These fundamental interactions seem able to account for all the physical processes and structures in the universe on all scales of size from atomic nuclei to galaxies of stars (Fig. 7-21). In order of decreasing strength these interactions are

1. The **strong interaction,** which holds protons and neutrons together to form atomic nuclei despite the mutual repulsion of the protons. The forces produced by this interaction have short ranges, only about 10^{-15} m, which is why nuclei are limited in size. Because the strong interaction is what its name suggests, nuclear binding energies are high. Electrons are not affected by the strong interaction.

2. The **electromagnetic interaction,** which gives rise to electric and magnetic forces between charged particles. This interaction is responsible for the structures of atoms, molecules, liquids, and solids. The force exerted when a bat hits a ball is electromagnetic. Although the electromagnetic interaction is about 100 times weaker than the strong interaction at short distances, electromagnetic forces are unlimited in range and, unlike strong forces, act on electrons.

3. The **weak interaction,** which affects all particles and, by causing beta decay, helps determine the compositions of atomic nuclei. The range of this interaction is even shorter than that of the strong interaction, and 10 trillion times less powerful.

4. The **gravitational interaction,** which is responsible for the attrac-

tive force one mass exerts on another. Because the strong and weak forces are severely limited in range and because matter in bulk is electrically neutral, the gravitational interaction dominates on a large scale. Planets, stars, and galaxies owe their existence to the gravitational interaction. This interaction is nevertheless extremely feeble on a small scale.

Before Newton, it was not clear that the gravity that pulls things down to the earth—which we might call terrestrial gravity—is the same as the gravity that holds the planets in their orbits around the sun. One of Newton's great accomplishments was to show that both terrestrial and astronomical gravity have the same nature. Another notable unification was made by Maxwell when he demonstrated that electric and magnetic forces can both be traced to a single interaction between charged particles.

Unifying the Interactions

What about the four fundamental interactions listed above? Are they all truly fundamental or are any of them, too, related in some way?

Studies made independently by Steven Weinberg and Abdus Salam in the 1960s indicate that the weak and electromagnetic interactions are really different aspects of the same basic phenomenon. Supported by experiment, this conclusion is now generally accepted (Fig. 7-22). A more recent development is a proposed link between the electroweak and strong interactions. Although such a grand unified theory is still far from its final form, it seems to be on the right track. One of the merits of the theory is that it can explain why the proton and the electron,

FIG. 7-22 One of the goals of physics is a single theoretical picture that unites all the ways in which particles of matter interact with each other. Much progress has been made, but the task is not finished.

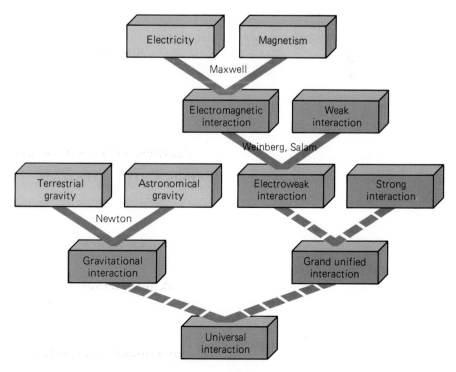

Timothy C. Miller (Bartol Research Institute, University of Delaware). I work on two astrophysics experiments located at the geographic South Pole.

A PHYSICIST AT WORK

The South Pole is a tremendously challenging place to work, and the environment is possibly the harshest anywhere on earth. With an average temperature of −55°F, polar air is drier than any desert, yet most of the world's fresh water is in Antarctica, frozen in the two-mile-thick ice sheet. The sun is up 24 hours per day in the summer but leaves the Pole in continuous darkness for six months each winter. There are no mountains within a hundred miles, so completely flat snow stretches all the way to the horizon.

A permanent scientific research station here is staffed year-round. There are up to 140 people at the station in the summer and about 25 who stay through the long, dark winter. Scientists fly to the Pole from all over the world to conduct research that cannot be done anywhere else, in fields such as glaciology, climatology, geology, atmospheric studies, and astrophysics. The high altitude, clean air, and long periods of darkness make the Pole uniquely suited for astronomical observations. During the summer U.S. Navy cargo planes fly in people and supplies a few times per day, weather permitting. With the nearest large Antarctic base over 800 miles away and the trip from the U.S. often taking several days, the Pole is one of the most isolated places on earth.

One project I work on at the Pole is called AMANDA, the Antarctic Muon and Neutrino Detector Array. One of its purposes is to detect particles called neutrinos that may originate near especially violent astronomical objects like black holes and quasars. Neutrinos have little or no mass, have no charge, and interact very feebly with ordinary matter, making them very difficult to detect (a typical neutrino will pass all the way through the earth without interacting). This also makes them very interesting astronomically because they travel to us from places that we cannot see with light since intervening matter blocks our view.

AMANDA consists of a huge array of very sensitive light detectors, buried a mile deep in the Antarctic ice and facing downward to look for neutrinos coming from the other side of the earth. Occasionally a neutrino will cause a flash of light that AMANDA will see through the very clear Antarctic ice. Since only neutrinos can pass unhindered through the earth, we essentially use the entire earth as a filter to eliminate all other particles.

The second experiment I work on is the South Pole Air Shower Experiment (SPASE). SPASE sits on the surface of the ice and looks for cosmic rays that strike the atmosphere. Cosmic rays consist of protons, electrons, and heavier nuclei that are created in the cosmos (see Sec. 18-5) and are accelerated to energies of up to 10 million times higher than particle accelerators on earth

which are very different kinds of particle (as discussed in the next section), have electric charges of exactly the same size.

What about gravitation? The final step in understanding how nature works is a single theory that ties together all the particles and interactions that are known—a "theory of everything." There are hints that this supreme goal is not beyond reach.

7-16 LEPTONS AND HADRONS

Ultimate Matter

All elementary particles fall into two broad categories that depend on their response to the strong interaction. **Leptons** (Greek for "light" or "swift") are not affected by this interaction and seem to be point particles with no size or internal structure. The electron is a lepton. **Hadrons** (Greek for "heavy" or "strong") are subject to the strong interaction and have definite sizes—they are about 10^{-15} m across—and internal structures. The proton and neutron are hadrons.

A very interesting lepton is the **neutrino,** which has no charge and little or no mass. The neutrino is associated with the weak interaction.

reach. At present we do not know how and where these particles are created. These are among the questions we hope to investigate with SPASE and AMANDA.

A typical research day at the South Pole is very different from a day doing physics homework in college. The material I learned in classes gives me the background to design these experiments and to understand the scientific goals, but making an experiment work at the Pole requires a lot of skills that I never learned in classes. Due to the limited facilities at the Pole, most of us sleep in heated tents on the snow. Before going outside to start the day, I have to put on several layers of cold weather gear, including parka, goggles, gloves,

and specially designed insulated "bunny boots." I then walk about half a mile out to the experiment site, a small building elevated on 20-foot-high stilts to protect it from the drifting snow, which would otherwise cover it within a few years. The day's activities can consist of many things, including installing or fixing equipment outside, debugging electronics indoors, writing software, or sometimes just dragging heavy cables or equipment from one place to another outside in the snow. Evenings are filled with conferences with colleagues back in the U.S. via e-mail. Most communications with the outside world occur via the Internet, since phone calls are limited and there is no TV reception.

With all of its hardships, field work at the Pole is exciting and rewarding. I think that the combination of the location and the possibility of scientific discoveries makes my work very special. When I was young, I had a keen interest in astronomy, science fiction, and stories of adventure. I often imagined myself exploring space, the ocean, or the far reaches of the earth. Carrying out research at the South Pole has allowed me to keep many of my childhood dreams as an adult, incorporating them into an exciting and challenging profession. Working at the South Pole for two months each year is hard and uncomfortable, but I do get to spend my life searching for new things about the universe, traveling all over the world, and visiting places where few people have ever been or even imagined being.

Electrons and positrons are given energies of 50 GeV in the two-mile-long Stanford Linear Accelerator in California. When the particles are brought together in head-on collisions, elementary particles of various kinds are produced.

Whenever a nucleus undergoes beta decay, a neutrino as well as an electron (or positron) is emitted. A neutrino can pass through vast amounts of matter—over 100 *light-years* of solid iron on the average—before interacting. (A light-year is the distance light travels in empty space in a year.) A vast number of neutrinos are produced in the sun in the course of the nuclear reactions that occur within it, and these neutrinos carry into space 6 to 8 percent of all the energy the sun generates. The energy the neutrinos from the sun and other stars carry is apparently lost forever in the sense that it cannot be changed into any other form. Neutrinos outnumber protons in the universe by about a billion to one.

Besides the proton and the neutron, the hadron family includes several hundred particles with extremely short lifetimes, less than a billionth of a second for some. These particles seem to play no role in the behavior of ordinary matter. They decay in various ways, often in a series of steps, and usually end up as protons, neutrons, or electrons; a few become gamma rays.

Quarks

The discovery that hadrons have internal structures was made with the help of experiments in which fast electrons were scattered by collisions with protons and neutrons. (We recall that the internal structure of the atom was revealed by the similar Rutherford experiment that used alpha particles as probes.) The particles that make up hadrons are the **quarks** mentioned earlier.

Only six kinds of quark are needed to account for all known hadrons. Those hadrons that are lighter than the proton consist of a quark and an antiquark. The proton, neutron, and heavier hadrons consist of three quarks. This is a welcome simplification, but quarks turn out to have two unprecedented properties. The first is that, unlike any other particle known, their electric charge is less than $\pm e$. Some quarks have a charge of $\pm\frac{1}{3}e$; others have a charge of $\pm\frac{2}{3}e$.

The second unusual aspect of quarks is that they do not seem able to exist outside of hadrons. No quark has ever been found by itself, even in experiments that ought to have been able to set them free. Thus there is no direct way to confirm that quarks have fractional charges, or even that they really exist.

Nevertheless there is a great deal of indirect evidence strongly in favor of quarks. For instance, every known hadron matches up with a particular arrangement of quarks, and predictions of hitherto unknown hadrons made on the basis of the quark model have turned out correct. Furthermore, a theory of the strong interaction based on quarks has been quite successful. This is the theory mentioned in the previous section that apparently can be linked to the theory of the electromagnetic-weak interaction to make a unified picture that accounts for all aspects of the behavior of matter except—thus far—gravitation.

The two quarks that make up the proton and neutron are called u (charge $+\frac{2}{3}e$) and d (charge $-\frac{1}{3}e$). As shown in Fig. 7-23, a proton consists of one d quark and two u quarks, and a neutron consists of one u quark and two d quarks. Thus all the properties of ordinary matter can be understood on the basis of just two leptons, the electron and the neutrino, and two quarks, u and d. Considering how diverse these properties are, this is an astonishing achievement. The other leptons and quarks are connected only with unstable particles created in high-energy collisions and seem to have nothing to do with ordinary matter.

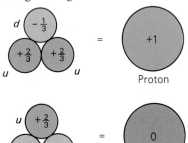

FIG. 7-23 Quark models of the proton and neutron. Electric charges are given in units of e.

IMPORTANT TERMS AND IDEAS

An **element** is a substance all of whose atoms have the same number of protons in their nuclei. This number is the **atomic number** of the element and equals the number of electrons that surround each nucleus of the element's atoms. The **isotopes** of an element have different numbers of neutrons in their nuclei. A **nucleon** is a neutron or proton; the **mass number** of a nucleus is the number of nucleons it contains. A **nuclide** is an atom whose nucleus has particular atomic and mass numbers.

In **radioactive decay,** certain atomic nuclei spontaneously emit **alpha particles** (helium nuclei), **beta particles** (electrons), or **gamma rays** (high-frequency electromagnetic waves). A **positron** is a positively charged electron emitted in some beta decays.

The **half-life** of a radionuclide is the time needed for half of an original sample to decay.

The mass of every nucleus is slightly less than the total mass of the same number of free neutrons and protons. The **binding energy** of a nucleus is the energy equivalent of the missing mass and must be supplied to the nucleus to break it up. Nuclei of intermediate size have the highest **binding energies per nucleon.** Hence the **fusion** of light nuclei to form heavier ones and the **fission** of heavy nuclei into lighter ones are both processes that liberate energy.

An **elementary particle** cannot be separated into other particles. The **antiparticle** of an elementary particle has the same mass and general behavior, but it has a charge of opposite sign and differs in certain other respects. A particle and its antiparticle can **annihilate** each other, with their masses turning entirely into energy. In the opposite process of **pair production,** a particle-antiparticle pair materializes from energy.

The four **fundamental interactions** are, in order of decreasing strength, the strong, electromagnetic, weak, and gravitational.

A **lepton** is an elementary particle that is not affected by the strong interaction and has no internal structure; the electron is a lepton. A **hadron** is an elementary particle that is affected by the strong interaction and is composed of **quarks,** particles with electric charges of $\pm\frac{1}{3}e$ or $\pm\frac{2}{3}e$ that have not been found outside hadrons as yet. Protons and neutrons are hadrons.

EXERCISES: MULTIPLE CHOICE

1. The basic idea of the Rutherford atomic model is that the positive charge in an atom is
   a. spread uniformly throughout its volume
   b. concentrated at its center
   c. readily deflected by an incoming alpha particle
   d. the same for all atoms

2. Nearly all the volume occupied by matter consists of
   a. electrons   c. neutrons
   b. protons   d. nothing

3. The atomic number of an element is the number of
   a. protons in its nucleus
   b. neutrons in its nucleus
   c. electrons in its nucleus
   d. protons and neutrons in its nucleus

4. The nuclei of the isotopes of an element all contain the same number of
   a. neutrons   c. nucleons
   b. protons   d. electrons

5. The chemical behavior of an atom is determined by its
   a. atomic number   c. binding energy
   b. mass number   d. number of isotopes

6. An alpha particle consists of
   a. two protons
   b. two protons and two electrons
   c. two protons and two neutrons
   d. four protons

7. An electron is emitted by an atomic nucleus in the process of
   a. alpha decay   c. gamma decay
   b. beta decay   d. nuclear fission

8. Gamma rays have the same basic nature as
   a. alpha particles   c. x-rays
   b. beta particles   d. sound waves

9. Radioactive materials do not emit
   a. electrons   c. alpha particles
   b. protons   d. gamma rays

10. Which of these types of radiation has the least ability to penetrate matter?
    a. alpha particles   c. gamma rays
    b. beta particles   d. x-rays

11. Which of these types of radiation has the greatest ability to penetrate matter?
    a. alpha particles   c. gamma rays
    b. beta particles   d. x-rays

12. When a nucleus undergoes radioactive decay, the number of nucleons it contains afterward is
    a. always less than the original number
    b. always more than the original number
    c. never less than the original number
    d. never more than the original number

13. The largest amount of radiation received by an average person in the United States comes from
    a. medical x-rays
    b. nuclear reactors
    c. fallout from past weapons tests
    d. natural sources

14. The half-life of a radionuclide is
    a. half the time needed for a sample to decay entirely
    b. half the time a sample can be kept before it begins to decay
    c. the time needed for half a sample to decay
    d. the time needed for the remainder of a sample to decay after half of it has already decayed

15. As a sample of a radionuclide decays, its half-life
    a. decreases
    b. remains the same
    c. increases
    d. any of the above, depending upon the nuclide

16. In a stable nucleus other than $^{1}_{1}\text{H}$ the number of neutrons is always
    a. less than the number of protons
    b. less than or equal to the number of protons
    c. equal to or more than the number of protons
    d. more than the number of protons

17. The electronvolt is a unit of
    a. charge   c. energy
    b. potential difference d. momentum

18. Relative to the sum of the masses of its constituent particles, the mass of an atom is
    a. greater
    b. the same
    c. smaller
    d. any of the above, depending on the element

19. The binding energy per nucleon is
    a. the same for all nuclei
    b. greater for very small nuclei
    c. greatest for nuclei of intermediate size
    d. greatest for very large nuclei

20. The splitting of an atomic nucleus, such as that of ^{235}U, into two or more fragments is called
    a. fusion   c. a chain reaction
    b. fission   d. beta decay

21. In a chain reaction
    a. protons and neutrons join to form atomic nuclei
    b. light nuclei join to form heavy ones
    c. neutrons emitted during the fission of heavy nuclei induce fusions in other nuclei
    d. uranium is burned in a type of furnace called a reactor

22. Enriched uranium is a better fuel for nuclear reactors than natural uranium because enriched uranium has a greater proportion of
    a. slow neutrons   c. plutonium
    b. deuterium   d. ^{235}U

23. Fusion reactions on the earth are likely to use as fuel
    a. ordinary hydrogen   c. plutonium
    b. deuterium   d. uranium

24. Of the following particles, the one that is not an elementary particle is the
    a. alpha particle   c. neutron
    b. beta particle   d. neutrino

25. An example of a particle-antiparticle pair is the
    a. proton and positron
    b. proton and neutron
    c. neutron and neutrino
    d. electron and positron

26. Atomic nuclei are stable despite the mutual repulsion of the protons they contain because of the action of the
    a. gravitational interaction
    b. electromagnetic interaction
    c. weak interaction
    d. strong interaction

27. The weakest of the four fundamental interactions is the
    a. gravitational interaction
    b. electromagnetic interaction
    c. strong interaction
    d. weak interaction

28. The mass of the neutrino is
    a. equal to that of the neutron
    b. equal to that of the electron
    c. equal to that of a quark
    d. very small or zero

29. Quarks are particles that
    a. have no mass
    b. have charges of less than e
    c. decay into protons
    d. decay into neutrinos

30. A particle that is believed to consist of quarks is the
    a. electron   c. neutron
    b. positron   d. neutrino

31. The number of protons in a nucleus of the boron isotope $^{11}_{5}$B is
    a. 5   c. 11
    b. 6   d. 16

32. The number of neutrons in a nucleus of the potassium nucleus $^{40}_{19}$K is
    a. 19   c. 40
    b. 21   d. 59

33. When the bromine isotope $^{80}_{35}$Br decays into the krypton isotope $^{80}_{36}$Kr, it emits
    a. a gamma ray   c. a positron
    b. an electron   d. an alpha particle

34. The product of the alpha decay of the bismuth isotope $^{214}_{83}$Bi is
    a. $^{210}_{79}$Au   c. $^{210}_{83}$Bi
    b. $^{210}_{81}$Tl   d. $^{218}_{85}$At

35. After 2 h has elapsed, one-sixteenth of the original quantity of a certain radioactive substance remains undecayed. The half-life of this substance is
    a. 15 min   c. 45 min
    b. 30 min   d. 60 min

36. The half-life of tritium is 12.5 years. If we start out with 1 g of tritium, after 25 years there will be
    a. no tritium left
    b. $\frac{1}{4}$ g of tritium left
    c. $\frac{1}{2}$ g of tritium left
    d. a total of 4 g of tritium

1. What is the difference between the ways in which the mass and the charge of an atom are distributed?

2. Alpha particle tracks through gases and thin metal foils are nearly always straight lines. To what conclusion regarding atomic structure does this observation lead?

3. The following statements were thought to be correct in the nineteenth century. Which of them are now known to be incorrect? For those that are incorrect, indicate why the statement is wrong and modify it to be in accordance with modern views. (a) Energy can be neither created nor destroyed. (b) The acceleration of an object is proportional to the force applied to it and inversely proportional to its mass. (c) Atoms are indivisible and indestructible. (d) All atoms of a particular element are identical.

4. In what ways are the isotopes of an element similar to one another? In what ways are they different?

5. Find the number of neutrons and protons in each of the following nuclei: $^{6}_{3}$Li; $^{13}_{6}$C; $^{31}_{15}$P; $^{94}_{40}$Zr.

6. Find the number of neutrons and protons in each of the following nuclei: $^{10}_{5}$B; $^{22}_{10}$Ne; $^{36}_{16}$S; $^{88}_{38}$Sr.

7. What limits the size of a nucleus?

8. How does the number of neutrons in a stable nucleus compare with the number of protons? Give the physical reason that underlies your answer.

9. (a) What is an alpha particle? A beta particle? A gamma ray? (b) How do they compare in general in ability to penetrate matter?

10. Radium undergoes spontaneous decay into helium and radon. Why is radium regarded as an element rather than as a chemical compound of helium and radon?

11. What happens to the atomic number and mass number of a nucleus when it emits an alpha particle?

12. What happens to the atomic number and mass number of a nucleus when it emits (a) an electron? (b) a positron? (c) a gamma ray?

13. (a) Under what circumstances does a nucleus emit an electron? A positron? (b) The oxygen nuclei $^{14}_{8}$O and $^{19}_{8}$O both undergo beta decay to become stable nuclei. Which would you expect to emit a positron and which an electron?

14. When the helium isotope $^{6}_{2}$He decays into the lithium isotope $^{6}_{3}$Li, is an electron or a positron emitted?

15. What happens to the half-life of a radionuclide as it decays?

16. If the half-life of a radionuclide is 1 month, is a sample of it completely decayed after 2 months?

17. Why is the $^{56}_{26}$Fe nucleus the most stable (that is, the most difficult to break apart) nucleus?

18. Suppose the strong interaction did not exist, so there were no nuclear binding energies. If the early universe contained protons, neutrons, and electrons, what kind or kinds of matter would eventually fill the universe?

19. What property of atomic nuclei makes it possible for nuclear fission and fusion to give off energy?

20. What are the differences and similarities between fusion and fission?

21. What is the function of the moderator in a uranium-fueled nuclear reactor?

22. What is the limitation on the fuel that can be used in a reactor whose moderator is ordinary water?

23. What fuel other than uranium can be used in a nuclear reactor?

24. Does a gamma ray require more or less energy to materialize into a neutron-antineutron pair than to materialize into a proton-antiproton pair?

25. Discuss the similarities and differences between the neutron and the neutrino.

26. What enables neutrinos to travel immense distances through matter without interacting?

27. Leptons and hadrons are the two classes of basic particle. How do they differ?

28. Which constituents of an atom consist of quarks and which do not?

29. No particle of fractional charge has yet been observed. If none is found in the future either, does this necessarily mean that the quark hypothesis is wrong?

30. The gravitational interaction alone governs the motions of the planets around the sun. Why are the other fundamental interactions not significant in planetary motion?

PROBLEMS

1. The polonium isotope $^{210}_{84}$Po undergoes alpha decay to become an isotope of lead. Find the atomic number and mass number of this isotope.

2. The carbon nucleus $^{11}_{6}$C decays radioactively by the emission of a positive electron. Find the atomic number, mass number, and chemical name of the resulting nucleus.

3. The thorium nucleus $^{223}_{90}$Th undergoes two successive negative beta decays. Find the atomic number, mass number, and chemical name of the resulting nucleus.

4. A $^{80}_{35}$Br nucleus can decay by emitting an electron or a positron and also by capturing an electron. What is the final nucleus in each case?

5. The uranium isotope $^{235}_{92}$U decays into a lead isotope by emitting seven alpha particles and four electrons. What is the symbol of the lead isotope?

6. If 1 kg of radium (half-life = 1600 years) is sealed into a container, how much of it will remain as radium after 1600 years? after 4800 years? If the container is opened after a period of time, what gases would you expect to find inside it?

7. After 10 years, 75 g of an original sample of 100 g of a certain radionuclide has decayed. What is the half-life of the nuclide?

8. One-sixteenth of a sample of $^{24}_{11}$Na remains undecayed after 60 h. Find the half-life of this radioisotope.

9. Find the kinetic energy (in eV) of an electron whose speed is 10^6 m/s.

10. Find the kinetic energy (in eV) of a potassium atom of mass 6.5×10^{-26} kg whose speed is 10^6 m/s.

11. Find the speed of an electron whose kinetic energy is 26 eV.

12. Find the speed of a neutron whose kinetic energy is 50 eV.

13. Atomic mass always refers to the mass of a neutral atom, not the mass of its bare nucleus. With this definition in mind, determine by how much the mass of a parent atom changes when its nucleus emits (a) an electron and (b) a positron. Ignore the kinetic energy of the emitted particle.

14. The binding energy of $^{20}_{10}$Ne is 160.6 MeV. Find its atomic mass.

15. The mass of $^{4}_{2}$He is 4.0026 u. Find its binding energy and binding energy per nucleon.

16. The mass of a $^{7}_{3}$Li nucleus is 0.042 u less than the sum of the masses of three protons and four neutrons. What is the binding energy (in MeV) per nucleon in $^{7}_{3}$Li?

17. (a) How much mass is lost per day by a nuclear reactor operated at a 1.0-GW power level? (b) If each fission releases 200 MeV, how many fissions occur per second to give this power level?

18. Old stars obtain part of their energy by the fusion of three alpha particles to form a $^{12}_{6}$C nucleus, whose mass is 12.0000 u. How much energy is given off in each such reaction?

19. The neutron decays in free space into a proton and an electron after an average lifetime of 15 min. What must be the minimum binding energy contributed by a neutron to a nucleus in order that the neutron not decay inside the nucleus? How does this figure compare with the observed binding energies per nucleon in stable nuclei?

20. Would you expect the gravitational attractive force between two protons in a nucleus to counterbalance their electrical repulsion? Calculate the ratio between the electric and gravitational forces acting between two protons. Does this ratio depend upon how far apart the protons are?

ANSWERS TO MULTIPLE CHOICE

1. b	**10.** a	**19.** c	**28.** d
2. d	**11.** c	**20.** b	**29.** b
3. a	**12.** d	**21.** c	**30.** c
4. b	**13.** d	**22.** d	**31.** a
5. a	**14.** c	**23.** b	**32.** b
6. c	**15.** b	**24.** a	**33.** b
7. b	**16.** c	**25.** d	**34.** b
8. c	**17.** c	**26.** d	**35.** b
9. b	**18.** c	**27.** a	**36.** b

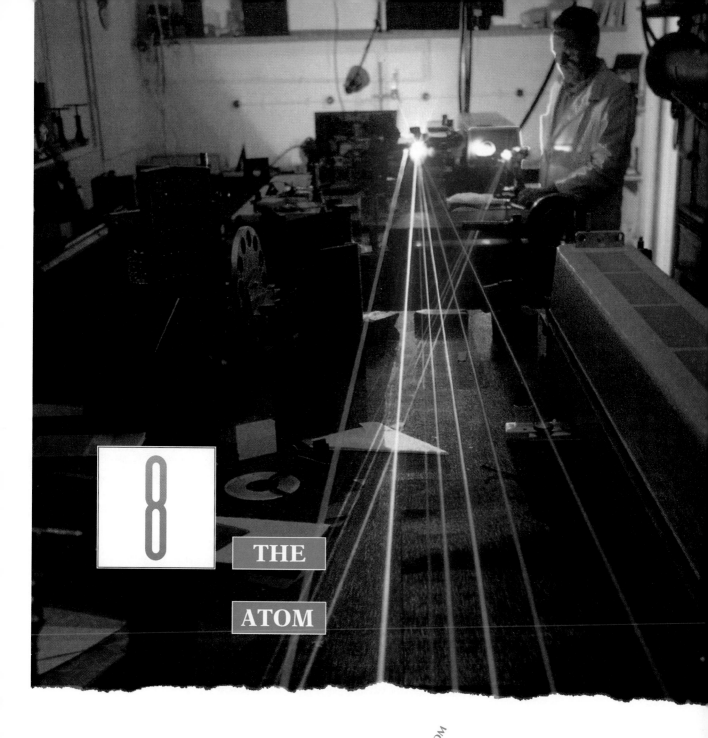

8

THE

ATOM

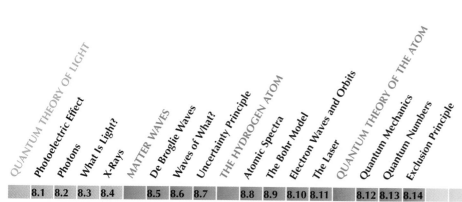

*E*very atom consists of a tiny, positively charged nucleus with negatively charged electrons some distance away. What keeps the electrons out there?

By analogy with the planets of the solar system, we might suppose that atomic electrons avoid being sucked into the nucleus by circling around it at just the right speed. This is not a bad idea, but it raises a serious problem. According to Maxwell's theory (Chap. 6), a circling electron should lose energy all the time by giving off electromagnetic waves. Thus the electron's orbit should become smaller and smaller, and soon it should spiral into the nucleus. However, atomic electrons do not behave like this. Under ordinary conditions atoms emit no radiation, and needless to say, they never collapse.

Whenever they have been tested outside the atomic domain, the laws of motion and of electromagnetism have always agreed with experiment—yet atoms are stable. In this chapter we shall see how the strange and radical concepts of the quantum theory of light and the wave theory of moving particles are needed to understand the world of the atom.

QUANTUM THEORY OF LIGHT

The concepts of "particle" and "wave" are clear enough to everybody. We regard a stone as a particle and the ripples in a lake as waves. A stone thrown into a lake and the ripples that spread out from where it lands seem to have in common only that both carry energy from one place to another. **Classical physics,** which refers to the physics covered in Chaps. 1 through 6, treats particles and waves as separate aspects of the reality we find in everyday life.

But the physical reality around us arises from the small-scale world of atoms and molecules, electrons and nuclei. In this world there are neither particles nor waves in our sense of these terms.

We think of electrons as particles because they have charge and mass and behave according to the laws of particle mechanics in such familiar devices as television picture tubes. However, there is plenty of evidence that makes sense only if a moving electron is a type of wave. We think of em waves as waves because they can exhibit such characteristic wave behavior as diffraction and interference. However, em waves also behave as though they consist of streams of particles. The wave-particle duality is central to an understanding of **modern physics,** which is the physics of the atomic world.

8-1 PHOTOELECTRIC EFFECT

Electrons Can Be Set Free from Atoms by Light, but How Does This Happen?

A century ago experiments showed that electrons are given off by a metal surface when light is directed onto it (Fig. 8-1). For most metals ultraviolet light is needed for this **photoelectric effect** to occur, but some metals, such as potassium and cesium, and certain other substances as well, also respond to visible light. The photoelectric cell that measures light intensity in a camera, the solar cell that produces electric current when sunlight falls on it, and the television

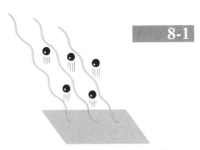

FIG. 8-1 In the photoelectric effect, electrons are emitted from a metal surface when a light beam is directed on it.

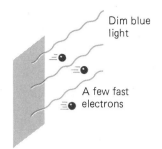

Dim blue light

A few fast electrons

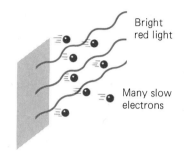

Bright red light

Many slow electrons

FIG. 8-2 The higher the frequency of the light, the more KE the photoelectrons have. The brighter the light, the more photoelectrons are emitted. Blue light has a higher frequency than red light.

camera tube that converts the image of a scene into an electric signal are all based upon the photoelectric effect.

Since light is electromagnetic in nature and carries energy, there seems to be nothing unusual about the photoelectric effect—it should be like water waves dislodging pebbles from a beach. But three experimental findings show that no such simple explanation is possible.

1. The electrons are always emitted at once, even when a faint light is used. However, because the energy in an em wave is spread out across the wave, a certain period of time should be needed for an individual electron to gather enough energy to leave the metal. Several months ought to be needed for a really weak light beam.

2. A bright light causes more electrons to be emitted than a faint light, but the average kinetic energy of the electrons is the same. The electromagnetic theory of light, on the contrary, predicts that the stronger the light, the greater the KE of the electrons.

3. The higher the frequency of the light, the more KE the electrons have. Blue light yields faster electrons than red light (Fig. 8-2). According to the electromagnetic theory of light, the frequency should not matter.

Until the discovery of the photoelectric effect, the electromagnetic theory of light had been completely successful in explaining the behavior of light. But no amount of ingenuity could bring experiment and theory together in this case. The result was the creation of the entirely new **quantum theory of light** in 1905 by Albert Einstein. The same year saw the birth of his equally revolutionary theory of relativity. All of modern physics has its roots in these two theories.

8-2   PHOTONS

Particles of Light

Einstein proposed that light consists of tiny bursts of energy called **photons.** He began with a hypothesis suggested 5 years earlier by the German physicist Max Planck to account for the spectrum of the light given off by hot objects. As we recall from Fig. 4-4, the color of this light varies with temperature, going from red to yellow to white as the object becomes hotter and hotter.

In order to explain this variation of color with temperature, Planck found it necessary to assume that hot objects contribute energy in separate units, or **quanta,** to the light they give off. The higher the frequency of the light, the more the energy per quantum. All the quanta associated with a particular frequency f of light have the same energy

$$E = hf \qquad quantum\ energy$$

Quantum energy = (Planck's constant)(frequency)

In this formula the quantity h, today known as **Planck's constant,** has the value

Planck's constant = $h = 6.63 \times 10^{-34}$ joule·second

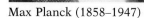

Max Planck (1858–1947)

All light-sensitive detectors, including the eye and the one used in this video camera, are based on the absorption of energy from photons of light by electrons in the atoms the light falls on.

Planck was not happy about this assumption, which made no sense in terms of the physical theories known at that time. He took the position that, although energy apparently had to be given to the light emitted by a glowing object in small bursts, the light nevertheless traveled with its energy spread out in waves exactly as everybody thought.

Einstein's Hypothesis

Einstein, however, felt that, if light is emitted in little packets, it should also travel through space and finally be absorbed in the same little packets. His idea fit the experiments on the photoelectric effect perfectly. He supposed that some minimum energy w is needed to pull an electron away from a metal surface. If the frequency of the light is too low—so that E, the quantum energy, is less than w—no electrons can come out. When E is greater than w, a photon of light striking an electron can give the electron enough energy for it to leave the metal with a certain amount of kinetic energy. Einstein's formula for the process is very simple:

$$hf = KE + w \qquad \textit{photoelectric effect}$$

where hf = energy of a photon of light whose frequency is f
  KE = kinetic energy of the emitted electron
  w = energy needed to pull the electron from the metal

Although the photon has no mass and always moves with the speed of light, it has most of the other properties of particles—it is localized in a small region of space, it has energy and momentum, and it interacts with other particles in more or less the same way as a billiard ball interacts with other billiard balls.

8-3 WHAT IS LIGHT?

Both Wave and Particle

The idea that light travels as a series of little packets of energy is directly opposed to the wave theory of light. And the latter, which provides the only way to explain such optical effects as diffraction and interference, is one of the best established of physical theories. Planck's suggestion that a hot object gives energy to light in separate quanta led to no more than raised eyebrows among physicists in 1900 since it did not apparently conflict with the picture of light as a wave. Einstein's suggestion in 1905 that light travels through space in the form of distinct photons, on the other hand, astonished most of his colleagues.

According to wave theory, light waves spread out from a source in the way ripples spread out on the surface of a lake when a stone falls into it. The energy carried by the light in this picture is spread out through the wave pattern (Fig. 8-3). According to the quantum theory, however, light travels from a source as a series of tiny bursts of energy, each burst so small that it can be taken up by a single electron. Curiously, the quantum theory of light, which treats light as a particle phenomenon, incorporates the light frequency f, a wave concept.

Which theory are we to believe? A great many scientific ideas have had to be changed or discarded when they were found to disagree with

FIG. 8-3 (a) The wave theory of light accounts for the diffraction of light into the shadow region when it passes through a narrow slit. (b) The quantum theory of light accounts for the photoelectric effect. Neither theory by itself can account for all aspects of the behavior of light. The two theories therefore complement each other.

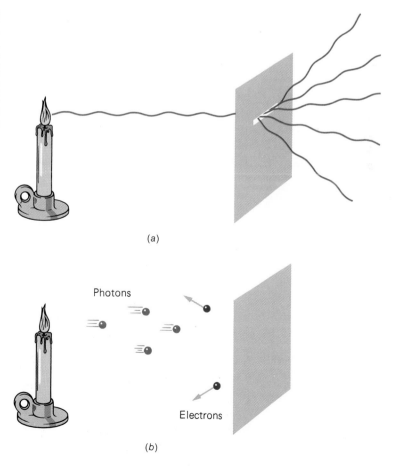

(a)

Photons

Electrons

(b)

experiment. Here, for the first time, two entirely different theories are needed to account for a single physical phenomenon.

In any particular event light exhibits *either* a wave nature or a particle nature, never both at the same time. This is an important point. The light beam that shows diffraction in passing the edge of an obstacle can also cause photoelectrons to be emitted from a metal surface, but these processes occur independently:

> **The wave theory of light and the quantum theory of light complement each other.**

Electromagnetic waves provide the only explanation for some experiments involving light, and photons provide the only explanation for all other experiments involving light. Light incorporates both wave and particle characters even though there is nothing in everyday life like that to help us form a mental picture of it.

8-4 X-RAYS

High-Energy Photons The photoelectric effect shows that photons of light can give energy to electrons. Is the inverse process also possible? That is, can part or all of the KE of an electron be turned into a photon? As it happens, the inverse photoelectric effect not only does occur but also had been discovered (though not understood) before the work of Planck and Einstein.

In 1895, in his laboratory in Germany, Wilhelm Roentgen accidentally found that a screen coated with a fluorescent salt glowed every time he switched on a nearby cathode-ray tube. (A cathode-ray tube is a tube with the air pumped out in which electrons are accelerated by an electric field. A TV picture tube is a type of cathode-ray tube.) Roentgen knew that the electrons themselves could not get through the glass walls of his tube, but it was clear that some sort of invisible radiation was falling on the screen.

The radiation was very penetrating. Thick pieces of wood, glass, and even metal could be placed between tube and screen, and still the screen glowed. Soon Roentgen found that his mysterious rays would penetrate

Wilhelm Roentgen (1845–1923)

FIG. 8-4 An x-ray tube. High-frequency electromagnetic waves called x-rays are emitted by a metal target when it is struck by fast-moving electrons. The cathode (negative electrode) is heated by a filament and emits electrons that are accelerated by an electric field between it and the positively charged target.

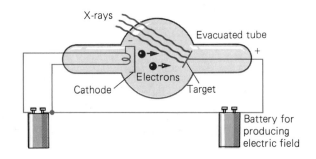

flesh and produce shadows of the bones inside. He gave them the name **x-rays** after the algebraic symbol for an unknown quantity. Roentgen refused to benefit financially from his work and died in poverty in the German inflation that followed World War I.

X-rays are given off whenever fast electrons are stopped suddenly. Figure 8-4 shows a cathode-ray tube especially designed to produce x-rays. In a television picture tube, an electron beam strikes the inside of the glass screen, which emits x-rays as a result. To prevent harm to viewers, the glass of the screen contains barium, a heavy metal that absorbs x-rays.

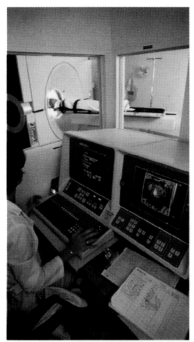

In a CAT scanner, a series of x-ray exposures of a patient taken from different directions are combined by a computer to give cross-sectional images of the part of the body being examined. In effect, the tissue is sliced up by the computer on the basis of the x-ray exposures, and any desired slice can be displayed. This technique enables an abnormality to be detected and its exact location established, which might be impossible to do from an ordinary x-ray picture.

What Are X-Rays?

After many attempts had been made to determine their nature, in 1912 Max von Laue was able to show by means of an interference experiment that they are electromagnetic waves of extremely high frequency. X-ray frequencies are much higher than those of ultraviolet light, but somewhat lower than those of the gamma rays produced by radioactive atomic nuclei.

The early workers with x-rays noted that increasing the voltage applied to the tube, which means faster electrons, gave rise to x-rays of greater penetrating power. The greater the penetrating ability, the higher the x-ray frequency turned out to be. Hence high-energy electrons produce high-frequency x-rays. The more electrons in the beam, the more x-rays were produced, but their energy depended only on the electron energy.

The quantum theory of light is in complete accord with these observations. Instead of photon energy being transformed into electron KE, electron KE is transformed into photon energy. The energy of an x-ray photon of frequency f is hf, and therefore the minimum KE of the electron that produced the x-ray should be equal to hf. This prediction agrees with experimental data.

The structure of a jet aircraft engine, laid bare in a single x-ray photograph.

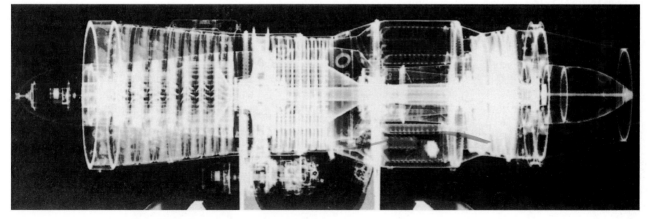

MATTER WAVES

As we have seen, light has both wave and particle aspects. In the topsy-turvy world of the very small, is it possible that what we normally think of as particles—electrons, for instance—have wave properties as well? So extraordinary is this question that it was not asked for two decades after Einstein's work. Soon after the question was raised came the even more extraordinary answer: yes.

8-5   DE BROGLIE WAVES

Matter Waves Are Significant Only in the Atomic World

In 1924 the French physicist Louis de Broglie proposed that moving objects act in some respects like waves. Reasoning by analogy with the properties of photons, he suggested that a particle of mass m and speed v behaves as though it is a wave whose wavelength is

$$\lambda = \frac{h}{mv}$$

$$\text{De Broglie wavelength} = \frac{\text{Planck's constant}}{\text{momentum}}$$

The more momentum mv a particle has, the shorter its **de Broglie wavelength** λ.

How can de Broglie's hypothesis be tested? Only waves can be diffracted and can reinforce and cancel each other by interference. A few years after de Broglie's work, experiments were performed in the United States and in England in which streams of electrons were shown to exhibit both diffraction and interference. The wavelengths of the electrons could be found from the data, and they agreed exactly with the formula $\lambda = h/mv$.

ELECTRON MICROSCOPES   The wave nature of moving electrons is the basis of the electron microscope. The resolving power of any optical instrument depends on the wavelength of whatever is used to illuminate the specimen being studied. In the case of a microscope that uses visible light, the highest useful magnification is about 500×. Higher magnifications give larger images but do not show more detail. Fast electrons, however, have wavelengths much shorter than those of visible light, and electron microscopes can produce useful magnifications of over 1,000,000×. X-rays also have short wavelengths, but it is not (yet?) possible to focus them adequately. In an electron microscope, an electron beam is directed at a thin specimen. Magnetic fields that act as lenses to focus the beam then produce an enlarged image of the specimen on a fluorescent screen or photographic film.

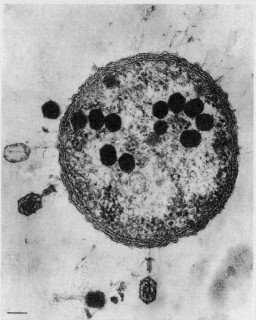

Electron micrograph showing bacteriophage viruses in a bacterium approximately 1 μm across. (Lee D. Simon/Photo Researchers)

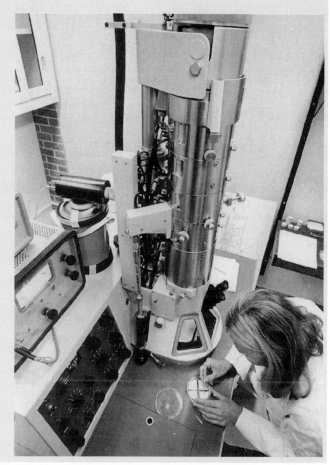

An electron microscope. (Guy Gillette/Photo Researchers)

There is nothing imaginary about these **matter waves.** They are perfectly real, just as light and sound waves are. However, they are not necessarily evident in every situation. A car moving at 80 km/h has a de Broglie wavelength of only about 10^{-38} m, which is so extremely small that no wave behavior can be detected. On the other hand, an electron whose speed is 10^7 m/s has a wavelength of 10^{-8} m, which is comparable in size to atomic dimensions. (A speed of 10^7 m/s is not much for an electron; the electrons in a television picture tube move faster.) It is not surprising, then, that the wave nature of moving electrons turns out to be crucial in understanding atomic structure and behavior.

Not only electrons but all other moving particles behave like waves under suitable circumstances. As with electromagnetic waves, the wave and particle aspects of moving bodies can never be observed at the same time. It therefore makes no sense to ask which is the "correct" description. All we can say is that in certain situations a moving body exhibits wave properties and in other situations it exhibits particle properties.

8-6 WAVES OF WHAT?

Waves of Probability   In water waves, the quantity that varies periodically is the height of the water surface. In sound waves, it is air pressure. In light waves, electric and magnetic fields vary. What is it that varies in the case of matter waves?

The quantity whose variations make up matter waves is called the **wave function,** symbol ψ (the Greek letter *psi*). The value of ψ^2 at a given place and time for a given particle determines the probability of finding the particle there at that time. For this reason ψ^2 is called the **probability density** of the particle. A large value of ψ^2 means the strong possibility of the particle's presence; a small value of ψ^2 means its presence is unlikely.

The de Broglie waves associated with a moving particle are in the form of a group, or packet, of waves, as in Fig. 8-5. This wave packet travels with the same speed v as the particle does. Even though we cannot visualize what is meant by ψ and so cannot form a mental image of matter waves, the agreement between theory and experiment means that we must take them seriously.

FIG. 8-5 (*a*) Particle description of a moving object. (*b*) Wave description of the same moving object. The packet of matter waves that corresponds to a certain object moves with the same speed v as the object does. The waves are waves of probability.

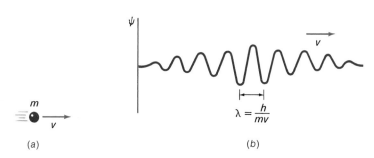

(a)   (b)

8-7 UNCERTAINTY PRINCIPLE

We Cannot Know the Future Because We Cannot Know the Present   To regard a moving particle as a wave packet suggests that there are limits to the accuracy with which we can measure such "particle" properties as position and speed. The particle whose wave packet is shown in Fig. 8-5 may be located anywhere within the packet at a given time. Of course, the probability density ψ^2 is a maximum in the middle of the packet, so the particle is most likely to be found there. But we may still find the particle anywhere that ψ^2 is not 0.

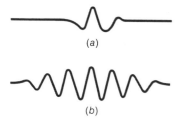

(a)

(b)

FIG. 8-6 (a) A narrow wave packet. The position of the particle can be precisely determined, but the wavelength (and hence the particle's momentum) cannot be established because there are not enough waves to measure λ accurately. (b) A wide wave packet. Now the wavelength can be accurately determined, but not the position of the particle.

The narrower its wave packet, the more precisely a particle's position can be specified (Fig. 8-6a). However, the wavelength of the waves in a narrow packet is not well defined. There are just not enough waves to measure λ accurately. This means that, since λ = h/mv, the particle's momentum mv and hence speed v are not precise quantities. If we make a series of momentum measurements, we will find a broad range of values.

On the other hand, a wide wave packet such as that in Fig. 8-6b has a clearly defined wavelength. The momentum that corresponds to this wavelength is therefore a precise quantity, and a series of measurements will give a narrow range of values. But where is the particle located? The width of the packet is now too great for us to be able to say just where the particle is at a given time.

Thus we have the **uncertainty principle:**

It is impossible to know both the exact position and the exact momentum of a particle at the same time.

This principle, which was discovered by Werner Heisenberg, is one of the most significant of physical laws.

Since we cannot know exactly both where a particle is right now and what its speed is, we cannot say anything definite about where it will be 2 s from now or how fast it will be moving then. *We cannot know the future for sure because we cannot know the present for sure.* But our ignorance is not total. We can still say that the particle is more likely to be in one place than another and that its speed is more likely to have a certain value than another.

All objects, of whatever size, are governed by the uncertainty principle, which means that their positions and motions likewise can be expressed only as probabilities. There is a chance that this book will

someday defy the law of gravity and rise up in the air by itself. But for objects this large—in fact, even for objects the size of molecules—such probabilities are so large as to be practically certainties. The likelihood that this book will continue to obey the law of gravity is so great that we can be quite sure it will stay where it is if left alone. Only in the behavior of electrons and other atomic particles do matter waves play an important part.

THE HYDROGEN ATOM

We now have what we need to make sense of atomic structures: the Rutherford model of the atom, the quantum theory of light, and the wave theory of moving particles. When linked together, these concepts give rise to a theory of the atom that agrees with experiment. Our starting point will be the hydrogen atom—the simplest of all—with its single electron outside a nucleus that consists of a single proton.

8-8 ATOMIC SPECTRA

Each Element Has a Characteristic Spectrum

When an electric current is passed through a gas, electrons in the gas atoms absorb energy from the current. The gas is said to be **excited.** An excited neon gas gives off a bright orange-red light; other excited gases give off light of other colors. We have all seen signs based on this effect, but not all of us are aware of how closely related the color of the light is to the way the electrons in the gas atoms are arranged.

Figure 8-7 shows an instrument called a **spectroscope** that disperses (spreads out) the light emitted by an excited gas into the different frequencies the light contains. Each frequency appears on the screen as a bright line, and the resulting series of bright lines is called an **emission spectrum** (Fig. 8-8a). Because some of the lines are more intense than the rest, the original undispersed light usually gives the impression of being a specific color, orange-red in the case of neon, even though other colors are present as well. An emission spectrum is different from

FIG. 8-7 An idealized spectroscope. Dispersion in the prism separates light of different frequencies.

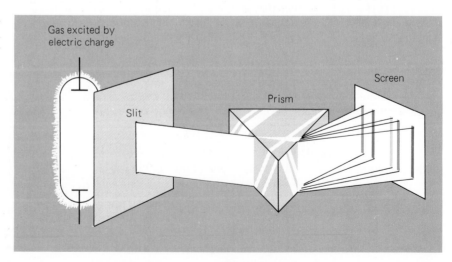

Gas excited by electric charge

Screen

Prism

Slit

FIG. 8-8 (*a*) Emission spectrum of sodium. (*b*) Absorption spectrum of sodium. Each dark line in the absorption spectrum corresponds to a bright line in the emission spectrum.

a **continuous spectrum,** which is the rainbow band produced when light from a hot object passes through a spectroscope (see photograph in Sec. 6-18). A continuous spectrum contains all frequencies, not just a few.

Absorption Spectra

Spectra of a different kind, **absorption spectra,** occur when light from a hot source passes through a cool gas before entering the spectroscope. The light source alone would give a continuous spectrum, but atoms of the gas absorb certain frequencies from the light that goes through it. Hence the original continuous spectrum is now crossed by dark lines, each line corresponding to one of the absorbed frequencies (Fig. 8-8*b*).

Some emission spectra in the visible region. These were produced by passing an electric current through gases or vapors, which caused them to radiate light of frequencies characteristic of the atoms or molecules present. From these frequencies a great deal can be learned of the electron structures of the atoms or molecules. *(a)* Molecular hydrogen. *(b)* Atomic hydrogen. *(c)* Sodium vapor. *(d)* Helium. *(e)* Neon. *(f)* Lithium vapor.

Gas atoms excited by electric currents in these tubes radiate light of wavelengths characteristic of the gas used.

If the emission spectrum of an element is compared with the absorption spectrum of the same element, the dark lines in the latter spectrum have the same frequencies as a number of the bright lines in the former spectrum. Thus a cool gas absorbs some of the frequencies of the light that it emits when excited. The spectrum of sunlight has dark lines in it because the luminous part of the sun, which radiates much like an object heated to 6000 K, has around it an envelope of cooler gas (Chap. 17).

Because the line spectrum of each element (either emission or absorption) contains frequencies that are characteristic of that element only, the spectrometer is a valuable tool in chemical analysis. Even the smallest traces of an element can be identified by the lines in a spectrum of an unknown substance. Helium was discovered in the sun through its spectrum 17 years before it was identified on the earth in 1895 (*helios* is Greek for "sun").

A century ago it was discovered that the frequencies in the spectrum of an element fall into sets called **spectral series** (Fig. 8-9). A simple formula relates the frequencies in each series. When the foundations of the modern picture of the atom had been laid, these spectral series provided the final clues for working out the details of atomic structure.

FIG. 8-9 The spectral series of hydrogen. The wavelengths (and hence frequencies) in each series can be related by simple formulas (1 nm = 1 nanometer = 10^{-9} m).

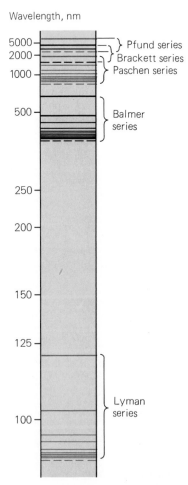

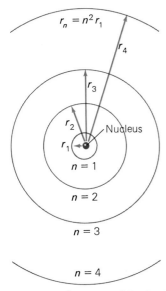

FIG. 8-10 Electron orbits in the Bohr model of the hydrogen atom (not to scale). The radius of each orbit is proportional to n^2, the square of the orbit's quantum number. The inner orbit is the electron's normal path, and the outer orbits represent states of higher energy. If the electron absorbs enough energy to jump to an outer orbit, it will return to the $n = 1$ orbit by a single jump or combination of jumps. Each inward jump is accompanied by the emission of a photon.

FIG. 8-11 Excitation by collision. When two atoms collide, some of the available energy is absorbed by one of the atoms, which goes into an excited energy state. The atom then emits a photon in returning to its ground (normal) state.

8-9 THE BOHR MODEL

Only Certain Electron Energies Are Possible in an Atom Niels Bohr, a Dane, put forward in 1913 a theory of the hydrogen atom that could account both for its stability and for the frequencies of the spectral lines of hydrogen. Bohr applied the then-new quantum ideas to atomic structure to come up with a model that, even though later replaced by a more complex picture of greater accuracy and usefulness, is still the mental image many scientists have of the atom.

Bohr began by proposing that an electron in an atom can circle the nucleus without losing energy only in certain specific orbits. Because these orbits are each a different distance from the nucleus, the energy of the electron depends on which orbit it is in. Thus Bohr suggested that atomic electrons can have only certain particular energies. An analogy is a person on a ladder, who can stand only on its steps and not in between.

An electron in the innermost orbit has the least energy. The larger the orbit, the more the electron energy. The orbits are identified by a **quantum number,** n, which is $n = 1$ for the innermost orbit, $n = 2$ for the next, and so on. Each orbit corresponds to an **energy level** of the atom.

Explaining Spectral Lines That atoms emit and absorb only light of certain frequencies, which we observe as spectral lines, fits Bohr's atomic model perfectly. An electron in a particular orbit can absorb only those photons of light whose energy will permit it to "jump" to another orbit farther out, where the electron has more energy. When an electron jumps from a particular orbit to another orbit closer to the nucleus, where it has less energy, it emits a photon of light. The difference in energy between the two orbits is hf, where f is the frequency of the absorbed or emitted light.

Figure 8-10 shows the possible orbits of the electron in a hydrogen atom. The circle nearest the nucleus represents the electron orbit under ordinary conditions, when the atom has the lowest possible energy. Such an atom is said to be in its **ground state.** The other circles represent orbits in which the electron would have more energy, since it would then be farther from the nucleus. (Similarly a stone on the roof of a building has more potential energy than it has on the ground, since it is farther from the earth's center when it is on the roof.)

Suppose an atom is in its ground state. If the atom is given energy—by strong heating, by an electric discharge, or by radiation—the electron may jump to a larger orbit (Fig. 8-11). This jump means that the atom

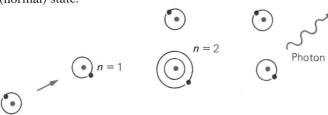

has absorbed energy. The atom keeps the added energy as long as it is in the **excited state,** that is, as long as the electron stays in the larger orbit. Because excited states are unstable, in a fraction of a second the electron drops to a smaller orbit, emitting a photon of light as it does so.

The energy (and hence the frequency) of the photon emitted from a hydrogen atom depends on the particular jump that its electron makes. If the electron jumps from orbit $n = 4$ to orbit $n = 1$ (Fig. 8-12), the energy of the photon will be greater than if the electron jumps from 3 to

FIG. 8-12 Spectral lines are the result of jumps between energy levels. The spectral series of hydrogen are shown in Fig. 8-9. When $n = \infty$, the electron is free.

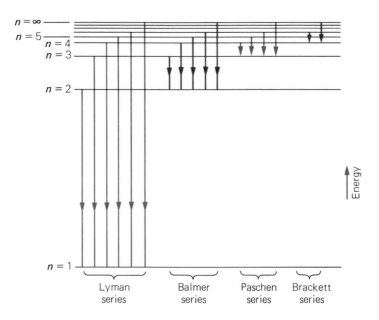

1 or 2 to 1. Starting from orbit 4, the electron may return to 1 not only by a single leap but also by stopping at 3 and 2 on the way. Corresponding to these jumps are photons with energies determined by the energy differences between 4 and 3, 3 and 2, 2 and 1.

Each electron jump gives a photon of a characteristic frequency and therefore appears in the hydrogen spectrum as a single bright line. The frequencies of the different lines are related since they correspond to different jumps in the same set of orbits. And the relations among the lines that Bohr predicted by this mechanism precisely matched the observed relations among the lines in the hydrogen spectrum.

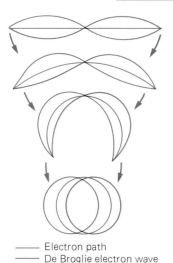

— Electron path
— De Broglie electron wave

FIG. 8-13 The condition for a stable electron orbit. The orbit of the electron in a hydrogen atom corresponds to a complete electron de Broglie wave joined on itself.

8-10   ELECTRON WAVES AND ORBITS

Standing Waves in the Atom

Why does an atomic electron follow certain orbits only? The answer comes from an analysis of the wave properties of an electron that circles a hydrogen nucleus. It turns out that the de Broglie wavelength of the electron is exactly equal to the circumference of its ground-state (that is, innermost) orbit. Thus the $n = 1$ orbit of the electron in a hydrogen atom corresponds to one complete electron wave joined on itself (Fig. 8-13).

This fact provides us with the final clue we need for a theory of the atom. If we consider the vibrations of a wire loop (Fig. 8-14), we find that their wavelengths always fit a whole number of times into the loop's circumference, so that each wave joins smoothly with the next. These are the only vibrations possible. Regarding electron waves in an atom as analogous to standing waves in a wire loop leads to an interesting concept:

> **An electron can circle a nucleus only in orbits that contain a whole number of de Broglie wavelengths.**

This idea is the decisive one for understanding the atom. It combines both the particle and the wave characters of the electron into a single statement, since the electron wavelength depends upon the orbital speed needed to balance the electrical attraction of the nucleus. These contradictory characters are basic aspects of the atomic world.

Now we see what the quantum number n of an orbit means—it is the number of electron waves that fit into the orbit (Fig. 8-15).

FIG. 8-14 Vibrations of a wire loop. In each case a whole number of wavelengths fit into the circumference of the loop.

Circumference = 2 wavelengths

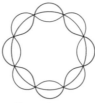

Circumference = 4 wavelengths

Circumference = 8 wavelengths

The presence of energy levels in an atom—which is true for all atoms, not just the hydrogen atom—is a further example of the basic graininess of physical quantities. In the world of our daily lives, matter, electric charge, energy, and so forth seem to be continuous and able to be cut up (so to speak) into chunks of any size at all. But in the world of the atom matter consists of elementary particles with definite rest masses; charge always comes in multiples of $+e$ and $-e$; em waves of frequency f appear as streams of photons each with the energy hf; and stable systems of particles, such as atoms, can have only certain energies.

Other quantities in nature are also grainy, or *quantized*. This quantization enters into every aspect of how electrons, protons, and neutrons interact to give the matter around us (and of which we are made) its familiar properties. In the case of an atom, the quantization of energy follows from the wave motion of moving bodies: the electron waves must be standing waves, hence the electrons can have only certain energies.

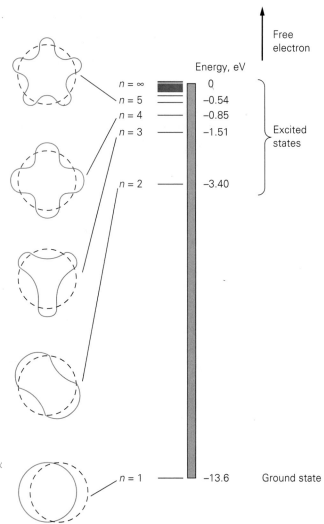

FIG. 8-15 Energy levels of the hydrogen atom. The energies are negative, which signifies that the electron is bound to its nucleus.

8-11 THE LASER

An Amplifier of Light That Produces Waves All in Step

A **laser** is a device that produces an intense beam of single-frequency, **coherent** light from the cooperative radiation of excited atoms. The light waves in a coherent beam are all in step with one another, as shown in Fig. 8-16. Ordinary light is incoherent since the atoms in light sources such as lamps and the sun emit light waves randomly.

A laser beam hardly spreads out at all. One sent from the earth to a mirror left on the moon by the Apollo 11 expedition remained narrow enough to be detected on its return to earth, a round-trip distance of over three-quarters of a million km. A light beam produced by any other means would have spread out too much to have been detected. The

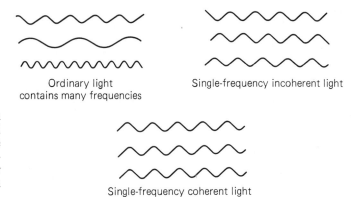

Ordinary light
contains many frequencies

Single-frequency incoherent light

Single-frequency coherent light

FIG. 8-16 A laser produces a beam of light whose waves all have the same frequency and are in step with one another (coherent). The beam is also very narrow and spreads out very little even over long distances.

word *laser* comes from *light amplification by stimulated emission of radiation*.

The key to the laser is that many atoms have one or more excited energy levels whose lifetimes are as much as 0.001 s instead of the usual 10^{-8} s. Such relatively long lived states are called **metastable.**

A laser uses atoms whose metastable states have some excitation energy E_1 (Fig. 8-17). The first step in laser operation is to bring as many of these atoms as possible to this metastable level. Often it is necessary to raise the atoms to a still higher state E_2, from which a number of them fall to the metastable level by emitting a photon of energy $E_2 - E_1$. Several ways exist to do this. In one of them, an external light source provides photons with the right energy. This method was used in the

FIG. 8-17 The principle of the laser. A metastable atomic state is one that lasts a much longer time than usual before a photon is emitted that brings the atom to a state of lower energy.

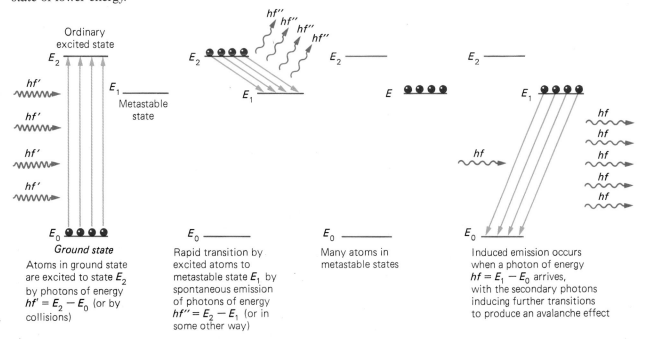

Ordinary
excited state

E_2

hf'

E_1
Metastable
state

hf'

hf'

hf'

E_0
Ground state

Atoms in ground state are excited to state E_2 by photons of energy $hf' = E_2 - E_0$ (or by collisions)

hf'' hf'' hf'' hf''

E_2

E_1

E_0

Rapid transition by excited atoms to metastable state E_1 by spontaneous emission of photons of energy $hf'' = E_2 - E_1$ (or in some other way)

E_2

E

E_0

Many atoms in metastable states

E_2

E_1

hf

E_0

hf
hf
hf
hf
hf

Induced emission occurs when a photon of energy $hf = E_1 - E_0$ arrives, with the secondary photons inducing further transitions to produce an avalanche effect

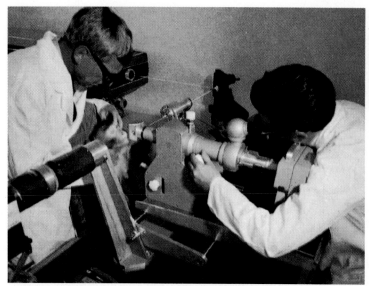

A laser produces an intense beam of single-frequency coherent light from the cooperative radiation of excited atoms or molecules. The light waves in a coherent beam are all in step, which greatly increases their effectiveness. A laser beam provides a way to "weld" a detached retina to the eye tissue behind it. Here the blue-green light from an argon laser is being directed into a monkey's eye for this purpose.

Laser beams at a laboratory in Madrid, Spain.

first lasers, in which xenon-filled flash lamps excited chromium ions in ruby rods to the required level E_2 (Fig. 8-18).

Another method is used in the helium-neon laser. Here an electric discharge in the gas mixture produces fast electrons whose impact on the gas atoms brings them to the required energy level. The advantage is that such a laser can operate continuously, whereas a ruby laser produces separate flashes of light.

With many atoms in the metastable state E_1, a few of them are likely to spontaneously emit photons of energy $hf = E_1 - E_0$ before the others, thereby falling to the ground state E_0. A typical laser is a transparent solid (such as a ruby rod) or a gas-filled tube with mirrors at both ends, one of them only partly silvered to allow some of the light inside to get out. The distance between the mirrors is made equal to a whole number of half-wavelengths of light of frequency f, so that the trapped light forms an optical standing wave (Fig. 6-12). This standing wave stimulates the other atoms in metastable states to radiate before they would

FIG. 8-18 A ruby laser. A ruby is a crystal that contains Cr^{3+} ions, which are chromium atoms that have lost three electrons each. A Cr^{3+} ion has a metastable level whose lifetime is about 0.003 s. The xenon flash lamp excites the Cr^{3+} ions to a level of higher energy from which they fall to the metastable level by losing energy to other ions in the crystal. Photons from the spontaneous decay of some Cr^{3+} ions cause other excited Cr^{3+} ions to radiate. The result is a large pulse of single-frequency, coherent red light from the partly silvered end of the rod.

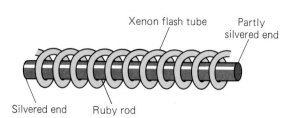

PRACTICAL LASERS Many kinds of lasers have been built and still others are being developed. In a compact disk player, the beam of a tiny solid-state semiconductor laser is focused to a spot a micrometer (a millionth of a meter) across to read data coded as pits that appear as dark spots on a reflective disk 12 cm in diameter. A compact disk can store over 600 megabytes of digital data, about 1000 times as much as the floppy disks used in computers. If the stored data is digitized music, the playing time can be over an hour. Similar lasers are used in transmission systems that carry telephone and television signals as flashes of infrared light along thin glass fibers.

Helium-neon gas lasers produce the narrow red beams that read bar codes in shops. More powerful carbon dioxide gas lasers are used in surgery, where a laser beam has the advantage of sealing small blood vessels it cuts through. Lasers find numerous applications in industry, from cutting fabric for clothing and making holes in nipples for babies' bottles to welding pipelines and heat-treating the surfaces of engine crankshafts to harden them.

normally do so. The result is an avalanche of photons, all of the same frequency f and all of whose waves are exactly in step, which greatly increases the power they can deliver.

QUANTUM THEORY OF THE ATOM

The theory of the hydrogen atom discussed above is basically that developed by Bohr in 1913 (although he did not have de Broglie's idea of electron waves to guide his thinking). It can account for much experimental data in a convincing manner. However, it has some severe limitations. For instance, although the Bohr theory correctly predicts the spectral series of hydrogen, it cannot do the same for the spectra of atoms that have two or more electrons each. Perhaps most important of all, it does not give what a really successful theory of the atom ought to: an understanding of how individual atoms interact with one another to form molecules, solids, and liquids.

These objections to the Bohr theory are not meant to be unfriendly, for it was one of those historic achievements that transform scientific thought, but rather to emphasize that a more general approach to the atom is required. Such an approach was developed in 1925–1926 by Erwin Schrödinger, Werner Heisenberg, and others, under the apt name of **quantum mechanics.** By the early 1930s the application of quantum mechanics to problems involving nuclei, atoms, molecules, and matter in the solid state made it possible to understand a vast body of otherwise puzzling data and—vital for any theory—led to predictions of remarkable accuracy.

8-12 QUANTUM MECHANICS

Probabilities, not Certainties

The real difference between newtonian mechanics and quantum mechanics lies in what they describe. The newtonian mechanics of Chap. 2 deals with the motion of an object under the influence of applied forces, and it takes for granted that such quantities as the object's position, mass, velocity, and acceleration can be measured. This assumption agrees completely with our everyday experience. Newtonian mechanics

BIOGRAPHY

Erwin Schrödinger (1887–1961) was born in Vienna, Austria and studied at the university there. Late in 1925, when he was a professor of physics in Zurich, Switzerland, Schrödinger wrote to a friend that he was "struggling with a new atomic theory. If only I knew more mathematics! I am very optimistic about this thing and expect that if I can only . . . solve it, it will be *very* beautiful." The struggle was successful, and early in 1926 Schrödinger published four papers on quantum mechanics that revolutionized physics and were indeed beautiful.

Later, while at Dublin's Institute for Advanced Study, Schrödinger became interested in biology, in particular the mechanism of heredity. He seems to have been the first to make definite the idea of a genetic code and to identify genes as long molecules that carry the code in the form of variations in how their atoms are arranged. Schrödinger's

1944 book *What Is Life?* was enormously influential and started James Watson on his search for "the secret of the gene," which he and Francis Crick discovered in 1953 to be the structure of the DNA molecule (see Sec. 12-16).

provides the "correct" explanation for the behavior of moving objects in the sense that the values it predicts for observable quantities agree with the measured values of those quantities.

Quantum mechanics, too, consists of relationships between observable quantities, but the uncertainty principle radically alters the meaning of "observable quantity" in the atomic realm. According to the uncertainty principle, the position and momentum of a particle cannot both be accurately known at the same time. (In newtonian physics, of course, such quantities are assumed to always have definite, measurable values.) What quantum mechanics explores are *probabilities*. Instead of saying, for example, that the electron in a normal hydrogen atom is always exactly 5.3×10^{-11} m from the nucleus, quantum mechanics holds that this is the *most probable* distance. In a suitable experiment, many trials would yield different values, but the one most likely to be found would be 5.3×10^{-11} m.

Quantum mechanics does not try to invent a mechanical model based on ideas from everyday life to represent the atom. Instead it deals only with quantities that can actually be measured. We can measure the mass of the electron and its electric charge, we can measure the frequencies of spectral lines emitted by excited atoms, and so on, and the theory must be able to relate them all. But we *cannot* measure the precise diameter of an electron's orbit or watch it jump from one orbit to another, and these notions therefore are not part of the theory.

Newtonian and Quantum Mechanics

Quantum mechanics abandons the traditional approach to physics in which models we can visualize are the starting points of theories. But although quantum mechanics does not give us a look into the inner world of the atom, it does tell us everything we need to know about the measurable properties of atoms. And there is something more: *newtonian mechanics is just an approximate version of quantum mechanics.* The certainties of Newton are an illusion. Their agreement with experiment is due to the fact that ordinary objects contain so many atoms that deviations from the most probable behavior are unnoticeable. Instead of two sets of physical principles, one for the world of the large and one for the world of the small, there is only a single set, and quantum mechanics represents our best effort to date at formulating it.

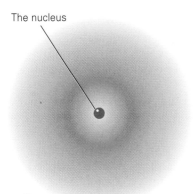

The nucleus

10^{-10}m

FIG. 8-19 Probability cloud for the ground state of the hydrogen atom. The denser the cloud, the more likely the electron is to be found there.

8-13   QUANTUM NUMBERS

An Atomic Electron Has Four in All

In the Bohr model of the hydrogen atom, the electron moves around the nucleus in a circular orbit. The only quantity that changes as the electron moves is its position on the circle. The single quantum number n is enough to specify the physical state of such an electron.

In the quantum theory of the atom, an electron has no fixed orbit but is free to move about in three dimensions. We can think of the electron as circulating in a **probability cloud** that forms a certain pattern in space. Where the cloud is most dense (that is, where ψ^2 has a high value), the electron is most likely to be found. Where the cloud is least dense (ψ^2 has a low value), the electron is least likely to be found. Figure 8-19 shows a cross section of the probability cloud for the ground (lowest-energy) state of the hydrogen atom.

Three quantum numbers determine the size and shape of the probability cloud of an atomic electron. One of them, the **principal quantum number,** is designated n as in the Bohr theory. This quantum number is the chief factor that governs the electron's energy (the larger n is, the greater the energy) and its average distance from the nucleus (the larger n is, the farther the electron tends to be from the nucleus).

The other two quantum numbers, l and m_l, together govern the electron's angular momentum and the form of its probability cloud. Angular momentum, as we learned in Chap. 3, is the rotational analog of linear momentum. According to quantum mechanics, angular momentum as well as energy is quantized—restricted to certain particular values—in an atom. The possible values of angular momentum for an atomic electron are determined by l, the **orbital quantum number.** An electron whose principal quantum number is n can have an orbital quantum number of 0 or any whole number up to $n - 1$. For instance, if $n = 3$, the values l can have are 0, 1, or 2.

The orbital quantum number l determines the *magnitude* of the electron's angular momentum. However, angular momentum, like linear momentum, is a vector quantity, and so to describe it completely requires that its *direction* be specified as well as its magnitude (Fig. 8-20). This is the role of the **magnetic quantum number** m_l.

What meaning can a direction in space have for an atom? The answer becomes clear when we reflect that an electron revolving about a nucleus is a current loop and has a magnetic field like that of a tiny bar magnet. In a magnetic field the potential energy of a bar magnet

FIG. 8-20 The right-hand rule for direction of angular-momentum vector.

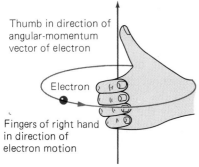

Thumb in direction of angular-momentum vector of electron

Electron

Fingers of right hand in direction of electron motion

TABLE 8-1

Quantum Numbers of an Atomic Electron

Name	Symbol	Possible Values	Quantity Determined
Principal	n	1, 2, 3, . . .	Electron energy
Orbital	l	0, 1, 2, . . . , $n-1$	Magnitude of angular momentum
Magnetic	m_l	$-l, . . . , 0, . . . , +l$	Direction of angular momentum
Spin magnetic	m_s	$-\frac{1}{2}, +\frac{1}{2}$	Direction of electron spin

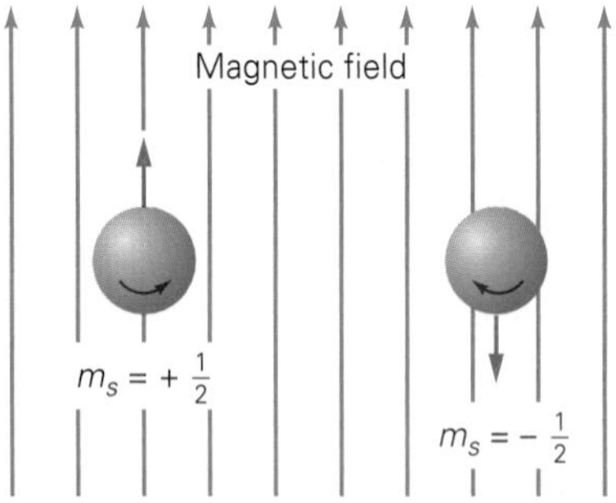

FIG. 8-21 The spin magnetic quantum number m_s of an atomic electron has two possible values, $+\frac{1}{2}$ and $-\frac{1}{2}$, depending upon how the electron aligns itself with a magnetic field.

depends both upon how strong the magnet is and upon its orientation with respect to the field. It is the direction of the angular-momentum vector (that is, the direction of the axis about which the electron may be thought to revolve) with respect to a magnetic field that is determined by m_l.

An electron whose orbital quantum number is l can have a magnetic quantum number that is 0 or any whole number between $-l$ and $+l$. For instance, if $l = 2$, the values m_l can have are $-2, -1, 0, +1,$ and $+2$.

Electron Spin

There is still another quantum number needed to describe completely an atomic electron. This is the electron **spin magnetic quantum number** m_s.

Electrons behave as though they were, in themselves, little bar magnets, which we can visualize as arising from electrons spinning on their axes. If we picture an electron as a charged sphere, such spinning means a circular electric current and hence magnetic behavior. A spinning electron can align itself either along a magnetic field, in which case m_s has the value $+\frac{1}{2}$, or opposite to the field, in which case $m_s = -\frac{1}{2}$ (Fig. 8-21). The concept of electron spin is essential for understanding many atomic phenomena.

Table 8-1 lists the quantum numbers of an atomic electron, their possible values, and their significance.

8-14 EXCLUSION PRINCIPLE

A Different Set of Quantum Numbers for Each Electron in an Atom

In an unexcited hydrogen atom, the electron is in its quantum state of lowest energy. What about more complex atoms? Are all 92 electrons of a uranium atom in the same quantum state, jammed into a single probability cloud? Many lines of evidence make this idea unlikely.

An example is the great difference in chemical behavior shown by certain elements whose atomic structures differ by just one electron. Thus the elements that have the atomic numbers 9, 10, and 11 are respectively the chemically active gas fluorine, the inert gas neon, and the metal sodium. Since the electron structure of an atom controls how the atom interacts with other atoms, it makes no sense that the chemical properties of the elements should change so sharply with a small change in atomic number if all the electrons in an atom were in the same quantum state.

In 1925 Wolfgang Pauli solved the problem of the electron arrangement in an atom that has more than one electron. His **exclusion principle** states that

> **Only one electron in an atom can exist in a given quantum state.**

Each electron in an atom must have a different set of quantum numbers n, l, m_l, m_s. In the next chapter we will see how the exclusion principle, together with the limits on the possible values of the various quantum numbers, determines the chemical behavior of the elements.

IMPORTANT TERMS AND IDEAS

The **photoelectric effect** is the emission of electrons from a metal surface when light shines on it. The **quantum theory of light** states that light travels in tiny bursts (or **quanta**) of energy called **photons.** The photoelectric effect can be explained only by the quantum theory of light, whereas the wave theory of light is needed to account for such other phenomena as interference; the two theories complement each other.

X-rays are high-frequency electromagnetic waves given off when matter is struck by fast electrons.

Moving objects have wave as well as particle properties; the smaller the object, the more conspicuous its wave behavior. The **matter waves** that correspond to a moving object have a **de Broglie** wavelength inversely proportional to its momentum. The quantity that varies in a matter wave is called the **wave function,** and its square is the object's **probability density.** The greater the probability density at a certain time and place, the greater the likelihood of finding the object there at that time. The **uncertainty principle** expresses the limit set by the wave nature of matter on finding both the position and the state of motion of a moving object at the same time.

An **emission** spectrum consists of the various frequencies of light given off by an excited substance. An **absorption** spectrum consists of the various frequencies absorbed by a substance when white light is passed through it.

According to the **Bohr model of the atom,** an electron can circle an atomic nucleus only if the electron's orbit is a whole number of de Broglie wavelengths in circumference. The number of wavelengths is the **quantum number** n of the orbit. Each orbit corresponds to a specific energy, and spectral lines originate in electron shifts from one orbit, and

hence **energy level,** to another. An atom in its **ground state** has the lowest possible energy; **excited states** correspond to higher energy levels.

A **laser** is a device that produces an intense beam of single-frequency light whose waves are all in step with one another, which greatly increases their effectiveness. Such light is said to be **coherent.**

Quantum mechanics is based on the wave nature of moving things; newtonian mechanics turns out to be an approximate version of quantum mechanics. Quantum mechanics shows that four quantum numbers are needed to specify the physical state of each atomic electron. One of these quantum numbers governs the direction of the **spin** of the electron. According to the **exclusion principle,** no two electrons in an atom can have the same set of quantum numbers.

IMPORTANT FORMULAS

Quantum energy of photon: $E = hf$

Photoelectric effect: $hf = KE + w$

De Broglie wavelength: $\lambda = \dfrac{h}{mv}$

EXERCISES: MULTIPLE CHOICE

1. When light is directed at a metal surface, the energies of the emitted electrons
   a. vary with the intensity of the light
   b. vary with the frequency of the light
   c. vary with the speed of the light
   d. are random

2. An increase in the brightness of the light directed at a metal surface causes an increase in the emitted electrons'
   a. wavelength   **c.** energy
   b. speed   **d.** number

3. The photoelectric effect can be understood on the basis of
   a. the electromagnetic theory of light
   b. the interference of light waves
   c. the special theory of relativity
   d. none of the above

4. In a vacuum, all photons have the same
   a. frequency   c. energy
   b. wavelength   d. speed

5. The mass of a photon
   a. is 0
   b. is the same as that of an electron
   c. depends on its frequency
   d. depends on its energy

6. When the speed of the electrons that strike a metal surface is increased, the result is an increase in
   a. the number of x-rays emitted
   b. the frequency of the x-rays emitted
   c. the speed of the x-rays emitted
   d. the size of the x-rays emitted

7. A phenomenon that cannot be understood with the help of the quantum theory of light is
   a. the photoelectric effect
   b. x-ray production
   c. the spectrum of an element
   d. interference of light

8. According to the theories of modern physics, light
   a. is exclusively a wave phenomenon
   b. is exclusively a particle phenomenon
   c. combines wave and particle properties
   d. has neither wave nor particle properties

9. According to the theories of modern physics,
   a. only stationary particles exhibit wave behavior
   b. only moving particles exhibit wave behavior
   c. only charged particles exhibit wave behavior
   d. all particles exhibit wave behavior

10. The speed of the wave packet that corresponds to a moving particle
    a. is less than the particle's speed
    b. is equal to the particle's speed
    c. is more than the particle's speed
    d. any of the above, depending on the circumstances

11. De Broglie waves can be regarded as waves of
    a. pressure   c. electric charge
    b. probability   d. momentum

12. The description of a moving body in terms of matter waves is legitimate because

    a. it is based upon common sense
    b. matter waves have actually been seen
    c. the analogy with electromagnetic waves is plausible
    d. theory and experiment agree

13. The narrower the wave packet of a particle is,
    a. the shorter its wavelength
    b. the more precisely its position can be established
    c. the more precisely its momentum can be established
    d. the more precisely its energy can be established

14. According to the uncertainty principle, it is impossible to precisely determine at the same time a particle's
    a. position and charge
    b. position and momentum
    c. momentum and energy
    d. charge and mass

15. If Planck's constant were larger than it is,
    a. moving bodies would have shorter wavelengths
    b. moving bodies would have higher energies
    c. moving bodies would have higher momenta
    d. the uncertainty principle would be significant on a larger scale of size

16. The emission spectrum produced by the excited atoms of an element contains frequencies that are
    a. the same for all elements
    b. characteristic of the particular element
    c. evenly distributed throughout the entire visible spectrum
    d. different from the frequencies in its absorption spectrum

17. The sun's spectrum consists of a bright background crossed by dark lines. This suggests that the sun
    a. is a hot object surrounded by a hot atmosphere
    b. is a hot object surrounded by a cool atmosphere
    c. is a cool object surrounded by a hot atmosphere
    d. is a cool object surrounded by a cool atmosphere

18. The classical model of the hydrogen atom fails because
    a. an accelerated electron radiates electromagnetic waves
    b. a moving electron has more mass than an electron at rest
    c. a moving electron has more charge than an electron at rest

d. the attractive force of the nucleus is not enough to keep an electron in orbit around it

19. An electron can revolve in an orbit around an atomic nucleus without radiating energy provided that the orbit
a. is far enough away from the nucleus
b. is less than a de Broglie wavelength in circumference
c. is a whole number of de Broglie wavelengths in circumference
d. is a perfect circle

20. According to the Bohr model of the atom, an electron in the ground state
a. radiates electromagnetic energy continuously
b. emits only spectral lines
c. remains there forever
d. can jump to another orbit if given enough energy

21. In the Bohr model of the atom, the electrons revolve around the nucleus of an atom so as to
a. emit spectral lines
b. produce x-rays
c. form energy levels that depend upon their speeds only
d. keep from falling into the nucleus

22. A hydrogen atom is said to be in its ground state when its electron
a. is at rest
b. is inside the nucleus
c. is in its lower energy level
d. has escaped from the atom

23. An atom emits a photon when one of its orbital electrons
a. jumps from a higher to a lower energy level
b. jumps from a lower to a higher energy level
c. is removed by the photoelectric effect
d. is struck by an x-ray

24. When an atom absorbs a photon of light, which one or more of the following can happen?
a. An electron shifts to a state of smaller quantum number
b. An electron shifts to a state of higher quantum number
c. An electron leaves the atom
d. An x-ray photon is emitted

25. Which of the following types of radiation is not emitted by the electronic structures of atoms?
a. ultraviolet light   c. x-rays
b. visible light   d. gamma rays

26. The operation of the laser is based upon
a. the uncertainty principle

b. the interference of de Broglie waves
c. stimulated emission of radiation
d. stimulated absorption of radiation

27. Which of the following properties is not characteristic of the light waves from a laser?
a. The waves all have the same frequency
b. The waves are all in step with one another
c. The waves form a narrow beam
d. The waves have higher photon energies than light waves of the same frequency from an ordinary source

28. The quantum-mechanical theory of the atom is
a. based upon a mechanical model of the atom
b. a theory that restricts itself to physical quantities that can be measured directly
c. less accurate than the Bohr theory of the atom
d. impossible to reconcile with Newton's laws of motion

29. A quantum number is not associated with an atomic electron's
a. mass
b. energy
c. spin
d. orbital angular momentum

30. The electrons in an atom all have the same
a. speed   c. orbit
b. spin magnitude   d. quantum numbers

31. According to the exclusion principle, no two electrons in an atom can have the same
a. spin direction
b. speed
c. orbit
d. set of quantum numbers

32. Light of wavelength 5×10^{-7} m consists of photons whose energy is
a. 1.1×10^{-48} J   c. 4×10^{-19} J
b. 1.3×10^{-27} J   d. 1.7×10^{-15} J

33. An x-ray photon has an energy of 6.6×10^{-15} J. The frequency that corresponds to this energy is
a. 4.4×10^{-48} Hz   c. 10^{15} Hz
b. 10^{-19} Hz   d. 10^{19} Hz

34. The de Broglie wavelength of an electron whose speed is 10^8 m/s is
a. 5.9×10^{-56} m   c. 7.3×10^{-12} m
b. 1.5×10^{-19} m   d. 1.4×10^{11} m

35. The speed of an electron whose de Broglie wavelength is 10^{-10} m is
a. 6.6×10^{-24} m/s   c. 7.3×10^6 m/s
b. 3.8×10^3 m/s   d. 10^{10} m/s

QUESTIONS

1. What differences can you think of between the photon and the electron?

2. The photon and the neutrino both have neither charge nor mass. What are the differences between them?

3. Compare the evidence for the wave nature of light with the evidence for its particle nature. Why do you think the wave nature of light was established long before its particle nature?

4. A certain metal surface emits electrons when light is shone on it. (a) How can the number of electrons per second be increased? (b) How can the energies of the electrons be increased?

5. Energy is carried in light by means of separate photons, yet even the faintest light we can see does not appear as a series of flashes. Explain.

6. How does the speed of a photon compare with the speed of an em wave?

7. When the speed of the electrons that strike a metal surface is increased, what happens to the speed, energy, and number per second of the x-ray photons that are emitted?

8. If Planck's constant were smaller than it is, would quantum phenomena be more or less conspicuous than they are now?

9. How does the speed of a matter wave compare with the speed of an em wave?

10. The uncertainty principle applies to *all* bodies, yet its consequences are significant only for such extremely small particles as electrons, protons, and neutrons. Explain.

11. Must a particle have an electric charge in order for matter waves to be associated with its motion?

12. What kind of experiment might you use to distinguish between a gamma ray of wavelength 10^{-11} m and an electron whose de Broglie wavelength is also 10^{-11} m?

13. A photon and a proton have the same wavelength. How does the photon's energy compare with the proton's kinetic energy?

14. A proton and an electron have the same de Broglie wavelength. How do their speeds compare?

15. What kind of spectrum is observed in (a) light from the hot filament of a lightbulb; (b) light from a sodium-vapor highway lamp; (c) light from a lightbulb surrounded by cool sodium vapor?

16. Most stars are hot objects surrounded by cooler atmospheres. What kind of spectrum does such a star give rise to?

17. In the Bohr model of the atom, the electron is in constant motion. How can such an electron have a negative amount of energy?

18. Why does the hydrogen spectrum contain many lines, even though the hydrogen atom has only a single electron?

19. Why is the Bohr theory incompatible with the uncertainty principle?

20. On the basis of the Bohr model of the atom, explain why the absorption lines in the spectrum of hydrogen have the same wavelengths as some of the emission lines of the same element.

21. What is meant by the ground state of an atom? What is the quantum number of the ground state of a hydrogen atom in the Bohr model?

22. What is coherent light? Is the light from a lightbulb coherent? The light from the sun?

23. What is a metastable atomic state?

24. How is quantum mechanics related to newtonian mechanics?

25. In what way does light from a laser differ from light from other sources?

26. Why is the optical length of a laser so important?

27. The Bohr theory permits us to visualize the structure of the atom, whereas quantum mechanics is very complex and concerned with such ideas as wave functions and probabilities. What reasons would lead to the replacement of the Bohr theory by quantum mechanics?

28. The four quantum numbers needed to describe an atomic electron are n, l, m_l, and m_s. What quantity is governed by each of them?

29. Under what circumstances do electrons exhibit spin?

30. Under what circumstances can two electrons share the same probability cloud in an atom?

PROBLEMS

1. Find the energy of a photon of ultraviolet light

whose frequency is 2×10^{16} Hz. Do the same for a photon of radio waves whose frequency is 2×10^5 Hz.

2. Find the energy of a photon of red light whose wavelength is 700 nm (1 nm = 1 nanometer = 10^{-9} m).

3. The eye can detect as little as 10^{-18} J of energy in the form of light. How many photons of frequency 5×10^{14} Hz does this amount of energy represent?

4. Yellow light has a frequency of about 5×10^{14} Hz. How many photons are emitted per second by a lamp that radiates yellow light at a power of 100 W?

5. The radiant energy reaching the earth from the sun is about 1400 W/m². If this energy is all green light of wavelength 5.5×10^{-7} m, how many photons strike each square meter per second?

6. A microwave oven operating at 2.5 GHz has a power output of 500 W. (a) What is the wavelength of the microwaves? (b) What is the energy of each photon? (c) How many photons per second does the over produce?

7. A detached retina is being "welded" back by using 20-ms pulses from a 0.50-W laser operating at a wavelength of 643 nm. How many photons are in each pulse?

8. The minimum frequency for photoelectric emission in calcium is 7.7×10^{14} Hz. Find the maximum KE of the electrons emitted when light of frequency 12.0×10^{15} Hz is directed on a calcium surface.

9. An energy of 4×10^{-19} J is required to remove an electron from the surface of a particular metal. What is the frequency of the light that will just dislodge electrons from the surface? What is the maximum energy of electrons emitted through the action of light of wavelength 2×10^{-7} m?

10. Electrons are accelerated through potential differences of approximately 10,000 V in television picture tubes. Find the maximum frequency of the x-rays that are produced when these electrons strike the screen of the tube.

11. Find the de Broglie wavelength of an electron whose speed is 2×10^7 m/s. How significant are the wave properties of such an electron likely to be?

12. Find the de Broglie wavelength of a 1-mg grain of sand blown by the wind at a speed of 20 m/s. How significant are the wave properties of such a grain of sand likely to be?

13. An electron microscope uses 40-keV (4×10^4 eV) electrons. Find its ultimate resolving power on the assumption that this is equal to the wavelength of the electrons.

14. Of the following transitions in a hydrogen atom (a) which emits the photon of highest frequency, (b) which emits the photon of lowest frequency, and (c) which absorbs the photon of highest frequency? $n = 1$ to $n = 2$, $n = 2$ to $n = 1$, $n = 2$ to $n = 6$, $n = 6$ to $n = 2$.

15. Calculate the speed of the electron in the innermost ($n = 1$) Bohr orbit of a hydrogen atom. [*Hint:* Begin by setting the centripetal force on the electron equal to the electrical attraction of the proton it circles around.]

ANSWERS TO MULTIPLE CHOICE			
1. b	10. b	19. c	28. b
2. d	11. b	20. d	29. a
3. d	12. d	21. d	30. b
4. d	13. b	22. c	31. d
5. a	14. b	23. a	32. c
6. b	15. d	24. b, c	33. d
7. d	16. b	25. d	34. c
8. c	17. b	26. c	35. c
9. b	18. a	27. d	

9

THE

PERIODIC

LAW

*A*lthough the line between physics and chemistry is hazy, with this chapter we are definitely across it. Chemistry began with a search for a way to change ordinary metals into gold. This fruitless task, called **alchemy** by the Arabs, was not abandoned until the seventeenth century. At that time John Mayow and Robert Boyle in England, Jean Rey in France, and Georg Stahl in Germany, among others, started to look systematically into the properties of matter and how they change in chemical reactions.

After a look at what is meant by chemical change, we go on to consider the periodic law, a natural classification of the elements into groups with similar characteristics. As we shall find, the periodic law has its roots in atomic structure. This is not surprising, since the way in which electrons are arranged in an atom is what determines how that atom interacts with other atoms—in other words, how it behaves chemically.

ELEMENTS AND COMPOUNDS

The properties of matter are altered in a number of processes. When a solid melts into a liquid or a liquid vaporizes into a gas, the cause is a change in the motions and separations of the molecules of the material. In other processes, however, the changes are in the molecules themselves. Examples are the rusting of iron, the burning of wood, and the souring of milk. Such processes are called **chemical reactions.**

9-1 CHEMICAL CHANGE

A Chemical Reaction Alters the Substances Involved

To begin our study of chemistry, let us examine a specific chemical reaction. Suppose we mix some powdered zinc metal with a somewhat larger volume of powdered sulfur on a ceramic surface and then ignite the mixture, say with a gas flame. The result is a small explosion with light and heat given off. When the fireworks have died down, we are left with a brittle white substance that resembles neither the original zinc nor the original sulfur (Fig. 9-1). What has happened?

Further experiments would show (1) that neither zinc nor sulfur alone gives such a reaction when heated; (2) that the explosion takes place just as well in a vacuum as in air; and (3) that the ceramic surface may be replaced by a metal or asbestos one without affecting the reac-

FIG. 9-1 Zinc and sulfur react chemically to give zinc sulfide, a substance whose properties are different from those of zinc and sulfur.

Zinc

Gray metal
Melts at 420°C
Density 7.1 g/cm³
Dissolves in dilute acids
Does not dissolve in carbon disulfide

Sulfur

Soft yellow solid
Melts at 113°C
Density 2.0 g/cm³
Does not dissolve in acids
Dissolves in carbon disulfide

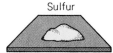

Zinc sulfide

Brittle white solid
Does not melt but decomposes into zinc and sulfur at 600°C
Density 3.5 g/cm³
Does not dissolve in either dilute acids or carbon disulfide

tion. Clearly the process involves both zinc and sulfur, but nothing else. We conclude that zinc and sulfur have joined chemically to form the new material, which is called *zinc sulfide*.

From Fig. 9-1 we can see that the properties of zinc, sulfur, and zinc sulfide are quite different from one another. Each material is a pure substance. Every particle of sulfur in the sulfur pile is like every other particle of sulfur, and the same is true for zinc and for zinc sulfide. However, if we simply mix zinc and sulfur together without heating them, the result is a **heterogeneous substance** whose properties vary from one particle to the next. With a microscope and tweezers we can separate particles of zinc from those of sulfur, which we cannot do in the case of zinc sulfide. There has been no change at all in the ingredients of the mixture of zinc and sulfur.

9-2 THREE CLASSES OF MATTER

Elements, Compounds, and Mixtures

Although the alchemists never reached their goal of turning ordinary metals into gold, their work did have an important result. What the alchemists discovered was that certain substances—such as zinc, sulfur, and gold—could be neither broken down nor changed into one another. Slowly the belief grew that only a limited number of such **elements** exist and that all other substances are combinations of them. A new material can be formed from other materials by chemical change only if the elements of the new material are present in the original ones. This observation, little more than two centuries old, marks the beginning of the science of chemistry.

Today more than 100 elements are known, most of them solids at room temperature and atmospheric pressure. About 75 percent (by mass) of the matter in the universe is a single element, hydrogen, and

---

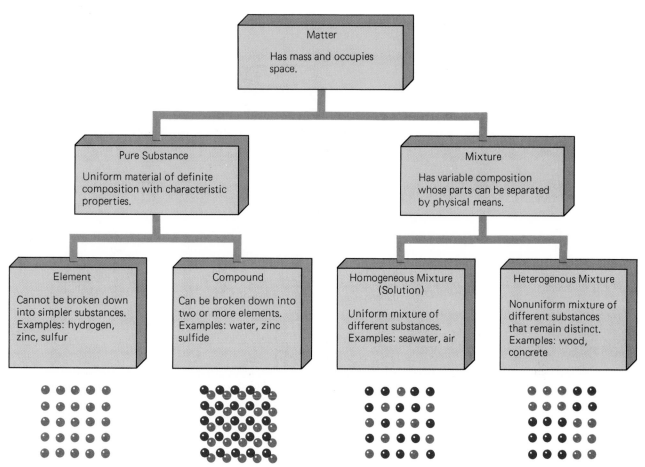

FIG. 9-2   Classification of matter.

nearly all the rest consists of one other element, helium. The other elements amount to less than 1 percent of the total. As for our planet, the four elements iron, oxygen, silicon, and magnesium make up 96 percent of the earth's mass. In the human body, oxygen is the most abundant element, followed by carbon, hydrogen, nitrogen, calcium, and phosphorus; no other element amounts to more than a fraction of a percent.

The matter around us contains elements by themselves and in a variety of combinations. Some materials consist of two or more elements joined together in chemical **compounds,** as in the case of zinc sulfide. Other materials are mixtures of elements or compounds or both. A mixture may be heterogeneous, with its components obvious to the eye and easy to separate. Wood is one example; our mixture of zinc powder and sulfur powder is another. When the components are so thoroughly mixed that the result is uniform, we have a **homogeneous mixture,** or **solution.** Thus seawater is a solution of various solids and gases dissolved in water. Figure 9-2 shows how matter is classified into its different forms.

Compound or Solution? How can we tell whether two elements have joined to make a compound or are just mixed to make a solution? A number of tests are available. Here are two:

FIG. 9-3 Elements and compounds have specific boiling and freezing points. A solution, here air, can therefore be separated into the elements or compounds it contains by boiling or freezing at an appropriate temperature, but a compound, here nitric oxide, cannot. Nitrogen boils at −196°C, oxygen boils at −183°C, and nitric oxide boils at −152°C.

1. See whether the new material can be separated into different substances by boiling or freezing. The changes of state we studied in Chap. 4 occur at specific temperatures for elements and compounds but not for mixtures. Air, for example, is a solution of several gases, mainly nitrogen and oxygen. Nitrogen boils at −196°C and oxygen boils at −183°C. If we heat liquid air to −196°C, most of the gas given off is nitrogen. The liquid left behind is richer in oxygen than the original sample. On the other hand, nitric oxide is a compound of nitrogen and oxygen that has a boiling point of −152°C. If we heat liquid nitric oxide to −152°C, all of it will boil away at this temperature, with no change in composition (Fig. 9-3).

2. Compare the relative amounts of the elements in different samples of the material. The elements in a given compound are always present in exactly the same proportions. However, the ingredients of a solution may be present in a range of proportions. At sea level, the mass ratio of the nitrogen and oxygen in air has an average value of 3.2:1, but there is more nitrogen than this at high altitudes. On the other hand, the mass ratio of these elements in nitric oxide is exactly 0.88:1 everywhere. If there is too much of either nitrogen or oxygen when nitric oxide is being made, the extra amount will not combine

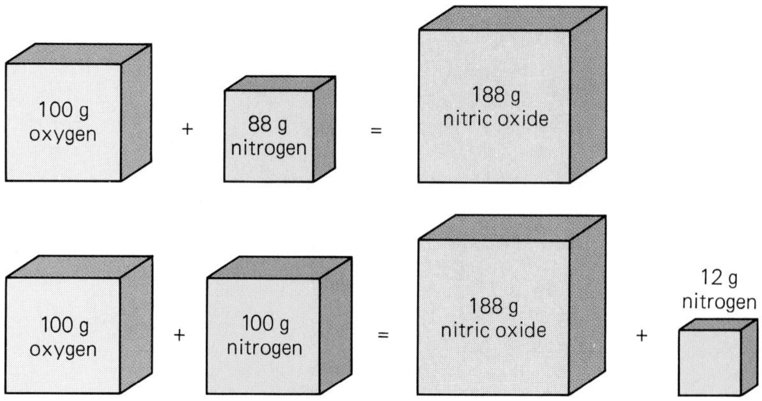

FIG. 9-4 An example of the law of definite proportions. Elements combine in a specific mass ratio when they form a compound. The mass ratio between the oxygen and nitrogen in nitric oxide is always 100:88.

but will be left over and can easily be separated (Fig. 9-4). The **law of definite proportions** is as basic to chemistry as the law of conservation of momentum is to physics.

9-3 THE ATOMIC THEORY

The Building Blocks of Matter

Two hundred years ago the structure of matter was still largely a mystery. Nobody knew what really happens when elements combine to form compounds. The explanation finally came from an English schoolteacher named John Dalton. Dalton began with the ancient Greek notion that all matter is built from basic particles called **atoms,** but he went much further. He proposed that all the atoms of each element were the same but were different from the atoms of other elements. Dalton was the first to establish the relative masses of atoms of the known elements, thus taking his ideas from the realm of philosophy and putting them into the realm of science. In Dalton's picture, compounds consist of atoms of different elements, with each compound having fixed ratios of the kinds of atoms present. Chemical reactions represent rearrangements of atoms, not changes in the atoms or the creation or destruction of atoms.

Molecules

Our modern picture of matter grew from Dalton's work. In the case of gases, the ultimate particles of a gaseous compound are its **molecules,** which in turn are made up of atoms of the elements in the compound. Some elemental gases, such as helium and neon, consist of individual atoms. Other elemental gases consist of molecules whose atoms are all the same. Thus each molecule of gaseous oxygen consists of a pair of oxygen atoms bound together.

Atoms and most molecules are very small, and even a tiny bit of matter contains huge numbers of them. If each atom in a penny were worth 1 cent, all the money in the world would not be enough to pay for it.

The molecules of a compound have fixed compositions, as Fig. 9-5 shows. This is the reason for the law of definite proportions. Each water molecule contains two hydrogen atoms and one oxygen atom, for exam-

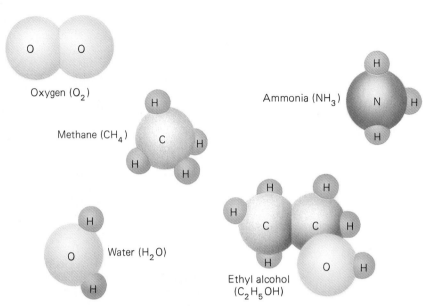

FIG. 9-5   Structures of several common molecules.

ple, and each ammonia molecule contains three hydrogen atoms and a nitrogen atom.

Two or more atoms linked into a molecule are represented by writing the symbols for their elements side by side. Thus a carbon monoxide molecule is CO and a zinc sulfide molecule is ZnS. When a molecule contains two or more atoms of the same kind, a subscript shows the number present. The familiar H_2O means that a molecule of water contains two H atoms and one O atom. A molecule of oxygen, with two O atoms, is written O_2; a molecule of nitrogen pentoxide, with two N atoms and five O atoms, is written N_2O_5. Each subscript number applies only to the symbol just in front of it. These expressions are called **chemical formulas.**

FIG. 9-6   Sodium chloride crystals consist of Na^+ and Cl^- ions rather than of neutral Na and Cl atoms or individual NaCl molecules.

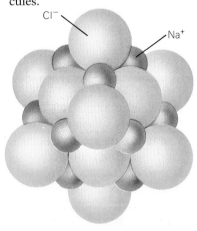

Not All Compounds Consist of Molecules

Elements in liquid and solid form are usually assemblies of individual atoms. Some liquid and solid compounds are also assemblies of individual molecules, but more often the situation is not so straightforward, as discussed in Chap. 10. For example, crystals of table salt, which is a compound of sodium and chlorine, consist of sodium and chlorine ions rather than of neutral atoms or molecules. The sodium ions are positively charged and the chlorine ions are negatively charged, as in Fig. 9-6. For every sodium ion Na^+, there is a chlorine ion Cl^-, so that the ratio between them is fixed, and the ions are firmly held together in a definite pattern. Sodium chloride is as much a compound as water, even though it is not composed of separate molecules, and its formula is NaCl.

BIOGRAPHY

John Dalton (1766–1844), the son of a Quaker weaver in England, began teaching at the age of 12, a year after his own formal education had ended. Besides having had no instruction in science, Dalton was an inept experimenter, a plodding, literal thinker, and poor at expressing his ideas. But these handicaps did not stop Dalton from placing the atomic theory of matter on a firm foundation. Indeed, by leading him to seek simple explanations for complex phenomena, they gave him an advantage over contemporaries whose minds were full of the misconceptions of the day.

Dalton's initial scientific interest was meteorology: he wrote a book on the subject and recorded weather data every day of his adult life. Studying the atmosphere started Dalton reflecting on the nature of gases and then on the nature of matter in general. He developed the concepts of atom and molecule, element and compound in detail, with numerical values derived from experiment. For instance, Dalton observed that in what is now called carbon monoxide gas the carbon:oxygen ratio by mass is 3:4 whereas in carbon dioxide it is 3:8. This led him to suggest that carbon monoxide molecules consist of one atom of carbon and one of oxygen (in symbols, CO), and that carbon dioxide molecules consist of one atom of carbon and two of oxygen (CO_2). Ratios such as these enabled Dalton to work out the relative atomic masses of many elements. Sometimes his figures were wrong (he assumed that water molecules contain one hydrogen atom for each oxygen atom, so the ob-

served mass ratio of 1:8 means that the atomic mass of oxygen is 8 times that of hydrogen; in fact, of course, there are two H atoms per O atom in water, H_2O, so the atomic mass ratio is actually 1:16), but on the whole he did very well. Soon after his book on this work appeared in 1808, most chemists accepted Dalton's ideas, and he became famous.

The atomic theory was not all that occupied Dalton, who always said he was too busy to marry. Among his other achievements were the first description of color blindness, from which he suffered, and the discovery that the warmer air is, the more water vapor it can hold.

THE PERIODIC LAW

The periodic law, now over a century old, was a giant step for chemists along the path toward understanding the nature and behavior of the elements. There cannot be many chemistry laboratories in the world that do not have a copy of the periodic table hanging on a wall. Before we examine the periodic law, we shall look at some of the ideas behind its discovery.

9-4 METALS AND NONMETALS

A Basic Distinction

The division between metals and nonmetals is a familiar one. All metals except mercury are solid at room temperature. Iron, copper, aluminum, tin, silver, and gold are examples. Nonmetals may be solid (carbon, sulfur), liquid (bromine), or gaseous (chlorine, oxygen, nitrogen) at room temperature. Metals outnumber nonmetals by more than 5:1.

A number of physical properties distinguish metals from nonmetals (Table 9-1). An obvious one is **metallic luster,** the characteristic sheen

TABLE 9-1

Some Physical Properties of Metals and Nonmetals

Property	Metals	Nonmetals
Metallic luster	Yes	No
Opaque to light	Yes	Only a few
Can be deformed without breaking	Yes	No
Conducts heat and electricity	Yes	No

of a clean metal surface. Related to this sheen is the fact that all metals are opaque—light cannot pass through even the thinnest sheet of a metal. Solid nonmetals do not show metallic luster and nearly all are transparent in thin sheets.

Another typical property of metals is their ability to be shaped by bending or hammering. One gram of gold can be beaten into a square meter of foil, and a copper rod can be pulled through a tiny hole in a steel plate to make a hair-thin wire. Solid nonmetals, though, are brittle and break instead of being deformed when enough force is applied. Metals are all good conductors of heat and electricity; nonmetals are insulators. Carbon is intermediate between metals and nonmetals. Carbon conducts heat and electricity better than other nonmetals do, and one form of it, graphite (familiar as the "lead" in pencils), is somewhat lustrous. However, all forms of carbon are brittle.

Gold leaf thinner than this page being glued to a statue. Metals differ from other solids in their ability to be rolled into thin sheets or be otherwise deformed without breaking.

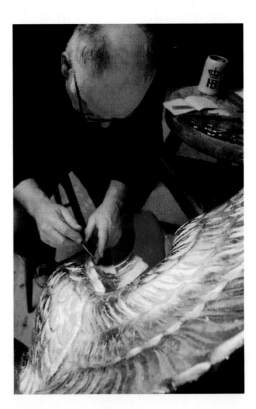

9-5 CHEMICAL ACTIVITY

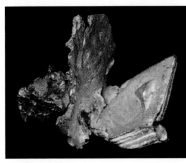

Gold occurs as the free metal because it is very inactive chemically, which is why gold objects such as rings do not tarnish or corrode. This gold nugget was found in Grass Valley, California.

The More Active an Element, the More Stable Its Compounds

Metals and nonmetals also differ in their chemical properties, but these differences are less clear-cut than the differences in physical properties because the elements in each category vary a great deal among themselves. In particular, some metals and nonmetals are very **active**, which means that they readily combine to form compounds. At the other extreme, **inactive** elements have little tendency to react chemically.

Sodium is an example of an active metal and gold is an example of an inactive one. A few seconds in the open air and sodium has lost its luster through chemical reactions, but a gold ring remains bright after a lifetime of exposure to perspiration as well as air. Sodium combines spectacularly with chlorine, giving off much heat and light. Gold combines with chlorine only sluggishly, with little energy set free. Sodium reacts with dilute acids and even with water. Gold is affected only by a mixture of concentrated hydrochloric and nitric acids.

Determining Activities

The relative activities of different elements can be established by measuring the amounts of heat given off in similar chemical reactions. Suppose we combine a given mass of chlorine with sodium and then the same mass of chlorine with gold. We would find that forming sodium chloride gives off more than 15 times as much heat as forming gold chloride. The conclusion is that sodium is much more active than gold.

Or we might start with similar compounds and ask how easily they can be separated into their component elements. In the case of gold chloride and sodium chloride, the results are that gold chloride breaks up when it is heated to about 300°C, but sodium chloride must be heated to well over 1000°C for this to happen. Gold chloride is accordingly considered to be a relatively unstable compound and sodium chloride to be a relatively stable compound. In general, the more active an element is, the more difficulty we have in decomposing its compounds.

Both metals and nonmetals can be arranged in order of their activities. In the partial listing of Table 9-2 the most active elements are at the top of each series and the least active are at the bottom.

TABLE 9-2

Relative Activities of Metals and Nonmetals

Metals	Nonmetals
Potassium	Fluorine
Sodium	Chlorine
Calcium	Bromine
Magnesium	Oxygen
Aluminum	Iodine
Zinc	Sulfur
Iron	
Lead	
Copper	
Mercury	
Silver	
Gold	

More active ↑ / Less active ↓

9-6 FAMILIES OF ELEMENTS

Members of Each Family Have a Lot in Common

Some elements resemble one another so much that they seem to be members of the same natural family. Three examples of such families are a group of active nonmetals called the *halogens*, a group of active metals called the *alkali metals*, and a group of gases that undergo almost no chemical reactions, the *inert gases*.

Halogens

The **halogens** (Table 9-3) are all highly active elements. In fact, fluorine is the most active element of all and can even corrode platinum, one of the most stable metals. The

TABLE 9-3

The Halogens

Element	Symbol	Atomic Number
Fluorine	F	9
Chlorine	Cl	17
Bromine	Br	35
Iodine	I	53
Astatine	At	85

Chlorine is a greenish-yellow gas with a strong odor at room temperature and atmospheric pressure. It is highly active chemically and is classed as a halogen.

halogens are responsible for some of the worst odors (*bromos* is Greek for "stink") and most brilliant colors (*chloros* is Greek for "green") to be found in the laboratory. The name halogen means "salt former," a token of the fact that these elements combine with many metals to give white solids that resemble table salt (which is NaCl, sodium chloride).

At room temperature fluorine is a pale-yellow gas, chlorine is a greenish-yellow gas, bromine is a reddish-brown liquid, iodine is a steel-gray solid, and astatine is a radioactive solid. About one part of chlorine per million parts of water is enough to kill any bacteria present, which is why chlorine is added to water supplies and to swimming pools.

What are the similarities among the halogens? For one thing, their molecules contain two atoms at ordinary temperatures: F_2, Cl_2, Br_2, I_2. (The half-life of astatine is too short for its chemical properties to be known.) Also, the compounds they form with metals have similar formulas. Here are three examples:

NaF	ZnF_2	AlF_3
NaCl	$ZnCl_2$	$AlCl_3$
NaBr	$ZnBr_2$	$AlBr_3$
NaI	ZnI_2	AlI_3

All the halogens react with hydrogen to form HF, HCl, HBr, and so on. These compounds can be dissolved in water to form acids, of which hydrochloric acid is a familiar example. The halogens dissolve readily in a liquid called carbon tetrachloride to give solutions colored in the same way as their vapors, but the halogens are only slightly soluble in water.

Alkali Metals

The **alkali metals** (Table 9-4) are all soft and very active chemically. Like sodium, the other alkali metals lose their lusters quickly in air, liberate hydrogen from water and dilute acids, and combine with active nonmetals to form very stable compounds. Formulas for their compounds follow similar patterns, for instance,

Bromides:	LiBr	NaBr	KBr	RbBr	CsBr	FrBr
Sulfides:	Li_2S	Na_2S	K_2S	Rb_2S	Cs_2S	Fr_2S
Hydroxides:	LiOH	NaOH	KOH	RbOH	CsOH	FrOH

All the alkali metals have rather low melting points for metals: cesium melts in a warm room, and even lithium, which has the highest melting point of the group, melts at only 186°C.

Inert Gases

The **inert gases** (Table 9-5), in contrast with the active halogens and alkali metals, are so inactive that they form only a handful of compounds with other elements. In fact, these elements are so inactive that their atoms do not even join together into molecules as the atoms of other gaseous elements do. All the inert gases are found in small amounts in the atmosphere, with argon making up about 1 percent of the air and the others much less. Their scarcity and inactivity prevented their discovery until the very end of the nineteenth century.

A volume of helium weighs much less than the same volume of air. Because it also cannot burn or explode, helium is ideal for lighter-than-

TABLE 9-4

The Alkali Metals

Element	Symbol	Atomic Number
Lithium	Li	3
Sodium	Na	11
Potassium	K	19
Rubidium	Rb	37
Cesium	Cs	55
Francium	Fr	87

TABLE 9-5		
The Inert Gases		
Element	Symbol	Atomic Number
Helium	He	2
Neon	Ne	10
Argon	Ar	18
Krypton	Kr	36
Xenon	Xe	54
Radon	Rn	86

The inert gas helium is used in these blimps because helium cannot burn or explode besides being less dense than air.

air craft such as balloons and blimps. We have already met radon, a radioactive product of radium decay, in Chap. 7. The other inert gases glow in various colors when excited by an electric current and are widely used in signs.

9-7 THE PERIODIC TABLE

A Pattern of Recurring Similarities among the Elements

A curious feature of the elements listed in Tables 9-3, 9-4, and 9-5 is that each halogen is followed in atomic number by an inert gas and then by an alkali metal. Thus fluorine, neon, and sodium have the atomic numbers $Z = 9$, $Z = 10$, and $Z = 11$, a sequence that continues through astatine (85), radon (86), and francium (87). When the properties of all the elements are checked to see what other regularities occur, the result is the **periodic law:**

> When the elements are listed in order of atomic number, elements with similar chemical and physical properties appear at regular intervals.

ATOMIC MASS As we recall from Chap. 7, nearly all elements have isotopes whose nuclei have different numbers of neutrons and hence whose atomic masses are different. The atomic mass of an element that chemists use is the *average* mass of the atoms of its various isotopes in the proportion in which they occur in nature. For instance, chlorine consists of 76 percent of the $^{35}_{17}Cl$ isotope, whose atomic mass is 34.97 u, and 24 percent of the $^{37}_{17}Cl$ isotope, whose atomic mass is 36.97 u. The average atomic mass of chlorine is 35.45 u, and this is the value given in Table 9-6.

The periodic law was first formulated in detail by the Russian chemist Dmitri Mendeleev about 1869. Although the modern quantum theory of the atom was many years in the future, Mendeleev was fully aware of the significance of his work. As he remarked, "The periodic law, together with the revelations of spectrum analysis, have contributed to again revive an old but remarkably long-lived hope—that of discovering, if not by experiment then at least by mental effort, the *primary matter.*"

A **periodic table** is a listing of the elements according to atomic number in a series of rows such that elements with similar properties form vertical columns. Table 9-6 is a simple form of the periodic table. Let us see how it organizes our knowledge of the elements.

TABLE 9-6

The Periodic Table of the Elements

The number above the symbol of each element is its atomic number, and the number below its name is its average atomic mass. The elements whose atomic masses are given in parentheses do not occur in nature but have been created in nuclear reactions. The atomic mass in such a case is the mass number of the most long-lived radioisotope of the element.

Group	1	2	3	4	5	6	7	8
Period								
1	1 H Hydrogen 1.008							2 He Helium 4.003
2	3 Li Lithium 6.941	4 Be Beryllium 9.012	5 B Boron 10.81	6 C Carbon 12.01	7 N Nitrogen 14.01	8 O Oxygen 16.00	9 F Fluorine 19.00	10 Ne Neon 20.18
3	11 Na Sodium 22.99	12 Mg Magnesium 24.31	13 Al Aluminum 26.98	14 Si Silicon 28.09	15 P Phosphorus 30.97	16 S Sulfur 32.07	17 Cl Chlorine 35.45	18 Ar Argon 39.95

Transition metals

Group	1	2	3	4	5	6	7	8	9	10	11	12	3	4	5	6	7	8
4	19 K Potassium 39.10	20 Ca Calcium 40.08	21 Sc Scandium 44.96	22 Ti Titanium 47.88	23 V Vanadium 50.94	24 Cr Chromium 52.00	25 Mn Manganese 54.94	26 Fe Iron 55.8	27 Co Cobalt 58.93	28 Ni Nickel 58.69	29 Cu Copper 63.55	30 Zn Zinc 65.39	31 Ga Gallium 69.72	32 Ge Germanium 72.59	33 As Arsenic 74.92	34 Se Selenium 78.96	35 Br Bromine 79.90	36 Kr Krypton 83.80
5	37 Rb Rubidium 85.47	38 Sr Strontium 87.62	39 Y Yttrium 88.91	40 Zr Zirconium 91.22	41 Nb Niobium 92.91	42 Mo Molybdenum 95.94	43 Tc Technetium (98)	44 Ru Ruthenium 101.1	45 Rh Rhodium 102.9	46 Pd Palladium 106.4	47 Ag Silver 107.9	48 Cd Cadmium 112.4	49 In Indium 114.8	50 Sn Tin 118.7	51 Sb Antimony 121.8	52 Te Tellurium 127.6	53 I Iodine 126.9	54 Xe Xenon 131.8
6	55 Cs Cesium 132.9	56 Ba Barium 137.3	57 La Lanthanum 138.9	72 Hf Hafnium 178.5	73 Ta Tantalum 180.9	74 W Tungsten 183.9	75 Re Rhenium 186.2	76 Os Osmium 190.2	77 Ir Iridium 192.2	78 Pt Platinum 195.1	79 Au Gold 197.0	80 Hg Mercury 200.6	81 Tl Thallium 204.4	82 Pb Lead 207.2	83 Bi Bismuth 209.0	84 Po Polonium (209)	85 At Astatine (210)	86 Rn Radon (222)
7	87 Fr Francium (223)	88 Ra Radium 226.0	89 Ac Actinium (227)	104 Unq Unnilquadium (261)	105 Unp Unnilpentium (262)	106 Unh Unnilhexium (263)	107 Uns Unnilseptium (262)	108 Uno Unniloctium (265)	109 Une Unnilennium (266)	110 Uun Ununnilium (269)								

Alkali metals

Halogens

Inert gases

Lanthanoids

57 La Lanthanum 138.9	58 Ce Cerium 140.1	59 Pr Praseodymium 140.9	60 Nd Neodymium 144.2	61 Pm Promethium (145)	62 Sm Samarium 150.4	63 Eu Europium 152.0	64 Gd Gadolinium 157.3	65 Tb Terbium 158.9	66 Dy Dysprosium 162.5	67 Ho Holmium 164.9	68 Er Erbium 167.3	69 Tm Thulium 168.9	70 Yb Ytterbium 173.0	71 Lu Lutetium 175.0

Actinoids

89 Ac Actinium (227)	90 Th Thorium 232.0	91 Pa Protactinium 231.0	92 U Uranium 238.0	93 Np Neptunium (237)	94 Pu Plutonium (244)	95 Am Americium (243)	96 Cm Curium (247)	97 Bk Berkelium (247)	98 Cf Californium (251)	99 Es Einsteinium (252)	100 Fm Fermium (257)	101 Md Mendelevium (258)	102 No Nobelium (259)	103 Lr Lawrencium (260)

How the Periodic Table Is Constructed

The first element in the table is hydrogen, which behaves chemically much like an active metal although physically it is a nonmetal. Next comes the inert gas helium, the alkali metal lithium, and the less active metal beryllium. Then follows a series of nonmetals of increasing nonmetallic activity: boron, carbon, nitrogen, oxygen, and finally the halogen fluorine. Thus from lithium to fluorine we have a complete sequence that goes from a highly active metal to a highly active nonmetal.

Following fluorine is neon, an inert gas like helium, and after neon is sodium, an alkali metal like lithium. Clearly it makes sense to break off the rows at helium and neon and start new rows with lithium and sodium under hydrogen. In the seven elements beyond neon, we find again a transition from active metals to active nonmetals.

After calcium, in the fourth row, complications appear. Scandium, the next element, is similar to aluminum in some properties but different in others. Titanium (Ti) is even less like carbon and silicon. Then come 10 metals (including iron, copper, and zinc) that are quite similar among themselves but conspicuously different from the nonmetals at the end of the first three rows. Only after the 10 metals do three relatives of these nonmetals appear, arsenic (As), selenium (Se), and bromine.

Between the gases helium and neon is a sequence of eight elements, and between neon and argon is another sequence of eight. However, between argon and krypton the sequence includes 18 elements. Beyond krypton is a second sequence of 18, including again a dozen metals with many properties in common. From xenon to the last inert gas, radon, is an even more complex sequence of 32 elements.

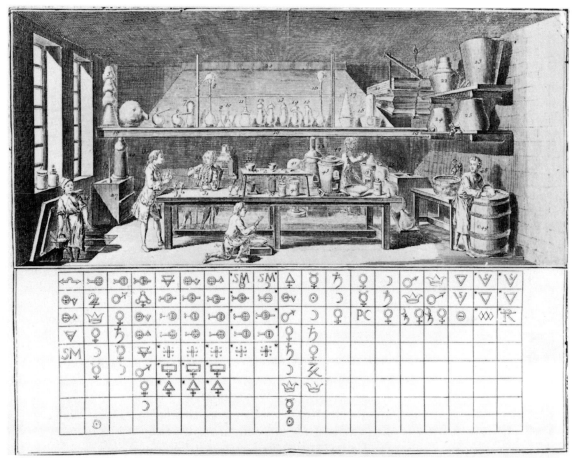

This 1780 table of chemical symbols is an early attempt to show relationships among the elements. A chemistry laboratory of the time is pictured in the engraving.

9-8 GROUPS AND PERIODS

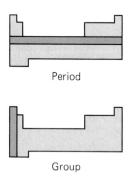

FIG. 9-7 The elements in a group of the periodic table have similar properties, whereas those in a period have different properties.

Elements in a Group Have Similar Properties; Elements in a Period Have Different Ones

The periodic table arranges families of similar elements in vertical columns called **groups.** The horizontal rows, called **periods,** contain elements with widely different properties (Fig. 9-7). Across each period is a steady change from an active metal through less active metals and weakly active nonmetals to highly active nonmetals and finally to an inert gas (Fig. 9-8). Within each column there is also a steady change in properties. Thus activity increases in the alkali metal family as we go from top to bottom down the group 1 column, and activity decreases in the halogen family as we go down the group 7 column.

Eight of the groups in Table 9-6 are numbered. The inert gases of group 8 are placed at the right since this puts them with the other nonmetals (Fig. 9-9). Each of the eight-element periods (periods 2 and 3) is

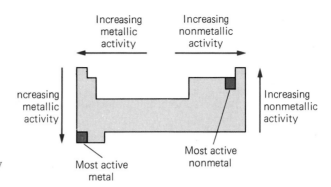

FIG. 9-8 How chemical activity varies in the periodic table.

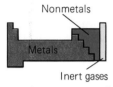

FIG. 9-9 The majority of the elements are metals.

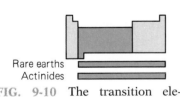

FIG. 9-10 The transition elements are metals.

broken after the second element in order to keep the members of the period in line with the most closely related elements of the long periods, which are 4 to 7.

The 10 **transition metals** in periods 4 and 5 are metals that resemble one another in chemical behavior but do not much resemble elements in the numbered groups (Fig. 9-10). Period 6 contains 32 elements, but 15 of them are brought out to a box below the table. These **rare-earth** metals are so much alike that they are hard to separate chemically and are all lumped together in the spot just to the right of barium, $Z = 56$. A similar group of closely related elements, the **actinides,** appears in the same position in period 7, and these elements are shown with the rare earths below the table.

The relationships brought out by the periodic table are a little vague in places, but on the whole the table brings together similar elements with considerable accuracy. Mendeleev's achievement is all the more remarkable when we recall that in 1869, when the periodic law was developed, the notion of atomic number had not been discovered and only 63 elements were known. Mendeleev used average atomic mass, not atomic number, to arrange the elements in the periodic table, and he and later chemists found it necessary to deviate from the strict sequence of atomic masses for certain elements. When atomic numbers were later determined for the elements, the values of Z were found to fit the sequence in the periodic table perfectly.

Mendeleev's Predictions Because so few elements were known in his time, Mendeleev had to leave gaps in his table in order to have similar elements fall in line. Sure of the correctness of his classification, he proposed that these gaps represented undiscovered elements. From the position of each gap, from the properties of the elements around it, and from the variation of these properties across the periods and down the columns, he went on to predict the properties of the unknown elements. His predictions included not only general chemical activity but also numerical values for boiling points, melting points, and so on.

As the unknown elements were discovered one by one and as their properties were found to agree with Mendeleev's predictions, the validity and usefulness of the periodic table became firmly established. Perhaps its greatest triumph came at the end of the last century, when the inert gases were discovered. Here were six new elements whose existence Mendeleev was not aware of, but they fitted perfectly as one more family of similar elements into the periodic table. The history of the periodic table is a beautiful example of the scientific method in action.

ATOMIC STRUCTURE

Now we return to the atomic theory of Chap. 8 to seek the basis of the periodic law. Two basic principles determine the structure of an atom that contains more than one electron:

1. The exclusion principle, which states that only one electron can exist in each quantum state of an atom. Thus each electron in a complex atom must have a different set of the four quantum numbers n, l, m_l, and m_s (see Table 8-1).

2. An atom, like any other system, is stable when its total energy is a minimum. This means that the various electrons in a normal atom are in the quantum states of lowest energy permitted by the exclusion principle.

9-9   SHELLS AND SUBSHELLS

They Contain Electrons with Similar Energies

Let us look into how electron energy varies with quantum state. In any atom, all the electrons with the same quantum number n are, on the average, about the same distance from the nucleus. These electrons therefore move around in nearly the same electric field and have similar energies. Such electrons are said to occupy the same atomic **shell.**

The energy of an electron in a particular shell also depends to some extent on the electron's orbital quantum number l. The higher the value of l, the higher the energy. Electrons that share a certain value of l in a shell are said to occupy the same **subshell.** All the electrons in a subshell have very nearly the same energy.

The subshells in a shell of given n can have any value of l from 0 to $n - 1$. Thus the $n = 1$ shell has only the single subshell $l = 0$; the $n = 2$ shell has the subshells $l = 0$ and $l = 1$; the $n = 3$ shell has the subshells $l = 0$, $l = 1$, and $l = 2$; and so on.

Closed Shells and Subshells

The exclusion principle limits the number of electrons that can occupy a given shell or subshell. A shell or subshell that contains its full quota of electrons is said to be **closed.**

The larger the orbital quantum number l, the more electrons the corresponding subshell can hold (see Sec. 8-13). When $l = 0$, the maximum number of electrons turns out to be 2; when $l = 1$, it is 6; when $l = 2$, it is 10; and so on. Adding up the electrons in its closed subshells gives the maximum number of electrons in a closed shell. Thus a closed $n = 1$ shell holds 2 $l = 0$ electrons; a closed $n = 2$ shell holds 2 $l = 0$ electrons plus 6 $l = 1$ electrons for a total of 8 electrons; a closed $n = 3$ shell holds these 8 electrons plus 10 more $l = 2$ electrons for a total of 18 electrons; and so on.

The concept of electron shells and subshells fits perfectly into the pattern of the periodic table, which turns out to mirror the atomic structures of the elements. Let us see how this pattern arises.

9-10 EXPLAINING THE PERIODIC TABLE

How an Atom's Electron Structure Determines Its Chemical Behavior

Table 9-7, which is illustrated in Fig. 9-11 (next page), shows the number of electrons in the shells of a number of elements. The table is arranged in the same manner as the periodic table to emphasize the relationship between the two tables.

Inert Gas Atoms

In order to interpret Table 9-7 we note that the electrons in a closed shell are all tightly bound to the atom, since the positive nuclear charge that attracts them is large relative to the negative charge of any electrons in inner shells. An atom that contains only closed shells or subshells has its electric charge uniformly distributed, so it does not attract other electrons and its electrons cannot be easily removed. We would expect such atoms to be passive chemically, like the inert gases—and the inert gases all turn out to have closed-subshell electron structures!

Hydrogen and Alkali Metal Atoms

Hydrogen and the alkali metals have single outer electrons. In the case of the hydrogen atom, the attractive force on the electron is due to a nuclear charge of only $+e$ and is not very great. In the case of the sodium atom, the total nuclear charge of $+11e$ acts on the two inner electrons, which are held very tightly. These two electrons shield part of the nuclear charge from the 8 electrons in the second shell, which are therefore attracted by a net charge of $+9e$. All 10 electrons in the first and second shells act to shield the outermost electron. This electron "sees" a net nuclear charge of only $+e$ and is held much less securely to the atom than any of the other electrons (Fig. 9-12a). This analysis also holds for the other alkali metals. As a result, atoms of hydrogen and of the alkali metals all tend to lose their outermost electrons in chemical reactions and therefore have similar chemical behavior.

Halogen Atoms

An atom whose outer shell lacks one electron from being closed tends to pick up such an electron through the strong attraction of the poorly shielded nuclear charge. The chemical behavior of the halogens is the result. In the chlorine atom, for instance, there are 10 electrons in the inner two shells, just as in the sodium atom. However, the nuclear charge of chlorine is $+17e$ as compared with only $+11e$ for sodium (Fig. 9-12b). Hence the net charge "felt" by each of the 7 outer electrons in chlorine is $+7e$, not the $+e$ in the case of the single outer electron of sodium, and the attractive force on an outer electron in chlorine is 7 times greater.

TABLE 9-7

Simplified Table of Electron Structures of Some Atoms (Subshells Are Filled When a Shell Has 2, 8, or 18 Electrons)

Electrons in	H							He
1st shell	1							2
Electrons in	**Li**	**Be**	**B**	**C**	**N**	**O**	**F**	**Ne**
1st shell	2	2	2	2	2	2	2	2
2d shell	1	2	3	4	5	6	7	8
Electrons in	**Na**	**Mg**	**Al**	**Si**	**P**	**S**	**Cl**	**Ar**
1st shell	2	2	2	2	2	2	2	2
2d shell	8	8	8	8	8	8	8	8
3d shell	1	2	3	4	5	6	7	8
Electrons in	**K**	**Ca**					**Br**	**Kr**
1st shell	2	2					2	2
2d shell	8	8					8	8
3d shell	8	8					18	18
4th shell	1	2					7	8
Electrons in	**Rb**	**Sr**					**I**	**Xe**
1st shell	2	2					2	2
2d shell	8	8					8	8
3d shell	18	18					18	18
4th shell	8	8					18	18
5th shell	1	2					7	8

FIG. 9-11 Electron structures of some atoms. In this schematic illustration of Table 9-7 the circles without dots represent closed inner shells.

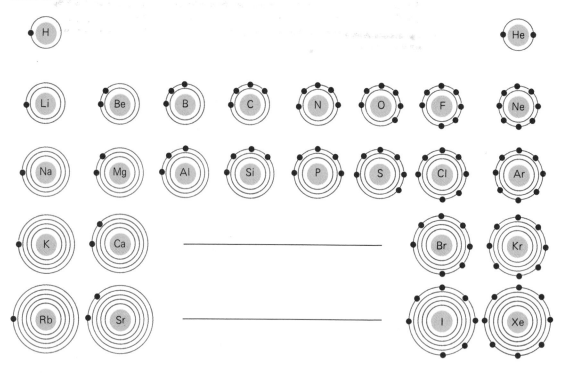

Metals and Nonmetals   The above considerations lead us to general descriptions of metal and nonmetal atoms:

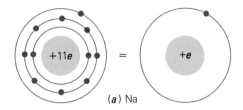

(a) Na

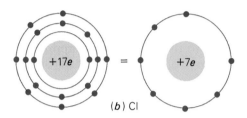

(b) Cl

FIG. 9-12 Electron shielding in sodium (a) and chlorine (b). Each outer electron in a Cl atom is acted upon by an effective nuclear charge 7 times greater than that acting upon the outer electron in a Na atom, even though the outer electrons in both cases are in the same shell.

A metal atom has one or several electrons outside closed shells or subshells. Such an atom combines chemically by losing these electrons to nonmetal atoms.

A nonmetal atom needs one or several electrons to achieve closed shells or subshells. Such an atom combines chemically by picking up electrons from metal atoms or by sharing electrons with other nonmetal atoms.

The inert gases are exceptions to these statements, of course, since their atomic structures make it hard for them to gain or lose electrons. As a result, they have almost no ability to react chemically.

The steady change in chemical properties as we go across a period from an alkali metal on the left to a halogen on the right is easy to account for. An atom of an element in group 2, for instance magnesium (Mg), has two electrons outside closed inner shells, as we see in Fig. 9-11. These electrons "feel" an effective nuclear charge of $+2e$ and so are more tightly held than the single outer electron in sodium, which "feels" an effective nuclear charge of only $+e$. Not surprisingly, the outer electrons in an Mg atom are harder to pull away than the outer electron in Na. Hence Mg is less active as a metal than Na. Aluminum (Al), with three outer electrons, holds them still more securely, which is why Al is less active than Mg.

In a nonmetal atom, the more the gaps in its outer shell, the weaker the electric field that attracts additional electrons to complete the shell. Sulfur (S), with two electrons missing from its outer shell, is therefore less active a nonmetal than chlorine, which is missing just one electron. Phosphorus (P), with three electrons missing, is even less active. We can now see why, in any period, metallic activity (losing electrons) decreases going to the right, while nonmetallic activity (gaining electrons) increases going to the right (Fig. 9-8).

Electron shells and subshells are not always filled in consecutive order in atoms with many electrons. The transition elements in any period have similar properties because their outer electron shells are the same and they add electrons successively to inner shells (see Table 9-7).

CHEMICAL BONDS

What is the nature of the forces that bond atoms together when compounds are formed? This question is of basic importance to the chemist. It is also important to the physicist because the quantum theory of the atom cannot be complete unless it provides a satisfactory answer. The ability of the quantum theory to explain chemical bonding is further testimony to the power of this approach.

9-11   TYPES OF BOND

Electric Forces Hold Atoms to One Another

Let us consider what happens when two atoms are brought closer and closer together. Three extreme situations may occur:

1. A covalent bond is formed. One or more pairs of electrons are shared by the two atoms. The shared electrons spend more time between the atoms than on their far sides, which produces an attractive force. An example is H_2, the hydrogen molecule, whose two electrons belong jointly to the two protons (Fig. 9-13).

2. An ionic bond is formed. One or more electrons from one atom shift to another atom, and the resulting positive and negative ions attract each other. An example is NaCl, where the bond exists between Na^+ and Cl^- ions and not between Na and Cl atoms (Fig. 9-14).

3. No bond is formed. The atoms do not interact to produce an attractive force.

In H_2 the bond is purely covalent and in NaCl it is purely ionic, but in many other molecules an intermediate type of bond occurs in which the atoms share electrons to an unequal extent. An example is the HCl

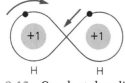

FIG. 9-13 Covalent bonding in hydrogen. The shared electrons spend more time on the average between their parent nuclei than on the far sides of the nuclei and therefore lead to an attractive internuclear force.

FIG. 9-14 Ionic bonding. Sodium and chlorine combine chemically by the transfer of electrons from sodium atoms to chlorine atoms. The resulting ions attract electrically.

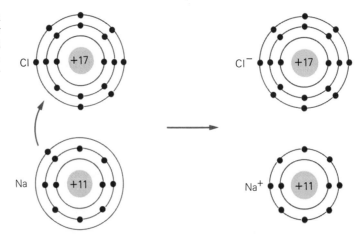

Much of the world's salt was once produced by evaporating seawater, as is still done in the Canary Islands.

molecule, where the Cl atom attracts the shared electrons more strongly than the H atom.

Ionic bonds usually do not result in the formation of molecules. Strictly speaking, a molecule is an electrically neutral group of atoms that is held together strongly enough to be experimentally observable as a particle. Thus the individual units that constitute gaseous hydrogen each consist of two hydrogen atoms, and we are entitled to regard them as molecules.

Crystals of table salt (NaCl), however, are aggregates of sodium and chlorine ions, as we saw in Fig. 9-6. Although arranged in a certain definite way, the ions do not pair off into individual molecules consisting of one Na^+ ion and one Cl^- ion. Salt crystals may in fact be of almost any size. There are always equal numbers of Na^+ and Cl^- ions in salt, so that the formula NaCl correctly represents its composition. Despite the absence of individual NaCl molecules in solid NaCl, the electric force between adjacent Na^+ and Cl^- ions makes NaCl as characteristic an example of chemical bonding as H_2.

9-12   COVALENT BONDING

Sharing Electron Pairs Produces an Attractive Force

As we saw in Fig. 9-13, two identical atoms, in this case hydrogen atoms, can bond together by sharing a pair of electrons. In some molecules more than one pair of electrons is shared. Examples are O_2, which has two shared electron pairs, and N_2, which has three. If we use a pair of dots to stand for a shared pair of electrons, the H_2, O_2, and N_2 molecules can be represented as follows:

H **:** H	O **::** O	N **:::** N
hydrogen molecule	oxygen molecule	nitrogen molecule

Substances whose atoms are joined by shared electron pairs are called **covalent.** In general, they are either nonmetallic elements or else

compounds of one nonmetal with another, although some compounds that contain metals belong to this class.

In some covalent compounds the shared electron pairs are closer to one atom than to the other. Two examples are HCl (hydrochloric acid) and H_2O:

$$H \quad \textbf{:} \quad Cl \qquad\qquad H \quad \textbf{:} \quad \overset{..}{O}$$
$$H$$

These substances are called **polar** covalent compounds, because one part of the molecule is relatively negative and another part is relatively positive. All gradations can be found between uniformly covalent molecules at one extreme, through polar covalent molecules, to ionic compounds at the other extreme. For example,

Covalent	Cl	**:**	Cl
Polar covalent	H	**:**	Cl
Ionic	Na	**:**	Cl

The carbon atom has four outer electrons to share with each other and with other atoms in covalent bonds. Covalent compounds that contain carbon are called **organic compounds.** Organic compounds are so important that all of Chap. 12 is devoted to them.

9-13 IONIC BONDING

Electron Transfer Creates Ions That Attract Each Other

The simplest example of a chemical reaction that involves electron transfer is the combination of a metal and a nonmetal. For a specific case, let us consider the burning of sodium in chlorine to give sodium chloride. From Fig. 9-11 it is clear that Na and Cl are perfect mates—one has an electron to lose, the other an electron to gain. In the process of combination, an electron goes from Na to Cl, as shown in Fig. 9-14.

The stability of the resulting closed electron shells in both ions is shown by the large amount of energy given off in the form of heat and light when this reaction takes place. The compound NaCl is quite unreactive because each of its ions has a stable electron structure. To break NaCl apart, which means to return the electron from Cl^- to Na^+, requires the same considerable energy that was set free when the compound was formed.

As we know, metal atoms tend to lose their outer electrons, like sodium in the above example. Nonmetal atoms, on the other hand, tend to gain electrons so as to fill in gaps in their outer shells. In most reactions of this sort the metal loses all its outer electrons, and the nonmetal fills all the gaps in its structure. When sodium combines with sulfur, for instance, each S atom has two spaces to fill for a closed outer shell (Fig. 9-15), but each Na atom has only one electron to give. Hence two Na atoms are needed for each S atom, and the resulting compound is Na_2S. When calcium combines with oxygen, each Ca atom contributes two electrons to each O atom, and the formula of the compound is CaO.

Compounds formed by electron transfer are called **ionic compounds.** Some are simple compounds like NaCl, Na_2S, and CaO. Others

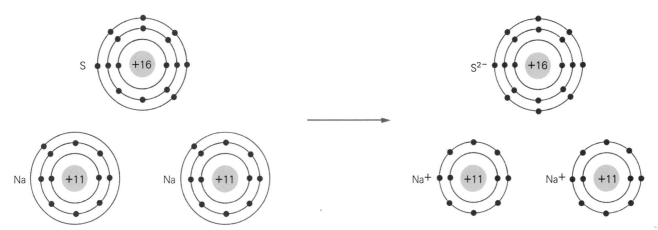

FIG. 9-15 Ionic bonding in Na_2S. Each sodium atom contributes one electron to the sulfur atom, and the resulting S^{2-} ion attracts the two Na^+ atoms.

have more complex formulas, such as Na_2SO_4, KNO_3, and $CaCO_3$. In these latter compounds electrons from the metal atoms have been transferred to nonmetal **atom groups** (SO_4, NO_3, CO_3) instead of to single nonmetal atoms.

Ionic compounds in general contain a metal and one or more nonmetals, and their crystal structures have alternate positive and negative ions. Most of them are crystalline solids with high melting points. We expect this, since melting involves separating the ions.

9-14   IONIC COMPOUNDS

Matching Up Ions   When a metal and a nonmetal combine to form an ionic compound, the atoms of the metal give up one or more electrons to atoms of the nonmetal. We can figure out the formula of the compound by knowing how many electrons the metal atoms tend to lose and how many electrons the nonmetal atoms tend to gain.

As we have already seen, Na tends to lose one electron to become Na^+ and Cl tends to gain one electron to become Cl^-. Hence the formula of sodium chloride is NaCl. Similarly sulfur tends to gain two electrons to become S^{2-}, so sodium sulfide must have the formula Na_2S in order that two electrons be available for each S atom. Calcium forms Ca^{2+} ions, hence calcium sulfide must have the formula CaS with two electrons shifting from each Ca atom to each S atom.

Table 9-8 shows the ions formed by some common elements when they enter into compounds. A few elements form different ions under different circumstances, for example copper (Cu^+, Cu^{2+}) and iron (Fe^{2+}, Fe^{3+}). In such cases the name of the element in its compound is followed by a Roman numeral to indicate the ionic charge. Thus $FeCl_2$ is called iron(II) chloride (in speech, "iron-two chloride") because it contains Fe^{2+} ions, and $FeCl_3$ is called iron(III) chloride because it contains Fe^{3+} ions.

With the help of Table 9-8 we can see what happens when a given metal combines with a given nonmetal. The positive and negative

TABLE 9-8

Ions of Some Common Elements

Element	Ion
Hydrogen	H^+
Lithium	Li^+
Sodium	Na^+
Potassium	K^+
Silver	Ag^+
Copper	Cu^+, Cu^{2+}
Mercury	Hg^+, Hg^{2+}
Magnesium	Mg^{2+}
Calcium	Ca^{2+}
Barium	Ba^{2+}
Zinc	Zn^{2+}
Iron	Fe^{2+}, Fe^{3+}
Aluminum	Al^{3+}
Tin	Sn^{2+}, Sn^{4+}
Lead	Pb^{2+}, Pb^{4+}
Fluorine	F^-
Chlorine	Cl^-
Bromine	Br^-
Iodine	I^-
Oxygen	O^{2-}
Sulfur	S^{2-}
Nitrogen	N^{3-}
Phosphorus	P^{3-}

charges on the ions must always balance out, and a little thought may be needed to find the right combination. Suppose we want the formula of aluminum oxide. We note that aluminum forms Al^{3+} ions and oxygen forms O^{2-} ions. The charges balance out for the formula Al_2O_3 since $2(+3) = +6$ for $2Al^{3+}$ and $3(-2) = -6$ for $3O^{2-}$.

9-15 ATOM GROUPS

They Act as Units in Chemical Reactions

Certain groups of atoms appear as units in many compounds and remain together during chemical reactions. An example is the group SO_4, which consists of a sulfur atom joined to four oxygen atoms. This **sulfate group,** whose ion has a charge of -2, is found in a number of compounds:

Sodium sulfate	Na_2SO_4
Potassium sulfate	K_2SO_4
Copper(II) sulfate	$CuSO_4$
Magnesium sulfate	$MgSO_4$

FIG. 9-16   When magnesium sulfate ($MgSO_4$) and barium chloride ($BaCl_2$) are dissolved in water, a precipitate of the insoluble compound barium sulfate is produced. The magnesium and chlorine ions remain in solution.

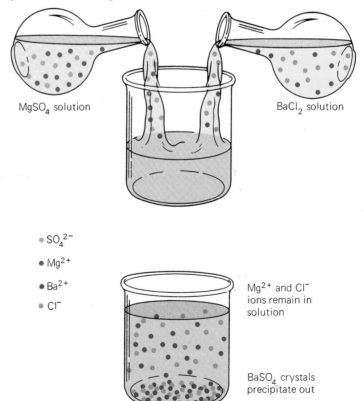

MgSO₄ solution

BaCl₂ solution

• $SO_4{}^{2-}$

• Mg^{2+}

• Ba^{2+}

• Cl^-

Mg^{2+} and Cl^- ions remain in solution

$BaSO_4$ crystals precipitate out

TABLE 9-9	
Ions of Some Common Atom Groups	
Atom Group	**Ion**
Ammonium	NH_4^+
Nitrate	NO_3^-
Permanganate	MnO_4^-
Chlorate	ClO_3^-
Hydroxide	OH^-
Cyanide	CN^-
Sulfate	SO_4^{2-}
Carbonate	CO_3^{2-}
Chromate	CrO_4^{2-}
Silicate	SiO_3^{2-}
Phosphate	PO_4^{3-}

How can we be sure that the sulfate group enters into chemical reactions as a unit? One way is to mix solutions of magnesium sulfate and barium chloride, $BaCl_2$. What happens is that a precipitate is formed, which analysis shows consists of barium sulfate, $BaSO_4$. (A **precipitate** is an insoluble solid that results from a chemical reaction in solution.) The solution left behind contains magnesium ions and chlorine ions (Fig. 9-16). The sulfate group has changed partners.

When two or more groups of a single kind are present in each molecule of a compound, the formula is written with parentheses around the group. An example is

Calcium nitrate $Ca(NO_3)_2$

The Ca^{2+} ion needs two NO_3^- ions to combine with in order that the charges balance out. Table 9-9 is a list of common atom groups and the charges their ions have.

9-16 NAMING COMPOUNDS

Language of Chemistry   Here are some of the rules chemists use to name compounds:

1. The ending *-ide* usually indicates a compound having only two elements:

Sodium chloride   NaCl
Calcium oxide   CaO

The hydroxides, which contain the OH^- ion, are the most common exceptions to this rule:

Barium hydroxide $Ba(OH)_2$

2. The ending *-ate* indicates a compound that contains oxygen and two or more other elements:

Sodium sulfate Na_2SO_4
Potassium nitrate KNO_3

3. When the same pair of elements occurs in two or more compounds, a prefix (*mono-* = 1, *di-* = 2, *tri-* = 3, *tetra-* = 4, *penta-* = 5, *hexa-* = 6, and so on) may be used to indicate the number of one or both kinds of atom in the molecule:

Carbon monoxide   CO
Carbon dioxide CO_2

4. When one of the elements in a compound is a metal that can form different ions, the scheme mentioned in Sec. 9-14 is used. In this scheme the ionic charge of the metal is given by a roman numeral:

Iron(II) chloride $FeCl_2$
Iron(III) chloride $FeCl_3$

The names of molecular compounds that contain hydrogen often follow tradition instead of a definite system. Thus

Methane CH_4

Water H_2O

Ammonia NH_3

9-17 CHEMICAL EQUATIONS

The Atoms on Each Side Must Balance

A **chemical equation** is a shorthand way to express the results of a chemical change. In a chemical equation the formulas of the **reactants** (reacting substances) appear on the left-hand side and the formulas of the products appear on the right-hand side. When charcoal (which is almost pure carbon) burns in air, for instance, what is happening is that carbon atoms are reacting with oxygen molecules in the air to form carbon dioxide molecules. The corresponding equation is therefore

$$C \; + \; O_2 \; \longrightarrow \; CO_2$$

carbon + oxygen $\longrightarrow$ carbon dioxide
atom   molecule   molecule

In order to correctly represent a chemical reaction, a chemical equation must be **balanced:** the number of atoms of each kind must be the same on both sides of the equation. Let us consider the decomposition (breaking down) of water that occurs when an electric current is passed through a water sample (Fig. 9-17). The process is called **electrolysis** and is written in words as

Water $\longrightarrow$ hydrogen + oxygen

Using the formulas for these substances, we might write

FIG. 9-17 Electrolysis of water. An electric current decomposes water into gaseous hydrogen and oxygen. The volume of the hydrogen evolved is twice that of the oxygen, since water contains twice as many hydrogen atoms as oxygen atoms. A trace of sulfuric acid is used to enable the water to conduct electricity.

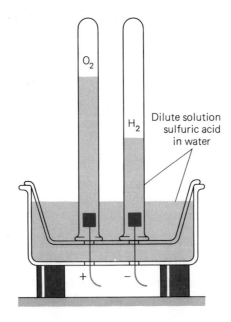

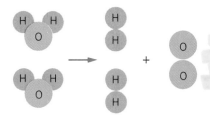

$$2H_2O \longrightarrow 2H_2 + O_2$$

FIG. 9-18 Schematic diagram of the electrolysis of water.

$$H_2O \longrightarrow H_2 + O_2 \quad \textit{unbalanced equation}$$

Here two atoms of oxygen are shown on the right-hand side but only one atom of oxygen on the left. The equation is therefore **unbalanced.** We cannot just write O instead of O_2 on the right because gaseous oxygen has the formula O_2. Nor can we write a subscript "2" after the O in H_2O because H_2O_2 is the formula for hydrogen peroxide, not water.

Balancing an Equation The first step toward balancing the equation is to show two molecules of H_2O on the left:

$$2H_2O \longrightarrow H_2 + O_2 \quad \textit{oxygen atoms balanced}$$

Now we have two O atoms on both sides of the equation, which means that these atoms are balanced. However, there are four H atoms on the left but only two H atoms on the right. The remedy is to put two H_2 molecules on the right:

$$\underset{\substack{\text{2 water}\\\text{molecules}}}{2H_2O} \longrightarrow \underset{\substack{\text{2 hydrogen}\\\text{molecules}}}{2H_2} + \underset{\substack{\text{1 oxygen}\\\text{molecule}}}{O_2} \quad \textit{balanced equation}$$

Since two O atoms and four H atoms appear on each side, the equation is balanced (Fig. 9-18).

An important point is that a number in front of a formula multiplies everything in the formula, whereas a subscript applies only to the symbol in front of it. Thus $2H_2O$ refers to two complete H_2O molecules, each one of which has two H atoms and one O atom.

Burning Propane A more complicated example is the burning of propane, C_3H_8, a gas widely used in cooking stoves and blowtorches. When it burns, propane combines with oxygen from the air to form carbon dioxide and water vapor. We begin by writing the unbalanced equation of the process:

$$C_3H_8 + O_2 \longrightarrow CO_2 + H_2O \quad \textit{unbalanced equation}$$

To balance the three carbon atoms on the left we need three on the right:

$$C_3H_8 + O_2 \longrightarrow 3CO_2 + H_2O \quad \textit{carbon atoms balanced}$$

Eight hydrogen atoms appear on the left. Hence we must have eight on the right, which means four H_2O molecules:

$$C_3H_8 + O_2 \longrightarrow 3CO_2 + 4H_2O \quad \textit{hydrogen atoms balanced}$$

The three CO_2 molecules have six oxygen atoms and the four H_2O molecules have four, for a total of ten oxygen atoms on the right. Five O_2 molecules on the left will provide the required ten oxygen atoms there:

$$C_3H_8 + 5O_2 \longrightarrow 3CO_2 + 4H_2O \quad \textit{balanced equation}$$

This is the balanced equation for the burning of propane. Since a propane stove or torch uses large amounts of oxygen and produces carbon dioxide, good ventilation is clearly necessary.

The combustion of propane produces carbon dioxide and water vapor and releases a great deal of energy.

It is worth noting that being able to balance the chemical equation for a certain reaction does not necessarily mean that the reaction can occur. And even if a reaction can take place, the balanced equation for the reaction does not tell us the particular conditions (of temperature and pressure, for instance) that might be needed.

IMPORTANT TERMS AND IDEAS

Elements are the simplest substances present in bulk matter. An element cannot be decomposed or changed into other elements by chemical means. Two or more elements may combine chemically to form a **compound,** a new substance whose properties are different from those of the elements it contains. According to the **law of definite proportions,** the elements that make up a compound are always combined in the same proportions by mass.

Other materials are **mixtures** of elements or compounds or both. The constituents of a mixture keep their characteristic properties. A **solution** is a uniform (or **homogeneous**) mixture.

The ultimate particles of any element are called **atoms.** The ultimate particles of gaseous compounds consist of atoms of the elements they contain joined together in separate **molecules.** Some compounds in the liquid and solid state also consist of molecules, but in many others the atoms are linked in larger arrays. In a given compound, however, the ratios between its various atoms are fixed.

The **periodic law** states that if the elements are listed in order of atomic number, elements with similar chemical and physical properties appear at regular intervals. Such similar elements form **groups.** The halogens, the alkali metals, and the inert gases are examples.

The electrons in an atom that have the same principal quantum number n are said to occupy the same **shell.** Electrons in a given shell that have the same orbital quantum number l are said to occupy the same **subshell.** Shells and subshells are **closed** when they contain the maximum number of electrons permitted by the exclusion principle. Atoms that contain only closed shells and subshells are extremely stable. The concept of shells and subshells is able to account for the periodic law.

A **metal** atom has one or several electrons outside closed shells or subshells. It combines chemically by losing these electrons to nonmetal atoms. A **nonmetal** atom lacks having closed shells or subshells by one or several electrons. It combines chemically by picking up electrons from metal atoms or by sharing electrons with other nonmetal atoms.

In a **covalent bond** between atoms, the atoms share one or more electron pairs. In an **ionic bond,** electrons are transferred from one atom to another and the resulting ions then attract each other. Many bonds in liquids and solids are intermediate between covalent and ionic.

Atom groups, such as SO_4^{2-} (the sulfate group), appear as units in many compounds and remain together during chemical reactions.

A **chemical equation** expresses the result of a chemical change. When the equation is **balanced,** the number of each kind of atom is the same on both sides of the equation.

EXERCISES: MULTIPLE CHOICE

1. A pure substance that cannot be decomposed by chemical means is
   a. an element   c. a solid
   b. a compound   d. a solution

2. Elements can be distinguished unambiguously by their
   a. hardnesses   c. atomic numbers
   b. colors   d. electrical properties

3. The number of known elements is approximately
   a. 50   c. 200
   b. 100   d. 500

4. At room temperature and atmospheric pressure, most elements are
   a. gases   c. metallic solids
   b. liquids   d. nonmetallic solids

5. Which of the following substances is a homogeneous mixture?
   a. iron   c. salt
   b. seawater   d. paper

6. Which of the following substances is a compound?
   a. iron   c. salt
   b. seawater   d. paper

7. The nonmetal whose chemical behavior is most like that of typical metals is
   a. hydrogen   c. chlorine
   b. helium   d. carbon

8. Compared with metals, nonmetals in general are
  a. better conductors of heat
  b. better conductors of electricity
  c. more active chemically
  d. less easily deformed (by bending, for instance)

9. Iodine is an example of
  a. an inert gas   **c.** a halogen
  b. an alkali metal   **d.** a compound

10. Of the following metals, the most active chemically is
  a. gold   **c.** iron
  b. aluminum   **d.** sodium

11. Of the following metals, the least active chemically is
  a. gold   **c.** lead
  b. calcium   **d.** sodium

12. Of the following nonmetals, the most active chemically is
  a. helium   **c.** oxygen
  b. fluorine   **d.** sulfur

13. At room temperature, chlorine is
  a. a colorless gas
  b. a greenish-yellow gas
  c. a reddish-brown liquid
  d. a steel-gray solid

14. Of the following nonmetals, the one not a halogen is
  a. fluorine   **c.** sulfur
  b. bromine   **d.** iodine

15. The place of an element in the periodic table is determined by its
  a. atomic number   **c.** density
  b. atomic mass   **d.** chemical activity

16. Each vertical column of the periodic table includes elements with chemical characteristics that are, in general,
  a. identical
  b. similar
  c. different
  d. sometimes similar and sometimes different

17. The periodic table of the elements does not
  a. permit us to make accurate guesses of the properties of undiscovered elements
  b. reveal regularities in the occurrence of elements with similar properties
  c. include the inert gases
  d. tell us the arrangement of the atoms in a molecule

18. The elements in group 1 of the periodic table (except for hydrogen) are

a. all metals
  b. all nonmetals
  c. both metals and nonmetals
  d. neither metals nor nonmetals

19. Of the elements in group 8 of the periodic table, at room temperature and atmospheric pressure
  a. all are gases
  b. all are liquids
  c. some are gases and the others liquids
  d. some are liquids and the others solids

20. An alkali metal atom
  a. has one electron in its outer shell
  b. has two electrons in its outer shell
  c. has a filled outer shell
  d. lacks one electron of having a filled outer shell

21. A halogen atom
  a. has one electron in its outer shell
  b. has two electrons in its outer shell
  c. has a filled outer shell
  d. lacks one electron of having a filled outer shell

22. An inert gas atom
  a. has one electron in its outer shell
  b. has two electrons in its outer shell
  c. has a filled outer shell
  d. lacks one electron of having a filled outer shell

23. The most important factor in determining the chemical behavior of an atom is its
  a. nuclear structure   **c.** atomic mass
  b. electron structure   **d.** solubility

24. An atom that loses its outer electron or electrons readily is
  a. an active metal
  b. an active nonmetal
  c. an inactive metal
  d. an inactive nonmetal

25. When they combine chemically with metal atoms, nonmetal atoms tend to
  a. gain electrons to become negative ions
  b. lose electrons to become positive ions
  c. remain electrically neutral
  d. any of the above, depending upon the circumstances

26. When atoms join to form a molecule,
  a. energy is absorbed
  b. energy is given off
  c. energy is neither absorbed nor given off
  d. any of the above, depending on the circumstances

27. In a covalent molecule,
  a. at least one metal atom is always present

   b. one or more electrons are transferred from one atom to another
   c. adjacent atoms share one or more electrons
   d. adjacent atoms share one or more pairs of electrons

28. An element that can form an ionic compound with chlorine is
    a. carbon   **c.** sulfur
    b. copper   **d.** neon

29. Sodium chloride crystals consist of
    a. NaCl molecules   **c.** Na^+ and Cl^- ions
    b. Na and Cl atoms   **d.** Na^- and Cl^+ ions

30. A compound whose name ends in -ate always contains
    a. hydrogen   **c.** hydrogen and oxygen
    b. oxygen   **d.** carbon

31. The number of atoms in a molecule of ammonium sulfide, $(NH_4)_2S$, is
    a. 3   **c.** 10
    b. 6   **d.** 11

32. The number of oxygen atoms in a molecule of aluminum sulfate, $Al_2(SO_4)_3$, is
    a. 3   **c.** 7
    b. 4   **d.** 12

33. The nitrate ion has the formula NO_3^-. The formula of mercury(II) nitrate is
    a. $HgNO_3$ **c.** $Hg(NO_3)_2$
    b. Hg_2NO_3 **d.** $Hg_2(NO_3)_2$

34. Which of the following chemical equations is balanced?
    a. $Fe_2O_3 + CO \longrightarrow Fe + 2CO_2$
    b. $Na_2SO_2 + S \longrightarrow S_2O_3 + S$
    c. $3CuO + 2NH_3 \longrightarrow 3Cu + 3H_2O + N_2$
    d. $4Al + 3Fe_3O_4 \longrightarrow 4Al_2O_3 + 9Fe$

35. Which of the following chemical equations is unbalanced?
    a. $2Hg + O_2 \longrightarrow 2HgO$
    b. $2H_2S + 3O_2 \longrightarrow 2H_2O + 2SO_2$
    c. $Na_2O + H_2O \longrightarrow 2NaOH$
    d. $SO_2 + H_2O \longrightarrow H_2SO_4$

36. The missing number in the equation $2Ca(NO_3)_2 \longrightarrow 2CaO + [\ ]NO_2 + O_2$ is
    a. 1   **c.** 3
    b. 2   **d.** 4

37. The missing number in the equation $4NH_3 + [\ ]O_2 \longrightarrow 4NO + 6H_2O$ is
    a. 1   **c.** 5
    b. 2   **d.** 10

QUESTIONS

1. The conversion of water to ice is considered a physical change, whereas the conversion of iron to rust is considered a chemical change. Why?

2. How can you show that water is a compound rather than a homogeneous mixture of hydrogen and oxygen?

3. Heating is a physical process. When mercuric oxide is heated, it becomes mercury and oxygen. Does this mean that mercuric oxide is a mixture rather than a compound?

4. Which of the following substances are homogeneous and which are heterogeneous? Blood, carbon dioxide gas, solid carbon dioxide, rock, steak, iron, rust, concrete, air, oxygen, salt, milk.

5. Which of the following homogeneous liquids are elements, which are compounds, and which are solutions? Alcohol, mercury, liquid hydrogen, pure water, seawater, beer.

6. How does the law of definite proportions help to distinguish between a compound of certain elements and a mixture of the same elements?

7. What is the most abundant element in the universe? In the human body?

8. From what physical and chemical characteristics of iron do we conclude that it is a metal? From what physical and chemical characteristics of sulfur do we conclude that it is a nonmetal?

9. Sodium never occurs in nature as the free element, and platinum seldom occurs in combination. How are these observations related to the chemical activities of the two metals?

10. Are the chemical properties of the elements in a vertical column or in a horizontal row of the periodic table similar to one another?

11. The element astatine (At), which appears at the bottom of the halogen column in the periodic table, has been prepared artificially in minute amounts but has not been found in nature. Using the periodic law and your knowledge of the halogens, predict the properties of this element, as follows:

    a. At room temperature, is it solid, liquid, or gaseous?
    b. How many atoms does a molecule of its vapor contain?
    c. Is it very soluble, moderately soluble, or slightly soluble in water?

d. What is the formula for its compound with hydrogen?

e. What are the formulas for its compounds with potassium and calcium?

f. Is its compound with potassium more or less stable than potassium iodide?

12. The following metals are listed in order of decreasing chemical activity: potassium, sodium, calcium, magnesium. How does this order agree with their positions in the periodic table? Where would you place cesium in the above list?

13. What is characteristic about the outer electron shells of the alkali metals? Of the halogens? Of the inert gases?

14. Group 2 of the periodic table contains the family of elements called the **alkaline earths.** How active chemically would you expect an alkaline earth element to be compared with the alkali metal next to it? Why?

15. Why do fluorine and chlorine exhibit similar chemical behavior?

16. Why do lithium and sodium exhibit similar chemical behavior?

17. Electrons are much more readily liberated from metals than from nonmetals when irradiated with visible or ultraviolet light. Can you explain why this is true? From metals of what group would you expect electrons to be liberated most easily?

18. The rare element selenium has the following arrangement of electrons: 2 in the first shell, 8 in the second, 18 in the third, and 6 in the fourth. Would you expect selenium to be a metal or a nonmetal? To what group in the periodic table would it belong?

19. At the end of the last century an entirely new group of elements, the inert gases, was discovered. Is it possible that, in the future, another as yet unknown group of the periodic table might be found?

20. Would you expect magnesium or calcium to be the more active metal? Explain your answer in terms of atomic structure.

21. What is the difference in atomic structure between the two isotopes of chlorine? How would you account for the great chemical similarity of the two isotopes?

22. Why are sodium atoms more chemically active than sodium ions?

23. Why are chlorine atoms more chemically active than chlorine ions?

24. What part of the atom is chiefly involved in each of the following processes?
a. the burning of charcoal
b. radioactive disintegration
c. the production of x-rays
d. the ionization of air
e. the emission of spectral lines
f. the explosion of a hydrogen bomb
g. the rusting of iron

25. More energy is needed to remove an electron from a hydrogen molecule than from a hydrogen atom. Why do you think this is so?

26. Illustrate with electronic diagrams (a) the reaction between a lithium atom and a fluorine atom, and (b) the reaction between a magnesium atom and a sulfur atom. Would you expect lithium fluoride and magnesium sulfide to be ionic or covalent compounds?

27. Why do the inert gas atoms almost never participate in covalent bonds?

28. The atoms in a molecule are said to share electrons, yet some molecules are polar. Explain.

29. What energy change would you expect when a molecule breaks up into its constituent atoms?

30. Which of the following compounds do you expect to be ionic and which covalent? IBr, NO_2, SiF_4, Na_2S, CCl_4, $RbCl$, Ca_3N_2.

31. What is the charge on alkali metal ions? On halogen ions? On oxygen ions?

32. The transition elements in any period have the same or nearly the same outer electron shells and add electrons successively to inner shells. How does this bear upon their chemical similarity?

33. The formula for liquid water is H_2O, for solid zinc sulfide ZnS, and for gaseous nitrogen dioxide NO_2. Precisely what information do these formulas convey? What information do they *not* convey?

34. How many atoms are present in a molecule of $Ca(NH_3)_4SO_4$? How many of them are hydrogen atoms?

35. How many atoms are present in a molecule of $C_3H_5(OH)_3$? How many of them are hydrogen atoms?

36. Name these compounds: $KMnO_4$, $HgBr_2$, Ca_3P_2, $FePO_4$, Na_2CrO_4.

37. Name these compounds: BaH_2, Li_3PO_4, PbO, $CuBr_2$, KOH.

PROBLEMS

1. What is the effective nuclear charge that acts on each electron in the outer shell of the calcium ($Z = 20$) atom? Would you think that such an electron is relatively easy or relatively hard to detach from the atom?

2. What is the effective nuclear charge that acts on each electron in the outer shell of the sulfur ($Z = 16$) atom? Would you think that such an electron is relatively easy or relatively hard to detach from the atom?

3. With the help of Tables 9-8 and 9-9 find the formulas of the following compounds: barium iodide; ammonium chlorate; tin(II) chromate; lithium phosphate.

4. With the help of Tables 9-8 and 9-9 find the formulas of the following compounds: potassium sulfide; sodium nitride; iron(III) hydroxide; calcium carbonate.

5. Which of the following equations are balanced?
a. $Zn + H_2SO_4 \longrightarrow H_2 + ZnSO_4$
b. $Al + 3O_2 \longrightarrow Al_2O_3$
c. $H_2CO_3 \longrightarrow H_2O + CO_2$
d. $3CO + Fe_2O_3 \longrightarrow 3CO_2 + 2Fe$

6. Which of the following equations are balanced?
a. $6Na + Fe_2O_3 \longrightarrow 2Fe + 3Na_2O$
b. $MnO + 4HCl \longrightarrow MnCl_2 + 2H_2O + Cl_2$
c. $C_4H_{10} + 9O_2 \longrightarrow 4CO_2 + 5H_2O$
d. $3H_2S + 2HNO_3 \longrightarrow 3S + 2NO + 4H_2O$

7. Insert the missing numbers in the following equations:
a. $Ca + [\ ]H_2O \longrightarrow Ca(OH)_2 + H_2$
b. $2Al + [\ ]H_2SO_4 \longrightarrow Al_2(SO_4)_3 + 3H_2$
c. $C_7H_{16} + 11O_2 \longrightarrow 7CO_2 + [\ ]H_2O$
d. $6H_3BO_3 \longrightarrow H_4B_6O_{11} + [\ ]H_2O$

8. Insert the missing numbers in the following equations:
a. $4Al + 3O_2 \longrightarrow [\ ]Al_2O_3$
b. $2HNO_3 + 3H_2S \longrightarrow 2NO + [\ ]H_2 + 3S$
c. $Ca(OH)_2 + 2NH_4Cl \longrightarrow CaCl_2 + [\ ]NH_3 + [\ ]H_2O$

d. $PCl_5 + [\ ]H_2O \longrightarrow H_3PO_4 + 5HCl$

Write balanced equations for the following reactions (Probs. 9 to 18):

9. Sulfur trioxide (SO_3) combines with water to form sulfuric acid (H_2SO_4).

10. Silver oxide decomposes into silver and gaseous oxygen.

11. Sodium reacts with water to give sodium hydroxide and gaseous hydrogen.

12. Potassium chlorate decomposes to give potassium chloride and gaseous oxygen.

13. Aluminum reacts with gaseous chlorine to give aluminum chloride.

14. Aluminum reacts with iron(III) oxide to give iron and aluminum oxide.

15. Aluminum reacts with a solution of hydrochloric acid (HCl) to give gaseous hydrogen and a solution of aluminum chloride.

16. Ethyl alcohol (C_2H_5OH) burns in air (that is, reacts with oxygen) to form carbon dioxide and water.

17. Acetylene gas (C_2H_2) burns in air to form carbon dioxide and water.

18. Sodium hydroxide reacts with copper(II) nitrate to give copper(II) hydroxide and sodium nitrate.

ANSWERS TO MULTIPLE CHOICE			
1. a	**11.** a	**21.** d	**31.** d
2. c	**12.** b	**22.** c	**32.** d
3. b	**13.** b	**23.** b	**33.** c
4. c	**14.** c	**24.** a	**34.** c
5. b	**15.** a	**25.** a	**35.** d
6. c	**16.** b	**26.** b	**36.** d
7. a	**17.** d	**27.** d	**37.** c
8. d	**18.** a	**28.** b	
9. c	**19.** a	**29.** c	
10. d	**20.** a	**30.** b	

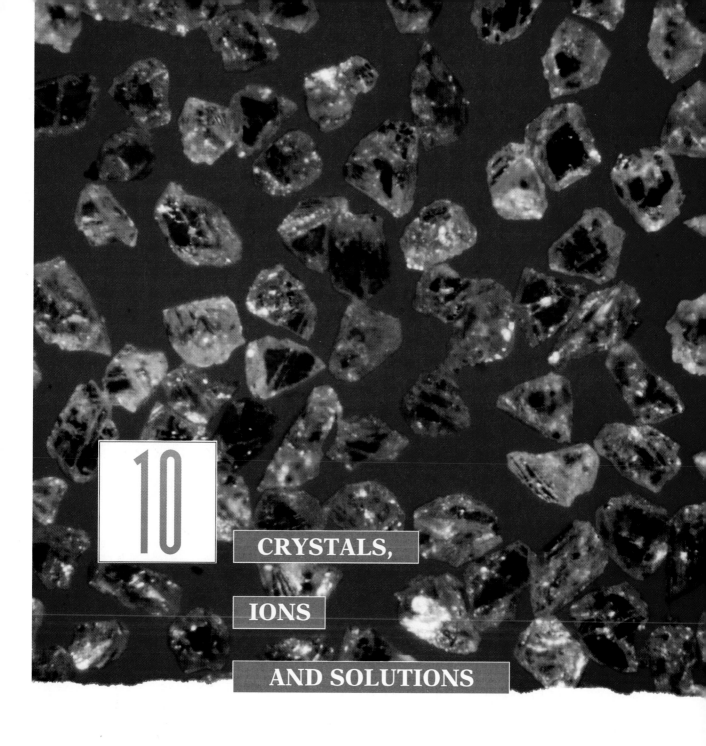

10

CRYSTALS,

IONS

AND SOLUTIONS

*T*he modern theory of the atom provides deep insights into many properties of matter. Exactly how are atoms held together in a solid? Why do metals conduct electricity but other solids do not? Why do some substances dissolve only in water and others dissolve only in liquids like alcohol or gasoline? In this chapter we shall look into the answers to these questions and others like them.

SOLIDS

A solid consists of atoms, ions, or molecules packed closely together and held in place by electric forces. Most solids are **crystalline,** which means that the particles they are made of are arranged in regular, repeated patterns. Every crystal of a given kind, whether large or small, has the same geometric form. The word *crystal* suggests salt and sugar grains, mineral samples, sparkling gemstones. But metals and snowflakes are crystalline, too, as are the fibers of asbestos and the clear, flat plates of mica. Clay is composed of tiny crystals that can trap water between them to give an easily shaped material.

Solids whose particles are irregularly arranged with no definite pattern are called **amorphous** (Greek for "without form"). Examples of amorphous solids are glass, pitch, and various plastics. One way to distinguish between the two kinds of solid is to see what happens when samples of each kind are heated. A crystalline solid melts at a specific temperature when the thermal energy of its particles is enough to break the bonds between them. An amorphous solid is really a very stiff liquid and softens gradually when heated because of the random nature of the bonds between its particles.

Crystalline solids fall into four classes, depending on how their particles are bonded together: **ionic, covalent, metallic,** and **molecular.** Let us look into how each type of bond arises.

GLASS

Glass is a transparent, amorphous solid that consists of silica (silicon dioxide, SiO_2, the chief constituent of most sands) combined with other oxides. Some glasses slowly crystallize with time and crack easily when that happens. Silica alone forms an excellent temperature-resistant glass ("quartz glass") that is transparent to ultraviolet light, unlike other glasses, but is too difficult to make for everyday use. Ordinary glass consists of about 75 percent SiO_2, 15 percent Na_2O, and 10 percent CaO. Pyrex glass, more resistant to temperature changes, is largely silica and B_2O_3 with small amounts of other oxides. Lead glass, a soft, highly refractive glass used in optical instruments and expensive glassware, is made up of SiO_2, PbO, and K_2O. Traces of certain metal oxides are responsible for most colored glass. The green glass of cheap bottles contains a little of the iron oxide FeO that was originally present as an impurity in its ingredients. Cobalt oxide gives glass a blue color, manganese oxide a violet color, and uranium oxide a yellow color. Red glass gets its hue from tiny particles of gold and copper.

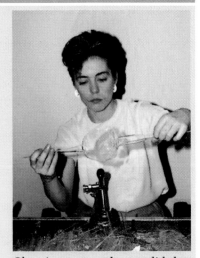

Glass is an amorphous solid that softens gradually when heated instead of melting at a specific temperature as crystalline solids do. For this reason hot glass is easily shaped.

10-1   IONIC AND COVALENT CRYSTALS

Electron Transfer and Electron Sharing in Solids

As we know, ionic bonds occur when metal atoms, which tend to lose electrons, interact with nonmetal atoms, which tend to pick up electrons. The result is a stable assembly of positive and negative ions. Ionic bonds are usually fairly strong and result in hard crystals with high melting points.

Figure 10-1 shows the arrangement of Na^+ and Cl^- ions in a sodium chloride crystal. The ions of each kind may be thought of as located at the corners and centers of the faces of a series of cubes, with the Na^+ and Cl^- cubes overlapping. Each ion thus has six nearest neighbors of the other kind. Such a crystal structure is called **face-centered cubic.**

A different structure is found in cesium chloride crystals, where each ion is located at the center of a cube at whose corners are ions of the other kind (Fig. 10-2). This structure is called **body-centered cubic, and each ion has eight nearest neighbors of the other kind.** Still other types of structures are found in ionic crystals.

The forces that hold covalent crystals together can be traced to electrons between adjacent atoms. Each atom involved in a covalent bond donates an electron to the bond, and these electrons are shared by both atoms.

Diamond is a crystalline form of carbon in which the atoms are linked by covalent bonds. Figure 10-3 shows the structure of a diamond crystal. Each carbon atom has four nearest neighbors and shares an electron pair with each of them. Since all the electrons in the outer shells of the carbon atoms participate in the bonding, it is not surprising that diamonds are extremely hard and must be heated to over 3500°C before they melt. (The name comes from the Greek *adamas,* which means "unconquerable.") Few purely covalent crystals are known. Besides diamond, some examples are silicon, germanium, and silicon carbide ("Carborundum").

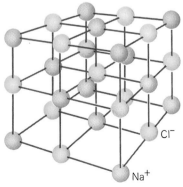

FIG. 10-1 The face-centered-cubic crystal structure of sodium chloride.

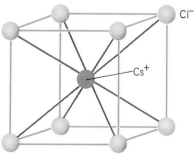

FIG. 10-2 The body-centered-cubic crystal structure of cesium chloride.

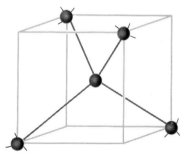

FIG. 10-3 The crystal lattice of diamond. The carbon atoms are held together by covalent bonds, which are shared electron pairs.

This apparatus uses x-rays to study the structure of a crystal. The atoms (or ions) in a crystal diffract a beam of x-rays at angles that vary with the wavelength of the x-rays, the spacing of the atoms, and their arrangement. From the pattern of diffracted x-rays, the corresponding pattern of atoms in the crystal can be found.

As in the case of molecules, it is not always possible to classify a given crystal as being wholly ionic or wholly covalent. Silicon dioxide (quartz) and tungsten carbide, for instance, contain bonds of mixed character.

GRAPHITE AND BUCKYBALLS Carbon can occur in two other forms besides diamond. One is the familiar **graphite,** a soft, black, lustrous solid that is a fair conductor of electricity—all properties very different from those of diamond. Coke, charcoal, and soot are composed mainly of small graphite crystals. Figure 10-4 shows the structure of graphite, which consists of sheets of carbon atoms in hexagonal arrays in which each atom is linked to three others. Weak van der Waals forces (see Sec. 10-3) bond the layers together. The layers can slide past each other readily and are easily flaked apart, which is why graphite is so useful as a lubricant and in pencils, where it is mixed with a clay binder.

Under ordinary conditions graphite is more stable than diamond, so crystallizing carbon produces only graphite. Because graphite is less dense than diamond, high pressures favor the formation of diamond. Natural diamond originates deep in the earth where pressures are enormous. To synthesize diamonds, graphite is dissolved in molten cobalt or nickel and the mixture is compressed at about 1900°C to about 60,000 atmospheres. The diamonds that form are less than 1 mm across and are widely used industrially for grinding and cutting tools. About 100 tons of synthetic diamonds are produced each year, 10 times the amount of natural diamonds mined.

The third form of carbon was accidentally discovered in 1985 at Rice University in Texas. The common-est version consists of 60 carbon atoms arranged in a cage structure of 12 pentagons and 20 hexagons whose geometry is the same as that of a soccer ball (Fig. 10-5). This extraordinary molecule was called "buckminster-fullerene" in honor of the American architect R. Buckminster Fuller, whose geodesic domes it resembles; the name is usually shortened to **buckyball.**

Buckyballs can be made in the laboratory from graphite, and are present in small quantities in ordinary soot and in a carbon-rich rock found in Russia. The original C_{60} buckyball is not the only form of fullerene known: C_{28}, C_{32}, C_{50}, C_{70} (also present in the Russian

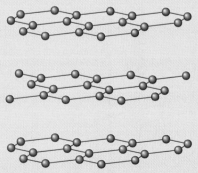

FIG. 10-4 Graphite is a form of carbon that consists of layers of carbon atoms in hexagonal arrays. The layers are held together by the weak van der Waals forces described in Sec. 10-3.

10-2 THE METALLIC BOND

The electron "gas" in a metal is responsible for the shiny surfaces of metal objects such as this silver teapot.

The Electron "Gas" That Bonds Metals Makes Them Good Conductors

A metal atom has only one or a few electrons in its outer shell, and these electrons are loosely attached. When metal atoms come together to form a solid, their outer electrons are given up to a common "gas" of electrons that move relatively freely through the assembly of metal ions. This negatively charged electron gas acts to hold together the positively charged metal ions.

The electron-gas picture of the metallic bond accounts nicely for the properties of metals. Metals conduct heat and electricity well because the free electrons can move about easily. In nonmetallic solids, all the electrons are bound firmly to particular atoms or pairs of atoms, which is why nonmetals are poor conductors (in other words, good insulators). Free electrons in metals respond readily to electromagnetic waves, which is why metals are opaque to light and have shiny surfaces. Since neighboring atoms in a metal are not linked to each other by specific bonds, most alloys—mixtures of different metals—do not obey the law of definite proportions discussed in Sec. 9-2. The copper and tin in bronze, for instance, need not be present in any exact ratio.

---

rock), C_{76}, C_{78}, and still larger ones have been made. Fullerene molecules are held together to form solids by weak van der Waals forces (like those that hold together the layers of C atoms in graphite); solid C_{60} is yellowish-brown and C_{70} is reddish-brown. Since their discovery, the fullerenes and their offshoots have shown some remarkable properties. For instance, solid C_{60} with potassium atoms in the spaces between the buckyballs ("potassium buckide") is a superconductor, and if buckytubes—cylinders of carbon atoms arranged in hexagons, like rolled-up graphite layers—can be made long enough, they will form exceedingly strong fibers, perhaps the strongest possible.

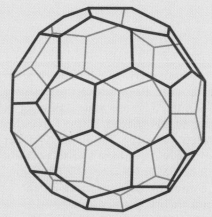

FIG. 10-5 In a buckyball, carbon atoms form a closed cagelike structure in which each atom is bonded to three others. Shown here is the C_{60} buckyball that contains 60 carbon atoms. The lines represent carbon-carbon bonds; their pattern of hexagons and pentagons closely resembles the pattern made by the seams of a soccer ball. Other buckyballs have different numbers of carbon atoms.

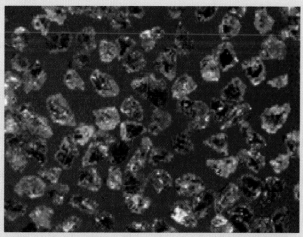

Extremely high temperatures and pressures are needed to change graphite, the ordinary form of carbon, into diamond. Only very small diamonds, such as these, have been produced artificially.

HOW SOLIDS DEFORM Few crystals have completely regular structures. A common defect is a missing line of particles, which is called a dislocation. Dislocations make it possible to permanently change the shape of a crystalline solid (a metal in particular) without breaking it. We might imagine that such a solid is deformed by the sliding of layers of its particles over one another, like sheets of paper in a pile. However, such sliding means that millions of bonds between particles have to be broken at the same time, which would require forces thousands of times stronger than those actually needed.

We can see what really happens in Fig. 10-6 when the forces shown act on a solid with a dislocation. Here the dislocation moves to the right as the particles in the layer under it shift their bonds with the particles in the layer on top one row at a time. In (c) the shape of the solid has become permanently altered. The forces needed for this step-by-step process are much smaller than those needed to deform a perfect crystal.

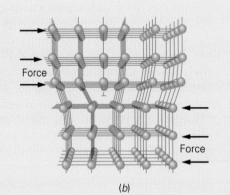
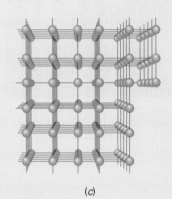

(a)   (b)   (c)

FIG. 10-6 The motion of a dislocation (line of missing particles) in a crystalline solid results in a permanent change in the shape of the solid.

10-3   MOLECULAR CRYSTALS

Van der Waals Forces Bond Molecules Together

Many molecules are so stable that they have no tendency to join together by transferring or sharing electrons. However, even these stable molecules can form liquids and solids through the action of what are called **van der Waals forces.** These forces are named after the Dutch physicist Johannes van der Waals, who suggested their existence nearly a century ago to account for the small but definite departures of actual gases from the ideal gas law. The explanation of how the forces come into being is more recent, of course, since it is based on the quantum theory of the atom.

We recall from Chap. 9 that molecules held together by polar covalent bonds behave as though they are negatively charged at one end and positively charged at the other end. An example is the H_2O molecule. In this molecule, the tendency for the shared electrons to favor the O atom makes the oxygen end of the molecule more negative than the end where the hydrogen atoms are (Fig. 10-7). Such **polar mole-**

FIG. 10-7 The electron distribution in a water molecule is such that the end where the H atoms are attached behaves as if positively charged and the opposite end behaves as if negatively charged. The water molecule is therefore polar.

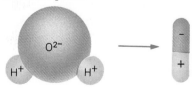

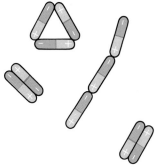

FIG. 10-8 Polar molecules attract each other.

Saran owes its clinging ability to polar molecules on its surface.

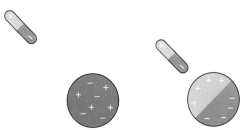

cules line up with the ends that have opposite charges adjacent, as in Fig. 10-8.

A polar molecule can also attract nonpolar molecules. Figure 10-9 shows a polar molecule approaching a nonpolar molecule. When the two molecules are close enough, the electric field of the polar molecule causes charges in the nonpolar molecule to separate. The two molecules now have charges of opposite sign facing each other, which produces an attractive force. The thin plastic sheets that stick so readily to whatever they touch do so because of polar molecules on their surfaces. The polar molecules cause molecules in the other material (the glass of the bowl you cover with cling film, for instance) that were originally nonpolar to become polar, and as a result, the plastic sheet is held firmly in place.

More remarkably, two nonpolar molecules can attract each other electrically. The electrons in a **nonpolar molecule** are distributed evenly *on the average*. However, the electrons are in constant motion, and so *at any moment* one part of the molecule has more than the usual number of electrons and the rest of the molecule has less. When two nonpolar molecules happen to get close together, their changing charge distributions tend to shift together, with adjacent ends always having opposite signs (Fig. 10-10). The result is an attractive force.

Van der Waals forces occur not only between all molecules but also between all atoms, including those of the inert gases that do not otherwise interact. Van der Waals bonds are much weaker than ionic, covalent, and metallic bonds. As a result, molecular crystals generally have low melting and boiling points and little mechanical strength. Ordinary ice and dry ice (solid CO_2) are examples of molecular solids.

Table 10-1 summarizes the characteristics of the four kinds of crystalline solids.

FIG. 10-9 Polar molecules attract normally nonpolar molecules.

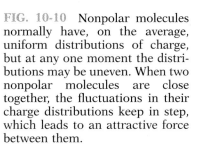

FIG. 10-10 Nonpolar molecules normally have, on the average, uniform distributions of charge, but at any one moment the distributions may be uneven. When two nonpolar molecules are close together, the fluctuations in their charge distributions keep in step, which leads to an attractive force between them.

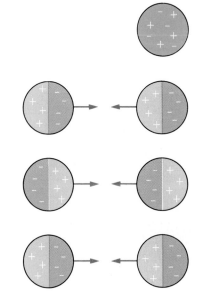

ICE   The molecular solid ice deserves an additional word. The crystal structure of ice is very open (Fig. 10-11) because an H_2O molecule can form bonds with only four other H_2O molecules. In other solids, each atom or molecule may have as many as 12 neighbors, which gives crystals that are much more compact than ice crystals. The molecules in liquid water are closer together on the average than those in ice. Water therefore expands when it freezes, which is why water pipes may burst on a cold winter's day. The expansion means that a given volume of ice weighs less than the same volume of water, which is why ice cubes float in a glass of water.

Since ice floats, a body of water outdoors freezes from the top down. Ice is a fair insulator of

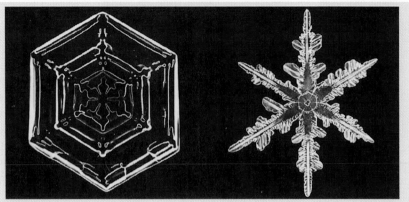

The water molecules in a snowflake are held together by van der Waals bonds.

heat, and so a layer of it on the surface of a body of water helps prevent further freezing. As a result, many lakes, rivers, and arms of the sea do not freeze solid in winter, which allows their plant and animal life to survive until the following spring.

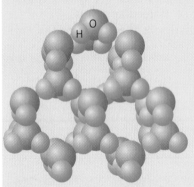

FIG. 10-11 Top view of an ice crystal, showing the open hexagonal arrangement of the H_2O molecules. The molecules in liquid water are randomly arranged; hence water is denser and ice floats.

Fishing through floating ice in the Arctic.

TABLE 10-1

Crystal Types (Bonds Are Strongest in Covalent Crystals, Weakest in Molecular Crystals)

Type		Bond	Example	Properties
Covalent	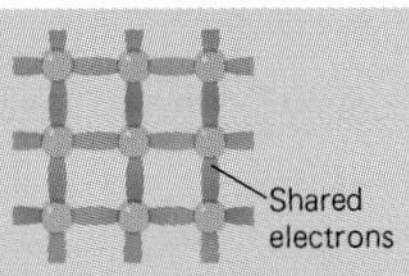 Shared electrons	Shared electrons	Diamond C	Very hard; high melting point
Ionic	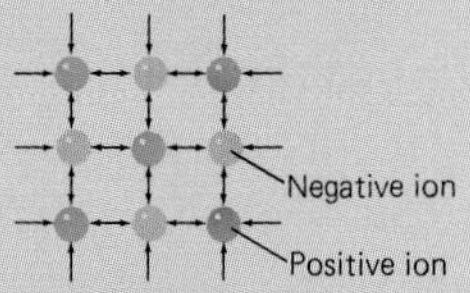 Negative ion / Positive ion	Electrical attraction	Sodium chloride NaCl	Hard; high melting point
Metallic	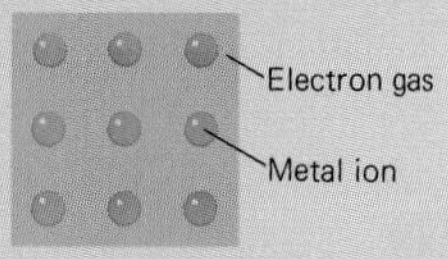 Electron gas / Metal ion	Electron gas	Copper Cu	Can be deformed; metallic luster; high electrical and thermal conductivity
Molecular	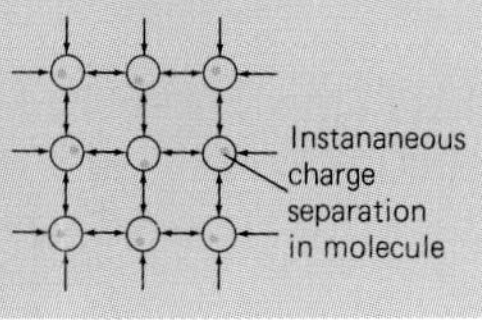 Instantaneous charge separation in molecule	Van der Waals forces	Ice H_2O	Soft; low melting point

SOLUTIONS

A solution is an intimate mixture of two or more substances. Solutions can be formed of any of the three states of matter. Thus air is a solution of several gases, seawater is a solution of various solids and gases in a liquid, and many alloys are "solid solutions" of two more metals. Here our concern will be with solutions in liquids.

10-4 SOLUBILITY

Solvent and Solute

In a solution that contains two substances, the substance present in the larger amount is called the **solvent** and the other substance is called the **solute.** When solids or gases dissolve in liquids, the liquid is always considered the solvent. When sugar is stirred into water, the sugar is the solute and the water is the solvent. Water is by far the most common and most effective of all solvents.

The **concentration** of a solution is the amount of solute in a given amount of solvent. Solutions, like compounds, are homogeneous, but unlike compounds, solutions do not have fixed compositions. To a sodium chloride solution whose concentration is 10 g of NaCl in 100 g of water, for example, we can add somewhat more NaCl or as much more water as we like. The concentration of the solution is altered, but it remains uniform.

Some pairs of liquids form solutions in all proportions. Any amount of alcohol may be mixed with any amount of water to form a homogeneous liquid, for instance. In general, however, a given liquid will dis-

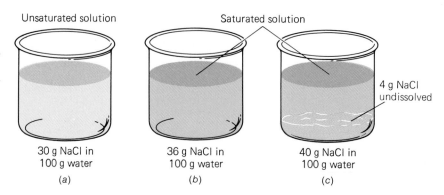

FIG. 10-12 The solubility of NaCl is 36 g per 100 g of water at 20°C. (a) 30 g of NaCl in 100 g of water produces an unsaturated solution. (b) 36 g of NaCl is the maximum amount that can dissolve, and it produces a saturated solution. (c) If 40 g of NaCl is added to 100 g of water, 4 g will remain undissolved.

Unsaturated solution

Saturated solution

4 g NaCl undissolved

30 g NaCl in 100 g water
(a)

36 g NaCl in 100 g water
(b)

40 g NaCl in 100 g water
(c)

solve only a limited amount of another substance. Sodium chloride can be stirred into water at 20°C until the solution contains 36 g of the salt for every 100 g of water. More salt will not dissolve, no matter how much we stir (Fig. 10-12). This figure, 36 g per 100 g of water, is called the **solubility** of NaCl in water at 20°C:

> **The solubility of a substance is the maximum amount that can be dissolved in a given quantity of a particular solvent at a given temperature and pressure.**

Saturated Solutions

A solution that contains the maximum amount of solute possible is said to be **saturated.** The solubilities of most solids increase with increasing temperature (Fig. 10-13). We all know that hot water is a better solvent than cold water; for example, hot tea can dissolve about twice as much sugar as iced tea. When a solution that is saturated at a high temperature is allowed to cool, some of the solute usually crystallizes out (Fig. 10-14).

Sometimes, if the cooling of a saturated solution is allowed to take place slowly and without disturbance, a solute may remain in solution even though its solubility is exceeded. The result is a **supersaturated solution.** Supersaturated solutions are often unstable, with the solute

FIG. 10-13 How the solubilities of various compounds in water vary with temperature. The higher the temperature, the greater the solubility.

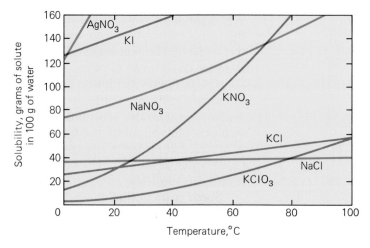

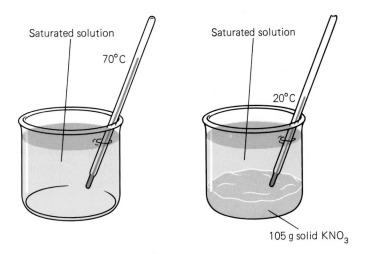

FIG. 10-14 The solubility of potassium nitrate, KNO_3, is 136 g per 100 g of water at 70°C and 31 g at 20°C. Cooling a saturated solution of KNO_3 from 70°C to 20°C causes 136 g − 31 g = 105 g of the salt per 100 g of water to crystallize out.

crystallizing out suddenly when the solution is jarred or otherwise disturbed.

The boiling point of a solution is usually higher than that of the pure solvent, and its freezing point is lower. Thus seawater, which contains about 3.5 percent of various salts (chiefly NaCl), boils at 100.3°C and freezes at −1.2°C. The more concentrated the solution, the greater the changes in boiling and freezing points. Ethylene glycol, $C_2H_4(OH)_2$, is often added to the water in the cooling system of a car to prevent the water from freezing in cold weather. A solution of 83 g of ethylene glycol per 100 g of water will not freeze until −25°C.

In contrast to the case of solids, the solubilities of gases in liquids *decrease* with increasing temperature. We all know that warming a glass of soda water, which is a solution of carbon dioxide gas in water, causes some of the gas to escape as bubbles. The solubility of a gas in a liquid depends on the pressure as well, increasing with increasing pressure. Soda water is bottled under high pressure, and when a bottle of it is opened, the drop in pressure causes some of the gas to leave the solution and form bubbles (Fig. 10-15).

FIG. 10-15 The higher the pressure and the lower the temperature, the greater the solubility of a gas in water.

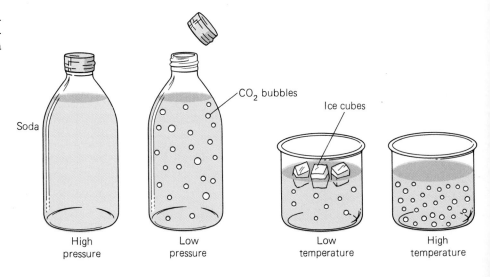

10-5 POLAR AND NONPOLAR LIQUIDS

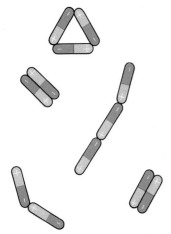

FIG. 10-16 Water molecules cluster together because of electric forces that arise from their polar character.

Like Dissolves Like   Some liquids are better solvents for some substances than for others. Water readily dissolves salt and sugar but not fats or oil. Gasoline, on the other hand, dissolves fats and oils but not salt or sugar.

The explanation for this behavior depends upon the electrical characters of the solvent and solute. Water is a **polar liquid** since its molecules behave as if negatively charged at one end and positively charged at the other (Fig. 10-5). Gasoline is a **nonpolar liquid** since the charges in its molecules are evenly distributed. Let us see what difference this makes.

Water and other polar liquids consist of groups of molecules rather than single, freely moving molecules. The molecules join together in clumps, positive charges near negative charges, as in Fig. 10-16. Water molecules can join together in a similar way with polar molecules of other substances, such as sugar (Fig. 10-17), so water dissolves these substances with ease.

Molecules of fats and oils are nonpolar and so do not interact with water molecules. If oil is shaken with water, the strong attraction of water molecules for one another squeezes out the nonpolar oil molecules from between them and the oil and water separate into layers. Oil or fat molecules mix readily, however, with the similarly nonpolar molecules of gasoline (Fig. 10-18).

A covalent substance, then, dissolves only in liquids whose molecules have similar electrical structures. In general, "like dissolves like."

Soaps and detergents are effective cleansing agents when dissolved in water because their molecules are polar at one end but nonpolar at the other. Dirt particles are nonpolar for the most part, and the nonpolar ends of soap molecules become attached to them (Fig. 10-19). The soap molecules are then pulled by their polar ends into the polar water with the dirt particles still attached to the nonpolar ends of the soap molecules. In this way dirt can be loosened and then washed away from skin, clothing, and other surfaces.

FIG. 10-17 Sugar dissolved in water. Polar compounds readily dissolve in water because their molecules can link up with water molecules.

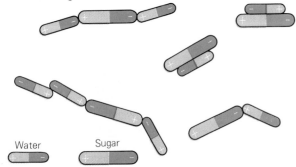

Water   Sugar

FIG. 10-18 Gasoline dissolves fat; water does not. Nonpolar compounds dissolve only in nonpolar liquids.

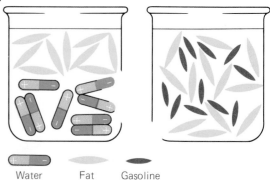

Water   Fat   Gasoline

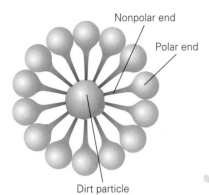

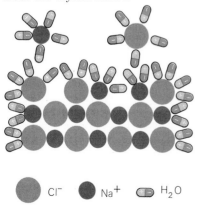

FIG. 10-19 Soap and detergent molecules are polar at one end and nonpolar at the other. The nonpolar ends become attached to a dirt particle to form a spherical cage around it called a micelle. Because the outside of a micelle is polar, it readily becomes suspended in water. By forming micelles around dirt particles, soap and detergent molecules remove them from dirty surfaces, and the micelles can then be rinsed away with water, taking the dirt with them.

Dissociation Ionic compounds consist of positive and negative ions and therefore dissolve only in highly polar liquids. Figure 10-20 shows how NaCl dissolves in water. At the surface of a salt crystal, water molecules are attracted to the ions, positive ends toward negative ions and negative ends toward positive ions. The pull of several water molecules is enough to overcome the electric forces that hold an ion to the crystal. The ion then moves off into the solution with its cluster of water molecules. As each layer of ions is removed, the next is attacked, until either the salt is completely dissolved or the solution becomes saturated. The separation of a compound into ions when it dissolves is called **dissociation.**

The ions released when an ionic compound dissolves are the same as those in its crystal structure. This is true not only for such simple compounds as NaCl, which dissociates into Na^+ and Cl^- ions, but also for more complex compounds that involve atom groups. An example is potassium nitrate, KNO_3, which dissociates into K^+ and NO_3^- ions.

Substances that separate into ions when dissolved in water are called **electrolytes.** Electrolytes include all ionic compounds soluble in water and some covalent compounds containing hydrogen (for example, hydrochloric acid, HCl) that form ions by reaction with water. Soluble covalent compounds, such as sugar and alcohol, that do not dissociate in solution are **nonelectrolytes.**

Electrolytes can be recognized by the ability of their solutions to conduct electric current. Hence their name. Conduction is possible because the ions are free to move, with positive ions migrating through the solution toward the negative terminal, negative ions migrating toward the positive terminal (Fig. 10-21).

FIG. 10-20 Solution of sodium chloride crystal in water. Water molecules exert electric forces on the Na^+ and Cl^- ions that are strong enough to remove them from the crystal lattice.

FIG. 10-21 (a) An electrolyte such as NaCl in solution conducts electric current through the motion of its ions. (b) Pure water is a nonelectrolyte, as are solutions of compounds that do not dissociate.

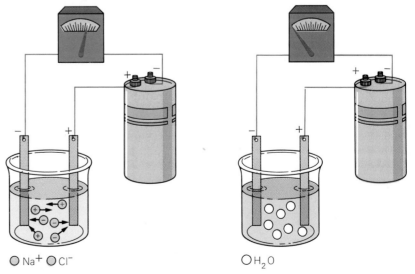

(a) NaCl solution   (b) Pure water

10-6 IONS IN SOLUTION

This device uses the color difference between chromic and dichromate ions to measure the alcohol concentration in a person's breath.

Ions Have Characteristic Properties of Their Own

One of the early objections to the theory of ionic solutions was that sodium chloride was supposed to break down into separate particles of sodium and chlorine, yet the solution remains colorless. Why, if chlorine is present as free ions, should we not find the greenish-yellow color of chlorine in the solution? The answer is that chloride ion Cl^- has altogether different properties from gaseous chlorine—a different color, a different taste, different chemical reactions.

We must regard a solution of sodium chloride not as a solution of NaCl or of Na and Cl atoms but as a solution of the two ions Na^+ and Cl^-. Each of these ions in solution has its own set of properties, properties quite different from those of NaCl crystals or of the active metal Na and the poisonous gas Cl_2. Each ion in an electrolytic solution is a new and separate substance.

By the "properties of an ion" we really mean the properties of solutions in which the ion occurs. A solution of a single kind of ion, all by itself, cannot be prepared; positive ions and negative ions must always be present together, so that the total number of charges of each sign will be the same. But each ion gives its own characteristic properties to all solutions containing it, and these properties can be recognized whenever they are not masked by other ions.

Let us look at some examples. A property of the copper ion Cu^{2+} is its blue color, and all solutions of this ion are blue (unless some other ion that has a stronger color is present). A characteristic of the hydrogen ion H^+ is its sour taste, and all solutions containing this ion (namely acids) are sour. The silver ion Ag^+ forms an insoluble white precipitate of AgCl when mixed with solutions of the chloride ion Cl^-. Any solution of an electrolyte that contains silver will give this precipitate when mixed with a solution of any chloride (Fig. 10-22).

To emphasize the differences between the properties of an ion and those of the corresponding neutral substance, Table 10-2 compares the chloride ion, Cl^-, and molecular chlorine, Cl_2. In general, Cl_2 is much more active chemically. This is to be expected, since Cl atoms have only seven outer electrons whereas Cl^- ions have closed outer shells with eight electrons.

For any ion we can list the properties common to all its solutions. In general, the properties of a solution of an electrolyte are the sum of the

TABLE 10-2

The Properties of Molecular Chlorine and Chloride Ion in Solution

Cl_2	Cl^-
Greenish-yellow color	Colorless
Strong, irritating taste and odor	Mild, pleasant taste
Combines with all metals	Does not react with metals
Combines readily with hydrogen	Does not react with hydrogen
Does not react with Ag^+	Forms AgCl with Ag^+

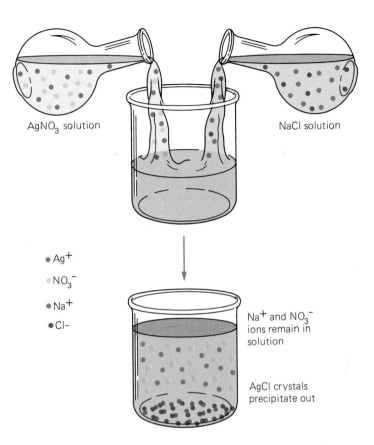

$\bullet$ Ag$^+$

$\bullet$ NO$_3^-$

$\bullet$ Na$^+$

$\bullet$ Cl$^-$

Na$^+$ and NO$_3^-$
ions remain in
solution

AgCl crystals
precipitate out

FIG. 10-22 When silver nitrate (AgNO$_3$) and sodium chloride (NaCl) are dissolved in water, a precipitate of the insoluble compound silver chloride is produced. The sodium and nitrate ions remain in solution.

properties of the ions that the solution contains. The properties of a sodium chloride solution are the properties of Na$^+$ plus those of Cl$^-$; the properties of a copper sulfate solution are the properties of Cu^{2+} plus those of SO$_4^{2-}$. Instead of learning the individual properties of hundreds of different electrolytes, we need only learn the properties of a few ions to be able to predict the behavior of any electrolytic solution that contains them.

10-7   EVIDENCE FOR DISSOCIATION

A Daring Idea a Century Ago   The hypothesis that many substances exist as ions in solution was proposed in 1887 by a young Swedish chemist, Svante Arrhenius. Today the idea of ions in solution follows naturally from our knowledge of the electrical structure of matter. We know that some compounds are formed by the shift of electrons from one kind of atom to another, so that some of the atoms become positive ions and the others negative ions. It is not hard for us to imagine that a polar liquid like water can separate these ions from a crystal. But in 1887 the modern picture of the atom was not even a dream. Without this knowledge Arrhenius's fellow chemists were hard to convince that neutral substances can break up into electrically charged fragments in solution.

Until the work of Arrhenius, Faraday's explanation for the ability of certain solutions to conduct electricity was generally accepted. Faraday held that the passage of a current caused the substance in solution to break up into ions. Arrhenius instead felt that ions are set free whenever an electrolyte dissolves, and he gave a number of reasons to support this notion.

One of the points Arrhenius made was that reactions between electrolytes take place almost instantaneously in solution, but occur very slowly or not at all if the electrolytes are dry. An example is the reaction between the silver nitrate and sodium chloride solutions shown in Fig. 10-22, which is very rapid. The speed with which the insoluble AgCl is formed suggests that its silver and chlorine components are already free in the original solutions and so are ready to combine at once. However, if dry $AgNO_3$ is mixed with dry NaCl, nothing happens because the components of each salt are held firmly in their respective crystals.

Freezing Points   Another piece of evidence cited by Arrhenius was the unexpectedly low freezing points of electrolyte solutions. The amount by which the freezing point of a solution is reduced (or its boiling point increased) depends on the concentration of solute particles present, not upon their nature. Equal numbers of sugar and of alcohol molecules dissolved in the same amount of water lower its freezing point by almost exactly the same amount. But the same number of NaCl units lowers the freezing point nearly *twice as much.* This suggests that there are no NaCl molecules as such and that solid NaCl breaks up into Na^+ and Cl^- ions when it dissolves. Similarly calcium chloride, $CaCl_2$, lowers the freezing point of water by nearly 3 times as much as sugar or alcohol, because each $CaCl_2$ unit dissociates in solution into three particles, one Ca^{2+} ion and two Cl^- ions.

FIG. 10-23 The composition of seawater. In the open ocean the total salt content varies about an average of 3.5 percent, but the relative proportions of the various ions are quite constant. (Percentages given are by mass.)

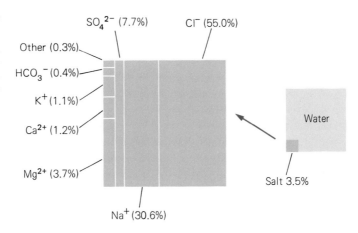

SO_4^{2-} (7.7%) Cl^- (55.0%)

Other (0.3%)

HCO_3^- (0.4%)

K^+ (1.1%)

Ca^{2+} (1.2%)

Water

Mg^{2+} (3.7%)

Salt 3.5%

Na^+ (30.6%)

10-8   # WATER

The Most Important Liquid

Although a minor constituent of the earth as a whole, water covers three-quarters of its surface. All the earth's water was once part of the rock of its interior and was freed as a result of geological processes. Water vapor is common in the gases that present-day volcanoes emit (Chap. 14). Life probably began in the early oceans, and water is essential to all living things.

Only 3 percent of the world's water is fresh, and of that 3 percent a third is trapped as ice in the Arctic and Antarctic. Seawater has a salt content (or **salinity**) that averages 3.5 percent. The composition of seawater is shown in Fig. 10-23. The ions Na^+ and Cl^- account for over 85 percent of the total salinity. The salinity of seawater varies around the world, but the proportions of the various ions are virtually the same everywhere because of mixing by currents both on and below the sea surface. Figure 10-24 shows where the ions found in seawater come from.

FIG. 10-24 Origins of seawater salts.

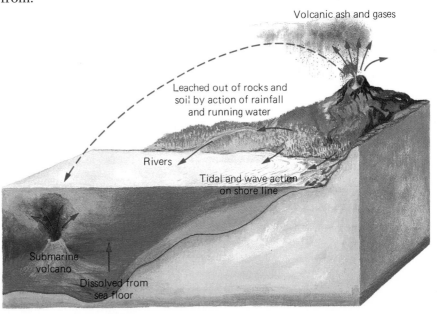

Volcanic ash and gases

Leached out of rocks and soil by action of rainfall and running water

Rivers

Tidal and wave action on shore line

Submarine volcano

Dissolved from sea floor

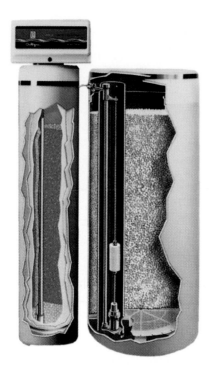

This household water softener uses an ion-exchange resin to remove calcium and magnesium ions from hard water.

"Hard" Water

Even freshwater is rarely free of ions in solution. "Hard" water contains dissolved minerals that prevent soap from forming suds and react with soap to produce a precipitate. When heated, hard water forms deposits in boilers, water heaters, and teakettles. These deposits are insoluble in water but not in acids, which provides a way to remove them. Calcium and magnesium ions are usually responsible for hard water. Groundwater often picks up the Ca^{2+} and Mg^{2+} ions from rocks such as limestone.

To "soften" water, the Ca^{2+} and Mg^{2+} ions must be removed, which can be done in several ways. In one common method, hard water is passed through a column filled with either a synthetic ion-exchange resin or a natural material called zeolite. Both materials absorb Ca^{2+} and Mg^{2+} ions into their structures while releasing an equivalent number of Na^+ ions. Since Na^+ ions do not affect soap or form compounds that precipitate out from hot water, the water is now "soft." When the ion-exchange column has reached its capacity of Ca^{2+} and Mg^{2+} ions, it can be flushed with a concentrated solution of NaCl to reverse the process and replace the accumulated Ca^{2+} and Mg^{2+} ions with Na^+ ions.

10-9   WATER POLLUTION

A Menace That Is Hard to Eliminate

Water can be polluted—that is, rendered unsuitable for a particular purpose, not necessarily human consumption only—in a variety of ways. Now that the dangers of water pollution are widely recognized, a great deal of effort is going into minimizing it. But so widespread are the existing sources of pollution, together with new ones appearing as patterns of industry and agriculture evolve, that this is a problem that will never go away.

Industry

A common source of pollution is industry, since such activities as metal refining, food processing, and the manufacture of bulk chemicals and paper require a lot of water: 300 tons of water is needed to produce a ton of steel.

Some industrial pollutants are especially dangerous because they are concentrated in the food chain of living things. A notorious example is mercury, which is widely used in the production of sodium hydroxide (NaOH) and chlorine gas from solutions of NaCl by electrolysis, as well as in other processes. Mercury-containing wastes have traditionally been dumped into the nearest body of water on the assumption that, since most mercury compounds are insoluble in water, no harm would result. Unfortunately some bacteria are able to convert mercury into the soluble compound dimethyl mercury, $(CH_3)_2Hg$, which is highly toxic. As lower forms of life are eaten by higher ones, the concentration of dimethyl mercury in the organisms increases. In Japan, thousands of people who regularly ate fish caught in Minamata Bay were afflicted with mercury poisoning due to wastes dumped by a nearby factory: brain damage, paralysis, blindness, and deformed children were the result. "Minamata disease" is only the best known of many similar occurrences of poisoning due to industrial pollution.

Agriculture

The fertilizers and pesticides used in agriculture are another major source of water pollution. Unlike the case of industry, where proper procedures can control the dispersion of harmful wastes, there is no way to keep chemicals deposited on the soil from spreading further. What can happen is illustrated by the potent and long-lasting insecticide DDT, now banned in the United States.

Water pollution from a steel mill. Public anger has forced governments to act against such abuse of the environment.

Washed from farmland into a body of water such as a lake, DDT enters the chain of life through tiny plant and animal organisms called plankton. With less than 0.1 part per million (ppm) of DDT in the lake water, the plankton may contain several ppm of DDT. The DDT concentration increases rapidly in the fish that eat plankton because DDT is retained in their fat and can reach over 1000 ppm in birds that eat the fish. One way that DDT affects birds is by weakening the shells of their eggs, which tend to break before the chicks hatch. This may wipe out whole species of birds in a DDT-polluted region.

Because pesticides are of great benefit to agriculture and, in many parts of the world, to disease control, it is not satisfactory merely to stop using them. The remedy is to develop pesticides that decompose rapidly, and much has already been done in this respect.

Dissolved Oxygen

Fertilizers contain the plant nutrients nitrogen and phosphorus. When a body of water becomes rich in these elements as a result of pollution by fertilizers, tiny organisms called algae cover its surface. As the algae die, aerobic bacteria (which need oxygen to live) use oxygen dissolved in the water to break down the algae into simple compounds, largely water and CO_2. Sewage and other organic wastes (such as those from food-processing plants and paper mills) are similarly attacked by these bacteria.

The oxygen needed to completely oxidize the organic debris in a given sample of water is called the **biochemical oxygen demand,** or **BOD,** of the water. BOD is a useful index of pollution because, when it is too great, the oxygen content of the water falls to the point where fish and other living things in the water die out. A sufficiently high BOD can lower the oxygen content so far that aerobic bacteria are unable to break down the wastes. Anaerobic bacteria (which do not need oxygen) then take over to produce such gases as methane (CH_4), which is flam-

Water pollution by fertilizers caused this algae raft in Florida.

mable, and hydrogen sulfide (H_2S), whose unpleasant odor is that of rotten eggs.

Another way in which the oxygen content of a body of water can be lowered is by heating, since the solubility of a gas in water decreases with increasing temperature. Because the heat output of an electric power plant is two or more times its electric output, such a plant has plenty of heat to get rid of. Using water from a nearby river or lake for cooling is the most common method, but the resulting thermal pollution does not benefit the local fish population.

ACIDS AND BASES

We continue our study of ions in solution by considering the three important classes of electrolytes: acids, bases, and salts. All are familiar in everyday life.

10-10   ACIDS

Hydrogen Ions Give Acidic Solutions Their Characteristic Properties

Acids are hydrogen-containing substances whose water solutions taste sour and change the color of the dye litmus from blue to red. In concentrated form such strong acids as the sulfuric acid used in storage batteries are poisonous, cause painful burns if allowed to remain on skin, and damage many materials. Hydrochloric acid, another strong acid, helps to digest food in our stomachs. Weak acids—such as the acetic acid of vinegar, the citric acid of lemons, and the lactic acid of yogurt—are far from being harmful and add a pleasant sour taste to foods and drinks.

What is it that underlies the behavior of acids? We have two clues:

1. All acids consist of hydrogen in combination with one or more nonmetals.

2. Solutions of acids conduct electricity and hence must contain ions.

It is therefore tempting to think that acids such as hydrochloric acid, HCl, and sulfuric acid, H_2SO_4, dissociate into hydrogen and nonmetal ions as follows:

$$HCl \longrightarrow H^+ + Cl^- \qquad \textit{dissociation of hydrochloric acid}$$

$$H_2SO_4 \longrightarrow 2H^+ + SO_4^{2-} \qquad \textit{dissociation of sulfuric acid}$$

From this it is natural to conclude that the characteristic properties of acid solutions are the properties of the hydrogen ion H^+.

This simple picture presents difficulties, however. For one thing, pure acids in liquid form do not conduct electric current, so a pure acid cannot be made up of ions. Because acids are covalent rather than ionic substances, they form ions not by the separation of ions already present but by reacting with water.

A second problem with the above simple picture of acids is that the ion H^+ is just a single proton—the nucleus of a hydrogen atom without its electron. All other ions are particles of the same general size as atoms, particles that consist of nuclei and electron clouds. The H⁺ ion would be entirely different, a naked proton a million billion times smaller than other ions. Such a particle cannot exist by itself in a liquid but must become attached immediately to some other atom or molecule.

Hydronium To avoid these difficulties, we could think in terms of reactions like this:

$$HCl + H_2O \longrightarrow H_3O^+ + Cl^-$$

Here the acid HCl is shown reacting with water instead of simply splitting up into ions, and the proton (the H⁺ ion) is shown attached to a water molecule rather than free in solution. The ion H_3O^+ is a combination of H⁺ with H_2O and is called the **hydronium ion** (Fig. 10-25). The characteristic properties of acids are described more correctly as properties of the hydronium ion than as properties of the simple hydrogen ion.

Nevertheless, chemists customarily write H⁺ for the characteristic ion of acids rather than H_3O^+ because it is more convenient to do so. For most purposes, then, we can say that

An acid is a hydrogen-containing substance that increases the number of H⁺ ions present when the substance is dissolved in water.

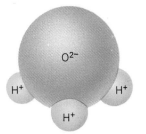

FIG. 10-25 A model of the hydronium ion, H_3O^+.

Although not strictly true, it is still legitimate to think of free hydrogen ions as being present in all acid solutions and giving these solutions their common properties. When we say that acid solutions taste sour, turn litmus pink, and liberate hydrogen gas by reaction with metals, we mean that the hydrogen ion does these things.

10-11 STRONG AND WEAK ACIDS

The More It Dissociates, the Stronger the Acid Acids differ greatly in how much they dissociate. Some acids, called **strong acids,** dissociate completely. The three most common strong acids are HCl (hydrochloric), H_2SO_4 (sulfuric), and HNO_3 (nitric). Other acids, called **weak acids,** dissociate only slightly.

The stronger an acid, the weaker the attachment of hydrogen in its molecules. In a strong acid like HCl the link is so weak that all the H⁺ and Cl⁻ ions split apart and go their separate ways in solution. In a weak acid like acetic acid, $HC_2H_3O_2$, on the other hand, the link is strong enough so that most of the molecules remain undissociated—in the case of acetic acid, 98 out of 100.

The greater dissociation of strong acids means that, in solutions of the same total concentration, a strong acid has a much larger proportion of hydrogen ions than a weak acid. It has a more sour taste, it is a better conductor of electricity, and if the two acids are poured on zinc, the evolution of hydrogen gas is much faster from the reaction with strong acid.

Why Carbon Dioxide Solutions Are Acidic

Certain substances that do not contain hydrogen still yield acidic solutions by reacting with water to liberate H^+ from H_2O. An interesting example is the gas CO_2, which when dissolved in water produces H^+ and HCO_3^- (hydrogen carbonate) ions:

$$CO_2 + H_2O \longrightarrow H^+ + HCO_3^-$$

Some HCO_3^- ions further dissociate into H^+ and CO_3^{2-} (carbonate) ions:

$$HCO_3^- \longrightarrow H^+ + CO_3^{2-}$$

It is customary to consider a solution of CO_2 in water (which is what soda water is) as containing "carbonic acid," although H_2CO_3 rarely exists as such. "Carbonic acid" is weak because relatively little of the dissolved CO_2 reacts with water to give H^+ ions. Rain and snow are slightly acidic because of the dissolved CO_2 they contain.

10-12 BASES

Hydroxide Ions Give Basic Solutions Their Characteristic Properties

Bases are familiar as substances whose solutions in water have a bitter taste, a slippery or soapy feel, and an ability to turn red litmus to blue. Their formulas, such as NaOH for sodium hydroxide and $Ba(OH)_2$ for barium hydroxide, show that bases consist of a metal together with one or more hydroxide (OH) groups. On dissolving in water, bases dissociate into ions according to reactions such as

$$NaOH \longrightarrow Na^+ + OH^- \qquad \textit{dissociation of sodium hydroxide}$$

$$Ba(OH)_2 \longrightarrow Ba^{2+} + 2OH^- \qquad \textit{dissociation of barium hydroxide}$$

Just as H^+ is the characteristic ion of acidic solutions, so OH^- is the characteristic ion in water solutions of bases. The properties of bases are properties of the OH^- ion. We may therefore say that

A base is a substance that contains hydroxide groups and whose solution in water increases the number of OH^- ions present.

The last part of the definition is needed because not all compounds that contain OH groups release them as OH^- ions in solution. An example is methanol (methyl alcohol), CH_3OH.

Like acids, bases may be classed as strong and weak according to how they dissociate in solution. Thus potassium hydroxide, KOH, is a strong base because it breaks up completely into K^+ and OH^- ions when it dissolves. The most common strong bases are KOH (caustic potash), NaOH (caustic soda, used in oven cleaners), and $Ca(OH)_2$ (slaked lime, used in the mortar that holds the bricks of a building together). Widely employed in industry, these bases are all poisonous and just as destructive to flesh and clothing as the strong acids.

Why Ammonia Solutions Are Basic

Bases differ from acids in that soluble weak bases are rare. However, many substances that do not contain OH in their formulas give basic solutions because they react with water to release OH^- ions from H_2O molecules. An example is the gas ammonia, NH_3, which reacts with water as follows:

$$NH_3 + H_2O \longrightarrow NH_4^+ + OH^-$$

The process is analogous to that by which CO_2 reacts with water to give an acidic solution. Ammonia solutions are often used in household cleansers. Two other compounds that are not bases but that give basic solutions are sodium carbonate (washing soda), Na_2CO_3, and sodium tetraborate (borax), $Na_2B_4O_7$, both also used as cleansing agents.

The name **alkali** is sometimes used for a substance that dissolves in water to give a basic solution. Alkali is an old Arabic word that referred originally to a bitter extract obtained from the ashes of a desert plant. Because NaOH and KOH are strong alkalis, sodium and potassium have become known as alkali metals. An alkaline solution is one that contains OH^- ions; the terms alkaline and basic mean the same thing.

10-13 THE pH SCALE

Less Than 7 Is Acidic; More Than 7 Is Basic

Even pure water dissociates to a small extent. The reaction can be written

$$H_2O \longrightarrow H^+ + OH^- \qquad \textit{dissociation of water}$$

The hydroxide ion OH^- attracts protons much more strongly than the neutral water molecule H_2O, and the reverse reaction

$$H^+ + OH^- \longrightarrow H_2O \qquad \textit{recombination of water}$$

This device determines the pH of a solution electrically. Its pH is a measure of how strongly acidic or basic a solution is.

FIG. 10-26   Typical pH values.

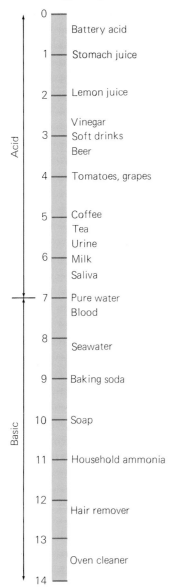

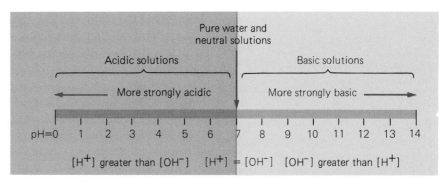

FIG. 10-27   The pH scale. The concentration of hydrogen ion is symbolized by $[H^+]$ and that of the hydroxide ion by $[OH^-]$. A neutral solution has a pH of 7. Litmus paper is red in an acidic solution, blue in a basic solution. An increase of 1 in pH corresponds to a decrease of a factor of 10 in H^+ concentration.

occurs readily. Thus we can write

$$H_2O \rightleftharpoons H^+ + OH^-$$

where the double arrow means that both reactions take place all the time in water.

The dissociation of water means that there are always some H^+ and OH^- ions in pure water, and the tendency for these ions to recombine keeps their concentration low. Only 0.0000002 percent of pure water is dissociated into ions on the average: 2 molecules out of every billion. In an acidic solution the concentration of H^+ is greater than in pure water, and the concentration of OH^- is lower. In a basic solution, the concentration of OH^- is greater than in pure water, and that of H^+ is lower.

The **pH scale** is a method for expressing the exact degree of acidity or basicity of a solution in terms of its H^+ ion concentration. This scale is so widely used that an acquaintance with it is worth having, but we need not concern ourselves here with its mathematical basis.

A solution that, like pure water, is neither acidic nor basic is said to be **neutral** and has, by definition, a pH of 7. Acidic solutions have pH values of less than 7; the more strongly acidic they are, the lower the pH. Basic solutions have pH values of more than 7; the more strongly basic they are, the higher the pH. A change in pH of 1 means a change in H^+ concentration by a factor of 10. Thus a solution of pH 4 is 10 times more acidic than a solution of pH 5 and 100 times more acidic than a solution of pH 6. Figure 10-26 shows typical pH values of some familiar solutions, and Fig. 10-27 illustrates the pH scale.

10-14   SALTS

An Acid plus a Base Gives Water and a Salt

When a sodium hydroxide solution is added slowly to hydrochloric acid, there is no visible sign that anything is happening. Both original solutions are colorless, and the resulting solution is also colorless. That a reaction does indeed occur can be shown in several ways, however. One sign is that the mixture becomes warm, which means that chemical energy is being liberated. In addition, if we measure the pH of the mixture as we add the NaOH, we would find that it gets closer and closer to 7 as the base is added—the concentration of H^+ ions is decreasing.

Evidently a base destroys, or **neutralizes,** the characteristic properties of an acid, and the reaction is accordingly called **neutralization.** In the same way the characteristic properties of a base can be neutralized by adding a strong acid.

Neutralization
What is the chemical change in the neutralization of HCl by NaOH? We could write simply

$$HCl + NaOH \longrightarrow H_2O + NaCl$$

However, we gain more insight into the process by considering the ions involved. HCl, a strong acid, dissociates completely in water to give H^+ and Cl^-; NaOH, a strong base, dissociates into Na^+ and OH^-; the product NaCl, also a soluble electrolyte, remains dissociated in solution. Of the four substances shown, only water is a nonelectrolyte, so it alone should appear intact in the equation. Hence we have

$$H^+ + Cl^- + Na^+ + OH^- \longrightarrow H_2O + Na^+ + Cl^-$$

Since Na^+ and Cl^- appear on both sides, we can omit them, leaving

$$H^+ + OH^- \longrightarrow H_2O \qquad \textit{neutralization}$$

This is the chemical change stripped of all nonessentials. The neutralization of a strong acid by a strong base in water solution is really a reaction between hydrogen ions and hydroxide ions to form water.

When an NaOH solution is neutralized with HCl, the resulting solution contains only the ions Na^+ and Cl^+. If the solution is evaporated to dryness, the ions combine to form the white solid NaCl. This substance, ordinary table salt, gives its name to an important class of compounds. Most **salts** are crystalline solids at ordinary temperatures and consist of a metal combined with one or more nonmetals.

How to Prepare a Salt
Any salt can be made by mixing the appropriate acid and base and evaporating the solution to dryness. Thus potassium nitrate, KNO_3, is formed when solutions of potassium hydroxide, KOH, and nitric acid, HNO_3, are mixed and evaporated; copper sulfate, $CuSO_4$, is formed when sulfuric acid, H_2SO_4, is poured on insoluble copper hydroxide, $Cu(OH)_2$, and the resulting solution is evaporated. In general, then, neutralization reactions give water and a solution of salt.

It is important to remember that a salt itself is not produced directly by a neutralization. Neutralization is a reaction between hydrogen ions and hydroxide ions. As a result of neutralization, ions may be left in solution that combine to form a salt when the solution is evaporated.

Table salt, NaCl, is not the only salt familiar to us in everyday life. Washing soda, Na_2CO_3, and borax, $Na_2B_4O_7$, have already been mentioned. Here are a few other examples. Baking soda, $NaHCO_3$, is used in baking powder, as a deodorizer, and in medicine as an antacid. Epsom salt, $MgSO_4 \cdot 7H_2O$ (this symbol means that seven water molecules are associated with each $MgSO_4$ unit in its crystals), has various medical uses. Saltpeter, KNO_3, is used to preserve meat and in gunpowder. Gypsum, $CaSO_4 \cdot 2H_2O$, is an ingredient of plaster. Many other salts can be added to this list.

IMPORTANT TERMS AND IDEAS

Solids that consist of particles arranged in repeated patterns are called **crystalline.** If the particles are irregularly arranged, the solid is called **amorphous.** The four types of bonds in crystals are **ionic, covalent, metallic,** and **molecular.**

The **metallic bond** arises from a "gas" of electrons that can move freely through a solid metal. These electrons are also responsible for the ability of metals to conduct heat and electricity well.

Van der Waals forces arise from the electric attraction between nonuniform charge distributions in atoms and molecules. They enable atoms and molecules to form solids without sharing or transferring electrons.

In a solution, the substance present in larger amount is the **solvent;** the other is the **solute.** When a solid or gas is dissolved in a liquid, the liquid is always considered the solvent. The **solubility** of a substance is the maximum amount that can be dissolved in a given quantity of solvent at a given temperature. A **saturated** solution is one that contains the maximum amount of solute possible.

Polar molecules behave as if negatively charged at one end and positively charged at the other; in **nonpolar molecules,** electric charge is uniformly distributed on the average. **Polar liquids** dissolve only ionic and polar covalent compounds, whereas **nonpolar liquids** dissolve only nonpolar covalent compounds. Water is a highly polar liquid, which is why it is so good a solvent.

Ionic compounds **dissociate** into free ions when dissolved in water; ions of a given kind in solution have properties that differ from those of the corresponding neutral substance. An **electrolyte** is any substance that separates into ions when dissolved in water.

Solutions of **acids** in water contain H^+ ions; solutions of **bases** in water contain OH^- ions. **Strong** acids and bases dissociate completely in solution; **weak** acids and bases only partially. The **pH** of a solution is a measure of its degree of acidity or basicity. Acid solutions have pH values of less than 7; basic solutions have pH values of more than 7. A **neutral** solution is neither acidic nor basic and has a pH of 7.

In acid-base **neutralization,** H^+ and OH^- ions join to form H_2O molecules.

Salts are usually crystalline solids that consist of positive metal ions and negative nonmetal ions. A salt can be formed by neutralizing the acid that contains the appropriate nonmetal ion with the base that contains the appropriate metal ion and then evaporating the solution to dryness.

EXERCISES: MULTIPLE CHOICE

1. An amorphous solid
   a. has its particles arranged in a regular pattern
   b. is held together by ionic bonds
   c. does not melt at a definite temperature but softens gradually
   d. consists of nonpolar molecules

2. Ionic crystals
   a. contain a "gas" of freely moving electrons
   b. consist of either positive or negative ions only
   c. dissolve only in polar liquids
   d. are soft and melt at low temperatures

3. A "gas" of freely moving electrons is present in
   a. ionic crystals   c. molecular crystals
   b. covalent crystals   d. metal crystals

4. The individual particles in a crystal held together by van der Waals forces are
   a. atoms   c. ions
   b. molecules   d. electrons

5. Van der Waals forces between atoms and between molecules arise from
   a. uniform charge distributions
   b. nonuniform charge distributions
   c. electron transfer
   d. electron sharing

6. Which solids have the lowest melting points in general?
   a. covalent   c. van der Waals
   b. ionic   d. metallic

7. Which solids are the best electrical conductors?
   a. covalent   c. van der Waals
   b. ionic   d. metallic

8. Diamond is an example of a
   a. covalent solid   c. metallic solid
   b. ionic solid   d. molecular solid

9. Suppose there were molecules that had no

attraction whatever for one another. A collection of such molecules would form a (an)

a. gas
b. liquid
c. amorphous solid
d. crystalline solid

10. A saturated solution is a solution that
   a. contains the maximum amount of solute
   b. contains the maximum amount of solvent
   c. is in the process of crystallizing
   d. contains polar molecules

11. A gas is dissolved in a liquid. When the temperature of the solution is increased, the solubility of the gas
   a. increases
   b. decreases
   c. remains the same
   d. any of the above, depending upon the nature of the solution

12. A solid is dissolved in a liquid. When the temperature of the solution is increased, the solubility of the solid
   a. increases
   b. decreases
   c. remains the same
   d. any of the above, depending upon the nature of the solution

13. Dissolving a solute in a liquid
   a. decreases the liquid's freezing point
   b. increases the liquid's freezing point
   c. does not change the liquid's freezing point
   d. any of the above, depending on the substances involved

14. A molecule that behaves as though positively charged at one end and negatively charged at the other is called a (an)
   a. hydronium ion
   b. acid molecule
   c. polar molecule
   d. nonpolar molecule

15. The most strongly polar liquid of the following is
   a. water
   b. alcohol
   c. a detergent
   d. gasoline

16. In general, ionic compounds
   a. are insoluble in all liquids
   b. dissolve best in polar liquids
   c. dissolve best in nonpolar liquids
   d. dissolve equally well in polar and nonpolar liquids

17. Nonpolar substances usually dissolve most readily in
   a. polar liquids
   b. nonpolar liquids
   c. acids
   d. bases

18. A substance that separates into free ions when dissolved in water is said to be
   a. polar
   b. saturated
   c. an electrolyte
   d. covalent

19. Which of the following would you expect to be a strong electrolyte in solution?
   a. H_2O
   b. HCl
   c. sugar
   d. alcohol

20. Which of the following has the least ability to conduct electric current?
   a. an acid solution
   b. a basic solution
   c. a salt solution
   d. a nonpolar liquid

21. The ions and atoms (or molecules) of an element
   a. have very nearly the same properties except that the ions are electrically charged
   b. may have strikingly different properties
   c. always exhibit different colors
   d. differ in that the ions are always more active chemically than the atoms or molecules

22. Dissociation refers to
   a. the formation of a precipitate
   b. the separation of a mixture of polar and nonpolar liquids (such as oil and water) into separate layers
   c. the separation of a solution containing ions into separate layers of + and − ions
   d. the separation of a substance into free ions

23. When an electrolyte is dissolved in water,
   a. the freezing point of the water is raised
   b. the solution is acidic
   c. the solution contains free ions
   d. the solution contains free electrons

24. The ions usually responsible for "hard" water are
   a. H^+ and OH^-
   b. Na^+ and Cl^-
   c. Ca^{2+} and Mg^{2+}
   d. SO_4^{2-} and NO_3^-

25. Acids invariably contain
   a. hydrogen
   b. oxygen
   c. chlorine
   d. water

26. The reason pure acids in the liquid state are not dissociated is that their chemical bonds are
   a. ionic
   b. covalent
   c. metallic
   d. van der Waals

27. While it is convenient to regard acidic solutions as containing H^+ ions, it is more realistic to describe them as containing
   a. hydronium ions
   b. hydroxide ions
   c. polar molecules
   d. hydrogen atoms

28. A common strong acid is
   a. acetic acid
   b. boric acid
   c. nitric acid
   d. citric acid

29. A base dissolved in water liberates
   a. H^- **c.** OH^+
   b. OH **d.** OH^-

30. A substance whose formula does not contain OH yet yields a basic solution when dissolved in water is
   a. NH_3 **c.** HCl
   b. CO_2 **d.** $NaCl$

31. The symbol of the ammonium ion is
   a. NH_3^- **c.** NH_4^-
   b. NH_3^+ **d.** NH_4^+

32. The net reaction between a strong acid and a strong base is
   a. $H_2O \rightarrow H^+ + OH^-$
   b. $H^+ + OH^- \rightarrow H_2O$
   c. $H_2O + H_2O \rightarrow H_3O^+ + OH^-$
   d. $H^+ + H_2O \rightarrow H_3O^+$

33. The formula for iron(III) hydroxide is
   a. $FeOH$ **c.** $Fe(OH)_3$
   b. Fe_3OH **d.** $Fe_3(OH)_3$

34. A strong acid or base in solution is completely
   a. dissociated   **c.** hydrolyzed
   b. neutralized   **d.** precipitated

35. Pure water contains
   a. only H^+ ions
   b. only OH^- ions
   c. both H^+ and OH^- ions
   d. neither H^+ nor OH^- ions

36. A pH of 7 signifies a (an)
   a. acid solution
   b. basic solution
   c. neutral solution
   d. solution of polar molecules

37. A concentrated solution of which of the following has the lowest pH?
   a. hydrochloric acid   **c.** sodium hydroxide
   b. acetic acid   **d.** ammonia

38. A concentrated solution of which of the following has the highest pH?
   a. hydrochloric acid   **c.** sodium hydroxide
   b. acetic acid   **d.** ammonia

QUESTIONS

1. State the four principal types of bonding in crystalline solids and give an example of each. What is the fundamental physical origin of all of them? What kind of particle is present in the crystal structure of each of them?

2. What kind of bonds hold H_2O molecules together in ice?

3. How could you tell experimentally whether a fragment of a clear, colorless material is glass or a crystalline solid?

4. What kind of solid contains a "gas" of freely moving electrons? Does this gas include all the electrons present?

5. Van der Waals forces are strong enough to hold inert gas atoms together to form liquids at low temperatures, but these forces do not lead to inert gas molecules at higher temperatures. Why not?

6. What ions would you expect to find in the crystal structures of MgO and K_2S?

7. What ions would you expect to find in the crystal structures of CaF_2 and KI?

8. Solids of which type have the lowest melting points? Solids of which type are the best conductors of heat and electric current?

9. Give two ways to tell whether a sugar solution is saturated or not.

10. Why is the solubility of one gas in another unlimited?

11. Why do bubbles of gas form in a glass of soda water when it warms up?

12. Ordinary tap water tastes different after it has been boiled. Can you think of the reason why?

13. What is the difference between a molecular ion and a polar molecule?

14. Give examples of polar and nonpolar liquids and state several substances soluble in each.

15. How could you distinguish experimentally between an electrolyte and a nonelectrolyte?

16. Contrast the properties and electron structures of Na and Na^+. Which would you expect to be more active chemically?

17. Name one property by which you could distinguish (a) Cl^- from NO_3^-, (b) Ag^+ from Na^+, (c) Cu^{2+} from Ca^{2+}.

18. Why do so many substances dissolve in water? Why do oils and fats not dissolve in water?

19. With the help of Fig. 10-13 predict what will happen when a concentrated solution of sodium nitrate at 50°C is added to a saturated solution of potassium chloride at the same temperature.

20. At 10°C, which is more concentrated, a saturated solution of potassium nitrate or a saturated solution of potassium chloride? At 60°C?

21. Do pure acids in the liquid state contain H^+ ions? If not, what do such acids consist of?

22. Is it correct to say that the only ions an acidic solution contains are H^+ ions and that the only ions a basic solution contains are OH^- ions? If not, what would correct descriptions of such solutions be?

23. Which is more strongly acidic, a solution of pH 3 or one of pH 5? Which is more strongly basic, a solution of pH 8 or one of pH 10?

24. In an acidic solution, why is the OH^- concentration lower than it is in pure water?

25. Justify the statement that water is both a weak acid and a weak base.

26. Which of the following are weak acids? Hydrochloric acid, nitric acid, acetic acid, sulfuric acid, citric acid.

27. Would you expect HBr to be a weak or strong acid? Why?

PROBLEMS

1. Give the ionic equation for the neutralization of potassium hydroxide by nitric acid. What chemical changes does this equation show?

2. What salt is formed when a solution of sodium hydroxide is neutralized by sulfuric acid? Give the equation of the process.

3. What salt is formed when a solution of calcium hydroxide is neutralized by hydrochloric acid? Give the equation of the process.

4. Boric acid (H_3BO_3) is a very weak acid. What would happen if solutions of Na_3BO_3 (sodium borate) and HCl were mixed?

5. The Al^{3+} ion tends to form $AlOH^{2+}$ ions in water solution. Would you expect a solution of $AlCl_3$ to be acidic, basic, or neutral? Explain your answer.

6. Give the equation of the reaction described below:
Johnny, finding life a bore,
Drank some H_2SO_4.
Johnny's father, an MD,
Gave him $CaCO_3$.
Now he's neutralized, it's true,
But he's full of CO_2.

ANSWERS TO MULTIPLE CHOICE			
1. c	**11.** b	**21.** b	**31.** d
2. c	**12.** a	**22.** d	**32.** b
3. d	**13.** a	**23.** c	**33.** c
4. b	**14.** c	**24.** c	**34.** a
5. b	**15.** a	**25.** a	**35.** c
6. c	**16.** b	**26.** b	**36.** c
7. d	**17.** b	**27.** a	**37.** a
8. a	**18.** c	**28.** c	**38.** c
9. a	**19.** b	**29.** d	
10. a	**20.** d	**30.** a	

11

CHEMICAL

REACTIONS

*T*he energy given off in chemical reactions powers our cars, airplanes, and ships, heats our homes, cooks most of our food, and is the energy source of the generating plants that produce most of our electricity. Our own bodies obtain the energy they need from chemical reactions in which the food we eat combines with oxygen from the air we breathe.

Not all chemical reactions liberate energy—some reactions must be supplied with energy in order to occur. Even those reactions that liberate energy may not take place unless some initial energy is furnished to start the process. Another aspect of chemical reactions is that they take time to be completed: a fraction of a second to many years, depending on a number of factors. Not all reactions even go to completion. Instead, an intermediate equilibrium situation often occurs with the products undergoing reverse reactions to form the starting substances just as fast as the primary reaction proceeds. These are some of the topics considered in this chapter.

COMBUSTION

The most spectacular chemical change our ancestors were familiar with was **combustion,** the process of burning. Early explanations of how wood turns into smoke and ashes amid dancing flames were based on demons and spirits. The fire god has a respected place in many religions.

The ancient Greeks made the first attempt at a nonsupernatural explanation, as recorded by Aristotle. Every flammable material was supposed to consist of "earth" and "fire." When the material burned, the fire escaped, leaving the earth behind as ashes. This idea persisted in various forms until the time of the French Revolution two centuries ago.

11-1 PHLOGISTON

Now It's There, Now It Isn't

The notion of fire as a basic substance was developed by two Germans, Johann Becher (1635–1682) and his student Georg Stahl (1660–1734), into the **phlogiston** hypothesis. The starting point was the same as Aristotle's, but Becher and Stahl showed how it could be extended to reactions other than burning. They used the word phlogiston (from the Greek word for "flame") for the substance that supposedly escaped during combustion. The story of the downfall of the phlogiston hypothesis and the growth of the modern picture of chemical change is a notable chapter in the history of ideas.

Today we never hear the word phlogiston, but once there was no more respected concept in chemistry. All substances that can be burned were supposed to contain phlogiston, which escapes as the burning takes place. Combustion requires air, but this is explained by assuming that phlogiston can leave a substance only when air is present to absorb it. When heated in air, many metals change slowly to soft powders: zinc and tin give white powders, mercury a reddish powder, iron a black scaly material. These changes, like the changes in ordinary burning, were ascribed to the escape of phlogiston.

A metal was assumed to be a compound of the corresponding powder plus phlogiston, and heating the metal simply caused the compound to decompose. Now, many of these powders can be changed back into metal by heating with charcoal. This observation was interpreted to mean that charcoal must be a form of phlogiston that simply reunited with the powder to form the compound (the metal). When hydrogen was discovered in 1766, its ability to burn without leaving any ash suggested that it was another form of phlogiston. One could predict, then, that heating one of the above powders with hydrogen would form a metal, and this prediction was confirmed by experiment.

So far so good, but soon the phlogiston hypothesis ran into serious trouble. When wood burns, its ashes weigh less than the original wood, and the decrease in mass can reasonably be explained as due to the escape of phlogiston. But when a metal is heated until it turns into a powder, the powder weighs *more* than the original metal! The believers in phlogiston were forced to assume that it sometimes could have negative mass, so that if phlogiston left a substance, the remaining material could weigh more than before. To us this notion of negative mass is nonsense, but in the eighteenth century it was taken quite seriously.

The Downfall of Phlogiston

The French chemist Antoine Lavoisier carried out a series of experiments in the latter part of the eighteenth century that overthrew the phlogiston hypothesis. Lavoisier knew that tin changed into a white powder when heated and that the powder weighed more than the original metal. To study the process in detail, he placed a piece of tin on a wooden block floating in water, as in Fig. 11-1. He covered the block with a glass jar and heated the tin by focusing sunlight on it with a magnifying glass—a common method of heating before gas burners and electric heaters were invented. The tin was partly changed into a white powder and the water level rose in the jar until only four-fifths as much air was left in the jar as there had been at the start. Further heating caused nothing more to happen.

FIG. 11-1 Lavoisier's experiment showed that tin, upon heating, combines with a gas from the air. (a) Before heating; (b) after heating. The tin is partly changed to a white powder, and the water level rises until only four-fifths as much air is left as there was at the start. Further heating causes no additional change.

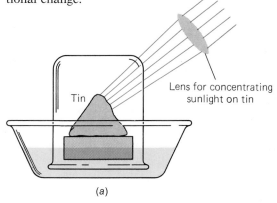

Lens for concentrating sunlight on tin

Tin

(a)

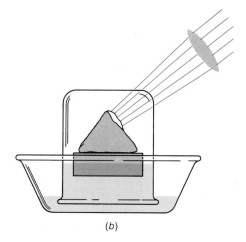

(b)

BIOGRAPHY

Antoine Lavoisier (1743–1794), the son of a wealthy French lawyer, studied law at first but soon became fascinated by chemistry and published his first paper, on the mineral gypsum, when he was 22. In that paper, as in his later work, Lavoisier stressed the importance of accurate measurements of mass in chemistry. This emphasis on reliable data marked the start of the modern era of chemistry: Lavoisier's influence on chemistry was much like that of Galileo on physics. With the help of a sensitive balance and a large magnifying glass for focusing sunlight, Lavoisier carried out a series of experiments that, among other achievements, demolished the notion of phlogiston and made clear the role of oxygen in combustion and other processes. In his research Lavoisier was greatly aided by his wife, Marie-Anne, who was 14 to his 28 when they were married. Besides work in the laboratory, Marie-Anne made engravings of the apparatus for publication and translated scientific books and papers from English to French. Benjamin Franklin and

Thomas Jefferson were among the visitors to the Lavoisier laboratory.

In 1787 Lavoisier and several colleagues systematized the language of chemistry in the book *Methode de Nomenclature Chimique*. They listed 55 substances they could not decompose, most of which (but not caloric or light, which were included) indeed proved to be elements, and named compounds after their constituents according to a standard scheme. Such familiar terms as oxide, nitrate, and sulfate were thereby introduced for the first time. Two years later Lavoisier published the first modern textbook on chemistry, an influential book in which the central place of the law of conservation of mass was made clear.

In order to support his research, while still a young man Lavoisier

had invested in the Ferme Générale, a private company that collected taxes for the French government, and he became an official of the company, as was his father-in-law. Brutal and corrupt, the Ferme Générale was the most hated institution in the country. In 1793, after the French Revolution, Lavoisier was arrested because of his association with the Ferme Générale; when he protested that he was a scientist, he was told that "the Republic has no need of scientists." In 1794 he and his father-in-law were guillotined. Lavoisier's widow later married Count Rumford, who proved that caloric did not exist 25 years after Lavoisier had done the same for phlogiston; but it was an unhappy union.

In another experiment, Lavoisier heated tin in a sealed flask until as much as possible was turned into powder. The flask was weighed before and after heating, and the two masses were the same. Then the flask was opened, and air rushed in. With the additional air, the mass of the flask was more than it had been at the start. The increase in mass was equal to the increase in mass of the tin.

To Lavoisier these results suggested that the tin had combined with a gas from the air. Since four-fifths of the original air was left after the reaction, he reasoned that one-fifth of air consisted of a gas that can combine with tin. Then the powder was a compound formed from this gas and tin, and the increase in mass of the powder over the tin was the mass of the gas. Water rose in the jar of Fig. 11-1 to take the place of the gas that had combined with the tin. When the sealed flask was opened in the second experiment, air rushed in to take the place of the gas that had combined with the tin. These explanations are simple and direct and involved substances that, unlike phlogiston, had definite masses.

11-2 OXYGEN

Joseph Priestley (1733–1804)

Combustion Is Rapid Oxidation

At about the time he was making these experiments, Lavoisier learned that Joseph Priestley had prepared a new gas with remarkable properties. Priestley was the poverty-striken minister of a small church in England, with only limited time and equipment, yet his scientific talents led him to a number of significant discoveries. The gas he had found caused lighted candles to flare up brightly and glowing charcoal to burst into flames. A mouse kept in a closed jar of the gas lived longer than one kept in a closed jar of air.

Lavoisier gave Priestley's new gas its modern name, **oxygen,** and found it to be involved not only in the changes that occur in metals when heated but in the process of combustion as well. The burning of candles, wood, and coal, according to Lavoisier, involves combining their materials with oxygen. When they burn, these materials seem to lose mass only because some of the products of the reactions are gases. Actually, as experiment shows, the total mass of the products in each case is more than the original mass of the solid material.

Under ordinary conditions, oxygen is a colorless, odorless, tasteless gas. Air owes its ability to support combustion to its oxygen content. Air cannot support combustion as well as pure oxygen because air consists of only about one-fifth oxygen (Fig. 11-2). The other four-fifths is mainly nitrogen, together with small amounts of other inactive gases.

Oxidation

When oxygen combines chemically with another substance, the process is called **oxidation,** and the other substance is said to be **oxidized.** In the experiments of Lavoisier, the tin reacted with oxygen in the air to become oxidized to a white powder. Rapid oxidation in which a lot of heat and light are given off is the process of combustion. As Priestley found, a lighted candle oxidizes rapidly in air, even more rapidly in pure oxygen.

Slow oxidation is involved in many familiar processes. One of them is the rusting of iron, in which iron is oxidized into the reddish-brown material we call rust. The energy to maintain life comes from the steady oxidation of food by oxygen taken in through our lungs and carried to all parts of our bodies by blood.

OZONE The molecules of ordinary oxygen consist of two oxygen atoms: O_2. A less stable form of oxygen is **ozone,** whose molecules consist of three oxygen atoms: O_3. The pungent odor of ozone is familiar near electrical discharges, such as sparks and lightning, which can produce O_3 from atmospheric O_2. Ozone is used industrially as a bleach and to purify water; it is toxic to living things and damages various materials, notably rubber. Two important processes in the atmosphere, one bad and one good, involve the small amount of ozone it contains. At low altitudes ozone contributes to the formation of smog, which is harmful to our health. At high altitudes, however, ozone provides an important service by absorbing dangerous ultraviolet radiation from the sun (see Sec. 13-1).

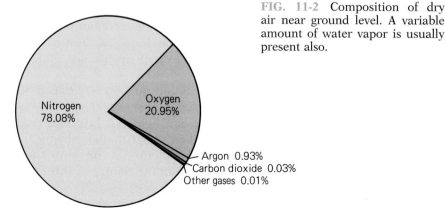

FIG. 11-2 Composition of dry air near ground level. A variable amount of water vapor is usually present also.

Nitrogen 78.08%

Oxygen 20.95%

Argon 0.93%
Carbon dioxide 0.03%
Other gases 0.01%

Rust forms when iron or steel reacts with oxygen and water. The formula for rust is often written $2Fe_2O_3 \cdot xH_2O$, which means that a variable number of water molecules are associated with every two iron(III) oxide units. Because rust is porous, oxygen and water vapor can penetrate it and continue to react with the metal underneath. A rusty object may therefore have very little strength left. On the other hand, the oxidation of aluminum to Al_2O_3 produces a hard, durable coating that prevents further corrosion. This is why steel cans are plated with tin to protect them, whereas aluminum cans need no treatment.

A fire can be put out in three ways. (1) Water can be sprayed on to cool the burning material below its ignition temperature. (2) The fire can be smothered by a heavier-than-air agent (such as carbon dioxide) that does not support combustion and keeps out air and hence oxygen. (3) A chemical (such as one of the "Halon" gases) that interferes with the combustion process can be sprayed on the fire.

A substance formed by the union of another element with oxygen is called an **oxide.** The white powder that Lavoisier obtained by heating tin is tin oxide. Rust is largely iron oxide. In general, oxides of metals are solids. Oxides of other elements may be solid, liquid, or gaseous. Thus one of the oxides of sulfur is the foul-smelling gas sulfur dioxide (SO_2). Carbon forms two gaseous oxides, carbon monoxide (CO) and carbon dioxide (CO_2). The oxide of silicon SiO_2 is found in nature as the solid called quartz, the chief constituent of ordinary sand and abundant in rocks. The oxide of hydrogen is water, H_2O.

Oxides of nearly all the elements can be prepared, most of them simply by heating the elements with oxygen. A few oxides (mercury oxide, lead oxide, barium peroxide) are easily decomposed by heating, which provides a convenient laboratory method for preparing oxygen. Other oxides, such as lime (calcium oxide), are not decomposed even at the temperature of an electric arc, 3000°C.

CHEMICAL ENERGY

Ever since our ancestors learned to control fire, people have been putting chemical energy to practical use. Today we transform it not only into heat and light but into mechanical energy and electric energy as well. Locked up in matter, chemical energy long remained a mystery. The modern picture of the atom and of the chemical bond, however, is able to explain the nature of this energy.

11-3 EXOTHERMIC AND ENDOTHERMIC REACTIONS

Some Reactions Liberate Energy; Others Absorb It Chemical changes that *give off* energy are called **exothermic reactions.** (In Greek, *exo* means "outside.") The burning of coal, which is largely carbon, and of hydrogen are both exothermic:

$$C + O_2 \longrightarrow CO_2 + E \qquad E = 9 \text{ kJ/g of } CO_2 \qquad \textit{formation of carbon dioxide}$$

$$2H_2 + O_2 \longrightarrow 2H_2O + E \qquad E = 13.6 \text{ kJ/g of } H_2O \qquad \textit{formation of water}$$

Chemical changes that take place only when heat or some other kind of energy is *absorbed* are called **endothermic reactions.** (In Greek, *endo* means "inside.") The decomposition of water into hydrogen and oxygen requires heating to very high temperatures or the supply of electric energy during electrolysis (see Fig. 9-17), so it is endothermic:

$$2H_2O + E \longrightarrow 2H_2 + O_2 \qquad E = 13.6 \text{ kJ/g of } H_2O \qquad \textit{decomposition of water}$$

The formation of nitric oxide (NO) from the elements N_2 and O_2 is an endothermic reaction that takes place only at high temperatures:

$$N_2 + O_2 + E \longrightarrow 2NO \qquad E = 3 \text{ kJ/g of NO} \qquad \textit{formation of nitric oxide}$$

The German airship *Hindenburg* was filled with hydrogen. In 1937, after crossing the Atlantic, it caught fire while landing at Lakehurst, New Jersey. The combustion of hydrogen is an exothermic reaction whose product is water.

Direct and Reverse Reactions From the law of conservation of energy we can predict that, if a given reaction is exothermic, the reverse reaction will be endothermic. Furthermore, the amount of heat liberated by the exothermic reaction must be the same as the amount of heat absorbed by the endothermic reaction. This prediction is borne out in the case of water, as we can see above, and is also verified in all other reactions where it can be tested. An example is sodium reacting with chlorine:

$$2Na + Cl_2 \longrightarrow 2NaCl + E \qquad E = 7 \text{ kJ/g of NaCl} \qquad \textit{formation of sodium chloride}$$

To break up NaCl takes the same amount of energy:

$$2NaCl + E \longrightarrow 2Na + Cl_2 \qquad E = 7 \text{ kJ/g of NaCl} \qquad \textit{decomposition of sodium chloride}$$

Dissociation and Neutralization The dissociation of most salts is an endothermic process. For example, when KNO_3 is dissolved in water, the container becomes cold, since the dissociation of the salt requires energy:

$$KNO_3 + E \longrightarrow K^+ + NO_3^- \qquad E = 0.36 \text{ kJ/g of } KNO_3 \qquad \textit{dissociation of potassium nitrate}$$

Neutralization, on the other hand, is an exothermic process. If concentrated solutions of NaOH and HCl are mixed, for instance, the mixture quickly becomes too hot to touch:

$$H^+ + OH^- \longrightarrow H_2O + E \qquad E = 3.2 \text{ kJ/g of } H_2O \qquad \textit{neutralization}$$

The Na^+ and Cl^- ions are omitted here because the principal chemical change in all neutralizations is simply the joining together of hydrogen ions and hydroxide ions. For the same reason the neutralization of any other strong acid by any other strong base liberates almost precisely the same amount of heat for each gram of water produced.

11-4 CHEMICAL ENERGY AND STABILITY

The Less PE Its Electrons Have, the More Stable the Compound The heat given off or absorbed in a chemical change is a measure of the stabilities of the substances (or mixtures of substances) involved. If a great deal of energy is needed to decompose a substance, the substance is (with rare exceptions) relatively stable. If the decomposition is either exothermic or weakly endothermic, the substance is normally unstable.

From the reactions given in Sec. 11-3 we can see at a glance that CO_2, H_2O, and NaCl are stable compounds, since the formation of each is strongly exothermic and its decomposition is endothermic. NO, on the other hand, is unstable, since its decomposition liberates heat. The combinations H_2 and O_2, Na and Cl_2, H^+ and OH^- are relatively unstable, since they react to give off energy. On the other hand, N_2 and O_2 form a stable mixture since energy must be supplied for them to react.

We can interpret chemical-energy changes in terms of atomic structure (Fig. 11-3). When sodium reacts with chlorine, for example, the outer electron of each Na atom is transferred to the outer shell of a Cl atom. In its new position the electron has less potential energy with respect to the atomic nuclei because it is held more firmly to the chlorine nucleus than it was to the sodium nucleus. The same is true for other ionic bonds.

When two hydrogen atoms react to form a hydrogen molecule, the atoms are joined by shared electrons. Such a covalent bond involves a decrease in the PE of the electrons because each is now attracted by two nuclei instead of a single nucleus. The same is true for other covalent bonds.

FIG. 11-3 (*a*) A book raised above a table has more PE than the same book lying on the table because the attractive force the earth exerts on the book is greater when the book is closer to the earth. (*b*) The outer electron in an Na atom has more PE than the same electron has when it is attached to a Cl atom to form a Cl⁻ ion because the electron is more strongly attracted to the nucleus of the Cl⁻ ion (see Fig. 9-12). (*c*) The electron in a H atom has more PE than the same electron has when it is part of a H molecule because in the H molecule the electron is attracted by two protons rather than one proton.

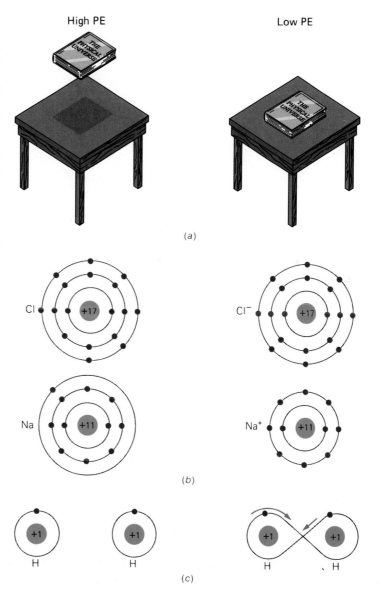

The Nature of Chemical Energy   Our conclusion is that

> **Chemical energy is electron potential energy.**

When electrons move to new locations during an exothermic reaction, some of their original PE is liberated. This freed energy may show itself in faster atomic or molecular motions that correspond to a higher temperature. Or the freed energy may excite outer electrons into higher energy levels from which they return to lower levels by giving off photons of light. In endothermic reactions, energy must be supplied to the atoms involved to enable some of their electrons to form bonds in which their PEs are greater than before.

11-5   ACTIVATION ENERGY

The Seed Energy Needed to Start an Exothermic Reaction   Wood burns in air to give off great quantities of heat. However, we can store a pile of firewood indefinitely without its catching fire. A mixture of hydrogen and oxygen can explode violently. However, hydrogen does not explode when mixed with air unless a flame or spark sets off the reaction. Why do not all exothermic reactions take place at once of their own accord?

Clearly, in order to begin, many exothermic processes must first be supplied with energy. A mixture of hydrogen and oxygen is like the car of Fig. 11-4, whose potential energy may be converted to kinetic energy if it moves down into the valley. However, the car cannot begin to go downward unless it is first given enough energy to climb to the top of the hill. Similarly the chemical energy of a mixture of hydrogen and oxygen can be freed only if the molecules have enough energy, or are sufficiently **activated,** to make the reaction start. The energy needed for activation, corresponding to the energy required to move the car up the hill, is called the **activation energy** of the reaction.

The electron picture of chemical combination suggests the reason for activation energy. The reaction of oxygen and hydrogen involves the formation of bonds between O and H atoms, a process that gives out energy. However, before these bonds can be formed, the covalent bonds

FIG. 11-4   Activation energy. The potential energy of the car will be converted into kinetic energy if it moves down into the valley. However, the car requires initial kinetic energy in order to climb the hill between it and the valley, analogous to the activation energy required in many exothermic reactions.

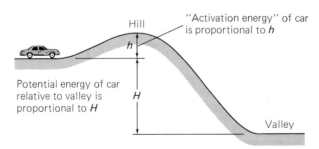

between the hydrogen atoms in H_2 molecules and the oxygen atoms in O_2 molecules must be broken. To break these bonds takes energy. Once the reaction starts, the energy already liberated can supply the needed energy, but in the beginning some outside energy must be supplied. Thus a mixture of hydrogen and oxygen need only be touched with a flame for the reaction to spread so rapidly that an explosion results. When a bed of coal is set on fire it continues to burn, since the heat liberated in one place is sufficient to ignite the coal around it.

Activated Molecules

A molecule with enough energy to react is called an **activated molecule.** In reactions that take place spontaneously at room temperature (for example, the reaction between hydrogen and fluorine), enough of the initial molecules have the required KE of thermal motion for their bonds to break during collisions without further activation. Ions in solution are, so to speak, already activated and react almost instantaneously. But a large number of exothermic reactions must have the preliminary activation of some molecules before the reactions can take place in a self-sustaining way.

FUELS

The first requirement of a good fuel is naturally that its combination with oxygen be strongly exothermic. Other desirable features are that it should be cheap, abundant, and easy to transport and store, and that the products of combustion should not endanger the environment. Many substances satisfy the first requirement, but none fulfills all the others as well.

Carbon and some of its compounds—which are found in nature as petroleum, natural gas, and coal—best fulfill the requirements for fuels. The chief products of their combustion are CO_2 and H_2O, which escape as gases into the air, and, in the case of coal, ashes. Some of the pros and cons of these fuels were discussed in Chap. 3. Here our interest is in their chemical aspects.

11-6   LIQUID FUELS

Pollution by Vehicle Exhausts Is a Major Problem

Petroleum, formed from the remains of tiny marine organisms buried under sediments millions of years ago, is the source of most liquid fuels. Nine million tons of petroleum is consumed every day worldwide. As described in Chap. 12, gasoline, kerosene, diesel fuel, and heating oil are mixtures of various hydrocarbons—compounds of carbon and hydrogen—derived from petroleum. All give off considerable heat when they are burned: 44 to 48 kJ/g (Fig. 11-5). In the engine of a car, the gaseous products of burning gasoline expand rapidly because of the intense heat generated by the reaction. This expansion forces down the pistons of the engine, which causes the crankshaft to turn and provide power to the car's wheels (see Fig. 4-34).

A mixture of gasoline and air is fed into the cylinders of a car's engine, one after another, to be ignited at just the right time by the spark plugs. Each gram of gasoline requires 15 g of air for complete combus-

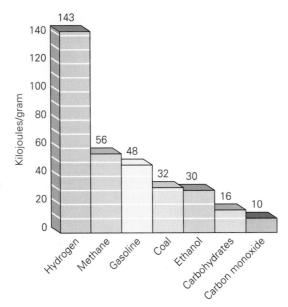

FIG. 11-5 Energy contents of various fuels. Shown are the number of kilojoules of energy liberated when 1 g of each fuel is burned. Carbohydrates provide much of the energy in our diets (see Sec. 12-11). All these fuels produce carbon dioxide and water when burned.

Traffic jam in Los Angeles. Car exhausts contain carbon monoxide and various hydrocarbons, which are poisonous, and nitrogen oxides, which contribute to acid rain through the nitric acid they form. More efficient engines, better fuels, and catalytic converters can reduce such emissions but not that of carbon dioxide, which contributes to global warming through the greenhouse effect described in Chap. 13.

tion. The CO_2 and H_2O that are produced are odorless and nonpoisonous. A "rich" mixture of gasoline and air contains a greater proportion of gasoline than this, and a "lean" mixture contains a smaller proportion.

A typical constituent of gasoline is the compound *octane*, C_8H_{18}, whose complete combustion is given by the equation

$$2C_8H_{18} + 25O_2 \longrightarrow 16CO_2 + 18H_2O \qquad \textit{complete combustion}$$

octane   oxygen   carbon   water
                       dioxide

Incomplete combustion, which occurs a little even under the best circumstances but much more when the gasoline-air mixture is rich, involves a reaction likely to be something like this:

$$2C_8H_{18} + 20O_2 \longrightarrow 11CO_2 + 15H_2O + \quad 3CO \quad + C_2H_6 \qquad \textit{incomplete}$$

octane   oxygen   carbon   water   carbon   ethane   *combustion*
                  dioxide   monoxide

Both carbon monoxide and ethane, which is one of various hydrocarbons that may be produced, are harmful substances. The hazard of CO comes from its tendency to combine permanently with the hemoglobin of the blood in place of O_2. This deprives the body of some of the oxygen it needs and leads to death if too much CO is inhaled.

We might think that the way to minimize pollution by car exhausts is to use a lean gasoline-air mixture, which increases the chance of complete combustion by providing more than enough O_2. But lean mixtures are hard to ignite. Worse, the leaner the mixture, the hotter it burns, and the higher temperatures promote the oxidation of the nitrogen present in the air. Nitrogen oxides, such as NO and NO_2, combine with water to form corrosive nitric acid and with unburned hydrocarbons to form a variety of toxic compounds, including potent carcinogens (cancer-producing agents).

Leaded Gasoline

Carbon monoxide, unburned hydrocarbons, and nitrogen oxides are not the only menaces to health in car exhausts. In the 1920s it was found that burning is more even in an engine when a lead compound called tetraethyllead is added to the gasoline. This prevents the "knocking" that sometimes happens when a car goes up a steep hill. The lead also helps engine performance by lubricating the valves that let fuel into each cylinder and exhaust gases out.

Almost all the gasoline used as fuel once contained tetraethyllead, and as a result engine emissions included poisonous lead compounds along with everything else. Not long ago millions of tons of lead per year was being dumped into the atmosphere in the United States (and still is elsewhere in the world). Lead that is breathed in tends to stay in the body, where it produces a variety of ailments, including mental disabilities in children. The problem was especially grave in inner cities, where heavy traffic made it inevitable that at least some children who grew up there absorbed enough lead from the air to affect their intelligence.

Reducing Pollution

There are several ways to reduce the amount of pollutants in vehicle exhausts. One is to build an engine in which just the right mixture of lead-free gasoline and air is completely burned (to minimize CO and unburned hydrocarbons) at just the right temperature (to minimize nitrogen oxides) throughout all the variations in speed and power output involved in driving a vehicle. Such an engine will obviously be very efficient as well. But it is extremely hard to produce an engine with these ideal properties, although much progress has been made, notably by computer control of engine operation.

Another approach is to change the polluting gases into harmless ones by passing the exhaust of a car through a device called a **catalytic converter.** Reactions that change polluting gases into harmless ones are promoted by the action of such metals as platinum and rhodium (catalysts are discussed later in this chapter), which are the active agents in a catalytic converter. Because lead inactivates the catalysts in a converter, only lead-free fuel can be used. Such fuel need not bring back knocking because other antiknock agents can be added to ordinary gasoline.

A third method is to change the composition of the fuel. "Reformulated gasoline" (RFG) consists of a mixture of hydrocarbons, different from that in ordinary gasoline, that is expected to reduce toxic emissions by around 15 percent. One advantage of RFG is that it evaporates less readily than ordinary gasoline, partly because it contains little of the volatile carcinogen benzene. Because RFG contains more oxygen, its combustion leaves less carbon monoxide and fewer unburned hydrocarbons. (In fact, simply adding ethanol to ordinary gasoline contributes oxygen to the fuel and thus helps it to burn more efficiently, and this is being done more and more often.) RFG costs more than ordinary gasoline, a small price to pay for cleaner air.

Construction of a catalytic converter.

11-7 GAS FUELS

The Least Polluting   When burned, gas fuels combine with oxygen to give CO_2 and H_2O as liquid fuels do. **Natural gas** consists of lighter hydrocarbons than petroleum, notably **methane,** CH_4. Natural gas is usually carried in pipelines, but sometimes in liquid form in special tankers.

Artificial gas fuels can be made from coal or from **coke,** which is largely carbon. (Coke is derived from coal by heating it in the absence of air to drive off many impurities as gases.) The most common artificial gas fuel is **water gas,** a mixture of carbon monoxide and hydrogen produced by passing steam over hot coke:

$$H_2O + C \longrightarrow CO + H_2 \qquad formation\ of\ water\ gas$$

When burned, the H_2 becomes H_2O and the CO becomes CO_2. Because of its carbon monoxide content, water gas is poisonous, unlike natural gas.

Hydrogen itself is an excellent fuel with the extremely high heat of combustion of 143 kJ/g, 3 times that of gasoline. Hydrogen is produced by the electrolysis of water and is too expensive at present for ordinary purposes, but it is widely used in welding and as a spacecraft fuel.

Proposals have been made to gasify coal while it is still underground. If the coal is ignited and just the right amount of compressed air is supplied, the combustion will be incomplete and carbon monoxide will be produced:

$$2C + O_2 \longrightarrow 2CO$$

The carbon monoxide can then be used as a fuel or can be reacted with steam to give hydrogen:

$$CO + H_2O \longrightarrow CO_2 + H_2$$

Natural gas, which is mainly methane, is carried in liquid form at low temperature (methane boils at −161°C) in tankers such as this. The tanks are spherical to minimize heat flow into them: a sphere has the least surface area for a given volume. The liquid natural gas is kept cold by being allowed to evaporate continuously, which absorbs the heat that passes through the tank walls. The gas that comes off is used to power the ship's engines, so it is not wasted.

11-8   SOLID FUELS

Burning Them Leads to Acid Rain   **Coal** was once the chief energy source under human control, but it has since been overtaken by petroleum. Since coal reserves far exceed those of other fossil fuels, it is quite possible that coal will eventually return to the lead. Most of the coal mined in the United States today is used to generate electricity.

The heat of combustion of coal varies, but a typical value is 32 kJ/g. Coke is used in place of coal where a hotter and less smoky flame is desired, for instance in steel mills. **Wood** is largely cellulose, one of the carbohydrates that plants produce by photosynthesis (Chap. 12). Its heat of combustion is much less than that of coal, ranging from 10 to 19 kJ/g.

If coal were pure carbon, it would be an excellent fuel. Unfortunately coal contains a number of other substances, one of which—sulfur—is largely responsible for the major ecological problem of acid rain.

Bituminous ("soft") coal, the most common variety, has an average sulfur content of 3 percent. Anthracite ("hard") coal and much of the furnace oil derived from petroleum contain somewhat smaller but still significant amounts of sulfur. Lignite, another type of coal, contains even more sulfur. When any of these fuels burns in a furnace, the sulfur is oxidized to sulfur dioxide, SO_2; the world total is 50 to 60 million tons per year released into the atmosphere. Some nitrogen from the air is oxidized in furnaces as well. Car exhausts are also major contributors of nitrogen oxides, as mentioned earlier. The sulfur and nitrogen oxides combine with atmospheric moisture to produce sulfuric and nitric acids. What goes up must come down, and the result is acid rain (and acid snow).

Coal, shown here piled next to an electric power plant in Newark, New Jersey, is the most abundant fossil fuel and is used to produce about half the electric energy generated in the United States. Opponents of nuclear energy often cite coal as a superior alternative, but coal mining is hazardous and the air pollution that results from burning coal has damaged the health of many people. Acid rain, too, is a result of burning coal. None of today's sources—fossil or nuclear—that is able to provide a substantial fraction of the vast amounts of energy modern civilization requires is free from serious objections of one kind or another.

Acid rain destroyed this forest in North Carolina.

Acid Rain   Normal rainwater is slightly acid, with a pH of about 5.6, because of dissolved CO_2. The presence of sulfuric and, to a lesser extent, nitric acid has increased the average acidity of rainwater in much of Canada, the United States, and Europe to pH values as low as 3.8, which is 60 times more acidic than normal rainwater. The record for a single rainfall, set in Scotland in 1970, is a pH of 2.4—more acidic than vinegar.

Acid rain has two main effects on soils. One is to dissolve and carry away valuable plant nutrients. The second is to convert ordinarily harmless aluminum compounds, which are abundant in many soils, to toxic varieties. As a result, forests are dying on a vast scale and fish are disappearing from tens of thousands of lakes and rivers due to aluminum washed into them. Drinking water has already been contaminated in a number of regions by metals released by acidified water, such as cadmium and copper in addition to aluminum. A long-term risk to the world's food supply exists as well since many farm soils are subject to damage by acid rain.

How can so serious a situation have been allowed to arise? Can nothing be done? In fact, several ways exist to reduce the sulfur and nitrogen oxide emissions of power plants, which are the cause of most acid rain. All of these methods cost money, which is why little was done in the past, and not much more is being done today, to stop acid rain. Pollution control adds over a third to both the building cost and the running costs of a coal-fired power station.

The usual method for coping with the pollution from power plants is to "scrub" the exhaust gases by mixing them with a lime ($CaCO_3$) solution. As much as 95 percent of the sulfur dioxide present can be converted to solid calcium sulfate ($CaSO_4$) in the reaction

$$2SO_2 + O_2 + 2CaCO_3 \longrightarrow 2CaSO_4 + 2CO_2$$

Disposing of the resulting useless sludge is a major job—a 1000-MW power plant may yield enough sludge each year to cover a square kilometer with a layer nearly a meter deep.

How the world's climates are affected by the CO_2 produced by burning fossil fuels is discussed in Chap. 13.

REACTION RATES   Some chemical changes are practically instantaneous. Thus in neutralization the acid and base react as soon as they are stirred together. Silver chloride is precipitated immediately when solutions that contain

TABLE 11-1		
Factors That Affect Reaction Rates		
Factor	**Effect**	
Temperature	The higher the temperature, the faster the reaction	
Concentration	The higher the concentration of the reactants, the faster the reaction	
Surface area	The greater the surface area of a solid reactant, the faster the reaction	
Catalyst	Can increase or decrease the reaction rate	

silver ions and chloride ions are mixed. The reaction involved in a dynamite explosion takes a fraction of a second. In contrast, other chemical changes, like the rusting of iron, take place slowly.

Reaction rates depend first of all on the nature of the reacting substances: iron corrodes faster than copper, for example. In any particular reaction the rate is influenced by four principal factors (Table 11-1). These are temperature, concentrations of the reacting substances, the exposed surface area in the case of reactions that involve solids, and the presence of an appropriate catalyst.

11-9 TEMPERATURE

Hotter Means Faster Reaction rates are always increased by a rise in temperature. This is why we put food in the refrigerator to retard its decay and why we use hot water rather than cold for washing. Reaction rates for many common processes that occur at or near room temperature are approximately doubled for every 10°C increase in temperature.

The kinetic theory of matter suggests one obvious reason for the increase of reaction rates with temperature. Most reactions depend on collisions between particles, and the number of collisions increases with rising temperature because molecular speeds are increased. But a 10°C rise is nowhere near enough to double the number of collisions in a particular sample. To find the real explanation we must go back to the idea of activation energy.

Why Reaction Rates Vary with Temperature If molecules must be activated before they can react, reaction rates should depend not on the total number of collisions per second but on the number of collisions between *activated* molecules. Activated molecules in a fluid (liquid or gas) are produced by ordinary molecular motion as a result of exceptionally energetic collisions. Such molecules remain activated only a short time before losing energy in further collisions (unless they react in the meantime).

In any fluid, then, a certain fraction of the molecules is activated at any time. The fraction may be very small at ordinary temperatures, but it increases rapidly as the temperature rises and molecular motion speeds up. Reaction rates increase with temperature chiefly because the number of activated molecules increases.

At room temperature, for instance, a mixture of hydrogen and oxygen contains very few molecules with sufficient energy to react. The reaction is so slow that the gases may remain mixed for years without anything happening. Even at 400°C the rate is small, but at 600° enough of the molecules are activated to make the reaction fast, and at 700° so many are activated that the mixture explodes.

This kind of behavior is typical of many reactions between molecules. At low temperatures the chemical changes are so slow that for all practical purposes they do not occur; in a range of intermediate temperatures the reactions are moderately rapid; and at high temperatures they become practically instantaneous. Reactions between ions, on the other hand, occur immediately even at room temperatures, since the ionic state itself is a form of activation.

11-10 CONCENTRATION AND SURFACE AREA

The Greater They Are, the Faster the Reaction   The effect of concentration on reaction speed is illustrated by rates of burning in air and in pure oxygen. The pure gas has almost 5 times as many oxygen molecules per cubic centimeter as air, and combustion in pure oxygen is correspondingly faster.

As a general rule, the rate of a simple chemical reaction is proportional to the concentration of each reacting substance. The number of collisions between activated molecules, which determines the reaction speed, depends on the total number of collisions and this, in turn, depends on how many molecules each cubic centimeter contains.

When a reaction takes place between two solids or between a fluid and a solid, the reaction speed depends markedly on the amount of solid surface exposed. A finely powdered solid presents vastly more surface than a few large chunks, and reactions of powders are accordingly much faster. Granulated sugar dissolves more rapidly in water than lump sugar; finely divided zinc is attacked by acid quickly, larger pieces are attacked slowly; ordinary iron rusts slowly, but the oxidation of iron powder is fast enough to produce a flame.

The explanation is obvious: the greater the surface, the more quickly atoms and molecules can get together to react. For a similar reason, efficient stirring speeds up reactions between fluids.

11-11 CATALYSTS

Faster or Slower by Magic (Almost)   **Catalysts** are substances that either speed up or slow down a reaction without being permanently changed themselves. As an example of catalytic action, let us consider the decomposition of hydrogen peroxide, H_2O_2. At ordinary temperatures solutions of hydrogen peroxide are unstable and slowly turn into water and oxygen:

$$2H_2O_2 \longrightarrow 2H_2O + O_2$$

If a little powdered manganese dioxide is added to the solution, the decomposition goes much faster, with oxygen bubbling up violently. At the end of the reaction the manganese dioxide remains unchanged. Commercial solutions of hydrogen peroxide usually contain a trace of the compound acetanilid, which acts as a catalyst of the opposite kind to retard their decomposition.

Catalysts accelerate reactions in different ways. In some cases the catalyst forms an unstable intermediate compound with one of the reacting substances, and this compound decomposes later in the reaction. Other catalysts, notably certain metals such as platinum, increase reaction rates by producing activated molecules at their surfaces. No adequate explanation is known for the action of some catalysts. A given

A CHEMIST AT WORK

Cynthia M. Friend (Harvard University). I spend my time trying to understand how surfaces of solids help transform molecules. Chemical reactions take place on a surface when a molecule originally in the gas or liquid state bonds to the surface. Such reactions are very important in technologies as disparate as fabrication of computer chips, lubrication in car engines, and prevention of pollution. When molecules bond to a surface, they may be changed so that specific reactions will occur quickly that otherwise might not occur in a lifetime of waiting.

In our current work my group is interested in how methane and related hydrocarbons can be converted to useful chemical building blocks or alternative fuels, such as methanol, by reaction with oxygen. Methane is abundant but not very reactive. By using a solid catalyst we can oxidize methane. Unfortunately this reaction generally occurs at high temperatures and all the methane and oxygen turns into carbon dioxide and water. The challenge is to find a way to induce methane to react but to stop the reaction before it is complete. Methanol is a fossil-fuel alternative, so a process that efficiently converts methane to methanol would be a source of fuel for cars. We are studying fundamental aspects of methane oxidation on metals such as rhodium and on some metal oxides.

Understanding how the structure and composition of a surface affects reactions at the surface is a central issue in our work with methane. Our aim is to be able to predict what materials would lead to specific types of reaction. Our approach is to pick apart the elementary steps of the reactions that involve methane. To do so we prepare molecules that may be produced from methane reaction in order to map out the steps that involve these potential intermediates. This approach is necessary because we cannot observe transient species that may react almost as fast as they are formed, perhaps in less than 10^{-9} s. This allows us to figure out the possible chemical pathways that can be taken by the methane.

Because we cannot observe the reactions directly, we examine how heat, light, and electrons affect the atoms and molecules on the surface. For example molecules there can be identified from the specific energies of infrared light that they absorb. Electron beams scattered by the atoms on the surface also give us information about structure and composition. Each result is like a piece of a jigsaw puzzle and finding its place in the big picture is very gratifying. As we add pieces, the puzzle begins to develop. At the same time, we look in detail at each piece to get a clearer picture so we can be sure the piece is in the right place. Ultimately we can figure out what missing pieces should look like; in other words, we predict the outcome of new experiments. We then see if our predictions are correct in the laboratory.

Did I always know I would be a scientist? Even though I have always been interested in science, I was not always sure I would be a scientist and I also did not always think I would be a chemist. Now I cannot imagine a more rewarding and interesting career. As a scientist I have limitless possibilities—there will always be new questions to ask and to answer.

reaction is usually influenced only by a few catalysts, and these may or may not affect other reactions. Catalysts are essential in many industrial processes.

The many chemical processes that take place in living things, for instance the digestion of food, are controlled by catalysts called **enzymes.** An enzyme is a protein molecule whose physical structure is such that it attracts specific molecules to its surface, promotes their reaction, and then releases the products. Each enzyme catalyzes a particular reaction, and thousands of different ones are present in the human body. The reacting molecules fit into the outside of the enzyme for a process much as a key fits into a lock: if their shapes do not match, nothing will happen.

11-12   CHEMICAL EQUILIBRIUM

One Step Forward; One Step Back

Most chemical reactions are reversible. That is, under suitable conditions the products of a chemical change can usually be made to react "backward" to give the original substances. We have seen many examples in other connections. Hydrogen burns in oxygen to form water, and water decomposes into these elements during electrolysis. Mercury and oxygen combine when heated moderately, and the oxide decomposes when heated more strongly. Carbon dioxide reacts with water to give hydrogen and hydrogen carbonate (HCO_3^-) ions, and these ions recombine all the time to form the original CO_2 and H_2O.

There is no reason why the forward and backward processes of a chemical change cannot take place at the same time. In a bottle of soda water, some of the CO_2 reacts with the water. But a number of the resulting H^+ and HCO_3^- ions then join together to give CO_2 and H_2O. The recombination rate increases with the ion concentration until, finally, as many ions recombine each second as are being formed. At this point the rates of the forward and backward reactions are the same, and the amounts of the various substances do not change. The situation can be represented by the single equation

$$H_2O + CO_2 \rightleftharpoons H^+ + HCO_3^-$$

The double arrow indicates that reactions in both directions occur together.

In soda water, dissolved carbon dioxide reacts with water to give hydrogen and hydrogen carbonate ions exactly as fast as other of these ions recombine to form carbon dioxide and water. This is an example of chemical equilibrium.

A State of Balance

A situation of the above kind is called **chemical equilibrium.** It is a state of balance determined by two opposing processes. The two processes do not reach equilibrium and stop, but instead continue indefinitely because each process constantly undoes what the other accomplishes.

As an analogy, we might imagine a person walking up an escalator while the escalator is moving down. If the person walks as fast in one direction as the escalator is moving in the other, the two motions will be in equilibrium and the person will remain in the same place indefinitely.

A great many chemical changes reach a state of equilibrium instead of going to completion. Equilibrium may be established when a reaction is nearly complete, or when it is only just starting, or when both products and reacting substances are present in comparable amounts.

The point at which equilibrium occurs depends on the rates of the opposing reactions. The initial reaction always dominates until the products are abundant enough for the reverse reaction to go at the same rate. Thus the extent to which an acid is dissociated depends on how fast its molecules break down into ions compared with how fast the ions recombine. HCl dissociates so rapidly in solution into H^+ and Cl^- that the reverse reaction has no chance to maintain a measurable amount of HCl. On the other hand, acetic acid dissociates slowly, and when only a small concentration of ions has been built up, the recombination occurs at the same rate as the dissociation.

11-13 ALTERING AN EQUILIBRIUM

How to Get Farther up the Down Escalator
Often a chemist wishing to prepare a compound finds that the reaction reaches equilibrium before very much of the compound has been formed. Once this happens, waiting for more of the product to form is useless, for its amount does not change after that. How can equilibrium conditions be altered to increase the yield of the product?

Since equilibrium represents a balance between two rates, what is needed to increase the yield is a way to change the speed of one reaction or the other. Speeding up or slowing down one of the reactions in an equilibrium is not as simple as changing the rate of a single reaction, but the same factors that affect reaction rates also influence equilibrium. The chemist has three chief methods available for shifting an equilibrium to favor one direction or the other. These are:

1. Change the concentration of one or more substances. For example, removing the gaseous product of a reaction will retard the reverse reaction. Thus opening a soda bottle allows CO_2 to escape, which decreases the rate of formation of H^+ and HCO_3^- and so lowers the acidity of the solution.

2. Change the temperature. If one reaction in an equilibrium is exothermic (gives off energy), the other is endothermic (absorbs energy). A rise in temperature, although it makes both reactions go faster, will favor the endothermic one.

3. Change the pressure. This is most effective in gas reactions where the number of product molecules differs from the number of initial molecules. Increasing the pressure favors the reaction that gives the fewest molecules. An example is the synthesis of ammonia from nitrogen and hydrogen in the reaction

$$N_2 + 3H_2 \rightleftharpoons 2NH_3$$

A rise in pressure increases the yield of ammonia because the ammonia occupies only half the volume of the gases that react to form it. Pressures of 500 to 1000 atm are used when this process is carried out industrially.

OXIDATION AND REDUCTION

Until now we have used the term **oxidation** to mean the chemical combination of a substance with oxygen. A related term is **reduction,** which refers to the removal of oxygen from a compound. When oxygen reacts with another substance (except fluorine), the oxygen atoms pick up electrons donated by the atoms of that substance. When the resulting compound is reduced, the atoms of the substances that had been oxidized regain the electrons initially lost to the oxygen atoms.

It is convenient to generalize oxidation and reduction to refer to *any* chemical process in which electrons are transferred from one element to another, regardless of whether or not oxygen is involved. Hence

> **Oxidation refers to the loss of electrons by the atoms of an element, and reduction refers to the gain of electrons.**

The oxidation of one element is always accompanied by the reduction of another. The two processes must take place together. For example, when zinc combines with chlorine, electrons are given up by the zinc atoms to the chlorine atoms. Thus the zinc is oxidized and the chlorine is reduced in the reaction:

$$Zn \longrightarrow Zn^{2+} + 2e^- \qquad oxidation$$

$$Cl_2 + 2e^- \longrightarrow 2Cl^- \qquad reduction$$

Reactions that involve electron transfer are called **oxidation-reduction reactions,** and they make up a large and important category of chemical reactions.

11-14   ELECTROLYSIS

A Chemical Reaction Caused by Electric Current

Electrolysis is an oxidation-reduction process. Let us consider the electrolysis of molten sodium chloride, which consists of the ions Na^+ and Cl^-. An **electrode** is a conductor through which electric current enters or leaves a solution. When electrodes in the molten NaCl are connected to the terminals of a battery, Na^+ ions are attracted to the cathode (negative electrode) and Cl^- ions to the anode (positive electrode), as in Fig. 11-6. At the anode each Cl^- is neutralized by giving up its extra electron and becomes a chlorine atom:

$$Cl^- \longrightarrow Cl + e^- \qquad oxidation$$

The Cl atoms pair off to form molecules of chlorine gas, Cl_2. At the cathode each Na^+ is neutralized by gaining an electron and becomes a sodium atom:

$$Na^+ + e^- \longrightarrow Na \qquad reduction$$

The net result of sending a current through molten salt, then, is to break up the compound NaCl into its constituent elements:

$$2NaCl \longrightarrow 2Na + Cl_2 \qquad electrolysis\ of\ sodium\ chloride$$

The sodium, a liquid at the temperature of molten salt, collects around the cathode, and chlorine gas bubbles up around the anode. This procedure is commonly used to prepare metallic sodium.

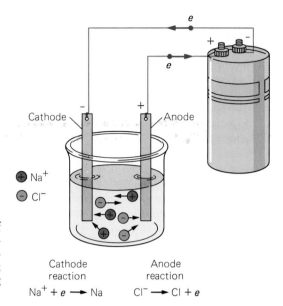

FIG. 11-6 The electrolysis of molten sodium chloride. The current in the liquid consists of moving Na^+ and Cl^- ions; the current in the wires consists of moving electrons.

Cathode reaction

$$Na^+ + e \longrightarrow Na$$

Anode reaction

$$Cl^- \longrightarrow Cl + e$$

Electroplating

Electrolysis is used in the process of **electroplating** in which a thin layer of one metal is deposited on an object made of another metal. Sometimes this is done because the plating metal is expensive, for instance gold or silver. In other cases the object is to protect the base metal from corrosion, as in the chromium plating of steel. Nonmetallic items can be plated by first coating them with a conducting material such as graphite.

Figure 11-7 shows how a spoon can be silver-plated. The procedure uses a solution of silver nitrate, which dissociates into Ag^+ and NO_3^- ions. Silver atoms lose electrons (which flow to the battery) at the anode and enter the solution as Ag^+ ions. These ions are attracted to the spoon, from which they pick up electrons (supplied by the battery) to become silver atoms again. In this way silver atoms are transferred from the

ALUMINUM Although aluminum is the third most abundant element in the earth's outer layer and the most abundant metal there, it was not discovered until 1827. Chemical methods for preparing aluminum were so expensive then that for some time its main use was in jewelry, and attempts to refine aluminum by electrolysis were balked by the difficulty of melting or dissolving aluminum ores.

Finally, in 1886, a 22-year-old American, Charles Martin Hall, found that cryolite, a mineral abundant in Greenland, when melted would dissolve the chief aluminum ore, Al_2O_3. Passing an electric current through a solution of Al_2O_3 in molten cryolite liberates aluminum at the negative electrode and oxygen at the positive one: this is the process used today to refine aluminum.

About 11 kJ of electric energy is needed to produce each gram of aluminum metal, which is a lot of energy. To recycle aluminum, notably the aluminum cans discarded by the billion every year, basically involves heating it to its melting point and then melting it, which takes less than 1 kJ/g. Hence recycling aluminum uses only 9 percent as much energy as refining it from its ores, a vast saving on the scale at which aluminum cans are used.

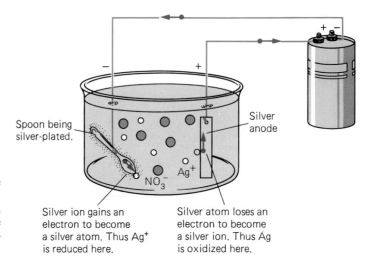

FIG. 11-7 Silver plating. The bath is a solution of silver nitrate, AgNO$_3$. The nitrate ions remain in solution because Ag atoms lose electrons at the anode more readily than NO$_3^-$ ions do.

Spoon being silver-plated.

Silver anode

Silver ion gains an electron to become a silver atom. Thus Ag$^+$ is reduced here.

Silver atom loses an electron to become a silver ion. Thus Ag is oxidized here.

Steel plated with tin to prevent corrosion is widely used for food and beverage containers.

anode to the spoon. Because an Ag atom loses an electron to become Ag$^+$ more readily than an NO$_3^-$ ion loses its extra electron, the NO$_3^-$ ions stay in solution and do not participate in the plating process.

11-15   ELECTROCHEMICAL CELLS

Turning Chemical Energy into Electric Energy

Oxidation-reduction reactions can produce electric currents. What we must do is arrange to have the electrons transferred in such a reaction pass through an external wire as they go from one reactant to the other. The dry-cell batteries of a flashlight, the storage battery of a car, and the fuel cell of a spacecraft are all

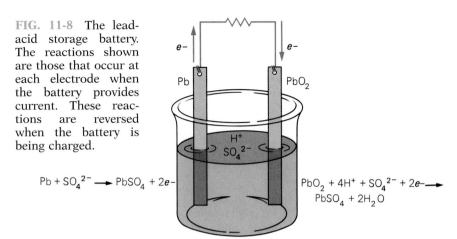

FIG. 11-8 The lead-acid storage battery. The reactions shown are those that occur at each electrode when the battery provides current. These reactions are reversed when the battery is being charged.

$$Pb + SO_4^{2-} \longrightarrow PbSO_4 + 2e-$$

$$PbO_2 + 4H^+ + SO_4^{2-} + 2e- \longrightarrow PbSO_4 + 2H_2O$$

FIG. 11-9 A hydrometer is often used to measure the density of the acid in a car's storage battery. By squeezing and then releasing the rubber bulb, the acid is drawn into a glass tube that contains a smaller sealed tube with a weight at its bottom. The higher the sealed tube floats, the denser the acid. A fully charged battery typically has an acid density of 1.28 g/cm³, which drops to about 1.15 g/cm³ when the battery is completely discharged.

based on oxidation-reduction reactions. They are known as **electrochemical cells.**

In the storage battery of a car, plates of lead and of lead dioxide, PbO_2, are in a solution of sulfuric acid which is dissociated into H^+ and SO_4^{2-} ions. The reactions that take place at each electrode when the battery is providing current are shown in Fig. 11-8. As the battery provides current, insoluble lead sulfate, $PbSO_4$, builds up on its plates. When there is no longer enough of the reactants to continue the electrode reactions, the battery is "dead."

To recharge the battery, a current is passed through it in the opposite direction. This current causes the two electrode reactions to proceed in reverse, which restores the plates and the acid bath to their original compositions (Fig. 11-9). The potential difference across a fully charged storage battery cell is 2.1 V; a "12-V" battery contains six cells connected in series.

Each cell of a storage battery contains alternate plates of lead and of lead dioxide, PbO_2, in a solution of sulfuric acid. The plates of each kind are connected together and to an outside terminal. The electric current such a battery produces consists of electrons that move from the lead plates through an outside circuit to the lead dioxide plates. The lower the temperature of the acid in a storage battery, the slower its ions move and the less current the battery can provide. In freezing weather, the current available for the starting motor of a car may be less than half that available on a warm day, and it may then be difficult or even impossible to start the car's engine. If a car's battery is too weak to start the car on a cold day, the remedy is to connect another battery to it by means of heavy jumper cables (+ terminal to + terminal, – terminal to – terminal), which doubles the available current.

ELECTRIC CARS   Because California has a severe air-pollution problem, a 1990 regulation calls for at least 2 percent of all cars sold there in 1998 to be emission-free. By 2003 the proportion must rise to 10 percent, which means over 100,000 cars per year. Other states and several European countries have shown interest in this approach to cleaner air.

In practice, emission-free cars will be battery-powered electric cars, and much effort has gone into developing them. The difficulty is not performance—the General Motors Impact shown here has a top speed of 120 km/h—but range and cost. The 400 kg of lead-acid batteries in the Impact can supply only as much energy as 5.7 L of gasoline, and under ideal conditions it can travel no more than perhaps 240 km before a lengthy recharge is needed. Using lights and heating or air-conditioning would cut this distance to an even smaller figure.

As for cost, electric cars at first will certainly be more expensive than comparable gasoline-powered cars, and the need to replace the batteries fairly often—every 40,000 km or so for the Impact—is an additional factor. Longer-lived batteries with greater energy content per kilogram than lead-acid batteries have been made, but they have their own bad features, usually an inability to provide energy fast enough for rapid acceleration. Although the first generation of electric cars will be adequate for short-range errands and commuting, a breakthrough in battery technology seems needed for such cars to rival gasoline-powered cars for general transportation.

Rechargeable batteries with a capacity of 13.6 kWh supply energy to the electric motors of this prototype General Motors car. Each front wheel has its own 57-hp motor.

Fuel Cells   In a **fuel cell,** the reacting substances are fed in continuously. As a result the cell can provide current indefinitely without having to be replaced or recharged. Fuel cells are used in spacecraft since they are very light in proportion to the electric power they can supply. In the future it is possible that fuel cells will be perfected to the point where they are economical sources of power for individual homes, electric cars, and large-scale electric plants.

Combining 1 kg of hydrogen and 8 kg of oxygen in a hydrogen-oxygen fuel cell produces over 2×10^8 J of electric energy, enough to power a 100-W lightbulb for 4 weeks. The overall reaction in such a cell is simply

$$2H_2 + O_2 \longrightarrow H_2O$$

and involves the flow of 4 electrons each time the reaction occurs. If a mixture of two volumes of hydrogen gas and one volume of oxygen gas is ignited, the result is a violent explosion with water as the product. In

The 446 fuel cells in each of these stacks produce a total of 100 kW df electric power. If trials are successful, such stacks may be the basis of a new generation of power stations.

a hydrogen-oxygen fuel cell the same chemical combination takes place, but the liberated energy is released in the form of electric current.

The ideal fuel cell would use readily available substances such as natural gas and air. Much progress has been made in recent years in developing such a cell, but practical difficulties remain to be overcome.

IMPORTANT TERMS AND IDEAS

Combustion is the rapid combination of oxygen with another substance during which heat and light are given off.

Endothermic reactions absorb energy and **exothermic reactions** liberate energy. Many exothermic reactions require initial **activation energy** in order to take place.

A **catalyst** is a substance that can change the rate of a chemical reaction without itself being permanently changed.

In a **chemical equilibrium,** forward and reverse reactions occur at the same rate, so the concentrations of the reactants and products remain constant.

Oxidation involves the loss of electrons by the atoms of an element in a chemical reaction, and **reduction** involves the gain of electrons. An example of an oxidation-reduction reaction is **electrolysis,** in which free elements are liberated from a liquid by the passage of an electric current. Batteries and fuel cells produce electric current by means of oxidation-reduction reactions.

EXERCISES: MULTIPLE CHOICE

1. When something burns,
   a. it combines with phlogiston

    b. it gives off phlogiston
    c. it combines with oxygen
    d. it gives off oxygen

2. The proportion of oxygen in air is about
    a. $\frac{1}{5}$ **c.** $\frac{1}{2}$
    b. $\frac{1}{3}$ **d.** $\frac{4}{5}$

3. A substance of unknown composition is heated in an open container. As a result,
    a. its mass decreases
    b. its mass remains the same
    c. its mass increases
    d. any of the above, depending on the nature of the substance and the temperature reached

4. When a piece of metal is oxidized, the resulting oxide is
    a. lighter than the original metal
    b. the same weight as the original metal
    c. heavier than the original metal
    d. any of the above, depending on the metal

5. Chemical energy is stored within atoms, molecules, and ions as
    a. activation energy
    b. electron kinetic energy
    c. electron potential energy
    d. thermal energy

6. A chemical reaction that absorbs energy is called
    a. endothermic
    b. exothermic
    c. activated
    d. oxidation-reduction

7. An example of an endothermic reaction is
    a. the dissociation of a salt in water
    b. the neutralization of an acid by a base
    c. the freezing of water
    d. combustion

8. If a given reaction is exothermic, the reverse reaction
    a. is exothermic
    b. is endothermic
    c. may involve no energy change
    d. any of the above, depending on the reaction

9. Reactions that give off energy
    a. never require activation energy
    b. sometimes require activation energy
    c. always require activation energy
    d. always require catalysts

10. The neutralization of a strong acid by a strong base
    a. absorbs energy

    b. liberates energy
    c. involves no energy change
    d. requires a catalyst to occur

11. The products of the complete combustion of hydrocarbon fuels are
    a. H_2 and CO   **c.** H_2O and CO
    b. H_2 and CO_2 **d.** H_2O and CO_2

12. Natural gas consists largely of
    a. hydrogen   **c.** octane
    b. methane   **d.** nitrogen oxides

13. Of the following substances, the one that gives off the most heat per gram when it is burned is
    a. hydrogen   **c.** gasoline
    b. methane   **d.** coal

14. A harmful product of the incomplete combustion of gasoline is
    a. carbon monoxide   **c.** octane
    b. carbon dioxide   **d.** hydrochloric acid

15. The impurity found in coal that contributes to acid rain is
    a. nitrogen   **c.** carbon
    b. sulfur   **d.** chlorine

16. Acid rain never contains dissolved
    a. carbon dioxide   **c.** sulfuric acid
    b. nitric acid   **d.** acetic acid

17. The least polluting of the following fuels is
    a. coal   **c.** diesel fuel
    b. gasoline   **d.** natural gas

18. Reaction rates increase with temperature primarily because
    a. dissociation into ions is more complete
    b. more collisions occur between the molecules involved
    c. more activated molecules are formed
    d. equilibrium does not occur at high temperatures

19. At ordinary temperatures, the temperature increase needed to double the rate of many common reactions is
    a. 1°C   **c.** 50°C
    b. 10°C   **d.** 100°C

20. The speeds of reactions between ions in solution
    a. depend critically upon temperature
    b. are essentially independent of temperature
    c. depend upon which catalyst is used
    d. are slow in general

21. Most chemical reactions are
    a. reversible   **c.** exothermic
    b. irreversible   **d.** endothermic

22. At equilibrium,
  a. both forward and reverse reactions have ceased
  b. the forward and reverse reactions are proceeding at the same rate
  c. the forward reaction has come to a stop, and the reverse reaction is just about to begin
  d. the mass of reactants equals the mass of products

23. The yield of the product C in the reversible reaction $A + B + \text{energy} \rightleftharpoons C$ can be increased by
  a. decreasing the temperature
  b. increasing the temperature
  c. increasing the surface area of the reactants
  d. changing the catalyst

24. When a gas reaction involves a decrease in the total number of molecules, the equilibrium can be shifted in the direction of higher yield by
  a. increasing the pressure
  b. decreasing the pressure
  c. increasing the temperature
  d. decreasing the temperature

25. Reduction occurs when a substance
  a. loses electrons   **c.** combines with oxygen
  b. gains electrons   **d.** reacts with an acid

26. A catalyst affects which one or more of the following?
  a. The energy needed for a chemical reaction to occur
  b. The energy a chemical reaction gives off
  c. The speed of a chemical reaction
  d. Whether a substance is oxidized or reduced in a chemical reaction

27. When an electric current is passed through molten sodium chloride,
  a. sodium metal is deposited at the positive electrode
  b. sodium ions are deposited at the positive electrode
  c. chlorine gas is liberated at the positive electrode
  d. chlorine ions are liberated at the positive electrode

28. The quantity actually stored in a "storage battery" is
  a. electric charge   **c.** voltage
  b. electric current   **d.** energy

29. Batteries and fuel cells employ
  a. oxidation reactions only
  b. reduction reactions only
  c. both oxidation and reduction reactions
  d. acid-base neutralization reactions

30. A fuel cell does not require
  a. a positive electrode
  b. a negative electrode
  c. oxidation-reduction reactions
  d. recharging

QUESTIONS

1. What role does air play in combustion?

2. If a given reaction is endothermic, what must be true of the reverse reaction?

3. Which of the following are exothermic reactions and which are endothermic?
  a. The explosion of dynamite
  b. The burning of methane
  c. The decomposition of water into its elements
  d. The dissociation of water into ions
  e. The burning of iron in chlorine
  f. The combination of zinc and sulfur to form zinc sulfide

4. From the observation that the slaking of lime [addition of water to CaO to form $Ca(OH)_2$] gives out heat, would you conclude that the following reaction is endothermic or exothermic?

$$Ca(OH)_2 \longrightarrow CaO + H_2O$$

5. In what fundamental way is the explosion of an atomic bomb different from the explosion of dynamite?

6. What is the origin of the energy liberated in an exothermic reaction?

7. What is the chief reason that reaction rates increase with temperature?

8. Why does an increase in temperature increase the rate of exothermic as well as endothermic reactions?

9. When carbon in the form of diamond is burned to produce CO_2, more heat is given off than when carbon in the form of graphite is burned. What form of carbon is more stable under ordinary conditions? What bearing does this conclusion have on the origin of diamonds?

10. Suggest three ways to increase the rate at which zinc dissolves in sulfuric acid.

11. Give two examples of reactions that are (a) practically instantaneous at room temperatures, (b) fairly slow at room temperatures.

12. Under ordinary circumstances coal burns slowly, but the fine coal dust in mines sometimes burns so rapidly as to cause an explosion. Explain the difference in rates. Would you expect the danger from spontaneous combustion to be greater in a coal pile containing principally large chunks or in one containing finely pulverized coal? Why?

13. Why is a reaction with a high activation energy slow at room temperature?

14. The solubility of a gas in a liquid decreases with increasing temperature. From this observation and what you know of how a change in temperature can affect an equilibrium, would you expect that dissolving a gas in a liquid is an exothermic or an endothermic process?

15. To what extent does the time needed for a strong acid to neutralize a strong base in solution depend on temperature?

16. Would you say that few, many, or most chemical reactions are reversible?

17. Can you think of the general condition for a reaction to go to completion instead of an equilibrium being established? Give examples of liquid-phase reactions that go to completion.

18. Ammonia gas dissolves in water and can react according to the equation

$$NH_3 + H_2O \rightleftharpoons NH_4^+ + OH^-$$

How would the amount of ammonium ion in solution be affected by
a. increasing the pressure of NH_3?
b. pumping off the gas above the solution?
c. raising the temperature?
d. adding a solution of HCl?

19. Hydrogen sulfide gas dissolves in water and ionizes very slightly:

$$H_2S \rightleftharpoons 2H^+ + S^{2-}$$

How would the acidity of the solution (concentration of H^+) be affected by
a. increasing the pressure of H_2S?
b. raising the temperature?
c. adding a solution of KOH?
d. adding a solution of silver nitrate [silver sulfide (Ag_2S) is insoluble]?

20. Limestone ($CaCO_3$) dissolves in carbonic acid to form calcium ion and acid carbonate ion:

$$CaCO_3 + H_2CO_3 \rightleftharpoons Ca^{2+} + 2HCO_3^-$$

How would this equilibrium be affected by
a. raising the temperature?
b. allowing the solution to evaporate?

c. increasing the pressure of CO_2, thereby increasing the concentration of H_2CO_3?
Under what natural conditions, then, is limestone most soluble? Under what conditions will it be precipitated from solution?

21. The reaction $2SO_2 + O_2 \longrightarrow 2SO_3$ is exothermic. How will a rise in temperature affect the yield of SO_3 in an equilibrium mixture of the three gases? Will an increase in pressure raise or lower this yield? In what possible way can the speed of the reaction be increased at moderate temperatures?

22. The three gases H_2, O_2, and H_2O are in equilibrium at temperatures near 2000°C. Write the equation for the equilibrium. Would the yield of H_2O be increased or decreased by raising the temperature? By raising the pressure?

23. What class of reactions involves the transfer of electrons between the reacting substances?

24. When an electric current is passed through molten NaCl, what substance is liberated at the anode (positive electrode)? At the cathode (negative electrode)?

25. In each of the following reactions, identify the element that is oxidized and the one that is reduced:

$$Zn + Cu^{2+} \longrightarrow Zn^{2+} + Cu$$
$$Fe + 2H^+ \longrightarrow Fe^{2+} + H_2$$
$$Cl_2 + 2Br^- \longrightarrow 2Cl^- + Br_2$$
$$Cl_2 + 2Fe^{2+} \longrightarrow 2Cl^- + 2Fe^{3+}$$

26. In each of the following reactions, identify the element that is oxidized and the one that is reduced:

$$Mg + 2H^+ \longrightarrow H_2 + Mg^{2+}$$
$$Ca + S \longrightarrow CaS$$
$$2Na + 2H_2O \longrightarrow 2Na^+ + 2OH^- + H_2$$
$$F_2 + 2Br^- \longrightarrow Br_2 + 2F^-$$
$$2Fe^{3+} + 3H_2S \longrightarrow 2FeS + S + 6H^+$$

27. Which loses electrons more easily, Na or Fe? Al or Ag? I^- or Cl^-? Which gains electrons more easily, Cl or Br? Hg^{2+} or Mg^{2+}?

28. In what part of the periodic table are the elements that are most easily reduced? In what part are those that are most easily oxidized?

29. How could you demonstrate that magnesium is

a better reducing agent (that is, more easily oxidized) than hydrogen?

30. What becomes of the electric energy provided in electrolysis? In what device is this energy transformation reversed?

31. Why must water be added periodically to a lead-acid storage battery when it is in normal operation?

32. A charging current is passed through a fully charged lead-acid storage battery. What happens?

33. In what basic way is a fuel cell different from a dry cell or a storage battery?

PROBLEMS

1. A cube of zinc 1 cm on each edge takes an hour to dissolve in an acid bath. If the zinc is first cut into eight cubes 0.5 cm on each edge, approximately how much time would be needed?

2. Lithium reacts with water to produce lithium hydroxide. What else is produced? Write the equation of the process. Which element is reduced and which is oxidized?

3. A hydrogen-oxygen fuel cell uses hollow, porous electrodes made of an inert conducting material. These electrodes enable the gases to interact gradually with the electrolyte. A typical electrolyte in such a cell is a solution of potassium hydroxide. At one electrode, hydrogen molecules combine with hydroxide ions to form water, and at the other electrode, oxygen molecules combine with water molecules to form hydroxide ions. Which electrode is negative and which is positive when the cell is in operation? What are the reactions that occur at each electrode? Does each reaction occur the same number of times as the other?

4. In the refining of iron, the iron(III) oxide, Fe_2O_3, in iron ore is reduced by carbon (in the form of coke) to yield metallic iron and carbon dioxide. Write the balanced equation of the process.

ANSWERS TO MULTIPLE CHOICE

1. c	**9.** b	**17.** d	**25.** b
2. a	**10.** b	**18.** c	**26.** c
3. d	**11.** d	**19.** b	**27.** c
4. c	**12.** b	**20.** b	**28.** d
5. c	**13.** a	**21.** a	**29.** c
6. a	**14.** a	**22.** b	**30.** d
7. a	**15.** b	**23.** b	
8. b	**16.** d	**24.** a	

12

ORGANIC

CHEMISTRY

*I*n many ways carbon is the most remarkable element. Hundreds of thousands of carbon compounds are known, far more than the number of compounds that do not contain carbon. Furthermore, carbon compounds are the chief constituents of all living things—hence the name **organic chemistry** to describe the chemistry of carbon and the name **inorganic chemistry** to describe the chemistry of all the other elements.

CARBON COMPOUNDS

At one time it was thought that carbon compounds—with the exception of the carbon oxides, the carbonates, and a few others—could be produced only by plants and animals (or indirectly from other compounds produced by them). Carbon was supposed to unite with other elements only under the influence of a mysterious "life force" possessed by living things. This ancient idea was disproved in 1828 by the German chemist Friedrich Wöhler, who prepared the organic compound urea by reacting the inorganic compounds lead cyanate and ammonia. Since Wöhler's time a great number of organic compounds have been made in the laboratory from inorganic materials, but the general distinction between the chemistry of carbon compounds and inorganic chemistry nevertheless remains useful.

12-1 CARBON BONDS

Carbon Atoms Can Form Covalent Bonds with Each Other

Let us see what the periodic table can tell us about carbon. Carbon is in period 2 at the top of group 4, which means it is halfway between the active metal lithium and the active nonmetal fluorine. Active metals tend to lose their outer electrons when they react chemically, and active nonmetals tend to gain electrons. Carbon, in the middle, does neither. Instead, it forms covalent bonds in which it shares four electron pairs.

As we saw in Table 9-7 and Fig. 9-11, the carbon atom has four electrons in its outer shell. For a carbon atom to achieve a closed outer shell, it can lose these four electrons, pick up four more for a total of eight, or share its four electrons with other atoms that contribute four other electrons so that eight electrons—four pairs—are shared.

The effective nuclear charge on the outer electrons in a carbon atom is $+4e$ (Fig. 12-1). The resulting force on the outer electrons is sufficient

FIG. 12-1 Electron shielding in carbon. Each outer electron is acted on by an effective nuclear force of $+4e$ because the inner electrons shield part of the actual nuclear charge of $+6e$.

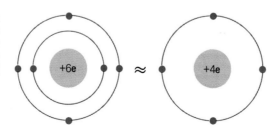

to keep them from being detached to leave a C^{4+} ion. However, the effective nuclear charge is not enough for a carbon atom to attract and hold four more electrons to give a C^{4-} ion. (See Fig. 9-12 for the reason why Na readily becomes Na^+ and Cl readily becomes Cl^-.) The result is that carbon atoms participate in four covalent bonds each when they form molecules with other atoms.

Why Carbon Forms Many Compounds   A carbon atom can bond strongly not only with many metallic and nonmetallic atoms but with *other carbon atoms* as well. This is the reason for the immense number and variety of carbon compounds, whose molecules have skeletons of linked carbon atoms. The strength of the bonds between carbon atoms is shown by the hardness of diamond, a crystalline form of carbon in which each atom is joined to four others by electron pairs (see Fig. 10-3). Atoms of a few elements near carbon in the periodic table, notably boron and silicon, are also able to bond with each other, but the range of their compounds is far more limited.

Because the bonds formed by carbon atoms are covalent, carbon compounds are mostly nonelectrolytes, and their reaction rates are usually slow. The affinity of carbon and hydrogen for oxygen makes many organic compounds subject to slow oxidation in air and to rapid oxidation if heated. Even in the absence of air most organic compounds are unstable at high temperatures; very few of them resist decomposition at temperatures over a few hundred degrees celsius.

12-2   ALKANES

The Hydrocarbons in Petroleum and Natural Gas   The simplest organic compounds are the **hydrocarbons** that contain only carbon and hydrogen. A group of hydrocarbons called the **alkanes** includes the familiar gases methane (CH_4), propane (C_3H_8), and butane (C_4H_{10}), all widely used as fuels in stoves, furnaces, and even cigarette lighters. Alkane molecules have single covalent bonds between their carbon atoms. The freezing points and boiling points of the alkanes all increase regularly as the molecular size increases (Table 12-1). Other series of organic compounds show similar regular changes in properties as the number of carbon atoms per molecule increases.

Natural gas and petroleum consist mainly of alkanes. About 80 percent of natural gas is methane, 10 percent ethane, and the rest mostly propane and butane. Alkanes with five or more carbon atoms are the main ingredients of petroleum ("crude oil"), whose exact composition varies from source to source. Methane is also one of the emissions from active volcanoes and a product of the bacterial decay of plant matter in the absence of oxygen. The "marsh gas" that bubbles up from the black ooze at the bottom of stagnant pools is largely methane, as is the "fire damp" that sometimes causes explosions in coal mines.

TABLE 12-1

The Alklane Series of Hydrocarbons*

Formula	Name	Freezing Point, °C	Boiling Point, °C	Commercial Name
CH_4	Methane	−183	−160	Fuel gases
C_2H_6	Ethane	−184	−89	Fuel gases
C_3H_8	Propane	−188	−42	
C_4H_{10}	Butane	−139	−1	
C_5H_{12}	Pentane	−130	36	Naphtha
C_6H_{14}	Hexane	−95	69	
C_7H_{16}	Heptane	−91	98	
C_8H_{18}	Octane	−57	126	Gasoline
C_9H_{20}	Nonane	−54	151	
$C_{10}H_{22}$	Decane	−30	174	Kerosene,
$C_{11}H_{24}$	Undecane	−27	197	jet fuel
$C_{16}H_{34}$	Hexadecane	18	287	
$C_{14}H_{30}$ to $C_{18}H_{38}$				Diesel fuel, heating oil
$C_{16}H_{34}$ to $C_{18}H_{38}$				Lubricating oil
$C_{16}H_{34}$ to $C_{32}H_{68}$				Petroleum jelly
$C_{20}H_{42}$ and up				Paraffin wax
$C_{36}H_{74}$ and up				Asphalt

*The data refer to the straight-chain compounds. Isomers of these hydrocarbons (see Sec. 12-5) have somewhat different properties.

12-3   PETROLEUM PRODUCTS

Fractional Distillation, Catalytic Cracking, and Polymerization

The separation of petroleum into its different alkanes is difficult because their properties are so similar. Suppose we want to separate pentane and hexane. Because pentane boils at 36°C and hexane at 69°C, it would seem that if we heated a mixture of the two, all the pentane would boil away first to leave pure hexane. The trouble is that hexane evaporates readily at 36°C, so the vapor produced by boiling the mixture would contain a certain amount of hexane as well as pentane. This procedure would thus give a vapor rich in pentane and a remaining liquid rich in hexane, but would not separate the two compounds completely.

Usually a complete separation of the alkanes in petroleum is not necessary, however. The basic process in petroleum refining is **fractional distillation,** in which crude oil is heated and its vapors are led off and condensed at progressively higher temperatures. What comes off at the lowest temperature is largely pentane and hexane, with minor amounts of both lighter and heavier hydrocarbons. The resulting colorless, volatile liquid, called naphtha, is used as a solvent and cleaning agent. The next fraction, consisting largely of alkanes from hexane to decane, is gasoline. Still heavier and less volatile fractions make up kerosene and diesel fuel. Higher temperatures give lubricating oil and still higher ones give the semisolid hydrocarbon mixtures called petro-

leum jelly and paraffin. Grease consists of oil to which a thickening agent has been added to prevent the mixture from running out from between the surfaces being lubricated.

Gasoline

Of all these products the most valuable, of course, is gasoline, needed to fuel the world's 400 million cars. Unfortunately the constituents of gasoline are present only to a minor extent in most petroleums. Two methods have been developed to increase the yield of gasoline. In one of them, the heavier hydrocarbons

Catalytic cracking unit at an oil refinery in New Jersey breaks down complex hydrocarbons into simpler ones.

Oil spills are common. Some are accidental, the result of shipwreck or a malfunction at an offshore oil well, but a great many are deliberate, the result of tankers illegally flushing out waste oil. Although the lighter hydrocarbons soon evaporate, the heavier ones remain floating on the surface or are washed ashore on adjacent coastlines.

Depending on the nature of the residues and the region, the effects of the residues on marine life may be drastic and immediate—dead plankton, dead fish, dead crustaceans, dead birds—or they may be gradual, taking the form of an altered balance of nature with declining populations. The lumps of tar that are one result of oil spills are a prominent feature on the surface of much of the world's oceans and are familiar sights on many beaches. Tankers and oil-well accidents are not the only source of hydrocarbon pollutants: land-based activities involving petroleum products pour a steady stream of them into lakes, rivers, seas, and oceans.

An especially severe oil spill occurred in 1989 when the tanker *Exxon Valdez* struck a well-marked reef outside the shipping lane in Alaska's Prince William Sound, an area of great natural beauty. Eleven million gallons of crude oil poured from its tanks. Exxon itself and both state and federal agencies reacted slowly and ineffectively to the spill, which allowed much of the oil to wash ashore along the Alaskan coast. The oil killed huge numbers of birds, fish, and other wildlife, and its traces will remain hazardous for many years.

Careless operation of the tanker *Exxon Valdez* led to its grounding on a reef off the Alaskan coast in 1989. Crude oil from the wreck devastated wildlife on a large scale. Nearly all of the thousands of oil spills that occur worldwide each year are either deliberate or the result of negligence and could be avoided.

are **cracked** into smaller molecules by heating them under pressure in the presence of catalysts. A typical cracking reaction is

$$C_{16}H_{34} \longrightarrow C_8H_{18} + C_8H_{16} \qquad \textit{cracking reaction}$$

Here hexadecane, one of the heavier alkanes in kerosene and diesel fuel, is broken down into lighter hydrocarbons that vaporize and burn more readily.

The second procedure is to **polymerize** lighter hydrocarbons, which means to join small molecules into larger ones under the influence of heat, pressure, and appropriate catalysts. An example is

$$C_3H_8 + C_4H_8 \longrightarrow C_7H_{16} \qquad \textit{polymerization reaction}$$

in which heptane, a liquid, is formed by the polymerization of two gases.

Alkane molecules have chains of carbon atoms linked together in line, as we shall see in Sec. 12-4. Such molecules, as we might expect, are nonpolar, with neither end much more positive or negative than the other. Because of this nonpolar character, the alkane hydrocarbons are insoluble in water. Chemically they are rather unreactive, and neither concentrated acids and bases nor most oxidizing agents affect them at moderate temperatures. Nor do biological agents such as bacteria attack them to any great extent. The combination of insolubility, relative inertness, and toxicity to living things is what makes the discharge of petroleum and its products into the sea such a serious matter.

STRUCTURES OF ORGANIC MOLECULES

12-4 STRUCTURAL FORMULAS

They Show How the Atoms in Organic Molecules Are Linked Together

Instead of a molecular formula such as CH_4 and C_2H_6, an organic compound is often represented by a **structural formula** in which the covalent bonds between the atoms in each molecule are shown by dashes. Each dash stands for a shared pair of electrons. Thus the structural formulas of the alkanes methane, ethane, and propane are

$$
\begin{array}{ccc}
\begin{array}{c}
\text{H} \\
| \\
\text{H}-\text{C}-\text{H} \\
| \\
\text{H} \\
\text{methane}
\end{array}
&
\begin{array}{c}
\text{H}\quad \text{H} \\
|\quad\; | \\
\text{H}-\text{C}-\text{C}-\text{H} \\
|\quad\; | \\
\text{H}\quad \text{H} \\
\text{ethane}
\end{array}
&
\begin{array}{c}
\text{H}\quad \text{H}\quad \text{H} \\
|\quad\; |\quad\; | \\
\text{H}-\text{C}-\text{C}-\text{C}-\text{H} \\
|\quad\; |\quad\; | \\
\text{H}\quad \text{H}\quad \text{H} \\
\text{propane}
\end{array}
\end{array}
$$

FIG. 12-2 Model of the methane molecule, CH_4.

A molecular formula tells us only how many atoms of each kind are present in each molecule of a compound. A structural formula tells us more. For instance, in the above three molecules we can see that each hydrogen atom is attached to a carbon atom and that in ethane and propane the carbon atoms are linked together. Figure 12-2 shows a three-dimensional model of the methane molecule.

The number of bonds an atom forms in an organic compound is the same as the number of electrons it has to gain or lose to achieve a closed outer shell. A carbon atom always participates in four bonds, as we have learned. A hydrogen atom always participates in a single bond, as does a chlorine atom; an oxygen atom participates in two bonds. Here are some examples:

$$
\begin{array}{ccc}
\begin{array}{c}
\text{H}\quad \text{H}\quad \text{H} \\
|\quad\; |\quad\; | \\
\text{H}-\text{C}-\text{C}-\text{C}-\text{Cl} \\
|\quad\; |\quad\; | \\
\text{H}\quad \text{H}\quad \text{H} \\
\text{propyl chloride}
\end{array}
&
\begin{array}{c}
\text{H} \\
| \\
\text{H}-\text{C}-\text{O}-\text{H} \\
| \\
\text{H} \\
\text{methanol}
\end{array}
&
\begin{array}{c}
\text{Cl} \\
| \\
\text{Cl}-\text{C}=\text{O} \\
\text{phosgene}
\end{array}
\end{array}
$$

12-5 ISOMERS

The Same Atoms but Arranged Differently

For methane, ethane, and propane the structural formulas given above are the only possible arrangements of carbon and hydrogen atoms that will satisfy the combination rules. Butane, on the other hand, may

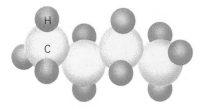

Normal butane

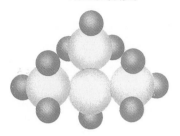

Isobutane

FIG. 12-3 The two isomers of butane, C_4H_{10}.

have its 4 C atoms and 10 H atoms arranged in two ways:

$$
\begin{array}{cccc}
\mathrm{H} & \mathrm{H} & \mathrm{H} & \mathrm{H} \\
| & | & | & | \\
\mathrm{H-C-C-C-C-H} \\
| & | & | & | \\
\mathrm{H} & \mathrm{H} & \mathrm{H} & \mathrm{H}
\end{array}
$$

normal butane

$$
\begin{array}{ccc}
\mathrm{H} & \mathrm{H} & \mathrm{H} \\
| & | & | \\
\mathrm{H-C-C-C-H} \\
| & | & | \\
\mathrm{H} & \mathrm{C} & \mathrm{H} \\
& /|\backslash \\
\mathrm{H} & \mathrm{H} & \mathrm{H}
\end{array}
$$

isobutane

These formulas show that there are two different compounds with the molecular formula C_4H_{10}. They differ in that one of the carbon atoms in isobutane is linked to three other carbon atoms, while in normal butane the carbon atoms are linked to only one or two others.

The physical properties of isobutane are different from those of normal butane because of this difference in molecular structure. The boiling point of isobutane, for instance, is –12°C, whereas that of normal butane, as listed in Table 12-1, is –1°C. Another difference is their densities (masses per unit volume): that of isobutane is 0.622 g/cm³ whereas that of normal butane is 0.604 g/cm³. Figure 12-3 shows three-dimensional models of the two kinds of butane.

Compounds that have the same molecular formulas but different structural formulas are called **isomers.** The number of possible isomers increases rapidly with the number of carbon atoms in the molecule; $C_{13}H_{28}$ has 813 theoretically possible isomers and $C_{20}H_{42}$ has 366,319. Only a few of the possible isomers have actually been prepared.

Merely flipping a structural formula end-for-end does not give the formula of an isomer. For instance, we can show the structure of methanol (methyl alcohol) in two ways:

$$
\begin{array}{cc}
\mathrm{H} & \mathrm{H} \\
| & | \\
\mathrm{H-C-OH} \qquad \mathrm{OH-C-H} \\
| & | \\
\mathrm{H} & \mathrm{H}
\end{array}
$$

However, the molecule is exactly the same in both cases.

12-6 UNSATURATED HYDROCARBONS

Double and Triple Carbon-Carbon Bonds

Hydrocarbons are not limited to the alkanes. A simple example of a nonalkane hydrocarbon is **ethene,** also called **ethylene,** whose formula is C_2H_4. The alkane with two C atoms is ethane, C_2H_6, whose structural formula is

$$
\begin{array}{cc}
\mathrm{H} & \mathrm{H} \\
| & | \\
\mathrm{H-C-C-H} \\
| & | \\
\mathrm{H} & \mathrm{H}
\end{array}
$$

ethane

Oxyacetylene cutting torch. The small cylinder contains acetylene and the large one contains oxygen. The flame temperature can reach 3000°C.

How can ethene, with two fewer H atoms, still have each C atom share four electron pairs? The answer is that there are *two* covalent bonds between the C atoms in ethene:

$$\underset{\text{H}}{\overset{\text{H}}{\diagdown}} \text{C} = \text{C} \underset{\text{H}}{\overset{\text{H}}{\diagup}}$$

ethene

Such a link between carbon atoms is called a **double bond** and involves the sharing of two electron pairs.

Triple bonds, with carbon atoms sharing three electron pairs, are also possible. The simplest case is that of acetylene, C_2H_2, a gas widely used in welding and metal-cutting torches. The structural formula of acetylene is

$$\text{H} - \text{C} \equiv \text{C} - \text{H}$$

acetylene

Multiple Bonds and Reactivity

Compounds with double and triple bonds are much more reactive than the alkanes, which have only single bonds. Both HCl and Cl_2 combine readily with ethene, for instance:

$$\underset{\text{H}}{\overset{\text{H}}{\diagdown}} \text{C} = \text{C} \underset{\text{H}}{\overset{\text{H}}{\diagup}} + \text{HCl} \longrightarrow \text{H} - \underset{\overset{|}{\text{H}}}{\overset{\overset{\text{H}}{|}}{\text{C}}} - \underset{\overset{|}{\text{Cl}}}{\overset{\overset{\text{H}}{|}}{\text{C}}} - \text{H}$$

$$H \underset{H}{\overset{H}{\diagdown}} C = C \underset{H}{\overset{H}{\diagup}} + Cl_2 \longrightarrow H - \underset{\underset{Cl}{|}}{\overset{\overset{H}{|}}{C}} - \underset{\underset{Cl}{|}}{\overset{\overset{H}{|}}{C}} - H$$

The other halogens and many other acids give similar reactions. Since compounds with multiple bonds are able to add other atoms to their molecules, they are called **unsaturated compounds.** The alkanes and similar compounds whose molecules have only single carbon-carbon bonds are called **saturated compounds** because they cannot add other atoms to their molecules.

12-7 BENZENE

Its Molecule Contains a Stable Ring of Six Carbon Atoms

Benzene, C_6H_6, is a clear liquid that does not mix with water and has a strong odor. Benzene is widely used as a solvent and in the manufacture of more complex organic compounds.

The six C atoms in benzene are arranged in a flat hexagonal ring, as shown in Fig. 12-4. What is especially interesting about this molecule is the manner in which its C atoms are attached to one another. In addition to single bonds between these atoms, six electrons are shared by the entire ring. The latter electrons belong to the molecule as a whole and not to any particular pair of atoms; these electrons are **delocalized.** (We recall from Chap. 10 that the outer-shell electrons in a metal are similarly delocalized.) The six delocalized electrons in benzene can be represented by an inner circle in its structural formula:

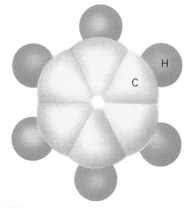

FIG. 12-4   Model of the benzene molecule, C_6H_6.

benzene

Aromatic Compounds

An **aromatic** compound is defined as one that contains a ring of six carbon atoms like that in benzene. The name arose because many of these compounds have strong odors. An example is toluene, which is a common solvent and paint thinner:

toluene

BIOGRAPHY

August Kekulé (1829–1896), a native of Germany, studied architecture before he became a chemist, and in fact his most notable work was on the architecture of molecules. Kekulé introduced the idea of structural formulas, proposed that a carbon atom forms four bonds when it combines, and suggested that carbon atoms could join with each other to form chains. All this came before the nature of chemical bonds was understood; indeed, even the electron was unknown until after Kekulé's death.

Kekulé's greatest achievement was to establish the structure of benzene, C_6H_6, a compound which had been discovered by Michael Faraday in 1826. Exactly how benzene's atoms were arranged was a mystery for the next 40 years until Kekulé (so he said) had a dream in which dancing snakes swallowed their tails. This led him to the concept of a flat ring of carbon atoms linked by alternate single and double bonds, with a hydrogen atom attached to each carbon atom:

Although this picture accounted for most of the properties of benzene and was the key to many of the extraordinary successes of organic chemistry, it had a flaw: experiments show that all of the carbon-carbon bonds in benzene are exactly the same, not alternatie single and double ones. Not until quantum theory was applied to molecular structure in this century did it become clear that six of the bonding electrons could be shared equally by the entire ring.

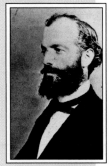

The C and H atoms that are part of a benzene ring are often omitted in representing the structures of aromatic molecules, as shown above.

Some aromatic compounds contain two or more benzene rings fused together, as in the case of naphthalene:

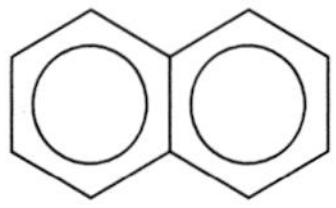

naphthalene

Naphthalene is familiar as the active ingredient in mothballs.

Organic compounds whose molecules do not contain ring structures are said to be **aliphatic.**

ORGANIC COMPOUNDS

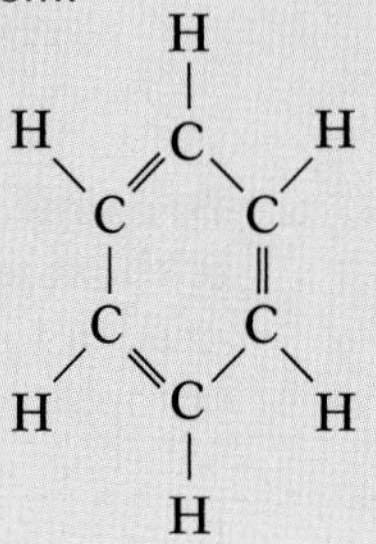

The remarkable range of organic compounds is hinted at in the hydrocarbons, which contain only carbon and hydrogen. Add just oxygen and the possibilities are multiplied many times over, giving compounds as diverse as they are numerous. Add still other elements and the result is staggering in variety and complexity. But regularities exist that permit the orderly classification of organic compounds and lead to an understanding of how their molecular structures govern their behavior. Given this understanding, the organic chemist can create compounds tailored to exhibit specific properties. Evidence of the success of this endeavor is found in the synthetic materials, from textile fibers to drugs, so widely used today.

12-8 ## HYDROCARBON GROUPS

A Handy Classification Scheme — In the classification system used for organic compounds that contain other elements besides carbon and hydrogen, the compounds are often regarded as **derivatives** of hydrocarbons—that is, as compounds obtained by substituting other atoms or atom groups for one or more of the H atoms in hydrocarbon molecules. Ordinarily such compounds are *not* prepared in this way, but their structural formulas suggest that they might be. For example, ethanol can be regarded as a derivative of ethane, with an OH group replacing an H atom:

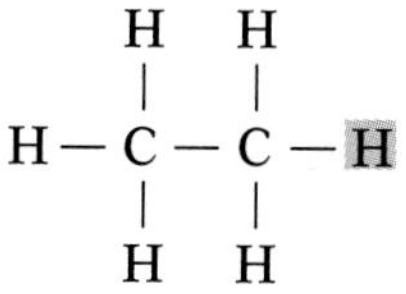

$$
\begin{array}{cccccc}
& H & H & & H & H \\
& | & | & & | & | \\
H- & C- & C- & H \qquad H- & C- & C- & OH \\
& | & | & & | & | \\
& H & H & & H & H \\
& & \text{ethane} & & & \text{ethanol}
\end{array}
$$

Similarly acetic acid can be regarded as a derivative of methane, with a COOH group replacing an H atom:

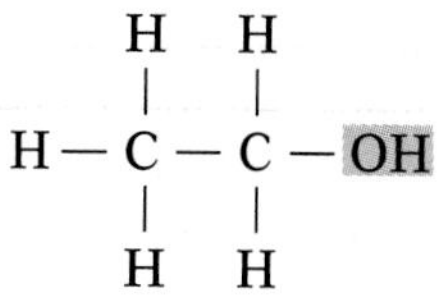

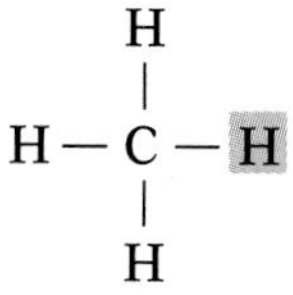

$$
\begin{array}{cc}
\begin{array}{c}
H \\ | \\ H-C-H \\ | \\ H \\ \text{methane}
\end{array}
&
\begin{array}{c}
H \quad\quad O \\ | \quad\; /\!/ \\ H-C-C \\ | \quad\quad \backslash \\ H \quad\quad OH \\ \text{acetic acid}
\end{array}
\end{array}
$$

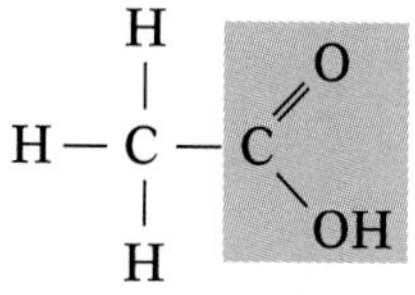

Hydrocarbon Groups — The carbon-hydrogen atom groups that appear in hydrocarbon derivatives are named from the hydrocarbons. Groups corresponding to the hydrocarbons methane, ethane, and propane are

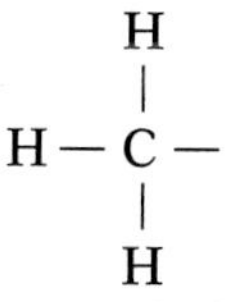

$$
\begin{array}{ccc}
\begin{array}{c}
H \\ | \\ H-C- \\ | \\ H \\ \text{methyl} \\ \text{group}
\end{array}
&
\begin{array}{c}
H \quad H \\ | \quad | \\ H-C-C- \\ | \quad | \\ H \quad H \\ \text{ethyl} \\ \text{group}
\end{array}
&
\begin{array}{c}
H \quad H \quad H \\ | \quad | \quad | \\ H-C-C-C- \\ | \quad | \quad | \\ H \quad H \quad H \\ \text{propyl} \\ \text{group}
\end{array}
\end{array}
$$

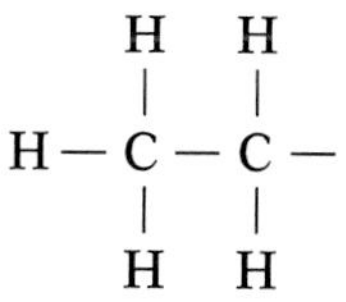

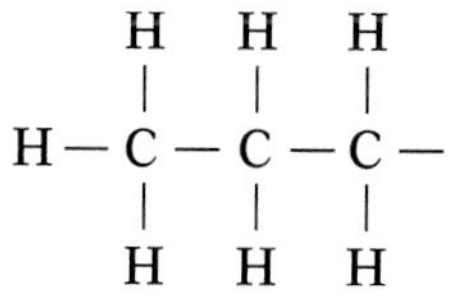

Thus the compound CH_3Cl is methyl chloride, C_3H_7I is propyl iodide, and $CH_3C_2H_5SO_4$ is methyl ethyl sulfate.

12-9 ## FUNCTIONAL GROUPS

Atom Groups with Characteristic Behaviors — Inorganic compounds that contain a particular atom group, such as OH or SO_4, have important aspects of their chemical behavior in common, as we know. The chemical behavior of many organic compounds is also determined to a large extent by the presence of certain atom groups, called **functional groups.** Table 12-2 shows some of the main functional groups found in organic molecules.

Alcohols   The hydroxyl (OH) group they contain makes many alcohol molecules somewhat polar, and the simpler alcohols are soluble in water. The polarity is not enough, however, to prevent them from mixing with many compounds less polar than water, which makes the alcohols useful as solvents. Ethanol (ethyl alcohol) is, of course, the active ingredient in wine, beer, and spirits.

The ethanol in beverages is produced by fermentation. In this process sugar is converted to ethanol and carbon dioxide, with yeast enzymes acting as catalysts. For example,

$$\underset{\substack{\text{glucose}\\\text{(a sugar)}}}{C_6H_{12}O_6} \xrightarrow[\substack{\text{yeast}\\\text{enzymes}}]{} \underset{\text{ethanol}}{2C_2H_5OH} + \underset{\substack{\text{carbon}\\\text{dioxide}}}{2CO_2} \qquad \textit{fermentation}$$

Wine is made by fermenting fruit juice, usually grape juice. Beer is made by fermenting grain, usually barley, and then adding a bitter extract of a plant called hop for flavor. Since yeast cells die when the alcohol concentration reaches about 15 percent, wine and beer cannot be stronger than this. In fact, fermentation generally stops somewhat earlier because the sugar runs out.

Distillation can produce stronger liquors. The fermented liquid is heated and the alcohol-rich vapor is then led off and condensed to give brandy (starting from fruit), whiskey (grain), rum (sugar cane), vodka (traditionally potatoes), and so on. The proof of an alcoholic beverage is twice its percentage content of ethanol. Thus 80 proof whiskey contains 40 percent ethanol, and pure ethanol is 200 proof.

Ethanol for industrial purposes is usually made more cheaply by reacting ethene, a by-product of petroleum refining, with water in the presence of an appropriate catalyst:

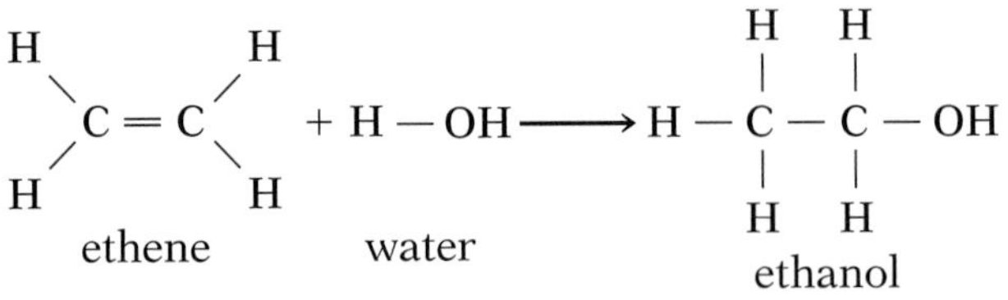

When an H atom is replaced by an OH group in an aromatic hydrocarbon, the result is a compound whose properties are different from those of ordinary alcohols. The simplest example is phenol ("carbolic acid"), C_6H_5OH, which was the first antiseptic and today is one of the raw materials for the plastic Bakelite and for the phenolic glues used in plywood:

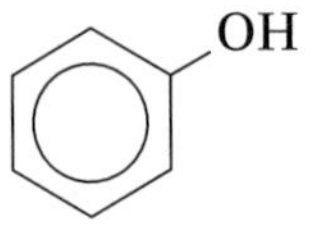

phenol

TABLE 12-2

Common Functional Groups

Name of Group	Structural Formula	Class of Compound	Example	Formula	Comments
Hydroxyl	—OH	Alcohol	Ethanol	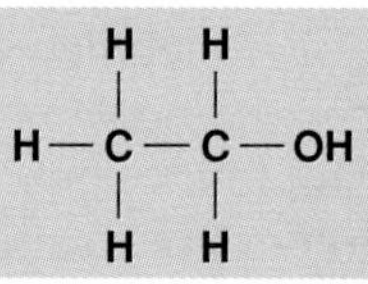 H—C—C—OH (with H's)	Used as a solvent and in beverages; prepared by fermenting sugar solution and synthetically from ethene and water.
Ether	—O—	Ether	Diethyl ether	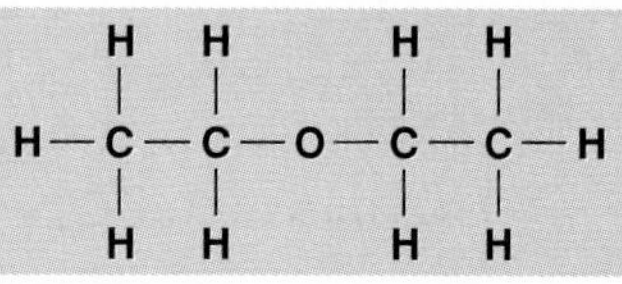 H—C—C—O—C—C—H (with H's)	Once widely used as an anesthetic, its side effects and flammability led to its replacement by safer compounds.
Aldehyde	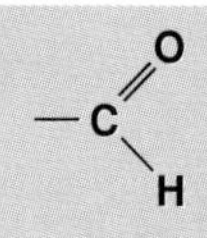 —C(=O)H	Aldehyde	Formaldehyde	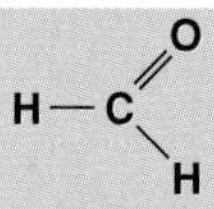 H—C(=O)H	A gas used to preserve biological specimens and as an embalming fluid when dissolved in water ("formalin").
Carbonyl	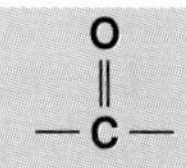 —C(=O)—	Ketone	Acetone	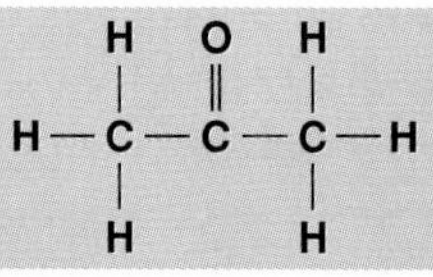 H—C—C(=O)—C—H (with H's)	A common solvent with a toxic vapor used in paints and as nail polish remover.
Carboxyl	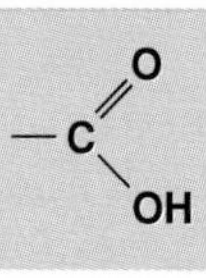 —C(=O)OH	Acid	Acetic acid	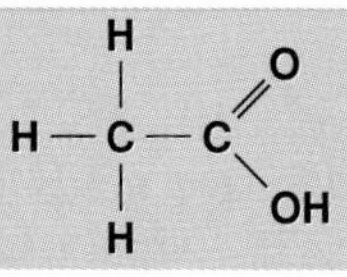 H—C—C(=O)OH (with H's)	Responsible for characteristic taste of vinegar; a weak acid like other organic acids.
Ester	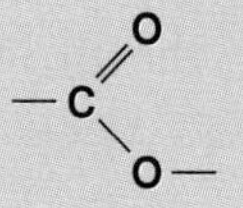 —C(=O)O—	Ester	Methyl acetate	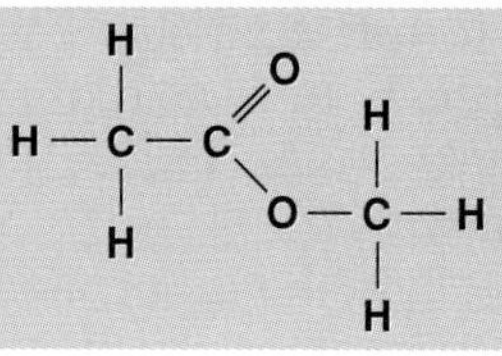 H—C—C(=O)O—C—H (with H's)	Formed by the reaction of methyl alcohol and acetic acid with water as the other product.

Familiar alcohols with more than one OH group are ethylene glycol and glycerol:

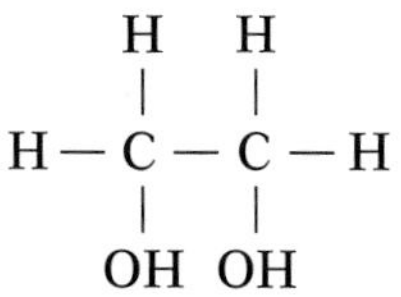

$$
\begin{array}{cc}
\text{H—C—C—H} & \text{H—C—C—C—H} \\
\text{OH OH} & \text{OH OH OH} \\
\text{ethylene glycol} & \text{glycerol}
\end{array}
$$

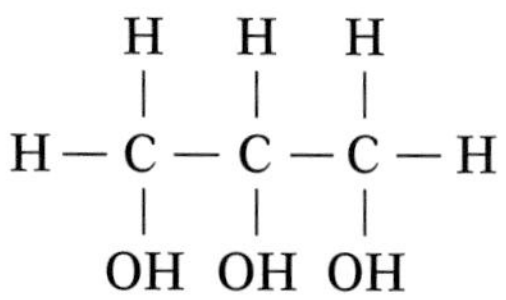

Ethylene glycol is used as an antifreeze in car engines. Glycerol, also known as glycerin, is a sweetish, viscous liquid used in many skin lotions and to prevent tobacco from drying out.

Ethers An ether has an oxygen atom bonded between two carbon atoms. Relatively inert chemically, ethers are widely used as solvents in organic processes since there is little or no danger they will interfere with the reactions.

Aldehydes and Ketones These compounds have similar chemical behavior because both contain the carbonyl atom group $\diagdown C = O$. In aldehydes the carbonyl group is at the end of a molecule with a hydrogen atom attached to the carbon atom, while in ketones the group is inside a molecule between two other carbon atoms. The double bond between C and O is highly polar, and as a result aldehydes and ketones are soluble in water.

Ethanol is oxidized in the liver into acetaldehyde:

$$\underset{\text{ethanol}}{H-\underset{\underset{H}{|}}{\overset{\overset{H}{|}}{C}}-\underset{\underset{H}{|}}{\overset{\overset{H}{|}}{C}}-OH} + \underset{\text{oxygen}}{O} \longrightarrow \underset{\text{acetaldehyde}}{H-\underset{\underset{H}{|}}{\overset{\overset{H}{|}}{C}}-C\overset{\diagup\!\!\!\diagup O}{\diagdown H}} + \underset{\text{water}}{H_2O}$$

This process is apparently responsible for many of the ill effects of drinking too much, such as nausea and hangovers. The acetaldehyde is oxidized further in the liver to acetic acid, and the acetic acid to CO_2 and H_2O. The liver oxidizes methanol (methyl alcohol) to the poisonous formaldehyde, which is believed to be the reason why methanol is so toxic.

The solvent acetone is the most familiar ketone.

Organic Acids Compounds that contain the carboxyl group, —COOH, are acids because the H atom is loosely held and can detach itself as H^+. Most organic acids are very weak. Familiar examples are the formic acid that causes insect bites to sting, the acetic acid of vinegar, the butyric acid of rancid butter and some cheeses, the citric acid of citrus fruits, the lactic acid of sour milk, and the acetylsalicylic acid of aspirin.

When an opened bottle of wine is stored for some time, the ethanol it contains gradually turns into acetic acid and the eventual result is vinegar. The conversion of ethanol to acetic acid is promoted by enzymes produced by bacteria in the wine:

$$\underset{\text{ethanol}}{H-\underset{\underset{H}{|}}{\overset{\overset{H}{|}}{C}}-\underset{\underset{H}{|}}{\overset{\overset{H}{|}}{C}}-OH} + \underset{\text{oxygen}}{O_2} \longrightarrow \underset{\text{acetic acid}}{H-\underset{\underset{H}{|}}{\overset{\overset{H}{|}}{C}}-C\overset{\diagup\!\!\!\diagup O}{\diagdown OH}} + \underset{\text{water}}{H_2O}$$

Esters Alcohols are, so to speak, organic hydroxides, but unlike their inorganic cousins they do not dissociate appreciably in water. They react slowly with acids to form compounds called esters, which are analogous to the salts of inorganic chemistry but are not electrolytes. An example is ethyl acetate, which is made by reacting ethanol with acetic acid:

The ester nitroglycerin is the active ingredient in the explosive dynamite. Dynamite was used to demolish this building in Indianapolis.

$$H-\overset{\overset{\displaystyle H}{|}}{\underset{\underset{\displaystyle H}{|}}{C}}-\overset{\overset{\displaystyle H}{|}}{\underset{\underset{\displaystyle H}{|}}{C}}-O-\overset{\overset{\displaystyle O}{\|}}{C}-\overset{\overset{\displaystyle H}{|}}{\underset{\underset{\displaystyle H}{|}}{C}}-H$$

ethyl acetate

This ester is an important commercial solvent. Several hundred thousand tons of it are used each year in the United States in manufacturing coatings of various kinds, from paint to nail polish.

Many esters have pleasant fruity or flowerlike odors and find extensive use in perfumes and flavors. Propyl acetate is responsible for the fragrance and taste of pears, octyl acetate for those of oranges, ethyl butyrate for those of apricots, and butyl butyrate for those of pineapples. The explosive nitroglycerin is an ester formed by the reaction of nitric acid with the alcohol glycerol. Animal and vegetable fats are all esters of glycerol as well.

12-10 POLYMERS

Molecules Linked into Giant Chains

Polymers are giant molecules that consist of hundreds or thousands of identical (or almost identical) subunits. Proteins, starch, cellulose, and rubber are natural polymers; polythene, polyvinyl chloride (PVC), Styrofoam, Teflon, nylon, and Dacron are synthetic polymers. For a long time polymers were thought to be merely assemblies of small molecules held together by van der Waals forces. Finally, in work that began in 1926, the German chemist Hermann Staudinger showed that polymers are true molecules of huge size held together with covalent bonds.

Hermann Staudinger (1881-1965)

Polythene We are already acquainted with the unsaturated hydrocarbon ethene:

$$\overset{\displaystyle H}{\underset{\displaystyle H}{\diagdown}}C=C\overset{\displaystyle H}{\underset{\displaystyle H}{\diagup}}$$

ethene

Because of the double bond, ethene molecules can, under the proper conditions of heat and pressure, polymerize to form chains thousands of units long whose formula we might write as

$$\cdots-\overset{\overset{\displaystyle H}{|}}{\underset{\underset{\displaystyle H}{|}}{C}}-\overset{\overset{\displaystyle H}{|}}{\underset{\underset{\displaystyle H}{|}}{C}}-\overset{\overset{\displaystyle H}{|}}{\underset{\underset{\displaystyle H}{|}}{C}}-\overset{\overset{\displaystyle H}{|}}{\underset{\underset{\displaystyle H}{|}}{C}}-\overset{\overset{\displaystyle H}{|}}{\underset{\underset{\displaystyle H}{|}}{C}}-\overset{\overset{\displaystyle H}{|}}{\underset{\underset{\displaystyle H}{|}}{C}}-\cdots$$

polythene

This material is polythene (or polyethylene), which is widely used as a packaging material because of its inertness and pliability. The ethene is called the **monomer** in the process, and polythene the polymer. (In Greek, *mono* means "alone" or "single," and *poly* means "many.") A train can be thought of as a polymer, with each of its cars as a monomer. Because of the large size of their molecules, polymers are usually solids.

Vinyls   One of the H atoms in ethene can be replaced by another atom or atom group to form the monomer for a polymer whose properties differ from those of polythene. Because the group

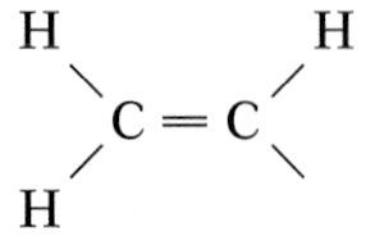

$$
\begin{array}{ccc}
\text{H} & & \text{H} \\
\diagdown & & \diagup \\
& \text{C} = \text{C} & \\
\diagup & & \diagdown \\
\text{H} & &
\end{array}
$$

vinyl group

is called the **vinyl group,** such polymers are classed as vinyls. Some familiar examples are shown in Table 12-3.

The benzene rings attached to alternate C atoms in polystyrene are relatively large and project like knobs from the polymer chain. This prevents adjacent chains from sliding past one another, and as a result poly-

TABLE 12-3

Some Common Vinyl Polymers

Monomer	Polymer	Uses
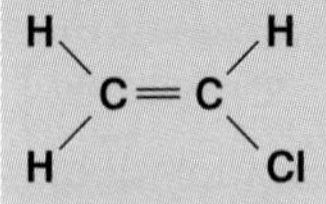 vinyl chloride	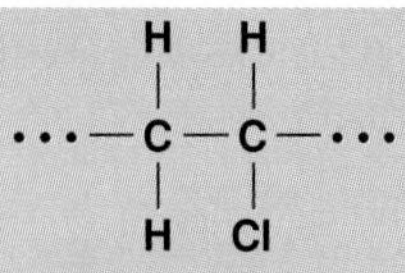 polyvinylchloride	Tubing, insulation, imitation leather, rain wear (PVC, Geon, Koroseal)
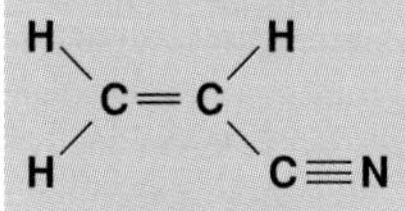 acrylonitrile	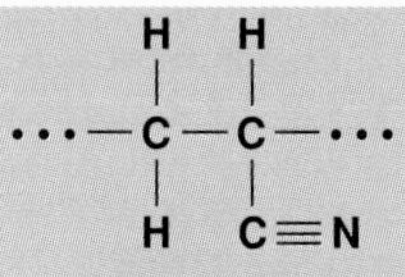 polyacrylonitrile	Textiles, carpets (Acrilon, Orlon)
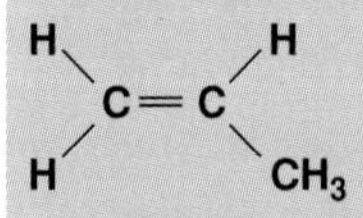 propene	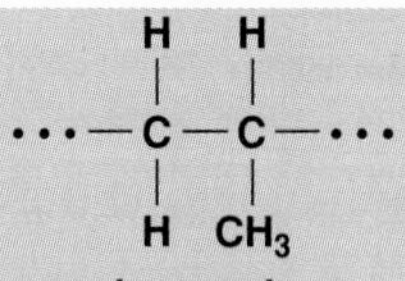 polypropylene	Carpets, ropes, molded objects, thermal underwear
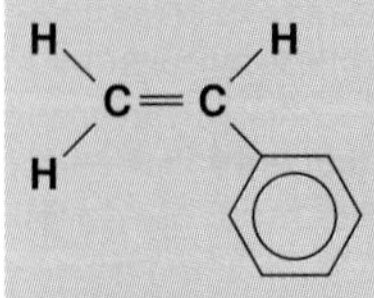 styrene	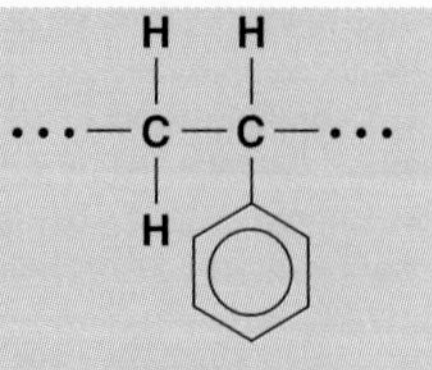 polystyrene	Molded objects, insulation, packing material (Styrofoam)

THE CUP AND THE ENVIRONMENT Styrofoam cups have nonrenewable petroleum as their ultimate raw material, their manufacture involves the carcinogen styrene, they cannot be recycled, and they are not biodegradable. Surely paper cups, with trees as their raw material and disposable in various ways, are more friendly to the environment? And are not china cups really the greenest of all, since they can be used over and over again?

Well, it all depends. Paper cups need more energy and are more polluting to produce than Styrofoam cups. Since they weigh 7 times as much as Styrofoam cups, their transportation uses more energy, too. China cups have to be washed, which involves energy, detergents, and water. Whether Styrofoam, paper, or china cups are best varies according to whether the criterion is energy use, nature and amount of raw materials, air pollution, water pollution, or volume of solid wastes. Overall, china cups win only if they are washed with the least water and detergent possible and are used several thousand times before breaking. Between the throwaway cups, Styrofoam seems less harmful than paper, although this is not entirely certain.

Life-cycle studies of the environmental impacts of Styrofoam, paper, and china cups give no firm answer as to which is best on an overall basis.

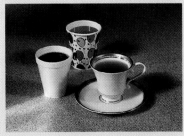

styrene is relatively stiff. If a substance that gives off a gas is added to the liquid monomer mixture, gas bubbles will form throughout the liquid as it polymerizes. The result is the familiar lightweight, rigid Styrofoam.

Lucite and Plexiglas In some monomers, such as methyl methacrylate, two of the H atoms in ethene are replaced by atom groups. Methyl methacrylate polymerizes to form the transparent plastics whose trade names are Lucite and Plexiglas:

methyl methacrylate

Lucite, Plexiglas

A feature of this material is that it is **thermoplastic,** which means that it softens and can be shaped when heated but becomes rigid again upon cooling.

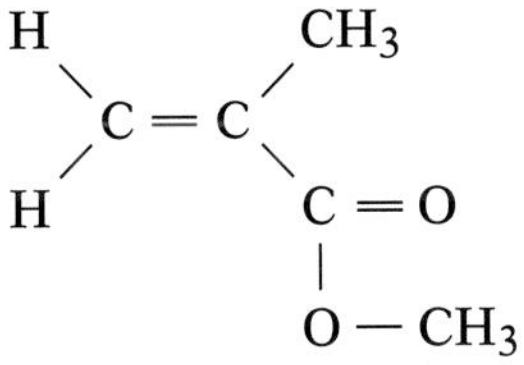

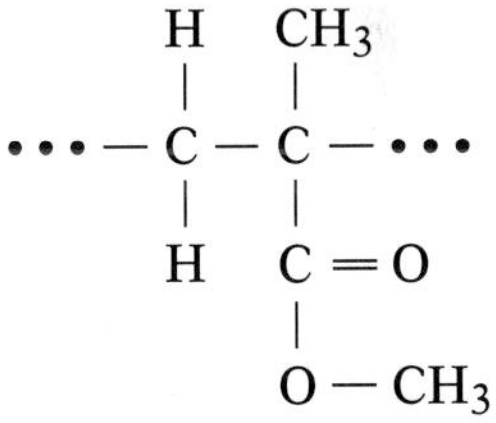

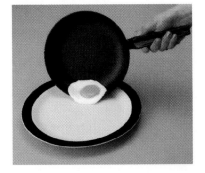

The strong bond between carbon and fluorine accounts for the durability and inertness of Teflon, which was used to coat this frying pan.

Teflon The monomer for Teflon is tetrafluorethene, which is ethene with all the H atoms replaced by fluorine atoms:

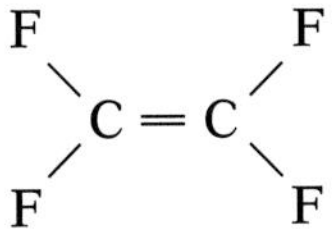

tetrafluorethene

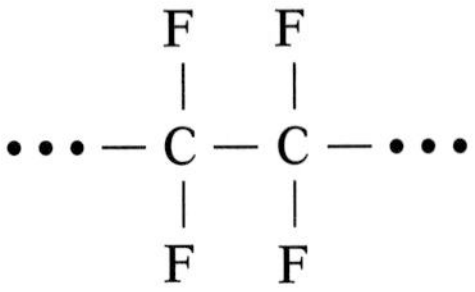

Teflon

The bond between fluorine and carbon is extremely strong, which makes Teflon tough and inert and able to withstand much higher temperatures than other polymers. Teflon has a very slippery surface, too. These properties make Teflon useful industrially for seals and bearings as well as for nonstick coatings for cooking utensils.

Copolymers Some polymers consist of two different monomers. An example of such a **copolymer** is Dynel, used among other things to make fibers for wigs, whose monomers are vinyl chloride and vinyl acetate. The kitchen wrap Saran is another copolymer.

Elastomers Certain monomers that contain two double bonds in each molecule form flexible, elastic polymers called **elastomers.** Rubber is a natural elastomer. A widely used synthetic elastomer is neoprene, a polymer of chloroprene:

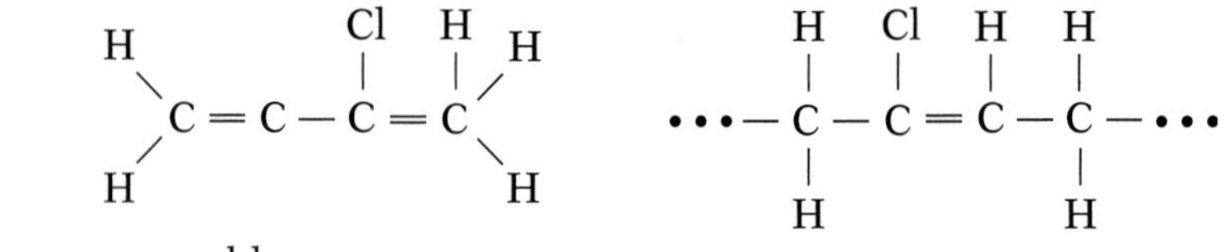

chloroprene   neoprene

A valuable property of neoprene is that liquid hydrocarbons such as gasoline affect it less than they do natural rubber. Another elastomer, nitrile rubber, is still more resistant to hydrocarbons and is used to line gasoline hoses.

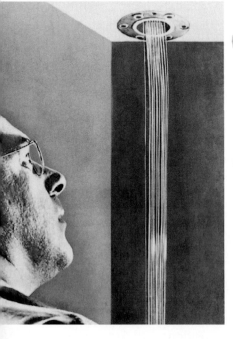

A die called a spinnaret is used to form nylon filaments from liquid polyamide.

Fibers Of the various kinds of synthetic fibers that have been developed, nylon and Dacron are the most familiar. Both are composed of chains of structural elements, just like polymers, but they are produced by chemical reactions rather than by the polymerization of monomer molecules. In the case of nylon, the result is a chain whose elements can be written

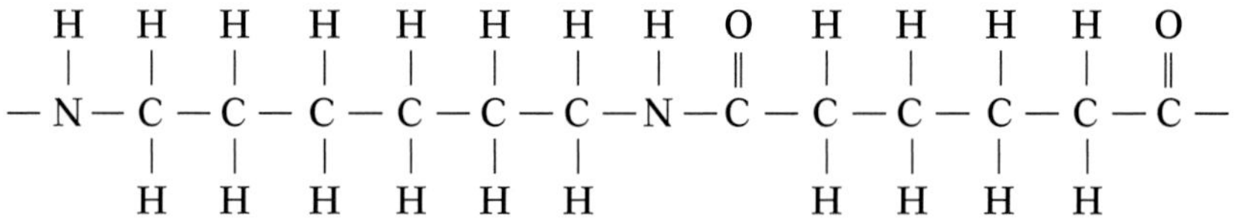

The atom group

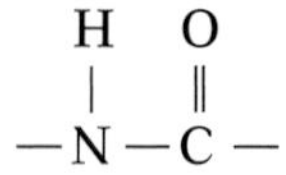

amide linkage

is known as an amide linkage, so nylon is called a **polyamide.** The N—H and C=O groups in nylon are polar, and their mutual attraction is what holds adjacent chains of molecules firmly together.

Dacron, whose structural elements are different from those of nylon, is a **polyester** because its elements are linked together by groups of the form

The hull of the yacht *Ardent Spirit* consists of fiberglass-reinforced polyester resin, and polyester fibers were woven to make the cloth for its sails.

$$\underset{\underset{-\text{C}-\text{O}-}{\overset{\overset{\text{O}}{\parallel}}{}}}{}$$ *ester linkage*

(see Table 12-2). Polyester resins reinforced with glass fibers are often used in boat hulls, truck bodies, and other large structures.

The strongest synthetic fiber yet developed—10 times as strong as steel for the same weight—is a form of polythene called Spectra. The molecules in Spectra are as much as 100 times as long as those in ordinary polythene, and great care is taken when they are formed into fibers to keep them aligned in the fiber direction.

CHEMISTRY OF LIFE

At one time the physical and biological worlds seemed two separate realms. They interacted with each other to be sure, but nevertheless they were thought to be distinct in that an intangible "life force" was thought to be present in living things but absent everywhere else.

Nowadays it is clear there is no life force. Instead there is a continuous chain of development from simple chemical compounds through more elaborate ones through viruses (which are neither "alive" nor "dead" by conventional definitions) through primitive one-celled organisms to complex plants and animals. Given the right chemical and physical conditions and plenty of time, there is every reason to believe that life will inevitably come into being from inorganic matter, as it has on our planet.

The four chief classes of organic compounds found in living matter are carbohydrates, lipids, proteins, and nucleic acids, which we shall examine in turn.

12-11 CARBOHYDRATES

The First Link in the Food Chain

Carbohydrates are compounds of carbon, hydrogen, and oxygen whose molecules contain two atoms of hydrogen for every one of oxygen. They are manufactured in the leaves of green plants from carbon dioxide and water in the process of **photosynthesis,** with energy for the reaction being provided by sunlight. Sugars, starches, and cellulose are all carbohydrates.

An important group of sugars consists of isomers that have the same molecular formula, $C_6H_{12}O_6$. These sugars exist both as straight chains of C atoms and as ring structures. The ring forms are more stable and so occur in nature most often, but a sugar molecule of this kind can shift back and forth between the two forms. Here are the straight-chain and ring forms of glucose, which is the sugar circulated by the blood to provide the body with energy:

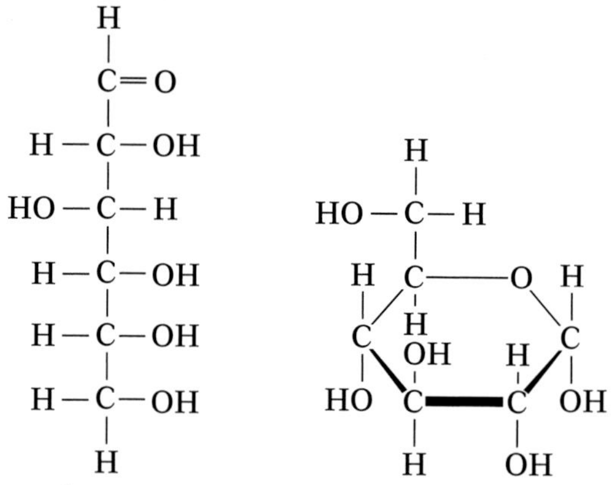

glucose
(straight-chain form)

glucose
(ring form)

The isomers of glucose—among them fructose (the sweetest-tasting of all), galactose, and mannose—have somewhat different structures and properties.

Simple sugars like those above are called **monosaccharides.** Two monosaccharide rings can link together to form a **disaccharide.** Thus sucrose (ordinary table sugar) consists of one ring of glucose and one of fructose; lactose (milk sugar) consists of one ring of glucose and one of galactose; and maltose (malt sugar) consists of two rings of glucose. The molecular formula of all three of these disaccharides is $C_{12}H_{22}O_{11}$, but of course their structures are different.

Polysaccharides

Polysaccharides are complex sugars that consist of chains of more than two simple sugars. They are naturally occurring polymers. In living things the polysaccharides serve both as structural components and as a medium of energy storage. In plants **cellulose,** which consists of a chain of about 1500 glucose rings, is the chief constituent of cell walls. Wood is mostly cellulose, as is cotton. In fact, cellulose is the most abundant organic compound on earth.

The cellulose in wood is extracted and made into paper at this mill in Maine.

Starch, whose 300 to 1000 glucose units are joined together in a slightly different way from those in cellulose, is a polysaccharide that plants use to store energy for later use. Starch occurs in grains that have an insoluble outer layer, and so remains in the cell in which it is formed until, when needed as fuel, it is broken down into soluble glucose molecules.

A polysaccharide found in animals is **chitin,** which forms the outer shells of insects and crustaceans such as lobsters and crabs. Chitin is much like cellulose in structure and is also very abundant. Another polysaccharide is **glycogen,** which is present in the liver and muscles of animals and is released when energy is required. Glycogen, the animal equivalent of starch, is soluble, but its molecules are so large that they cannot pass readily through cell walls. When glucose is needed by an animal, its stored glycogen is split into the much smaller glucose molecules.

Plant and Animal Energy

Living things obtain the energy they need by the oxidation of nutrient molecules. Generally the nutrient molecule most directly involved is glucose, and its oxidation is an exothermic reaction that yields carbon dioxide and water as products:

$$C_6H_{12}O_6 + 6O_2 \longrightarrow 6CO_2 + 6H_2O + \text{energy}$$

glucose   oxygen   carbon   water
                          dioxide

oxidation of glucose

This reaction takes place not all at once, as this equation would indicate, but in a complex series of steps that involve a number of other substances. However, the net effect is the oxidation of glucose. The oxidation of glucose is evidently the reverse of photosynthesis and is the final

Microorganisms in their digestive systems enable cattle to convert the cellulose in plants to glucose.

process by which the energy in sunlight is turned into the energy used by living things.

Usually the carbohydrates in the food we eat are in the form of disaccharides and polysaccharides. In digestion these are **hydrolyzed with the help of water to monosaccharides.** Hydrolysis is promoted by enzymes, which are the specialized protein molecules that act as catalysts in most biochemical processes.

Although many animals can hydrolyze starch to glucose, very few can hydrolyze cellulose. Some plant-eating animals, for instance cattle, have microorganisms such as yeasts, protozoa, and bacteria in their digestive tracts, and enzymes from these microorganisms hydrolyze cellulose in the plants that are eaten. The resulting glucose can then be used by the animal.

After digestion, glucose passes into the bloodstream to be circulated throughout the body. Glucose not immediately needed by the cells is converted into glycogen in the liver and elsewhere. If there is too much glucose to be stored as glycogen, the excess is synthesized into fats.

12-12   PHOTOSYNTHESIS

How the Sun Powers the Living World

As mentioned above, plants combine carbon dioxide from the air with water absorbed through their roots to form carbohydrates in photosynthesis. Photosynthesis is highly endothermic, with the necessary energy coming from sunlight:

$$6CO_2 + 6H_2O + energy \longrightarrow C_6H_{12}O_6 + 6O_2 \qquad photosynthesis$$

The energy is absorbed not directly by the CO_2 and H_2O but instead by a substance called **chlorophyll,** which is part of the green coloring matter of leaves. Chlorophyll acts as a catalyst that passes solar energy to the reacting molecules in a complicated way.

Perhaps 70 billion tons of carbon dioxide is cycled through plants each year (Fig. 12-5). Photosynthesis is only about 1 percent efficient on the average in utilizing the sunlight that reaches plants. A few plants have much higher efficiencies—as much as 11 percent for sugar cane, which is why the ethanol sometimes used to replace or supplement gasoline is made from this source of sugar.

Photosynthesis not only maintains the oxygen content of the atmosphere but seems to have been responsible for it in the first place. The early atmosphere of the earth, which is thought to have consisted of gases emitted during volcanic action, contained oxygen only in combination with other elements in compounds such as water (H_2O), carbon dioxide (CO_2), and sulfur dioxide (SO_2). Primitive organisms, which probably obtained their own energy by fermentation, eventually began to produce free oxygen by photosynthesis. In time the oxygen content of the atmosphere increased to the point where more complex organisms could evolve. Besides the oxygen now in the atmosphere, photosynthesis is believed to account for much of the oxygen that is combined with

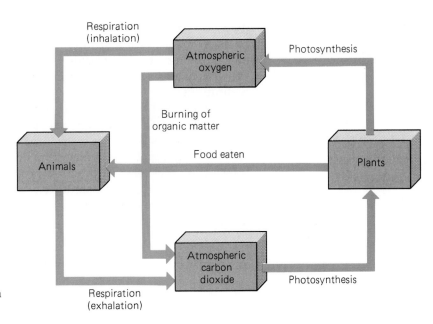

FIG. 12-5 The oxygen-carbon dioxide cycle in the atmosphere.

other elements in the oxides, carbonates, and sulfates found in sediments and sedimentary rocks.

12-13 LIPIDS

Where the Calories Are Fats and such fatlike substances as oils, waxes, and sterols are collectively known as **lipids.** Like carbohydrates, lipids contain only the elements C, H, and O. This is natural since lipids are synthesized in plants and animals from carbohydrates. The proportions and arrangements of these elements are different in lipids, though.

A fat molecule consists of a glycerol molecule with three **fatty acid** molecules attached to it. The hydrocarbon chains in solid fats have only single bonds between their carbon atoms; hence they are saturated. Most animal fats are saturated. Liquid fats, such as vegetable oils, are unsaturated, with double bonds linking one or more carbon atoms. Polyunsaturated fats have more than one double bond per molecule. The double bonds introduce bends in such molecules, which prevents them from being closely packed together. As a result the interactions between nearby molecules are weaker than in the case of saturated fat molecules, and unsaturated fats are liquid at room temperature whereas saturated fats are solid.

Adding hydrogen atoms to the double-bonded carbon atoms in a liquid fat saturates the chains and gives a solid fat. Margarine is produced by such a **hydrogenation** process, with vegetable oils such as soybean and cottonseed oils being heated with hydrogen in the presence of a catalyst.

Fats are used for energy storage and other purposes, such as insulation against cold and mechanical protection, in most living things. The digestion of a fat molecule involves the breaking of the ester links between its glycerol and fatty acid parts. The oxidation of the glycerol and fatty acids then proceeds in a fairly complicated way and releases nearly twice as much energy as in the case of the same mass of carbohydrate.

Cholesterol is a lipid found in the bloodstream. **Atherosclerosis** ("hardening of the arteries") is a serious condition caused by deposits, largely of cholesterol, that restrict the flow of blood. Heart attacks occur when not enough blood reaches the heart muscles due to partly blocked arteries for the heart to function properly. These attacks are a leading cause of death. Eating unsaturated rather than saturated fats has been found to keep the cholesterol level in the blood low and so helps prevent atherosclerosis.

12-14   PROTEINS

The Building Blocks of Living Matter

Proteins are the principal constituent of living cells. They are compounds of carbon, hydrogen, oxygen, nitrogen, and often sulfur and phosphorus; some proteins contain still other elements.

The basic chemical units of which protein molecules are composed are 20 **amino acids.** The simplest amino acid is glycine,

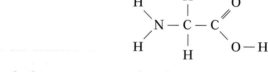

in which we recognize the characteristic carboxyl group —COOH of organic acids. Typical protein molecules consist of several hundred amino acids joined together in chains, and their structures are accordingly quite complex. The formula of one of the proteins found in milk is $C_{1864}H_{3012}O_{576}N_{468}S_{21}$, which gives an idea of the size of some protein molecules.

Plant and animal tissues contain proteins both in solution, as part of the fluid present in cells and in other fluids such as blood, and in insoluble form, such as the skin, muscles, hair, nails, horns, and so forth of animals. Silk is an almost pure protein. The human body contains about 100,000 different proteins, all of which it must make from the 20 amino acids it obtains from the digestion of the food proteins it takes in. One of the great successes of modern biochemistry was the discovery of how living cells build the complex arrangements of amino acids in their proteins.

In a protein molecule, the links between the amino acids consist of **peptide bonds** that are like the amide bonds in nylon. These chains of amino acids, called **polypeptide chains,** are usually coiled or folded in intricate patterns (Fig. 12-6). An important aspect of the patterns is the cross-linking that occurs between different chains and between different parts of the same chain.

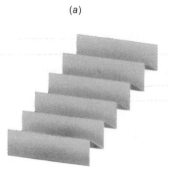

FIG. 12-6 (a) The alpha helix form of a protein molecule. Each amino acid unit in the helix is linked by hydrogen bonds (a type of van der Waals bond) to other units above and below it. Most proteins have molecules like this. (b) The pleated sheet form of a protein molecule. Two or more chains of amino acid units are linked side-to-side along the sheet by hydrogen bonds. Fibrous proteins—such as those in hair, silk, cartilage, and horn—are of this kind. Some proteins have still other forms.

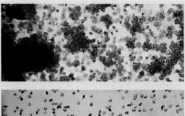

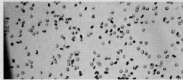

Matching blood types before a transfusion. *Top:* The blood types of donor and recipient are incompatible, which causes the red cells to clump together. *Bottom:* When the blood types are compatible, no clumping occurs.

TISSUE MATCHING The carbohydrates and the lipids do not share the specificity of the proteins. Glucose, for instance, is a carbohydrate found in all plants and animals, but no protein is similarly widespread. Even individuals of the same species have some proteins that are not quite identical, so that tissues cannot ordinarily be transplanted because of the danger of "rejection" of the graft. The matching of blood types before a transfusion is to ensure that the proteins in the blood of the donor are the same as those in the blood of the recipient. Drugs have been developed that can prevent transplanted organs such as kidneys and hearts from being rejected. Unfortunately such drugs also weaken the body's ability to defend itself from infection since both responses of the body to foreign proteins involve the same mechanism.

The sequence of amino acids in a protein is just as important as which ones they are. The amino acid units in even a small protein molecule, such as insulin with 51 units, can be arranged in a great many different ways. However, only one arrangement has the biological effects associated with insulin. A parallel is with the formation of a word from the 26 letters of the alphabet. *Run* and *urn* have the same letters but mean different things because the order of the letters is different. The alphabet of the proteins has only 20 letters, corresponding to the various amino acids, but the words may contain hundreds of letters whose relative positions in three dimensions are significant. The extraordinary number of different proteins, each serving a specific biological need in an organism, is not surprising in view of this picture of protein structure.

Dietary Protein The human body can synthesize only some of the 20 amino acids it requires. The others must be present in our diets or else our bodies will not be able to manufacture the various proteins essential to life. A proper diet must therefore include not just an adequate total amount of proteins but also the right ones.

Most proteins of animal origin—such as those in meat, fish, eggs, and milk—contain all the needed amino acids, but plant proteins do not. The important amino acid lysine is missing in corn, wheat, and rice; isoleucine and valine are missing in wheat; threonine is missing in rice; and so on. Although it is certainly possible to live without eating meat or other animal products, a vegetarian diet not only must be sufficiently varied to include all the required amino acids but must provide all of these acids every day since they are not stored in the body and are needed together for protein manufacture.

Pure ammonia (NH₃) is some-times applied directly to soil to supply it with the nitrogen that plants (here soybeans) need to manufacture proteins.

12-15 SOIL NITROGEN

A Vital Component Because the amino acids of which proteins are composed are nitrogen compounds, nitrogen is one of the most important elements to us. The ultimate source of all our protein is plants, although much of it comes to us secondhand in such animal proteins as those in meat, eggs, and milk. Plants manufacture their proteins from simpler nitrogen compounds that enter their roots from the soil in which they grow.

Green plants cannot draw upon the stable molecules of free nitrogen in the air around them. All their nitrogen, and therefore all the nitrogen that goes into animal bodies as well, comes from nitrogen compounds in the soil. The nitrogen molecules we breathe can do us no good either, for the atoms in these molecules are held together by strong bonds that our body processes are unable to break. Like a shipwrecked sailor surrounded by seawater but dying of thirst, we are surrounded by an ocean of nitrogen but would perish except for the combined nitrogen that plants can absorb through their roots.

The formation of plant proteins steadily removes nitrogen compounds from the soil. Just as steadily, nitrogen compounds are returned to the soil by the decay of animal wastes and of dead plants and animals. The nitrogen of proteins is converted by decay into ammonia and ammonium salts which are then oxidized to nitrates by soil bacteria. But the replenishment is never complete. Some nitrogen is lost permanently from the soil when nitrates and ammonium salts dissolve in streams and rainwash, and when bacteria decompose nitrates into free nitrogen.

Nature makes good these losses in two ways. Another kind of soil bacteria, the "nitrogen-fixing" bacteria, have the ability to break down

FIG. 12-7   The nitrogen cycle on land.

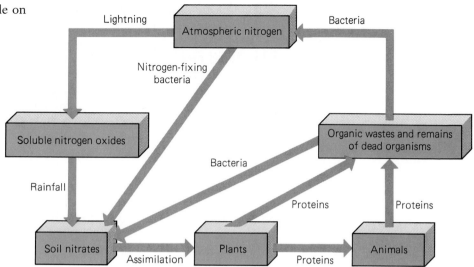

the stable nitrogen molecules of the air and to manufacture nitrates from the atoms. Also, lightning causes atmospheric nitrogen and oxygen to combine into nitrogen oxides, which are brought down to the earth in rainwater. So in nature nitrogen goes through a continuous cycle (Fig. 12-7) that keeps the amount of fixed nitrogen in the soil approximately constant.

We have drastically disturbed this natural cycle. Much of the protein that enters our bodies is not returned to the soil but instead is dumped as sewage into the oceans. The use of plant material and manure for cooking fires in primitive areas further contributes to the conversion of combined nitrogen to free nitrogen. To be sure, manure is still used as fertilizer in some regions, and legumes, which are plants (such as beans) on which nitrogen-fixing bacteria grow, are widely cultivated, but artificial fertilizers have become essential as sources of nitrogen for a large part of the world's agriculture. Over 125 million tons of such fertilizers is used each year, much of it ending up in groundwater, lakes, and rivers. Some of the consequences of this pollution are described in Sec. 10-8; high concentrations of nitrates in drinking water are also a health hazard.

12-16   NUCLEIC ACIDS

The Genetic Code   The **nucleic acids** are very minor constituents of living matter from the point of view of quantity. However, because they control the processes of heredity by which cells and organisms reproduce their proteins and themselves, these acids are extremely important. If anything may be said to be the key to the distinction between living and nonliving matter, it is the nucleic acids.

Nucleic acid molecules consist of long chains of units called **nucleotides.** As in the case of the amino acids in a polypeptide chain,

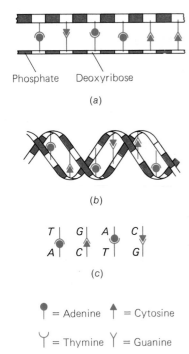

Phosphate   Deoxyribose

(a)

(b)

$$T \downarrow \quad G \downarrow \quad A \downarrow \quad C \downarrow$$
$$A \downarrow \quad C \downarrow \quad T \downarrow \quad G \downarrow$$

(c)

🍡 = Adenine   ↑ = Cytosine

Y = Thymine   Y = Guanine

FIG. 12-8   The structure of DNA. (a) The nitrogen bases link a double chain of alternate phosphate and deoxyribose groups. Adenine and thymine are always paired, and cytosine and guanine are always paired. (b) The chains are not flat but form a double helix. (c) The four "letters" of the genetic code.

both the kinds of nucleotide present and their arrangement govern the biological behavior of a nucleic acid.

Each nucleotide has three parts, a **phosphate group** (PO_4), a **pentose sugar,** and a **nitrogen base.** A pentose sugar is one that contains five carbon atoms. In **ribonucleic acid** (RNA) the sugar is **ribose,** $C_5H_{10}O_5$, and in **deoxyribonucleic acid** (DNA) the sugar is **deoxyribose,** $C_5H_{10}O_4$, which has one O atom less than ribose. Nitrogen bases have characteristic ring structures of nitrogen and carbon atoms. Four nitrogen bases are found in DNA: adenine, guanine, cytosine, and thymine. The nitrogen bases in RNA are the same except that uracil replaces thymine.

The structure of a DNA molecule is shown in Fig. 12-8. Pairs of nitrogen bases form the links between a double chain of alternate phosphate and deoxyribose groups. Adenine and thymine are always coupled together, as are cytosine and guanine. The chains are not flat but spiral around each other in a double helix, as in Fig. 12-8b. The double helix structure of DNA was discovered in 1953 by the American biologist James D. Watson and the English physicist Francis H. C. Crick, who were working together at Cambridge University.

Figure 12-8c shows the four "letters" of the genetic code. There may be hundreds of millions of such letters in a DNA molecule, and their precise sequence governs the properties of the cell in which the molecule is located. DNA molecules thus represent the biological blueprints that are translated into the processes of life.

The complexity of living things is mirrored in the complexity of DNA molecules, which are the largest known to science. DNA molecules are normally folded and coiled into microscopic packages called **chromosomes.** If the 23 human chromosomes were stretched out, they would

James D. Watson (1928–) *at left* and Francis H. C. Crick (1916–)

Model of a small part of a DNA molecule, which has the form of a double helix. The development and functioning of every living organism are ultimately controlled by the DNA in its cells. When the organism reproduces, copies of its DNA are passed on to the new generation.

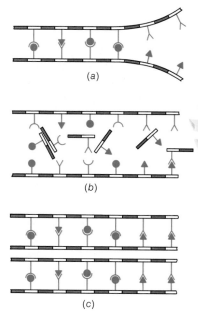

(a)

(b)

(c)

FIG. 12-9 Simplified model of DNA replication. (*a*) When a cell reproduces, each double DNA chain it contains breaks into two single ones, much like a zipper opening. (*b*) The single chains then pick up from the cell material the nucleotides needed to complete their structures. (*c*) The result is two identical DNA chains.

total about a meter in length. If DNA were as thick as a strand of spaghetti, a chromosome would be over 10 km long.

What DNA and RNA Do

DNA controls the development and functioning of a cell by determining the proteins the cell makes. This is only one aspect of the role of DNA in the life process. Another follows from the ability of DNA molecules to reproduce themselves, so that when a cell divides, all the new cells have the same characteristics (that is, the same **heredity**) as the original cell (Fig. 12-9). Finally, changes in the sequence of bases in a DNA molecule can occur under certain circumstances, for example, during exposure to x-rays. These changes will be reflected in alterations in the cell containing the DNA molecule. If such a **mutation** occurs in the reproductive cells of an organism, the result may be that the descendants of the original organism will be different in some way from their ancestor.

Thus three fundamental attributes of life can be traced to DNA: the structure of every organism, its ability to reproduce, and its ability to evolve into different forms in later generations.

The other type of nucleic acid, RNA, differs from DNA in a number of respects. RNA molecules are much smaller than DNA molecules, for example, and usually consist of only single strands of nucleotides. One type of RNA carries instructions for the synthesis of specific proteins from the DNA in a cell's nucleus to the place where the synthesis occurs. The instructions are in the form of a code in which each successive group of three nucleotides determines the particular amino acid to be added next to the protein polypeptide chain being formed. For example, the group GCA (guanine-cytosine-adenine) corresponds to the amino acid alanine, and GGA corresponds to glycine.

The Human Genome

Every cell in a plant or animal contains in its DNA coded instructions for making all the proteins the organism needs. The set of instructions for each protein is called a **gene.** Genes make up about 3 percent of the human **genome,** the 3 billion or so base pairs present in our chromosomes. The purpose of the other 97 percent of the genome (once dismissed as "junk DNA") is not known. Perhaps it helps control the action of the genes; perhaps it provides raw material for new genes; perhaps it consists of genes that evolution has discarded; perhaps its function is something else entirely.

The human genome contains about 100,000 genes. Preparing a detailed map that shows not only each gene but also its sequence of base pairs is an immense task. The result would be nothing less than a blueprint of human life. An international project to map the human genome was started in 1988; no one is sure how long the task will take.

A knowledge of the genome has practical importance apart from its intrinsic interest. Many diseases have a specifically genetic basis (for instance, cystic fibrosis and muscular dystrophy), and most, if not all, others have some genetic linkage. An example of a genetic linkage is the increased tendency members of some families have to contract a particular disease, such as cancer. A complete genetic map will be a step toward better health for everybody.

Fumaroles are vents in the earth's surface associated with volcanic activity that give out a mixture of gases, mainly water vapor, carbon dioxide, and nitrogen and sulfur compounds. The earth's oceans and atmosphere are believed to be the result of similar outgassing on a large scale early in the earth's history. These fumaroles are on the Italian island of Vulcano.

12-17 ORIGIN OF LIFE

An Inevitable Result of Natural Processes

Whatever the earth's beginnings may have been, it is safe to assume that, at some time in the remote past, the surface was considerably warmer than it is at present. The atmosphere of the young earth almost certainly contained compounds of hydrogen, oxygen, carbon, and nitrogen; the most likely were water, methane, ammonia, carbon dioxide, and hydrogen cyanide (HCN).

Eventually the earth's surface cooled, and torrents of rain began to fall. The rain carried down with it some of the atmospheric gases, so that the infant oceans had a certain proportion of these gases dissolved in them. Weathering and erosion began at this time also, and the oceans acquired their salt and mineral content early since these processes, along with volcanic activity on the ocean floor, must have been exceptionally rapid at first. Life may have had its origin in or around these oceans.

Chemical reactions occur most readily in liquids, and furthermore water is the best solvent. Hence the early oceans must have been home to chemical processes of all sorts, with ample energy available in sunlight and lightning discharges. Of the great many compounds that must have been formed, five classes have particular biological significance: the sugars, glycerol, the fatty acids, the amino acids, and the nitrogen bases.

It is naturally not proper to assume that, because a certain compound contains certain elements, merely bringing together these elements will yield the compound. However, the reaction sequences that are necessary to go from the primitive ingredients of the atmosphere and oceans to the specific compounds listed above seem straightforward and likely to have occurred—or, rather, there is no known reason why

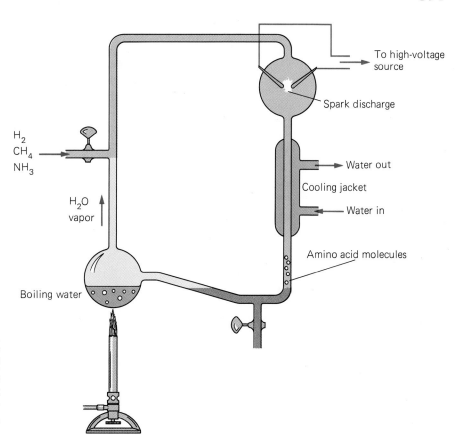

FIG. 12-10 In this experiment, first performed in 1952 by Stanley Miller under the direction of Harold Urey, amino acids were created by passing sparks that simulate lightning through a mixture of water vapor, hydrogen, methane, and ammonia.

they should *not* have occurred. And this is one of the rare hypotheses about the early earth that can be directly verified in the laboratory. When an electric discharge simulating lightning is passed through a mixture of water, hydrogen, methane, and ammonia, amino acids and other compounds of biological importance are created (Fig. 12-10). Even fatty acids and the basic structural parts of the chlorophyll molecule have been produced in this way. Later work has shown that atmospheres with other plausible compositions, notably with carbon dioxide predominating, can also yield amino acids.

Confirming the laboratory studies was the discovery of amino acids and other complex organic molecules in meteorites and comets, which are relics of the youth of the solar system. It seems quite likely that there was no shortage of the building blocks of life on the earth's surface long ago.

Given molecules of the five classes of significant organic compounds, further reactions of equal plausibility (not all of them verified in the laboratory as yet) lead to more complex compounds such as the fats, proteins, and nucleic acids that are directly involved in living matter. And given these latter compounds, especially the nucleic acids that govern protein synthesis and are able to replicate themselves, the emergence of primitive cells, the basic biological units, becomes inevitable.

The progression from an ocean with dissolved gases, salts, and minerals to living organisms certainly did not occur with the neatness and dispatch with which, say, a baker combines certain ingredients, puts the mixture in an oven, and takes out a cake an hour later. About a billion years elapsed between the formation of the earth and the formation of the first cells. Although biochemists do not agree on exactly which of several proposed routes from primitive molecules to primitive cells was the one most likely to have been followed, this immense span of time seems sufficient for any of them.

IMPORTANT TERMS AND IDEAS

Organic chemistry concerns the chemistry of carbon compounds. The number and variety of organic compounds result from the ability of carbon atoms to form covalent bonds with each other as well as with other atoms.

Structural formulas show not only the numbers of each kind of atom in a molecule but also how the atoms are joined together. **Isomers** are compounds with the same molecular formulas but different structural formulas, corresponding to different arrangements of the same atoms.

Double bonds and **triple bonds** are possible between carbon atoms in addition to single ones. The molecules of a **saturated** organic compound contain only single carbon-carbon bonds; those of an **unsaturated** compound contain one or more double or triple bonds. Unsaturated compounds react more readily than saturated ones.

A **functional group** is a group of atoms whose presence in an organic molecule determines its chemical behavior to a large extent. Thus the hydroxyl (OH) group characterizes the alcohols, the carboxyl (COOH) group characterizes the organic acids, and so on.

A **polymer** is a long chain of simple molecules (**monomers**) linked together. Plastics, synthetic fibers, and synthetic elastomers are polymers.

Carbohydrates are compounds of carbon, hydrogen, and oxygen manufactured in green plants from carbon dioxide and water by **photosynthesis** with sunlight providing the needed energy. Sugars, starches, and cellulose are carbohydrates. **Lipids** are fats and fatlike substances such as oils and waxes synthesized from carbohydrates by plants and animals. **Proteins,** the principal constituents of living matter, consist of long chains of **amino acid** molecules. The sequence of amino acids in a protein molecule together with the shape of the molecule determines its biological role.

The **nucleic acid** molecules DNA and RNA consist of long chains of **nucleotides,** atom groups whose precise sequence governs the structure and function of cells and organisms. DNA has the form of a double helix and carries the genetic code; one type of RNA has the form of a single helix and acts as a messenger in protein synthesis.

EXERCISES: MULTIPLE CHOICE

1. The science of organic chemistry has as its subject
   a. compounds produced by plants and animals
   b. carbon compounds
   c. compounds with complex molecules
   d. the determination of structural formulas

2. Compared with inorganic compounds in general, most organic compounds
   a. are more readily soluble in water
   b. are more easily decomposed by heat
   c. react more rapidly
   d. form ions more readily in solution

3. The number of covalent bonds each carbon atom has in organic compounds is usually
   a. one   c. four
   b. two   d. six

4. As a class, the alkanes are
   a. highly reactive
   b. soluble in water
   c. unaffected by acids and bases
   d. harmless to living things

5. In general, in the alkane series of hydrocarbons, a high molecular mass implies a (an)
   a. low boiling point
   b. high boiling point
   c. low freezing point
   d. artificial origin

6. Gasoline is a mixture of
   a. alkanes
   b. isomers of octane
   c. hydrocarbon derivatives
   d. unsaturated hydrocarbons

7. Compounds that have the same molecular formulas but different structural formulas are called
   a. hydrocarbons   c. polymers
   b. isomers   d. derivatives

8. Unsaturated hydrocarbon molecules are characterized by
   a. double or triple bonds between carbon atoms, so that additional atoms can be added readily
   b. the ability to absorb water
   c. the ability to dissolve in water
   d. benzene rings in their structural formulas

9. Hydrocarbon molecules with only single bonds between carbon atoms are
   a. fairly reactive   c. extremely dense
   b. fairly unreactive   d. unsaturated

10. Hydrocarbon molecules with double or triple bonds between carbon atoms are
    a. fairly reactive   c. extremely dense
    b. fairly unreactive d. saturated

11. The number of bonds between the carbon atoms in acetylene, C_2H_2, is
    a. one   c. three
    b. two   d. four

12. Which of the following compounds can exist?
    a. C_2H_3 c. C_2H_5
    b. C_2H_4 d. C_2H_7

13. The benzene molecule is notable for having a
    a. ring of six carbon atoms
    b. ring that can consist of any number of carbon atoms
    c. straight chain of six carbon atoms
    d. helix of carbon atoms

14. Hydrocarbon derivatives in general are
    a. formed by burning hydrocarbons in air
    b. alcohols
    c. hydrocarbons that have had H atoms replaced by other atoms or atom groups
    d. hydrocarbons that have more than one bond between carbon atoms

15. All alcohols
    a. are safe to drink
    b. can be made only by fermentation
    c. contain just one OH group
    d. contain one or more OH groups

16. The presence of a COOH group is characteristic of an
    a. alkane   c. aldehyde
    b. organic acid   d. ester

17. The conversion of sugar to ethanol and carbon dioxide with enzymes acting as catalysts is called
    a. fermentation   c. photosynthesis
    b. polymerization   d. digestion

18. Organic acids are
    a. strong and highly corrosive
    b. rather weak
    c. not found in nature but must be artificially made
    d. characterized by simple molecules

19. The reaction of ethanol with acetic acid produces ethyl acetate, which is a (an)
    a. aldehyde
    b. ester
    c. aromatic compound
    d. polymer

20. A long chain of identical simple molecules linked together is called a (an)
    a. monomer   c. isomer
    b. polymer   d. elastomer

21. Living cells consist mainly of
    a. carbohydrates   c. proteins
    b. lipids   d. nucleic acids

22. Living things differ most from one another in their constituent
    a. carbohydrates   c. amino acids
    b. lipids   d. proteins

23. The most abundant organic compound on earth is
    a. methane   c. glucose
    b. benzene   d. cellulose

24. Carbohydrates are produced in green plants by
    a. polymerization   c. photosynthesis
    b. fermentation   d. respiration

25. Animals store energy in the form of
    a. starch   c. chitin
    b. cellulose   d. glycogen

26. Plants store energy in the form of
    a. starch   c. chitin
    b. cellulose   d. glycogen

27. Cellulose is not
    a. a carbohydrate
    b. the chief constituent of wood
    c. present in all plants
    d. easily digested by most animals

28. Fats and oils are
    a. proteins   c. nucleic acids
    b. carbohydrates   d. lipids

29. Lipids are synthesized in plants and animals from
  a. proteins   **c.** enzymes
  b. carbohydrates   **d.** nucleic acids

30. A given mass of fat compared with the same mass of carbohydrate can provide the body with about
  a. half as much energy
  b. the same amount of energy
  c. twice as much energy
  d. 10 times as much energy

31. It is healthier to eat unsaturated rather than saturated fats because unsaturated fats
  a. are easier to digest
  b. contain more energy
  c. keep the cholesterol content of the blood low
  d. keep the glucose content of the blood low

32. Proteins consist of combinations of
  a. amino acids
  b. nucleic acids
  c. esters of glycerin with organic acids
  d. DNA and RNA molecules

33. The number of amino acids important to life is
  a. 2   **c.** 20
  b. 8   **d.** 69

34. Which of the following statements is correct?
  a. The human body can synthesize all the amino acids it requires
  b. The human body can synthesize some of the amino acids it requires
  c. The human body can synthesize none of the amino acids it requires
  d. Different people require different groups of amino acids

35. The matching of blood types before a transfusion is to ensure that the blood of donor and recipient both have the same
  a. carbohydrates   **c.** lipids
  b. proteins   **d.** nucleic acids

36. Most biochemical processes in living matter are catalyzed by
  a. enzymes   **c.** lipids
  b. glycogen   **d.** DNA

37. The structure of a DNA molecule resembles a
  a. single helix   **c.** pleated ribbon
  b. double helix   **d.** straight chain

38. Each three-nucleotide group in a DNA molecule corresponds to a particular
  a. amino acid   **c.** enzyme
  b. nucleic acid **d.** protein

1. Why are there more carbon compounds than compounds of any other element?

2. In what ways do organic compounds, as a class, differ from inorganic compounds?

3. What is the principal bonding mechanism in organic molecules?

4. What kind of carbon-carbon bonds are found in alkane molecules?

5. What procedure can be used to separate the different alkanes present in petroleum?

6. Is gasoline a compound? If not, what is it?

7. Explain why structural formulas are more important in organic chemistry than in inorganic chemistry.

8. In general, how do the reactivities of hydrocarbon molecules that contain only single bonds compare with the reactivities of hydrocarbon molecules that contain double or triple bonds as well?

9. Why are substances whose molecules contain triple carbon-carbon bonds relatively rare?

10. The isomers of a compound have the same chemical formula. In what way do they differ from one another?

11. How many covalent bonds are present between the carbon atom and each oxygen atom in carbon dioxide, CO_2?

12. Distinguish between unsaturated and saturated hydrocarbons, giving examples of each.

13. What is the difference between aromatic and aliphatic compounds?

14. Why are all aromatic compounds unsaturated?

15. Is it possible for a molecule with the formula C_4H_2 to exist? If not, why not?

16. Is it possible for a molecule with the formula C_2H_3 to exist? If not, why not?

17. Is it possible for a molecule with the formula C_2H_6 to exist? If not, why not?

18. In which of the compounds C_2H_2, C_2H_4, and C_4H_{10} are the carbon-carbon bonds single, in which are they double, and in which are they triple?

19. The aliphatic compound pentene has the molecular formula C_5H_{10}. Are all the bonds in the pentene molecule single ones?

20. The aliphatic compound heptane has the molecular formula C_7H_{16}. Are all the bonds in the heptane molecule single ones?

21. To what class of organic compounds does the compound belong whose structure is shown below?

$$H - \underset{\underset{H}{|}}{\overset{\overset{H}{|}}{C}} - \underset{\underset{H}{|}}{\overset{\overset{H}{|}}{C}} - \overset{\diagup O}{\underset{\diagdown H}{C}}$$

22. What do you think is the name of the compound whose structural formula is shown below?

$$H - \underset{\underset{H}{|}}{\overset{\overset{H}{|}}{C}} - O - \underset{\underset{H}{|}}{\overset{\overset{H}{|}}{C}} - \underset{\underset{H}{|}}{\overset{\overset{H}{|}}{C}} - H$$

23. What do you think is the name of the compound whose structural formula is shown below?

$$\underset{Cl}{\overset{Cl}{\diagdown}} C = C \underset{\diagdown Cl}{\overset{\diagup H}{}}$$

24. The alcohols are the organic equivalents of what class of inorganic compounds?

25. Give an example of an ester, an organic acid, an alcohol, a sugar, and a methane derivative.

26. Which of the following (a) dissolve in water, (b) are acids, (c) react with ethyl alcohol to give esters, (d) react with acetic acid to give esters?

C_2H_5COOH C_3H_8
C_2H_4 C_2H_5OH
HCl $C_3H_5(OH)_3$

27. Compare the properties of a simple ester, for instance, methyl acetate, with those of a salt, for instance, sodium chloride.

28. Trace the energy you use in lifting this book back through the various transformations it undergoes to its ultimate source.

29. How does a plant obtain its carbohydrates and fats? An animal?

30. What are the products of the oxidation of glucose? Is the process endothermic or exothermic?

31. What is believed to be the origin of atmospheric oxygen?

32. When sugar undergoes fermentation to produce ethanol, what other compound is also formed?

33. Can you think of any function other than energy storage that body fat might have?

34. Why do plants need nitrogen? Why can they not use nitrogen from the air? Where do nitrogen compounds in the soil come from?

35. What are the basic structural units of proteins? How does the human body obtain them?

36. How many different "letters" are found in the genetic code? What is the nature of the "letters"?

PROBLEMS

1. Each molecule of butene, C_4H_8, has a double bond between two of its carbon atoms. Give the structural formula(s) for butene and its isomers, if any.

2. Each molecule of propene, C_3H_6, has a double bond between two of its carbon atoms. Give the structural formula(s) for propene and its isomers, if any.

3. Give structural formulas for the two isomeric propyl alcohols that share the molecular formula C_3H_7OH.

4. Xylene molecules consist of benzene molecules in which two of the hydrogen atoms have been replaced by CH_3 groups. Give structural formulas for the three isomers of xylene.

5. Give structural formulas for the three isomers of pentane, C_5H_{12}.

6. The carbon atoms in normal hexane, C_6H_{14}, form a straight chain. All the bonds are single. In cyclohexane the six carbon atoms are arranged in a ring. Give the structural formula of cyclohexane. Are all the bonds single?

7. Bromopropane is propane with one H atom replaced by a Br (bromine) atom. How many isomers does bromopropane have? What are their structural formulas?

8. Use structural formulas to show the reaction between methyl alcohol and acetic acid to produce methyl acetate.

ANSWERS TO MULTIPLE CHOICE

				17. a	**23.** d	**29.** b	**35.** b
				18. b	**24.** c	**30.** c	**36.** a
1. b	**5.** b	**9.** b	**13.** a	**19.** b	**25.** d	**31.** d	**37.** b
2. b	**6.** a	**10.** a	**14.** c	**20.** b	**26.** a	**32.** a	**38.** a
3. c	**7.** b	**11.** c	**15.** d	**21.** c	**27.** d	**33.** c	
4. c	**8.** a	**12.** b	**16.** b	**22.** d	**28.** d	**34.** b	

13

ATMOSPHERE

AND

HYDROSPHERE

The earth's **atmosphere** is an invisible envelope of gas we hardly notice except when the wind blows or when rain or snow falls. The atmosphere is also responsible for the blue of the sky, for the colors of sunrise and sunset, and for the rainbow, as we learned in Chap. 6. Less obvious but more important is the role of the atmosphere in the living world. Its oxygen, nitrogen, and carbon dioxide are essential for life. It screens out deadly ultraviolet and x-rays from the sun. It carries energy and water over the face of the earth. And, by weathering away rocks, it helps form the soil in which plants grow.

All the water of the earth's surface is included in the **hydrosphere.** Oceans, seas, rivers, and lakes cover about three-quarters of the surface area of our planet. The oceans, of course, make up by far the greatest part of the hydrosphere, and they are a major factor in shaping the environment of life on earth.

THE ATMOSPHERE

13-1   REGIONS OF THE ATMOSPHERE

Four Layers   The chief gases of the atmosphere and their average abundances are given in Table 13-1. Water vapor is also present but to a variable extent, from nearly none to about 4 percent. In addition, the lower atmosphere contains a great many small particles of different kinds, such as soot, bits of rock and soil, salt grains from the evaporation of seawater droplets, and spores, pollen, and bacteria.

Those of us who have been among mountains know that the higher up we go, the thinner and colder the air becomes. In the lower atmosphere, air temperature falls an average of 6.5°C per km of altitude. At an elevation of only 5 km (about 16,400 ft) the pressure is down to half what it is at sea level (Fig. 13-1) and the temperature is about −20°C. At about 11 km (36,000 ft) the pressure is only one-fourth its sea-level value, which means that 75 percent of the atmosphere lies below. The temperature at 11 km is about −55°C, which is cold but no colder than it sometimes is at ground level in Siberia and northern Canada. The atmosphere stays that cold for 14 km more.

A passenger in an airplane climbing past 11 km would notice a marked change in the atmosphere. Above that level there are almost no clouds, no storms, not even dust. Since the character of the atmosphere changes so abruptly at the 11-km level, this is taken as the boundary between the two lowest layers of the atmosphere. The dense part near the ground is called the **troposphere,** and the clear layer above it is called the **stratosphere.** Such features of the weather as clouds and storms, fog and haze, belong to the troposphere. Figure 13-2 shows how the atmosphere is divided into regions.

Ozone   The most striking aspect of the stratosphere is the presence of ozone, O_3, which was described in Sec. 11-2. Ozone is produced in the stratosphere when solar radiation breaks up O_2 molecules into O atoms. Some O atoms then join O_2 molecules to

TABLE 13-1

The Composition of Dry Air near Ground Level

Gas	Average Percentage by Volume
Nitrogen	78.08
Oxygen	20.95
Argon	0.93
Carbon dioxide	0.03
Neon	0.0018
Helium	0.00052
Methane	0.00015
Krypton	0.00011
Hydrogen, carbon monoxide, xenon, ozone,	<0.0001

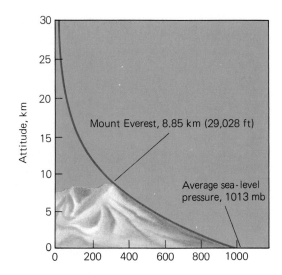

FIG. 13-1 The variation of atmospheric pressure with altitude. The millibar (mb) is the pressure unit used in meteorology, where 1 mb = 100 Pa = 1 hPa.

give ozone molecules: $O + O_2 \rightarrow O_3$. As O_3 molecules are being formed, others are combining with O atoms to give O_2 molecules: $O + O_3 \rightarrow 2O_2$. Hence the situation is one of equilibrium, with ozone being formed and destroyed at the same rates.

Ozone is an excellent absorber of ultraviolet radiation. It is so good, in fact, that the relatively small amount of ozone in the stratosphere is able to filter out nearly all the dangerous short-wavelength ultraviolet radiation reaching the earth from the sun. Because this radiation is so harmful to living things, it is believed that life did not leave the protection of the sea to become established on land until the ozone layer had come into being. The maximum concentration of ozone occurs at 22

FIG. 13-2 The variation of temperature with altitude. The altitudes of the boundaries between regions of the atmosphere are averages.

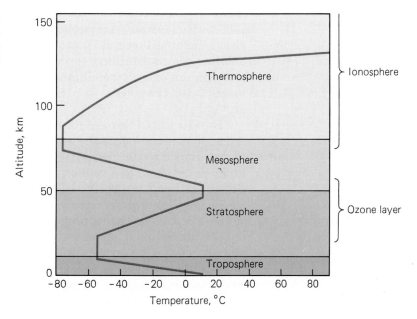

SMOG IN LOS ANGELES On an average day 9000 tons of carbon monoxide, various hydrocarbons, nitrogen and sulfur oxides, and small particles such as soot are emitted in the Los Angeles basin, mainly by vehicles and industry. Sunlight acting on this already nasty mixture adds toxic ozone to it. The result of breathing such polluted air is a high rate of respiratory diseases such as bronchitis and asthma and an increased risk of cancer. The air in Los Angeles is unhealthful on half the days each year.

Los Angeles is not alone in having dirty air: all 20 of the world's largest cities exceed World Health Organization limits in at least one pollutant. The situation in the Los Angeles basin is made worse by the frequent temperature inversions that occur there; Mexico City is another victim of this phenomenon. Ordinarily air temperature falls steadily with increasing altitude in the lower atmosphere. Sometimes, however, a temporary situation arises in which a layer of air aloft is warmer than the air below it. This is called a **temperature inversion.** Polluted air cannot rise past an inversion because when it reaches the inversion, the density of the polluted air is greater than that of the warm air layer on top. Hence the inversion acts to trap the polluted air, which results in a persistent smog.

Smog in Los Angeles. When cool air from the Pacific blows into the Los Angeles basin, it forms a dense layer under warm air above. The warm layer stops polluted air from rising upward. The result of such a temperature inversion is heavy smog, made worse by nearby mountains that prevent the smog from escaping inland.

km, where less than one molecule in 4 million is O_3—hardly an impressive amount for so efficient an absorber.

Because the ozone in the stratosphere is so valuable, pollutants that attack it are highly undesirable. Among the worst such pollutants is a group of gases called the **chlorofluorocarbons** (CFCs); an example of a CFC is CCl_2F_2, whose trade name is Freon 12. The chlorine in CFCs catalyzes the breakdown of O_3 molecules, and once in the stratosphere, the chlorine can remain there for a very long time. CFCs are widely used in refrigerators and air conditioners, as cleaning solvents, and in making foam-plastic containers. In each application some CFCs escape into the atmosphere.

By now the CFCs have measurably weakened the ozone shield. This is a serious matter since the additional ultraviolet radiation that reaches the earth as a result increases skin cancer and cataract rates, depresses immune systems, and reduces crop yields, among other effects. Over Antarctica weather patterns have enabled the CFCs to punch a huge seasonal "hole" of low O_3 concentration in the ozone layer, and something similar seems to be happening in the Arctic.

By international agreement production of CFCs is being phased out and other gases, less harmful to the ozone layer, are replacing them. But because the changeover is taking place gradually and because CFCs in the atmosphere take an average of 75 years to release their chlorine, the ozone layer will continue to deteriorate for a long time to come.

BIOGRAPHY

Evangelista Torricelli (1608–1647)

was born near Ravenna, Italy, and studied mathematics in Rome. He revered Galileo and was his secretary in Florence for the last few months of Galileo's life. Torricelli then became court mathematician to the Grand Duke of Tuscany, Galileo's old position. Galileo had found it odd that, on the upstroke of its piston, a pump like that shown in Fig. 4-11 could lift water no more than about 10 m, and he suggested that Torricelli investigate the matter. As others did in those days, Galileo thought that pumps were able to lift water because "nature abhors a vacuum," so there seemed no reason for the observed limit of 10 m.

Torricelli concluded that the behavior of a lift pump made sense if air has mass, so that the atmosphere presses on the water feeding the pump and thereby pushes it upward in the cylinder as the piston is raised. The limit of 10 m meant that the atmosphere was finite in extent and did not fill the entire universe, despite the belief of many scientists then. To test his ideas, in 1643 Torricelli filled a glass tube, closed at one end, with mercury, whose density is nearly 14 times that of water. Holding his thumb over the open end of the tube, Torricelli turned it upside down and placed the open end in a dish of mercury. When he removed his thumb, the mercury level dropped in the tube but stopped when its height was about 760 mm above the mercury in the dish, $\frac{1}{14}$ of the maximum height that water could be lifted by a pump. This was the first barometer, with the mercury height directly proportional to the pressure of the atmosphere.

Torricelli observed that the mercury height varied slightly from day to day and interpreted this to mean that atmospheric pressure varied similarly. Three years later the French mathematician and physicist Blaise Pascal realized that, if atmospheric pressure is due to the weight of the atmosphere, then it should decrease with altitude. Himself not very robust, Pascal gave his brother-in-law two barometers to take up a mountain. The mercury columns dropped by several centimeters, which verified Torricelli's insights. A pressure unit equal to the pressure exerted by a mercury column 1 mm high is called the **torr** in honor of Torricelli; blood pressures in medicine are conventionally expressed in torr.

Mesosphere and Thermosphere

The ozone of the stratosphere causes a rise in temperature to 10°C or so in the vicinity of 50 km. At this altitude the atmosphere is $\frac{1}{1000}$ of its density at sea level. The temperature then falls once more to another minimum of about −75°C at 80 km. The portion of the atmosphere between 50 and 80 km is known as the **mesosphere.**

Above 80 km the properties of the atmosphere change radically, for now ions become abundant. The **thermosphere** extends from 80 km to about 600 km, with the temperature increasing to about 2000°C. (We must keep in mind that the density of the thermosphere is extremely low, so that despite the high temperatures a slowly moving object there would not get hot if shielded from direct sunlight.) The thermosphere is the home of the ionosphere, which was described in Sec. 6-12.

13-2 ATMOSPHERIC MOISTURE

Another Vital Cycle

Water vapor consists of water molecules that have escaped, or **evaporated,** from a body of water at a temperature below the boiling point of water. The moisture content, or **humidity,** of air refers to the amount of water vapor that it contains. Most of the atmosphere's water vapor comes from the evaporation of seawater. A little also comes from evaporation of

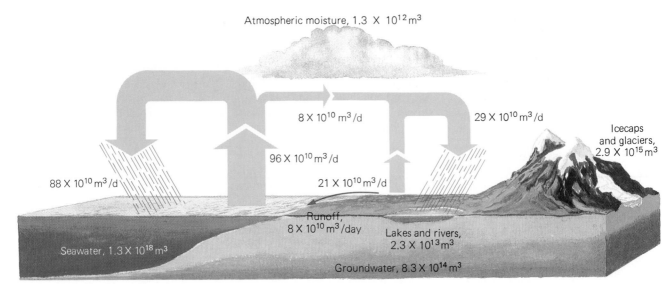

Atmospheric moisture, 1.3 × 10^{12} m^3

8 × 10^{10} m^3/d

29 × 10^{10} m^3/d

Icecaps and glaciers, 2.9 × 10^{15} m^3

96 × 10^{10} m^3/d

88 × 10^{10} m^3/d

21 × 10^{10} m^3/d

Runoff, 8 × 10^{10} m^3/day

Lakes and rivers, 2.3 × 10^{13} m^3

Seawater, 1.3 × 10^{18} m^3

Groundwater, 8.3 × 10^{14} m^3

FIG. 13-3 The world's water content and its daily cycle. Upward arrows indicate evaporation; downward arrows indicate precipitation. If all the water vapor in the atmosphere were condensed, it would form a layer only about 2.5 cm thick.

water in lakes, rivers, moist soil, and vegetation (Fig. 13-3). Since water vapor is continually being added to air by evaporation and periodically removed by condensation as clouds, fog, rain, and snow, the humidity of the atmosphere varies a lot from day to day and from one region to another. If it were not for the ability of water to evaporate, to be carried by winds, and later to fall to the ground, all the earth's water would be in its oceans and the continents would be lifeless deserts.

We can regard air as a sort of sponge for water vapor, and like an ordinary sponge, air at a given temperature can absorb only so much water and no more. Air is said to be **saturated** when it holds this maximum amount of water vapor. (To be accurate, air has nothing to do with evaporation. Even if there were no air, vapor would still escape from bodies of water. But, since moving air carries water vapor from one region to another and since air is the medium in which water vapor condenses as clouds, fog, rain, or snow, it is convenient to think of the air as "taking up" and "holding" different amounts of vapor.)

We usually describe air as humid if it is saturated or nearly saturated, and as dry if it is highly unsaturated. Humid weather is uncomfortable because little moisture can evaporate from the skin into saturated air, and so perspiration does not produce its usual cooling effect. Very dry air is harmful to the skin and mucous membranes because their moisture evaporates too rapidly to be replaced.

13-3   CLOUDS

Some Are Water Droplets, Others Are Ice Crystals

When air that contains water vapor is cooled past its saturation point, some of the vapor condenses to a liquid. Dew forms because the temperature of the ground falls at night, which cools the nearby air. Fogs result when large volumes of air come in contact with cold land or water. Clouds come into being when air

RELATIVE HUMIDITY

Meteorologists express the moisture content of air in terms of **relative humidity,** a percentage that indicates the extent to which air is saturated with water vapor. A relative humidity of 100 percent means that the air is completely saturated with water vapor, 50 percent means that the air contains half of the maximum it could hold, and 0 percent means perfectly dry air.

The amount of moisture that air can hold increases with temperature (Fig. 13-4). As a result, a sample of air becomes less saturated when it is heated (its relative humidity goes down) and more saturated when it is cooled (its relative humidity goes up).

Suppose outside air at, say, 10°C and 70 percent relative humidity is taken inside a house and heated to 20°C. The relative humidity indoors will then drop to only about 35 percent, even though the actual vapor density stays the same. Such a low relative humidity is not desirable, so a way to add more moisture to heated air in winter should be provided.

On the other hand, if summer air at, say, 30°C and 70 percent relative humidity is cooled, it will reach saturation (100 percent relative humidity) at only 24°C. Further cooling will cause water to condense out. An

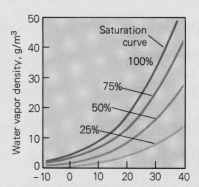

FIG. 13-4 How the mass of water vapor per cubic meter of air varies with temperature for various relative humidities. Each curve corresponds to a different relative humidity. The curve for 100% relative humidity represents saturation, when the air can hold no more water vapor.

air-conditioning system should therefore include a way to remove water vapor from the air being cooled to keep the relative humidity at a comfortable level.

is cooled by expansion when it rises. Clouds cover about half the earth's surface at any time.

Compressing a gas causes it to heat up, as anyone who has used a tire pump knows. The opposite effect is the cooling of a gas when it expands. When a warm, moist air mass moves upward, it expands because the pressure decreases, and it becomes cooler. A cooling rate of about 0.65°C for each 100 m of rise is normal. If the temperature drop is large enough, the air becomes saturated and some of its water vapor condenses into clouds. The condensation occurs around salt and dust particles in the air. The result is tiny water droplets or ice crystals, depending on the temperature, that are small enough to remain suspended aloft indefinitely. High clouds consist of ice crystals, low clouds of water droplets.

Three processes in the atmosphere can cause clouds to form:

1. A warm air mass moving horizontally meets a land barrier such as a mountain and rises (Fig. 13-5). Coastal mountains that lie in the paths of moisture-laden ocean winds may have permanent cloud caps over them.

2. An air mass is heated by contact with a warm part of the earth's surface. The air mass expands, and its buoyancy causes it to rise. This process, called **convection,** is discussed later (see Fig. 13-11, page 416).

3. A warm air mass meets a cooler air mass and, being less dense, is

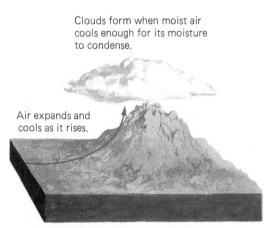

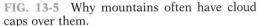

Clouds form when moist air cools enough for its moisture to condense.

Air expands and cools as it rises.

FIG. 13-5 Why mountains often have cloud caps over them.

Cloud cap over the mountainous island of Saba in the West Indies.

forced upward over the cooler mass. This process is also discussed later (see Fig. 13-20, page 424).

Rain, Snow, Sleet, and Hail

Rain falls when a cloud (or part of one) is cooled suddenly, so that condensation is rapid. Some of the water droplets in the cloud become larger than others and because air resistance affects them less, they move within the cloud faster than the smaller droplets. The smaller droplets tend to stick to the larger ones when they come in contact, and the result is larger and larger droplets. Finally the droplets become drops that average 2 mm in diameter, 100 times larger than typical cloud droplets. These drops are too heavy to stay aloft and the result is rain.

If a cloud is cold enough, it consists of ice crystals rather than water droplets, and these crystals can grow into snowflakes. As the photographs on page 312 show, snowflakes have very open structures. As a result, 10 cm of snow is equivalent to only about 1 cm of rain.

Sleet consists of raindrops that have frozen on the way to the ground. Hail occurs when cloud particles alternately freeze and thaw as they grow into rocklike lumps of ice that may be larger than golfballs.

Clouds are sometimes **seeded** with silver iodide to induce rain to fall from them. The crystal structure of silver iodide resembles that of ice; hence water molecules and droplets in a cloud can readily become attached to a silver iodide crystal. Such crystals are thus efficient condensation nuclei and so promote rain from a cloud. "Dry ice" (solid carbon dioxide) can also be used to seed clouds. However, the usual reason for not enough rain is not enough clouds of the right kind in the first place, so cloud seeding is seldom of much help in a drought.

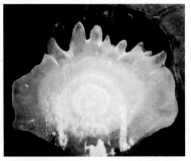

Cross section of a hailstone the size of a grapefruit that fell in Kansas in 1970.

Cirrostratus

Cumulus

Cumulonimbus

Cirrocumulus

Altocumulus

Nimbostratus

Cirrus clouds occur at high altitudes and are wispy or featherlike. Stratus clouds occur in broad layers. The cirrostratus clouds shown here combine features of both cloud types and consist of ice crystals. Cumulus clouds are puffy or heaped upward and are normally associated with fair weather. Clouds associated with rain or snow have the word *nimbus* (Latin for "rain") in their names. Shown are cumulonimbus clouds. A "mackerel sky" consists of high-altitude cirrocumulus clouds. A cloud that occurs at a higher altitude than usual for its type is given the prefix *alto,* as is the case of the altocumulus clouds shown at bottom left. Nimbostratus clouds are thick, dark, and relatively shapeless. Their presence is usually accompanied by steady rain or snow.

Meteorology, the study of weather and weather patterns, is concerned with what we can think of as a vast air-conditioning system. Our spinning planet is heated strongly at the equator and weakly at the poles, and its water content is concentrated in the great ocean basins. From our point of view, it is the task of the atmosphere to redistribute this heat and moisture so that large areas of the land surface are habitable.

But air-conditioning by the atmosphere is far from perfect. It fails miserably in deserts, on mountaintops, in the polar regions. On sultry midsummer nights or on bitter January mornings we may question its efficiency even in our favored part of the world. Still, the atmosphere does succeed in making a surprisingly large part of the globe fit for people to live in.

Besides regulating air temperature and humidity, we expect the atmosphere to provide us with water in the form of rain or snow. The weather and climate of a region describe how effectively these things are done. **Weather** refers to the temperature, humidity, air pressure, cloudiness, and rainfall (or snowfall) at any given time. **Climate** is a summary of weather conditions over a period of years, including how temperature and rainfall vary with the seasons. For instance, a notable feature of the climate of North Dakota is its extreme warmth in summer and extreme cold in winter, whereas comfortable year-round temperatures with rainfall concentrated in the winter months characterize the climate of southern California. The speeds and directions of winds are often significant in describing weather and climate.

13-4 ATMOSPHERIC ENERGY

A Giant Greenhouse in the Sky The energy that warms the air, evaporates water, and drives the winds comes to us from the sun. Solar energy arriving at the upper atmosphere is called **insolation** (for *in*coming *sol*ar rad*iation*).

Every object gives off electromagnetic waves whose intensity and predominant wavelength depend on the temperature of the object (Fig. 13-6). The hotter the object, the more energy it emits and the shorter the average wavelength. The sun, whose surface temperature is about 5700°C, gives off a great deal of energy and its radiation is mainly visible light. The earth, whose surface temperature averages about 15°C, is a feebler source of energy and its radiation is mainly in the long-wavelength infrared part of the spectrum to which the eye is not sensitive.

About 30 percent of the insolation is reflected back into space, chiefly by clouds (Fig. 13-7). The atmosphere absorbs perhaps 19 percent of the insolation, with ozone, water vapor, and water droplets in clouds taking up most of this amount. Slightly over half of the total insolation therefore reaches the earth's surface, where it is absorbed and becomes heat. A little of this heat is given to the atmosphere through contact with the warm surface, somewhat more by means of water evaporated from the oceans.

The warm earth also radiates energy back into the atmosphere, but the energy now is in the form of long-wavelength infrared radiation. These long waves are readily absorbed by atmospheric carbon dioxide

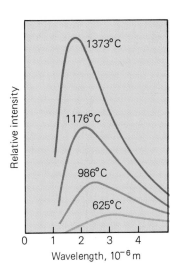

FIG. 13-6 Every object gives off electromagnetic (em) radiation. These curves show how the intensity of the radiation varies with wavelength for objects at the temperatures indicated. The predominant wavelength decreases as temperature increases. Em radiation from the sun is mostly visible light; em radiation from the earth is mostly infrared light.

and water vapor, whose molecules then transfer energy to the rest of the atmosphere. Thus a major source of atmospheric energy is radiation from the earth, not direct sunlight.

The way in which the atmosphere is heated from below rather than from above is often called the **greenhouse effect.** The interior of a greenhouse is warmer than the outside air because sunlight can enter through its windows, but the infrared radiation that the warm interior gives off cannot go through glass. As a result much of the incoming energy is trapped inside. The atmosphere thus behaves like a giant greenhouse, with atmospheric gases that absorb infrared radiation acting like the windows of a greenhouse.

Factors That Govern Air Temperatures

If the earth had no atmosphere, it would grow intensely hot during the day and unbearably cold at night, as the airless moon does. The earth's atmosphere prevents these extremes. The constant movement of air around the world keeps daytime temperatures in any one place from climbing very high, and the ability of moist air to absorb the earth's infrared radiation prevents the rapid escape of heat by night.

How hot the atmosphere becomes over any particular region depends on a number of factors. Air near the equator is on the average much warmer than air near the poles because the sun's rays are more effective in heating the surface when they come from overhead than when they come at a slanting angle (Fig. 13-8). Air over a mountaintop may become warm at midday but cools quickly because it is thinner and contains less carbon dioxide and water vapor than air lower down. A region covered with clouds usually has lower air temperatures than a nearby region in sunlight.

Because the temperature of water changes more slowly than that of rocks and soil, the atmosphere near large bodies of water is usually cooler by day and warmer by night than the atmosphere over regions far from water. Desert regions commonly show abrupt changes in air temperature between day and night because so little water vapor is there to

FIG. 13-7 The greenhouse effect. Much of the energy in the short-wavelength visible light from the sun that is absorbed by the earth's surface is in turn radiated by the earth as long-wavelength infrared light that is absorbed by CO_2 and H_2O in the atmosphere. Some energy reaches the atmosphere by contact with the earth and by means of water evaporated from the sea. Thus the atmosphere is heated mainly from below by the earth rather than from above by the sun. On the average, the total energy the earth and its atmosphere radiate into space equals the total energy they receive from the sun.

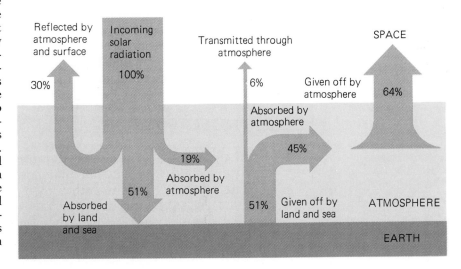

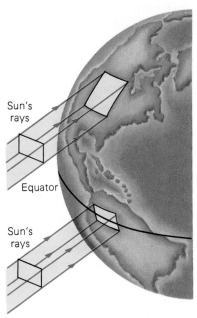

FIG. 13-8 The equatorial regions of the earth are on the average warmer than the polar regions because at the equator the sun's rays are spread over a smaller surface.

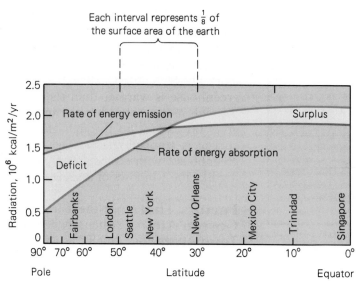

FIG. 13-9 The annual balance between incoming solar radiation and outgoing radiation from the earth. More energy is gained than lost in the tropical regions, and more energy is lost than gained in the polar regions. The latitude scale is spaced so that equal horizontal distances on the graph correspond to equal areas of the earth's surface.

absorb infrared radiation. The atmospheric temperatures of some regions are influenced profoundly by winds and by ocean currents.

The earth's average temperature does not change very much with time, hence there must be a balance between incoming and outgoing energy. That such a balance does indeed occur can be seen with the help of Fig. 13-9, which shows how the rates at which radiant energy enters and leaves the earth vary with latitude.

More energy arrives at the tropical regions than is lost there, and the opposite is true at the polar regions. Why then do not the tropics grow warmer and warmer while the poles grow colder and colder? The answer is to be found in the motions of air and water that shift energy from the regions of surplus to the regions of deficit. About 80 percent of the energy transport around the earth is carried by winds in the atmosphere, and the remainder is carried by ocean currents. We shall look at both of these mechanisms in the rest of this chapter.

13-5   THE SEASONS

**They Are
due to the
Tilt of the
Earth's Axis**

The average distance between the earth and the sun is about 150 million km. Because the earth's orbit is an ellipse rather than a circle, the earth-sun distance varies during the year from about 2.4 million km closer than the average to the same amount farther away. The earth is nearest the sun early in January and farthest from the sun early in July.

We might be tempted to attribute the seasons to the shape of the earth's orbit, especially if we happen to live in the southern hemisphere where January is a summer month and July a winter month. But this cannot be the reason, if only because the seasons are reversed in the northern hemisphere. In fact, the sunlight that reaches the earth varies in intensity by too little between the orbital extremes to give rise to the difference between summer and winter. After all, the earth's orbit differs from a perfect circle by only ±1.6 percent.

The tilt of the earth's axis, not the shape of its orbit, is what causes the seasons. As a result of this tilt, for half of each year one hemisphere receives more direct sunlight than the other hemisphere, and in the other half of the year it receives less (Fig. 13-10). A beam of light that arrives at an angle to a surface delivers less energy per m^2 than does a similar beam that arrives perpendicularly, as we can see in Fig. 13-8.

The noon sun is at its highest in the sky in the northern hemisphere on about June 22 when the north pole is tilted most toward the sun. The period of daylight in the northern hemisphere is longest on this date. The noon sun is at its lowest 6 months later when the north pole is tilted most away from the sun. The period of daylight is shortest on this date. In the southern hemisphere the situation is, of course, reversed. On about March 21 and September 23 the sun is directly overhead at noon at the equator. The periods of daylight and darkness are then equal everywhere on the earth.

FIG. 13-10 The seasons are caused by the tilt of the earth's axis together with its annual orbit around the sun. As a result, the daylight side of the northern hemisphere is tilted away from the sun in January, which means that sunlight strikes this hemisphere at a glancing angle and delivers less energy to a given area than in June. The seasons are reversed in the southern hemisphere.

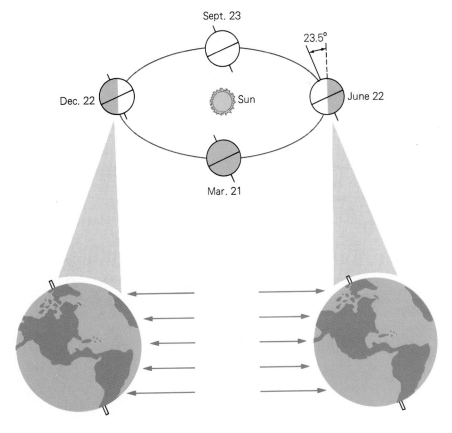

| 13-6 | **WINDS** |

Currents of Air Driven by Temperature Differences

Winds are horizontal movements of air caused by pressure differences in the atmosphere. The greater the pressure difference between two regions, the faster the air between them moves. All pressure differences between places on the earth's surface can be traced to temperature differences.

When a region is warmer than its surroundings, the air above it is heated and expands (Fig. 13-11). The hot air rises, leaving behind a low-pressure zone into which cool air from the high-pressure neighborhood flows. The flow toward the heated region at low altitudes is balanced by an outward flow of the air that has risen. This air then cools and sinks to replace the air that has moved inward. Air movements of this kind, produced as the result of unequal heating of the earth's surface, are called **convection currents.**

Coriolis Effect

The rotation of the earth affects the path of something that moves above its surface as this path is seen from the surface. This phenomenon is called the **coriolis effect,** and it has these results:

> **In the northern hemisphere, a path that would be a straight line over a stationary earth instead is curved to the right. In the southern hemisphere the curvature is to the left.**

Only motion along the equator is not affected by the coriolis effect.

FIG. 13-11 Convection currents are produced by unequal heating. The temperature of a land surface rises more rapidly in sunlight than the temperature of a water surface. The resulting convection produces the **sea breeze** found on sunny days near the shores of a body of water. At night, the land cools more than the sea, and the convection current is reversed to give a **land breeze** that blows offshore.

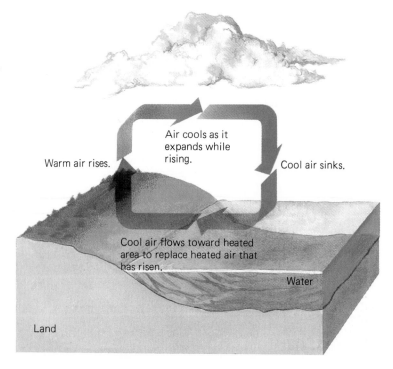

Air cools as it expands while rising.

Warm air rises.

Cool air sinks.

Cool air flows toward heated area to replace heated air that has risen.

Water

Land

FIG. 13-12 (a) Because of the coriolis effect, which is a consequence of the earth's rotation, winds in the northern hemisphere are deflected to the right. As a result, air does not flow directly toward the center of a low-pressure region but spirals inward in a counterclockwise direction. (b) Similarly, air flows away from the center of a high-pressure region in a clockwise spiral. In the southern hemisphere these directions are reversed. An **isobar** is a line of constant pressure on a weather map; it corresponds to a contour line of constant altitude on an ordinary map.

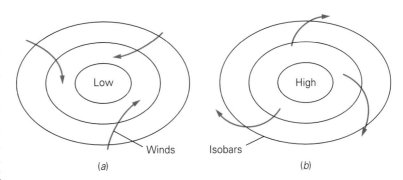

Because of the coriolis effect, winds are deflected from straight paths into curved ones. Thus the air rushing into a low-pressure region does not move directly inward but instead follows a spiral path that is counterclockwise in the northern hemisphere and clockwise in the southern (Fig. 13-12). Examples of such spiral motion are, in order of

MONSOONS The seasonal winds called **monsoons** are large-scale sea and land breezes modfied by the coriolis effect. During the summer, a continent is warmer than the oceans around it. The rising air over the continent creates a low-pressure region that pulls in moisture-laden sea breezes that bring rain. In winter the situation is reversed, with dry air moving seaward.

Monsoons are most pronounced in certain parts of Africa and Asia. Figure 13-13 shows how the summer and winter monsoons of India and southeast Asia arise.

In summer, the motion of air around the low-pressure center on land is counterclockwise, and it is clockwise around the high-pressure center in the Indian Ocean. The result is a wet southwest monsoon that blows from May to September. Every few years this monsoon is weaker than usual, less rain falls, crops suffer, and there may be widespread famine. From October to April dry winds blow from the northeast, as in Fig. 13-13b. About half the world's population depends on summer monsoons to provide the rain needed for agriculture.

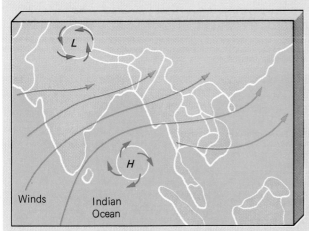

(a) Summer monsoon

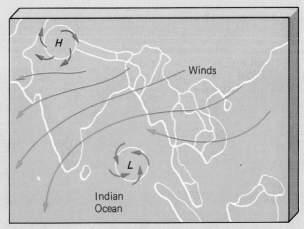

(b) Winter monsoon

FIG. 13-13 (a) The summer monsoon of India and south Asia. Heating of the land produces a low-pressure region (L) centered inland and a high-pressure region (H) centered in the Indian Ocean that together cause southwest winds to occur. Rice cultivation in this part of the world depends on the warm, moist air brought by the summer monsoon. (b) The winter monsoon. Now the land is cooler than the ocean, so the low- and high-pressure regions are reversed to give dry northeast winds.

decreasing size (but increasing violence), middle-latitude weather systems, hurricanes and other tropical storms, and tornadoes.

13-7 GENERAL CIRCULATION OF THE ATMOSPHERE

Alternate Belts of Wind and Calm

The earth is heated most at the equator and least at the poles. We therefore expect to find convection currents as part of the general atmospheric circulation.

Suppose for the moment that our planet did not rotate and that its surface was made up entirely of either land or water. On such an earth, air circulation would depend only on the difference in temperature between equator and poles. Air would rise along the heated equator, flow at high altitudes toward the poles, and at low altitudes return from the poles toward the equator (Fig. 13-14). We in the northern hemisphere would experience a steady north wind. (Winds are named for the direction they come from, so a north wind blows from the north.) Around the equator would be a belt of relatively low pressure, and near each pole a region of high pressure.

Because the earth does rotate, however, the north and south winds coming from the poles are deflected by the coriolis effect into large-scale eddies that lead to a generally eastward drift in the middle of each hemisphere and a westward drift in the tropics. The main features of the general circulation of the atmosphere are shown in Fig. 13-15.

The various wind zones were important to shipping in the days of sail, as their names indicate. Thus the steady easterlies on each side of the equator became known as the **trade winds** because they could be relied upon by sailing ships. The region of light, erratic wind along the equator, where the principal movement of air is upward, constitutes the

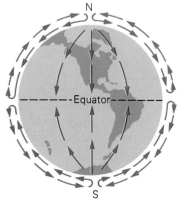

FIG. 13-14 The convectional circulation that would occur if the earth did not rotate. The arrows in the center of the diagram indicate surface winds.

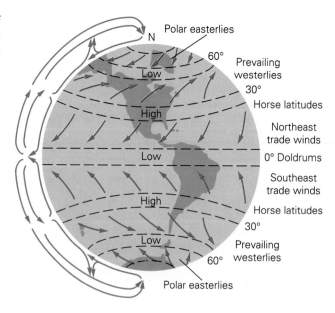

FIG. 13-15 Simplified pattern of horizontal and vertical circulation in the actual atmosphere. Regions of high and low pressure are indicated.

doldrums. The **horse latitudes** that separate the trade winds in both hemispheres from the **prevailing westerlies** poleward of them are also regions of calm and light winds. Their name is supposed to have come from the practice of throwing overboard horses that were being carried on sailing ships when the ships were becalmed and ran short of drinking water.

Jet Streams   With increasing altitude the belts of westerly winds broaden until almost the entire flow of air is west to east at the top of the troposphere. The westerly flow aloft is not uniform but contains narrow cores of high-speed winds called **jet streams,** usually only a few hundred kilometers wide but with winds of up to 500 km/h. The jet streams form zigzag patterns around the earth that change continuously and affect the variable weather of the middle latitudes by their influence on the paths of air masses closer to the surface. They are important to aircraft—an hour or more can be saved on a long west-east flight by using a route with a jet stream as a tail wind, and a corresponding addition to the flight time can be prevented in the other direction by avoiding a jet stream as much as possible.

At any given time the circulation near the earth's surface is more complicated than the pattern given in Fig. 13-15. An important factor is the presence of large seasonal low- and high-pressure regions caused by unequal heating due to the irregular distribution of land masses and sea masses (Fig. 13-16). Smaller, short-lived cells also occur and profoundly affect local weather conditions, as we shall see next.

FIG. 13-16 Average January sea-level pressures (in millibars) and wind patterns. High- and low-pressure systems are indicated. Isobars connecting points of equal pressure are shown in white.

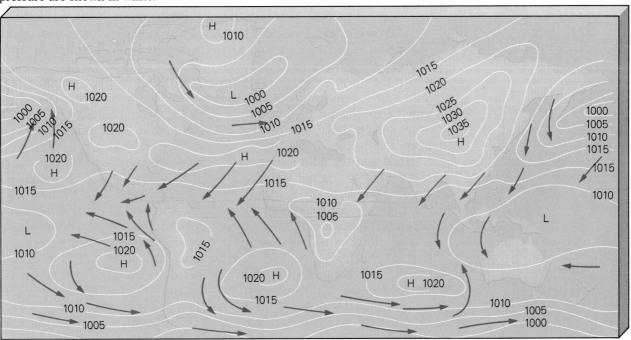

EL NIÑO   The winds of the tropical Pacific usually blow westward from the Americas. These are the trade winds shown in Fig. 13-15. Together with the ocean currents they drive, the trade winds create a huge pool of warm water in the western side of the Pacific basin. The heated moist air above this pool rises to give an area of low pressure and brings abundant rain to Indonesia and northern Australia.

Every few years, however, a different pattern takes over. For reasons that are not completely understood, the trade winds weaken and the warm water pool shifts to the central Pacific. Now droughts parch Indonesia and Australia while torrential rains soak Peru and Ecuador on South America's west coast. The upwelling of nutrient-rich cold water along this coast when the surface waters were blown westward stops and the coastal water becomes warm. As a result the fish population drops sharply in the eastern Pacific and the economies of Peru and Ecuador suffer accordingly. The entire phenomenon is called **El Niño,** Spanish for "the Christ child," because the warm water appears around Christmas.

The influence of an El Niño can extend well beyond the Pacific, in part by disrupting the jet streams. The severe El Niño of 1982–1983 brought droughts that devastated agriculture in India, southern Africa, and Brazil, and abnormally heavy rains gave rise to floods in southern China and California. All the continents were affected in some way.

In the past, 3 to 7 years would pass after an El Niño before another appeared. The El Niño that started in 1990, however, returned a record 4 years in a row. Was this a once-in-a-lifetime oddity or does it indicate a long-lasting change in the wind systems of the world? Is global warming (see Sec. 13-12) the reason for more frequent El Niños? Nobody knows the answers as yet.

13-8 MIDDLE-LATITUDE WEATHER SYSTEMS

Why Our Weather Is So Fickle Day-to-day weather is more variable in the middle latitudes than anywhere else on earth. If we visit central Mexico or Hawaii, in the belt of the northeast trades, we find that one day follows another with hardly any change in temperature, humidity, or wind direction. On the other hand, in nearly all parts of the continental United States, drastic changes in weather are commonplace. The reason for this variability lies in the movement of warm and cold air masses and of storms derived from them through the belts of the westerlies.

In the northern part of the westerly belt an irregular boundary separates air moving generally northward from the horse latitudes and air moving southward from the polar regions. Great bodies of cold air at times sweep down over North America, and at other times warm air from the tropics extends far northward. The cold air is ultimately warmed and the warm air cooled, but a large volume of air can maintain nearly its original temperature and humidity for days or weeks. These huge tongues of air, or isolated bodies of air detached from them, are the **air masses** of meteorology. The kind of air in an air mass depends on its source (Fig. 13-17). A mass formed over northern Canada is cold and dry, one from the North Atlantic or North Pacific is cold and humid, one from the Gulf of Mexico warm and humid, and so on. Weather prediction in the United States depends largely on following the movements of air masses from these various source areas.

Cyclones and Anticyclones Weather systems associated with air masses are usually several hundred to a thousand or more km across and move from west to east. A **cyclone** is an air mass

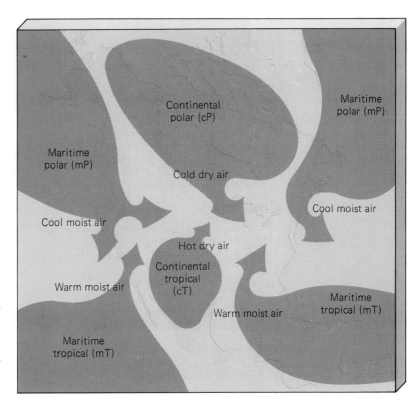

FIG. 13-17 The air masses that affect weather in North America. The importance of the various air masses depends upon the season. In winter, for instance, the continental tropical air mass disappears and the continental polar air mass exerts its greatest influence.

in which the pressure is low at the center. As air rushes in toward the center of a cyclone, the moving air is deflected toward the right in the northern hemisphere and toward the left in the southern because of the coriolis effect. As a result cyclonic winds blow in a counterclockwise spi-

Cyclonic weather systems are responsible for the variable weather of the middle latitudes. A typical system is 1500 km in diameter and moves eastward at 40 km/h; its characteristic winds usually do not exceed 65 km/h. The system here was centered about 2000 km north of Hawaii when it was photographed from a spacecraft.

A tornado, such as this one in Kansas in March 1990, is a narrow but extremely violent cyclonic storm.

ral in the northern hemisphere and in a clockwise spiral in the southern hemisphere (see Fig. 13-12).

An **anticyclone** is centered on a high-pressure region from which air moves outward. The coriolis effect therefore causes anticyclonic winds to blow in a clockwise spiral in the northern hemisphere and in a counterclockwise spiral in the southern hemisphere. These spirals are conspicuous in cloud formations photographed from earth satellites.

A cyclone is a region of low pressure, and air flowing into one rises in an upward spiral (Fig. 13-18). The rising air cools and its moisture

FIG. 13-18 (a) In a cyclone, surface air spirals counterclockwise in toward the low-pressure center and rises. The air cools as it moves upward, which causes its moisture to condense out as clouds and rain. (b) In an anticyclone, cold air from above spirals outward from the high-pressure center as it sinks. The falling air warms, which decreases its relative humidity to give a clear sky.

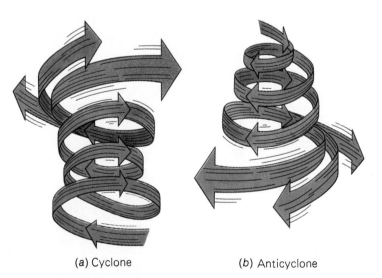

(a) Cyclone   (b) Anticyclone

content condenses into clouds. As a rule, cyclones bring unstable weather conditions with clouds, rain, strong shifting winds, and abrupt temperature changes. An anticyclone is a region of high pressure, and air flows out of it in a downward spiral. The descent warms the air and its relative humidity accordingly drops, hence condensation does not occur. The weather associated with anticyclones is usually settled and pleasant with clear skies and little wind.

Fronts

Middle-latitude cyclones originate at the **polar front,** which is the boundary between the cold polar air mass and the warmed air mass next to it. It is common for a kink to develop in this front with a wedge of warm air protruding into the cold air mass. This produces a low-pressure region that moves eastward as a cyclone. The eastern side of the warm-air wedge is a **warm front** since warm air moves in to replace cold air in its path; the western side is a **cold front** since cold air replaces warm air (Fig. 13-19).

As warm air rises along an inclined front, it is cooled and part of its moisture condenses out. Clouds and rain are therefore associated with both kinds of fronts (Fig. 13-20). A cold front is generally steeper, since cold air is actively burrowing under warm air. The temperature difference is greater as well, so rainfall on a cold front is heavier and of shorter duration than on a warm front. A cold front with a large temperature difference is often marked by violent thundersqualls.

A warm front typically moves eastward at about 150 km/day. The cold front behind it moves faster, up to twice as fast, and eventually it

FIG. 13-19 Weather maps show pressure patterns, winds, rain, and snow. This is a weather map of the eastern United States one April morning. A cold air mass on the west and north (polar continental air) is separated from a warm air mass (tropical maritime air) by a cold front extending from Louisiana to Michigan and by a warm front from Michigan to Virginia. Where the north end of the warm air mass lies between the two fronts a cyclone has formed, bringing rain (colored area) to the Great Lakes region. The unit of pressure in this map is the millibar. The small circles indicate clear skies; solid dots indicate cloudy skies. The small lines show wind direction, which is toward the circle or dot, and wind strength; the greater the number of tails, the faster the wind.

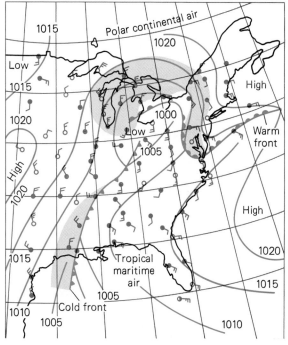

FIG. 13-20 Cross-section diagrams of (*a*) a warm front and (*b*) a cold front. In each case the front is moving to the right. Photographs of the various cloud forms appear on p. 411.

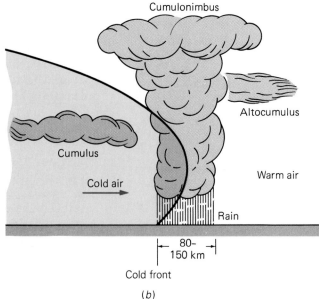

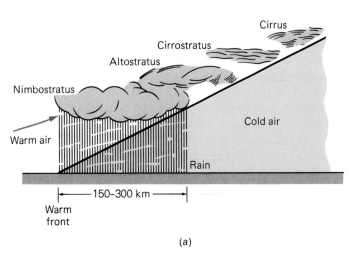

(a)

(b)

overtakes the warm front to force the wedge of warm air upward (Fig. 13-21). The resulting **occluded front** is the last stage in the evolution of a cyclone, which soon afterward disappears. The total life span of a middle-latitude cyclone may be as little as a few hours or as much as a week, though the usual range is 3 to 5 days.

WEATHER FORECASTING   In order to predict the weather for a certain region, it is necessary to know present and past conditions in detail over a much larger area. Data on air pressure, temperature, humidity, cloud cover, precipitation, and winds are collected every 3 hours at about 10,000 weather stations around the world and are available to forecasters everywhere in a standard code. Additional information is provided by weather satellites that look at the atmosphere from above and by balloon-borne instruments that make measurements at various altitudes. In both cases the observations are radioed to ground stations.

The first step in preparing a forecast is to draw a weather map, called a **synoptic** map, that shows the current situation. By comparing this map with the synoptic maps for the past day or two, a meteorologist can see how the various weather systems are developing and moving and may be able to spot new ones in their early stages. Then a **prognostic** map is prepared that shows for a future time the weather pattern the meteorologist thinks will have evolved from the current situa-

tion. The prognostic map indicates the weather that can be expected at that time.

Nowadays computers are used to digest the vast amounts of data that go into synoptic maps and also to help predict what will happen next. People are still essential, though: an educated guess by someone familiar with local weather may be more reliable than the calculations of a supercomputer. Forecasts for a day ahead are usually quite accurate, those for longer periods in the future less so. Even with all the resources available to meteorologists today, some phenomena, such as hurricanes, do not always behave as expected.

Are more accurate long-range forecasts a realistic prospect? The trouble is that the atmosphere is a very complex system and it interacts with land and sea in complicated ways. Although weather exhibits many regularities, even small differences in initial conditions may lead to entirely different outcomes. Because, to give a famous example, a typhoon in Bangladesh could, in principle, be triggered by the fluttering of a butterfly's wings in Iowa, some element of uncertainty will always exist in weather prediction.

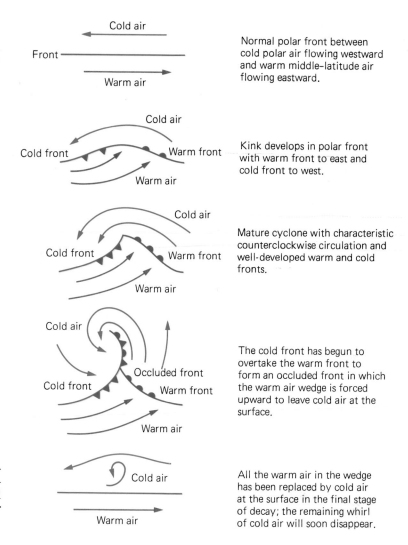

Cold air

Front

Warm air

Normal polar front between cold polar air flowing westward and warm middle–latitude air flowing eastward.

Cold air

Cold front   Warm front

Warm air

Kink develops in polar front with warm front to east and cold front to west.

Cold air

Cold front   Warm front

Warm air

Mature cyclone with characteristic counterclockwise circulation and well-developed warm and cold fronts.

Cold air

Cold front   Occluded front

Warm front

Warm air

The cold front has begun to overtake the warm front to form an occluded front in which the warm air wedge is forced upward to leave cold air at the surface.

Cold air

Warm air

All the warm air in the wedge has been replaced by cold air at the surface in the final stage of decay; the remaining whirl of cold air will soon disappear.

FIG. 13-21 Life cycle of a middle-latitude cyclone in the northern hemisphere. Conventional weather-map symbols are used for cold, warm, and occluded fronts.

Tropical cyclones are different in character from those of the middle latitudes. The tropical ones are small but extremely violent, with very low pressures at their centers and winds of 120 km/h or more. The hurricanes that from time to time batter the West Indies and the eastern coast of the United States are tropical cyclones, as are the typhoons of the Pacific Ocean. Figure 13-22 shows the track of Hurricane Emily, which devastated Bermuda in 1987.

CLIMATE The climate of a region refers both to its average weather over a period of years and to the typical variations in its weather elements during each day and during each year. The most significant weather elements in determining climate are temperature and precipitation. Climates differ considerably around the world, ranging from the tropics, where there is no winter, to the polar regions, where summer is brief.

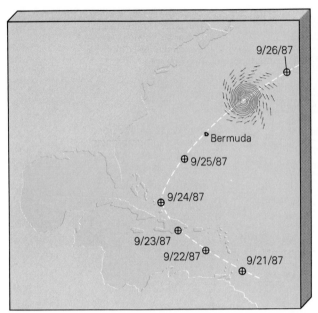

FIG. 13-22 The track of Hurricane Emily in September 1987, showing the positions of the storm center at 24-h intervals. Wind speeds of up to 200 km/h (124 mi/h) were recorded. This track is typical of North Atlantic hurricanes, which are born near the bulge of Africa and grow in strength by absorbing energy from the warm tropical water as they are swept westward by the trade winds. They then turn north to curve around the permanent high-pressure region of the mid-Atlantic, and finally are blown eastward by the westerlies of the higher latitudes until they lose energy and die out. When such a hurricane swings over land, its winds can do immense damage. Seawater surging ashore to flood low-lying coastal area is even more destructive and causes most hurricane deaths.

A hurricane is a large, violent tropical storm typically 160 km in diameter whose winds spiral inward and upward at velocities of 120 km/h or more around an eye of low pressure. Heavy rainfall accompanies the passage of a hurricane except in the eye, which may be 15 to 30 km across. Most hurricanes occur on the western sides of the Pacific, Indian, and North Atlantic oceans during the late summer and early fall, and their most vigorous phases last for a few days to a week or so. Hurricanes usually move at 15 to 50 km/h but may move faster or remain stationary for a day or more. This photograph of Hurricane Gladys was taken from the spacecraft Apollo 9 when the eye of the hurricane was southwest of Florida. Cuba is in the background.

13-9 **TROPICAL CLIMATES**

Hot and Wet or Hot and Dry

The equatorial belt of the doldrums, with its rapid evaporation and strong rising air currents, provides an ideal situation for abundant rain. Throughout the year the weather is hot with almost daily rains and light, changeable winds. The steaming rain forests of Africa, South America, and the East Indies occur where this belt crosses land.

The horse latitudes, roughly 30° north and south of the equator, are also belts of little wind, but their climate is anything but humid. Air in these belts moves downward to become warmer and hence less saturated with water vapor. Thus the climate is dry, with clouds and rain only at long intervals. In a few areas, however, such as the Gulf Coast of the United States, the prevailing dryness is modified by moisture-laden winds of local origin.

Year-round high temperatures and abundant rainfall are characteristic of the equatorial regions and encourage plant growth. This rain forest is in Trinidad.

Air returning to the warm equatorial belt from the horse latitudes tends to keep the little moisture it contains, except where a mountain range or strong convection currents force the air upward. Hence the trade-wind belts are for the most part dry regions. Seasonal shifts of the trade-wind belts give rainfall during part of the year to the equatorial margins of these belts. The poleward portions of the trade-wind belts, together with the adjacent horse latitudes, are the regions of the world's great deserts—the Sahara, the deserts of South Africa, the dry districts of Mexico and northern Chile, the dry interior of Australia.

13-10 MIDDLE-LATITUDE CLIMATES

Variety Is the Rule
The belts of prevailing westerlies generally have moderate average temperatures, although continental interiors show great variations with the seasons. On the other hand, oceanic islands and the western coasts of continents in these belts have even temperatures throughout the year.

In the northern hemisphere the huge land masses of North America and Eurasia bring complications that are well illustrated by climates in the United States. Damp winds from the Pacific Ocean are forced upward by mountain ranges along the West Coast, and the western sides of the mountains therefore receive abundant rainfall. Once across the mountain barriers the westerlies have little remaining moisture, so that the region from the mountains east to the Great Plains is largely dry.

If the westerlies kept their direction as steadily as do the trade winds, dry conditions would continue across the continent to the East Coast. Instead, the cyclonic storms characteristic of this belt often bring moisture-laden air from the Gulf of Mexico and the Atlantic Ocean into the Mississippi Valley and the eastern states. Rainfall increases eastward across the country, becoming very abundant along the Gulf of Mexico.

Temperatures on the West Coast are conditioned by the prevailing wind from the ocean and change relatively little from season to season. In most other parts of the country, however, the difference between summer and winter is very marked. The fine climates of Florida and southern California owe their mildness to nearby warm oceans and to their locations near the junction of the belt of the westerlies and the horse latitudes.

In the bleak arctic and antarctic regions, summers are short, winters long and cold. Moderate winds are the rule, although violent gales occur at times. The total amount of snow during the year is small simply because the low temperatures prevent the accumulation of much water vapor in the air.

13-11 CLIMATIC CHANGE

An Icy Past
Weather we expect to vary, both from day to day and from season to season. Nor are we surprised when one year has a colder winter or a drier summer than the one before. Less familiar are changes in climate. Even though climate represents

The Viking Erik the Red established a colony at this site in Greenland in the tenth century, which was a time of warm climate. His son Leif Eriksson sailed westward from here and landed in what is now Newfoundland, the first European to set foot in the Americas. (Another Viking, Bjarni Herjolfsson, had reached the same region earlier but had not gone ashore.) The Greenland colony died out several centuries later when the Arctic climate deteriorated.

averages in weather conditions over periods of, say, 20 or 30 years, there is plenty of evidence that it, too, is not constant but instead fluctuates markedly over long spans of time. The most dramatic fluctuations were the **ice ages** of the distant past.

The last ice age reached its peak about 18,000 years ago when huge ice sheets as much as 4 km thick covered much of Europe and North America (see Chap. 15). The vast amount of water locked up as ice lowered sea level nearly 100 m below what it is today. Then the ice began to melt and climates became progressively less severe; in a period of 12,000 years the average annual temperature of central Europe rose from –4°C to +9°C. By about 4000 B.C. average temperatures were a few degrees higher than those of today. A time of declining temperatures then set in, reaching a minimum in Europe between 900 and 500 B.C.

A gradual warming-up followed and came to a peak between A.D. 800 and 1200. So generally fine were climatic conditions then that the Vikings established flourishing colonies in Iceland and Greenland from which they went on to visit North America. But then came cooler summers and exceptionally cold winters with extensive freezing of the Arctic Sea. So extreme was the weather from the fifteenth to the nineteenth centuries in Europe that it has been called the "Little Ice Age." Greenland became a less attractive place than formerly and the colony there disappeared; the coast of Iceland was surrounded by ice for several months per year (in contrast to a few weeks per year today); and glaciers advanced farther across mountain landscapes than ever before or since in recorded history. The average temperature in Europe then was around 1°C less than it is these days. This does not seem like much, but it made a considerable difference in climates.

Recent Climates

Late in the nineteenth century a trend toward higher temperatures led to a marked shrinkage of the world's

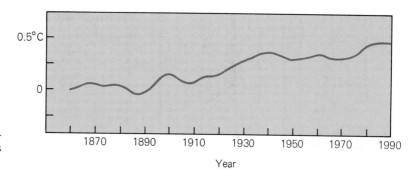

FIG. 13-23 Changes in world-wide average annual temperatures since 1860.

glaciers. In the first half of the present century especially pronounced temperature increases took place whose most noticeable consequences were milder winters in the higher latitudes. These balmy conditions peaked about 1940, after which average worldwide temperatures did not change by much for 30 years (Fig. 13-23). Then temperatures again started upward, with the 1980s and 1990s being the warmest decades since records began to be kept.

The total increase in average temperature since 1860 has been about 0.5°C, enough to raise sea level by over 12 cm. Most of the rise is due to thermal expansion—when any substance is heated, as we know, it expands. Climates have changed, too, with the regions nearest the poles being affected most strongly.

If temperatures continue to increase, climatic patterns will shift even more. Some plants and animals will have a hard time keeping up with the new environments, and a number of species may disappear. Others will thrive. Eventually the polar ice caps will melt on a large enough scale to flood coastal areas everywhere, leaving hundreds of millions of people to be resettled on higher ground. It is not surprising that more and more attention is being given to searching out the sources of climatic change and to trying to estimate the effects of these sources on climates in the near and far futures.

13-12 ORIGINS OF CLIMATIC CHANGE

Why Do Climates Change?

A number of factors can influence climate, and it is not always clear which one is responsible for which change.

The most obvious factor is the sun's energy output. This is not constant but increases and decreases during the 11-year sunspot cycle (Sec. 17-6). Variations in a number of weather phenomena (for instance, in the paths of winter storms across the North Atlantic Ocean) parallel the sunspot cycle. Although the connections with the sunspot cycle seem to be real, nobody yet knows exactly how they come about. Longer-term rises and falls in solar radiation occur as well; the Little Ice Age may have been caused by one of them.

The most promising explanation for the large-scale climatic changes, however, relates them to periodic changes that occur in the tilt of the earth's axis, the shape of its orbit, and the time of year when the

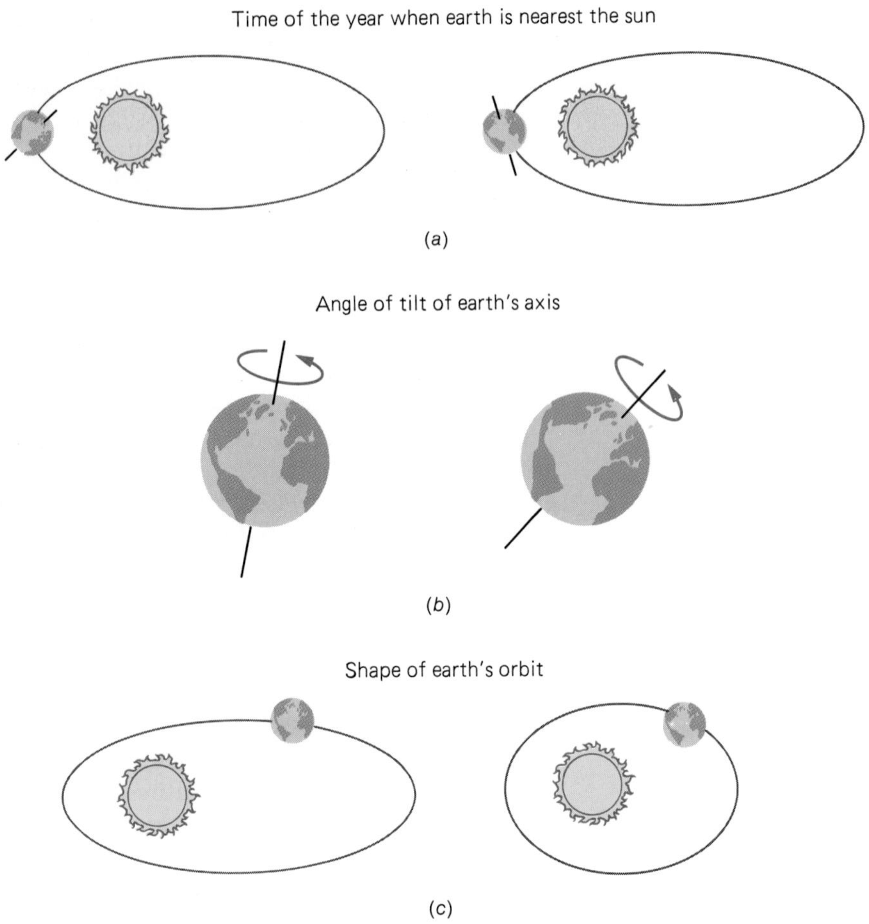

Time of the year when earth is nearest the sun

(a)

Angle of tilt of earth's axis

(b)

Shape of earth's orbit

(c)

FIG. 13-24 Three variations in the earth's motion that may be responsible for causing ice ages. (a) The time of year when the earth is nearest the sun varies with a period of about 23,000 years. (b) The angle of tilt of the earth's axis of rotation varies with a period of about 41,000 years. (c) The shape of the earth's elliptical orbit varies with a period of about 100,000 years. These variations have relatively little effect on the total sunlight reaching the earth but a considerable effect on the sunlight reaching the polar regions in summer.

earth is closest to the sun (Fig. 13-24). Seventy years ago Milutin Milankovitch (1879–1958), a Yugoslav mathematician and physicist, worked out how these changes might affect climate by altering the amount of sunlight (that is, insolation) received by the earth.

The differences in insolation on a global basis are small, 0.3 percent at most, but Milankovitch argued that what really counts is not the total insolation but the insolation in the polar regions in summer, which varies by up to 20 percent. Too little summer sunshine would not melt all the snow that fell during the winter before, and in time the accumulated snow would turn into great sheets of ice. In the southern hemisphere the ice would melt when it left the Antarctic continent and fell into the sea, but in the northern hemisphere the ice would move down across North America and Eurasia to produce an ice age.

Major ice ages have occurred every 100,000 years or so, with smaller cycles of cold and warm at closer intervals. The strongest evidence in favor of Milankovitch's hypothesis is that the periods of advance and retreat of the ice sheets are in accord with the various periods of the earth's orbital variations. Additional support comes from current theoretical models of the earth's climate, which respond to the known insolation changes with a prediction of regular ice ages.

Global Warming

What about the warming trend shown in Fig. 13-23? Apparently this warming results from an enhancement of the greenhouse effect, mainly due to an increase in the carbon dioxide content of the atmosphere. Analyzing air bubbles trapped in Antarctic ice shows that the CO_2 concentration in the atmosphere was approximately constant from 10,000 years ago until the last century. Huge amounts of CO_2 were then being exchanged between plants and the atmosphere and between the atmosphere and the oceans, but always in roughly equal amounts. In recent times the balance seems to have been upset by a rapidly increasing source of CO_2: the burning of coal, oil, and natural gas to heat our homes and to power industry and transportation. Ninety percent of the world's energy comes from carbon-based fuels. Each kilogram of carbon burned yields almost 4 kg of CO_2, and at present our chimneys and exhaust pipes pour out 20 trillion kg of CO_2 per year.

As we can see from Fig. 13-25, the CO_2 content of the atmosphere has gone up by over 20 percent since 1860 and is today increasing at the rate of 0.3 percent per year. This increase represents about half the CO_2 from combustion. The other half is absorbed by plants and by the oceans. As fossil fuels continue to be burned at a high rate, the greenhouse "window" of CO_2 becomes a better trap for heat and the atmosphere will continue to warm up. Other gases released by modern civilization also contribute to the greenhouse effect. The CFCs mentioned in Sec. 13-1 that weaken the ozone layer are especially bad—one molecule of a common CFC is equivalent to 10,000 molecules of CO_2 in its ability to trap heat.

Interestingly enough, the warming due to increased CO_2 has been canceled out to a certain extent in some regions—notably North America, Europe, and China—by the presence of sulfate particles, which are also given off when fossil fuels are burned. These particles form a thin haze that partly blocks sunlight. But sulfate particles stay aloft for only a few days before falling to the ground; hence they do not build up in the atmosphere or get very far from where they are produced. On the other hand, the lifetime in the atmosphere of emitted CO_2 is a century

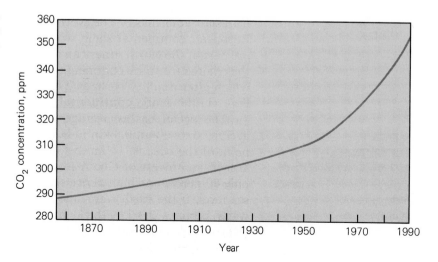

FIG. 13-25 Carbon dioxide concentration in the atmosphere since 1860, in parts per million (ppm).

or more, and as the CO_2 continues to accumulate its warming effect will increase to overwhelm the cooling due to the sulfate particles.

Even with the sulfates, the steady rise in CO_2 levels is increasing global temperatures by about 0.2°C per decade. At this rate, by 2100 the earth's average temperature will be higher than it has been for at least 100,000 years. Exactly how life on earth will be affected can only be guessed at, but it is sure to mean profound changes; most estimates point to an overall drop in food production. We are in the midst of a gigantic experiment in climate modification, one that our descendants may well regret. To stop the experiment means a drastic cut in our use of fossil fuels, not an easy thing to do as we saw in Chap. 3. International conferences have agreed that CO_2 emissions should be reduced, but they continue to rise.

THE HYDROSPHERE

Over 70 percent of the earth's surface is covered by the oceans and the seas that join with them. These waters are filled with plant and animal life—indeed, the early oceans were home to living things for more than a billion years before they began to move ashore. In addition, the oceans influence continental life in a variety of indirect ways. The oceans provide the reservoir from which water is evaporated into the atmosphere, later to fall as rain and snow on land. The oceans participate in the oxygen–carbon dioxide cycle both through the life they support and through the vast quantities of these gases dissolved in them. And the oceans help determine climates by their ability to absorb solar energy and transport it around the world.

13-13   OCEAN BASINS

Water, Water Everywhere   Each of the world's oceans lies in a vast basin bounded by continental land masses. Typically an ocean bottom slopes gradually downward from the shore to a depth of 130 m or so before starting to drop more rapidly (Fig. 13-26). The average width of this **continental shelf** is 65 km, but it ranges from less than a kilometer (off the mountainous western coast of South America) to over 1000 km (off the low arctic coasts of the Eurasian land mass). The North, Irish, and Baltic Seas are part of the European continental shelf, while the Grand Banks off Newfoundland are part of the North American shelf. A sharp change in steepness marks the change from the continental shelf to the steeper **continental slope,** which after a fall of perhaps 2 km joins the **abyssal plain** of the ocean floor via the gentle **continental rise.**

The ocean basins average 3.7 km in depth, while the continents average only about 0.8 km in height above sea level. The deepest known point of the oceans, 11 km below the surface, is found in the Marianas Trench southwest of Guam in the Pacific. By contrast, Mt. Everest is only 80 percent as high above sea level as the Marianas Trench is below sea level. If the earth were smooth, it would be covered with a layer of water perhaps 2.4 km thick, but it seems likely that the oceans have always been confined to more or less distinct basins and presumably will continue to be.

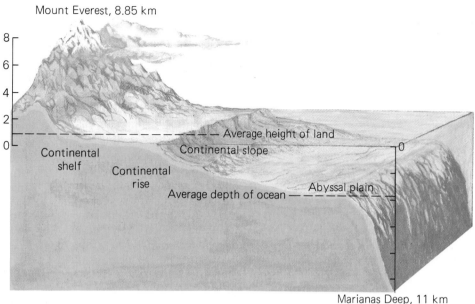

Mount Everest, 8.85 km

Average height of land

Continental shelf

Continental slope

Continental rise

Average depth of ocean

Abyssal plain

Marianas Deep, 11 km

FIG. 13-26 Profile of the earth's surface. The vertical scale is greatly exaggerated. Heights and depths are in km.

The ocean floor, like the continents, has mountain ranges and valleys, isolated volcanic peaks, and vast plains, many of them rivaling or exceeding in size their counterparts on land. The Hawaiian Islands, for instance, are volcanoes that rise as much as 9000 m above the ocean floor, about half of their altitude being above sea level. Less conspicuous from the surface is the Mid-Atlantic Ridge, an immense submarine mountain range that extends from Iceland past the tip of South America before swinging into the Indian Ocean. Such islands as the Azores, Ascension Island, and Tristan da Cunha are all that protrude from the ocean of this ridge.

Seawater and Freshwater

By far the greatest part of the earth's surface water is in the form of seawater, as Fig. 13-3 shows. Much of the freshwater is stored as ice in the form of the glaciers and ice caps that cover one-tenth the land area of the earth. About 90 percent of this ice is located in the Antarctic ice cap, about 9 percent in the Greenland ice cap, and the remaining 1 percent in the various glaciers of the world. If suddenly melted, the ice would raise the sea level by perhaps 75 m. (By comparison, if all the water vapor in the atmosphere were condensed, the sea level would go up by only about 2.5 cm.)

Most of the earth's surface water probably appeared about 4 billion years ago when the young earth assumed its present internal structure. The water came from the rocks of the interior and took with it the same ions found in seawater today: the oceans have always been salty. Additional salts have been added to the oceans in the various ways illustrated in Fig. 10-24. However, seawater salinity has not changed very much because of various mechanisms that remove salt from the oceans. One of these mechanisms is quite direct: the loss of salt to the atmosphere when wind blows spray off wave tops. The resulting salt particles serve

OCEAN WAVES AND TSUNAMIS   Nearly all the waves in the ocean are produced by winds blowing over its surface. The stronger the wind, the longer it blows, and the greater the distance over which the wind has been in contact with the water, the higher the waves that result. These three factors govern the amount of energy transferred to the water and thus determine how violent the resulting disturbance will be.

A **tsunami** consists of ocean waves caused by an earthquake on the seabed. In the open ocean tsunamis may be less than a meter high and, since their crests are many kilometers apart, ships at sea do not notice them. When tsunami waves reach shallow water, however, they steepen in height and can go on to do immense damage on land.

Because the location and magnitude of a major earthquake can be established soon after it occurs, even in midocean, warnings of a possible tsunami can usually be given a few hours before the first wave arrives. Tsunamis used to be called "tidal waves," but since they have nothing to do with tides, scientists now use the Japanese word for them. Nearly all tsunamis occur in the Pacific Ocean.

An earthquake that occurred near Unimak Island, Alaska, on April 1, 1946 gave rise to a tsunami that caused great damage in the Hawaiian Islands 3800 km away. This photograph shows the wave breaking over a pier in the harbor of Hilo, Hawaii, where the maximum rise of water was about 8 m. The man in the foreground was never seen again.

as nuclei for water molecules to stick to, and a substantial amount falls on land in rain and snow. Another mechanism that reduces salinity is the incorporation of various compounds in the shells of marine organisms, which end up in the sediments that coat the ocean floors.

13-14   OCEAN CURRENTS

Four Great Whirlpools   The oceans affect climate in two ways. First, they act as heat reservoirs that moderate seasonal temperature extremes; more heat is stored in the upper few meters of the oceans than in the entire atmosphere. In spring and summer the oceans are cooler than the regions they border, since the insolation they receive is absorbed in a greater volume than on solid, opaque land. The heat stored in the ocean depths means that in fall and winter the oceans are warmer than nearby land areas. Heat flows readily between moving air and water. With enough of a temperature difference, the rate of energy transfer from warm water to cold air (or from warm air to cold water) can exceed the rate at which solar energy arrives at the earth.

Lacking such an adjacent heat reservoir, continental interiors experience lower winter temperatures and higher summer temperatures than those of coastal districts. In Canada, for instance, temperatures in the city of Victoria on the Pacific coast range from an average January minimum of 9°C to an average July maximum of 20°C, whereas in Winnipeg, in the interior, the corresponding figures are −13°C and 27°C.

The second way oceans influence climate is through surface currents, which are produced by the friction of wind on water. Such currents are much slower than movements in the atmosphere, with the fastest normal surface currents having speeds of about 10 km/h.

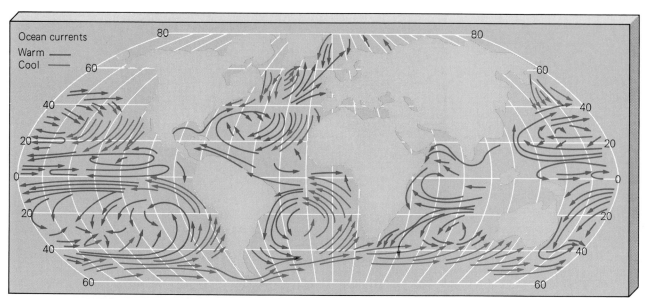

FIG. 13-27 Principal ocean currents of the world.

The wind-driven surface currents parallel to a large extent the major wind systems (Fig. 13-27). The northeast and southeast trade winds drive water before them westward along the equator, forming the **equatorial current.** In the Atlantic Ocean this current runs into South America and in the Pacific the current runs into the East Indies. At each of these contacts the current divides into two parts, one flowing south and the other north. Moving away from the equator along the continental edges, these currents then come under the influence of the westerlies, which drive them eastward across the oceans. Thus gigantic whirlpools are set up in both the Atlantic and the Pacific on either side of the equator. Many complexities are produced in the four great whirls by islands, continental projections, and undersea mountains and valleys.

The Gulf Stream   The western side of the North Atlantic whirl—a warm current moving partly into the Gulf of Mexico, partly straight north along our southeastern coast—is the famous **Gulf Stream.** Forced away from the coast in the latitude of New Jersey by the westerlies, this current moves northeastward across the Atlantic. The current splits on the European side into one part that moves south to complete the whirl and another part that continues northeastward past Great Britain and Norway into the Arctic Ocean. To compensate for the addition of water into the polar sea, the cold **Labrador Current** moves southward along the east coast of North America as far as New York. Down the west coast of North America moves the **Japan Current,** the southward-flowing eastern part of the North Pacific whirl.

Since ocean currents retain the temperatures of the latitudes from which they come for a long time, they exert a direct influence on the temperatures of neighboring lands. The influence is greatest, of course, where prevailing winds blow shoreward from the sea. The warm Gulf Stream has a far greater effect in tempering the climate of northwestern

Surface temperatures of the North Atlantic Ocean are shown in this computer simulation. Temperatures range from about 28°C (dark red) in the tropics to 0°C (dark blue) near Greenland. The ribbon of dark red off the southern part of the American coast is the Gulf Stream, whose continuation across the Atlantic is responsible for the relatively warm water shown in yellow and green.

Europe than that of eastern United States, since prevailing winds in these latitudes are from the west. Cyclonic storms bring east winds to the Atlantic Seaboard often enough, however, for the Gulf Stream to help raise temperatures in the South Atlantic states, and the Labrador Current is in part responsible for the rigorous climate of New England and eastern Canada.

Thus the oceans, besides acting as water reservoirs for the earth's atmosphere, play a direct part in temperature control—both by preventing abrupt temperature changes in lands along their borders and by aiding the winds, through the motion of ocean currents, in their distribution of heat and cold over the surface of the earth.

IMPORTANT TERMS AND IDEAS

The earth's **atmosphere** is its envelope of air, which consists mainly of nitrogen and oxygen. The lowest part of the atmosphere is the **troposphere,** in which such weather phenomena as clouds and storms occur. Next comes the **stratosphere,** whose content of **ozone** (O_3) absorbs most of the ultraviolet radiation from the sun. Above the stratosphere is the **mesosphere** and still higher is the **thermosphere,** which contains most of the ions that make up the **ionosphere.** The ionosphere reflects radio waves.

Air at a given temperature is said to be **saturated** when it contains the maximum amount of water vapor possible without liquid water condensing out. The **relative humidity** of a volume of air is the ratio between the water vapor it contains and the amount that would be present at saturation. Low-altitude **clouds** consist of tiny water droplets; high-altitude clouds consist of ice crystals.

The energy that powers weather phenomena is **insolation,** which is **in**coming **sol**ar radi**ation.** Most of the insolation passes through the atmosphere and is absorbed by the earth's surface. The warm earth then radiates energy back into the atmosphere which absorbs some of it. This indirect heating of the atmosphere is called the **greenhouse effect.**

The **seasons** occur because the earth's axis is tilted, so that in half of each year one hemisphere receives more direct sunlight than the other hemisphere.

Convection currents are due to the uneven heating of a fluid: the warmer parts of the fluid expand and rise while the cooler parts sink. Such currents occur where air in the equatorial regions is heated strongly, expands and rises, and moves toward the poles. In the polar regions the air cools and sinks and then flows back on the surface toward the equator. In this way energy is shifted from the tropics to the higher latitudes.

The **coriolis effect** is the deflection of winds to the right in the northern hemisphere, to the left in the southern, as a consequence of the earth's rotation. Because of the coriolis effect, the convection currents in the atmosphere follow curved paths. A **cyclone** is a weather system centered on a low-pressure region. In the northern hemisphere the coriolis effect deflects winds moving inward in a cyclone into a counterclockwise spiral, in the southern hemisphere into a clockwise spiral. An **anticyclone** is a weather system centered on a high-pressure region; winds blowing outward from an anticyclone spiral clockwise in the northern hemisphere and counterclockwise in the southern hemisphere.

A **front** is the boundary between a mass of warm air and a mass of cold air; clouds and rain usually occur at a front.

Climate refers to averages in weather conditions over a period of years. The **ice ages** were times of severe cold in which ice sheets covered much of Europe and North America. Such long-term climatic changes are probably caused by changes in the earth's motions around the sun. The recent warming of the atmosphere may be due to an increased amount of CO_2 in the atmosphere, which contributes to the greenhouse effect.

EXERCISES: MULTIPLE CHOICE

1. Arrange the following gases in the order of their abundance in the earth's atmosphere:
   a. oxygen   **c.** nitrogen
   b. carbon dioxide   **d.** argon

2. Much of Tibet lies in altitudes above 5.5 km (18,000 ft). At such altitudes the Tibetans are above approximately
   a. 10 percent of the atmosphere
   b. 50 percent of the atmosphere
   c. 90 percent of the atmosphere
   d. 99 percent of the atmosphere

3. The region of the atmosphere closest to the earth's surface is the
   a. thermosphere   **c.** troposphere
   b. ionosphere   **d.** stratosphere

4. A characteristic feature of the stratosphere is the presence of
   a. clouds   **c.** ozone
   b. dust   **d.** ions

5. The atmosphere constituent chiefly responsible for absorbing ultraviolet radiation from the sun is
   a. carbon dioxide   **c.** ozone
   b. water vapor   **d.** helium

6. Chlorofluorocarbon (CFC) gases, which are widely used in refrigeration and in foam plastics, catalyze the breakdown in the upper atmosphere of
   a. water vapor   **c.** oxygen
   b. carbon dioxide   **d.** ozone

7. Dry air has a relative humidity of
   a. 0   **c.** 50 percent
   b. 1 percent   **d.** 100 percent

8. The higher the temperature of a volume of air, the
   a. more water vapor it can hold
   b. less water vapor it can hold
   c. greater its possible relative humidity
   d. lower its possible relative humidity

9. When saturated air is cooled,
   a. it becomes able to take up more water vapor
   b. some of its water content condenses out
   c. the relative humidity goes down
   d. convection currents result

10. Clouds consist of
    a. water droplets at all altitudes
    b. ice crystals at all altitudes
    c. water droplets at low altitudes and ice crystals at high altitudes
    d. ice crystals at low altitudes and water droplets at high altitudes

11. If the atmosphere contained fewer salt crystals and dust particles than it now does,
    a. clouds would form less readily
    b. clouds would form more readily
    c. the formation of clouds would be unaffected
    d. snow would never fall

12. Clouds occur when moist air is cooled by
    a. expansion when it rises
    b. expansion when it falls

c. compression when it rises
d. compression when it falls

13. The atmosphere is heated chiefly by
    a. infrared radiation from the sun
    b. ultraviolet radiation from the sun
    c. infrared radiation from the earth
    d. ultraviolet radiation from the earth

14. The term *greenhouse effect* is used to describe the
    a. condensation of moisture to form dew
    b. conversion of carbon dioxide to oxygen by green plants
    c. heating of the atmosphere by direct solar radiation
    d. heating of the atmosphere by infrared radiation from the earth

15. Energy is transported from the tropics to the polar regions chiefly by
    a. winds   c. carbon dioxide
    b. ocean currents   d. ozone

16. The seasons occur as a result of
    a. variations in the sun's energy output
    b. variations in the distance between the earth and the sun
    c. variations in the orbital speed of the earth
    d. the tilt of the earth's axis

17. Because of the coriolis effect, a wind in the northern hemisphere is deflected
    a. upward   c. toward the right
    b. downward   d. toward the left

18. Because of the coriolis effect, a wind in the southern hemisphere is deflected
    a. upward   c. toward the right
    b. downward   d. toward the left

19. On a summer day sunlight warms coastal land until its temperature is higher than that of the adjacent sea. The result is a "sea breeze" that blows
    a. from sea to land
    b. from land to sea
    c. parallel to the shore
    d. any of the above, depending on the part of the world

20. The flow of air in the upper atmosphere is largely from
    a. east to west
    b. west to east
    c. north to south
    d. south to north

21. The middle latitudes usually experience winds from the

a. north   c. east
b. south   d. west

22. The general direction of the trade winds in both hemispheres is from the
    a. north   c. east
    b. south   d. west

23. An airplane flies at the same speed relative to the air at a high altitude from New York to Paris and back.
    a. The New York to Paris flight takes less time.
    b. The Paris to New York flight takes less time.
    c. The two flights take the same time on the average.
    d. Any of the above, depending on the season.

24. The trade-wind belts are regions of generally
    a. little rainfall   c. low temperatures
    b. much rainfall   d. westerly winds

25. A cyclone is a weather system centered about a
    a. region of low pressure
    b. region of high pressure
    c. hurricane
    d. cold front

26. The winds in an anticyclone
    a. blow directly toward its center
    b. spiral toward its center
    c. blow directly away from its center
    d. spiral away from its center

27. Unstable weather is associated with
    a. cyclones
    b. anticyclones
    c. trade winds
    d. the greenhouse effect

28. The most pleasant weather is found in
    a. a cold front   c. a cyclone
    b. a warm front   d. an anticyclone

29. The chief reason why the equatorial regions are warmer than the polar regions is that
    a. the equator is closer to the sun
    b. sunlight falls more nearly vertically on the equatorial regions
    c. sunlight is reflected by ice and snow in the polar regions
    d. there is more CO_2 in the air over the equatorial regions

30. The greatest seasonal variations in temperature occur in
    a. the west coasts of the continents
    b. the east coasts of the continents
    c. continental interiors
    d. isolated islands

31. Rain is most abundant in the
    a. prevailing westerlies of the middle latitudes
    b. horse latitudes between the middle latitudes and the trade-wind belts
    c. trade-wind belts
    d. doldrums of the equatorial regions

32. Ice ages
    a. cover the entire earth with a sheet of ice
    b. freeze all the oceans
    c. occurred seldom in the past
    d. occurred frequently in the past

33. The approximate percentage of the earth's surface covered by water is
    a. 10 percent   **c.** 70 percent
    b. 50 percent   **d.** 90 percent

34. Compared with the average height of the continents above sea level, the average depth of the ocean basins below sea level is
    a. smaller
    b. greater
    c. about the same
    d. sometimes smaller and sometimes greater, depending upon the tides

35. The deepest known point of the oceans is found in the
    a. Atlantic Ocean   **c.** North Sea
    b. Pacific Ocean   **d.** Panama Canal

36. The Hawaiian Islands are
    a. part of a sunken continent
    b. floating on the surface of the ocean
    c. located in shallow water
    d. volcanic peaks

37. Tsunamis are caused by
    a. monsoons
    b. typhoons
    c. icebergs
    d. undersea earthquakes

38. Most surface ocean currents are due to
    a. melting glaciers
    b. rivers
    c. winds
    d. differences in the altitude of the ocean surface

39. The Gulf Stream is a
    a. warm current in the North Atlantic
    b. cool current in the North Atlantic
    c. river that flows into the Gulf of Mexico
    d. warm wind blowing over the North Atlantic

40. The climate of northwestern Europe is greatly affected by the

    a. Gulf Stream   **c.** Labrador Current
    b. abyssal zone   **d.** trade winds

QUESTIONS

1. In what two important processes does the carbon dioxide in the atmosphere have an essential role?

2. Suppose you are climbing in an airplane that has no altimeter. How could you tell when you are approaching the top of the troposphere?

3. The tropopause, stratopause, and mesopause are respectively the upper boundaries of the troposphere, stratosphere, and mesosphere. What is characteristic of the air temperature at each of these boundaries?

4. What would happen if ozone were to disappear from the upper atmosphere?

5. Why are chlorofluorocarbon (CFC) gases, which are widely used in refrigeration and in foam plastics, harmful when released into the atmosphere?

6. What causes ionization to occur in the upper atmosphere?

7. What does Fig. 13-3 tell us about the relative amounts of freshwater present on or near the earth's surface as groundwater, in lakes and rivers, in ice caps and glaciers, and as moisture in the atmosphere? How does the total amount of freshwater compare with the amount of seawater?

8. What does it mean to say that a certain volume of air has a relative humidity of 50 percent? Of 100 percent?

9. The air in a closed container is saturated with water vapor at 20°C. (a) What is the relative humidity? (b) What happens to the relative humidity if the temperature is reduced to 10°C? (c) If the temperature is increased to 30°C?

10. Why does the air in a heated room tend to be dry?

11. Why does dew form on the ground during clear, calm summer nights?

12. What do high-altitude clouds consist of? Low-altitude clouds?

13. Cumulus clouds form when warm air rises vertically by convection, and stratus clouds form when a warm air mass moving horizontally encounters a cooler mass and is forced upward on top of the cooler mass. Which kind of clouds would you expect to be characteristic of a warm front?

14. How do dust and salt particles in the atmosphere affect weather?

15. From time to time a gigantic volcanic explosion sends a large amount of dust into the atmosphere, where it may remain for some years. How many consequences of such an event can you think of?

16. What is insolation? How does it affect the atmosphere?

17. What is the greenhouse effect and how is it related to the absorption of solar energy by the earth's atmosphere?

18. Why do you think temperatures in the troposphere decrease with altitude?

19. What is the chief source of atmospheric heat?

20. Why does air near the equator average higher temperatures than air near the poles?

21. What are the two mechanisms by which energy of solar origin is transported around the earth? Which is more important?

22. Account for the abrupt changes in temperature between day and night in desert regions.

23. In the northern hemisphere, the longest day is in June and the shortest day is in December, but the warmest weather occurs in July and August and the coldest weather in January and February. What is the reason for these time lags?

24. If the earth's axis were tilted less than it is today, what would be the effect on the difference in average temperature between summer and winter?

25. How many times per year is the noon sun directly overhead at the equator?

26. What is the direction of the prevailing winds of the middle latitudes in each hemisphere?

27. Where in the atmosphere do the jet streams occur? What is their general direction?

28. A crew is planning to sail from the United States to England and later to sail home. What routes should they follow across the Atlantic on the trips out and back in order to have their course downwind as much of the time as possible?

29. An airplane flies at the same speed relative to the air at a high altitude from Chicago to London and then back to Chicago. How do the flight times for each leg compare?

30. A wind in the northern hemisphere begins to blow from the north. In what approximate direction is the wind deflected because of the coriolis effect?

31. What is the name of weather systems centered about regions of high pressure? In what direction do winds in the northern hemisphere spiral around such a region? In the southern hemisphere?

32. Why are clouds and rain more likely to be associated with a cyclone than with an anticyclone?

33. How does the weather associated with a cyclone differ from that associated with an anticyclone?

34. What is the usual lifetime of a cyclone in the middle latitudes?

35. When you face a wind associated with a cyclone in the northern hemisphere, in what approximate direction will the center of low pressure be? In what direction will the center of low pressure be if you do this in the southern hemisphere?

36. What is the approximate sequence of wind directions when the center of a cyclone passes north of an observer in the northern hemisphere?

37. What is the approximate sequence of wind directions when the center of an anticyclone passes south of an observer in the northern hemisphere?

38. What is the difference between the rainfall that accompanies the passage of a warm front and that which accompanies the passage of a cold front?

39. The salinity of seawater varies with location, but the relative proportions of the various ions in solution are almost exactly the same everywhere regardless of local circumstances. What is the significance of the latter observation?

40. How does the average depth of the ocean basins below sea level compare with the average height of the continents above sea level?

41. A wind begins to blow over the surface of a calm body of deep water. What factors govern the height of the waves that are produced?

42. The giant whirls of the oceans involve clockwise

flows in the northern hemisphere and counter-clockwise flows in the southern. Why?

43. (a) If you were planning to drift in a raft across the North Atlantic from the United States to Europe by making use of ocean currents, what would your route be? (b) If you were planning to drift from Europe to the United States, what would your route be?

44. England and Labrador are at about the same latitude on either side of the North Atlantic Ocean, but England is considerably warmer than Labrador on the average. Why?

45. The California Current along the California coast is cooler than the ocean to the west. How does this fact explain the numerous fogs on this coast?

46. The island of Oahu (one of the Hawaiian Islands) is at latitude 21°N and is crossed by a mountain range trending roughly northwest to southeast. Account for the more abundant rainfall on the northeastern side of the range.

47. The northeast and southeast trade winds meet in a belt called the doldrums. What is the characteristic climate of the doldrums and why does it occur?

48. Why are most of the world's deserts found in the horse latitudes, which separate the trade winds from the prevailing westerlies in both hemispheres?

49. The Milankovitch theory of ice ages relates them to variations in the tilt of the earth's axis, the shape of its orbit, and the time of year when the earth is closest to the sun. However, these variations affect the total amount of solar energy

reaching the earth by no more than 0.3 percent. How did Milankovitch account for this apparent contradiction?

50. Water vapor is the most important "greenhouse gas" in the atmosphere, yet more attention is paid in this respect to carbon dioxide. Why do you think this is so?

51. What has kept certain regions of the earth from sharing fully in global warming?

52. Why is it believed that seawater has always been salty?

53. Why does the equatorial current flow westward?

54. Why does the Gulf Stream affect climate in Europe to a greater extent than climate in North America even though it flows close to the eastern coast of North America?

ANSWERS TO MULTIPLE CHOICE			
1. c, a, d, b	11. a	21. d	31. d
2. b	12. a	22. c	32. d
3. c	13. c	23. a	33. c
4. c	14. d	24. a	34. b
5. c	15. a	25. a	35. b
6. d	16. d	26. d	36. d
7. a	17. c	27. a	37. d
8. a	18. d	28. d	38. c
9. b	19. a	29. b	39. a
10. c	20. b	30. c	40. a

14

THE

ROCK

CYCLE

Soil, vegetation, and rock fragments form a thin surface layer on most land areas, but solid rock is always underneath. Rock underlies the sediments on the ocean floors as well. The deepest oil wells, which go down over 8 km, are drilled through rock similar to that at the surface. Some of the rock now out in the open was once buried several km inside the earth, and the material that makes up some volcanic rock probably rose in molten form from still greater depths, perhaps as much as 100 km down. These samples of rock from well below the earth's surface also turn out to be very much like rock that formed close to the surface.

Such direct observation tells us that the outer part of the earth, called its **crust,** is composed almost entirely of rock. However, the thickness of the crust is only 0.5 percent of the earth's 6400-km radius. There is no firsthand information about the rest of our planet, but its interior can be probed by indirect methods. After we have learned something about the structure of our planet and about the rocks that clothe it, we shall turn to the processes whose action has produced the landscapes around us.

14-1 COMPOSITION OF THE CRUST

Oxygen and Silicon Are the Most Abundant Elements

The average composition of the earth's crust is shown in Fig. 14-1. Only a few elements are abundant in the crust, while others are present in quite small amounts. Oxygen makes up nearly half the mass of the crust, most of it combined with silicon. Silicon and the two metals iron and aluminum account for three-fourths of the rest of the crust's mass. Lumped together in the 1.4 percent of "all others" are the carbon, hydrogen, and nitrogen present in all living things and such familiar metals as copper, lead, and silver.

Silicon never occurs by itself in nature, but its compounds make up about 87 percent of the rock and soil of the earth's crust. In the chemistry of the earth, silicon has the same sort of central role that carbon has in the chemistry of living things.

FIG. 14-1 Average chemical composition of the earth's crust. Percentages are by mass.

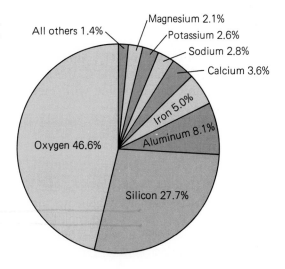

443

Asbestos is a fibrous mineral whose silicon-oxygen units are arranged in chains. Asbestos was once widely used as insulation in buildings and around steam pipes, for fireproof theater curtains, and in brake linings. When inhaled, asbestos fibers can cause serious lung diseases such as cancer. For many years after their discovery, the dangers of asbestos were deliberately hidden from the general public. It is believed that hundreds of thousands of people around the world have died prematurely or will do so as a result of breathing in asbestos fibers at work or in their daily lives.

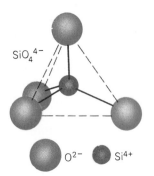

FIG. 14-2 The silicon-oxygen tetrahedron is the fundamental unit in all silicate structures. Dashed lines show the tetrahedral form of SiO_4^{4-}; solid lines are bonds between the ions.

Nearly all the earth's silicon is combined either with oxygen in silicon dioxide (SiO_2), sometimes called **silica,** or with oxygen and one or more metals in the **silicates.** The differences in composition and structures of the silicates are reflected in a variety of colors, hardnesses, and crystal forms. The softness of talc and the hardness of zircon and beryl, the transparency of topaz and the deep color of garnet, the platy crystals of mica and the fibrous crystals of asbestos give some idea of the range of silicate properties. Glass is a mixture of silicates.

The basic structural unit of all silicates is the SiO_4^{4-} tetrahedron (a pyramid with a triangular base) shown in Fig. 14-2. In some silicates these units occur as single ions linked by positive metal ions. In more complex silicates the units form continuous chains, as in asbestos, or sheets, as in mica, with metal ions lying between them. Three-dimensional networks of SiO_4^{4-} units also occur. The number and variety of silicate minerals are due to the many different ways in which the basic SiO_4^{4-} unit can combine with metal ions to form stable crystal structures.

14-2   MINERALS

What Rocks Are Made Of   Most rocks are heterogeneous solids. The different homogeneous materials, called **minerals,** in a coarse-grained rock like granite are obvious to the eye. For a fine-grained rock, a microscope may be needed to distinguish the minerals that compose it.

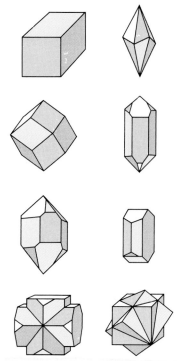

FIG. 14-3 Some crystal forms found in minerals.

A mineral is a crystalline inorganic solid found in nature that has a fairly specific chemical composition. Minerals are not usually identified by chemical names for two reasons. First, the same compound may occur in different forms: for instance, the minerals calcite and aragonite are both largely calcium carbonate ($CaCO_3$), but they differ in crystal form, hardness, density, and so on. Second, most minerals vary somewhat in composition from sample to sample, whereas the compositions of chemical compounds are invariable. More than 2000 minerals are known, but the majority are rare. The number of minerals important in ordinary rocks is so small that knowing something about only half a dozen is enough for an introduction to geology.

Crystal Form and Cleavage

In describing minerals, two important properties are **crystal form** and **cleavage.** Minerals are crystalline solids, which means that their atoms are arranged in lattice structures with definite geometric patterns. When a mineral hardens in a location where its crystals can grow freely, as in a cavity, perfect crystals are formed with smooth faces that meet each other at sharp angles. Each mineral has crystals of a distinctive shape, so that well-formed crystals make it easy to recognize a mineral (Fig. 14-3). Unfortunately good mineral crystals are rare, since neighboring crystals usually interfere with one another's growth.

When well-developed crystals are not present, the characteristic crystal structure of a mineral may still reveal itself in the property called cleavage. This is the tendency of a substance to split along certain planes, which are determined by the arrangement of atoms in its lattice. When a mineral sample is struck with a hammer, its cleavage planes are revealed as the preferred directions in which it breaks.

Even in its original state, a mineral may show cleavage by flat, parallel faces and minute parallel cracks. The flat surfaces of mica flakes, for instance, and the ability of mica to be peeled apart in thin sheets show that this mineral has almost perfect cleavage in one direction (Fig. 14-4). Some minerals (for example, quartz) have no cleavage at all. When struck they shatter, like glass, along random curved surfaces.

Six Common Minerals

Here are details of six common minerals and mineral groups.

Quartz Well-formed quartz (SiO_2) crystals are six-sided prisms and pyramids that show no cleavage. They are colorless or milky (often gray, pink, or violet because of impurities), have a glassy luster, and are hard enough to scratch glass. They occur in many kinds of rock, sometimes appear as long, narrow deposits called **veins,** and often form

FIG. 14-4 (a) Mica has one direction of cleavage and fractures irregularly if broken across its cleavage plane. (b) Feldspar has two perpendicular cleavage planes and fractures irregularly if broken across them. (c) Calcite has three directions of cleavage that are not perpendicular to each other.

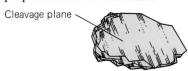

Cleavage plane

(a)

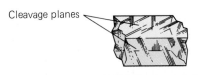

Cleavage planes

(b)

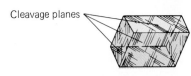

Cleavage planes

(c)

Quartz consists of silicon dioxide crystals and is found in many kinds of rock.

Feldspar is the most abundant mineral in the earth's crust. Shown is a sample of orthoclase feldspar.

assemblies of crystals inside cavities. Clear quartz (rock crystal) is used in jewelry and in optical instruments. Smoky quartz, rose quartz, and amethyst are colored varieties used in jewelry. Quartz sand is the chief raw material for making glass.

Feldspar This is the name of a group of minerals with very similar properties. The two classes of feldspar are a silicate of K and Al called orthoclase and a series of silicates of Na, Ca, and Al collectively called plagioclase. The crystals are rectangular, with blunt ends; they show good cleavage in two directions approximately at right angles. They are sometimes clear; if not, their color is white or light shades of gray and pink. Feldspar is slightly harder than glass but not as hard as quartz. It is the most abundant single constituent of rocks, making up about 60 percent of the total weight of the earth's crust. Pure feldspar is used in the making of porcelain and as a mild abrasive.

The micas are aluminum silicates having a sheet-type cleavage. As fine flakes, mica is the "shiny" mineral in some metamorphic rocks. Shown is white mica.

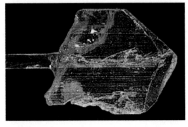

Mica The two chief varieties of this familiar mineral are white mica—a silicate of H, K, and Al—and black mica—a silicate of H, K, Al, Mg, and Fe. Mica is easily recognized by its perfect and conspicuous cleavage in one plane. It is a very soft mineral, only a trifle harder than a fingernail. Large sheets of white mica free from impurities are used as insulators in electrical equipment.

Olivine is an olive-green ferromagnesian mineral (a magnesium-iron silicate) that occurs mainly in igneous rocks.

Ferromagnesian Minerals This name refers to a large group of minerals with varied properties. All of them are silicates of iron and magnesium, and most of them contain other metallic elements as well (for instance, calcium). Black mica belongs to this group; its composition includes H, K, and Al in addition to Fe and Mg. Nearly all ferromagnesian minerals are dark green to black, but apart from composition and color the members of this group differ greatly from one another. The most abundant dark-colored constituents of common rocks belong to this group.

Clay Minerals This is a group of closely related minerals that are the chief constituents of clay. All are silicates of aluminum, some with a

Water molecules fit readily into the layered structures of clay minerals such as kaolin, shown here. When baked, wet clay loses its water content and becomes hard due to the formation of silicates that bind together the clay particles. The transformation of clay from a soft, easily shaped material into a permanent one is the basis of such ceramic products as bricks, pottery, and porcelain.

little Mg, Fe, and K. They consist of microscopic crystals, white or light-colored when pure, often discolored with iron compounds. They have a dull luster and are very soft, forming a smooth powder when rubbed between the fingers. Clay minerals have a low density and absorb water readily. They are distinguished from chalk by softness. Kaolin, one of the clay minerals, is an important ingredient in the manufacture of ceramics, paper, paint, and certain plastics.

Calcite, the chief constituent of limestone and marble, consists of calcium carbonate crystals.

Calcite. Calcite ($CaCO_3$) crystals are hexagonal, somewhat like those of quartz. Unlike quartz they show perfect cleavage in three directions at angles of about 75°, so that fragments of calcite have a characteristic rhombic shape. They are colorless or light in color, with a glassy luster. They are hard enough to scratch mica or a fingernail, but can be scratched by glass or by a knife blade. Like quartz, calcite is a common mineral of veins and crystal aggregates in cavities. It is the chief constituent of the common rocks limestone and marble and commercially serves as a source of lime for glass, mortar, and cement. Eggshells are mainly calcite, incidentally.

14-3   IGNEOUS ROCKS

Once Molten, Now Solid   There seems no limit to the variety of rocks on the earth's surface. We find coarse-grained rocks and fine-grained rocks, light rocks and heavy rocks, soft rocks and hard rocks, rocks of all sizes, shapes, and colors. But if we look closely, we will find order in this diversity, and a straightforward scheme for classifying rocks according to their origin has been developed.

→ 1. **Igneous rocks** are rocks that have cooled from a molten state. The formation of some igneous rocks can actually be observed when molten lava cools on the side of a volcano. For others an igneous origin inside the earth is inferred from their composition and structure. Two-thirds of crustal rocks are igneous.

→ 2. **Sedimentary rocks** consist of materials derived from other rocks and deposited by water, wind, or glacial ice. Some consist of separate rock fragments cemented together; others contain material precipitated from solution in water. Although sedimentary rocks make up only about 8 percent of the crust, three-quarters of surface rocks are of this kind.

3. **Metamorphic rocks** are igneous or sedimentary rocks that have been changed, or metamorphosed, by heat and pressure deep under the earth's surface. The changes may involve the formation of new minerals or simply the recrystallization of minerals already present.

Obsidian is a glassy rock of volcanic origin.

Properties of Igneous Rocks

The minerals in igneous rocks usually appear in the form of irregular grains that consist of interlocking crystals. This is to be expected when crystals grow together and interfere with one another's development. The principal minerals of these rocks contain silicon: quartz, feldspar, mica, and the ferromagnesians.

The siliceous liquids from which igneous rocks form are thick and viscous, much like molten glass. Sometimes, in fact, molten lava has the right composition and cools rapidly enough to form a natural glass—the black, shiny rock called **obsidian.**

More often cooling is slow enough to allow mineral crystals to form. If cooling is fairly rapid and if the molten material is highly viscous, the resulting rock may consist of tiny crystals or partly of crystals and partly of glass. If cooling is very slow, mineral crystals grow large and a coarse-grained rock is formed. Table 14-1 lists some common igneous rocks according to composition and grain size.

Grain size usually tells us not only the rate of cooling but also the environment in which a rock was cooled. Sufficiently fast cooling to give fine-grained rocks is common when molten lava reaches the earth's surface from a volcano and spreads out in a thin flow. Coarse-grained rocks, on the other hand, have cooled slowly, which must have occurred well beneath the surface. Such rocks are now exposed only because erosion has carried away the material that once covered them.

The faces of four American presidents (Washington, Jefferson, Theodore Roosevelt, and Lincoln) are carved in the granite of Mt. Rushmore, South Dakota. Granite is a coarse-grained igneous rock, generally light in color, in which quartz and feldspar are abundant.

Basalt is a dark, fine-grained rock that emerged molten from the earth's interior and hardened on the surface. Shown is pahoehoe basalt whose smooth, ropy surface resulted from the rapid cooling of a very fluid lava.

TABLE 14-1

Some Igneous Rocks

Mineral Composition	Coarse-Grained Rocks	Fine-Grained Rocks
Quartz Feldspar Ferromagnesian minerals	Granite	Rhyolite
No quartz Feldspar predominant Ferromagnesian minerals	Diorite	Andesite
No quartz Feldspar Ferromagnesian minerals predominant	Gabbro	Basalt

14-4 SEDIMENTARY ROCKS

Compacted Sediments or Precipitates from Solution

Sediments laid down by wind, water, or ice can become rock through the pressure of overlying deposits and by the gradual cementing of their grains with material deposited from underground water. The resulting rocks usually have distinct, somewhat rounded grains that have not grown together like the crystals of igneous rocks. A few sedimentary rocks, however, consist of intergrowing mineral grains that precipitated from solution in water.

Since sediments are normally deposited in layers, most sedimentary rocks have a banded appearance owing to slight differences in color or grain size from one layer to the next. Sedimentary rocks may often be recognized at a glance by the presence of fossils—remains of plants or animals buried with the sediments as they were laid down.

TABLE 14-2

Some Sedimentary Rocks

Group	Type	Constituents
Fragmental rocks	Conglomerate	Rock fragments
	Sandstone	Quartz usually most abundant
	Shale	Clay minerals
Chemical and biochemical precipitates	Chert	Microcrystalline quartz
	Limestone	Calcite

Sedimentary rocks may be divided into two groups according to the nature of the original sediments, as in Table 14-2. The three **fragmental rocks** are distinguished by their grain size. **Conglomerate** is cemented gravel whose fragments may have any composition and any size, from pebbles to boulders. Conglomerate becomes **sandstone** as fragment size decreases. Sand grains may consist of many different minerals, but quartz is generally the most abundant. The hardness of sandstone and conglomerate depends largely on how well their grains are cemented together. Some varieties crumble easily. Others, especially those with silica as the cementing material, are among the toughest of rocks. **Shale** is consolidated mud or silt; it is a soft rock usually composed of thin layers.

Limestone is a fine-grained rock that consists chiefly of calcite. It may be formed either as a chemical precipitate or by the consolidation

An outcrop of conglomerate in New York State that contains many fragments of limestone. The large chunk is about 30 cm long.

Shale is a soft sedimentary rock that has consolidated from mud deposits. These shale layers in the Hudson River Valley of New York were formed over 400 million years ago.

Limestone deposit in western Texas. Some limestones originate as precipitates from solution; others have consolidated from the shells of marine organisms.

CORAL REEFS   A   common feature of shallow waters in the tropics is limestone coral reefs. These reefs are produced by tiny creatures called coral polyps, which extract calcium carbonate from seawater to form hard external skeletons. When the polyps die, their skeletons remain behind, with new polyps growing on top. Coral reefs harbor plants and animals of many kinds, a profusion of life comparable with that in a rain forest. The Great Barrier Reef off Australia's east coast is a series of coral reefs that extends for about 2000 km, and Bermuda, the Bahamas, and the Florida Keys are coral islands. The Pacific Ocean contains many atolls, which are rings or horseshoes of coral reefs that enclose lagoons. Limestone deposits that were once coral reefs are found in Wisconsin, Illinois, Indiana, and Texas.

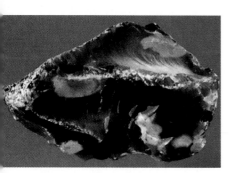

Flint, a type of chert, consists largely of microcrystalline quartz and hence is hard and durable. Many Native American arrowheads were made from flint.

of shell fragments. Small amounts of impurities may give limestone almost any color. **Chalk** is a loosely consolidated variety of limestone, often made up mainly of the shells of tiny one-celled animals.

Many Native American arrowheads are made of either the igneous rock obsidian or the sedimentary rock **chert** (microcrystalline quartz). Both rocks are hard and have sharp edges when broken. Two familiar varieties of chert are **flint** and **jasper.** Fragments of chert show the same sharp edges and smooth, concave surfaces as broken quartz or obsidian, but the surfaces have a characteristic waxy luster. Chert may have almost any color; often a single specimen shows bands and pockets of several different colors. Not nearly as abundant as the other sedimentary rocks just described, chert is nevertheless common in pebble beds and gravel deposits because its hardness and resistance to chemical decay enable it to survive rough treatment from streams, waves, and glaciers.

14-5   METAMORPHIC ROCKS

Formed from Other Rocks by Heat and/or Pressure

The enormous pressures and high temperatures below the earth's surface can profoundly change sedimentary and igneous rocks that become deeply buried. Minerals stable at the surface are often unstable when crushed and baked and may react to form different substances. Other minerals keep their identities but their crystals increase in size. Hot liquids may add some new materials and dissolve out others. So many kinds of change are possible that no general rules can be set down to tell metamorphic rocks from others.

Many metamorphic rocks are characterized by a property called **foliation,** which means the arrangement of flat or elongated mineral grains in parallel layers. This effect is caused by extreme pressure in one direction, with the mineral grains growing out sideward as the rock is

Slate results from the metamorphism of shale under pressure. This scene is in southern New York State.

The metamorphic rock gneiss is foliated and has bands of different material.

squeezed. Foliation gives a rock a banded or layered appearance, and when it is broken, the rock tends to split along the bands. Layering is also characteristic of sedimentary rocks, but in them the layering is caused by slight variations in color or grain size; layering in metamorphic rocks is due to the lining up of mineral grains.

Some Common Metamorphic Rocks

The commoner metamorphic rocks may be classified according to the presence or absence of foliation, as in Table 14-3.

Slate is produced by the low-temperature metamorphism of the sedimentary rock shale, whose clay minerals form tiny flakes of mica. Although the individual flakes are too small to be seen, mica is responsible for the shiny surfaces seen whenever slate is split along its foliation. Slate is harder than shale, finely foliated, and usually black or dark gray but sometimes light-colored.

Schist is formed from shale at higher temperatures than those that give slate, or from fine-grained igneous rocks. In it the mineral grains responsible for the foliation are large enough to be visible, giving the foliation surfaces a characteristic spangled appearance. Schist does not split as easily along the foliation as slate does, and its surfaces are rougher.

Gneiss is a coarse-grained rock formed under conditions of high temperature and pressure from almost any other rock except pure limestone and pure quartz sandstone. Its composition naturally depends on the nature of the original rock, but quartz, feldspar, and mica are the commonest minerals. In appearance gneiss resembles granite, except for its banding and foliation.

The metamorphism of pure limestone and of pure quartz sandstone are relatively simple processes. Since each consists of a single mineral of simple composition, heat and pressure can produce no new substances but instead cause the growth and interlocking of crystals of calcite and quartz. Thus limestone becomes **marble,** a rock composed of calcite in crystals large enough to be easily visible, and sandstone becomes the hard rock **quartzite.** Quartzite sometimes looks like sandstone, but its grains are so firmly intergrown that it splits across separate grains when

TABLE 14-3

Some Metamorphic Rocks			
Group	**Type**	**Constituents**	**Origin**
Foliated rocks	Slate	Mica and usually quartz, both in microscopic grains	Shale
	Schist	Mica and/or a ferromagnesian mineral, usually quartz also	Shale or fine-grained igneous rock
Foliated and banded rocks	Gneiss	Quartz, feldspar, mica	Various
Unfoliated rocks	Marble Quartzite	Chiefly calcite Chiefly quartz	Limestone Sandstone

The metamorphic rock marble is often used for statues, such as this one of a warrior on the Capitol Building, Washington, D.C.

broken to give smooth fracture surfaces in contrast to the rough surfaces of sandstone.

WITHIN THE EARTH

The earth's solid crust together with the atmosphere and oceans above is directly accessible to our instruments, and we may legitimately hope one day to understand their structures and behavior in detail. The interior of the earth, however, is beyond our direct reach. What we need to study it is some kind of indirect probe, and the waves sent out by earthquakes have turned out to be ideal for this purpose. Largely through the analysis of earthquake waves we now know a great deal about the earth's interior, which is hardly less remote than the most distant star, and we are continually learning more.

14-6   EARTHQUAKES

When Our Planet Trembles

An **earthquake,** the most destructive of natural phenomena, consists of rapid vibrations of rock near the earth's surface. A single shock usually lasts no more than a few seconds, though severe quakes may last for as much as 3 min. Even in such brief times the damage done may be immense. Widespread fires often follow earthquakes in inhabited regions since broken water mains hinder their control, and landslides are common. Usually the first shock is the most severe, with weaker and weaker disturbances following from time to time for days or months

THE RICHTER SCALE Earthquake magnitudes are often expressed on the **Richter scale.** Each step of 1 on this scale represents an increase in vibration amplitude of a factor of 10 and an increase in energy release of a factor of about 30. Thus an earthquake of magnitude 5 produces vibrations 10 times larger than a quake of magnitude 4 and releases 30 times more energy. An earthquake of magnitude 0 is barely detectable; if the energy given off by such an earthquake could be concentrated, it would be just about enough to blow up a tree stump. An earthquake of magnitude 3 would be felt by people living near the location of the quake, and some damage to structures would occur when the magnitude is 4 or 5. Significant destruction is likely if the magnitude is 6 or more. The energy given off in a magnitude 8.7 earthquake, the strongest observed thus far, is about double the energy content of the world's yearly production of coal and oil. The Alaska earthquake of 1964 was nearly this severe, the San Francisco earthquake of 1906 somewhat less so (Fig. 14-5).

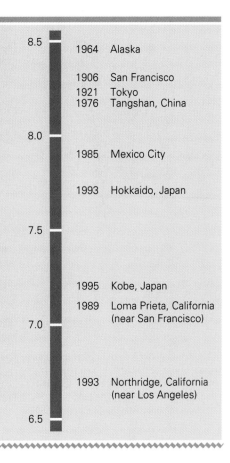

FIG. 14-5 Some major earthquakes and their magnitudes on the Richter scale. The San Francisco and Tokyo earthquakes, together with the fires that followed them, led to the almost complete destruction of those cities. Over a quarter of a million people died in the Tangshan earthquake.

afterward. A major earthquake may be felt over many thousands of square kilometers, but its destructiveness is limited to a much smaller area.

Of the million or so earthquakes per year strong enough to be noticed, only a few liberate enough energy to do serious damage. About

FIG. 14-6 How an earthquake occurs. (*a*) The sides of a fault between crustal blocks have stuck together, preventing movement even though the forces (arrows) on them remain active. (*b*) As time goes on, stresses build up in the adjacent blocks, which deform as a result. (*c*) When the locked-in stresses become too great, the blocks suddenly shift to release them. The stored-up elastic energy powers the vibrations that constitute an earthquake. The **focus** of an earthquake is the place where the crustal blocks moved; the **epicenter** of the quake is the point on the surface directly over the focus.

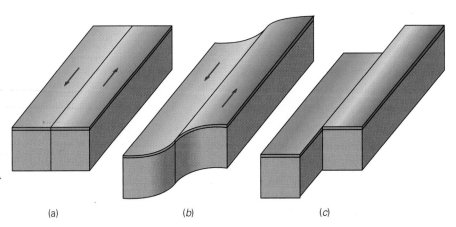

(a)   (b)   (c)

A GEOLOGIST AT WORK

Andrea Donnellan (Jet Propulsion Laboratory and Seismological Laboratory, CalTech). I am a geologist who uses the Global Positioning System (GPS) [see Fig. 2-30], a satellite navigation system, to monitor motions of the earth's crust and to learn more about tectonic and earthquake processes. For the first time we can actually measure slow quiet motions of the earth's crust that occur as a result of plate tectonics. While seismometers detect earthquake vibrations, GPS can be used to measure strain buildup before earthquakes as well as motions that occur as a result of and after earthquakes. I particularly enjoy my work because it encompasses a wide variety of skills ranging from technical field work to mathematics and computer skills.

I may start my day by driving into the field and installing a GPS receiver at a particular site. Sometimes this involves carrying the receiver and antenna up a hill to its location where it is set up precisely over a certain mark. The movements of the earth's crust that we measure are so tiny that we must set the antenna up over the same mark to within a millimeter. The receiver is then left to track for several days while I return to the office to analyze data. We must secure the antenna and receiver carefully so that the antenna does not move during the survey. We've learned that cows like to scratch their backs on the antenna and that bears are quite curious about the receiver box!

In analyzing the GPS data I compare precise positions of each station with positions found six months to one year earlier and see if the station has moved. Between earthquakes the stations may move as little as just a few millimeters per year. When an earthquake occurs, however, a station may move significantly. Following the Northridge earthquake one of my stations moved upward 40 centimeters, indicating that the mountain on which the station was located had risen. If the stations are located close to the epicenter of an earthquake they continue to move rapidly for up to a few years following the earthquake. When we find that a station moves we must first rule out that it shifted locally (for example, slid down a hill) or that there was a problem with the data analysis that produced an incorrect position that implied station movement.

As the stations move we can calculate their velocities and then the amount of strain buildup in the earth's crust. From this information we can begin in construct models of faults that might be producing the surface movements. We try to determine which faults are responsible for the motions and how much they may be moving. We then estimate how large an earthquake a particular fault may produce. Very little is known about the earthquake process, so with data collected following earthquakes we hope to learn more about this process and how the earth's crust is composed and deforms.

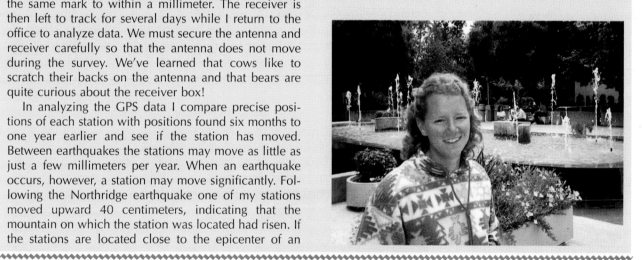

15 really violent earthquakes occur each year on the average. When one of them happens to involve a densely populated urban area, the effects can be appalling: over 250,000 people died in the 1976 earthquake in Tangshan, China, as buildings collapsed around them.

The great majority of earthquakes are caused by the sudden movement of large blocks of the earth's crust less than 70 km from the surface along fracture lines called **faults.** When the stresses that develop in the crust in a certain region become too great for the rock there to support, one side of a fault slips past the other side (Fig. 14-6). This movement causes vibrations that send out waves that may travel for long distances from their origin.

FIG. 14-7 Earthquakes are expected during the next 100 years in the parts of the United States shown in color. The darker the color, the greater the likely damage. The earthquake probability is especially high in California because locked-in stresses have built up in many of the thousands of faults that riddle the region. The most severe earthquakes in U.S. history were a trio that occurred in the winter of 1811–1812 at New Madrid in southeastern Missouri, which is in the dark spot south of the Great Lakes in this map. The quake was felt over most of the country east of the Rockies; the vibrations set church bells ringing as far away as Boston.

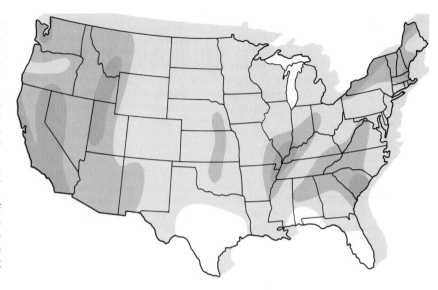

Regions in which severe earthquakes are comparatively frequent include the mountain chains that fringe the Pacific and a broad belt that extends from the Mediterranean basin across southern Asia to China (see Fig. 14-19). Major earthquakes have occurred now and then elsewhere, but by far the greatest number have been concentrated in these zones. In or near the earthquake belts lie most of the world's active volcanoes—which, as we shall see in Chap. 15, is no coincidence. Figure 14-7 is a map prepared by the U.S. Coast and Geodetic Survey that shows where earthquakes may be expected in the United States in the next 100 years.

Earthquakes give no obvious warning, though a variety of subtle effects show promise as signs that a quake is about to occur. Neverthe-

The upper deck of this highway in Oakland, California, collapsed in the magnitude 7.1 earthquake that shook San Francisco and its vicinity on October 17, 1989. Horizontal vibrations of the ground are responsible for most of the damage an earthquake causes.

less, despite intensive monitoring of the region, the earthquakes that rocked California in 1989 and 1993 came as surprises. There is a long way to go before severe earthquakes can be predicted with any degree of reliability.

14-7 STRUCTURE OF THE EARTH

Core, Mantle, and Crust   When an earthquake occurs at a fault in the crust, the rocks on both sides of the fault vibrate and send out waves that travel both through the earth's interior—"body waves"—and along its surface—"surface waves." The two kinds of body waves and the two kinds of surface waves are shown in Fig. 14-8.

FIG. 14-8 (a) Earthquake body waves travel through the earth's interior. P waves are longitudinal, like sound waves. S waves are transverse, like waves in a stretched string. P waves can move through a liquid; S waves cannot. (b) Earthquake surface waves travel on the earth's surface. Love waves are transverse, with their vibrations parallel to the surface. Rayleigh waves involve orbital motions, like water waves. All the waves shown here are moving from left to right. The arrows show the directions in which the rock particles move.

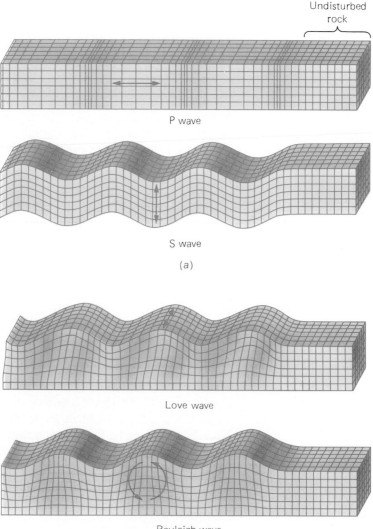

FIG. 14-9 Earthquake waves are detected by instruments called **seismographs.** This is a record of waves from an earthquake that occurred about 5000 km from the location of the seismograph.

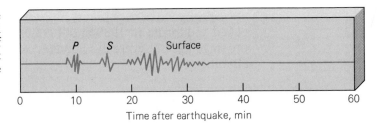

An easy way to remember the difference between P and S waves is to think of P waves as "push-pull" vibrations and of S waves as "shakes." P waves are the fastest and so arrive first at a distant point when an earthquake occurs somewhere (Fig. 14-9). The S waves, which are slower, come next. The surface waves, which have to travel along the ground rather than the shorter distance through the earth, appear last. However, the surface waves may produce the strongest vibrations, particularly when the distance is no more than a few thousand kilometers. The Love surface waves are responsible for most earthquake damage to buildings and other structures.

The internal structure of the earth was discovered with the help of P and S earthquake waves. These waves do not travel in straight lines inside the earth (except directly downward toward the earth's center) because of refraction. As we learned in Chap. 6, refraction refers to the change in direction of a wave when its speed changes. The speeds of P and S waves change with depth in the earth in two ways, which have different effects on these waves.

One change in speed is a gradual increase with depth. This causes the P and S waves to travel in curved paths within the earth, as shown in Fig. 14-10. Figure 6-8 shows a similar effect in ocean waves as they approach the shore.

The other change in speed is more abrupt and occurs because the earth's interior consists of layers of different materials. When an earthquake body wave crosses the boundary between two layers, its speed changes, and the result is a sharp change in direction. Figure 6-24 shows a similar effect in light waves as they go from air into glass. As we see in Fig. 6-26, waves that move perpendicularly through a boundary are not deflected.

Shadow Zones   Let us suppose an earthquake occurs somewhere. We consult various seismograph stations and find that most of them—never all—have detected P waves from this event. Curiously, the stations that P waves did not reach lie along a large band on the side of the earth opposite the quake. We would find, if we looked at the records of other earthquakes, that no matter where they took place, similar **shadow zones** for their P waves existed. This is the chief clue that confirmed an early suspicion that the earth's interior is not uniform.

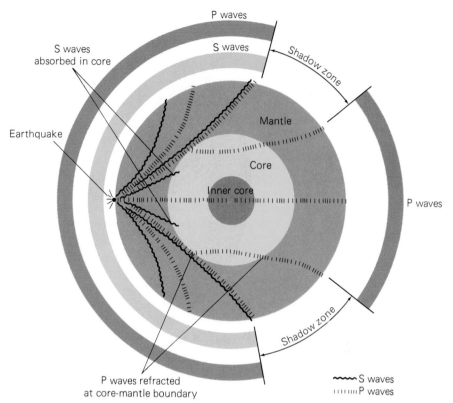

S waves
absorbed in core

Earthquake

P waves

S waves

Shadow zone

Mantle

Core

Inner core

P waves

Shadow zone

P waves refracted
at core-mantle boundary

~~~ S waves
ⅠⅠⅠⅠⅠⅠ P waves

FIG. 14-10 How earthquake waves travel through the earth. The existence of a shadow zone where neither P nor S waves arrive is evidence for a central core. The inability of S waves to get through the core suggests that it is liquid.

Figure 14-10 shows why this conclusion is necessary. In the picture, the earth below its thin outer crust is divided into a central **core** and a surrounding **mantle.** P waves from an earthquake that move only through the mantle are refracted so that they reach only a little over half the earth's surface. Those P waves that enter the core are bent sharply toward the center of the earth, and as a result the only part of the earth's surface they reach is a region opposite the quake. The shadow zone separates P waves that have reached the surface through the mantle only from P waves that have also passed through the core.

Analyzing the data shows that the mantle is 2900 km thick, which means that the core has a radius of 3470 km, over half the earth's total radius. However, the core makes up less than 20 percent of the earth's volume.

Supporting the above finding and giving further information about the nature of the core is the behavior of the S waves. It is found that these cannot get through the core at all (see Fig. 14-10). Because they are transverse, S waves cannot pass through a liquid, so the conclusion is that the earth's core is liquid! A liquid core accounts not only for the behavior of S waves but also for the marked changes in the speed of P waves when they enter and leave the core.

Sensitive seismographs have detected faint traces of P waves in the shadow zones, which suggests that within the liquid core is a smaller

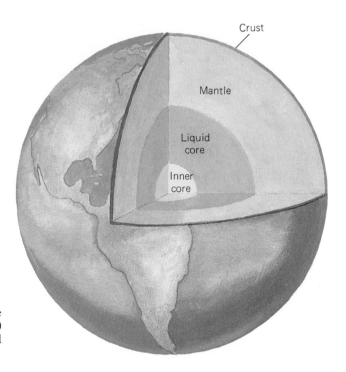

FIG. 14-11 Structure of the earth. The mantle constitutes 80 percent of the earth's volume and about 67 percent of its mass.

solid inner core. This inner core is believed to have a radius of 1390 km. Thus the earth's interior has the onionlike structure shown in Fig. 14-11.

The Moho   From observations of the waves from a 1909 earthquake it became clear that there is a distinct difference between the outer shell of the earth and the denser mantle below it. The surface between them is known as the **Mohorovicic discontinuity** (or just **Moho**), after its Croatian discoverer, and is considered to be the lower boundary of the crust. Under the oceans the crust is seldom much more than 5 km thick. Under the continents, though, the crust averages about 35 km, as much as 70 km under some mountain ranges (Fig. 14-12).

FIG. 14-12   The Mohorovicic discontinuity separates the earth's crust from the mantle below it. The crust is thicker and has a different composition under the continents than under the oceans.

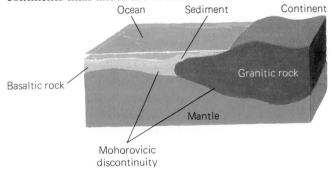

14-8   THE EARTH'S INTERIOR

A Mantle of Rock, a Core of Molten Iron, All Heated Mainly by Radioactivity
In the absence of a hole over 6000 km deep, anything said about the composition of the earth's interior can be no more than an educated guess, but a great deal of evidence supports the guesses that have been made.

In the case of the upper mantle, most studies point to igneous rocks composed mainly of ferromagnesian minerals. Deep in the mantle enormous pressures squeeze minerals into crystal structures that are the most compact possible. Olivine, for instance, is known to occur in two forms, the normal one of the crust and another whose particles are packed together in an especially tight arrangement. In the upper mantle the more open variety is probably the main component, the denser variety lower down. In the innermost part of the mantle, minerals have probably separated into very dense oxides of silicon, iron, and magnesium that occupy even less space.

The Core
Now we come to the liquid outer core. The average density of the earth is about twice the average density of the rocks at the earth's surface. The material of the mantle is only moderately denser than surface rocks, so that the core must be very heavy indeed.

Several clues point to iron as the logical candidate for the core material. It has almost the right density, it is liquid at the estimated pressure and temperature of the core, and it is abundant in the universe generally. Furthermore, iron is a good conductor of electricity, which is necessary in order to explain the earth's magnetism (Sec. 14-9). Because those meteorites that contain iron also contain some nickel, a good guess is that there is nickel in the core also. As for the solid inner core, the kernel of the earth, many geophysicists believe it to be crystalline iron or iron-nickel because of the pressures there.

The Earth's Heat
Whether or not the fires of Hell lie beneath the earth's surface, it is certainly hot enough there. The rate at which heat flows outward from the interior is immense, about 100 times greater than the energy involved in such geological events as volcanoes and earthquakes. There is plenty of heat to spare to account for mountain building and other deformations that occur in the crust. In fact, the geological history of the earth is mainly a consequence of the steady heat streaming through its outer layers. Temperatures inside the earth are believed to go from perhaps 375°C at the top of the mantle to at least 3000°C in the center of the core, and possibly much more.

Part of the earth's heat is a relic of its early history, but most of it comes from radioactive uranium, thorium, and potassium isotopes. The earth is believed to have come into being 4.5 billion years ago as a cold clump of smaller bodies of metallic iron and silicate minerals that had been circling the sun. Heat due to radioactivity accumulated in the interior of the infant earth and in time caused partial melting. The influence

Diamond-bearing rock about to be blasted out in a mine several kilometers deep near Pretoria in South Africa. Such rock originates in the mantle where temperatures and pressures are high enough for diamonds to be formed. Diamonds are thought to be very old, having come into being as much as 3.3 billion years ago. An average of 5 tons of rock must be mined for each carat of rough diamonds found.

of gravity then caused the iron to migrate inward to form the core while the lighter silicates rose to form the mantle and crust.

Today most of the earth's radioactivity is concentrated in the crust and upper mantle, where the heat it produces escapes through the surface and cannot collect to remelt the rest of the mantle or the inner core.

14-9   GEOMAGNETISM

FIG. 14-13 The earth's magnetic field originates in electric currents in its core of molten iron. The magnetic axis is tilted by 11° from the axis of rotation, so magnetic compasses do not point to true north.

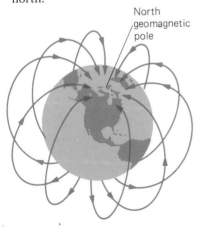

North geomagnetic pole

Electric Currents in the Core Seem to Be Responsible Although the earliest known description of the compass and its use in navigation was published by Alexander Neckham in 1180, knowledge of the compass seems to have been widespread even further back in antiquity. Until 1600, however, it was believed that a force exerted by Polaris, the North Star, is what attracted magnetized needles. In that year Sir William Gilbert wrote of experiments he had performed with spherical pieces of lodestone, a naturally magnetized mineral. By comparing the direction of the magnetic force on an iron needle at various positions near a lodestone sphere with similar measurements made in other parts of the world by explorers, Gilbert concluded that the earth behaves like a giant magnet— *"magnus magnes ipse est globus terrestris."*

The earth's magnetic field is much like the field that would come from a giant bar magnet located a few hundred km from the earth's center and tilted by 11° from the direction of the earth's axis of rotation (Fig. 14-13). No such magnet can possibly exist, since iron loses its magnetic properties above about 770°C and most of the earth's interior is much hotter than that. Instead, the magnetic field is believed to arise

BIOGRAPHY

William Gilbert (1544–1603)

practiced medicine in London and became the personal physician to Queen Elizabeth I in 1600. At about that time he published *De Magnete*, a book that described the results of his years of research on magnetism. An early advocate of experiment and observation in science—Galileo admired his work—Gilbert found that iron objects can be magnetized by rubbing them with a lodestone (an iron oxide mineral that is naturally magnetic); that heating an iron magnet causes it to lose its magnetism; that like magnetic poles repel and unlike poles attract (he introduced the word pole for the end of a magnet); and, most significant of all, that the earth itself is a giant magnet. In addition he showed that a number of popular beliefs were incorrect, for instance that garlic destroys magnetism and that magnetism cures headaches. Not until two centuries after *De Magnete* were any major additions made to our knowledge of magnetism.

Gilbert also studied static electricity, which the ancient Greeks had discovered by rubbing amber with silk, and found that other substances, too, can be charged by rubbing. Gilbert called all such substances *electrics*,

after the Greek for "amber," and clearly distinguished between magnetic and electric attractions. In astronomy, Gilbert was the first prominent Englishman to support the copernican model of the universe, and he believed that the stars were at different distances from the earth, perhaps circled by habitable planets. The **gilbert** is a magnetic unit named after this remarkable man.

from coupled fluid motions and electric currents in the liquid iron of the core. We recall from Fig. 5-22 that an electric current in the form of a loop has around it a magnetic field of the same kind as that of a bar magnet.

In many places, rock specimens of different ages have opposite magnetic polarities. The only explanation in most cases is that the earth's magnetic field has reversed itself periodically while these rocks were being formed. In the past 76 million years, 171 field reversals seem to have occurred. Such flip-flops seem consistent with the notion that the field is due to electric currents in the core, since changes may well take place in the patterns of flow in the liquid iron there from time to time.

EROSION

It is not obvious that this solid earth under us, made up largely of hard, strong rock, is in a state of constant change. But rocks, hills, and mountains are permanent only by comparison with the brief span of human life, and the long history of the earth goes back not scores of years but billions of years. In this immense stretch of time continents have shifted across the globe, mountain ranges have been thrust upward and then leveled, and broad seas have appeared and disappeared.

All the processes by which rocks are worn down and by which the debris is carried away are included in the general term **erosion**. The underlying cause of erosion is gravity. Such agents of erosion as running water and glaciers derive their destructive energy from gravity, and gravity is responsible for the transport of the removed material.

14-10 WEATHERING

How Exposed Rocks Decay

We have all seen the rough, pitted surfaces of old stone buildings. This kind of disintegration, brought about by rainwater and the gases in the air, is called **weathering.** Weathering contributes to erosion by preparing rocks for easy removal by the more active erosional agents, such as running water.

Some of the minerals in igneous and metamorphic rocks are especially susceptible to **chemical weathering,** since they were formed under conditions very different from those at the earth's surface. Ferromagnesian minerals are readily attacked by atmospheric oxygen, aided by carbon dioxide dissolved in water (which gives an acid solution) and by organic acids from decaying vegetation. The result is the red and brown colors that commonly appear as stains on the surface of rocks containing these minerals. Feldspars and other silicates containing aluminum are broken down to clay minerals. Among common sedimentary rocks limestone is most readily attacked by chemical weathering because calcite dissolves in weak acids. Exposures of limestone can often be identified simply from the pitted surfaces and enlarged cracks that solution produces.

Quartz and white mica resist chemical attack and usually remain as loose grains when the rest of a rock is thoroughly decayed. Rocks consisting wholly of silica, like chert and most quartzites, are practically immune to chemical weathering.

Mechanical weathering is often aided by chemical attack. Not only is the structure of a rock weakened by the decomposition of its minerals, but fragments of it are wedged apart because the chemical changes in a mineral grain usually result in an increased volume. The most effective process of mechanical disintegration that does not require chemical

After standing in the clean, dry air of Egypt for about 36 centuries, the carvings on Cleopatra's Needle were still sharp and clear. In 1881 the granite obelisk was moved to New York City, where the combination of climate and atmospheric pollution has almost erased the carvings. Acid rain running down the obelisk is the reason the damage increases toward the base.

Underground limestone gradually dissolves in acid water that seeps through it. When the limestone is near the surface, its disappearance can lead to a collapse of the ground above to leave a **sinkhole.** This sinkhole occurred in Winter Park, Florida. When limestone farther underground is dissolved, the result is a cave.

The expansion of water as it freezes into ice in cracks in rock has carved the sharp peak of the Matterhorn, a mountain on the border between Switzerland and Italy.

action is the freezing of water in crevices. Just as water freezing in a car's engine on a cold night may split the engine block, so water freezing in tiny cracks is effective in disrupting rocks. Plant roots help in rock disintegration by growing in cracks.

Weathering processes coat the naked rock of the earth's crust with a layer of debris made up largely of clay mixed with rock and mineral fragments. The upper part of the weathered layer, in which rock debris is mixed with decaying vegetable matter, is the **soil.** The formation of soil is an important result of weathering.

SOIL

Though the bulk of the earth's crust is solid rock, what we see on that part of the surface not covered by water is mainly soil with only occasional outcrops of bedrock. Soil originates in the weathering of rock, a complex disintegration process whose result is a coat of rock fragments and clay minerals mixed with varying amounts of organic matter.

Any type of rock may form the parent material of a soil. The particles of rock typically vary in size down to microscopic fineness and are intimately mixed with dark, partly decomposed plant debris called humus. The humus content decreases with depth. A great many factors are involved in the production of soil, including microorganisms such as bacteria and fungi that are responsible for the decay of plant and animal residues and are important in maintaining the nitrogen content of soil. A significant fraction of the organic matter in soil, in fact, consists of the bodies, living and dead, of these microorganisms. Even so lowly a creature as the worm plays a vital role in mixing together the various soil constituents.

Cross section of a typical soil. The darkening toward the top is due to the presence of humus.

The running water of a stream may accomplish more erosion during a few hours of heavy rain than in months or years of normal flow.

14-11 STREAM EROSION

Running Water Is the Chief Agent of Erosion

By far the most important agent of erosion is the running water of streams and rivers. The work of glaciers, wind, and waves is impressive locally, but compared with running water, these other erosion agents play only minor roles in shaping the earth's landscapes. Even in deserts, mountainsides are carved with the unmistakable forms of stream-made valleys.

In a young landscape, streams and rivers are just starting their work and flow downhill swiftly with many rapids and waterfalls. At first a river carves a narrow V-shaped channel (Fig. 14-14a). As time goes on,

FIG. 14-14 Successive stages in the development of a river valley.

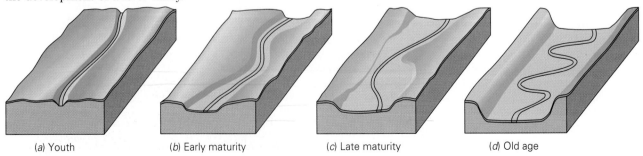

(a) Youth   (b) Early maturity   (c) Late maturity   (d) Old age

The floodplain of the South Platte River in Colorado. The wandering, shifting channel is typical of the old age of a river. The bends grow larger as their outer banks (where the water flows faster) are eroded, while sediments are deposited on the inner banks. Sometimes a bend becomes so extreme that its ends join together at a time of high water, which cuts a straight connection that leaves behind an oxbow lake. Such a lake can be seen to the left of the river near the top of this photograph.

the channel deepens until the river's lower end is near the level of a nearby valley or perhaps nearly at sea level. As the rest of the channel becomes less steep, its sides begin to be cut away to broaden it (Fig. 14-14b). The eventual result is a **floodplain** with a flat floor (Fig. 14-14c). In dry weather the river wanders over its plain in a meandering course; in very wet weather it overflows its channel and spreads across the plain. The floodplain grows wider and wider, the river becomes more and more sluggish, and the sides of its valley become lower and lower (Fig. 14-14d).

The treelike drainage pattern of the Wadi Hadramawt in Aden is conspicuous from the air.

FIG. 14-15 Parallel ridges and valleys produced by stream erosion in tilted layers of hard and soft rocks. Soft layers underlie the valleys; hard layers, the ridges. Landscapes and rock structures of this sort are typical of the Appalachian Mountains.

During this development of the major valley, secondary streams extend their smaller valleys on either side. Soon a characteristic treelike pattern develops, separated from the patterns of adjacent rivers by sharp divides of high ground. As floodplains widen along the main streams, divides are lowered by attack from the streams on either side. In the final stages of valley growth, when floodplains are wide and rivers broadly meandering, most of the divides are erased and those remaining are low and rounded.

Actual landscapes seldom conform exactly to the simple valley shapes and patterns just described. One reason is the presence of rocks of different hardness: hard rocks usually remain as cliffs and high ridges, while the more easily eroded soft rocks wear away (Fig. 14-15). Many of the striking landforms produced by erosion are due simply to differences in resistance from one rock layer to the next.

Whatever the valley shapes produced in various stages of landscape development, whatever different kinds of rock may be present, the goal of stream erosion is to reduce the land surface to a flat plain almost at sea level. However, this goal is never reached because, as we shall learn, other geologic processes continually counteract the effects of running water. There are very few regions in which geologic processes involving uplift do not occur at the same time as stream erosion. Thus most landscapes are the result of a complex of factors and reflect a balance among them rather than the action of stream erosion alone.

14-12   GLACIERS

Rivers and Seas of Ice   In a cold climate with abundant snowfall, the snow of winter may not completely melt during the following summer, and a deposit of snow accumulates from year to year. Partial melting and the continual increase in pressure cause the lower part of a snow deposit to change gradually into ice. If the ice is sufficiently thick, its weight forces it slowly downhill. A moving mass of ice formed in this manner is called a **glacier.** About 10 percent of the earth's land area is covered by glacial ice at the present time.

Today's glaciers are of two main kinds:

1. **Valley glaciers**—found, for instance, in the Alps, on the Alaskan coast, in the western United States—are patches and tongues of ice lying in mountain valleys. These glaciers move slowly down their valleys and melt at their lower ends. The combination of downward movement and melting keeps their ends in roughly the same position from year to year. Valleys carved by glaciers have U-shaped cross sections instead of the V shapes produced by stream erosion. Movement in the faster valley glaciers (a meter or more per day) is enough to keep their lower ends well below the timber line.

2. Glaciers of another type cover most of Greenland and Antarctica. These huge masses of ice thousands of square km in area that engulf hills as well as valleys are called **continental glaciers** or **ice caps.** They, too, move downhill, but the "hill" is the slope of their upper surfaces. An ice cap has the shape of a broad dome with its surface

The U-shaped cross section of this valley in Alaska suggests that it was cut by the ancestor of the shrunken glacier at its head at a time when the climate there was colder than at present.

sloping outward from a thick central portion of greatest snow accumulation. Its motion is outward in all directions from its center. The icebergs of the polar seas are fragments that have broken off the edges of ice caps. Similar sheets of ice extended across Canada and northern Eurasia during the ice ages.

Glacial Erosion   As a glacier moves, rock fragments held firmly by the ice at its bottom are dragged along. These fragments scrape and polish the underlying rock and are themselves ground down. Smooth, grooved rock surfaces and deposits of debris that contain boulders with flattened sides are common where the lower end of a valley glacier has melted back to reveal some of its bed. When such evidence of ice erosion is found far from present-day glaciers, we can infer that glaciers must have been active there in the past.

Glacial erosion is locally very impressive, particularly in high mountains. The amount of debris and the size of the boulders that a glacier can carry or push ahead of itself are often startling. But overall, the erosion accomplished by glaciers is small. Only rarely have they gouged rock surfaces deeply, and the amount of material they carry long distances is little compared with that carried by streams. Most glaciers of today are only feeble descendants of mighty ancestors, but even these ancestors succeeded only in modifying landscapes already shaped by running water.

14-13   GROUNDWATER

Water, Water Everywhere (Almost)   Most of the water that falls as rain does not run off at once in streams but instead soaks into the ground. All water that thus penetrates the surface is called **groundwater.** There is more groundwater than all the freshwater of the world's lakes and rivers, though less than the water locked up in ice caps and glaciers (see Fig. 13-3).

The soil, the layer of weathered rock under it, and any porous rocks below act together as a sponge that can absorb huge quantities of water. During and just after a heavy rain all empty spaces in the sponge may be filled, and the ground is then said to be **saturated** with water. When the rain has stopped, water slowly drains away from hills into the adjacent valleys. A few days after a rain porous material in the upper part of a hill contains relatively little moisture, while that in the lower part may still be saturated. Another rain would raise the upper level of the saturated zone, prolonged drought would lower it. The fluctuating upper surface of the saturated zone is called the **water table.**

Beneath valleys the water table is usually closer to the surface than beneath nearby hills (Fig. 14-16). Groundwater in the saturated zone seeps slowly downward and sideward into streams, lakes, and swamps. The motion is rapid through coarse material like sand or gravel, slow through fine material like clay. It is this flow of groundwater that maintains streams when rain is not falling; a stream goes dry only when the water table drops below the level of its bed. A **spring** is formed where groundwater comes to the surface in a more or less definite channel.

An **aquifer** is a body of porous rock through which groundwater moves. Aquifers underlie more than half the area of the continental United States.

Because groundwater moves slowly, it can accomplish little mechanical wear, but its prolonged contact with rocks and soil allows it to dissolve much soluble material. Some of the dissolved material remains in solution (hence the "hardness" of water from many wells), and some is redeposited elsewhere. In regions underlain by limestone, the most soluble of ordinary rocks, caves are produced when water moving through tiny cracks enlarges the cracks by dissolving and removing the adjacent limestone.

FIG. 14-16 Cross sections through a landscape underlain by porous material. The position of the water table is shown (a) just after a heavy rain, (b) several days later, and (c) after a prolonged drought. The spring, the stream, and the upper well would be dry during the drought.

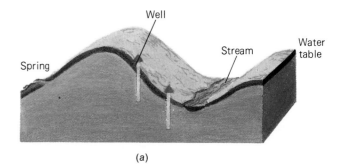

(a)

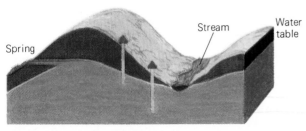

(b)

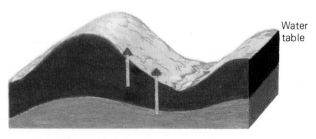

(c)

14-14   SEDIMENTATION

What Becomes of the Debris of Erosion

Most of the material transported by the agents of erosion is eventually deposited to form **sediments** of various sorts. The ultimate destination of erosional debris is the ocean, and the most widespread sediments collect in shallow parts of the ocean near continental edges. But much sedimentary material is carried to the sea in stages, deposited first in thick layers elsewhere—in lakes, in desert basins, in stream valleys. Each erosional agent has its own ways of giving up its load, and these ways leave their stamp on the deposits formed.

Rivers and streams, the chief agents of erosion, lose some of the abundant debris they carry whenever their speeds drop or their volumes of water decrease. Four sites of deposition are common:

1. Debris carried in time of flood is deposited in gravel banks and sandbars on the stream bed when the swiftly flowing waters begin to recede.

2. The floodplain of a meandering river is a site of deposition whenever the river overflows its banks and loses speed as it spreads over the plain. In Egypt, for example, before construction of the Aswan Dam, the fertility of the soil was maintained for centuries by the deposit of black silt left each year when the Nile was in flood.

3. A common site of deposition, especially in the western United States, is the point where a stream emerges from a steep mountain valley and slows down as it flows onto a plain. Such a deposit, usually taking the form of a low cone pointing upstream, is called an **alluvial fan.**

4. A similar deposit is formed when a stream's flow is stopped abruptly as the stream enters a lake or sea. This kind of deposit, built largely

Alluvial fan in Death Valley, California, consists of debris deposited by a mountain stream when it slowed down on reaching the plain.

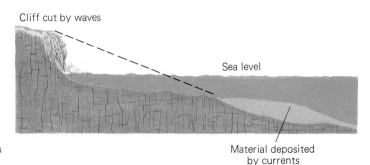

Cliff cut by waves

Sea level

Material deposited
by currents

FIG. 14-17 Cross section of a steep, rocky shoreline.

underwater and with a surface usually flatter than that of an alluvial fan, is called a **delta.**

Some of the material a glacier scrapes from its channel is spread as a layer of irregular thickness under the ice, and some is heaped up at the glacier's lower end where the ice melts. The pile of debris around the end and along the sides of the glacier, called a **moraine,** is left as a low ridge when the glacier melts back. Moraines in mountain valleys and in the North Central states are part of the evidence for a former wide extent of glaciation.

All the material deposited directly by ice goes by the name of **till,** an indiscriminate mixture of fine and coarse material. Till includes huge boulders that are often embedded in the fine, claylike material a glacier produces by its polishing action. Typically, most of the boulders are angular, but a few are rounded and show the flat scratched faces produced as they were dragged along the bed of the glacier.

Most important of the agents of deposition, because they handle by far the largest amount of sediment, are ocean currents. Currents deposit not only the materials eroded from coastlines by wave action but also the abundant debris brought to the ocean by rivers, wind, and glaciers. Visible deposits of waves and currents include beaches and sandbars, but the great bulk of the sediments brought to the ocean are laid down underwater (Fig. 14-17).

Groundwater deposits material in the pore spaces of sediments, a process that helps to convert the sediments to rock. Much dissolved material precipitates in cracks to form veins, which are found in all kinds of rocks. Quartz and calcite are common in veins, and the ores of various metals are also found there. Spectacular examples of groundwater deposition are the **stalactites** that hang from the roofs of limestone caves, the **stalagmites** that rise from their floors, and the colorful deposits often found around hot springs and geysers.

Lithification Sediments buried beneath later deposits are gradually hardened into rock, a process called **lithification.** Lithification can be a complex process, taking thousands or even millions of years. One important change in the sediment is compaction, the squeezing together of its grains under the pressure of overlying deposits.

Carlsbad Caverns in New Mexico. Caves such as this occur when groundwater dissolves limestone and carries it away. The stalactites that hang like icicles from the roof of the cave and the stalagmite columns that grow from its floor consist of material that has precipitated from solution.

Some recrystallization may accompany compaction. The calcite crystals of limy sediments, in particular, grow larger and interlock with one another.

Chemical changes brought about by circulating groundwater contribute to the hardening of many sediments. The grains of coarse sediments are cemented by material precipitated from solution in groundwater, and some sediments have much of their original material dissolved away and replaced by other substances. In petrified wood the original organic compounds have been removed, molecule by molecule, and replaced by silica. The whole process takes place so gradually that the finest details of the wood structure may be preserved.

Sedimentary rocks are especially important in geology because they contain material that was deposited at or near the earth's surface and so record the changing surface conditions of past time. If we read the evidence with sufficient insight and imagination, we find before us a panorama of earth history: seas that once spread widely over the land, the advance and retreat of immense glaciers, the shifting sand dunes of ancient deserts, and much more. Revealed are the living creatures that inhabited lands and seas of the past, for many sedimentary rocks contain fossil remains of plants and animals. Igneous and metamorphic rocks tell us by their structures something about conditions in the earth's interior; sedimentary rocks tell us about the history of surface landscapes.

Petrified wood in Arizona.

These sandstone beds in Zion National Park, Utah, were once sand dunes that had been deposited by winds.

VULCANISM

Erosion and sedimentation are leveling processes through which the higher parts of the earth's surface are worn down and the lower parts are filled with the resulting debris. If their work could be completed, the continents would disappear and the earth would become a smooth globe covered with water. The fact that the continents still exist, not to mention mountain ranges on them, testifies to the action of two other processes:

1. Processes of **vulcanism,** which involve movements of molten rock
2. Processes of **diastrophism** (or **tectonism**), which involve movements of the solid materials of the crust

Vulcanism and diastrophism often occur together. The flow of molten rock may cause adjacent rock structures to be distorted, and it is common for major crustal shifts to be accompanied by volcanic eruptions and subsurface migrations of molten rock.

14-15   VOLCANOES

Rivers of Lava, Clouds of Gas and Dust

A **volcano** is an opening in the earth's crust through which molten rock, usually called **magma** while underground and **lava** above ground, pours forth. Because lava accumulates near their openings, most volcanoes in the course of time build up mountains with a characteristic conical shape that steepens toward the top, with a small depression, or **crater,** at the summit. Lava escapes almost continuously from a few volcanoes, but the majority are active only at intervals.

A volcanic eruption is one of the most awesome spectacles in all nature. Earthquakes may provide warning of an eruption a few hours or a few days beforehand—minor shocks probably caused by the move-

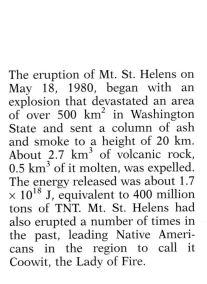

The eruption of Mt. St. Helens on May 18, 1980, began with an explosion that devastated an area of over 500 km² in Washington State and sent a column of ash and smoke to a height of 20 km. About 2.7 km³ of volcanic rock, 0.5 km³ of it molten, was expelled. The energy released was about 1.7 × 10¹⁸ J, equivalent to 400 million tons of TNT. Mt. St. Helens had also erupted a number of times in the past, leading Native Americans in the region to call it Coowit, the Lady of Fire.

ment of gases and liquids underground. Eruptions follow a variety of patterns. Usually an explosion or a series of explosions comes first, sending a great cloud billowing upward from the crater. In the cloud are various gases, dust, and fragments of rock. The exceptionally violent eruption of the volcano Tambora in Indonesia in 1815 is thought to have blown over 150 km³ of pumice and ash into the atmosphere. Enough stayed aloft to keep a substantial amount of sunlight from reaching Europe and North America the following year—the "year without a summer."

Gas may continue to come out in large quantities after the first explosion, and further bursts may recur at intervals. The cloud may persist for days or weeks with its lower part glowing red at night. Activity gradually slackens, and soon a tongue of white-hot (1100°C or so) lava spills over the edge of the crater or pours out of a fissure on the mountain slope. Other flows may follow the first, and explosions may continue, though they become weaker and weaker. Slowly the volcano quiets down, with only a small cloud of water vapor above to suggest its activity.

Volcanic Violence

The chief factors that determine whether an eruption will be a quiet lava flow or a furious explosion are the **viscosity** of the magma and the amount of gas it con-

FIG. 14-18 Cross section of a volcano. During explosive eruptions much liquid rock is blown into fragments when it emerges. Deposits of the finer material may form the rock tuff, and deposits of the coarser material may form a kind of conglomerate called volcanic breccia. In the volcano shown, lava flows (solid color) alternate with beds of tuff and volcanic breccia.

A river of red-hot lava pours from the crater of this volcano in Hawaii.

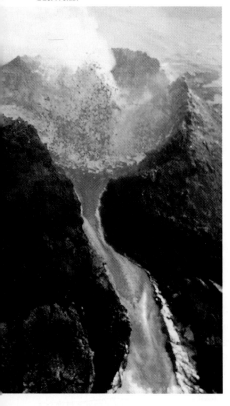

tains. (The greater the viscosity of a liquid, the less freely it flows; honey is more viscous than water.) Magma is a complex mixture of various metal oxides and silica and usually has a lot of gas dissolved in it under pressure. Like most molten silicates, lava is extremely viscous and, with rare exceptions, creeps downhill slowly, like thick syrup or tar. The viscosity depends upon chemical composition: magmas with high percentages of silica are the most viscous.

Gas also affects viscosity: magmas with little gas in them are the most viscous. If the magma feeding a volcano happens to be rich in both gas and silica, the eruption will be explosive. A magma with modest gas and silica contents results in a quiet eruption.

Volcanic gases include water vapor, carbon dioxide, nitrogen, hydrogen, and various sulfur compounds. The most prominent is water vapor. Some of it comes from groundwater heated by magma, some comes from the combination of hydrogen in the magma with atmospheric oxygen, and some was formerly incorporated in rocks deep in the crust and carried upward by the magma. Much of the water vapor condenses when it escapes to give rise to the torrential rains that often accompany eruptions.

Lava hardens into one or another of the various volcanic rocks (Fig. 14-18), which are all fine-grained because lava cools too rapidly for large mineral crystals to grow. Basalt is by far the most common volcanic rock and, when molten, is fluid enough to spread out over a wide area. The more viscous andesite usually produces steep, conical mountains and generally is associated with more rugged landscapes than basalt. Rhyolite, the most siliceous of ordinary lavas, forms small, thick flows and domes. Rhyolitic lava is sometimes so viscous and cools so rapidly that crystallization does not take place, leaving the natural glass obsidian.

Volcanic rocks of all kinds frequently have rounded holes due to gas bubbles that were trapped during the final stages of solidification. Viscous lavas may harden with so many cavities that the light, porous rock pumice results. Pumice is so light that it floats in water.

Volcanic Regions

About 500 volcanoes are active today around the borders of the Pacific Ocean, on some of the Pacific Islands, in Iceland, in the Mediterranean region, in the West Indies, and in East Africa (Fig. 14-19). Fifty or so of them erupt in an average year.

In many other parts of the world volcanoes were active in earlier times. Where volcanoes became extinct in the recent geological past, their former splendor is suggested by isolated conical mountains, by solidified lava flows, and by hot springs, geysers, and steam vents. Some of the great mountains in the western United States are old volcanoes, scarred here and there with lava flows so recent that vegetation has not yet gained a foothold on them. In regions where volcanoes have been inactive for still longer, erosion may have removed all evidence of the original mountains and left only patches of volcanic rocks to indicate former igneous activity.

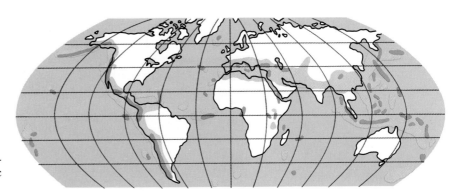

FIG. 14-19 The principal earth-quake (light color) and volcanic (dark color) regions of the world.

14-16 INTRUSIVE ROCKS

They Have Hardened Under-ground from Magma
Molten rock that rises through the earth's crust but does not reach the surface solidifies to form intrusive bodies (often called **plutons**) of various kinds. Because these bodies cool slowly, intrusive igneous rocks are coarser-grained than volcanic rocks that cool rapidly on the surface. We find intrusive rocks exposed only where erosion has uncovered them after they hardened.

The igneous origin of volcanic rocks is clear enough, for we can actually watch lava harden to solid rock. But no one has ever seen an intrusive rock like granite in a liquid state in nature. The belief that granite was once molten follows from indirect evidence such as the following:

1. Granite shows the same relations among its minerals that a volcanic rock shows. The separate grains are intergrown, and those with higher melting points show by their better crystal forms that they crystallized a little earlier than the others.

2. Some small intrusive formations show a continuous change between coarse granite and a rock indistinguishable from the volcanic rock rhyolite, whose igneous origin is clear.

3. Granite is found in masses that cut across layers of sedimentary rock and from which small irregular branches penetrate into the surrounding rocks. Sometimes blocks of the sedimentary rocks are found completely engulfed by the granite.

4. That granite was at a high enough temperature to be molten is shown by the baking and recrystallization of the rocks that it intrudes.

These four types of evidence apply equally well to the other intrusive rocks.

Plutons
A **dike** is a wall of igneous rock that cuts across exist-ing rock layers (Fig. 14-20). The largest dikes are hun-dreds of meters thick, but more often their thickness is between a few

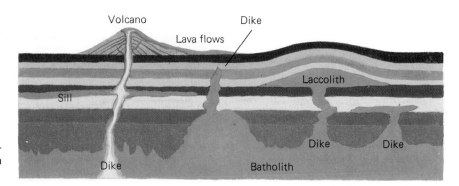

FIG. 14-20 A batholith and associated dikes and sills; a laccolith and a volcano are also shown.

tenths of a meter and a few meters. The distinction between dikes and veins is that a dike is molten rock that has filled a fissure and solidified, whereas a vein consists of material deposited along a fissure from solution in water.

Any kind of igneous rock may occur in a dike. Rapid cooling in small dikes may give rocks similar to those of volcanic origin, and slow cooling in larger dikes gives coarse-grained rocks. Dikes may cut any other kind of rock. They are often associated with volcanoes: some of the magma forces its way into cracks instead of moving upward through the central orifice. In regions of intrusive rocks, dikes are often found as offshoots of larger masses, as in Fig. 14-20. Also shown in the figure are **sills** and **laccoliths,** intrusive bodies that lie parallel to the strata in which they are found.

Batholiths are very large plutons that extend downward as much as several km. Visible exposures of batholiths cover hundreds of thousands of square km. The great batholith that forms the central part of the Sierra Nevada in California, for example, is about 800 km long and, in

A dike of igneous rock intruded in sedimentary beds in Arizona's Grand Canyon. The beds on either side of the dike do not match, which suggests that the dike was intruded along a fault.

places, over 160 km wide. Granite is the principal rock in batholiths, although many have local patches of diorite and gabbro. Batholiths are always associated with mountain ranges, either mountains of the present or regions whose rock structure shows evidence of mountains in the distant past. The intrusion of a batholith is evidently one of the events that occurs in the process of mountain building.

14-17 THE ROCK CYCLE

Rocks Are Not Necessarily Forever

As we have seen, rocks can change from one kind to another in a variety of ways. An igneous rock, for instance, can be broken up by erosion into fragments that eventually end up in a deposit of sediments that in time lithifies into sedimentary rock. Heat and pressure can later transform this rock into a metamorphic counterpart, which may later be melted underground into magma and still later harden into an igneous rock—probably not the same as the one the cycle began with, but perhaps a cousin to it. Other life histories are also possible, as shown in Fig. 14-21.

FIG. 14-21 The rock cycle. Depending upon circumstances, different paths are possible, including the conversion of one kind of metamorphic rock into another.

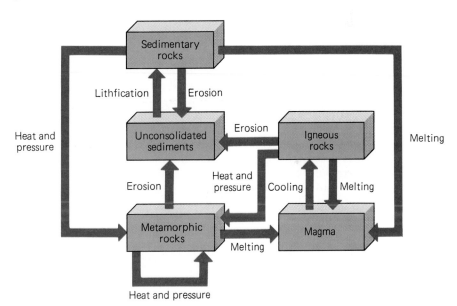

IMPORTANT TERMS AND IDEAS

The earth's outer shell of rock is called its **crust.** Oxygen and silicon are the most abundant elements in the crust. Crystalline **silicates** consist of continuous structures of oxygen and silicon ions, usually with metal ions as well. **Minerals** are the substances of which rocks are composed. The most abundant minerals are silicates; also common are carbonates and oxides. Quartz, feldspar, mica, calcite, and the ferromagnesian and clay minerals are six important kinds of minerals.

Igneous rocks (such as granite and basalt) have cooled from a molten state. **Sedimentary rocks** (such as sandstone and limestone) have consolidated from materials derived from the disintegration or solution of other rocks and deposited by water, wind, or glaciers. **Metamorphic rocks** (such as slate and marble) have been altered by heat and pressure beneath the earth's surface. **Soil** consists of rock fragments mixed with varying amounts of organic material.

Nearly all **earthquakes** are due to the sudden movement of solid rock along fracture surfaces called **faults** that are near the earth's surface. Earthquakes send out **body waves,** which pass through the earth's interior, and **surface waves,** which travel along the earth's surface. The two kinds of body waves are **P waves,** which are longitudinal, and **S waves,** which are transverse. The two kinds of surface waves are **Love waves,** which are transverse, and **Rayleigh waves,** which are orbital, like water waves.

The analysis of earthquake wave records shows that the earth's interior contains a **core,** probably consisting mainly of molten iron with a small solid inner core, and a solid **mantle** of ferromagnesian silicates. The temperature of the earth's interior increases with depth.

The **earth's magnetic field** resembles the magnetic field that would be produced by a huge bar magnet located near the earth's center. The field is due to electric currents in the molten iron core, and the direction of the field has reversed itself many times in the past.

The processes by which rocks are worn down and the debris carried away are included in the term **erosion.** The chief agent of erosion is running water, although **weathering** (the gradual disintegration of exposed rocks), **glaciers** (rivers and seas of ice formed from accumulated snow), and **groundwater** (subsurface water) also contribute. Most of the eroded material is deposited to form sediments, the bulk of which are laid down on ocean bottoms. **Lithification** refers to the gradual hardening into rock of sediments buried under later deposits.

A **volcano** is an opening in the earth's crust through which molten rock (called **magma** while underground, **lava** above ground) comes out. Intrusive bodies called **plutons** are formed by the solidification of magma under the surface.

EXERCISES: MULTIPLE CHOICE

1. The most abundant element in the earth's crust is
   a. oxygen   **c.** silicon
   b. nitrogen   **d.** carbon

2. The second most abundant element is
   a. iron   **c.** carbon
   b. silicon   **d.** aluminum

3. Minerals are
   a. silicon compounds
   b. common types of rock
   c. homogeneous solids of which rocks are composed
   d. always compounds

4. Cleavage, the tendency of certain minerals to split along certain planes, is conspicuous in
   a. quartz   **c.** chalk
   b. mica   **d.** chert

5. The most abundant mineral in the earth's crust is
   a. quartz   **c.** feldspar
   b. mica   **d.** calcite

6. The crystalline form of silicon dioxide (SiO_2) is
   a. quartz   **c.** feldspar
   b. mica   **d.** calcite

7. A mineral that is not a silicate is
   a. quartz   **c.** feldspar
   b. mica   **d.** calcite

8. The ferromagnesian minerals are usually
   a. transparent   **c.** bluish
   b. white or pink   **d.** dark green or black

9. Rocks that have been formed by cooling from a molten state are called
   a. igneous rocks   **c.** metamorphic rocks
   b. sedimentary rocks   **d.** precipitated rocks

10. Rocks that have been altered by heat and pressure beneath the earth's surface are called
    a. igneous rocks   **c.** metamorphic rocks
    b. sedimentary rocks   **d.** precipitated rocks

11. The majority of surface rocks are
    a. sedimentary
    b. metamorphic
    c. igneous
    d. found in the mantle as well

12. A general characteristic of rocks of volcanic origin is
    a. the presence of shale
    b. a light color
    c. coarse-grained structure
    d. fine-grained structure

13. Foliation occurs in
    a. sedimentary rocks   **c.** igneous rocks
    b. metamorphic rocks   **d.** all of the above

14. An example of a foliated rock is
    a. marble   **c.** slate
    b. sandstone   **d.** granite

15. Of the following rocks, the one that does not originate in sediments laid down by water, wind, or ice is

a. marble   c. shale
b. conglomerate   d. sandstone

16. Limestone may be metamorphosed into
   a. marble   c. gneiss
   b. quartzite   d. schist

17. Shale may be metamorphosed into
   a. marble   c. slate
   b. sandstone   d. granite

18. Mica is not found in
   a. slate   c. gneiss
   b. schist   d. marble

19. Fossils are most likely to be found in
   a. granite   c. gneiss
   b. shale   d. basalt

20. Most earthquakes are caused by
   a. volcanic eruptions
   b. landslides
   c. lightning strikes
   d. shifts of crustal rocks

21. Regions in which earthquakes are frequent are also regions in which
   a. the geomagnetic field is strong
   b. hurricanes are common
   c. volcanoes occur
   d. petroleum is found

22. Relative to an earthquake of magnitude 5 on the Richter scale, an earthquake of magnitude 6 releases
   a. 2 times more energy
   b. 10 times more energy
   c. 30 times more energy
   d. 100 times more energy

23. Earthquake P waves
   a. are longitudinal vibrations like sound waves
   b. are transverse vibrations like waves in a taut string
   c. cannot pass through the earth's core
   d. always travel in straight lines through the earth's interior

24. The earth's crust
   a. has very nearly the same thickness everywhere
   b. varies irregularly in thickness
   c. is always thinnest under the continents
   d. is always thinnest under the oceans

25. The part of the earth with the greatest volume is the
   a. inner core   c. mantle
   b. outer core   d. crust

26. The radius of the earth's core is roughly
   a. $\frac{1}{10}$ the earth's radius
   b. $\frac{1}{4}$ the earth's radius
   c. $\frac{1}{2}$ the earth's radius
   d. $\frac{3}{4}$ the earth's radius

27. The rocks of the mantle are believed to consist largely of
   a. feldspar
   b. quartz
   c. clay minerals
   d. ferromagnesian minerals

28. The reasons why the earth's core is believed to be largely molten iron do not include the
   a. magnetic properties of iron
   b. electrical conductivity of iron
   c. density of iron
   d. relative abundance of iron in the universe

29. Most of the heat of the earth's interior is believed to
   a. be left over from its formation
   b. come from radioactive materials
   c. be due to chemical reactions in the core
   d. be provided by solar radiation absorbed by the crust

30. The earth's magnetic field
   a. never changes
   b. has reversed itself many times
   c. is centered exactly at the earth's center
   d. originates in a permanently magnetized iron core

31. A rock readily attacked by chemical weathering is
   a. limestone   c. granite
   b. obsidian   d. chert

32. The principal agent of erosion is
   a. groundwater   c. ice
   b. running water   d. wind

33. Which of the following is not produced by rivers?
   a. flood plains   c. alluvial fans
   b. deltas   d. moraines

34. In their early stages, river valleys exhibit characteristic
   a. U-shaped cross sections   c. water tables
   b. V-shaped cross sections   d. moraines

35. The last stage in the erosion of a river is the production of a (an)
   a. delta   c. moraine
   b. alluvial fan   d. floodplain

36. The approximate percentage of the earth's land area covered by ice today is
a. 1 percent   **c.** 10 percent
b. 2 percent   **d.** 30 percent

37. A fairly fast valley glacier might have a speed of
a. 1 m/h   **c.** 1 m/month
b. 1 m/day   **d.** 1 m/year

38. Most of the groundwater present in soil and underlying porous rocks comes from
a. streams and rivers   **c.** springs
b. melting glaciers   **d.** rain

39. A body of porous rock through which groundwater moves is called a (an)
a. water table   **c.** moraine
b. aquifer   **d.** batholith

40. The largest amounts of sediment are deposited
a. by glaciers
b. on river beds
c. on the ocean floors
d. by chemical precipitation

41. Minerals deposited by groundwater in rock fissures form
a. dikes   **c.** sills
b. veins   **d.** moraines

42. Most caves are produced by the solvent action of groundwater on
a. limestone   **c.** granite
b. sandstone   **d.** schist

43. The chief constituent of volcanic gases is
a. nitrogen   **c.** carbon dioxide
b. oxygen   **d.** water vapor

44. Molten rock underneath the earth's surface is called
a. magma   **c.** obsidian
b. lava   **d.** till

45. The most common volcanic rock is
a. granite   **c.** limestone
b. basalt   **d.** shale

46. The holes found in most volcanic rocks are due to
a. gases trapped in solidifying lava
b. erosion
c. marine organisms
d. rapid cooling

47. Active volcanoes are not found
a. in the West Indies
b. in the Mediterranean region
c. on the rim of the Pacific Ocean
d. in eastern Canada

48. A batholith is a
a. fissure from which groundwater emerges
b. natural rock pillar
c. large body of intrusive rock
d. volcanic cone

QUESTIONS

1. What is the relationship between rocks and minerals?

2. Which of the following naturally occurring substances are minerals? Diamond, calcite, petroleum, ice, soil, wood, salt, coal.

3. Both cleavage and crystal form are characteristic mineral properties. What is the difference between the two?

4. List the three classes of rock in the order of their abundance in the earth's crust.

5. What kind of rocks consist of random arrangements of irregular mineral grains that have grown together?

6. What kind of rocks are most likely to contain fossils?

7. Obsidian is a rock that resembles glass, in particular by sharing the property that its structure is closer to that of a liquid than to that of a crystalline solid. What does this observation suggest about the manner in which obsidian is formed?

8. Granite and rhyolite have similar compositions, but granite is coarse-grained whereas rhyolite is fine-grained. What does the difference in grain size indicate about the environments in which each rock formed?

9. Diorite is an igneous rock that has hardened slowly underground, and andesite, whose composition is similar, is an igneous rock that has hardened on the earth's surface. How can they be distinguished from one another?

10. Shale is a sedimentary rock that consolidated from mud deposits. What are the various metamorphic rocks that shale can become under progressively increasing temperature and pressure?

11. What happens to the density of a rock that undergoes metamorphism?

12. Gneiss is formed at greater depths than slate. Which rock would you expect to have the greater density?

13. The mineral grains of many metamorphic rocks are flat or elongated and occur in parallel layers. (a) What is this property called? (b) How does it originate?

14. What are the three most common types of cemented-fragment (or *clastic*) sedimentary rocks? What distinguishes them from one another?

15. What is the nature of chert and why is it so resistant to chemical and mechanical attack?

16. (a) What is the origin of limestone? (b) What rock is formed by the metamorphism of limestone? (c) What is the difference in structure that the metamorphism produces?

17. How could you distinguish (a) granite from gabbro? (b) basalt from limestone? (c) schist from diorite?

18. How could you distinguish (a) chert from obsidian? (b) conglomerate from gneiss? (c) quartz from calcite?

19. Name the following rocks: (a) a fine-grained, unfoliated rock with intergrowing crystals of quartz, feldspar, and black mica; (b) a finely foliated rock with microscopic crystals of quartz and white mica; (c) a fine-grained rock consisting principally of kaolin.

20. Name the following rocks: (a) a rock consisting of intergrown crystals of quartz; (b) the rock resulting from the metamorphism of limestone; (c) an intrusive igneous rock with the same composition as andesite.

21. What are the similarities and differences between seismic P and S waves?

22. What can be said about an earthquake whose magnitude is 0 on the Richter scale? Whose magnitude is 8 or more?

23. Why is the mantle thought to be solid?

24. (a) Why is it considered likely that the earth's outer core is liquid? (b) Why is the liquid thought to be largely iron? (c) Why is nickel believed to be present as well?

25. How does the radius of the earth's core compare with the total radius of the earth?

26. Where is the earth's crust thinnest? Where is it thickest?

27. What evidence is there in favor of the idea that the earth's interior is very hot?

28. What is the principal cause of the high temperatures believed to exist in the earth's interior?

29. Why is it unlikely that the earth's magnetic field originates in a huge bar magnet located in its interior?

30. Why does a compass needle in most places not point due north?

31. In what way is the weathering of rock important to life on earth?

32. What common rocks are almost immune to chemical weathering?

33. Both marble and slate are metamorphic rocks. Would you expect a marble tombstone or a slate one to be more resistant to weathering?

34. Granite consists of feldspars, quartz, and ferromagnesian minerals. (a) What becomes of these minerals when granite undergoes weathering? (b) What kinds of sedimentary rocks can the weathering products form?

35. What is the source of energy that makes possible the erosion of landscapes?

36. Why are streams and rivers the principal agents of erosion on the earth's surface?

37. Is there a limit to the depth to which streams can erode a particular landscape? Is there a limit in the case of glaciers?

38. What agent of erosion produces valleys with a V-shaped cross section? A U-shaped cross section?

39. Under what circumstances does a glacier form?

40. Glaciers grind away rock with far more force than streams and rivers, yet running water has had more influence in shaping landscapes around the world than glaciers. Why?

41. How is it possible for glaciers to wear down rocks that are harder than glacial ice?

42. What is an aquifer?

43. What is the eventual site of deposition of most sediments?

44. What kind of material is found in an alluvial fan? In a moraine?

45. Why are clay minerals and quartz particles abundant in sediments that have not been chemically deposited?

46. In sand derived from the attack of waves on

granite, what mineral would you expect to be most abundant?

47. What is the probable origin of the following sedimentary rocks?
   a. A thick limestone
   b. A conglomerate with well-rounded boulders and numerous thin beds of sandy and clayey material
   c. A sandstone consisting of well-sorted, well-rounded grains of quartz

48. Why are hot-spring deposits thicker than the deposits found around ordinary springs?

49. What kinds of rocks are likely to be found in lava flows? What is the most common volcanic rock?

50. What factors determine the viscosity of a magma? What kinds of landscapes are produced by volcanoes whose lavas have relatively high and relatively low viscosities?

51. What is the cause of the holes found in many volcanic rocks?

52. What is the main constituent of volcanic gases?

53. What characteristic landscape features do active volcanoes produce? From what features could you conclude that volcanoes were once active in a region where eruptions have long since ceased?

54. Distinguish between a dike and a vein.

55. (a) Why are metamorphic rocks often found near plutons? (b) Where would you expect to find the wider zone of thermal metamorphism, near a dike or near a batholith?

56. Suppose you find a nearly vertical contact between granite and sedimentary rocks, the sedimentary beds ending abruptly against the granite. How could you tell whether the granite had intruded into the sedimentary rocks or after solidifying had moved against the sedimentary rocks by tectonic movement along a fault?

57. Distinguish between the foliation of a metamorphic rock and the stratification of a sedimentary rock.

58. An experiment is performed to determine the lowest temperature at which a certain magma can exist within the earth by melting a sample of rock that has hardened from this magma in a furnace. How meaningful are the results of this experiment?

ANSWERS TO MULTIPLE CHOICE			
1. a	**13.** b	**25.** c	**37.** b
2. b	**14.** c	**26.** c	**38.** d
3. c	**15.** a	**27.** d	**39.** b
4. b	**16.** a	**28.** a	**40.** c
5. c	**17.** c	**29.** b	**41.** b
6. a	**18.** d	**30.** b	**42.** a
7. d	**19.** b	**31.** a	**43.** d
8. d	**20.** d	**32.** b	**44.** a
9. a	**21.** c	**33.** d	**45.** b
10. c	**22.** c	**34.** b	**46.** a
11. a	**23.** a	**35.** d	**47.** d
12. d	**24.** d	**36.** c	**48.** c

15

THE

EVOLVING

EARTH

Nothing about the earth is fixed, permanent, unchanging. What is today a great mountain that pierces the sky may tomorrow be nibbled down into a mere hill, while elsewhere an undersea mass of sediments may be thrust upward into a lofty plateau. How do we know that such things can happen? After all, though plenty of geologic activity is taking place around us, for the most part the pace is exceedingly slow. Only after millions of years can the processes now at work yield large-scale changes in the pattern of the continents and in their landscapes. What justifies the belief that the earth's crust never stops evolving is the record of the past, a record that can be read in the rocks of the present.

During the past few decades a major advance has occurred in our understanding of the large-scale forces that shape and reshape the earth's crust. The notion that the continents are slowly drifting relative to one another—a notion going back three-quarters of a century but largely scorned for most of that period—has turned out to be the only way to explain a variety of striking observations. These same observations also provide clues as to what makes the continents drift. So far-reaching are the implications of the new dynamic picture of the crust, and so suddenly did they come to light, that it is legitimate to speak of a revolution in geologic thought.

TECTONIC MOVEMENT

Terra firma, the solid earth, is a symbol of stability and strength. On foundations of rock we anchor our buildings, our dams, our bridges. The massive rock of mountain ranges seems strong enough to withstand any force.

Yet even casual observation shows how naive such notions of the earth's stability are. High up on mountainsides we find shells of marine animals, shells that can be there only if rock formed beneath the sea has been lifted far above sea level. Sedimentary rocks, which must have been deposited originally in horizontal layers, are found tilted at steep angles or folded into arches and basins. Other layers have broken along cracks, and the fractured ends have moved apart. Gigantic forces must occur in the crust in order to lift, bend, and break even the strongest rocks. Such forces and the changes they cause are called **tectonic** (from the Greek word for "carpenter").

15-1   TYPES OF DEFORMATION

Faults and Folds, Tilts and Warps, Rises and Falls

Cracks are found in rock formations of all kinds, some due to molten rock contracting as it cools and others due to mechanical stresses in the crust. A fracture surface along which motion has taken place is called a **fault.** In a rock outcrop a fault appears as a fairly straight line against which sedimentary layers and other structures end abruptly. Near the fault, layers may be bent or crumpled, and along the fault streaks of finely powdered material may have developed from friction during movement. Three important kinds of fault are shown in Fig. 15-1.

Movement along faults usually takes place as a series of small, sudden displacements, with intervals of years or centuries between succes-

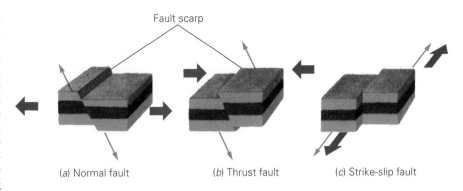

Fault scarp

(a) Normal fault   (b) Thrust fault   (c) Strike-slip fault

FIG. 15-1 (*a*) A normal fault is an inclined surface along which a rock mass has slipped downward; it is the result of forces that tend to stretch the earth's crust. (*b*) A thrust fault is an inclined surface along which a rock mass has moved upward to override the neighboring mass; it is the result of forces that tend to compress the crust. (*c*) A strike-slip fault is a surface along which one rock has moved horizontally with respect to the other; it is the result of oppositely directed forces in the crust that do not act along the same line. Erosion modifies the fault scarps left by normal and thrust faults.

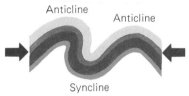

Anticline

Anticline

Syncline

FIG. 15-2 Cross section showing effect of folding in horizontal strata. Folds always shorten the crust and hence are produced by compressional forces. An anticline is an arch (a fold convex upward), and a syncline is a trough (a fold convex downward). In regions of intense folding, anticlines and synclines follow one another in long series.

sive jerks. An immediate effect of displacement along a thrust fault or normal fault is a small cliff. Erosion then attacks the cliff and may level it before the next movement. If successive movements follow one another fast enough, erosion may not be able to keep up, with a high cliff as the result. Cliffs of this sort are called **fault scarps.** Good examples of scarps produced by normal faults are the steep mountain fronts of many of the desert ranges in Utah, Nevada, and eastern California. A more deeply eroded scarp produced by thrust faulting is the eastern front of the Rocky Mountains in Glacier National Park.

Folding takes place by slow, continuous movement, in contrast to the sudden displacements along faults (Fig. 15-2). Sometimes folding produces hills and depressions in the landscape directly, but more often erosion keeps pace with folding. Indirectly folds affect landforms by exposing tilted beds of varying degrees of resistance to the action of streams, so that long, parallel ridges and valleys develop, like those of the Appalachian Mountains (see Fig. 14-15). In these mountains, as in

Thrust fault in sandstone near Klamath Falls, Oregon. A thrust fault occurs when a rock mass moves upward relative to adjacent rock.

Folded shale beds near Palmdale, California.

FIG. 15-3 Drowned valleys on the Atlantic Coast of the United States.

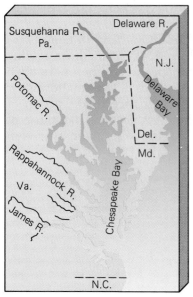

many others, the folding is very ancient. The present ridges are due to deep erosion after successive uplifts of the stumps of the old folds.

Large-scale crustal movements may involve whole continents or large parts of them, which may rise or fall, tilt or warp. Such events are occurring today as they have in the past. Coastal features in many parts of the world provide obvious evidence. For instance, a wave-cut cliff and terrace high above the present shore means that the coast there has been raised in fairly recent times, or stream erosion would have eaten these features away.

The sinking of land with respect to sea level is shown by long, narrow bays that fill the mouths of large stream valleys. A body of water like Chesapeake Bay (Fig. 15-3) could not be formed by wave erosion, since wave attack normally straightens a coastline rather than deeply indenting it. Instead, the shape of the bay suggests that it lies in a stream-carved valley whose lower part has been submerged beneath the sea. Elsewhere in the landscapes and sedimentary rocks of the United States a multitude of regional movements are recorded.

15-2   MOUNTAIN BUILDING

The Rock Record Tells a Complicated Story   Mountains can form in a number of ways. Some mountains are accumulations of lava and fragmental material ejected by a volcano. Others are small blocks of the earth's crust raised along faults. But the great mountain ranges of the earth, like the Appalachians, the Rockies, the Alps, and the Himalayas, have a much longer and more complex history involving sedimentation, folding, faulting, igneous activity, repeated uplifts, and deep erosion.

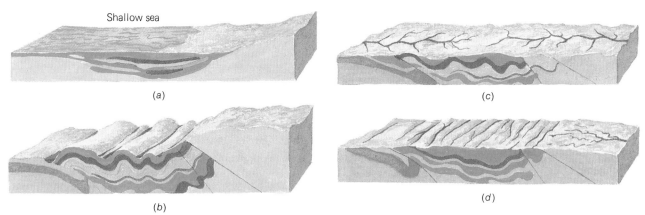

(a)

(b)

(c)

(d)

FIG. 15-4 Successive stages in the evolution of the Appalachian Mountains. (a) Sediments accumulating in the Appalachian basin; (b) folding and thrust faulting of rocks in the basin; (c) original mountains worn down to a nearly level plain by stream erosion; (d) renewed erosion of the folded strata following vertical uplift, producing the parallel ridges and valleys of the present landscape.

The layers of sedimentary rock in a mountain range are usually thicker than layers of similar age found under adjacent plains. Most of the layers in both mountains and plains, as shown by their sedimentary structures and their fossils, were formed from deposits that accumulated in shallow seas or on low-lying parts of the land—that is, on surfaces not far above or below sea level. As deposition continued, the surface must have been slowly sinking at the same time. The greater thickness of the strata in the present mountain area means that this part of the earth's surface was sinking more rapidly than nearby areas.

Besides their thickness, another conspicuous feature of the sedimentary rocks in major mountain ranges is their complex structure (Fig. 15-4). They are crumpled by intense folding and locally broken along huge thrust faults and minor normal faults. Thus in the development of a mountain range the next step must be a period in which the piled-up sediments are subjected to intense compressional forces. (Later in this

The Teton mountain range in Wyoming came into being when the eastern end of a crustal block was raised along a fault. As a result the Teton range slopes gradually toward the west from a steep eastern front (shown here).

chapter we will see how these forces arise.) The compression raises the folded layers high above the sea, and erosion begins to wear down the exposed beds as folding continues. The appearance of most present-day mountain landscapes is not due directly to the compressions that folded and faulted their rocks but instead to erosion between periodic uplifts.

15-3   CONTINENTAL DRIFT

An Evolving Jigsaw Puzzle   A glance at a map of the world suggests that at some time in the past the continents may have been joined together in one or two giant supercontinents. If the margins of the continents are taken to be on their continental slopes (see Fig. 13-26) at a depth of 900 m instead of their present sea-level boundaries, the fit between North and South America, Africa, Greenland, and western Europe is remarkably exact, as Fig. 15-5 shows.

But merely matching up outlines of continents is not by itself proof that the continents have migrated around the globe. The first really detailed theory of continental drift was proposed early in this century by Alfred Wegener, who based his argument on biological and geologic evidence.

FIG. 15-5 How some of the continents fit together. The boundary of each continent is taken at a depth of 900 m on its continental slope; the tan regions represent land above sea level at present, and the light orange regions represent submerged land on the continental shelf and slope. Overlaps are shown in dark orange and gaps in blue.

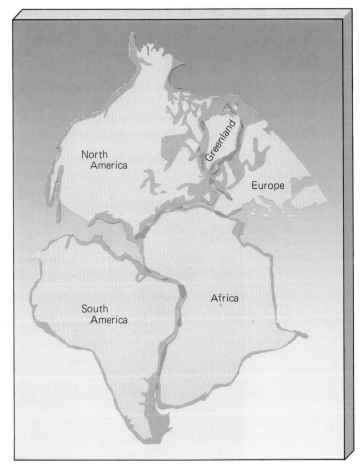

BIOGRAPHY

Alfred Wegener (1880–1930) was a German meteorologist with a special interest in the Arctic. He participated in four expeditions to Greenland and died on the last of them. Like others before him, Wegener was struck by the apparent fit of the continents shown in Fig. 15-5, but unlike them he went on to develop biological and geological arguments to support the notion that the continents had once been united. Wegener did not stop with tracing the subsequent movements of the continents but also considered other effects of their motions. He believed that mountain ranges were forced up by pressure on the leading edges of the drifting continents, with the Rockies and Andes as examples. The trailing edges of the drifting continents, in Wegener's view, left behind fragments that ended up as such island arcs as the West Indies.

Wegener first presented his ideas in 1912 and continued to refine them until his death. Some geologists were immediately attracted by the boldness and comprehensiveness of his scheme. Others were strongly opposed, at least partly because Wegener, a meteorologist, seemed to them an outsider heedlessly trying to overthrow established geological concepts. Doubters of Wegener explained the similarity of patterns of early life around the world by postulating a series of land bridges linking the continents. But no traces of such bridges were ever found. As for the mountains, it was widely believed that the earth was in the process of shrinking, with its surface becoming wrinkled the way a baked apple does as its water content evaporates. But the massive earth cannot contract as a baked apple does: the wrinkles of its continents can only have been produced by horizontal forces. A more serious objection to continental drift was Wegener's inability to come up with a really convincing mechanism for pushing the continents around. Today such a mechanism is known, and even though he got some details wrong,

Wegener now can be seen as a brilliant pioneer and not as the "pseudo-scientist" his opponents called him.

Wegener was troubled by the parallel evolution of living things. Going back through the ages, the fossil record shows that, until about 200 million years ago, whenever a new species appeared, it did so in many now-distant regions where suitable habitats existed. Evolution, in other words, proceeded at the same rate and in the same way in continents and oceans that today are widely separated. Only in the last 200 million years have plants and animals in the different continents developed in markedly different ways.

What Wegener suggested instead was that the continents were once all part of one huge landmass he called **Pangaea** ("all earth"). Then Pangaea broke up, and the continents slowly drifted to their present locations. This model found additional support in geologic data regarding prehistoric climates. Somewhat over 200 million years ago South Africa, India, Australia, and part of South America were burdened with great ice sheets, while at the same time a tropical rain forest covered North America, Europe, and China. At various other times, there was sufficient vegetation in Alaska and Antarctica for coal deposits to have resulted, and so currently frigid a place as Baffin Bay was a desert.

Wegener and his followers examined what was known about the climates of the distant past and tried to arrange the continents in each geologic period so that the glaciers of that period were near the poles and the hot regions were near the equator. Their efforts were, in general, quite successful, and in some cases startlingly so. Deposits of glacial debris and fossil remains of certain distinctive plant species follow each other in the same succession in Argentina, Brazil, South Africa, Antarctica, India, and Australia, for example. A recent discovery of this kind

FIG. 15-6 The landmasses of the earth as they may have appeared in the past and as they are today. The breakup of Pangaea into Laurasia and Gondwana began about 200 million years ago.

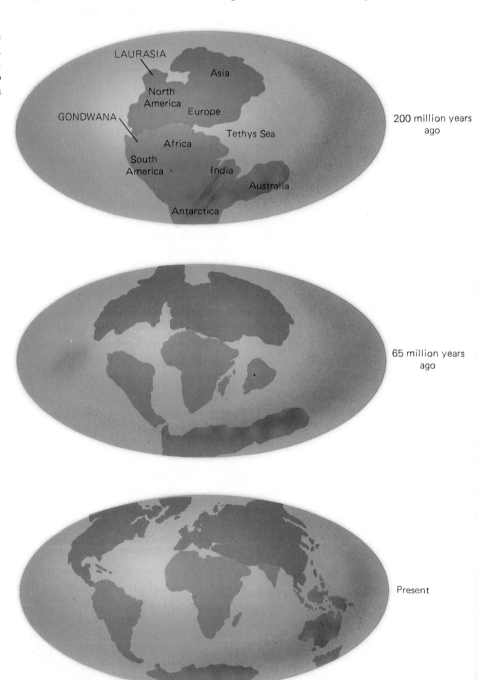

200 million years
ago

65 million years
ago

Present

was the identification of a skull of the reptile Lystrosaurus in a sandstone layer in the Alexandra mountain range of Antarctica. This creature, which was about 3 ft long, flourished long ago in Africa.

Laurasia and Gondwana

Today it seems clear that Pangaea did exist. About 200 million years ago it began to break apart into two supercontinents, **Laurasia** (which consisted of what is now North America, Greenland, and most of Eurasia) and **Gondwana** (South America, Africa, Antarctica, India, and Australia). Laurasia and Gondwana were almost equal in size (Fig. 15-6). The separation of Pangaea into these supercontinents is supported by detailed geologic and biological evidence, for instance the differences between Laurasian and Gondwana fossils of the same age after the breakup.

Laurasia and Gondwana were separated by a body of water called the **Tethys Sea.** Today a little of the Tethys Sea survives as the Mediterranean, Caspian, and Black Seas. Its original extent can be gauged from the sediments that were subsequently uplifted to form the mountain ranges that stretch from Gibraltar eastward to the Pacific. The Pyrenees, Alps, and Caucasus of Europe, the Atlas Mountains of North Africa, and the Himalayas of Asia all were once part of the Tethys Sea.

Not long after Pangaea divided into Laurasia and Gondwana, the supercontinents themselves started to break up. The North Atlantic and Indian Oceans were the first to open, followed by the South Atlantic. Perhaps 80 million years ago Greenland began to move away from North America; 45 million years ago Australia split off from Antarctica and India finished its journey north to Asia; and 20 million years ago Arabia separated from Africa.

PLATE TECTONICS

"Continental drift" is not too accurate a description because the continents turn out to be merely passengers on a number of rigid rock plates that are continually moving across the face of the earth. Before we examine how this motion occurs and what its consequences are, we must first see why it is possible in the first place.

15-4 LITHOSPHERE AND ASTHENOSPHERE

A Hard Layer over a Soft Layer

At the heart of all current explanations for major crustal changes is the idea that the mantle of the earth is not a stiff structure, like the crust, but rather contains a layer near the top that is able to flow.

The crust and the outermost part of the mantle together make up a shell of hard rock 50 to 100 km thick called the **lithosphere.** The lithosphere has no sharp boundary, as the crust does, but gradually turns into the softer **asthenosphere,** a region about 100 km thick. (*Lithos* means "rock" in Greek and *astheno* means "weak.") The asthenosphere is soft because its material is close to the melting point at the temperature and pressure found at such depths. Above the asthenosphere the temperature is too low and below it the pressure is too high for the material to deform easily.

WHY CONTINENTS HAVE DEEP ROOTS

The notion of a relatively soft asthenosphere makes it possible to understand both why the continents are raised above the rest of the crust and why the crust is much thicker beneath them than beneath the oceans (see Fig. 14-12). Suppose that we place several blocks of wood of different sizes in a pool of water. The larger blocks float higher than the smaller while simultaneously extending down farther into the water (Fig. 15-7). Thus, if the lithosphere with its continents is imagined as floating on a denser asthenosphere, we can refer for guidance to the floating wooden blocks. This analogy suggests that exceptionally high regions—mountain ranges and plateaus—have roots that extend far downward. Such is actually the case; in fact, its discovery led the British scientist George Airy a century ago to propose the floating of the entire lithosphere.

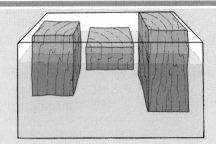

FIG. 15-7 Large blocks of wood float higher and extend farther downward than smaller blocks of wood. This is why the thicker continental crust extends lower into the mantle than the oceanic crust, as shown in Fig. 14-12.

The speeds of earthquake waves in the crust are different from their speeds in the mantle, which suggests that the two regions differ in their compositions or crystal structures or both. The difference between the lithosphere and the asthenosphere instead lies in their degrees of rigidity. Insofar as brief, suddenly applied forces are concerned, the asthenosphere behaves like a solid and so can transmit the transverse S waves of earthquakes, for example. But when forces act on the asthenosphere over long periods of time, it responds like a thick, viscous fluid.

15-5 THE OCEAN FLOORS

The Youngest Part of the Crust

The mountains and valleys, plains and plateaus of the continents were mapped long ago, and few surprises are in store for future explorers. But the continents occupy less than 30 percent of the area of the earth's crust, while the rest lies hidden in darkness thousands of meters under the seas and oceans. Only in the past few decades have the floors of the oceans been studied and their physical characteristics determined. It is largely these findings that have clarified the evolution of the crust.

Methods used to investigate the ocean floors are not particularly subtle—the real problem is the vastness of the area to be covered. These days depths are charted by means of echo sounders. An instrument of this kind sends out a pulse of high-frequency sound waves and the time needed for the pulse to reach the sea floor, be reflected there, and then return to the surface is a measure of how deep the water is (Fig. 15-8). A variant of this method reveals something of the structure of the sea floor itself. What is done is to set off an explosive charge in the water and study the returning echoes—one echo will come from the top of the sediment layer, and a later one from the hard rock underneath.

FIG. 15-8 The principle of echo sounding. (a) A pulse of high-frequency sound waves is sent out by a suitable device on a ship. (b) The time at which the pulse returns to the ship is a measure of the sea depth.

(a)

(b)

Samples of the sea floor can be obtained by dropping a hollow tube to the bottom on a long cable and then pulling it up filled with a core of the sediments into which it sank. Longer sediment cores can be obtained by drilling. These sediments can be examined later in the laboratory for their composition, their age (by radioactive dating), their fossil content, their magnetization, and so forth. Another important technique is to tow a magnetometer behind a survey ship to obtain an idea of the direction and intensity of the magnetization of the rocks of the ocean floor over wide areas.

Four Important Findings

Four findings about ocean floors have proved of crucial importance:

1. The ocean floors are, geologically speaking, very young. The oldest oceanic crust dates back only about 200 million years, in contrast to continental rocks that date back as much as 3800 million years. Many parts of the ocean floor

Samples from a core of sediments extracted from the ocean floor are removed for examination. The record of the past is most complete in marine sediments.

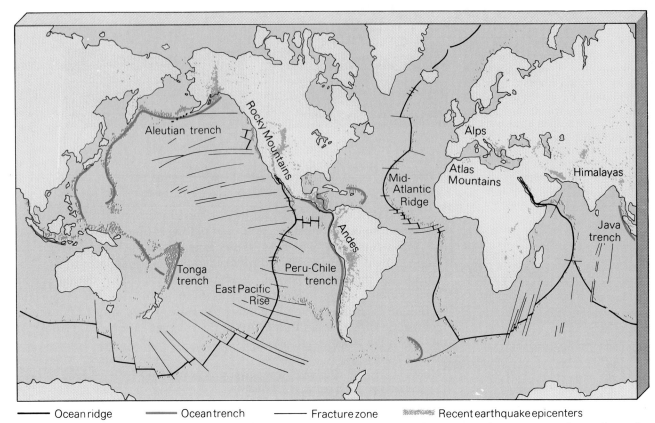

— Ocean ridge   — Ocean trench   — Fracture zone   ⋯ Recent earthquake epicenters

FIG. 15-9 The worldwide system of oceanic ridges, rises, and trenches. The ridges and rises are offset by transverse fracture zones.

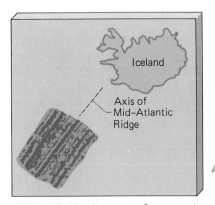

FIG. 15-10 Pattern of magnetization along the Mid-Atlantic Ridge southwest of Iceland. Sea-floor rocks whose directions of magnetization are the same as that of today's geomagnetic field are shown in dark blue; the intervening spaces represent rocks whose magnetization is in the opposite direction.

are much younger still, so that about one-third of the earth's surface has come into existence in $1\frac{1}{2}$ percent of the earth's history.

2. A worldwide system of narrow **ridges** and somewhat broader **rises** runs across the oceans (Fig. 15-9). An example is the Mid-Atlantic Ridge, which runs down the Atlantic Ocean from north to south. Iceland, the Azores, Ascension Island, and Tristan da Cunha are some of the higher peaks in this ridge. These ridges are offset at intervals by fracture zones that indicate sideways shifts of the ocean floors.

3. There is also a system of **trenches** several kilometers deep that rims much of the Pacific Ocean. These trenches parallel the belts in which most of today's earthquakes and volcanoes occur. Some of the trenches have **island arcs** on their landward sides that consist largely of volcanic mountains projecting above sea level.

4. The direction in which ocean-floor rocks are magnetized is the same along strips parallel to the midocean ridges, but the direction is reversed from strip to strip going away from a ridge on either side (Fig. 15-10).

15-6 OCEAN-FLOOR SPREADING

Alternate Magnetization Is the Proof

The first step toward understanding the above observations was taken in the early 1960s by the American geologists Harry H. Hess and Robert S. Dietz, who independently proposed that the ocean floors are spreading. (A similar hypothesis was put forward in 1928 by Arthur Holmes in England, but it remained practically unnoticed because supporting data were lacking.)

The basic idea of ocean-floor spreading is that molten rock is continually rising up along the midocean ridges (Fig. 15-11). The parts of the lithosphere on either side of a ridge are pushed apart by the molten rock at speeds of a few centimeters per year—about as fast as fingernails grow—with the new material filling the gap as it hardens.

The fourth key observation mentioned earlier, which concerns the magnetization of rocks on either side of an ocean ridge, confirms the hypothesis of sea-floor spreading in a convincing way. As Fig. 15-10 shows in the case of a portion of the Mid-Atlantic Ridge southwest of Iceland, successive strips of rock lying parallel to the ridge are magnetized in alternate directions. To interpret this pattern, we draw upon the fact that the earth's magnetic field has periodically reversed itself many times in the past (Sec. 14-9). What must have been happening is clear. As molten rock, unmagnetized in its liquid state, comes to the surface of the crust at a ridge, it hardens and the iron content of its minerals becomes magnetized in the same direction as that of the geomagnetic field at the time. When the direction of the geomagnetic field reverses, the new molten rock that cools then becomes magnetized in the opposite direction. Thus strips of alternate magnetization follow one another going away on both sides from a ridge.

The reversals of the earth's magnetic field have been dated by measurements made on magnetized lava flows on land using methods based on radioactivity. This information can be used to find the ages of the magnetized strips that make up the sea floor, and thereby to establish the speeds with which the lithosphere plates have been moving. These speeds vary from about 1 cm per year in the North Atlantic Ocean to about 10 cm per year in the Pacific Ocean.

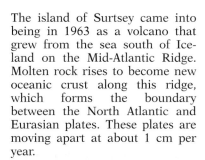

The island of Surtsey came into being in 1963 as a volcano that grew from the sea south of Iceland on the Mid-Atlantic Ridge. Molten rock rises to become new oceanic crust along this ridge, which forms the boundary between the North Atlantic and Eurasian plates. These plates are moving apart at about 1 cm per year.

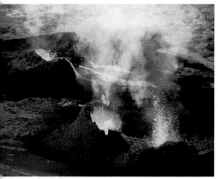

FIG. 15-11 Ocean-floor spreading. A midocean ridge forms where molten rock rises from the asthenosphere along a crack in the ocean floor and pushes apart the lithosphere on both sides.

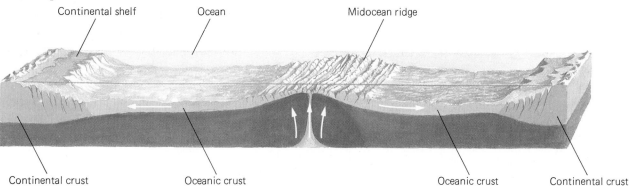

Continental shelf   Ocean   Midocean ridge

Continental crust   Oceanic crust   Oceanic crust   Continental crust

15-7 PLATE TECTONICS

How the Continents Drift

Although the ocean floors are spreading at the mid-ocean ridges, the earth as a whole does not expand. The spread of the ocean floors must therefore be balanced by other large-scale processes in the lithosphere. The study of these processes and their consequences has come to be known as **plate tectonics.**

The starting point of plate tectonics is the observation that the lithosphere is split not only along the ridges of the ocean floors but also along the trenches and fracture zones of Fig. 15-9. These cracks divide the lithosphere into seven huge **plates** and a number of smaller ones, all of which float on the plastic asthenosphere (Fig. 15-12).

New lithosphere is created where plates move apart at the midocean ridges, as we saw in Fig. 15-11. Where plates come together, on the other hand, lithosphere may be destroyed: the edge of one of the plates may slide under the edge of the other and partially melt upon reaching the hot asthenosphere. Figure 15-14 shows the three possible kinds of colli-

FIG. 15-12 The chief lithospheric plates whose motion results in continental drift. The plates are bounded by ridges, rises, trenches, and faults. The arrows show the directions of plate motions. Dashed lines indicate uncertain boundaries. The African plate is thought to be stationary. Because Japan lies on or near the intersections of four plates, one-tenth of the world's earthquakes rock its islands.

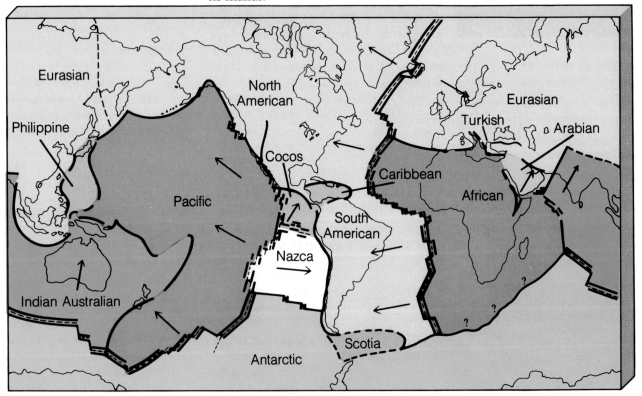

THE HAWAIIAN ISLANDS The Hawaiian Islands were formed as the Pacific plate moved northwestward over a hot spot in the mantle under the plate (Fig. 15-13). From time to time a plume of molten rock erupted from this hot spot to create a volcano above it that rose above the ocean floor. These volcanoes, now the islands of the Hawaiian chain, became extinct after they passed the hot spot. The only volcanic activity in that region today is on the island of Hawaii, which is over the hot spot.

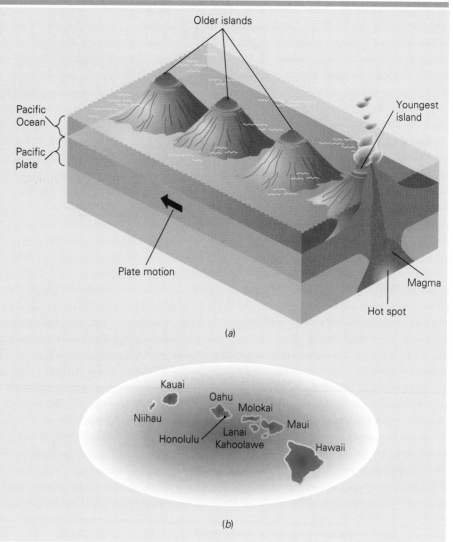

FIG. 15-13 (*a*) Origin of the Hawaiian Islands. (*b*) The islands today.

sion. To understand what is happening, we must keep in mind that continental crust (largely granitic rock) is less dense than oceanic crust (largely basaltic rock).

1. **Oceanic-continental plate collision.** When a plate whose edge is covered with oceanic crust moves against a plate whose edge is covered with continental crust, the denser oceanic slab slides underneath the continental slab at a steep angle (Fig. 15-14*a*). The region of contact is called a **subduction zone,** and a trench is formed there. Some of the oceanic slab melts in the asthenosphere, and magma from the lighter materials in the slab is carried upward by its buoyancy. The magma rises through cracks in the continental plate above to produce volcanoes at the surface and bodies of intrusive rock below the surface. An example of such a collision occurs at

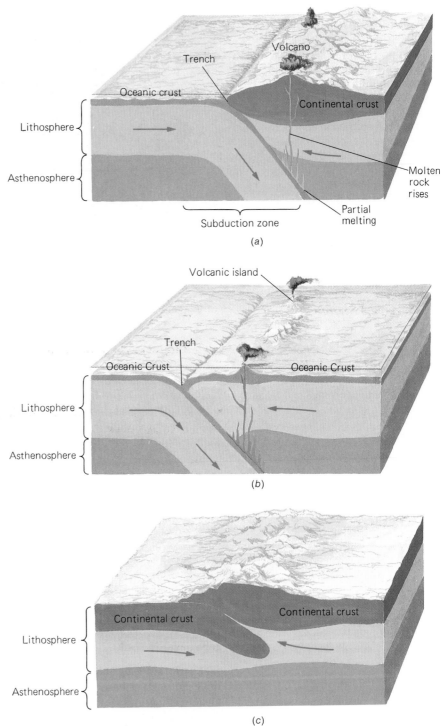

FIG. 15-14 Three situations can occur when lithospheric plates come together. (*a*) Oceanic-continental plate collision. The Andes Mountains of South America are the result of such a collision. (*b*) Oceanic-oceanic plate collision. The islands of the West Indies originated in this way. (*c*) Continental-continental plate collision. A collision of this kind thrust up the Himalaya Mountains between India and the rest of Asia.

the western edge of South America, where the oceanic Nazca plate (which is moving eastward) meets the continental South American plate (which is moving westward). The result is a trench along the coasts of Peru and Chile and a range of volcanic mountains, the

The Andes are a young mountain range thrust upward along the western edge of South America where the eastward-moving Nazca plate is forced under the westward-moving Atlantic plate. This landscape is in Peru.

Andes (Fig. 15-15). About 400 of the world's 500 or so active volcanoes are found above subduction zones.

2. **Oceanic-oceanic plate collision.** When two plates whose edges are covered with oceanic crust collide, one of them slides underneath the other in a subduction zone (Fig. 15-14*b*). Volcanoes are again formed, this time on the ocean floor. Where these volcanoes are high enough to rise above the ocean, they appear as the chains of islands called **island arcs.** The island arcs that border the Asiatic side of the Pacific—the Aleutians, Japan, the Philippines, Indonesia, the Marianas—are believed to have come into being in this way. The West Indies in the Atlantic form another example of an island arc.

FIG. 15-15 The sea floor spreads apart at midocean ridges where molten rock rises to the surface of the lithosphere. At a trench, one lithosphere plate is forced under another into the asthenosphere, where it melts. Mountain ranges, volcanoes, and island arcs are found where plates collide. The vertical scale is greatly exaggerated.

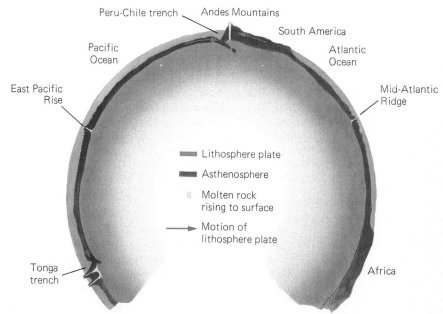

Mt. Everest, the highest mountain on earth, is part of the Himalaya range in Asia that was forced upward when the Indian plate collided with the Eurasian plate starting about 45 million years ago. The Indian plate is still moving north at 2 cm per year.

FIG. 15-16 A transform fault separates two plates that are sliding past each other. An example is the San Andreas Fault in California.

Transform fault

3. **Continental-continental plate collision.** Here both plates are too light relative to the underlying asthenosphere and too thick for either one to be forced under. The result is that the plate edges are pushed together and buckle, forming a mountain range (Fig. 15-14c). The massive Himalayas that divide India from the rest of Asia were thrust upward in this manner. The geologic evidence for this event is backed up by the fossil record. The oldest mammal fossils in India date back only 45 million years, just when the Indian and Eurasian plates came together, and these fossils are similar to those found in Mongolia, the part of Asia that India joined. The Ural Mountains between Europe and Asia are a much older range that formed in the same manner.

Because the earth's crust under the oceans is being continuously created at midocean ridges and destroyed at subduction zones, all of it is relatively young. On the other hand, most continental crust is quite old, as much as 3.8 billion years old as compared with less than 180 million years for oceanic crust.

Transform Faults

Another type of plate boundary is a **transform fault,** which occurs where the edges of two plates slide past each other (Fig. 15-16). Earthquakes are common along transform faults, as we would expect. Since most transform faults

Part of the San Andreas Fault near Taft, California. The earthquake of October 17, 1989 that rocked San Francisco originated about 100 km south of the city in a zone of this fault whose sides shifted by about 2 m relative to each other. The slippage occurred underground.

are in ocean basins, the quakes associated with them usually are unnoticed except by geologists. A conspicuous exception is the 1200-km-long San Andreas Fault in California, which is one of the faults that lie along the boundary between the Pacific and North American plates. The Pacific plate is moving northward relative to the North American plate at an average of about 6 cm/year. Movement along the fault occurs continuously in some regions but elsewhere in sudden jerks that release accumulated stresses. The San Francisco earthquakes of 1906 and 1989 were caused by such abrupt slippages, and further earthquakes in that area are sure to happen in the future.

When present trends in the evolution of the earth's crust are projected 30 million years into the future, the result is a picture like that shown in Fig. 15-17. The Atlantic Ocean has grown wider, the Pacific narrower. Part of California has broken off from the rest of North America, and the Arabian peninsula has been forced around to become an integral part of Asia. The islands of the West Indies have grown into a land bridge between the Americas, and the western Pacific islands have also increased markedly in extent.

FIG. 15-17 How the earth may appear 30 million years from now if present plate motions continue.

And after that? All that can be said is that the face of the earth will probably continue to change in the future, just as it has been changing as far back in the past as we have any evidence.

METHODS OF HISTORICAL GEOLOGY

Two kinds of events in the history of the crust are significant. In one category are physical changes such as the drift of the continents, the upthrust and wearing away of mountains, the spread and retreat of glaciers and ice sheets. In the other are changes in the living things that populate the earth, from primitive one-celled organisms to the complex plants and animals of today. Although the rock record is less complete the farther back we go, it is still possible to trace the physical and biological evolution of the crust for billions of years into the past. Let us see how this is done.

15-8   PRINCIPLE OF UNIFORM CHANGE

"No Vestige of a Beginning, No Prospect of an End"
The modern view of geologic evolution had its start two centuries ago. Before then a major obstacle to scientific thinking about the earth's history was the Bible. The Book of Genesis tells about the earth's beginning with beauty and simplicity. A seventeenth-century theologian, Bishop James Ussher, used the stories in the Bible to pinpoint the moment when our planet was created from a formless void: 9 o'clock in the morning of October 12, 4004 B.C. However, even the smallest knowledge of geology shows that the events recorded in today's landscapes cannot possibly have occurred in only a few thousand years. Early investigators nevertheless devoted their efforts to trying to fit what they found in the rocks around them into a literal interpretation of Genesis.

Even when freed from biblical shackles, geologists of the past usually went far astray in their arguments. One reason was their habit of generalizing from limited evidence. Another was their readiness to postulate tremendous events to explain particular findings. For instance, a comprehensive theory was formulated by Abraham Gottlob Werner on the basis of the geologic structures he found near his home in Freiberg, Germany. What he saw was granite overlain by folded, somewhat metamorphosed rocks, with these in turn overlain by flat sedimentary beds. Untraveled and deaf to the reports of others, Werner considered this sequence to be the same worldwide. Each of the three types of rocks, he presumed, was deposited by a universal ocean, with granite precipitating first and the upper beds last. All rocks, in Werner's system, were sedimentary rocks, and the geologic history of the earth consisted of three sudden precipitations from an ancient ocean that were followed by the disappearance of most of the water.

The French biologist Georges Cuvier greatly influenced geology at the beginning of the nineteenth century by studying **fossils,** which are the remains or traces of organisms preserved in rocks. In successive rock layers around Paris, Cuvier found distinct groups of animal fossils, different from one another and from the present animals of that region. He concluded that each group appeared on the earth as a result of a special creation and that each was destroyed by a universal disaster before the next creation. Thus Cuvier also regarded the earth's history as a succession of catastrophes that were separated by intervals of stable conditions.

Hutton, Lyell, and Darwin

James Hutton, a Scot, based his thinking on a much larger body of observational evidence than Werner or Cuvier. By accepting that the earth is very old, Hutton found no need to invoke special mechanisms since he could account for what he saw in terms of processes under way in the present-day world: "In the phenomena of the earth," he concluded, "I see no vestige of a beginning, no prospect of an end."

Hutton's ideas, outlined in his 1785 book *Theory of the Earth*, were soon taken up by others who modified and extended them. Chief among Hutton's followers was Charles Lyell, whose goal was "to explain the former changes of the earth's surface by forces now in operation." This he was largely able to do, and his guiding idea became known as the **principle of uniform change.** Of course, the vigor of geologic processes has varied from time to time and from place to place, so that uniform is not the best word. However, with a few exceptions, such variations are quite different from the catastrophic happenings earlier thinkers were so fond of.

An important link in the chain of ideas that was forming was supplied by Lyell's friend Charles Darwin in 1859. Darwin's **theory of evolution** showed that changes in living things as well as those in the inorganic world of rocks could be explained in terms of processes operating

EVOLUTION Evolution is the "view of life" (in Darwin's words) that has turned out to be the only perspective from which all the different aspects of biology make sense. "I had two distinct objects in view: firstly to show that species had not been separately created, and secondly, that natural selection has been the chief agent of change." Starting from the simplest organisms, "endless forms most beautiful and most wonderful have been, and are being, evolved."

Evolution operates through two mechanisms. The first is reproduction with variation: the members of each generation are not necessarily the same as each other or as their parents. Darwin did not know how variation comes about, but today we can trace it to changes (called mutations) in the DNA of which the genes of living things are composed.

The second mechanism is natural selection: the individual plants and animals that are best adapted to their environments are most likely to thrive and produce similarly successful offspring. In simple terms, "survival of the fittest." When living conditions alter drastically, which has happened often in the history of the earth, the pace of evolution is faster than during periods of stability. At such times previously dominant species (such as the dinosaurs) may even disappear entirely, while previously marginal species (such as the mammals) become favored.

A huge body of evidence supports the basic concepts of evolution. Thus the fossil record shows the gradual development of new forms of life from older ones, with a time scale established by radiometric methods that is long enough (billions of years) for this development to have taken place. The distribution of life forms around the world shows the effects of evolution: Australia was long isolated from the other continents, and many plants and animals there are distinct from those elsewhere. The anatomies of different species often show great similarities (human arms have much in common with whale flippers), which suggests descent from a common ancestor. Certain body parts that today have no function can be traced back to forebears that needed them: an example is the coccyx at the base of the spine, the remnant of a tail. And, of course, even before Darwin farmers speeded up evolution by choosing plant and animal variations to breed that were especially desirable for their purposes. Even this very brief list shows the power of evolution to account for the history, variety, and relatedness of life on earth.

Charles Darwin
(1809–1882)

all around us. The fossil groups found by Cuvier near Paris did not come from special creations but were stages in a continuous line of development. Lyell understood at once the significance of Darwin's work and became one of his most active supporters.

How far back does the principle of uniform change hold? It must break down eventually because the earth in its early stages as a planet was certainly different from today's earth. All geology can say for certain is that the oldest rocks now exposed contain a clear record of processes very similar to those of the present. Beyond that lies the realm of hypothesis, and in Chap. 18 we shall see how theories of the origin of the earth connect up with geologic history.

15-9   ROCK FORMATIONS

**History
under
Our Feet**

The crustal events of the past are recorded in the rocks and landscapes of the present. It is usually possible to reconstruct these events in terms of processes

The Colorado River carved Arizona's Grand Canyon from sedimentary layers, most of them deposited at times when the region was covered by shallow seas. Exposed rocks are mainly limestone, shale, and sandstone.

still at work reshaping the face of the earth. Thus from moraines, lakes, and U-shaped valleys we learn of the spread and retreat of ancient glaciers. Wave-cut cliffs and terraces above the sea suggest recent elevation of the land. Hot springs and isolated, cone-shaped mountains signify past volcanic activity.

Other past events have left their traces as well. A geologist finds a bed of salt or gypsum buried beneath other strata, and he or she knows that the region must once have had a desert climate in which a lake or an arm of the sea evaporated. A layer of coal implies an ancient swamp in which partly decayed vegetation accumulated. A limestone bed with numerous fossils suggests a clear, shallow sea in which lived clams, snails, and other hard-shelled organisms. As the long history is carried further and further back, the evidence becomes more shadowy and the geologist's reconstruction of the earth's surface similarly imprecise.

In trying to figure out how the earth's crust evolved, geology is faced with two fundamental problems: to arrange in order the events recorded in the rocks of a single outcrop or small region, and to correlate events in various regions of the world to give a connected history of the earth as a whole.

FIG. 15-18 Schematic cross section showing folded sedimentary rocks that were displaced along a fault and then intruded by granite.

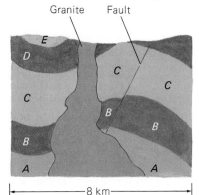

Reading the Rocks

Some of the principles used by geologists to figure out the history of a small area are straightforward:

1. In a sequence of sedimentary rocks, the lowest bed is the oldest and the highest bed is the youngest. In Fig. 15-18 bed *A* must have been deposited before the others and bed *E* last.

2. Sedimentary beds were originally deposited in approximately horizontal layers.

3. Tectonic movement took place after the deposition of the youngest

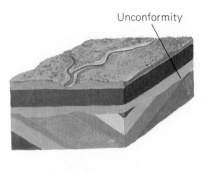

Unconformity

FIG. 15-19   An unconformity is an irregular eroded surface that separates one set of rock layers from an earlier set. Shown is an unconformity above tilted lower layers; the layers above and below an unconformity can also be parallel.

This angular unconformity in Scotland consists of sandstone layers from the Devonian Period that overlie layers of Silurian rock that have been tilted almost vertically.

bed affected. Thus the layers of Fig. 15-18 were not folded until after bed *D* was laid down, and the fault must be younger than bed *C*.

4. An igneous rock is younger than the youngest bed it intrudes. The granite pluton shown in Fig. 15-18 is younger than bed *D*. (The age of an igneous rock refers to the time at which it solidified.)

Obvious as these rules are, much ingenuity may be needed to apply them to regions of heavily folded and faulted layers. The problem is especially difficult in regions where much of the rock structure is hidden by later sediments or vegetation.

A structure like that shown in Fig. 15-19 requires further attention. Here the lower, tilted beds are cut off by an uneven surface on which rest the upper horizontal beds. An irregular surface of this sort, separating two series of rocks, is called an **unconformity.**

An unconformity is a buried surface of erosion. It always involves at least four geologic events: the deposition of the oldest strata; tectonic movement that raises and perhaps tilts the existing strata; erosion of the elevated strata to produce an irregular surface; and finally a new period of deposition that buries the eroded surface. Usually this last event involves the lowering of the eroded surface either beneath the sea or to a level where stream deposition can occur.

15-10   RADIOMETRIC DATING

A Clock Based on Radioactivity

To trace the sequence of geologic events in various places is not enough. We also need to know which events happened at the same time in different places if we are to understand how the earth's crust evolved.

We could get the data we need by following rock layers from one region to another around the globe, but in fact after a shorter or longer distance the layers are always either cut off by erosion or concealed by later deposits.

Two methods can be used to figure out the worldwide sequence of the events that have shaped the earth's surface: radioactive dating and fossil identification. As we shall find, each method has different advantages and disadvantages.

Methods based on radioactive decay make it possible to establish the ages of many rocks on an absolute rather than a relative time scale. Because the decay of any particular radionuclide proceeds at a steady

TABLE 15-1

Radionuclides Used for Dating Rocks		
Parent Nuclide	Stable Daughter Nuclide	Half-Life, Billion Years
Potassium 40	Argon 40	1.3
Rubidium 87	Strontium 87	47
Thorium 232	Lead 208	13.9
Uranium 235	Lead 207	0.7
Uranium 238	Lead 206	4.5

rate regardless of external conditions, the ratio between the amounts of that nuclide and its stable daughter in a rock sample gives an indication of the age of the rock. The greater the proportion of the daughter nuclide, the older the rock. Radioactive decay was discussed in Chap. 7; see especially Fig. 7-9.

Table 15-1 lists the radionuclides that have been found most useful in dating rocks. Age measurements of fossil-bearing rocks with the radioactive clock reveal that relatively large-brained humanlike creatures walked erect in Africa about 2 million years ago; that primitive mammals existed about 200 million years ago; that sea animals with hard shells first became abundant about 600 million years ago. The most ancient rocks whose ages have been determined are found in northern Canada and are nearly 4 billion years old; zircon samples have been discovered in Australia that seem to have crystallized even earlier.

Radiocarbon Another radionuclide permits the dating of more recent remains of living things. The carbon isotope ^{14}C, called **radiocarbon,** is beta radioactive with a half-life of 5700 years. Radiocarbon is produced in small quantities in the earth's atmosphere by the action of cosmic rays (discussed in Chap. 18) on nitrogen atoms, and the carbon dioxide of the atmosphere accordingly contains a small proportion of radiocarbon. Green plants take in carbon dioxide in

This mass spectrometer at the University of Arizona measures the proportion of ^{14}C atoms in carbon from a sample of plant or animal material. Because ^{14}C is radioactive whereas the more abundant ^{12}C isotope is not, their ratio indicates how long ago a plant or animal died.

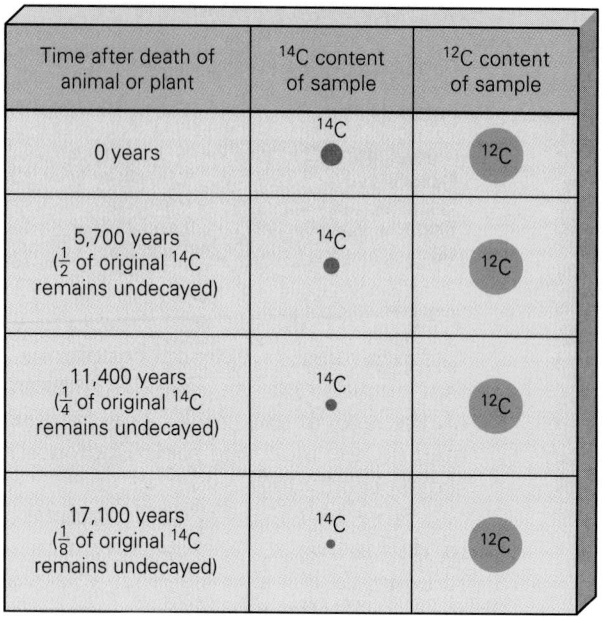

Time after death of animal or plant	^{14}C content of sample	^{12}C content of sample
0 years	^{14}C	^{12}C
5,700 years ($\frac{1}{2}$ of original ^{14}C remains undecayed)	^{14}C	^{12}C
11,400 years ($\frac{1}{4}$ of original ^{14}C remains undecayed)	^{14}C	^{12}C
17,100 years ($\frac{1}{8}$ of original ^{14}C remains undecayed)	^{14}C	^{12}C

FIG. 15-20 The principle of radiocarbon dating. The radioactive ^{14}C content of a sample of dead animal or plant tissue decreases steadily, while its ^{12}C content remains constant. Hence the ratio of ^{14}C to ^{12}C contents indicates the time that has elapsed since the death of an organism. The half-life of ^{14}C is 5700 years.

order to live, and so every plant contains radioactive carbon that it absorbs along with intake of ordinary carbon dioxide. Animals eat plants and thus become radioactive themselves. As a result, every living thing on earth is slightly radioactive because of the radiocarbon it takes in. The mixing of radiocarbon with ordinary carbon is very efficient and so living plants and animals all have the same proportion of radiocarbon to ordinary carbon (^{12}C).

After death, however, the remains of living things no longer absorb radiocarbon, and the radiocarbon they contain keeps decaying away to nitrogen. After 5700 years, then, they have only half as much radiocarbon left—relative to their total carbon content—as they had as living matter, after 11,400 years only one-fourth as much, and so on. Determining the proportion of radiocarbon to ordinary carbon thus makes it possible to evaluate the ages of ancient objects and remains of organic origin (Fig. 15-20). This elegant method permits the dating of archeological specimens such as mummies, wooden implements, cloth, leather, charcoal from campfires, and similar remains of ancient civilizations that are between 1000 and 70,000 years old.

15-11   FOSSILS

Tracing the History of Life

Perhaps the most fascinating technique at the geologist's command for establishing relationships among rocks of different regions and for arranging beds in sequence makes use of **fossils.** The most common fossils, of course, are the hard parts of animals, such as shells, bones, and teeth. On rare occasions an entire animal may be preserved: ancient insects have been trapped in amber, and immense woolly mammoths have been found frozen in the Arctic.

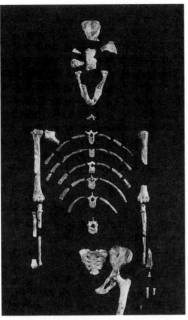

Fossil skeleton of "Lucy," a humanlike creature who lived 3 million years ago in what is today Ethiopia.

Fossil fern found in a coal deposit in Pennsylvania.

Plant fossils are relatively scarce since plants do not have durable hard parts. The structure of tree trunks is sometimes beautifully shown in petrified wood. The incomplete decay of buried leaves and wood fragments produces black, carbonaceous material that may keep the original organic structures—coal is a thick deposit of such material. Occasionally fine sediments show impressions of delicate structures like leaves, feathers, and skin fragments. Some fossils are merely trails or footprints left in soft mud and covered by later sediments.

Conditions necessary for preservation have been much the same throughout geologic history. Chemical decay, bacteria, and scavengers have quickly disposed of most of the organisms that have lived on the earth, and only special conditions of burial permit the survival of fossil groups. These conditions most often occur on the floor of shallow seas, where life is abundant and sediments are sometimes deposited rapidly. Our picture of marine life in the past is accordingly far more complete than our picture of the organisms that lived on land, but even the marine record is fragmentary.

What Fossils Tell Us An important conclusion from the fossil record is that groups of organisms show a progressive change from those buried in ancient rocks to those of the most recent strata. In general, the change is from simple forms to more complex forms, from forms very different from those in the present world to creatures much like those around us today. These observations are part of the foundation for Darwin's theory that life has evolved by a steady development from simple organisms to complex ones.

Because plants and animals have changed continuously through the ages, rock layers from different periods can be recognized by the kinds of fossils they contain. This fact makes possible the arrangement of beds in a relative time sequence, even when their relationships are not directly apparent, and also provides a means of correlating the strata of different localities. If, for example, fossil snail shells and clamshells are found in a rock layer in New York that are similar to fossil shells from a layer in the Grand Canyon, the two layers are likely to be approximately the same age.

Fossils are useful not only in tracing the development of life and in correlating strata but in helping us to reconstruct the environment in which the organisms lived. Some creatures, like barnacles and scallops, live only in the sea, and it is likely that their close relatives in the past were similarly restricted to salt water. Other animals can exist only in freshwater. On land some organisms prefer desert climates, others cold climates, others warm and humid climates. Evidently many details about the conditions in which a rock was formed can be revealed by its fossil organisms.

15-12   GEOCHRONOLOGY

Precambrian, Paleozoic, Mesozoic, Cenozoic

Fossils enable us to arrange in sequence geologic events over the entire earth. Enough of these events can be dated accurately by radioactive methods for good estimates to be made for the dates of the others.

The most recent 570 million years of the earth's history have been divided by geologists into three major divisions called **eras** (Table 15-2):

Cenozoic ("recent life") Era; began 65 million years ago
Mesozoic ("intermediate life") Era; began 225 million years ago and lasted 160 million years
Paleozoic ("ancient life") Era; began 570 million years ago and lasted 345 million years

The record of events before the Paleozoic Era is so dim that geologists are not agreed about the proper division of this early time into eras. Although the nearly 4 billion years before the Paleozoic makes up seven-eighths of the history of our planet, it is usually considered as a single long division, **Precambrian time.**

The divisions of geologic time shown in Table 15-2 are based on dramatic changes that occurred from time to time in the earth's history—changes in landscapes, in climates, in types of organisms. In particular, the fossil record reveals a number of occasions, called **extinctions,** when animal and plant life became sharply reduced in both number and variety, to be followed in each case by the rapid evolution of new forms (Fig. 15-21). The times of extinction are used to divide geologic history into **periods.** During a typical period an expansion of living things is followed by an interval in which biological change is more gradual; then a time of more sudden change ends the period. Periods are similarly

TABLE 15-2

Geologic Time. (The Earth Came into Existence about 4500 Million Years Ago)

Millions of Years Ago	Era	Period	Epoch	Duration, Million Years	The Biologic Record	The Geologic Record
65	Cenozoic	Quaternary	Holocene	0.01	Humans become dominant	
			Pleistocene	2.5	Rise of humans; large mammals abundant	Ice Age
		Tertiary	Pliocene	4.5	Flowering plants abundant	Atlantic Ocean widens; Alps and Himalayas form; Red Sea opens
			Miocene	19	Grasses abundant; rapid spread of grazing mammals	India collides with Asia
			Oligocene	12	Apes and elephants appear	Australia separates from Antarctica
			Eocene	16	Primitive horses, camels, rhinoceroses	Norwegian Sea and Baffin Bay open
			Paleocene	11	First primates	
225	Mesozoic	Cretaceous		71	Spread of flowering plants; dinosaurs die out	Laurasia and Gondwana begin to break up
		Jurassic		54	First birds; dinosaurs at their peak	Laurasia separates from Gondwana
		Triassic		35	Dinosaurs and first mammals appear	Pangaea complete
	Paleozoic	Permian		55	Rise of reptiles; large insects abundant	Laurasia and Gondwana come together
		Carboniferous		65	Large nonflowering plants in enormous swamps; large amphibians; extensive forests; sharks abundant	Coal being formed; Africa moves against Europe and North America
		Devonian		50	First forests and amphibians; fish abundant	Greenland and North America join Europe
		Silurian		35	First land plants and large coral reefs	
		Ordovician		70	First vertebrates (fish) appear	
570		Cambrian		70	Marine shelled invertebrates (earliest abundant fossils)	Early supercontinent breaks up
4,000	Precambrian time	Late Precambrian			Marine invertebrates, mainly without shells: bacteria, sponges, worms	Early supercontinent forms
4,500		Early Precambrian			Primitive bacteria and algae	

513

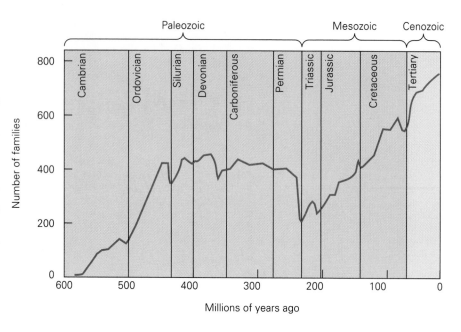

FIG. 15-21 How the number of families of living things has varied throughout history. A family is a group of related species, and the greater the number of families, the greater the diversity of life. Each dip in the curve represents a biological extinction. The most severe extinction marks the end of the Paleozoic Era; the dinosaurs disappeared in the extinction at the end of the Mesozoic Era. Life is at its most diverse today, although human activity has begun to turn the graph downward.

THE GREAT DYING

As Fig. 15-21 shows, the most severe mass extinction yet marks the divide between the Paleozoic and Mesozoic Eras. In all, 96 percent of the earth's plant and animal species vanished a quarter of a billion years ago. What happened?

Because the Great Dying (as it has been called) took place so long ago, geological processes have inexorably erased most potential evidence. But some clues remain, and they point to a series of immense volcanic eruptions that covered much of Siberia with lava and sent vast clouds of dust and sulfurous gases billowing into the atmosphere. With the sky veiled against sunlight for many years, the earth's surface cooled. Sea level dropped as much as 200 m as the polar ice caps grew, and the marine life that had been flourishing in the shallow waters of the world could not survive drying out. Life on land suffered at the same time from the cold, from acid rain due to the sulfur compounds in the air, and possibly from a shortage of atmospheric oxygen, a lot of which must have been absorbed in the decay of marine organisms left stranded by the receding oceans. Eventually the volcanic activity subsided, conditions more favorable to living things returned, and the biological world began to thrive again.

subdivided into shorter **epochs;** only the epochs of the Cenozoic Era are shown in Table 15-2.

The division into eras was based on exceptionally marked worldwide extinctions and subsequent expansions, and correlations were found between large-scale events in the earth's crust and the biological record. Not all these correlations have turned out to be as clear-cut as they once appeared, but the traditional organization of geologic time is still convenient and is universally used by scientists.

EARTH HISTORY

The earth came into being about 4.5 billion years ago. Current ideas on its origin are discussed in Chap. 18. With the oldest surface rocks the story of the earth leaves the realm of speculation. These rocks, which date back 4 billion years, consist of sedimentary and metamorphic rocks intruded by igneous rocks, which suggests that the cycle of erosion and diastrophism was already well established as far back as the visible record of the earth's history extends.

15-13   PRECAMBRIAN TIME

Precambrian Time
(4500–570)
Late Precambrian
(2500–570)
marine invertebrates, mainly without shells; bacteria, sponges, worms
Early Precambrian
(4500–2500)
primitive bacteria and algae
(Dates are millions of years ago)

Long Ago but Not Far Away

Nothing is known for certain about the locations of the continents in Precambrian time. The splitting of Pangaea into Laurasia and Gondwana and their subsequent breakup into today's continents began less than 200 million years ago, a mere 4 percent of the earth's age. Plate movement seems to have been going on farther back in the past, with the Precambrian continents differently arranged across the face of the earth.

Although practically unaltered Precambrian sedimentary and igneous rocks are sometimes found, more often Precambrian rocks show considerable metamorphism and in many places have been greatly deformed. Both stream deposits and marine deposits are found in the sedimentary beds. At least two glacial periods have been identified in the Precambrian. The volcanic rocks include all types, with basalt flows then as now the most common. Intrusive rocks are represented in great abundance and variety. Evidently geologic processes a billion years ago were not very different from those in the modern world. Precambrian rocks are exposed at the surface over a broad area covering most of eastern Canada and nearby parts of the United States.

Early Life

Precisely when and where life began on earth nobody knows, because primitive organisms seldom leave traces. Nevertheless bacterialike structures have been found that suggest living things flourished at least as far back as 3.6 billion years ago. Indirect evidence is abundant in the form of beds of marble, whose ultimate source is typically the shells of aquatic creatures, and of layers of graphite (a form of pure carbon), which may well have originated in organic debris since all living things contain much carbon. Both marble and graphite are found in rocks up to 3 billion years old.

Sample of Precambrian rock from northern Canada that shows ripples and cracks characteristic of fine sediments deposited in shallow water and occasionally dried by exposure to the sun. Similar conditions occur along the shores of present-day lakes and seas.

During the billion or so years that followed the appearance of life, primitive cyanobacteria apparently grew widely in the oceans. Joined together in sticky bluish-green mats, these bacteria could survive solar ultraviolet light deadly to most other organisms, and their photosynthetic activity began to provide the atmosphere with its oxygen content. This oxygen, together with the ozone layer it made possible that screens out solar ultraviolet light (and is now endangered), was the final condition needed for advanced forms of life. Evidence of multicellular algae and fungi has been discovered in the Gunflint Formation in Ontario, Canada, that dates from 1.9 billion years ago. At this time there was already a good deal of oxygen in the atmosphere, though less than there is today.

15-14   THE PALEOZOIC ERA

PALEOZOIC ERA
Permian (290–225)
rise of reptiles, large insects abundant
Carboniferous (345–290)
large nonflowering plants in enormous swamps; large amphibians; extensive forests; sharks abundant
Devonian (395–345)
first forests and amphibians; fish abundant
Silurian (430–395)
first land plants and large coral reefs
Ordovician (500–430)
first vertebrates (fish) appear
Cambrian (570–500)
marine invertebrates with shells (earliest abundant fossils)
(dates are millions of years ago)

Plants and Animals in Variety and Abundance

The history of the Paleozoic Era is remarkably complete. No longer are there the doubts and vagueness that characterize Precambrian events, for Paleozoic strata are widely exposed, and their wealth of fossils makes possible correlation of rocks and events from one side of the earth to the other.

The oldest fossils of the Paleozoic Era are those of marine **invertebrates,** creatures without internal skeletons but with external shells of various kinds. All the major groups of the invertebrates are represented. Clams and snails increased in number and evolved considerably in the Paleozoic seas. Tiny coral polyps built widespread reefs in the middle Devonian Period. Starfish and sea urchins were not common, but some of their distant relatives that today are rare or extinct were more numerous then. Fishes were many and varied.

In late Paleozoic rocks there is for the first time much evidence of land-dwelling organisms. In the coal swamps of Carboniferous times grew dense forests of primitive plants—huge fernlike trees, enormous horsetails, primitive conifers. A modern person wandering through such a forest would find no bright-colored flowers, no grasses, few plants at all familiar except possibly some of the ferns and mosses. In and near these primeval forests lived a great variety of animals: scorpions, land snails, primitive insects. The land-living vertebrates of the late Paleozoic are early members of our own family tree. In fact, the basic body plans that appeared in the Paleozoic have served for all animals ever since.

Fossil amphibians, oldest of the land vertebrates, appear first in Devonian rocks. These are relatives of modern frogs and salamanders, sluggish creatures that laid their eggs and spent the early part of their lives in water. Fossils of reptiles begin to appear in late Carboniferous rocks. These animals looked at first much like their amphibian ancestors but had the great advantage of being able to lay their eggs on land. The dry climate at the end of this era was hard on the amphibians, but the reptiles, not needing water to hatch their eggs, multiplied rapidly and developed many different species. During the Permian Period reptiles became the dominant creatures of the land.

Trilobites, distant cousins of insects, were abundant in Paleozoic seas and oceans. Their fossils are common because they were among the first animals to have hard shells. Trilobites ranged in length from a few millimeters to 50 cm; the larger trilobite fossil shown here is 75 mm long.

The Paleozoic Era ended with a time of intense tectonic activity, which affected many parts of the world including the North American continent. The sediments that had accumulated for more than 300 million years in the Appalachian trough were crumpled, fractured, and uplifted into a mountain chain that must have rivaled any modern range in height and grandeur. It is not hard to connect these dramatic geologic events with the crushing together of the continental masses that formed the supercontinent Pangaea at this time.

15-15   COAL AND PETROLEUM

Both Came from Once-Living Matter
Coal is plentiful in Paleozoic rocks. Coal was formed from plant matter that accumulated under conditions where complete decay was prevented. A bed of coal nearly always implies an ancient swamp. Coal has been formed in swamps from the Devonian to the present day, but seldom have conditions been so favorable as in the Carboniferous Period. Apparently there were broad swamps almost at sea level that became periodically submerged so that partly decayed vegetation was covered with thin layers of marine sediments.

The formation of coal begins with the slow bacterial decay of the cellulose content of plants. Taking place underwater and in the absence of air, this decay results in a gradual removal of oxygen and hydrogen from the cellulose to leave a residue that is largely carbon. Also contributing to coal formation was the action of heat and pressure resulting from burial beneath later sediments.

The origin of petroleum is more obscure, for two reasons: fossils are not preserved in a fluid, and petroleum often migrates long distances

About half the coal extracted in the United States comes from underground mines such as this one. The rest is gouged in the open from deposits that lie near the surface after the overlying soil has been stripped away. Most of the underground mines are in the eastern part of the country; most of the surface mines are in the western part. The coal currently consumed in the world each year took about 2 million years to accumulate.

Oil drilling rig in the Gulf of Mexico off the Louisiana coast. Oil furnishes about half the energy needs of today's industrial societies as well as being a valuable raw material for the chemical industry. Demand may exceed supply in the next century.

from where it forms. Because petroleum hydrocarbons can be detected in modern marine sediments, because oils resembling petroleum can be prepared artificially from organic material, and because petroleum is associated with rocks formed from sediments deposited in shallow seas, there seems to be little doubt that marine life such as algae and plankton is the source of petroleum. Most petroleum seems to have been formed more recently than coal: over half in the Cenozoic Era, about a quarter in the Mesozoic, and the rest in the Paleozoic.

Three steps seem to have been involved in producing petroleum. The first was bacterial decay in the absence of oxygen, an ideal site being the floor of a shallow sea. Then, as the organic debris was buried under later sediments, it was further modified by low-temperature chemical reactions. The final step was the "cracking" of complex hydrocarbons to straight-chain alkane hydrocarbons (Chap. 12) under the influence of temperatures of 70 to 130°C deep underground. If the temperatures became higher, the result was natural gas rather than oil.

Both gas and petroleum, like groundwater, can migrate freely through such porous rocks as loosely cemented sandstones and conglomerates. Wherever formed, they often find their way into porous beds, and it is from these beds that they are obtained by drilling. Since both gas and petroleum are lighter than water, they may be displaced by groundwater and so move upward to the surface to form oil seeps.

Petroleum reservoirs consist of porous sandstones or carbonates that lie under layers of impermeable clays or shales. The most common kind of reservoir is one formed by an anticline, as in Fig. 15-22a. The world's largest producers of petroleum are Saudi Arabia, the United States, and Russia, in that order; smaller but still substantial amounts come from a number of other countries.

FIG. 15-22 Two common types of structural traps in which petroleum accumulates: (*a*) a trap formed by an anticline. (*b*) a trap formed by a fault. In both cases petroleum in a porous reservoir rock (such as sandstone) is prevented from migrating upward by an impermeable cap rock (such as shale). A well drilled at A would strike petroleum, one drilled at B would strike gas, and one at C only water. About 80 percent of known petroleum deposits are found in anticline traps.

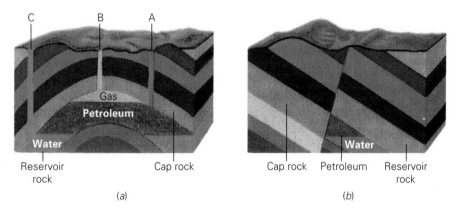

(a)   (b)

15-16   THE MESOZOIC ERA

MESOZOIC ERA
Cretaceous (136–65)
spread of flowering plants; dinosaurs die out
Jurassic (190–136)
first birds; dinosaurs at their peak
Triassic (225–190)
dinosaurs and first mammals appear
(dates are millions of years ago)

The Age of Reptiles   The earliest Mesozoic sediments were laid down about 225 million years ago, a long time by ordinary reckoning. But the earth was already very old. Some 345 million years had elapsed since the beginning of the Paleozoic and $3\frac{1}{2}$ billion years since the oldest known rocks of the Precambrian. All the time that we include in the Mesozoic and Cenozoic Eras is only one-sixteenth of the history recorded in rocks of the earth's crust.

The Mesozoic Era saw Pangaea split into Laurasia and Gondwana, and this division was followed by their own breakup (see Fig. 15-6). Early in the Mesozoic Era North America began to part from Europe, and somewhat later, perhaps 120 million years ago, South America and Africa began to drift apart. By the end of the era Gondwana no longer existed. Australia, New Zealand, and India had all left Africa, though Arabia still remained attached. Africa itself was in the process of a shift northeastward, thus closing the western end of the Tethys Sea, while India, well on its way toward Asia, was moving into the eastern end. The Mid-Atlantic Ridge was already a prominent feature of the floor of the infant Atlantic Ocean.

On Mesozoic lands developed a group of reptiles that included some of the largest animals the earth has seen—the **dinosaurs,** descendants of the primitive reptiles that had survived from the Paleozoic. (The word dinosaur comes from the Greek for "terrible lizard.") Some dinosaurs were carnivores with bodies designed for pursuing and eating other animals. Some were herbivores with jaws and digestive organs adapted for a vegetarian diet. Active species lived in open plains, more sluggish ones in swamps. Some had bony armor for protection, others depended on speed to escape their enemies. The fastest dinosaurs could outrun today's sprinters, though not today's racehorses. The really large dinosaurs, though, moved sedately at a pace slower than that of a person walking. Not all the dinosaurs by any means were large in size, but the biggest ones were truly enormous, 40-ton beasts over 25 m long; the largest modern land animal, the African elephant, weighs in at a mere 5 tons.

Meanwhile other land organisms were developing. Flowering plants appeared in mid-Mesozoic and with them a host of modern-looking

Dinosaur forms can be inferred by comparing their skeletons with the skeletons of living reptiles. Impressions of dinosaur skins sometimes occur in the rocks in which their bones are found, although the skin colors can only be guessed at. This painting shows some of the dinosaurs that lived in western North America 65 to 70 million years ago at the end of the Mesozoic Era. The dinosaurs here were all plant-eaters except *Tyrannosaurus* at upper right, which had a mass of perhaps 7000 kg; its teeth were 15 cm long. By contrast, the largest modern land predator is the polar bear, whose mass is at most around 700 kg.

insects suited for helping in the pollination of flowers. The first true birds, with feathered wings rather than membranes, arose from dinosaur ancestors in the Jurassic. Sometime in the Triassic appeared the first **mammals,** creatures that probably descended from a group of small Permian reptiles. All during the Mesozoic mammals remained inconspicuous, never much larger than a cat. However, in several respects mammals represented an evolutionary advance over reptiles: they were warm-blooded (constant body temperatures), hence better able to cope with changes of temperature; they had bigger brains relative to their body size; and they cared for their young after birth, so some of the experience of one generation could be passed on to the next.

What Happened to the Dinosaurs? For over 100 million years dinosaurs large and small roamed the earth. Then, 65 million years ago, all of them disappeared. Not a single dinosaur fossil has ever been found in rocks formed after the end of the Mesozoic Era. Dinosaurs were not the only victims: as many as 70 percent of the plant and animal species of the world were wiped out at about the same time. (Insects were more durable: most of the insect species present at the end of the Mesozoic are still with us. Today there are 200 million insects for each person on the earth.) This mass extinction is what divides the Mesozoic from the Cenozoic, and its cause has become the subject of much controversy.

One point of view starts from the fact that extinctions of organisms are by no means unusual in geologic history. New species have evolved, flourished, and then died out as far back as the fossil record exists. There is no shortage of past events that could have led to these extinctions. The shifting of the continents and surges in volcanic activity associated with such shifts certainly led to repeated changes in climate, in sea level, in the amount of sunlight reaching the surface (which affects the photosynthesis that is at the base of the chain of life), in the carbon

EXTINCTIONS AND UNIFORM CHANGE The widespread biological extinctions of the past, for instance the drastic one in which the dinosaurs disappeared, lead to an interesting question. How can the events that caused them be squared with the principle of uniform change, according to which the earth's geologic history can be explained, in Lyell's words, by "forces now in operation"?

Certainly the event that caused not only the dinosaurs but a large proportion of the world's other living things to perish is a catastrophe in the usual definition of the word. But we should note that the ideas proposed to account for this catastrophe are all in accord with the same laws of nature we find in today's universe. Such is not the case with Werner's idea that all rocks were formed in three precipitations from a universal ocean or with Cuvier's idea of the periodic destruction and re-creation of life by divine edict. Their catastrophes make no sense in terms of what we know about physics, chemistry, and biology. On the other hand, volcanoes, comets, and asteroids do exist, and even though they may affect the earth on a large scale only rarely, they still fit into the pattern of the principle of uniform change.

dioxide content of the atmosphere, and so forth. In particular, the end of the Mesozoic was a time of worldwide tectonic and volcanic activity, and it would have been surprising if a large-scale extinction had not occurred then. The late Mesozoic extinction is famous because the dinosaurs disappeared so suddenly and completely, but in fact, as Fig. 15-21 shows, other extinctions, notably the Great Dying that ended the Paleozoic, saw more species disappear.

Until 1979 there seemed no reason to suppose that the fate of the dinosaurs was unusual in any fundamental way. In that year came the discovery of exceptionally large traces of the element iridium in a thin clay layer in Gubbio, Italy, that separated marine limestones of the late Mesozoic from younger limestones. Iridium, a metal similar to platinum, is rare in the earth's crust but relatively abundant in the meteorites that bombard the earth from space. Other late-Mesozoic deposits rich in iridium were soon found elsewhere. The obvious inference was that a comet or asteroid, perhaps 10 or 15 km across, struck the earth 65 million years ago. The impact could have sent up a vast cloud of dust and sulfur compounds that remained in the stratosphere for months, perhaps years, blocking sunlight and thereby wiping out much of the life on our planet.

Supporting the impact theory was the discovery in 1991 of the remains of a huge crater about 180 km across that is centered on the edge of Mexico's Yucatán peninsula (Fig. 15-23). The crater's size and age are about right. Furthermore, the clay layer that marks the end of the Mesozoic seems to have originated in both ocean-floor and continental rocks that were pulverized by the impact, which fits in with the crater's location at the rim of an ocean basin.

There are yet other aspects to the disappearance of the dinosaurs. For one thing, they seem to have been on the way out even before the Yucatán impact. For another, the most intense volcanic activity in the earth's history took place at the end of the Mesozoic, leaving 300,000 km^2 of India covered with several km of lava. Certainly the gas and ash

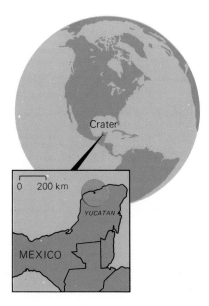

FIG. 15-23 A gigantic crater whose traces were found in Mexico's Yucatán Peninsula may have been formed by the impact of the comet or asteroid that led to the extinction of the dinosaurs and many other forms of life at the end of the Mesozoic Era.

Volcanic eruptions at the end of the Mesozoic flooded a huge area of what is now India with basalt to form landscapes like this. Debris flung into the air by the eruptions could have partly blocked sunlight for a long period and thus contributed to the biological extinction at that time. The impact of a comet or asteroid on the other side of the globe in Yucatán may have set off the volcanic activity.

thrown into the atmosphere by the eruptions did life on our planet no good. Furthermore, iridium is abundant in the mantle and could have been brought to the surface in the volcanic magma. Could the volcanic activity, centered opposite Yucatán on the globe, have been triggered by seismic waves from the cosmic jolt in Yucatán? It seems quite possible.

Whatever the reason (or reasons) that dinosaurs are no longer with us, it is clear that the history of our planet is far from being a dull subject.

15-17 THE CENOZOIC ERA

The Age of Mammals
In many ways the Cenozoic Era of today has been different from preceding eras. During the Cenozoic the continents have stood for the most part well above sea level. No longer do shallow seas spread widely. In North America, beds originating from marine deposits are found only in narrow strips along the Pacific Coast and on the Atlantic Coast from New Jersey south to Yucatán. The thick Tertiary beds east and west of the Rocky Mountains are river, lake, and wind deposits made in continental basins. And climates during much of the Cenozoic have had a diversity like those of the present. The distribution of plants and animals shows that, instead of having widespread moderate climates like those of other eras, Cenozoic continents have had zones of distinct hot, cold, humid, and dry climates.

A characteristic of Cenozoic times has been widespread volcanic activity. From the Rockies to the Pacific Coast, lava flows and tuff beds testify to former volcanoes, some of which have only recently become extinct. In the mid-Tertiary, immense flows of basalt inundated an area

CENOZOIC ERA
Quaternary {
Holocene (0.01–0)
humans are the most complex and successful form of life
Pleistocene (2.5–0.01)
rise of humans; large mammals abundant
}
Tertiary {
Pliocene (7–2.5)
flowering plants abundant
Miocene (26–7)
grasses abundant; rapid spread of grazing mammals
Oligocene (38–26)
apes and elephants appear
Eocene (54–38)
primitive horses, camels, rhinoceroses
Paleocene (65–54)
first primates
(dates are millions of years ago)
}

of nearly a half-million km^2 in Oregon, Idaho, and Washington. Some of these flows today form the somber cliffs of the Columbia River Gorge.

Tectonic Activity The Cenozoic has also been a time of almost continuous tectonic disturbance, in contrast with the long periods of crustal stability in previous eras. Movements associated with the mountain-building episodes that divide the Cenozoic from the Mesozoic lasted well into the Tertiary. In mid-Tertiary the Alps and Carpathians of Europe and the Himalayas of Asia were folded and uplifted. Toward the end of the Tertiary the Cascade range of Washington and Oregon was formed, and other mountain ranges that have been active to the present day began to form around the border of the Pacific. Mountain ranges that had been folded earlier—the Appalachians, the Rockies, the Sierra Nevada-were repeatedly uplifted during the Cenozoic, and erosion following these uplifts has created their present topography.

It is not hard to associate this reshaping of continental landscapes with the spreading of sea floors and the grinding together of lithospheric plates that are still in action today. In the Cenozoic the continents continued their earlier drifts. In addition Greenland parted from Norway, Australia parted from Antarctica (New Zealand had done so earlier), and the Bay of Biscay opened up. More recently the Arabian peninsula broke off from Africa, the Gulf of California opened to separate Baja California from mainland Mexico, and Iceland rose above the surface of the Atlantic Ocean.

Mammals Thanks to their great adaptability, some mammals managed to survive the mass extinction that ended the Mesozoic Era. Without the dinosaurs to keep them in check, these mammals multiplied and evolved rapidly. Carnivores like cats and wolves, armored beasts like rhinoceroses, agile creatures like deer and rabbits—ancestors of all these modern forms roamed the early Tertiary landscape. A few mammals, like the whales and porpoises, took to life in the sea; another line, the bats, developed wings.

By the middle Tertiary, mammals dominated the earth as reptiles had before them. Side by side with the mammals developed modern birds and the trees of modern forests. As the end of the Tertiary approached, both the physical and the biological worlds assumed more and more closely their present aspects.

15-18 THE ICE AGE

One of Many During the Pleistocene Epoch great ice caps formed in Canada and northern Europe, and valley glaciers advanced in high mountains elsewhere. This was only the latest in a series of glacial periods that have punctuated earth history, but since these particular glaciers took part so directly in shaping present landscapes, the Pleistocene is often referred to simply as the Ice Age.

Glacial deposits in North America show that ice spread outward from three centers of accumulation in Canada, the ice front in its far-

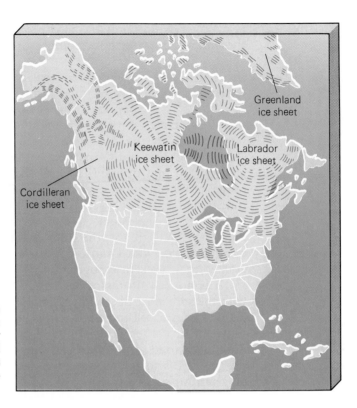

FIG. 15-24 The maximum extent of Pleistocene glaciers in North America. There were four major ice advances that covered up to 30 percent of the earth's land surface. Today about 10 percent lies under ice.

thest advance reaching the Missouri River on the west and the Ohio River to the east (Fig. 15-24). Many different times of ice advance can be distinguished, with long interglacial periods between. During at least one of these interglacial times the ice disappeared completely, and the climate became warmer than it is today. The changing climates of the Pleistocene proved a severe ordeal for living things. Nowadays mammals are still dominant, but in numbers and diversity of species they have declined markedly since the later Tertiary.

Human History   According to the fossil record, our own species had its infancy in the Great Rift Valley of eastern Africa. Roughly 6 million years ago, biochemical evidence indicates that the descendants of a common ancestor split into two branches, one that evolved into some of today's apes and the other into today's humans (Fig. 15-25). The earliest of the humanlike creatures walked on two legs but still had small brains. There does not seem to have been a continuous development from then on but rather a succession of species, sometimes overlapping, with each flourishing for a time and then either disappearing or turning into another. By 2.5 million years ago the line that would become modern humans had larger brains than before and were making and using stone tools (hence their species name Homo habilis, "handy man"). Farther into the Pleistocene, with still larger brains, our forebears began to spread out from Africa into Europe and Asia. Finally, a few hundred thousand years ago, modern humans (Homo sapiens, "wise man") emerged. The later record is most

This painting is one of many that were made in a cave at Lascaux, France, during the most recent Ice Age 15,000 years ago.

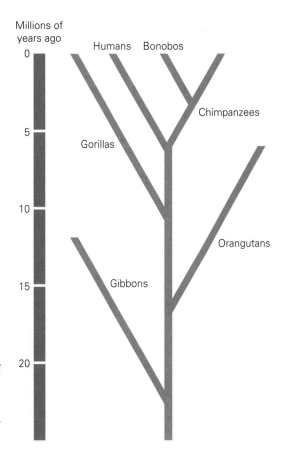

FIG. 15-25 Evolutionary tree of humans and apes, based on DNA evidence. Monkeys split off earlier. Chimpanzees, bonobos, and humans have 98 percent of their genes in common.

complete in Europe, where stone implements, burial sites, drawings on cave walls, and skeletal fragments give a fairly connected history.

Humans crossed a broad land bridge across the Bering Strait from Asia to Alaska between 30,000 and 12,000 years ago, when heavy glaciation had lowered sea level worldwide by 100 or 150 m below what it is today. At that time animal life in North America—which included wooly mammoths, saber-toothed tigers, giant ground sloths, and zebralike horses—was spectacularly varied and abundant. Did hunting by the newly arrived humans wipe out the large animals or was it some environmental change that did so? No one knows for sure.

15-19 **POPULATION AND THE FUTURE**

The Biggest Problem of All   In the past, biological changes were brought about largely by changes in the physical environment, so the biological narrative follows the physical one through the ages. The modern world seems to be witnessing a reversal of this sequence. Not only are we able by virtue of our ingenuity to flourish in the environment that nature has provided, but we are able to alter that environment in a number of ways. These alterations have permitted a vast number of people to live relatively free (by histor-

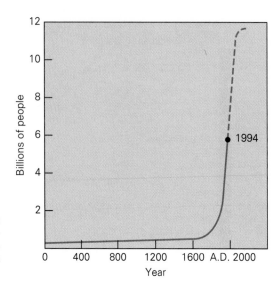

FIG. 15-26 World population according to United Nations figures. Estimates for future population vary widely; shown is the midrange projection.

ical standards) of starvation and disease. Today deaths due to ailments brought on by diets of plenty compete in number with those due to famine.

If continued into the future and swollen in scale by population growth, however, our present patterns of industry and agriculture menace the interplay between people and environment that has been so successful in the past. The indiscriminate use of pesticides and artificial fertilizers has already destroyed the ecological balance in many areas; numerous inland waters have been poisoned by industrial waste; noxious gases fill the air in densely populated regions; acid rain due to power-plant emissions has damaged forests and lakes on a large scale; vast areas of forest have had their trees cut and their soil washed away as a result; the ozone layer is weakening and global warming due to carbon dioxide emissions seems already to have begun; deposits of radioactive wastes lie waiting for chance catastrophe to disperse them; and so on. Although a remarkable amount has been done in developed countries to limit environmental damage, in much of the world little or no effort is being made in this direction.

Most ominous of all is that, whatever we do in the future, a lot more of us will be doing it. From perhaps 500 million in 1650, 1 billion in 1800, 2.5 billion in 1950, and 5.6 billion in 1994, the world's population is still climbing (Fig. 15-26). The current rate of population increase is about 250,000 *per day*—another Mexico every year, another India every decade.

How will all these people be fed? Where will the natural resources come from to accommodate their other needs? How will the environment respond to the increased assaults on it when it is deteriorating even now? And even if these problems can be solved, what kind of lives will our children and grandchildren have if the earth becomes a human anthill?

Since safe, efficient, and inexpensive means of family planning exist, there is no reason why the world's population cannot be stabilized at a reasonable level. In the developed world, the average number of children per woman is already 2 or less (only 1.3 in Italy). In much of the underdeveloped world, however, this average is 6 or more (8.3 in Rwanda). Ignorance and poverty are not alone responsible: forces of tradition and of religion that oppose family planning have been able to obstruct the wishes of many people who want to limit their fertility. A lot more will have to be done if the graph of Fig. 15-26 is not to continue its climb toward disaster.

IMPORTANT TERMS AND IDEAS

Movements of the solid materials of the earth's crust are called **tectonic.** The principal kinds of tectonic movement are faulting, folding, and regional uplift, sinking and tilting. A **fault** is a fracture surface along which motion has taken place. **Mountain building** begins with the deposition of sediments in a sinking area, followed by tectonic movement and erosion.

The crust and the upper part of the mantle make up a shell of hard rock called the **lithosphere.** Below the lithosphere is a layer of hot, soft rock called the **asthenosphere.** According to the theory of **plate tectonics,** the lithosphere is divided into seven huge **plates** and a number of smaller ones. The plates can move relative to one another in four ways: two plates can move apart with molten rock rising to form new ocean floor at the gap; one plate can slide under another and melt; two plates can collide and buckle; and adjacent plates can slide past each other.

Continental drift occurs because of plate motion. Today's continents were once part of two supercontinents called **Laurasia** (North America, Greenland, Europe, and most of Asia) and **Gondwana** (South America, Africa, Antarctica, India, and Australia) that were separated by the **Tethys Sea.**

According to the **principle of uniform change,** geologic processes in the past were the same as those in the present. An **unconformity,** which is a buried surface of erosion, indicates that tectonic uplift, erosion, and sedimentation have occurred in that order.

Radionuclides and their decay products in rocks make it possible to date geologic formations. The remains of living things can be dated with the help of **radiocarbon,** the radioactive carbon isotope ^{14}C. **Fossils,** the remains of organisms preserved in rocks, are useful in correlating strata, in tracing the development of living things, and in reconstructing ancient environments.

Geologic time is divided into **Precambrian time** and the **Paleozoic, Mesozoic,** and **Cenozoic Eras.** Eras are subdivided into **periods** and the periods into **epochs.**

EXERCISES: MULTIPLE CHOICE

1. A crack in the earth's crust along which movement has taken place is called
   a. a fault   c. an earthquake
   b. a fold   d. a moraine

2. A long, narrow bay with an irregular outline, such as Chesapeake Bay, was formed by
   a. wave action
   b. stream erosion and later submersion
   c. glacier erosion and later submersion
   d. faulting

3. The rugged character of mountain landscapes is largely the result of
   a. folding   c. stream erosion
   b. faulting   d. glacier erosion

4. The Tethys Sea once separated
   a. North and South America
   b. South America and Africa
   c. Europe and Asia
   d. Laurasia and Gondwana

5. A mountain range that was not once part of the Tethys Sea is the
   a. Alps   c. Andes
   b. Pyrenees   d. Himalayas

6. North America, Greenland, and most of Eurasia once made up the supercontinent of
   a. Pangaea   c. Gondwana
   b. Laurasia   d. Atlantis

7. The thickness of the lithosphere varies between
   a. 2 and 5 km   c. 50 and 100 km
   b. 5 and 35 km   d. 500 and 1000 km

8. The layer of soft rock beneath the lithosphere is called the
   a. stratosphere   c. asthenosphere
   b. thermosphere   d. mantle

9. As compared with the continents, the ocean floors are
   a. much younger

b. much older

c. about the same age

d. in some places older and in others younger

10. The ocean floor near a midocean ridge
    a. has the same constant magnetization on both sides
    b. is magnetized in one direction on one side and in the opposite direction on the other side
    c. has strips of alternate magnetization on both sides
    d. has no consistent pattern of magnetization

11. According to the hypothesis of sea-floor spreading, molten rock is rising up along the
    a. trenches that rim the Pacific Ocean
    b. ridges on midocean floors
    c. location of the Tethys Sea
    d. equator

12. The number of large plates into which the lithosphere is divided is
    a. 3   **c.** 20
    b. 7   **d.** 50

13. A typical speed for a lithospheric plate is
    a. 3 mm/year   **c.** 3 m/year
    b. 3 cm/year   **d.** 3 km/year

14. A region where an edge of a lithospheric plate slides under an edge of another plate is called a
    a. transform fault **c.** moraine
    b. fault scarp   **d.** subduction zone

15. The Indian subcontinent
    a. was always part of Asia
    b. came into being 45 million years ago as the result of extensive vulcanism
    c. collided with Asia 45 million years ago
    d. began to move away from Asia 45 million years ago

16. The Andes seem to be the result of
    a. an oceanic plate sliding under a continental plate
    b. two continental plates colliding
    c. two oceanic plates colliding
    d. an enormous earthquake

17. If present trends continue,
    a. California will be detached from the rest of North America
    b. the Atlantic will become narrower
    c. The Pacific will become wider
    d. The West Indies will sink below the ocean surface

18. Scientific observation does not support
    a. the principle of uniform change
    b. Darwin's theory of evolution
    c. continental drift
    d. the origin of the earth in 4004 B.C.

19. An uneven surface on which a horizontal upper bed rests is called a (an)
    a. stratum   **c.** dike
    b. fault   **d.** unconformity

20. Radiocarbon dating is based upon the fact that
    a. ^{14}C is continually being formed in the remains of living things after their death
    b. ^{14}C is not radioactive
    c. the ^{14}C content of the remains of living things depends upon the time in the past when they came into being
    d. the ^{14}C content of the remains of living things depends upon the time in the past when they died

21. A suitable object for radiocarbon dating would be one whose age is
    a. 10,000 years   **c.** 1 million years
    b. 100,000 years   **d.** 10 million years

22. The oldest rocks that have been dated are about
    a. 4 million years old   **c.** 1 billion years old
    b. 4.5 million years old   **d.** 4 billion years old

23. Most fossils are found in
    a. volcanic rocks
    b. metamorphic rocks
    c. sedimentary rocks that formed from deposits on the floors of shallow seas
    d. sedimentary rocks that formed from stream deposits

24. The division of geologic time into eras and periods is based upon
    a. the coming of ice ages
    b. the disappearance of continents
    c. biological extinctions
    d. worldwide flooding

25. The earth was formed
    a. in 4004 B.C.
    b. about 2 million years ago
    c. about 4.5 billion years ago
    d. about 10 billion years ago

26. Arrange the following divisions of geologic time in their proper sequence, starting with the oldest:
    a. Paleozoic Era   **c.** Cenozoic Era

b. Precambrian time   **d.** Mesozoic Era

27. Precambrian rocks are
    a. never found
    b. extremely rare
    c. exposed in a number of regions
    d. the most common rocks

28. In rocks of which one or more of the following eras does evidence of life appear?
    a. Precambrian   **c.** Mesozoic
    b. Paleozoic   **d.** Cenozoic

29. Ancient geologic processes as revealed in Precambrian strata were
    a. primarily volcanic
    b. primarily glacial
    c. primarily erosion and sedimentation
    d. of the same kinds as those of the present time

30. Large exposures of Precambrian rocks are to be found in
    a. eastern Canada and adjacent parts of the United States
    b. the Rocky Mountains
    c. Texas and New Mexico
    d. the Middle West

31. Coal is composed of
    a. petrified wood
    b. buried plant material that has partially decayed
    c. buried animal material that has partially decayed
    d. a variety of igneous rock

32. A bed of coal usually implies that the region was once a
    a. desert   **c.** swamp
    b. coniferous forest **d.** river bed

33. Most coal deposits were formed in the
    a. Cenozoic   **c.** Paleozoic
    b. Mesozoic   **d.** Precambrian

34. Amphibians, fishes, and marine invertebrates were the dominant form of life in the
    a. Cenozoic   **c.** Paleozoic
    b. Mesozoic   **d.** Precambrian

35. Dinosaurs were abundant in the
    a. Cenozoic   **c.** Paleozoic
    b. Mesozoic   **d.** Precambrian

36. Dinosaurs were the dominant form of animal life for a period of about
    a. 1000 years   **c.** 100 million years
    b. 1 million years **d.** 1 billion years

37. Unusually large traces of the rare element iridium are found in sediments deposited at the end of the
    a. Cenozoic   **c.** Paleozoic
    b. Mesozoic   **d.** Precambrian

38. The ancestors of the birds were
    a. reptiles   **c.** amphibians
    b. mammals   **d.** insects

39. Mammals were the dominant form of life in the
    a. Cenozoic   **c.** Paleozoic
    b. Mesozoic   **d.** Precambrian

40. Laurasia and Gondwana broke up in the
    a. Cenozoic   **c.** Paleozoic
    b. Mesozoic   **d.** Precambrian

41. The Cenozoic Era represents a period
    a. of almost continuous tectonic activity
    b. of relative stability, with erosion and sedimentation the chief geologic processes
    c. of relatively uniform climate around the world
    d. in which the reptile was the most advanced form of life

42. During the Ice Age
    a. there was a single glacial advance
    b. there were several glacial advances and retreats
    c. the entire earth was covered with an ice sheet
    d. all animal life perished and had to start over again afterward

QUESTIONS

1. The energy source of erosional processes is the sun. Where does the energy involved in tectonic activity come from?

2. List all the evidence you can for each of the following statements:
    a. Granite is an igneous rock.
    b. Mica schist is a rock that has been subjected to nonuniform pressure.
    c. Compressional forces exist in the earth's crust.
    d. Tectonic movement is going on at present.

3. What landscape features are associated with faults?

4. When stream erosion has been active for a long time in a region underlain by folded strata, what determines the position of the ridges and valleys?

5. What geologic process is chiefly responsible for the landscape of a mountain range?

6. Why is it believed that the region where the Rocky Mountains now stand was once near or below sea level?

7. What is the difference between the earth's crust and its lithosphere? How is it possible for a plastic asthenosphere to occur between a rigid lithosphere and a rigid mantle? If the asthenosphere is plastic, how can transverse seismic waves travel through it?

8. When continental drift was first proposed three-quarters of a century ago, it was assumed that the continents move through soft ocean floors. Why is this hypothesis no longer considered valid? How does continental drift actually occur?

9. How do the ages of the ocean floors compare with the ages of continental rocks? What is the reason for the difference, if any?

10. What kind of biological evidence supports the notion that all the continents were once part of a single supercontinent? What kind of climatological evidence supports the concept of continental drift?

11. The eastern coast of South America is a good fit against the western coast of Africa. What sort of evidence would you look for to confirm that the two continents were once part of the same landmass?

12. How does the density of granitic continental crust compare with the density of basaltic oceanic crust?

13. What body of water separated the ancient supercontinents of Laurasia and Gondwana? Are there any remnants of this body of water in existence today? If so, what are they?

14. What mountain ranges of today were once part of the Tethys Sea? What kind of evidence indicates that the region where these mountains are was once below sea level?

15. North America, Greenland, and Eurasia fit quite well together in reconstructing Laurasia, but there is no space available for Iceland. Why is the omission of Iceland from Laurasia reasonable?

16. How does the origin of the Himalayas differ from that of the oceanic mountains that constitute the Mid-Atlantic Ridge?

17. Which are younger, the Rocky Mountains or the Himalayas?

18. Is the Atlantic Ocean becoming narrower or wider? The Pacific Ocean?

19. The San Andreas Fault in California is a strike-slip fault that lies along the boundary between the Pacific and American plates. What does this indicate about the nature of the boundary?

20. How would you expect the ages of the South Pacific islands far away from the East Pacific Rise to compare with those near the rise?

21. The Andes are a young, still-growing mountain range on the west coast of South America. Why is there no similar mountain range on the east coast of South America?

22. The distance between the continental shelves of the eastern coast of Greenland and the western coast of Norway is about 1300 km. If Greenland separated from Norway 65 million years ago and their respective plates have been moving apart ever since at the same rate, find the average speed of each plate.

23. The oldest sediments found on the floor of the South Atlantic Ocean, which are 1300 km west of the axis of the Mid-Atlantic Ridge, were deposited about 70 million years ago. What rate of plate movement does this finding suggest?

24. In Fig. 15-27, beds A to F consist of sedimentary rocks formed from marine deposits and rocks G and H are granite. What sequence of events must have occurred in this region?

25. What is an unconformity? If one is shown in Fig. 15-27, where is it?

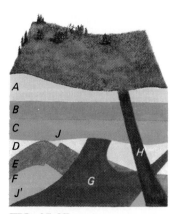

FIG. 15-27

26. What is the age of the oldest rocks known? How was this age determined?

27. What is the basis of the radiocarbon dating procedure?

28. What steps would you take to determine the age of an ancient piece of wood by radiocarbon dating?

29. List as many different kinds of fossils as you can.

30. Why are fossils still useful in dating rock formations despite the development of radioactive methods?

31. Why are fossils never found in igneous rocks and only rarely in metamorphic rocks?

32. Why are most fossils found in beds that were once the floors of shallow seas?

33. What is the oldest division of geologic time? In what division are we living today?

34. During what divisions of geologic time have living things existed on the earth?

35. The earth's history is sometimes divided into two **eons,** Cryptozoic ("hidden life") and Phanerozoic ("visible life"), with the first corresponding to Precambrian time and the second extending from the beginning of the Paleozoic Era to the present day. What do you think is the reason for this division?

36. The early atmosphere of the earth probably consisted of carbon dioxide, water vapor, and nitrogen, with little free oxygen. What is believed to be the source of oxygen in the present-day atmosphere? What bearing has this question on the relatively rapid development of varied and complex forms of life that marks the start of the Paleozoic Era?

37. Precambrian rocks include sedimentary, igneous, and metamorphic varieties. What does this suggest about the geologic activity in Precambrian time?

38. Precambrian rocks are exposed over a large part of eastern Canada. What does this suggest about the geologic history of this region since the end of Precambrian time?

39. What conspicuous difference is there between Precambrian sedimentary rocks and those of later eras?

40. What are the chief kinds of organisms that have left traces in Precambrian sedimentary rocks?

41. Paleozoic sedimentary rocks derived from marine deposits are widely distributed in all the continents. What does this indicate about the height of the continents relative to sea level in the Paleozoic Era?

42. Why is it believed that large parts of the United States were once covered by shallow seas?

43. About 200 million years ago today's continents were all part of the supercontinent Pangaea. During what geologic era did Pangaea break apart into Laurasia and Gondwana? During what era did Laurasia break up into North America, Greenland, and Eurasia?

44. Under what circumstances is coal formed?

45. What is believed to be the origin of petroleum? of natural gas?

46. What are some of the events associated with continental drift that could produce a widespread biological extinction?

47. What kind of animals were the dinosaurs? Were they mostly small, mostly large, or were they of all sizes?

48. What is the evidence that a comet or asteroid impact caused the great dinosaur extinction at the end of the Mesozoic? Are any other explanations possible?

49. During what geologic era did birds develop? From what type of animal did they evolve?

50. What are some of the chief differences between reptiles and mammals?

51. The same reptiles were present on all continents during the Mesozoic Era, but the mammals of the Cenozoic Era are often different on different continents. Why?

52. In rocks of what era or eras would you expect to find fossils of (a) horses; (b) ferns; (c) clams; (d) insects; (e) apes?

53. What were the Ice Ages? When did they occur?

54. Minnesota has a great many shallow lakes. How do you think they originated?

55. The Scandinavian landmass has been rising since the end of the most recent ice advance; the current rate is about 1 cm/year. Can you think of any reason for this rise?

ANSWERS TO MULTIPLE CHOICE			
1. a	**6.** b	**11.** b	**16.** a
2. b	**7.** c	**12.** b	**17.** a
3. c	**8.** c	**13.** b	**18.** d
4. d	**9.** a	**14.** d	**19.** d
5. c	**10.** c	**15.** c	**20.** d

21. a	**27.** c	**33.** c	**39.** a
22. d	**28.** a, b, c, d	**34.** c	**40.** b
23. c	**29.** d	**35.** b	**41.** a
24. c	**30.** a	**36.** c	**42.** b
25. c	**31.** b	**37.** b	
26. b, a, d, c	**32.** c	**38.** a	

16

THE

SOLAR

SYSTEM

Whether we look at them with the naked eye or with the help of the largest telescope, the stars are just points of light. On the other hand, most of the planets appear as disks in a telescope of even modest power. This does not mean, of course, that the planets are larger than the stars, only that the planets are much closer to us. The sun, like other stars, glows brightly because it is extremely hot. The planets are too cool to shine by themselves and we see them by the sunlight they reflect. The sun together with its accompanying planets, their satellites, and other smaller bodies make up the **solar system.** The members of the solar system dwell in emptiness and are separated by vast distances from everything else in the universe.

THE FAMILY OF THE SUN

Until the seventeenth century the solar system was thought to consist of only five planets besides the sun, the earth, and the moon. In 1609, soon after having heard of the invention of the telescope in Holland, Galileo built one of his own. With his telescope, Galileo found four additional members of the solar system: the brighter of the moons (or **satellites**) that circle Jupiter. Since Galileo's time, improved telescopes have led to the discovery of many more members of the sun's family.

The list of planets now includes nine. In order of average distance from the sun, they are Mercury, Venus, Earth, Mars, Jupiter, Saturn, Uranus, Neptune, and Pluto (Fig. 16-1). All except Mercury and Venus have satellites. Thousands of small objects called **asteroids,** all less than 1000 km in diameter, follow separate orbits about the sun in the region between Mars and Jupiter. Comets and meteors, in Galileo's time thought to be atmospheric phenomena, are now recognized as also belonging to the solar system.

FIG. 16-1 The solar system. The orbits of Mercury and Venus are too small to be shown on this scale. Pluto's orbit is by far the most elliptical. Diameters of sun and planets are exaggerated.

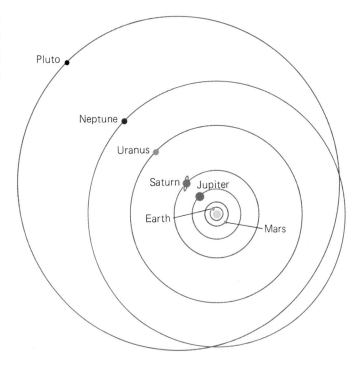

In recent years our knowledge of the solar system has been greatly increased by the voyages of spacecraft, most of them from the United States but with some notable Soviet ones as well. Spacecraft have landed on Venus and Mars, and astronauts have walked on the surface of the moon. So remarkable is modern technology that signals from the tiny 8-W transmitter on Pioneer 10 were still being picked up on the earth a dozen years after it left the solar system, over 20 years after it left Florida.

16-1   THE SOLAR SYSTEM

X . 6

The Inner Planets Are Mostly Rock, the Outer Planets Are Mostly Liquefied Gas

Not only is the solar system isolated in space, but each of its principal members is far from the others. From the earth to our nearest neighbor, the moon, is about 384,000 km; from the earth to the sun is about 150 million km. It took the Apollo 11 spacecraft 3 days to reach the moon. Traveling at the same speed, Apollo 11 would take more than 3 years to reach the sun.

If we use a golf ball to represent the sun, a grain of sand 4 m away would represent the earth on the same scale. The moon would be a dust speck about 1 cm from the sand grain. The largest planet, Jupiter, would be a small pebble 18 m from the golf ball, would be another dust speck 150 m from the golf ball. With three more sand grains, and a few dozen more wide orbit of Pluto, the model is com-

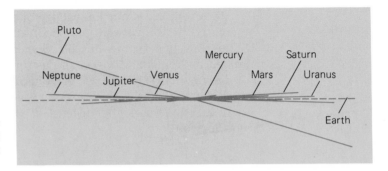

FIG. 16-3   The orbits of the planets seen edgewise. Except for that of Pluto, the orbits all lie nearly in the same plane.

1. Nearly all the revolutions and rotations are in the same direction. Only the rotation of Venus and the revolutions of a few minor satellites about their parent planets run contrary to the general motion. (Uranus is an exception of a different kind, since it rotates about an axis only 8° from the plane of its orbit.)

2. All the orbits except those of Pluto and comets lie nearly in the same plane (Fig. 16-3).

The principal data about the planets and their orbits are summarized in Table 16-1. The **inner planets** of Mercury, Venus, Earth, and Mars are relatively small, have similar densities, and are composed largely of rocky material. They Only one satellite them is only one satellit Mars are

TABLE 16-1

The Planets

Planet	Symbol	Mean Distance from Sun, Earth = 1[a]	Diameter, Thousands of km	Mass, Earth = 1[b]	Mean Density, Water = 1[c]	Surface Gravity, Earth = 1[d]	Escape Speed, km/s[e]	Period of Rotation on Axis	Period of Revolution around Sun	Eccentricity of Orbit[h]	Inclination of Orbit to Ecliptic[i]	Known Satellites
Mercury	☿	0.39	4.9	0.055	5.4	0.38	4.3	59 days	88 days	0.21	7°00'	0
Venus	♀	0.72	12.1	0.82	5.25	0.90	10.4	243 days[f]	225 days	0.01	3°34'	0
Earth	⊕	1.00	12.7	1.00	5.52	1.00	11.2	24 h	365 days	0.02	—	1
Mars	♂	1.52	6.8	0.11	3.93	0.38	5.0	24.5 h	687 days	0.09	1°51'	2
Jupiter	♃	5.20	143	318	1.33	2.6	60	10 h	11.9 yr	0.05	1°18'	16
Saturn	♄	9.54	120	95	0.71	1.2	36	10 h	29.5 yr	0.06	2°29'	23
Uranus	⛢	19.2	51	15	1.27	1.1	22	16 h[g]	84 yr	0.05	0°46'	15
Neptune	♆	30.1	50	17	1.70	1.2	24	16 h	165 yr	0.01	1°46'	8
Pluto	♇	39.4	2.4	0.03	1.99	0.43	3.2	6 days	248 yr	0.25	17°12'	1

[a]The mean earth-sun distance is called the *astronomical unit*, where 1 AU = 1.496×10^8 km.

[b]The earth's mass is 5.98×10^{24} kg.

[c]The density of water is 1 g/cm³ = 10^3 kg/m³.

[d]The acceleration of gravity at the earth's surface is 9.8 m/s².

[e]Speed needed for permanent escape from the planet's gravitational field.

[f]Venus rotates in the opposite direction from the other planets.

[g]The axis of rotation of Uranus is only 8° from the plane of its orbit.

[h]The difference between the minimum and maximum distances from the sun divided by the average distance.

[i]The ecliptic is the plane of the earth's orbit.

The solid nucleus of Halley's comet, photographed from the European spacecraft *Giotto* in March 1986. The nucleus is irregular, about 16 km long and 8 km across, and its mass is about 100 billion tons. Gases and dust stream outward from the sunlit side at left. The nucleus is composed of ice and various frozen gases, stony materials, and what are apparently polymerized organic compounds.

The comet West as it appeared in the sky in March 1976.

organic molecules. In addition, smaller amounts of various frozen gases are present, notably carbon monoxide, carbon dioxide, methane, ammonia, and hydrogen cyanide.

The paths followed by comets are quite different from the nearly circular planetary orbits. A comet approaches the sun from far out in space beyond the orbit of Pluto, swings around the sun, and then retreats. The orbit is a long, narrow ellipse, and the comet returns at regular intervals. Although most orbits are so large that their periods range up to millions of years, a few are smaller. Halley's comet, for instance, reappears every 76 years; it is the brightest one known. This notable comet has returned 28 times since the first sure record of its observation was made in 239 B.C. It was named after the English astronomer Edmund Halley, a contemporary of Newton, who in 1705 predicted that a comet last seen in 1682 would reappear in 1758, as it did. Halley's comet came within 93 million km of the earth in November 1985 on its most recent visit, when it was studied at close range by European, Russian, and Japanese spacecraft, and will return next in 2061.

FIG. 16-4 The tail of a comet always points away from the sun because of pressure from the sun's radiation and from the solar wind of ions. The tail is longest near the sun and is probably absent far away from the sun.

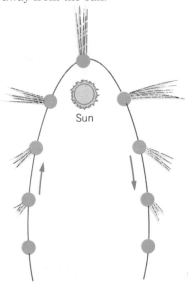

Sun

Heads and Tails

In the far reaches of the solar system comets are fairly small, only a few km across. Near the sun the frozen gases vaporize and take with them dust to form the huge but thin clouds that make up the comet heads and tails we see. Comets are visible only when close to the sun. This is partly because of sunlight scattered by cometary material but mainly because the gases are excited to luminescence by solar radiation.

Comets appear as small, hazy patches of light, often accompanied by long, filmy tails—hence their name, which comes from the Greek word for "hairy." Most comets are visible only with the help of a telescope, but from time to time one becomes conspicuous enough to be seen with the naked eye. Watched for a few weeks or months, a comet at first grows larger and its tail longer and more brilliant. Then it fades gradually, loses its tail, and eventually disappears.

A comet's tail always points away from the sun regardless of which way the comet is heading (Fig. 16-4). One reason is the **solar wind,** the stream of ions that constantly flows outward from the sun, which tends to sweep the comet's gases with it. Another reason is the pressure of solar radiation on the dust particles that are part of a comet. The two components of the tail, gas and dust, can be distinguished in photographs of Halley's comet taken from the European spacecraft *Giotto*, which passed near the comet in 1986.

Comet tails may stretch for tens of millions of km across the sky. Whenever a comet passes near the sun, the comet loses a little of its material to space permanently, and as a result is fainter each time it returns. Once past the sun, a comet begins to contract again into a relatively small body.

16-3   METEORS

"Shooting Stars" Usually Smaller Than a Grain of Sand

Meteoroids are small fragments of matter that the earth meets as it travels through space. Most meteoroids are smaller than grains of sand. The majority are believed to be the result of asteroid collisions; others are the debris of comets.

Moving swiftly through the atmosphere, meteoroids are heated rapidly by friction. Usually they burn up completely about 100 km above the earth, appearing as bright streaks in the sky—**meteors,** or "shooting stars." Sometimes, though, they are so large to begin with that a substantial portion may get through the atmosphere to the earth's surface. The largest known fallen meteoroids, called **meteorites,** weigh several tons. The smallest meteoroids are so light that they float through the atmosphere without burning up. Many tons of these fine, dustlike **micrometeorites** reach our planet daily.

A keen observer on an average clear night can spot 5 to 10 meteors an hour. Most of these meteors are random in occurrence and follow no particular pattern either in time or place in the sky. At several specific times of year, however, great meteor showers occur, with 50 or more

Meteor streak in the night sky. Few meteoroids are larger than a grain of sand, and those that enter the earth's atmosphere usually burn up before they can reach the ground. On a clear dark night, 5 to 10 meteors per hour can be seen.

TABLE 16-2

Some Major Meteor Showers

Shower	Maximum Display	Typical Hourly Rate
Quadrantid	Jan. 3	30
Eta Aquarid	May 4	5
Perseid	Aug. 12	40
Orionid	Oct. 22	13
Taurid	Nov. 1	5
Leonid	Nov. 17	6
Geminid	Dec. 14	55

meteors sometimes visible per hour that appear to come from the same region of the sky. The showers come about when the earth moves through a swarm of meteoroids that all follow the same orbit about the sun.

When the meteoroids of a shower are spread out along their common orbit, the number of meteors seen is about the same each year. This is the case with the Perseid showers that occur between August 10 and 14. When the meteoroids are bunched together, the number of meteors seen varies from year to year. The Leonid showers of mid-November are an example, with intense displays every 33 years. The most recent one was in 1966 when thousands of meteors per hour were seen. Other conspicuous meteor showers are listed in Table 16-2. The Eta Aquarid and Orionid showers are due to dust particles from Halley's comet that are spread out in its orbit. The earth crosses this orbit twice a year with these showers as a result.

Meteoroid speeds range up to 72 km/s. This limit is significant, because a higher speed would mean an object arriving from outside the solar system. The conclusion is that all meteoroids, random as well as shower, are members of the solar system that follow regular orbits around the sun until they collide with the earth or another planet.

Most meteorites that have been examined fall into two classes: stony meteorites, the great majority, which consist of silicate minerals much like those in common rocks, and iron meteorites, which consist largely of iron with a small percentage of nickel. A few meteorites are intermediate in character. All meteorites are sufficiently different from terrestrial rocks to be recognized as such, although the iron ones, which are in the minority, are easier to identify.

The earth's neighbors in space—the moon, Mercury, Venus, Mars, and the satellites of Mars—are all deeply pitted with many huge meteor

Discovered in Greenland in 1894, this iron meteorite is 3.3 m long.

THE PHASES OF VENUS

Galileo, the first person to study Venus with a telescope, found that it has phases like those of the moon and that its apparent size changes cyclically. He correctly interpreted this evidence as supporting the copernican thesis that Venus revolves around the sun. A good pair of binoculars can reveal the phases of Venus.

Figure 16-5 shows how these effects arise. A year on Venus is only 225 earth days long, so the positions of Venus and the earth relative to each other and to the sun change continually. When Venus is between the sun and the earth, as at 1, it appears dark to an observer on earth. If it could be seen, it would have its largest apparent diameter. At 2 and 8 Venus appears as a crescent, but because it is still fairly close to the earth, it is quite bright. At 3 and 7 we see half of Venus illuminated, and at 4 and 6 more

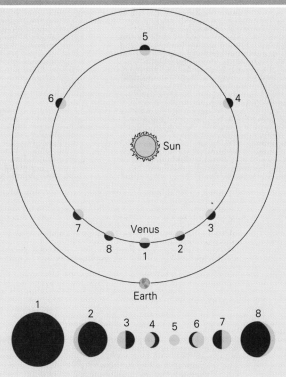

FIG. 16-5 The appearance and apparent size of Venus as seen from the earth depend upon the relative positions of the two planets and the sun. The "new Venus" at 1 is 6 times larger than the "full Venus" at 5. Venus appears brightest between 1 and 2 and between 8 and 1. To an observer on the earth, Venus never appears very far from the sun and either rises just before sunrise or sets just after sunset.

than half. Although the sun gets in the way of our seeing the full Venus at 5, it is so far from the earth that it would not be especially brilliant even if it were visible.

the sky from the sun and appears alternately as a "morning star" and an "evening star" (Fig. 16-5). Venus is usually farther from the sun than Mercury, however, and so is visible for longer periods than Mercury. Venus was named after the Roman goddess of love and beauty and is represented by the traditional symbol of a mirror.

Venus has the distinction of spinning "backward" on its axis. That is, looking downward on its north pole, Venus rotates clockwise, whereas the earth and the other planets rotate counterclockwise. The rotation of Venus is also extremely slow, so that a day on that planet represents 243 of our days.

In size and mass Venus is more like the earth than any other member of the sun's family. Mountains, craters, and fault-like cracks mar its surface, which has two major highland regions, the larger with about the area of the United States. These "continents" cover only about 5 percent of Venus in contrast to the 30 percent extent of the earth's continents, and no water laps their edges. Some of the mountains of Venus are quite high, one of them topping Everest. Evidence of volcanic activity is abundant, including volcanic peaks, extensive lava flows, and a huge volcanic crater about 100 km across. But there is no evidence that the crust of Venus, like that of the earth, consists of huge shifting plates.

 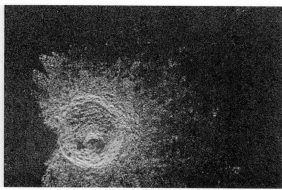

In 1990 the United States space probe Magellan began to circle Venus and to radio back radar images of its rugged surface. Features as small as 120 m across could be detected. The picture at left shows a ridge and valley network formed by intersecting faults that resembles the landscape of a region in the western United States. The picture at right shows an impact crater 12.5 km in diameter probably formed by a meteoroid moving at a shallow angle with the surface when it hit. The central peak in the crater consists of material that rebounded after the impact. Volcanic mountains and rivers of hardened lava were also found.

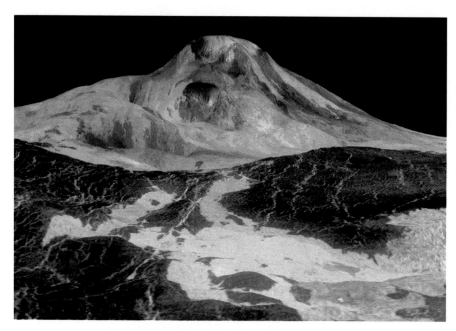

Radar images from the Magellan spacecraft were combined to give this false color perspective view of Maat Mons, the second highest mountain on Venus. Lava flows extend for hundreds of kilometers across the fractured plains in the foreground.

Life on Venus

Is there life on Venus? Bernard de Fontanelle thought so when he wrote in 1686 that the inhabitants of Venus must "resemble the Moors of Granada; a small, dark people, burned by the sun, full of wit and fire, always in love, writing verse, fond of music, arranging festivals, dances, and tournaments every day."

Alas, the truth is less romantic. American spacecraft have passed close to Venus and Soviet ones have parachuted instruments to its sur-

6 months long and at least as pronounced as ours. Mars has long fascinated astronomers and laypeople alike because it is in so many ways similar to the earth, which leads to the question of whether life of some kind is present there.

16-7   IS THERE LIFE ON MARS?

Maybe Once, but Probably Not Now   If life does exist on Mars, it is adapted to an environment that would soon destroy most earthly organisms. For a start, Martian climates are severe by our standards. Over half again farther from the sun than the earth, Mars receives much less solar energy per m² than we do. Its atmosphere, largely carbon dioxide, is extremely thin—equivalent to the earth's atmosphere at an altitude of nearly 40 km—so little of the heat from the sun is retained after nightfall. Daytime temperatures in summer at the equator rise to perhaps 20°C, but at night drop to perhaps −70°C. The average surface temperature of the entire planet is about −30°C. The scanty Martian atmosphere is also unable to screen out harmful solar ultraviolet radiation, a function carried out in the earth's atmosphere by the ozone present at high altitudes (see Chap. 13).

Another difficulty life would have to face on Mars is the lack of liquid water. Water in other forms is certainly there, as water vapor in the atmosphere and frozen (along with carbon dioxide) in the ice caps of the polar regions. In earlier times, though, liquid water does seem to have been present on Mars. Some surface features photographed by spacecraft strongly suggest dried-up riverbeds and erosion by running water. The earth's surface water probably was vented from volcanoes early in its history, and there seems no reason why the same process should not have occurred on Mars, whose surface is dotted with extinct volcanoes. Seas and oceans quite possibly were part of the Martian landscape, as they are on the earth today. The early Martian atmosphere would have enabled the greenhouse effect to keep the surface warm enough for the water to remain liquid.

The craters we now see in the eroded regions of Mars must have been formed by meteoroid impacts after the surface water disappeared. From the number of these craters it seems clear that there has been no erosion to speak of for at least several hundred million years, perhaps for a few billion.

The Likelihood of Life Today   The fact that most terrestrial life requires liquid water and oxygen plus protection from solar ultraviolet radiation does not necessarily mean that life of some kind could not exist in their absence. The life processes of certain bacteria on the earth require carbon dioxide rather than oxygen, so an oxygen-containing atmosphere is not indispensable, at least for primitive forms of life. Conceivably there could have evolved on Mars organisms that can thrive on traces of water gleaned from the minerals in surface rocks. And shells of some sort might protect Martian life from ultraviolet radiation. Or life there could exist underground, its energy source heat from the interior rather than sunlight.

In 1976 two Viking spacecraft landed on Mars and sent back data and photographs such as this one. Martian rocks are porous and jagged, like basaltic lavas of the earth and moon, and the soil resembles weathered lava. Iron oxides are responsible for the red color. No trace of life was found.

Since conditions there long ago may have been comparable to those on the earth, life of some kind could have come into being on Mars. Then the carbon dioxide of the atmosphere gradually became incorporated in Martian rocks, as happens on the earth today. But Mars lacks the tectonic and volcanic processes that recycle the earth's carbon dioxide, so this gas, vital for the greenhouse effect, all but disappeared.

Where did the water go? Nobody knows, but much of it may remain permanently frozen below the Martian surface. These changes were very gradual, and it is not entirely absurd to speculate that living things there could have adapted to the progressively harsher environment and have survived in some form to the present.

The absence of any sign of life in photographs taken from thousands of km away from the Martian surface means nothing. At such distances terrestrial life would probably not be apparent to a visitor from elsewhere (and a closer look might well suggest that the car is the most conspicuous type of life on earth).

In 1976 two American Viking spacecraft landed on Mars. Among their various tasks were several sensitive experiments able to detect life in Martian soil. No evidence for present life or chemical traces left by past life was found. Even worse for the hypothesis of life on Mars, the soil turned out to be self-sterilizing. Solar ultraviolet radiation turns the soil minerals into strong oxidizing agents that quickly destroy organic compounds. Of course, the experiments are not conclusive since they were limited to two landing sites, and such essential elements for life as nitrogen and phosphorus were indeed discovered. The best that can be said at present is that conditions suitable for life possibly once existed on the surface of Mars but probably do not today. A new series of visits to Mars to search for signs of life is scheduled for the last years of this century.

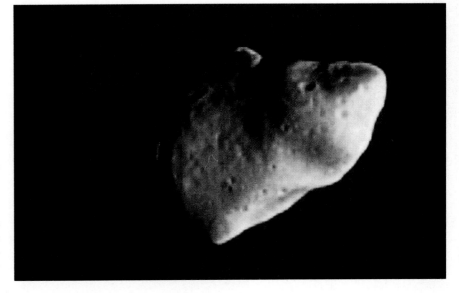

The first close-up photograph of an asteroid was taken in 1991 from the Galileo spacecraft. Called Gaspra, the asteroid is about 20 km long and 13 km wide and is pocked with craters, some over a kilometer across. Gaspra seems to have a magnetic field as strong as the earth's.

16-8   ASTEROIDS

Millions of Tiny Planets between Mars and Jupiter

The asteroids are small, rocky objects, nearly all of which circle the sun in a belt between Mars and Jupiter. A few pass inside the earth's orbit, and one, Icarus, gets even closer to the sun than Mercury.

The largest asteroid, Ceres, is about 1000 km across, roughly the extent of Texas. It was the first to be discovered, on January 1, 1801, by the Italian astronomer Giuseppe Piazzi, who named it after the patron goddess of his native Sicily. For a long time asteroids were thought to be fragments from the breakup of a planet that was supposed to have once existed between Mars and Jupiter. Nowadays asteroids are considered to be bits of matter from the early solar system that never became part of a larger body because of the gravitational influence of the nearby giant planet Jupiter.

The larger asteroids, several thousand in all with relatively few over 100 km across, have been tracked and named. Although there are millions more, the asteroid belt is so vast that they are usually far apart. But not always: most of the meteoroids that reach the earth are believed to be debris from asteroid collisions. Most asteroids are irregular in shape, too small for gravity to have pulled them into spheres. At least one asteroid, Ida, has a satellite. Called Dactyl (after the Dactyli, mythical creatures supposed to live on Mount Ida in Greece), it is 1.5 km in diameter.

The Asteroid Danger

Because some asteroids pass inside the earth's orbit, there is a chance that one of them might someday collide with the earth. What would happen? In fact, at least two such events seem to have occurred already. One took place on the morning of June 30, 1908, in the Tunguska River region of Siberia. An object was seen falling from the sky followed by a brilliant flash and a great blast that devastated over 2000 km^2 of forest. No large crater was

An incoming asteroid that exploded over Siberia in 1908 had this effect on a huge area of forest. Here is how the event appeared to a Siberian farmer 200 km away: "When I sat down to have my breakfast beside my plow, I heard sudden bangs, as if from gunfire. My horse fell to its knees. From the north side above the forest a flame shot up. Then I saw that the fir forest had been bent over by the wind, as I thought of a hurricane. I seized hold of my plow with both hands so it would not be carried away. The wind was so strong it carried soil from the surface of the ground, and then the hurricane drove a wall of water up the Angora River."

found, and only microscopic iron and silicate particles were found in the soil afterward. The object exploded at an altitude of about 8 km, which is what would be expected of an asteroid but not of the head of a comet, an alternative explanation once favored.

The Tunguska asteroid was probably about 30 m across, and although lots of such small asteroids are out there, they are too small to pose a large-scale threat to our planet. But a really big asteroid is another story. As we saw in Sec. 15-16, an asteroid perhaps 10 to 15 km across struck Mexico 65 million years ago, an event whose consequences apparently led to the extinction of the dinosaurs as well as many other forms of life. If another asteroid of such size headed for the earth could be detected in time, which seems possible, a missile with a nuclear warhead could be sent to intercept it whose explosion might be able to divert the asteroid. Such a defense is under serious study.

THE OUTER PLANETS

16-9   JUPITER

Almost a Star
The mammoth planet Jupiter is fittingly named after the most important of the Roman gods. The planet Jupiter is represented by a stylized lightning bolt because lightning was supposed to come from the god Jupiter. The symbolism has turned out to be appropriate since electric discharges regularly occur in Jupiter's atmosphere that produce bursts of radio waves detectable on the earth.

Jupiter, like Venus, is shrouded in clouds, but those of Jupiter are more spectacular. The clouds are responsible for Jupiter's relatively high reflectivity, which together with its size makes it a bright object in the

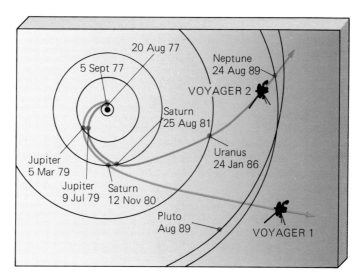

Nor can the rings be gaseous—they reflect sunlight and radar signals far too well. The only question left is the size and nature of the particles of which they consist. In 1980 and 1981 two Voyager spacecraft reached Saturn after 4-year journeys (Fig. 16-7). They radioed back vast amounts of data on the planet, its satellites, and, of course, its rings. The findings about the rings confirmed what astronomers had suspected earlier: the particles are chunks of rock and ice that range from the size of buildings down to fine dust. In addition, the rings we see from the earth are not uniform but are split into thousands of narrow ringlets. Two of the ringlets in one of the outer, faint rings seem to be braided around each other, a peculiarity that may result from the gravitational pulls of two small satellites whose orbits are just inside and just outside the ringlets.

16-11 URANUS, NEPTUNE, PLUTO

Far Out The three most distant planets—Uranus, Neptune, and Pluto—owe their discovery to the telescope. Uranus was found in 1781 by the great English astronomer William Herschel. In Greek mythology, Uranus personifies the heavens and is the father of Saturn and the grandfather of Jupiter.

Herschel at first suspected Uranus to be a comet because, through his telescope, it appeared as a disk rather than as a point of light like a star. Observations made over a period of time showed its position to change relative to the stars, and its orbit, found from these data, revealed it to be a planet. Neptune was discovered in 1846 as the result of predictions based on its gravitational effects on the orbit of Uranus (see Chap. 1). Pluto was discovered in 1930 by the American astronomer Clyde Tombaugh, who was looking for a new planet whose existence was suspected on the basis of irregularities in the orbits of Uranus and Neptune. As it happens, the total mass of Pluto and its satellite is not enough to account for these irregularities, and it seems to have been

BIOGRAPHY

William Herschel (1738–1822) was born in Germany (his name was originally Friedrich Wilhelm) and went to England in 1757 to avoid military service. In England Herschel prospered as an organist and music teacher, and composed 24 symphonies. He also found time for his hobby of astronomy. Good telescopes were expensive, so Herschel, with the help of his sister Caroline, built his own. By 1774 Herschel had made the finest telescope in the world, a reflector whose mirror he and his sister had ground themselves. With this telescope he systematically searched the sky, a labor rewarded in 1781 by the discovery of a seventh planet, Uranus. This created a great stir everywhere, both because Uranus was the first new planet to be identified for many centuries and because its distance from the sun was twice that of Saturn, thus doubling the size of the solar system. Six years later Herschel found two satellites circling Uranus and called them Titania and Oberon; eventually the total was raised to 15.

In 1788 Herschel married a rich widow, which enabled him to become a full-time observer of the heavens. The list of his achievements is unparalleled in astronomy. Among his other discoveries were double stars, and he established that each member of a pair revolved around the other. Using a thermometer to measure the heat produced, he studied the intensity of sunlight in various parts of its spectrum. To his surprise, the heating continued past the red end, which suggested to him that the sun emits invisible (today called infrared) radiation besides visible light. Most significantly, Herschel's studies of the Milky Way led him to conclude that it is a vast disk-shaped galaxy of stars, one of which is the sun. He speculated—correctly—that a number of fuzzy objects in the sky were other galaxies far away in space.

Herschel's renown became such that King George III provided the money for a huge new telescope with a mirror 1.2 m across. The first time the telescope was used, Herschel discovered two more satellites of Saturn to add to the five already known, and he was soon able to show that Saturn's rings are not stationary but revolve about their mother planet. He remained active almost until his death at 84 (the same span as the orbital period of Uranus). Herschel's work was carried on by his son John, who was a pioneer in astronomical photography.

pure accident that Tombaugh found Pluto. Neptune was named after the Roman god of the sea, and Pluto after the Greek god of the underworld, the home of the dead.

Uranus has the distinction of rotating about an axis only 8° from the plane of its orbit, so we can think of it as spinning on its side. The most likely reason is a collision with a large object, perhaps the size of the earth, early in the history of the solar system. Because of this tilt, summer at a particular place on Uranus is a 42-year period of continuous sunshine, and winter is a 42-year period of continuous darkness.

The spacecraft Voyager 2 passed near Uranus in 1986 and near Neptune 3 years later (Fig. 16-7). Photographs radioed back to the earth show that both planets have several rings around them. Like the rings of Saturn, these rings consist of small particles but are too narrow to be seen from the earth.

When the moon is farthest from the earth, its apparent diameter is less than that of the sun. The moon cannot then block the entire solar disk even when the moon lies directly between the sun and the earth. In this situation the result is an **annular** eclipse of the sun, with a ring of sunlight appearing around the rim of the moon.

16-14 THE LUNAR SURFACE

Mountains and Maria, but No Water and No Atmosphere

To look at the moon is to wonder. What is it made of? What is its origin? What is the nature of its landscape? What geologic processes occur on its surface and in its interior? Is there life on the moon? What is the ultimate destiny of the earth-moon system?

Until July 20, 1969 the study of the moon was more notable for the questions asked than for the answers available. On that day Neil Armstrong set foot on the moon, the first person ever to do so, after a 3-day voyage aboard the spacecraft Apollo 11 with two companions. Four days after that they returned to earth, bringing with them samples of the lunar surface. That historic expedition and the others that followed it answered a great many questions about our companion in space. But questions still remain, the chief one being: Where did the moon come from?

The moon was not entirely a mystery even before the voyages of Apollo 11 and of the other manned spacecraft that followed it there. Even a small telescope reveals the chief features of the lunar landscape: wide plains, jagged mountain ranges, and innumerable craters of all sizes. Each mountain stands out in vivid clarity, with no clouds or haze to hide the smallest detail. Mountain shadows are black and sharp-edged. When the moon passes in front of a star, the star remains bright and clear up to the moon's very edge. From these observations we con-

The moon rotates on its axis at the same rate that it orbits the earth. As a result, the moon always presents this face to us. The dark areas are the maria, lava flows that have been broken up by meteoroid bombardment.

clude that the moon has little or no atmosphere. Water is likewise absent, as indicated by the complete lack of lakes, oceans, and rivers.

But there is still no substitute for direct observation and laboratory analysis. Each spacecraft that landed on the moon and returned to earth, whether piloted by human beings or not, has brought back information and samples of the greatest value. The lack of a protective atmosphere and of running water to erode away surface features means there is much to be learned on the moon about our common environment in space, both past and present. And from the composition and internal structure of the moon hints can be gleaned of its history, hints that bear upon the history of the earth as well. Thus the study of the moon is also a part of the study of the earth, doubly justifying the effort of its exploration.

Landscape Features

With the help of no more than binoculars it is easy to distinguish the two main kinds of lunar landscape, the dark, relatively smooth **maria** (the singular is **mare**) and the lighter, ruggedly mountainous highlands. The mountains of the moon are thousands of meters high, which means that the moon's surface is about as irregular as the earth's.

Mare means "sea" in Latin, but the term is still used even though it has been known for a long time that these regions are not covered with water. The largest of the maria is Mare Imbrium, the Sea of Showers, which is over 1000 km across. The maria are circular depressions covered with dark, loosely packed material—not solid rock. They are not perfectly smooth but are marked by small craters, ridges, and cliffs. The maria consist of lava flows similar to basalt that have been broken up by meteoroid impacts. It is curious that nearly all the maria are on the lunar hemisphere that faces the earth.

The lunar highlands are scarred by innumerable craters, some with mountain peaks at their centers. Certain craters, such as Tycho and

The far side of the moon, which always faces away from the earth, was photographed from the Galileo spacecraft in 1990. The dark markings are basaltic lava flows formed over 3 billion years ago. The dark region at lower left is a huge basin that was probably caused by the impact of a giant meteoroid.

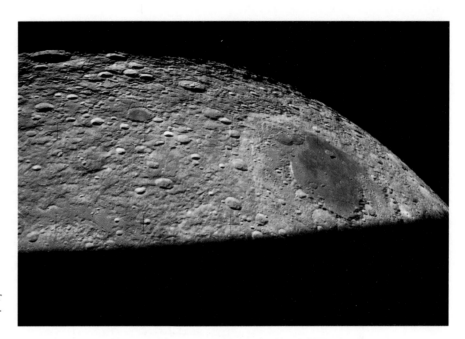

Cratered landscapes of the lunar highlands were caused by meteoroid impacts.

Copernicus, have conspicuous streaks of light-colored matter radiating outward. These **rays** may extend for hundreds or thousands of km, and seem to consist of lunar material sprayed outward after the meteoroid impacts that caused the craters. The impacts melted this material, and it cooled quickly in flight into glassy particles that reflect light well.

The **rilles** of the highlands are especially intriguing. These are narrow trenches up to 250 km long that look like dried-up riverbeds. The rilles were probably created by the collapse of subsurface channels through which molten rock once flowed from active volcanoes.

Instruments placed on the lunar surface show that moonquakes are few and mild compared with earthquakes. This implies that the interior of the moon is relatively cool, since heat from inside the earth powers its tectonic activity. As in the case of the earth, the data show the moon to consist of a rigid crust, a thick solid mantle, and a small, dense core. Evidence in lunar rocks of a magnetic field in the past suggests that the core may once have consisted of molten iron, as the earth's core does today. Later the core cooled, which stopped the motions that generated the magnetic field. Because the moon's average density is only 60 percent of the earth's average density, the moon's core must be much smaller, relative to its size, than the earth's core.

16-15 EVOLUTION OF THE LUNAR LANDSCAPE

A Violent Past, a Quiet Present
The analysis of lunar rock and soil samples has led to a number of conclusions about the history of our satellite. On the basis of such samples the moon's story can be taken farther back than the earth's because, since the moon lacks an atmosphere, weathering did not occur

Astronaut Charles M. Duke, Jr., collecting lunar samples during the Apollo 16 expedition in 1972. The crater at left is 40 m in diameter and 10 m deep. Behind it is the Lunar Roving Vehicle that permitted Duke and his fellow astronaut John W. Young to explore the lunar surface some distance from the landing craft.

there. Even particles of pure iron have been found on the moon's surface, whereas only iron compounds occur on the earth's surface. Rocks have been found on the moon that radioactive dating reveals crystallized much earlier than the most ancient terrestrial rocks, which are nearly 4 billion years old. Some lunar samples apparently solidified very soon after the solar system came into being. These rocks are older than any found on the earth.

As the molten outer part of the early moon gradually hardened into a light-colored crust, meteoroids of all sizes rained down. Some, as large as Rhode Island and better described as asteroids, smashed great basins

This 2-cm rock fragment collected during the Apollo 11 expedition to the moon in 1969 resembles certain volcanic rocks on the earth, although it is different chemically.

hundreds of km across. About 4 billion years ago, as the bombardment moderated, radioactivity inside the moon produced enough heat to melt rock there. Lava that reached the surface flowed into the impact basins to form the dark maria that today cover about 20 percent of the moon's surface.

The youngest rocks found on the moon are 3 billion years old, so all igneous activity there must have stopped at that time. Meteoroids continued to crater the landscape and to pulverize surface rock into the powdery debris that today coats the moon.

16-16   ORIGIN OF THE MOON

Not an Easy Problem   Until recently theories of the origin of the moon fell into three categories (Fig. 16-10):

1. The moon was initially part of the earth and split off to become an independent body.

2. The moon was formed elsewhere in the solar system and was later captured by the earth's gravitational field.

3. The moon and the earth came into being together as a double-planet system.

Each of these approaches once seemed quite attractive, but strong arguments against all of them eventually appeared.

FIG. 16-10 Four theories of the moon's origin: (1) The moon split away from the earth. (2) The moon was captured as it approached the earth from elsewhere. (3) The earth and moon were formed together from different clouds of particles. (4) Another early planet struck the earth and formed a larger earth plus the moon. The last theory has the fewest objections to it.

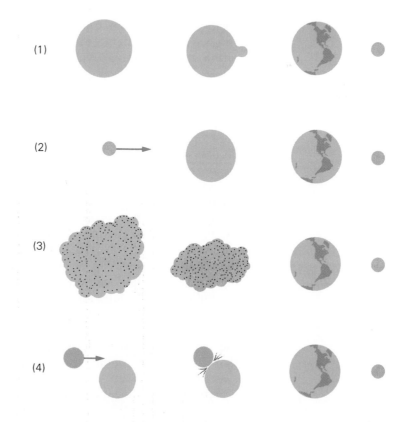

The first hypothesis assumes that the original earth was spinning so fast that it became unstable and broke in two. If we add together the earth's angular momentum of rotation on its axis and the moon's angular momentum of revolution around the earth, the original earth would have completed a turn every $5\frac{1}{2}$ h or so. This is quite fast, but only half the speed required for such a body to break up. Furthermore, even if the body had broken up, the mass ratio of the fragments ought to be perhaps 8:1, not the 81:1 ratio of the actual earth and moon. And if the moon had originated in this way, it would have escaped from the earth's gravitational field altogether instead of going into orbit around it. Finally, the compositions of lunar rocks are too different from those on earth for the moon to be its daughter.

The second hypothesis must overcome the problem that a body approaching the earth from far away will, if nothing else happens, either crash into the earth or else simply swing past it and move off again. But if such a body lost just the right amount of energy in just the right way near the earth (perhaps by friction in a dust cloud or by being struck by a large meteoroid), it could have become permanently trapped in an orbit. Chance plays too many roles in this hypothesis for such a celestial marriage to be very likely.

In the third idea, the earth and moon are regarded as sister planets. Since the solar system almost certainly originated in the gradual fusing together of small particles of matter in a cloud around the sun (Sec. 18-11), there could have been two centers of accumulation rather than one. However, nobody has been able to come up with a mechanism whereby a double planet instead of two separate ones could have evolved. And if the moon and earth came from the same material, why are their compositions so different today? Well, maybe the moon grew later from another cloud of matter near an already formed earth. But this modification, too, has serious weaknesses.

The Collision Hypothesis

A fourth proposal, suggested not long ago, may have less against it than the others. Early in the history of the solar system, when the bits of matter in a particular region began to collect together, they could have produced more than one large body. Suppose another planet a little larger than Mars and with a composition slightly different from that of the infant earth developed nearby, and this planet crashed into the earth. The mantle of the other planet and some of the earth's would have been thrown off into orbit around the earth by the impact, later to form the moon. In this picture of the moon's origin, the other planet's iron core was added to the earth's core. The orbital angular momentum that the moon has today can be accounted for if the collision were a glancing one rather than head-on.

Pulling so dramatic an event as a collision between two planets out of a hat always arouses suspicion among scientists. Why should we take this idea any more seriously than, say, the capture of the moon by the earth in the second hypothesis described above? The answer is that, if similar secondary planets also occurred elsewhere in the solar system, then a number of other long-standing oddities can also be understood. For instance, we have already noted that a large-scale collision could have knocked Uranus into its sideways rotation. So it is possible that the moon's origin will prove to be just one example of a normal phenomenon, and at last we will know where our companion in space came from.

IMPORTANT TERMS AND IDEAS

Comets and **meteoroids** are relatively small objects that pursue regular orbits in the solar system. Comets glow partly by the reflection of sunlight but mainly through the excitation of their gases by solar ultraviolet radiation. **Meteors** are the flashes of light that meteoroids produce when they enter the earth's atmosphere. Fallen meteoroids are called **meteorites.**

In order from the sun, the planets are **Mercury, Venus, Earth, Mars, Jupiter, Saturn, Uranus, Neptune,** and **Pluto.** All but Mercury and Venus have satellites. In addition, thousands of **asteroids** are in orbits that lie between those of Mars and Jupiter. The planets and their satellites are visible by virtue of the sunlight they reflect.

The **inner planets** (Mercury, Venus, Earth, and Mars) are considerably smaller, less massive, and denser than the **outer planets,** which (except for Pluto) apparently consist largely of hydrogen, helium, and hydrogen compounds such as methane, ammonia, and water vapor. Jupiter, Saturn, Uranus, and Neptune are circled by **rings** of particles.

The **moon** shines by virtue of reflected sunlight. The **phases** of the moon occur because the area of its illuminated side visible to us varies with the position of the moon in its orbit. A **lunar eclipse** occurs when the earth's shadow obscures the moon, a **solar eclipse** when the moon obscures the sun.

The moon has neither an atmosphere nor surface water. Although geologically inactive at present, its surface shows signs of having once melted and of having experienced many volcanic eruptions long ago. Meteoroid bombardment has been a more recent factor in shaping its landscapes.

EXERCISES: MULTIPLE CHOICE

1. Comets
   a. follow orbits around the earth
   b. follow orbits around the sun
   c. move randomly through the solar system
   d. are the tracks left in the atmosphere by meteoroids

2. Comets consist largely of
   a. hydrogen   **c.** ice and frozen gases
   b. iron   **d.** rocks

3. The tail of a comet
   a. is constant in size
   b. always points away from its direction of motion
   c. always points toward the sun
   d. always points away from the sun

4. Meteors occur in the night sky
   a. only during spring and fall
   b. only during specific shower periods
   c. in approximately the same numbers every night
   d. every night, but with greater frequency during shower periods

5. On an average clear night, the number of meteors that can be seen per hour is
   a. 1 or 2   **c.** 100 to 200
   b. 5 to 10   **d.** 1000 to 2000

6. Meteoroids
   a. come from the sun
   b. come from the moon
   c. come from outside the solar system
   d. are members of the solar system

7. The majority of meteoroids consist of
   a. silicate minerals   **c.** iron
   b. calcite   **d.** nickel

8. The planets shine because they
   a. emit light
   b. reflect sunlight
   c. reflect moonlight
   d. reflect starlight

9. The planet nearest the sun is
   a. Mercury   **c.** Saturn
   b. Venus   **d.** Pluto

10. The smallest of the following planets is
   a. Mars   **c.** Saturn
   b. Jupiter   **d.** Uranus

11. The largest of the following planets is
   a. Mars   **c.** Saturn
   b. Jupiter   **d.** Uranus

12. The planet closest to the earth in size and mass is
   a. Mercury   **c.** Mars
   b. Venus   **d.** Uranus

13. The length of the year is shortest on
 a. Mercury   **c.** Earth
 b. Venus   **d.** Pluto

14. A planet with virtually no atmosphere is
 a. Mercury   **c.** Jupiter
 b. Mars   **d.** Saturn

15. The planet that appears brightest in the sky is
 a. Venus   **c.** Jupiter
 b. Mars   **d.** Saturn

16. The planet whose surface most resembles that of
 the moon is
 a. Mercury   **c.** Jupiter
 b. Venus   **d.** Saturn

17. An astronaut would weigh least on the surface of
 a. Venus   **c.** Mars
 b. Earth   **d.** Jupiter

18. A dense atmosphere of carbon dioxide is found
 on
 a. Mercury   **c.** Mars
 b. Venus   **d.** the moon

19. A planet with no known satellites is
 a. Venus   **c.** Saturn
 b. Jupiter   **d.** Uranus

20. The largest of Jupiter's satellites is about the size
 of
 a. a large ship   **c.** Mercury
 b. the moon   **d.** the earth

21. A planet that has phases like those of the moon
 is
 a. Venus   **c.** Jupiter
 b. Mars   **d.** Saturn

22. Compared with the earth, Mars has
 a. a denser atmosphere
 b. more surface water
 c. a lower average surface temperature
 d. a shorter year

23. The spacecraft that traveled to Mars did not find
 a. heavily cratered regions
 b. lava flows
 c. dust storms
 d. liquid water

24. The rings of Saturn
 a. are gas clouds
 b. are sheets of liquid
 c. are sheets of solid rock
 d. consist of separate particles

25. The planet with the most mass is
 a. Earth   **c.** Saturn
 b. Jupiter   **d.** Pluto

26. A planet that consists largely of hydrogen and
 helium is
 a. Mercury   **c.** Mars
 b. Venus   **d.** Jupiter

27. Most asteroids lie in a belt between
 a. Mercury and the sun
 b. Earth and Mars
 c. Earth and the moon
 d. Mars and Jupiter

28. The asteroids have diameters of up to about
 a. 1 km   **c.** 1000 km
 b. 10 km   **d.** that of the moon

29. The asteroids probably consist largely of
 a. ice   **c.** silicate minerals
 b. frozen methane   **d.** metallic iron

30. Relative to the time the moon takes to circle the
 earth, the period of its rotation on its axis is
 a. shorter
 b. the same
 c. longer
 d. any of the above, depending on the time of the
 year

31. The moon's diameter is about
 a. $\frac{1}{10}$ that of the earth   **c.** $\frac{1}{2}$ that of the earth
 b. $\frac{1}{4}$ that of the earth   **d.** $\frac{3}{4}$ that of the earth

32. At new moon, the moon is
 a. between the earth and the sun
 b. on the opposite side of the earth from the sun
 c. on the opposite side of the sun from the earth
 d. to the side of a line between the earth and the
 sun

33. An eclipse of the moon occurs when the
 a. moon passes directly between the earth and
 the sun
 b. sun passes directly between the earth and the
 moon
 c. earth passes directly between the sun and the
 moon
 d. moon's dark side faces the earth

34. Eclipses of the sun occur at
 a. new moon   **c.** full moon
 b. first or last quarter   **d.** anytime

35. Eclipses of the moon occur at
 a. new moon   **c.** full moon
 b. first or last quarter   **d.** anytime

36. The average density of the moon is
 a. lower than that of the earth
 b. about the same as that of the earth
 c. higher than that of the earth
 d. unknown

37. The moon's surface is
  a. perfectly smooth
  b. irregular but relatively smoother than the earth's surface
  c. about as irregular as the earth's surface
  d. much more irregular than the earth's surface

38. The moon's maria are
  a. bodies of water like the earth's oceans
  b. solid lava flows
  c. lava flows pulverized by meteoroid impacts
  d. hardened sediments

39. Most of the craters on the moon are probably the result of
  a. volcanic action
  b. meteoroid bombardment
  c. erosion
  d. collisions with asteroids

40. The moon's surface shows no signs of
  a. volcanic action
  b. meteoroid bombardment
  c. ever having melted
  d. glacier erosion

41. The time needed for the Apollo spacecraft to reach the moon was about
  a. 3 h   **c.** 3 weeks
  b. 3 days   **d.** 3 months

42. Relative to the oldest rocks that have been found on the earth, the oldest known lunar rocks are
  a. very much younger
  b. somewhat younger
  c. about the same age
  d. older

43. The interior of the moon is probably
  a. liquid   **c.** gaseous
  b. solid   **d.** hollow

QUESTIONS

1. If the earth had no atmosphere, would comets still be visible from its surface? Would meteors?

2. Why do comets have tails only in the vicinity of the sun? Why do these tails always point away from the sun, even when the comet is receding from it?

3. When a comet is close enough to the sun to be seen from the earth, stars are visible through both the comet's head and tail. What does this imply about the danger to the earth from a collision with a comet?

4. The Perseid meteor shower appears early every August. Does this mean that the orbits of the meteoroids in the Perseid swarm all have periods of exactly 1 year?

5. Over 90 percent of the meteorites found after a known fall are stony, yet most of the meteorites in museums are iron. Why do you think this is so?

6. The moon has virtually no atmosphere. How does the likelihood of being struck by a meteoroid on the moon's surface compare with that on the earth's surface?

7. How is it possible to distinguish the planets from the stars by observations with the naked eye? By observations with a telescope?

8. Which planets are visible to the naked eye?

9. Why do the planets shine?

10. Which planet is the brightest? How does its brightness compare with that of the brightest star? Which planet is the second brightest?

11. Why is Venus a brighter object in the sky than Mars?

12. Which planets would you expect to show phases like those of the moon?

13. Suppose you were on Mars and watched the earth with the help of a telescope. What changes would you see in the earth's appearance as it moves around the sun in its orbit?

14. Why are Mercury and Venus always seen either around sunrise or around sunset?

15. Which is the largest planet? The smallest? Which planet is nearest the sun?

16. On which planet is the length of the year shortest? On which is it longest?

17. On which planets would a person weigh less than on the earth? On which planets would a person weigh more?

18. Which planet resembles the earth most in size and mass? In surface conditions?

19. Is there any evidence that planets other than the earth today have crusts that consist of huge moving plates?

20. Venus is brighter when it appears as a crescent than when we can see its full disk. Why?

21. About how much time elapses between sunrise and sunset on Mars? On Jupiter?

22. Which planets have crusts of rock?

23. Why is it very unlikely that there is life on Mercury?

24. Why is the average temperature on Mars lower than that on the earth? Why is there a greater variation in Martian temperatures?

25. Mars has surface features that seem to be the result of erosion by running water. Why does the presence of many meteoroid craters in these regions suggest that the running water disappeared long ago?

26. What is the nature of Saturn's rings?

27. Which planets besides Saturn have rings? What is the nature of these rings?

28. Would you expect the particles in each of Saturn's rings to revolve with the same period, like parts of the same phonograph record?

29. What are thought to be the chief constituents of the giant planets Jupiter, Saturn, Uranus, and Neptune?

30. Jupiter radiates more energy than it receives from the sun. What is the most likely origin of the additional energy?

31. Distinguish between asteroids and meteoroids.

32. We always see the same hemisphere of the moon. Why?

33. What is wrong with the statement that the moon is more useful to us than the sun because the moon provides illumination at night when it is needed most?

34. What are the relative positions of the earth, moon, and sun at full moon?

35. Approximately how much time elapses between new moon and full moon?

36. Eclipses of the sun and of the moon do not occur every month. Why not?

37. If the moon were smaller than it is, would total eclipses of the sun still occur? Would total eclipses of the moon still occur?

38. In what phase must the moon be at the time of a solar eclipse? At the time of a lunar eclipse?

39. To what approximate length of time on the earth does the length of "day" at a given place on the moon correspond? The length of "night"?

40. Is the moon the largest satellite in the solar system? The smallest?

41. If the moon circled Jupiter in an orbit the same size as its present one around the earth, would its period of revolution be the same, shorter, or longer? Why?

42. What has been the chief influence that shaped the lunar landscape during the past 3 billion years?

43. The moon's maria are dark, relatively smooth regions conspicuous even to the naked eye. What is their nature?

44. The moon's surface is about as irregular as that of the earth. What does this imply about temperatures in the moon's interior?

45. Moonquakes are weaker and occur much less often than earthquakes. What do these facts imply about temperatures in the moon's interior?

46. Why is it believed that the moon's interior is different in composition from the earth's interior?

47. The densities of rocks on the moon's surface are about the same as the density of the moon as a whole. What does this observation suggest about the thermal history of the moon?

48. Why is it believed that large-scale igneous activity ceased on the moon about 3 billion years ago?

ANSWERS TO MULTIPLE CHOICE

1. b	**12.** b	**23.** d	**34.** a
2. c	**13.** a	**24.** d	**35.** c
3. d	**14.** a	**25.** b	**36.** a
4. d	**15.** a	**26.** d	**37.** c
5. b	**16.** a	**27.** d	**38.** c
6. d	**17.** c	**28.** c	**39.** b
7. a	**18.** b	**29.** c	**40.** d
8. b	**19.** a	**30.** b	**41.** b
9. a	**20.** c	**31.** b	**42.** d
10. a	**21.** a	**32.** a	**43.** b
11. b	**22.** c	**33.** c	

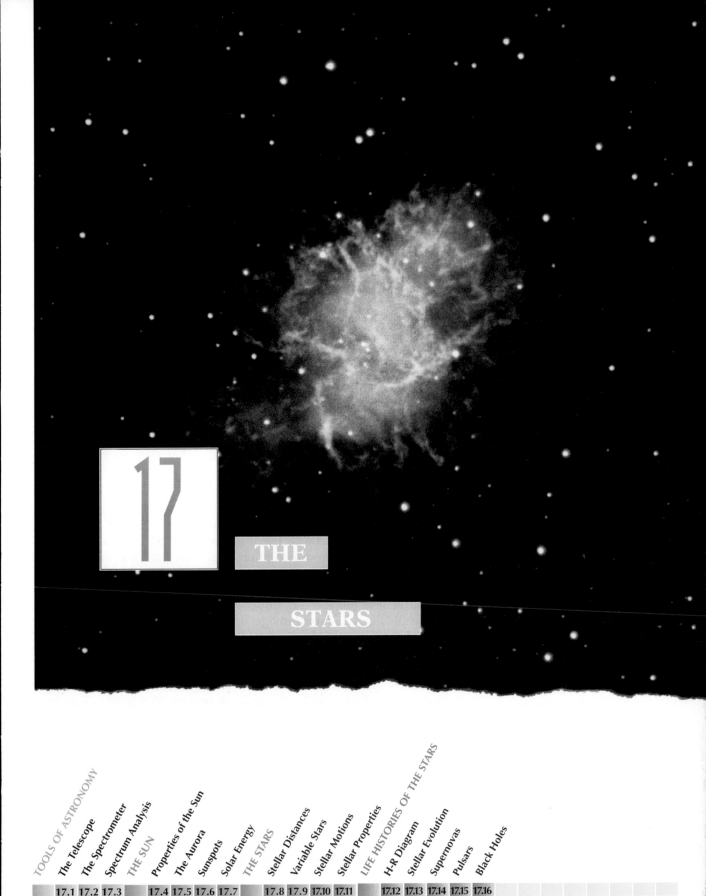

17

THE

STARS

*T*he study of the stars began in earnest toward the end of the eighteenth century with the work of William Herschel. Herschel sought among the stars some kind of order, something as profound as the regularities Copernicus, Kepler, and Newton had found in the solar system. Like a pioneer in any other branch of science, Herschel began with observation and spent many years cataloging stars and measuring their apparent motions. From this study he was able to arrive at a structure for the universe that is not far from the one that today's astronomers believe to be correct.

Of the billions of stars in the universe, none (besides the sun) appears as more than a point of light to even the most powerful telescope. As recently as the last century most scientists despaired of ever knowing the physical nature of the stars. Today, however, thanks to spectroscopic analysis, we not only have a great deal of detailed information on thousands of stars but also are able to trace the evolution of a star from its birth through maturity to its last agonies and eventual "death."

TOOLS OF ASTRONOMY

Light is the messenger that brings us information about the universe. Because the information arrives in code, so to speak, the astronomer must decipher it before being able to assess its significance. The tools of astronomy are devices that collect light and sort it into its component wavelengths, which are the elements of the code.

17-1 THE TELESCOPE

All Modern Ones Are Reflectors

In Herschel's time, as in ours, the telescope was the basic astronomical instrument. Much of his success was due to the improvements he introduced in telescope construction. Herschel was the first to build and use a large reflecting telescope, an instrument in which light is reflected from a concave mirror instead of being refracted through a lens. A large lens tends to sag under its own weight, with the change in shape producing a distorted image, whereas a mirror of any size can be adequately supported from behind. In addition, there is no problem of the dispersion of light of different wavelengths with a mirror (Sec. 6-18).

All modern astronomical telescopes are reflectors. The largest with a single mirror is located in the former Soviet Union and has a mirror 6 m in diameter. Technical problems have prevented this telescope from reaching its full potential. Next in size is a 5-m reflector at Mt. Palomar in California that has been enlarging the horizons of astronomy since 1948.

The latest generation of astronomical telescopes does not rely on single large mirrors, which practical difficulties limit to at most 8 m. Instead, a number of individual mirrors are linked to produce a single image. In one approach, the Keck Telescope in Hawaii uses hexagonal segments to give a collecting surface 10 m across. In another approach, separate circular mirrors are used. A new American telescope of this type has four 7.5-m mirrors mounted in one structure to give the equiv-

The new Keck Telescope in Hawaii, the world's largest, has a mirror 10 m in diameter consisting of 36 hexagonal segments that can be continuously adjusted to compensate for any distortions that may arise.

alent of a 15-m mirror. A similar telescope being developed in Europe for installation in a Chilean desert where the air is especially dry, stable, and clear should be able to see an astronaut on the moon.

Advantages of Size In stellar astronomy the purpose of a big telescope is not magnification, for the stars are too distant to ever appear as more than points of light. One virtue of large mirrors and lenses lies in their light-gathering power, which enables more light from a given object to be collected. Thus faint objects that would otherwise be invisible are revealed by a large telescope, and more light from brighter objects is available for study.

The second advantage of a large telescope is its ability to distinguish, or **resolve,** small details. As mentioned in Sec. 6-20, the diffraction of light waves causes every optical image to be blurred to a certain extent. The larger the lens or mirror, the less the blurring and the sharper the image.

Nowadays the light collected by a telescope is directed to a photographic plate or electronic sensor rather than to the astronomer's eye. These devices have the advantage that they respond to the *total amount* of light that falls on them over the period of time they are exposed. The eye, on the other hand, responds only to the *brightness* of the light that reaches it. A telescope with a camera or sensor attached can be trained on the same area of the sky for hours or, if necessary, for several nights to detect objects too faint for the eye to pick up. Photographic plates provide permanent records that enable positions and properties of stars as they appear today to be compared with what they were years ago and with what they will be in years to come.

17-2   THE SPECTROMETER

Without It, Little Would Be Known about the Stars

By itself, a telescope is of limited use in studying the stars. What is needed is a combination of a telescope and a spectrometer, the same instrument that contributed so much to our knowledge of atomic structure (Chap. 8). A spectrometer breaks light up into its separate wavelengths, as shown in Fig. 8-7. The resulting band of colors, with each wavelength separate from the others, is the spectrum that is recorded on a photographic plate.

The spectrum of a star does not seem impressive. If photographed in natural colors, it generally consists of a rainbow band crossed by a multitude of fine dark lines. Ordinarily color film is not used, so the spectrum shows simply black lines on a light gray background.

At first glance it does not seem that a few black lines on a photographic plate can get us very far in understanding the stars. But each of those lines has its own story to tell about how it was produced, and a specialist can piece together data from different lines into a comprehensive picture of a star. Some types of information obtainable from spectra are outlined in the next section.

A serious problem in astronomical spectroscopy is the absorption of light in the earth's atmosphere. Spectrometers mounted in sounding rockets, high-altitude balloons, satellites, and spacecraft are needed to study those parts of solar and stellar spectra that cannot reach the earth's surface.

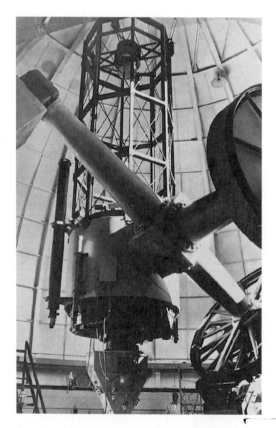

A spectrometer mounted on the 1.8-m-diameter reflecting telescope at Lowell Observatory, Arizona. Spectra obtained with such instruments have provided the clues needed to understand the nature and evolution of the universe. The mounting of the telescope enables it to examine all parts of the sky.

17-3 SPECTRUM ANALYSIS

Spectra Can Tell Us a Surprising Amount

Structure A spectrum of dark lines on a continuous colored background is an **absorption spectrum;** it is produced when light from a hot object passes through a cooler gas (Chap. 8). Atoms and molecules of the gas absorb light of certain wavelengths—wavelengths that they would emit if they themselves were hot—and so leave narrow gaps in the band of color. Thus a star that has this kind of spectrum (and nearly all of them do) reveals at once something of its structure: it must have a hot, glowing interior surrounded by a relatively cool gaseous atmosphere.

Temperature From the continuous background of a star's spectrum, astronomers can find the temperature of its surface. What they need to know is where in the spectrum the star's radiation is brightest. Since the wavelength of maximum intensity decreases as the temperature rises, the point of maximum intensity in the spectrum is a measure of temperature (Fig. 17-1). Thus the hottest stars are blue-white (maximum intensity at the short-wavelength end of the spectrum), stars of intermediate temperature are orange-yellow, and the coolest visible stars are red. This relation holds for materials on the earth as well as for the stars, as we know from experience (Fig. 4-4).

Composition Each element has a spectrum of lines with characteristic wavelengths. The elements present in a star's atmosphere can therefore be identified from the dark lines in its spectrum. In principle, all we have to do is measure the wavelength of each line in the spectrum and compare these wavelengths with those produced by various elements in the laboratory.

Condition of Matter In practice the identification of lines in a star's spectrum is not quite so easy. The wavelengths and intensities of the lines characteristic of a given element depend not only on the ele-

FIG. 17-1 Relative intensity of the wavelengths of light emitted by bodies with the temperatures indicated. The wavelength of greatest intensity is shorter for hot bodies than for cooler ones. The red curve represents measurements of the sun's photosphere. (1 nm = 10^{-9} m)

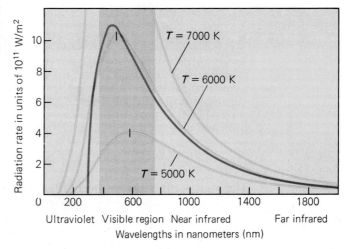

$T = 7000$ K

$T = 6000$ K

$T = 5000$ K

Radiation rate in units of 10^{11} W/m^2

0   200   600   1000   1400   1800

Ultraviolet Visible region Near infrared   Far infrared

Wavelengths in nanometers (nm)

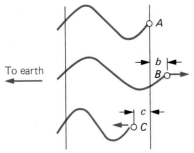

FIG. 17-2 The doppler effect in stellar spectra. Star *A* is stationary with respect to the earth. Star *B* is moving away from the earth; it moves the distance *b* during the emission of one light wave, whose wavelength is therefore increased by *b*. Star *C* is approaching the earth; it moves the distance *c* during the emission of one wave, whose wavelength is thus decreased by *c*. Hence stars receding from the earth have spectral lines shifted toward the red (long-wavelength) end, while stars approaching the earth have spectral lines shifted toward the blue (short-wavelength) end.

ment but also on such conditions as temperature, pressure, and degree of ionization. These difficulties prove to be blessings in disguise, however, for once the lines are identified, they tell us not only which elements are present in a star's atmosphere but also something about the physical conditions in which the elements exist.

Chemical compounds also have spectral lines of recognizable wavelengths, so spectra provide a means of determining how much of the matter in a star's atmosphere is in the form of molecules rather than atoms.

Magnetic Fields   The presence of a magnetic field causes individual energy levels within atoms to divide into several sublevels. When such atoms are excited and radiate, their spectral lines are accordingly split, each into a number of lines close to the original one. This phenomenon is called the **Zeeman effect** after its discoverer, the Dutch physicist Pieter Zeeman. With the help of the Zeeman effect the magnetic nature of sunspots has been established, and a large number of stars and clouds of matter in space have been discovered that appear to be strongly magnetized.

Motion   As we learned in Chap. 6, the doppler effect causes sounds produced by vehicles moving toward us to seem higher pitched than usual, whereas sounds produced by vehicles receding from us seem lower pitched than usual. Similarly a star moving toward the earth has a spectrum whose lines are shifted toward the violet (high-frequency) end, and a star moving away from the earth has a spectrum in which each line is shifted toward the red (low-frequency) end, as in Fig. 17-2. From the amount of the shift we can calculate the speed with which the star is approaching or receding.

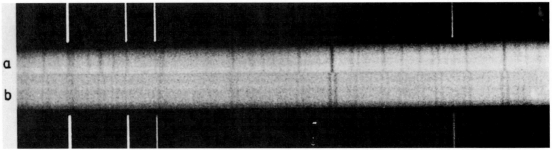

Spectra of the double star Mizar, which consists of two stars that circle each other, taken 2 days apart. In (a) the stars are in line with no motion toward or away from the earth, so their spectral lines are superimposed. In (b) one star is moving toward the earth and the other is moving away from the earth, so the spectral lines of the former are doppler shifted toward the blue end of the spectrum and those of the latter are shifted toward the red end.

THE SUN

The sun is the glorious body that dominates the solar system, and the origin and destiny of the earth are closely connected with its life cycle. The astronomer has another reason for studying the sun closely, for it is in many ways a typical star, a rather ordinary member of the assembly of perhaps 10^{20} stars that make up the known universe. Thus the properties of the sun that we can observe by virtue of its relative closeness are interesting not only in themselves but also because they give us information about stars in general that would otherwise be out of reach.

17-4 PROPERTIES OF THE SUN

The Nearest Star

The sun's mass can be found from the characteristics of the earth's orbital motion around it. The result is a mass of 1.99×10^{30} kg, more than 300,000 times the earth's mass. The sun's radius of 6.96×10^8 m can be established by simple geometry from the fact that its angular diameter as seen from the earth is 0.53°. The volume of the sun is such that 1,300,000 earths would fit into it.

The gases in the sun's interior are very hot, so they emit light copiously, and they are very dense, so the light is reabsorbed almost at once. Because both temperature and density fall off with distance from the sun's center, eventually there is a region in which the gases are still hot enough to radiate a great deal of light but not dense enough to prevent the light from escaping. This region, called the **photosphere,** is what we see as the "surface" of the sun. The temperature of the photosphere, about 6000 K, is found in two ways: from the shape of its spectrum (see Fig. 17-1) and from the rate at which it gives off energy.

Of the thousands of lines in the sun's spectrum, all can be identified with those of elements known on the earth. Lines of only a few extremely stable compounds are found since temperatures in the sun are high enough to break down nearly all molecules into atoms.

Although conditions on the sun are very different from those on the earth, the basic matter of the two bodies appears to be the same. Even the relative amounts of different elements are similar, except that there is much more of the light elements hydrogen (72 percent) and helium (27 percent) on the sun. At the relatively low temperatures here on the earth, most elements have combined to form compounds; in the hot sun the elements are present mostly as individual atoms, the majority of them ionized.

Above the Photosphere

A rapidly thinning atmosphere, mainly of hydrogen and helium, extends past the photosphere. From this atmosphere great flamelike **prominences** sometimes project into space, much like sheets of gas standing on their sides. Prominences occur in a variety of forms; a typical example is about 200,000 km long, 10,000 km wide, and 50,000 km high. Prominences are often associated with sunspots and, like sunspots, seem to have magnetic fields associated with them.

During a total eclipse of the sun, when the moon obscures the sun's disk completely (see photograph on page 560), a wide halo of pearly light can be seen around the dark moon. This halo, or **corona,** may extend out as much as a solar diameter and seems to have a great number of fine lines extending outward from the sun immersed in its general luminosity. The corona consists of ionized atoms, mainly pro-

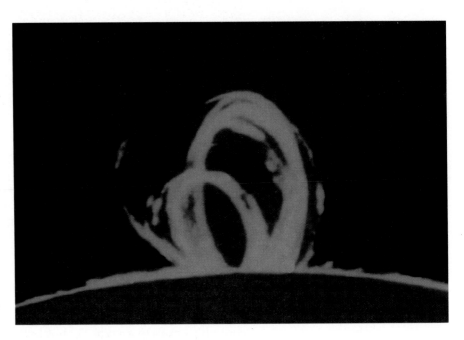

These solar prominences follow magnetic field lines.

tons, and electrons in extremely rapid motion; its temperature can reach 1 million K.

Although the corona we see during eclipses is relatively near the sun, it is found in very diffuse form much farther out, well beyond the earth's orbit. The outward flow of ions and electrons in this extension of the sun's atmosphere is the **solar wind.** The solar wind has been detected by spacecraft and helps deflect comet tails away from the sun as well as producing auroras in the earth's upper atmosphere.

17-5 THE AURORA

Fire in the Sky

The aurora (or "northern lights") is one of nature's most awesome spectacles. In a typical auroral display, colored streamers seem to race across the sky, and glowing curtains of light pulsate as they change their shapes into weird forms and images. In the climax of the display the heavens seem on fire, with silent green and red flames dancing everywhere. Then, after a while, the drama fades away, and only a faint reddish arc remains. Auroras are most common in the far north and far south. **Aurora borealis** is the name given to this phenomenon in the northern hemisphere, and **aurora australis** in the southern.

Here is how an aurora that occurred in Russia in 1370 was described at the time: "During the autumn there were many signs in the sky. For many nights, people saw pillars in the sky and the sky itself was red, as if covered with blood. So red was the sky that even on the earth covered with snow all seemed red like blood, and this happened many times."

Auroras are caused by the solar wind. The protons and electrons take about a day to reach the earth. When they enter the upper atmos-

Aurora in Alaska. Auroras are caused by streams of fast protons and electrons from the sun that excite gases in the upper atmosphere to emit light.

phere, they interact with the nitrogen and oxygen there so that light is given off. The process is similar to what occurs in a neon-filled glass tube when electricity is passed through it. The gas molecules are excited by charged particles moving near them, and this energy is then radiated as light in the characteristic wavelengths of the particular element. The green hues of an auroral display come from oxygen, the blues from nitrogen, and the reds from both oxygen and nitrogen.

The incoming streams of solar protons and electrons are affected by the earth's magnetic field in a complicated way. As a result most auroral displays occur in doughnutlike zones about 2000 km in diameter cen- geomagnetic north and south poles. So though, is so immer isi-

varies with solar activity, but it is always present to some extent. Auroras and the airglow occur during the day as well as the night but are too dim to be visible in the daytime.

SUNSPOTS

They Come and Go in an 11-Year Cycle

Dark patches called **sunspots** at times mar the intense luminosity of the sun's surface. Sunspots are cooler areas that appear dark only because we see them against a brighter background. A spot whose temperature is 5000 K is hot enough to glow brilliantly but is considerably cooler than the rest of the solar surface.

Sunspots change continually in form, each one growing rapidly and then shrinking, with lifetimes of from 2 to 3 days to more than a month. The largest sunspots are many thousands of km across, large enough to swallow several earths. Galileo, one of the first to study sunspots, noted that they moved across the sun's disk. He interpreted this finding, as we do today, as a sign that the sun rotates on its axis. Solar rotation is confirmed by doppler shifts in the spectral lines of radiation from the edges of the sun's disk. The sun rotates faster at its equator, where a complete turn takes 27 days, than near its poles, where a complete turn takes about 31 days.

Sunspots generally appear in groups, each with a single large spot together with a number of smaller ones. Some groups contain as many as 80 separate spots. They tend to occur in two zones on either side of the solar equator and are rarely seen either near the equator or at lati-

Magnetic storms can have serious consequences. The fluctuating magnetic fields of a major storm in 1989 induced currents in electric transmission lines in eastern Canada strong enough to trip protective devices. This triggered power failures over a wide area by a domino effect: as one section of the electric grid became overloaded and shut down, load was automatically shifted to another section that in turn became overloaded and shut down, and so on. The solar streams can also affect orbiting satellites by damaging the solar cells that provide their energy and by erasing control software stored in the satellites' computer memories.

tudes on the sun higher than 35°. Strong magnetic fields are associated with sunspots, and there is little doubt these fields are involved in sunspot formation.

The number of spots on the sun increases and decreases with time in a regular cycle that covers about 11 years. There is evidence that other stars also have cool spots, some of which come and go periodically like those on the sun. Many of these starspots are much larger than sunspots.

Sunspots and the Earth

The sunspot cycle has aroused much interest because a number of effects observable on the earth—such as disturbances in the terrestrial magnetic field (called **magnetic storms**), shortwave radio fadeouts, changes in cosmic-ray intensity, and unusual auroral activity—follow this cycle. It seems likely that the ionosphere changes that affect radio transmissions are due to intense bursts of ultraviolet and x-radiation that are more frequent during sunspot maximum. The magnetic, cosmic-ray, and auroral effects are due to vast streams of energetic protons and electrons that shoot out of the sun from the vicinity of sunspot groups.

Some aspects of weather and climate seem to be synchronized with sunspot activity. For instance, very few sunspots appeared between 1645 and 1715, a period during which temperatures worldwide were lower than usual—the "Little Ice Age" mentioned in Chap. 13 occurred at about that time. Apparently the events in the sun that cause sunspots are correlated with a slightly higher energy output. Even a small change in the sun's energy output would be enough to alter climates on the earth to the observed extent.

17-7 SOLAR ENERGY

It Comes from the Conversion of Hydrogen to Helium

Here on the earth, 150 million km from the sun, an area of 1 m^2 exposed to the vertical rays of the sun receives energy at a rate of about 1.4 kW. Adding up all the energy received over the earth's surface gives a staggering total, although this is only a tiny fraction of the sun's total radiation. And the sun has been emitting energy at this rate for billions of years. Where does all this energy come from?

We might be tempted to think of combustion, for fires give off what seems like a lot of heat and light. But a moment's thought shows that the sun is too hot to burn. Burning involves oxygen reacting with other substances to form compounds, but the sun is so hot that compounds cannot exist there except in its atmosphere. And even if burning were possible, the heat even the best fuels could give would be much too little to maintain the sun's temperatures.

Solar energy can come only from processes that take place inside the sun. Calculations based on reasonable assumptions lead to an estimate of 14 million K for the temperature and 1 billion atm for the pressure near the sun's center. The density of the matter there is nearly 10 times that of lead on the earth's surface. Under these conditions atoms of the lighter elements have lost all their electrons, and atoms of the heavier elements retain only their inmost electron shells. Thus matter in the sun's interior consists of atomic debris—free electrons and positive nuclei surrounded by a few electrons or none at all.

These atomic fragments move about far more rapidly than gas molecules at ordinary temperatures. Such speeds mean that two atomic nuclei may get close enough to each other—despite the repulsive electric force due to their positive charges—to react and form a single large nucleus. When this occurs among the light elements, the new nucleus has a little *less* mass than the combined masses of the reacting nuclei, as we saw in Chap. 7. The missing mass is converted to energy according to Einstein's formula $E = mc^2$. So huge an amount of energy is given off in nuclear fusion reactions of this kind that there is no doubt they are responsible for solar energy.

Fusion Reactions in the Sun

Most solar energy comes from the conversion of hydrogen into helium. This takes place both directly by collisions of hydrogen nuclei (protons) and indirectly by a series of steps in which carbon nuclei absorb a succession of hydrogen nuclei (Figs. 17-3 and 17-4). Each step

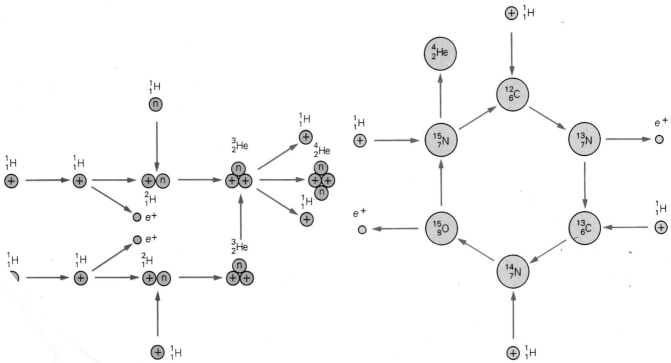

FIG.   proton-proton cycle. This is the chief nuclear   equence that takes place in stars like the sun an   rs. Energy is given off at each step. The net resu   ombination of four hydrogen nuclei to form a heli   cleus and two positrons.

FIG. 17-4   The carbon cycle also involves the combination of four hydrogen nuclei to form a helium nucleus with the evolution of energy. The $^{12}_{6}C$ nucleus is unchanged by the series of reactions. This cycle predominates in stars hotter than the sun.

BIOGRAPHY

Hans Bethe (1906–) studied and taught at various German universities until Hitler came to power, when he moved first to England and then to the United States. He was professor of physics at Cornell University from 1937 to 1975 and has remained active in research and public affairs even though officially retired. Notable among Bethe's many and varied contributions is his 1938 account of the sequences of nuclear reactions that power the sun and stars. During World War II he directed the theoretical physics division of the laboratory at Los Alamos, New Mexico where the atomic bomb was developed. A strong believer in nuclear energy—"it is more necessary now than ever before because of global warming"—Bethe has also been an effective advocate of nuclear disarmament.

can be duplicated in the laboratory and the energy released can be measured. The sun's interior is ideal for such energy-producing events. For the entire process by either fusion mechanism, every kilogram of helium formed means that about 0.007 kg—about the mass of a teaspoon of water—of matter disappears. The corresponding energy release is

$$E = mc^2 = (0.007 \text{ kg})(3 \times 10^8 \text{ m/s})^2 = 6.3 \times 10^{14} \text{ J}$$

About 20 million kg of coal would have to be burned to obtain this amount of energy!

The relative likelihoods of the carbon and proton-proton cycles depend on temperature. In the sun and other stars like it, which have interior temperatures up to about 14 million K, the proton-proton cycle predominates. Most of the energy of hotter stars comes from the carbon cycle.

In every second the sun converts more than 4 billion kg of matter to energy, and it has enough hydrogen to be able to release energy at this rate for billions of years to come. In fact, the amount of matter lost in all of geologic history is not enough to have changed the sun's radiation appreciably. This confirms other evidence that the earth's surface temperature has not changed by very much during this period.

Origin of the Heavy Elements

The reactions that convert hydrogen to helium are not the only ones that take place in the sun and other stars. With hydrogen and helium as raw materials and high temperatures and pressures to make things happen, most of the other elements are formed as well. The heaviest ones require still more extreme conditions to be produced, conditions that occur during the supernova explosions of heavy stars that are discussed later in this chapter. Such explosions also serve to scatter into space the various elements already created in the parent star.

Once scattered, these elements, heavy and light, mix with the hydrogen and helium of interstellar space and in turn become incorporated in new stars and new planets. We are all made of stardust.

THE STARS

17-8   STELLAR DISTANCES

Friedrich Bessel (1784–1840)

Not Easy   Aristotle pointed out long ago that, if the earth
to Measure   revolves around the sun, the stars should appear to
shift in position, just as trees and buildings shift in
position when we ride past them. Since he could detect no such shifts,
Aristotle concluded that the earth must be stationary.

Another interpretation of the apparent lack of movement among the
stars, suggested by some of the Greeks and later by Copernicus, is that
the stars are simply too far away for such movement to be easily
detected. An undoubted shift for one star was finally discovered in 1838
by the German astronomer Friedrich Bessel, and others were found
later. These shifts are so very small that the long failure to detect them
is not surprising.

Bessel's discovery made possible the direct measurement of distances to the nearer stars. The method is simple (Fig. 1-10). The position
of a star is determined twice, at times 6 months apart. From the measured change in the angle of the telescope, together with the fact that
the telescope was moved by the 300-million-km diameter of the earth's
orbit during the 6 months, the distance to the star can be calculated.

The **parallax,** as this apparent shift in position is called, is large
enough to be measurable only for a few thousand of the nearer stars.
The parallax of the closest star is equivalent to the diameter of a dime
seen from a distance of 6 km. The distance from the earth to this star,
Proxima Centauri, is about 4×10^{16} m.

A unit sometimes used to express stellar distances is the **light-year,**
which is the distance light travels in a year and is equal to 9.46×10^{12}
km. Thus Proxima Centauri is a little over 4 light-years away—which
means that we see the star not as it is today but as it was 4 years ago.
Only about 40 stars are within 16 light-years of the solar system. Such
distances between stars are typical for much of the visible universe.
Space is almost completely empty, far more empty even than the solar
system with its tiny isolated planets.

Apparent   Parallax measurements are possible only for distances
and Intrinsic   up to about 300 light-years. However, several indirect
Brightnesses   methods are available to find the distances of stars
farther away than this. A very useful one is based on
a comparison of the apparent and intrinsic brightnesses of stars.

The **apparent brightness** of a star is its brightness as we see it from
the earth. This quantity expresses the amount of light that reaches us
from the star. Its **intrinsic brightness,** on the other hand, is the true
brightness of the star, a figure that depends upon the total amount of
light it radiates into space.

The apparent brightness of a star depends on two things: its intrinsic brightness and its distance from us. A star that is actually very bright
may appear faint because it is far away, and a star that is actually faint
may have a high apparent brightness because it is close.

Walter Adams (1876–1956)

If both the apparent and the intrinsic brightness of a star are known, we can calculate its distance by finding out how far away an object with this intrinsic brightness must be located in order to send us the amount of light we observe. Such a calculation is not hard, and so we can find the distance to any star whose intrinsic brightness can be established.

A way to find the intrinsic brightness of a star was discovered by the American astronomer Walter Adams. Studying the spectra of the nearer stars, for which intrinsic brightnesses are known, Adams observed that the spectra of stars with high intrinsic brightness showed certain relationships among the strengths of their lines. The spectra of stars with low intrinsic brightness showed somewhat different relationships. Adams thus could establish the intrinsic brightness of a star simply by looking at its spectrum. Assuming that the relationship holds for more distant stars, Adams then was able to use their spectra to find their intrinsic brightnesses and hence their distances. With this method stellar distances have been determined up to several thousand light-years.

17-9 VARIABLE STARS

Stars Whose Brightness Changes, Usually in Regular Cycles

An extension of the above method for finding stellar distances is based upon the properties of a certain type of **variable star.** A variable star is one whose brightness varies continually. Some variables show wholly irregular fluctuations, but most repeat a fairly definite cycle of change.

A typical variable grows brighter for a time, then fainter, then brighter once more, with irregular minor fluctuations during the cycle. Cycles range in length all the way from a few hours to several years. Maximum brightness for some variables is only slightly greater than minimum brightness, but for others it is several hundred times as great. Since the sun's radiation changes slightly during the sunspot cycle, we may think of it as a variable star with a very small range in brightness (a few percent at most) and a long period (11 years).

The light changes in some variable stars are easy to explain. These stars are actually double stars whose orbits we see edgewise, so that one member of each pair periodically gets in the way of the other. In other variables the appearance of numerous spots at regular intervals may be what is dimming their light. Still others seem to be pulsating, swelling and shrinking so that their surface areas change periodically. Perhaps

Superimposed photographs taken at different times of the region of the sky in which the variable star WW Cygni appears. Only the brightness of this star has changed.

the irregular variables are passing through or behind ragged clouds of gas and dust that absorb some of their light.

Cepheid Variables

A particular class of variable stars, called **Cepheid variables,** helps astronomers find out how far from us distant star groups that contain them are. Cepheid variables are fairly young, very bright yellow stars 5 to 10 times as massive as the sun. Their name comes from a typical example discovered in 1784 in the constellation Cepheus. Polaris, the North Star, is a Cepheid variable whose brightness varies by 10 percent with a period of 4 days.

Early in this century the American astronomer Henrietta Leavitt happened to be studying the Cepheid variables in a nearby galaxy called the Small Magellanic Cloud. She noticed that the brighter a Cepheid was, the longer its cycle took. Since all the stars in this galaxy are very nearly the same distance from the earth, Leavitt concluded that the average intrinsic brightness of a Cepheid anywhere in the sky can be found just by measuring its period. Comparing this calculated intrinsic brightness with the Cepheid's apparent brightness then gives its distance, as we know. This method can be used for greater distances than spectroscopic determinations because the period of a Cepheid can be determined even when it is very faint; the present record is 49 million light-years.

Unfortunately the individual stars in really distant galaxies appear smeared together, even in the largest telescopes, so the Cepheids they may contain cannot be picked out for analysis. More recently other methods have been developed for finding how far away such galaxies are. One method is based on the observation that the faster a galaxy rotates, the greater its intrinsic brightness. Nearby galaxies whose distances are known from the Cepheids in them are used to calibrate the rotation-brightness relationship.

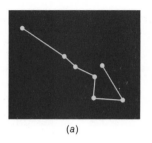 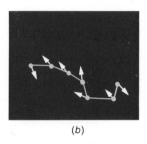 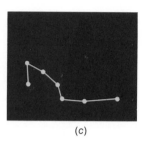

(a) (b) (c)

FIG. 17-5 The Big Dipper (a) as it was 200,000 years ago, (b) as it is today (arrows show directions of motion of the stars), and (c) as it will be 200,000 years from now.

17-10 STELLAR MOTIONS

The Stars Are Not Fixed in Space

As mentioned earlier, the speeds of stars that move toward or away from the earth can be found from the doppler shifts in their spectral lines. Motion across the line of sight can be followed by direct observation.

The great distances of the stars make their apparent movements so slow that it is easy to think of them as being fixed in space. Nevertheless most stars are moving at speeds of several km per second relative to the earth (Fig. 17-5).

What about the sun? If the sun is moving toward a certain part of the sky, stars in that direction, on the average, should appear to be approaching us and to be spreading apart, just as trees in a forest seem to approach and spread apart when we drive toward them. Average stellar motions of this sort are indeed found near the constellation Cygnus, and in the opposite part of the sky stars are apparently receding and coming closer together. A study of these motions indicates that the sun and its family of planets are moving toward Cygnus at a speed of 200 to 300 km/s.

17-11 STELLAR PROPERTIES

Mass, Temperature, and Size

We now turn to the properties of the stars. Many different types of stars are known, most of which fit into a pattern that can be understood in terms of a regular evolutionary sequence. Some stars, however, are still puzzles to the astronomer.

Mass The points of light that appear to the eye as single stars are often actually double, two stars close together. The members of such a star pair attract each other gravitationally, and each circles around the other. From the characteristics of the orbits the masses of the stars can be calculated. Although this method is limited to double stars, double stars are common and stars of all kinds are found as members of such pairs.

What the measurements show is that stellar masses range from $\frac{1}{40}$ to 100 times that of the sun—a smaller span than planetary masses. It is not hard to see why normal stars should be limited in mass. In a body whose mass is smaller than a certain limit, gravity cannot squeeze its

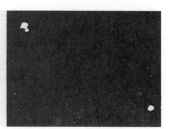

The double star shown here makes a complete rotation every 50 years.

matter sufficiently to produce the temperatures needed for nuclear fusion reactions. At the other extreme, a very heavy star would become so hot due to speeded-up nuclear reactions that gravity could not keep it together against the resulting outward pressure.

Temperature   The temperature of a star is determined by finding the part of its spectrum in which the radiation is most intense (Fig. 17-1). This tells us the temperature of the star's surface—the photosphere from which radiation is emitted.

The surface temperatures of a few very hot stars range up to 40,000 K, but the great majority are between 3000 and 12,000 K. Probably many stars are cooler than 3000 K, which is near the boiling point of iron. However, unless they are relatively close to the earth, their radiation is then too feeble for us to detect. Like the sun, other stars must have enormously high internal temperatures to maintain their surface radiation. As mentioned earlier, the hottest stars are blue-white, those of intermediate temperature are orange-yellow, and the coolest are red.

Size   If we know a star's surface temperature and its intrinsic brightness, we can find its size. The temperature tells us how much radiation is emitted from each square meter of the star's surface: the hotter it is, the more intense the radiation given off. The intrinsic brightness is a measure of the total radiation from the star's entire surface. We need only divide the total radiation by the radiation per square meter to find the number of square meters in the star's surface, and from this area the diameter and volume can be calculated.

There is also a more direct method of measuring stellar diameters based on the interference of light that can be used on the larger stars. Results obtained in this way agree with estimates from temperatures and intrinsic brightness.

The diameters of stars, unlike their masses, have an enormous range (Fig. 17-6). The smallest stars, composed almost entirely of neutrons, are only about 20 km across. The largest, like the giant red star Antares in the constellation Scorpio, have diameters 500 or more times that of the sun. Antares is so huge that, if the sun were placed at its center, the four inner planets could pursue their normal orbits inside the star with plenty of room to spare.

Giant stars like Antares have densities less than one-thousandth that of ordinary air—densities that correspond to a fairly good vacuum here on earth.

FIG. 17-6 The range of stellar sizes, from Antares (at the bottom) through the sun to a large white dwarf (black dot). Neutron stars are even smaller.

LIFE HISTORIES OF THE STARS

Looking around us, we see human beings who seem quite different from one another: babies, children, young men and women, the middle-aged, the old. If we had just arrived from another world, we might think these kinds of people are all different species, not individuals of the same kind in different stages of development. Because stars have such long lives, it is easy for us to make the same mistake and think of them as belonging to separate categories. In fact, stars, like people, are born, mature, grow old, and die, so that the various kinds of stars we see fit into regular patterns of evolution.

17-12 H-R DIAGRAM

Most Stars Belong to the Main Sequence

Early in this century two astronomers, Ejnar Hertzsprung in Denmark and Henry Norris Russell in America, independently discovered that the intrinsic brightnesses of most stars are related to their temperatures. This relationship is shown in the graph of Fig. 17-7, which is called a **Hertzsprung-Russell** (or **H-R**) **diagram.** Each point on this graph represents a particular star.

About 90 percent of all stars belong to the **main sequence,** with most of the others in the **red giant** class at the upper right and in the **white dwarf** class at the lower left. The names giant and dwarf refer, as we might expect, to very large and very small stars respectively. The most abundant stars in the main sequence are the **red dwarfs** at its lower end, all of them too faint to be seen by the naked eye. Proxima Centauri, the nearest star to the sun, is a red dwarf.

The position of a star on the H-R diagram is related to its physical properties. Stars at the upper end of the main sequence are large, hot, massive bodies. Stars at the lower end are small, dense, and reddish, cool enough so that chemical compounds form a considerable part of their atmosphere. In the middle are average stars like our sun,

Ejnar Hertzsprung (1873–1967)

Henry Norris Russell (1877–1957)

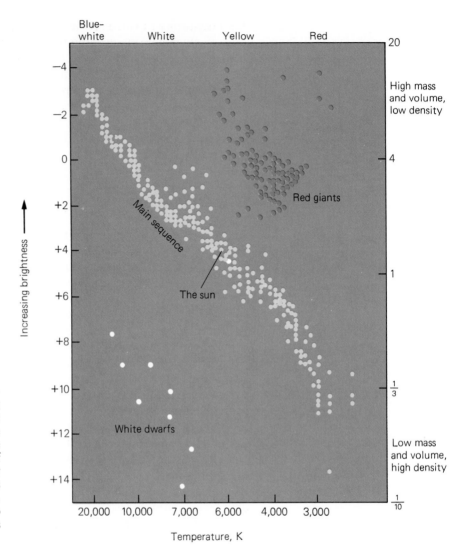

FIG. 17-7 The Hertzsprung-Russell diagram plots stars according to temperature and intrinsic brightness. The numbers at the left express absolute magnitude (the astronomical measure of intrinsic brightness) with low numbers indicating bright stars and high numbers, faint stars. Masses at right correspond to main-sequence stars. Star colors are indicated at the top.

with moderate temperatures, densities, and masses and rather small diameters.

To the red giant class belong the huge, diffuse stars like Antares, with low densities and diameters often larger than that of the earth's orbit. Many of these stars have low surface temperatures, as their reddish color indicates, but their enormous surfaces make them very bright.

White Dwarfs
The position of the white dwarfs in the H-R diagram reflects a combination of intensely hot surface and small total radiation. These properties suggest that such stars must be small, in fact comparable in size to the earth. However, their masses are all close to that of the sun, so that the density of a white dwarf is about 10^6 g/cm^3! A pinhead of such matter would weigh nearly a pound here on earth, and a cupful would weigh many tons.

Densities like this seem hard to believe, but they have been checked by enough methods to leave little doubt of their correctness. The only possible explanation is that atoms in these stars have collapsed. Instead of ordinary atoms with electrons relatively far from their nuclei, white dwarfs must have electrons and nuclei packed closely together. Matter in this state does not exist on earth, but its properties can be calculated from theories whose predictions have turned out correct in other situations.

Only a few hundred white dwarfs are known. Their scarcity seems to be more apparent than real, since they are so faint that only the nearer ones can be seen even in large telescopes. Enough of them have been found in recent years to suggest that the universe contains great numbers of these remarkable objects. Probably about 10 percent of the stars in the Milky Way are white dwarfs.

17-13 STELLAR EVOLUTION

Life History of a Star The relationships revealed by the H-R diagram cannot have occurred through chance alone. Are the stars in different parts of the diagram perhaps in various stages of development? Does the mass of a star control its temperature and the composition of its atmosphere? The answers to these questions seem to be yes, and the H-R diagram fits in well with modern ideas of the life history of a star.

Stars are believed to originate in gas clouds in space, clouds that consist largely of hydrogen. If a part of a gas cloud is dense enough, gravity will begin to pull it together into a still denser clump. The contraction heats the clump, much as the gas in a tire pump is heated by

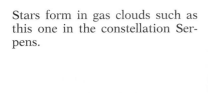

Stars form in gas clouds such as this one in the constellation Serpens.

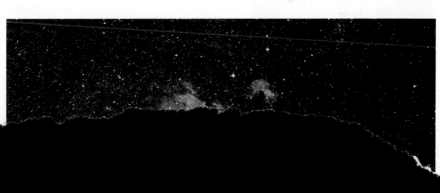

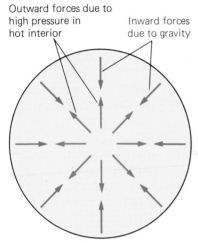

Outward forces due to high pressure in hot interior

Inward forces due to gravity

FIG. 17-8 The tendency of a star to contract gravitationally is balanced by the tendency of its hot interior to expand.

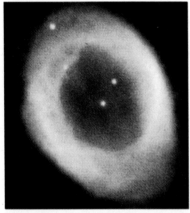

This planetary nebula in Lyra is a shell of gas moving outward from the star in the center, which is in the process of becoming a white dwarf. For a few th̶ this dying st̶ heat the̶ but̶ o̶

compression, and the clump glows as a result. Such an infant star appears among the cooler giants in the H-R diagram.

Some thousands or millions of years later, the star's temperature will rise to the point where the nuclear reactions of Fig. 17-3 begin to occur, which convert its hydrogen into helium. The increase in temperature shifts the position of the star on the H-R diagram downward and to the left. From this time on the star's tendency to contract is opposed by the pressure of its hot interior (Fig. 17-8). Such a star is a stable member of the main sequence and maintains a constant size as long as its hydrogen supply holds out.

A star does not shine because some special force causes it to do so—it shines because it has a certain mass and a certain composition.

The temperature a star reaches depends on its mass. Gravity in a large mass is more powerful than in a small mass and leads to more intense energy production to balance the resulting inward forces. For stars with abundant hydrogen, calculations show that the relationship between mass and temperature should be exactly that shown by stars in the H-R main sequence. The large, heavy stars at the upper end of the main sequence have high temperatures and shine brightly. The red dwarfs at the lower end are relatively cool and only faintly luminous.

A heavy star consumes its hydrogen rapidly, so its lifetime in the main sequence is shorter than that of a less massive star. Accordingly, fewer than 1 percent of main-sequence stars are giants. The supply of hydrogen in a fairly modest star like our sun might last for 10 billion years. Probably the sun is now about halfway through this part of its life cycle. The heaviest stars may stay only as little as 10 million years in the main sequence and the lightest ones may stay a trillion years.

Old Age

When the hydrogen supply at last begins to run low in a star like the sun, the life of the star is by no means over but instead enters its most spectacular phase. Further gravitational contraction makes its core still hotter, and other nuclear reactions become possible—in particular those in which nuclei larger than helium are built up, for instance, the combination of three helium nuclei to form a carbon nucleus. The outer part of the star is heated and expands to as much as 100 times its former diameter. The expansion produces a cooling so that the result is a very large, cool star with a hot core. Such

For the sun, the star in which we have the most personal interest, we can expect about 5 billion more years without much change from present conditions. Then the sun's temperature will gradually increase, and life on earth will ultimately become impossible—not, as was once thought, because the sun will cool off but rather because it will grow too hot. In time the sun will expand into a giant, perhaps as large around as the orbit of Mars, and still later it will collapse into a white dwarf. A dismal end, but one that our most remote descendants are not likely to survive to witness.

Eventually the new energy-producing reactions run out of fuel too, and again the star shrinks, this time all the way down to the white dwarf state. A shell of gas from the outer part of the star goes out into space to form a bubblelike **planetary nebula.** (Nebula is Latin for "cloud.") As a slowly contracting dwarf the star may glow for billions of years more with its energy now coming from the contraction, from nuclear reactions that involve elements heavier than helium, and from proton-proton reactions in a very thin outer atmosphere of hydrogen. Ultimately the star will grow dim and in time cease to radiate at all. It will now be a **black dwarf,** a lifeless lump of matter. The universe is not thought to be old enough for any white dwarfs to have become black dwarfs as yet.

17-14   SUPERNOVAS

Exploding Stars   A heavy star—more than 8 times the sun's mass—at the upper end of the main sequence has a rather different later history.

Such a star does not proceed in the usual way from red giant to white dwarf. Instead, after a relatively short lifetime of some millions of years, the star's great mass causes it to collapse abruptly when its fuel has run out and then to explode violently. The explosion, which takes less than a second, flings into space much of the star's mass. An event of this kind, called a **supernova,** is billions of times brighter than the original star ever was. Nuclear reactions during the explosion create the heaviest elements which, together with the other elements formed during the earlier part of the star's life, are flung into space. A supernova occurs once every 20 or 30 years on the average in the Milky Way galaxy of which the sun is a member.

What is left after a supernova explosion is a dwarf star of extraordinary density. Its mass is typically 1.5 times that of the sun but its diameter is only about 20 km, the size of a city. The matter of such a star weighs billions of tons per teaspoonful. If the earth were this dense, it would fit into a large apartment house. Under the pressures inside such a star, the most stable form of matter is the neutron. Until not long ago the notion of **neutron stars** was purely speculative, but several hundred have been identified thus far. Figure 17-10 shows the life history of a heavy star, and Fig. 17-11 shows how neutron stars and white dwarfs compare in size with the earth and the sun.

FIG. 17-11 A comparison of a white dwarf and a neutron star with the sun and the earth. Both white dwarfs and neutron stars are thought to have masses similar to that of the sun.

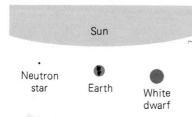

Sun

Neutron
star   Earth

White
dwarf

FIG. 17-10 Life histor⁓   ⁓up Such a star
spen⁓⁓ ⁓⁓ ⁓⁓ ⁓⁓

17-15 ·PULSARS

Spinning Neutron Stars

In 1967 unusual radio signals were discovered by Jocelyn Bell, then a graduate student at Cambridge University in England, that were coming from a source in the constellation Vulpecula. The signals fluctuated with an extremely regular period, exactly 1.33730113 s. Since then several hundred more **pulsars** have been discovered with periods between 0.001 and 4 s. At first only radio emissions from pulsars were observed. Later, however, flashes of visible light were detected from several pulsars that were exactly synchronized with the radio signals.

The power output of a pulsar is about 10^{26} W, which is comparable with the total power output of the sun. So strong a source of energy cannot possibly be switched on and off in a fraction of a second, nor can it be the size of the sun. Even if the sun were to suddenly stop radiating, it would take 2.3 s before light stopped reaching us, because all parts of the solar surface that we see are not at the same distance away. Hence we conclude that a pulsar must have the mass of a star, in order to be able to emit so much energy, but it must be very much smaller than a star, in order that its signals fluctuate so rapidly.

From the above and other considerations it seems likely that pulsars are neutron stars that are spinning rapidly. Conceivably a pulsar has a strong magnetic field whose axis is at an angle to the axis of rotation, and this field traps tails of ionized gases that do the actual radiating.

This wispy object in space, called the Crab Nebula, has a pulsar at its heart. The nebula is the remnant of a supernova that was seen in A.D. 1054, and it has been expanding rapidly and glowing brightly ever since. Until recently it was impossible to understand where all the energy of the nebula is coming from, since the explosion itself occurred so long ago. However, the light and radio flashes from the Crab pulsar seem powerful enough to furnish the entire nebula with its energy. This pulsar flashes 30 times per second and is slowing down rapidly, 1 part in 2400 per

Whatever the mechanism, though, a pulsar is apparently like a lighthouse whose flashes are due to a rotating beam of light. The identification of pulsars with neutron stars is supported by evidence that the periods of pulsars are very gradually decreasing, which would be expected as they continue to lose energy.

The closest known pulsar is 280 light-years away. It is very dim, which suggests that there may be many—half a million?—other dim pulsars in our galaxy that are too faint to be seen because they are farther away.

17-16   BLACK HOLES

Even Light Cannot Escape from Them   Does a neutron star represent the ultimate in compression? Apparently not. After it becomes a supernova, a very heavy star leaves behind a remnant too massive to be stable as a neutron star. The remnant continues to contract until it is only a few km across. Such an object is called a **black hole** for a most interesting reason.

As we learned in Chap. 3, one of the results of Einstein's general theory of relativity is that light is affected by gravity. Thus starlight passing near the sun is bent by a small but measurable extent.

The heavier a star and the smaller it is, the stronger the pull of gravity at its surface and the higher the escape speed needed for something to leave the star. In the case of the earth, the escape speed is 11.2 km/s. In the case of the sun, 617 km/s. If the ratio M/R between the mass and radius of a star is large enough, the escape speed is more than the speed of light, and nothing, not even light, can ever get out. The star cannot radiate and so is invisible, a black hole in space. For a star with the sun's mass, the critical radius is 3 km, a quarter of a million times smaller than the sun's present radius. Anything passing near a black hole will be gobbled up, never to return to the outside world.

Detecting Black Holes   Since it is invisible, how can a black hole be detected? A black hole that is a member of a double-star system (double stars are quite common) will reveal its presence by its gravitational pull on the other star. In addition, the intense gravitational field of the black hole will suck matter from the other star, which will be compressed and heated to such high temperatures that x-rays will be emitted profusely. One of a number of invisible objects that astronomers believe on this basis to be black holes is known as Cygnus X-1. Its mass is perhaps 16 times that of the sun, and its radius may be only about 10 km. Enormous black holes are believed to be at the centers of galaxies of stars, as discussed in the next chapter.

Only very heavy stars end up as black holes; lighter ones evolve into white dwarfs and neutron stars. But as time goes on, the strong gravitational fields of both white dwarfs and neutron stars attract more and more cosmic dust and gas. When they have gathered up enough additional mass, they too will become black holes. If the universe lasts long enough (see Sec. 18-9), then everything in it may be in the form of black holes.

IMPORTANT TERMS AND IDEAS

A star's **spectrum** contains dark lines that correspond to particular frequencies of light absorbed by the ions, atoms, and molecules in the atmosphere around the star. These lines are superimposed on a bright background emitted by the star's surface. The spectrum and the background can be analyzed to provide information on the structure, temperature, composition, condition of matter, magnetism, and motion of the star.

The **sun** is a typical star whose temperature is 6000 K at the surface and perhaps 14 million K near the center. **Solar energy** comes from the conversion of hydrogen to helium, which can take place through both the **proton-proton** and **carbon cycles.** **Sunspots** are regions slightly cooler than the rest of the solar surface and have strong magnetic fields associated with them. Surrounding the sun is a diffuse **corona** of ions and electrons whose outward flow is the **solar wind** that extends through much of the solar system.

Auroras result from the excitation of gases in the earth's upper atmosphere by streams of protons and electrons from the sun.

The distance to a nearby star can be found from its apparent shift in position, or **parallax,** as the earth moves in its orbit. The distance to a star far away can be found by comparing its **intrinsic** and **apparent brightnesses,** which are its actual brightness and its brightness as seen from the earth. Stellar distances are often expressed in **light-years,** where a light-year is the distance light travels in a year.

Variable stars fluctuate continually in brightness. Notable are the **Cepheid variables,** whose intrinsic brightness and periods of variation are related, which permits their distances to be found. Other variable stars are actually pairs of stars that revolve about their common centers of gravity and periodically block each other's light.

The **Hertzsprung-Russell (H-R)** diagram is a graph on which the intrinsic brightnesses of stars are plotted versus their temperatures. Most stars belong to the **main sequence,** which appears as a diagonal band on the diagram, but there are also cool, large, **red giant** stars and hot, small, **white dwarfs** that lie outside the main sequence. Heavy stars eventually explode into **supernovas** and later subside into **neutron stars** whose interiors consist entirely of neutrons. A **pulsar** is a spinning neutron star that emits flashes of light and radio waves. A **black hole** is a "dead" star that has contracted so much that its gravitational field is strong enough for the escape speed to be greater than the speed of light. Nothing, not even light, can escape from a black hole.

EXERCISES: MULTIPLE CHOICE

1. Which one or more of the following are reasons why telescopes with large mirrors are useful in astronomy?
   a. Their ability to resolve nearby objects in the sky
   b. Their ability to gather light
   c. Their magnifying ability
   d. Their ability to disperse light of different wavelengths

2. The fact that the spectra of most stars consist of dark lines on a bright background means that these stars
   a. have cool interiors surrounded by hot atmospheres
   b. have hot interiors surrounded by cool atmospheres
   c. have hot interiors surrounded by hot atmospheres
   d. have cool interiors surrounded by cool atmospheres

3. The examination of starlight with a spectrometer cannot provide information about an isolated star's
   a. temperature   c. mass
   b. structure   d. magnetic field

4. The sun is a (an)
   a. unusually small star
   b. unusually large star
   c. unusually hot star
   d. rather ordinary star

5. The surface temperature of the sun is about
   a. 600 K   c. 6 million K
   b. 6000 K   d. 14 million K

6. The sun's atmosphere
   a. extends out into the solar system
   b. consists mainly of oxygen and nitrogen
   c. consists of burning hydrogen
   d. is relatively cool

7. Auroras are caused by
   a. streams of colored gases from the sun
   b. streams of charged particles from the sun
   c. comets
   d. micrometeorites

8. Auroras occur mainly in the polar regions because of the effect of
   a. the moon
   c. the geomagnetic field
   b. low temperatures there
   d. the earth's equatorial bulge

9. Sunspots are
   a. dark clouds in the sun's atmosphere
   b. regions somewhat cooler than the rest of the sun's surface
   c. regions somewhat hotter than the rest of the sun's surface
   d. of unknown nature

10. The duration of the sunspot cycle is approximately
    a. 27 days
    c. 3 years
    b. 6 months
    d. 11 years

11. Sunspot activity does not affect
    a. shortwave radio communication
    b. the earth's magnetic field
    c. the aurora
    d. volcanic eruptions

12. The temperature of the sun's interior is believed to be about
    a. 600 K
    c. 14 million K
    b. 6000 K
    d. 14 billion K

13. Most of the matter in the sun is in the form of
    a. neutral atoms
    c. molecules
    b. ions
    d. liquids

14. The sun's energy comes from
    a. nuclear fission
    b. radioactivity
    c. the conversion of hydrogen to helium
    d. the conversion of helium to hydrogen

15. Proxima Centauri, the closest star to the sun, is at a distance of about
    a. 4 light-years
    c. 4 million light-years
    b. 400 light-years
    d. 4 billion light-years

16. If we know both the intrinsic and the apparent brightnesses of a star, we can find its
    a. mass
    c. distance
    b. temperature
    d. age

17. Cepheid variable stars are valuable to astronomers because
    a. they are very brilliant, easily seen stars
    b. their intrinsic brightnesses and periods of variation are directly related
    c. their periods of variation are short
    d. they all lie in the constellation Cepheus

18. In order to determine the mass of a star, it must
    a. belong to the main sequence
    b. be a Cepheid variable
    c. be relatively close
    d. be part of a double-star system

19. The size of a star can be found from its
    a. temperature and mass
    b. temperature and intrinsic brightness
    c. temperature and apparent brightness
    d. mass and apparent brightness

20. The temperature of a star can be determined from
    a. the wavelength at which its radiation is brightest
    b. the wavelength at which its radiation is dimmest
    c. its apparent brightness
    d. the strength of its magnetic field

21. The stellar property that varies least is
    a. distance from the sun
    c. size
    b. mass
    d. temperature

22. The color of a relatively cool star is
    a. white
    c. blue
    b. yellow
    d. red

23. The reason stars more than about 100 times as massive as the sun are not found is that
    a. they would split into double-star systems
    b. they would be black holes from which no light can escape
    c. the gravity of such a star would not hold it together against the pressure produced by nuclear reactions in its interior
    d. the high internal pressures would prevent nuclear reactions from taking place

24. Stars belonging to the main sequence
    a. have the same mass
    b. have the same temperature
    c. radiate at a steady rate
    d. fluctuate markedly in brightness

25. The greater the mass of a main-sequence star, the
    a. younger it is
    b. longer it will remain part of the main sequence
    c. less hydrogen it contains
    d. hotter it is

26. The main sequence on the H-R diagram includes
    a. the sun
    c. red giants
    b. white dwarfs
    d. pulsars

27. A typical white dwarf star is about the size of
  a. a large building   **c.** the earth
  b. the moon   **d.** Jupiter

28. When the hydrogen supply in a typical main-sequence star begins to run out, other nuclear reactions occur and the star becomes a
  a. red giant   **c.** supernova
  b. white dwarf   **d.** neutron star

29. At the end of its life cycle, a typical star becomes a
  a. red giant   **c.** supernova
  b. white dwarf   **d.** neutron star

30. A star that explodes as a supernova
  a. has a mass much greater than that of the sun
  b. has a mass much smaller than that of the sun
  c. has a mass about equal to that of the sun
  d. may have any mass

31. The brightest of the following types of stars is a
  a. red giant   **c.** supernova
  b. white dwarf   **d.** neutron star

32. The heaviest elements are created in
  a. stellar interiors
  b. stellar atmospheres
  c. black holes
  d. supernova explosions

33. A pulsar is not
  a. the remnant of a supernova explosion
  b. rotating rapidly
  c. composed largely of neutrons
  d. as large as the earth

34. A "black hole" appears black because
  a. it is too cool to radiate
  b. it is surrounded by an absorbing layer of gas
  c. its gravitational field is too strong to permit light to escape
  d. its magnetic field is too strong to permit light to escape

35. Black holes are remnants of
  a. stars with small masses
  b. stars with large masses
  c. white dwarfs
  d. black dwarfs

36. The sun will eventually become a
  a. pulsar   **c.** white dwarf
  b. supernova   **d.** black hole

QUESTIONS

1. Why are large telescopes valuable in astronomy?

2. A photograph of a star cluster shows many more stars than can be seen by looking by eye at the cluster through the same telescope. Why is there a difference?

3. Why do you think it is useful to put an astronomical telescope in a satellite?

4. What part of a star is responsible for the continuous background of its spectrum? What part is responsible for the dark absorption lines?

5. Arrange the following types of stars in order of decreasing surface temperature: yellow stars, blue stars, white stars, red stars.

6. If the earth were moving toward a star instead of the star toward the earth, would lines in the star's spectrum be shifted? If so, toward which end of the spectrum?

7. Suppose you examine the spectra of two stars receding from the earth and find that the lines in one are displaced farther toward the red end than those in the other. What conclusion can you draw?

8. Suppose the earth's magnetic field were to disappear. What effect would this have on the aurora?

9. Why do sunspots appear dark if their temperatures are over 4500 K?

10. Give two methods for determining how fast the sun rotates on its axis. Is the rotation speed the same for the entire sun?

11. What aspect of sunspots changes during a sunspot cycle?

12. Why is the sun's corona ordinarily not visible? How do we know it exists?

13. What evidence suggests that the sun is almost wholly gaseous?

14. What evidence can you suggest to support the hypothesis that the sun's radiation rate has not changed appreciably in the past 2 billion years or so?

15. Why do you think astronomers believe that the sun's energy originates in its interior rather than in its surface layers?

16. Why is it impossible for combustion to be the source of the sun's energy?

17. How can the composition of the sun be determined?

18. Helium was discovered in the sun before it was found on the earth (hence its name, which comes from *helios,* the Greek word for "sun"). How can this sequence have come about?

19. What aspect of the formation of helium from hydrogen results in the evolution of energy?

20. How are stellar masses determined?

21. Which varies more, the masses of the stars or their sizes?

22. What data are needed to determine a star's average density? How would you expect the density to change from the surface layers to the interior?

23. Which stars do you think have the highest densities? The lowest?

24. A red star and a white star of the same apparent brightness are the same distance from the earth. Which is larger? Why?

25. Explain how the distance to a star cluster that contains Cepheid variables is determined.

26. What methods can be used to determine the intrinsic brightness of a star?

27. How is a star's diameter estimated from measurements of temperature and intrinsic brightness?

28. The stars Betelgeuse and Deneb have similar intrinsic brightnesses, but Betelgeuse is red and Deneb is white. Which is larger? Which has the greater density? Which is hotter?

29. What information can be gained by comparing the intrinsic and apparent brightnesses of a star?

30. What information can be gained from a knowledge of the surface temperature and intrinsic brightness of a star?

31. The spectrum of a certain star shows a doppler shift that varies periodically from the red to the blue end of the spectrum. What kind of star is this?

32. Must a star be spherical?

33. Is it possible for an object with the mass and composition of the sun to exist without radiating energy?

34. Why is the sun considered to be a star?

35. Why are most stars part of the main sequence on the H-R diagram?

36. Main-sequence stars are supposed to evolve into red giants, but relatively few stars lie between the main sequence and the group of red giants on the H-R diagram. Why?

37. Why do main-sequence stars have masses that lie between about $\frac{1}{40}$ and 100 times the mass of the sun?

38. A giant star is much redder than a main-sequence star of the same intrinsic brightness. How does this observation indicate that the giant star is larger than the main-sequence star?

39. What are the chief characteristics of an average star in the upper left of the H-R diagram? In the lower left? In the upper right? In the middle of the main sequence?

40. The sun is near the middle of the main sequence. Where would a star be located in the main sequence if its mass is 10 times that of the sun? Would it remain in the main sequence for a longer or shorter time than the sun? Would it be hotter or cooler than the sun?

41. Sirius, the brightest star in the sky apart from the sun, is a blue-white star of great intrinsic brightness. What does this suggest about its temperature? About its average density? About its position in the H-R diagram?

42. As a main-sequence star evolves, what happens to its position on the main sequence?

43. Why are there relatively few giant stars in the main sequence?

44. Which of the following types of star is the smallest? The largest? The most common? Neutron stars, white dwarfs, red dwarfs, black dwarfs.

45. Which of the following types of star is the most common in the sky? Least common? Main-sequence stars, white dwarfs, red giants, supernovas, double stars, Cepheid variables.

46. Into what kind of star will the sun eventually evolve?

47. In what part of its life cycle is a white dwarf star?

48. Why are white dwarfs considered to have such enormous densities?

49. After a very long time, a white dwarf will cool down and become a black dwarf. What will the corresponding evolutionary path of the star be on the H-R diagram?

50. What is left behind after a supernova explosion?

51. What is the characteristic behavior of a pulsar? What is believed to be its nature? From what does a pulsar originate?

52. What prevents light from escaping from a black hole?

53. How large are black holes? Can any star evolve into a black hole?

ANSWERS TO MULTIPLE CHOICE

1. a, b	**3.** c	**5.** b	**7.** b
2. b	**4.** d	**6.** a	**8.** c

9. b	**16.** c	**23.** c	**30.** a
10. d	**17.** b	**24.** c	**31.** c
11. d	**18.** d	**25.** d	**32.** d
12. c	**19.** b	**26.** a	**33.** d
13. b	**20.** a	**27.** c	**34.** c
14. c	**21.** b	**28.** a	**35.** b
15. a	**22.** d	**29.** b	**36.** c

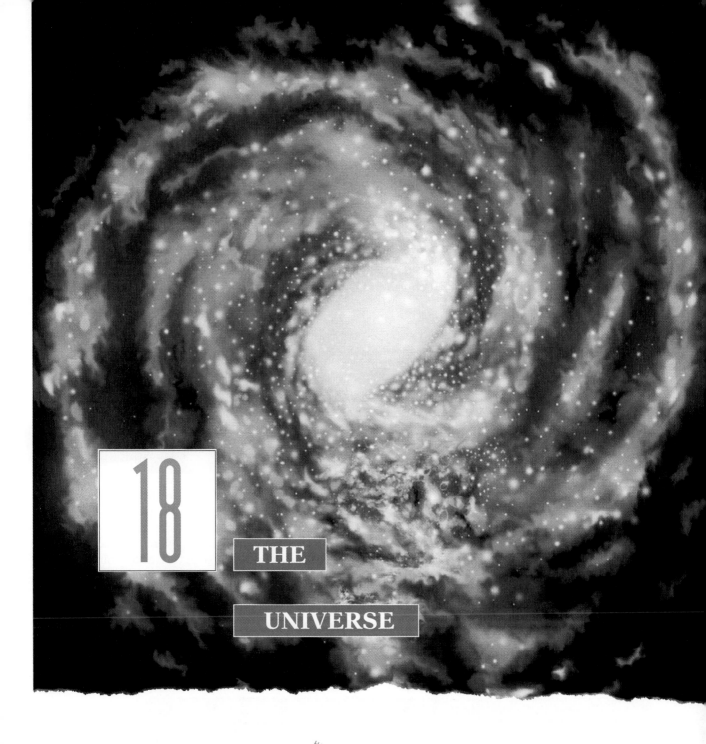

18

THE

UNIVERSE

Stars are not scattered at random throughout the universe but instead occur in immense swarms called **galaxies.** Each galaxy is separated from the others by vast reaches of nearly empty space. The stars of the galaxy to which our sun belongs appear in the sky as the Milky Way.

Of the many remarkable properties galaxies have, one stands out: they are moving apart from one another, so that the universe as a whole is expanding at a steady rate. If we project this expansion backward, we find that it began about 15 billion years ago. Can it be that the entire universe was born in a cosmic **big bang** at that time and has been evolving ever since into the galaxies of today? As we shall see in this chapter, several lines of evidence support such a picture, which has been filled out in considerable detail.

GALAXIES

A galaxy is an island universe of stars. Just as studying the sun tells us a lot about stars in general, so studying our galaxy, the Milky Way, tells us a lot about galaxies in general.

18-1 THE MILKY WAY

A Spinning Disk of Stars

The great band of misty light we see in the sky on a clear night is called the **Milky Way** and forms a continuous band around the heavens. When we look at it with a telescope, the Milky Way is an unforgettable sight. Instead of a dim glow, we now see countless individual stars, stars as numerous as sand grains on a beach. In other parts of the sky the telescope also reveals stars too faint for the naked eye, but nowhere else that many. Clearly the stars are not distributed evenly in space—a basic observation that implies much about the structure and evolution of the universe.

The appearance of the Milky Way tells us something about how the stars in our galaxy are arranged. Most of them are concentrated in a relatively thin disk with the sun near its central plane. When we look toward the rim of the disk, we see a great many stars, so many that they seem to form a continuous band of light. When we look above or below

A mosaic of several photographs of the Milky Way between the constellations Sagittarius and Cassiopeia.

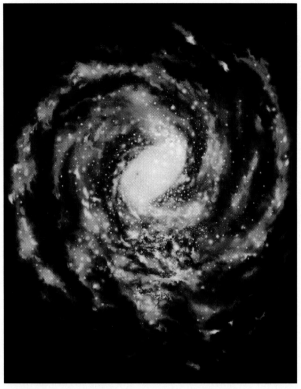

Computer simulation of our galaxy. The concentration of stars in the disk of the galaxy appears in the night sky as the Milky Way. The galaxy seems to have two main spiral arms and several smaller arms. The sun is a member of one of the latter and in this picture is located about two-thirds of the radius of the galaxy below its center.

the disk, far fewer stars are to be seen. The disk of stars has a thicker central nucleus, so that it is shaped something like a fried egg.

The disk of the galaxy is roughly 100,000 light-years in diameter and 10,000 light-years thick near the center. It is one of the larger galaxies of the universe. The sun is about two-thirds of the distance from the center, which lies in the direction of the densest part of the Milky Way in the constellation Sagittarius. The 100 billion stars of the galaxy are chiefly located in spiral arms that extend from the nucleus. Surrounded by so many similar bodies, the sun is not unusual in position, size, mass, or temperature. It is probably not even unusual in possessing a family of planets. There may well be millions of planets in our galaxy, many of them no doubt inhabited by some form of advanced life.

The stars of our galaxy revolve about its center, which is what they must do if the galaxy is not to gradually collapse because of the gravitational attraction of its parts. (The planets do not fall into the sun because of their orbital motion, too.) The orbital speed of the sun and nearby stars around the galactic center is about 220 km/s. At this rate the sun makes a complete circuit once every 240 million years.

Clouds of gas and dust surround the center of our galaxy and prevent us from seeing it. However, radio waves, infrared and ultraviolet light, x-rays, and gamma rays from near the center have been detected, and they suggest a giant black hole there. Its mass seems to be several million times the sun's mass, and it sucks in material blown off by nearby supernova explosions. It is this material, on its way to disappearing into the black hole, that gives off the observed radiations.

18-2 STELLAR POPULATIONS

They Provide a Clue to the History of Our Galaxy

The globular cluster in Hercules contains perhaps a million stars.

The stars in our galaxy fall into two categories. **Population I** stars are those in the central disk of the galaxy. These stars are of all ages, from those just coming into being to old ones that must have been formed early in the life of the galaxy.

Population II stars make up the **globular clusters** that surround the galaxy above and below its central disk. To the naked eye the largest of these assemblies of stars are just visible as faint patches of light. Through a telescope they are more spectacular, dense with stars near their centers and thinning out toward the edges. About 125 of these clusters have been discovered, all orbiting the center of the galaxy.

In photographs of the globular cluster in the constellation Hercules, one of the largest, more than 50,000 stars have been counted. These are only the brightest stars, since the cluster is so far away that faint ones cannot be seen. The total number of stars may be close to a million. Small clusters may have only 10,000 or so stars. The nearest clusters are

---

GALACTIC NEBULAS

Galactic nebulas are irregular masses of diffuse material within our galaxy. Some appear as small glowing rings or disks surrounding stars, some take the form of lacy filaments, and many are wholly irregular in outline. The brightest, the Great Nebula in the constellation Orion, is barely visible with the naked eye, but most of them are much fainter. Probably these nebulas consist of gas and dust and shine only because they reflect light from nearby stars or are excited to luminescence by stellar radiation.

Clouds of gas and dust similar to galactic nebulas but without any luminosity sometimes reveal their presence as dark patches that obscure the light of stars beyond them. Such dark nebulas may be fairly abundant but are difficult to find because bright stars shine through them except where they are especially dense. In fact, our entire galaxy is filled with rarefied nebular material

The Great Nebula in Orion is a gas cloud excited to luminescence by hot stars in its center.

with a density somewhere near one atom or molecule per cubic centimeter. The dark nebulas are only local concentrations of this interstellar matter. Empty space is not nearly as empty as it was once thought to be, but in most places the amount of interstellar material is so small that starlight passes right through it.

about 20,000 light-years away from us and the farthest more than 100,000 light-years away. Light from the great Hercules cluster travels 33,000 years before reaching our eyes, so we see it as it appeared toward the end of the Ice Age. The average distance between stars in a globular cluster is about 1 light-year, which means they are much more closely packed than those near the sun.

The stars in the globular clusters of Population II are all very old, almost as old as the galaxy. What seems to have happened is that all the matter of the galaxy was originally a spherical cloud of gas from which stars were forming. In time the cloud concentrated in a central disk, leaving behind those stars that had already formed. New stars continued to come into being from the gas clouds of the disk, which is why the stars there are of all ages.

This picture is supported by the compositions of the stars of the two populations. Population II stars are extremely rich in hydrogen and helium. This follows from an early origin since the materials of the young universe are thought to have been these elements. Population I stars, on the other hand, contain heavier elements in some abundance. Such elements are produced in stars by the nuclear reactions that are responsible for stellar energy, and are thrown into space during the explosions that occur in the final stages of a star's active life. Compared with Population II stars, those of Population I are richer in heavy elements because they were formed from material that already contained such elements, material that was the debris of dying stars from Population II.

18-3   RADIO ASTRONOMY

Another Messenger from the Sky

A **radio telescope** is a directional antenna connected to a sensitive radio receiver. Usually a metal dish, like the concave mirror of a reflecting telescope, gathers radio waves over a large area and concentrates them on the antenna itself. With such an arrangement the direction from which a particular radio signal arrives can be established.

The largest steerable-dish radio telescope is 100 m in diameter and is located in West Germany. Still larger is a fixed dish 305 m across in Arecibo, Puerto Rico, that consists of wire mesh fitted into a bowl-shaped hollow in the landscape. As the earth turns, a band in the sky can be surveyed; the great sensitivity of this telescope makes up for its restriction to this band. A number of radio telescopes can be linked together electronically to precisely locate the source of radio waves by interference methods.

Radio waves from space seem to originate in three ways. A common source is the random thermal motion of ions and electrons in a very hot gas, such as the atmosphere of a young star or the remnant thrown out by a supernova explosion. Another source consists of high-speed electrons that move in a magnetic field. The strongest radio sources, called **quasars,** are of this kind and may emit more energy as radio waves than as light waves. Quasars are discussed later in this chapter.

A radiotelescope is a directional antenna designed to pick up radio waves from space. Shown are several of the 27 dish antennas, each 25 m in diameter, that make up the Very Large Array near Socorro, New Mexico. The data obtained from these antennas can be combined to reveal details about radio sources in space that would otherwise require a single dish many kilometers across (see Sec. 6-20).

Molecules in Space

A third source of radio waves in space is hydrogen atoms and molecules of various kinds. These waves are spectral lines that happen to lie in the radio-frequency part of the spectrum rather than in the optical part. In particular, the hydrogen line whose wavelength is 21 cm has proved invaluable to astronomers because most interstellar material consists of cool hydrogen. Dark nebulas in our galaxy can be mapped accurately with the help of radio telescopes tuned to receive 21-cm radiation.

A surprisingly large number of chemical compounds have been detected in galactic space by the radio waves their molecules emit when excited by collisions. Most of the molecules are fairly simple, such as carbon monoxide (CO), ammonia (NH_3), and water (H_2O); the hydroxide radical OH is also common. In addition some fairly complex organic molecules have been found, including formaldehyde (H_2CO), acetaldehyde (CH_3CHO), acrylonitrile (CH_2CHCN), and the alcohols methanol (CH_3OH) and ethanol (CH_3CH_2OH). Experiment and calculation have brought to light plausible reactions by which many of these compounds can be formed in space, but there is much still to be learned.

18-4   SPIRAL GALAXIES

Island Universes of Stars

Our Milky Way galaxy is only one of hundreds of millions of galaxies in the universe. Although galaxies have a variety of shapes, most appear as flat spirals with two curving arms that radiate from a bright nucleus. The telescope shows us spiral galaxies from different angles: some full face, some obliquely, and some edgewise, as in the accompanying photographs. Spiral galaxies contain from 1 to 100 billion stars.

The Great Galaxy in Andromeda, the spiral galaxy nearest the earth, closely resembles our own Milky Way galaxy. The Andromeda galaxy contains about 100 billion stars, and its period of rotation is about 200 million years. The two bright objects nearby are dwarf galaxies held as satellites by the gravitational pull of their huge neighbor.

The spiral galaxy in Coma Berenices appears edge-on to us.

Our galaxy is among the largest known, with so much mass that 10 smaller galaxies revolve around it.

Other galaxies are so far away from ours that in early telescopes they were merely faint, fuzzy objects. William Herschel, the first to study these objects intensively, suggested that they were galaxies like the Milky Way, "island universes" in the sea of space. In his time, this proposal was little more than a guess, but today even galaxies too far away for their separate stars to be resolved have had their character revealed by spectroscopic studies. Doppler shifts have been detected in galaxies that show their stars to be revolving around the galactic centers, just as happens in our galaxy.

Most galaxies are concentrated in space in groups of up to a few hundred. Our galaxy is one of the two dozen or so members of the **Local Group.** The largest galaxy of this group is the one in the constellation Andromeda, with our galaxy next in size. Groups of galaxies are further assembled into immense clusters that contain thousands of galaxies—and thousands of such clusters have been detected.

Through all this vast array of stars and galaxies runs a uniformity of material and structural pattern. The elements of the earth are the elements of the galaxies, the sun generates energy by a process repeated in billions of other stars, and the form of our galaxy appears again and again in the rest of the universe. Enormous black holes are believed to lie at the hearts of several nearby galaxies, just as has been proposed in the case of our galaxy, and indirect evidence suggests that this is true in

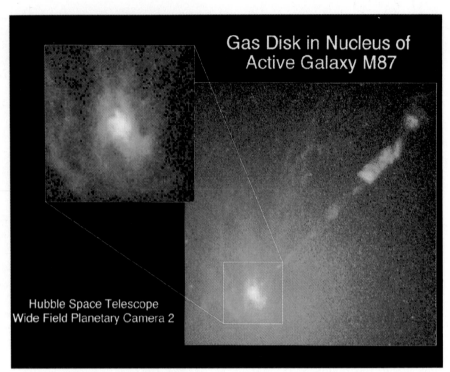

The galaxy M87, which is 50 million light-years from the earth, as seen from the Hubble Space Telescope. A disk of hot gas orbits the center of this galaxy at a speed (750 km/s) that suggests the center is a black hole with a mass of 2 to 3 billion times that of the sun.

general. We can examine at first hand only a tiny fragment of the universe, yet so consistent is the whole that from this fragment we can extend our knowledge to as far as our instruments can reach.

18-5   COSMIC RAYS

Atomic Nuclei Speeding through the Galaxy

The story of cosmic rays began early in this century, when it was discovered that the ionization in the atmosphere increased with altitude. At that time most scientists thought that the small number of ions always present in the air was due to radioactivity. After all, naturally radioactive substances such as radium and uranium are found everywhere on the earth. If this were the correct explanation, then when we go high into the atmosphere away from the earth and its content of radioactive materials, the proportion of ions that we find should drop. Instead it increases, as a number of balloon-borne experimenters learned between 1909 and 1914.

Finally Victor Hess, an Austrian physicist, suggested the correct explanation. From somewhere *outside* the earth ionizing radiation is continually bombarding our atmosphere. This radiation was later called **cosmic radiation** because of its extraterrestrial origin.

Primary cosmic rays, which are the rays as they travel through space, are atomic nuclei, mainly protons, that move nearly as fast as light. Their origin is uncertain, but they may have been shot out during supernova explosions in our galaxy and then trapped there by magnetic fields. The most energetic primaries may come from outside the galaxy.

Victor Hess (1883–1964)

About as much energy reaches the earth in the form of cosmic rays as in the form of starlight, a hint of their significance. Over a billion billion primaries arrive at the earth each second and carry with them energy equivalent to the output of a dozen large power plants. The most energetic primary ever detected had an energy of about 20 J, enough to heat a teaspoon of water by 10°C. This is a billion times as much energy as any particle accelerated in a laboratory on the earth.

When a primary cosmic ray enters the earth's atmosphere, it disrupts atoms in its path to produce a shower of secondary particles. On the average more than one of these secondaries passes through each square centimeter at sea level per minute. Among the secondary particles are neutrons, which interact with nitrogen nuclei to produce the radioactive carbon 14 that is the basis of the radiocarbon dating method discussed in Chap. 16. Secondary cosmic rays cause some of the mutations in living things that are part of the process of evolution.

THE EXPANDING UNIVERSE

Now we turn to the evidence that led to the idea that the entire universe is growing larger and larger.

18-6 RED SHIFTS

The Galaxies Are All Moving Away from One Another

The spectra of galaxies share the curious feature that the lines in nearly all of them are shifted toward the red. Furthermore, the amount of shift increases with the distance of the galaxy from us. We can see this in Fig. 18-1, which shows two of the absorption lines of calcium in the spectra of several galaxies. Each galactic spectrum is shown between two comparison spectra, so that the shift of the lines toward the red (to the right in these pictures) is clear.

BIOGRAPHY *E d w i n H u b b l e* (1889–1953) was born in Missouri and, although always interested in astronomy, pursued a variety of other subjects at the University of Chicago. He then went to Oxford University in England, where he concentrated on law and Spanish. After two years of teaching at an Indiana high school, Hubble realized what his true vocation was and returned to the University of Chicago to study astronomy. At Mount Wilson Observatory in Pasadena, California, Hubble made the first accurate measurements of the distances of spiral galaxies, which showed that they are far away in space from our own Milky Way galaxy. It had been known for some time that such galaxies have red shifts in their spectra that indicate motion away from the Milky Way, and Hubble joined his distance figures with the observed red shifts to conclude that the recession speeds were proportional to distance. This implies that the universe is expanding, a remarkable discovery that has led to the modern picture of the evolution of the universe. In his later work Hubble tried to determine the structure of the universe by finding how the concentration of remote galaxies varies with distance, a very difficult task that even today has not been fully accomplished.

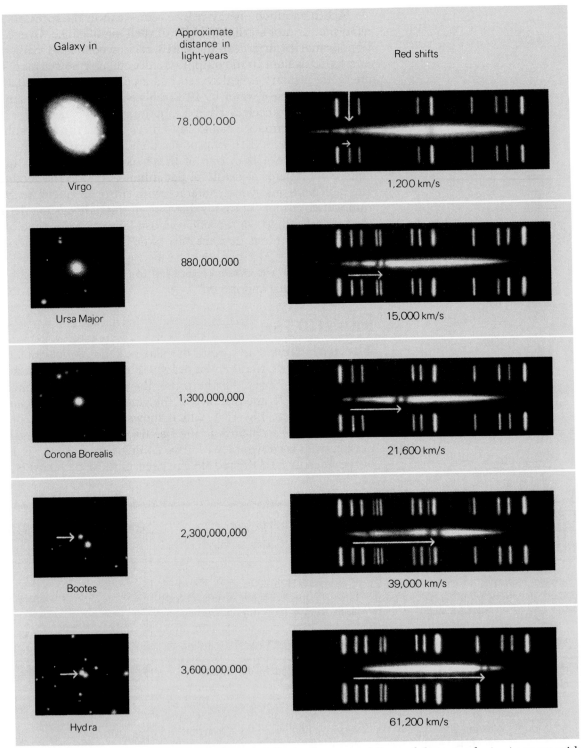

FIG. 18-1   The red shifts in the spectral lines of distant galaxies increase with increasing distance. The indicated lines occur in the spectrum of calcium. Reference spectra are shown above and below each galactic spectrum. The red shifts are interpreted as doppler shifts that come about because of the expansion of the universe.

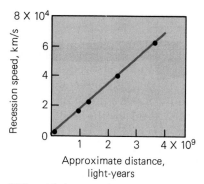

FIG. 18-2 Graph of recession speed versus distance from the earth for the galaxies of Fig. 18-1. The speed of recession averages about 17 km/s per million light-years.

In Chap. 17 we saw that red shifts in stellar spectra result from motion away from the earth. We must therefore conclude that all the galaxies in the universe (except the few nearby ones in the Local Group) are receding from us. The recession speeds can be computed from the extent of the red shifts, and the results are startling: several hundred kilometers per second for the nearer galaxies, more than 200,000 km/s for the farthest ones. By comparison, the speed of light—the maximum speed anything can have—is 300,000 km/s.

Hubble's Law

If we plot the recession speeds of the galaxies shown in Fig. 18-1 versus their distances from the earth, as in Fig. 18-2, we find that these quantities are proportional. The greater the distance, the faster the galaxies are traveling. The speed increases by about 17 km/s per million light-years. When this graph is extended to cover the still faster and more distant galaxies that have been studied, the experimental points fall on the same line. The proportionality between galactic speed and distance was discovered in 1929 by the astronomer Edwin Hubble and is known as **Hubble's law.**

At first it might seem as though our galaxy had some strange repulsion for all other galaxies, forcing them to move away from us with ever-increasing speed. But it is hardly likely that we occupy so important a place in the universe. A more reasonable conclusion is that *all* the galaxies are moving away from one another, spreading apart like the fragments of a bursting rocket. An observer on our galaxy *or on any other* would then get this impression that the neighbors are fleeing in all directions.

The universe, in other words, seems to be expanding, with its component galaxies moving ever farther apart (Fig. 18-3).

The discovery that the universe is expanding should not really have come as a surprise. One of the findings of Einstein's general theory of relativity, which we recall from Chap. 3 interprets gravity as a warping of space and time, was that the universe cannot stand still, so to speak, because all its contents attract one another. The universe must either expand or contract. However, when Einstein came to this conclusion in 1917, astronomers were unanimous in thinking that the universe was static and unchanging. Einstein therefore felt his theory must be wrong as it was and arbitrarily altered it to suit this view—the biggest mistake of his scientific career, as he later commented. Today it is clear that Einstein's original ideas were right, and they fit in well with the observed expansion of the universe.

FIG. 18-3 Two-dimensional analogy of the expanding universe. As the balloon is inflated, the spots on it become farther apart. A bug on the balloon would find that the farther away a spot is from it, the faster the spot seems to be moving away. This is true no matter where the bug is.

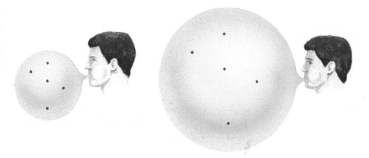

18-7 QUASARS

Brilliant, Tiny, and Far Away

The most remarkable objects in the sky whose spectra show red shifts are **quasars.** In a telescope, a quasar appears as a sharp point of light, just as a star does, not a fuzzy dot as it would if it were a distant galaxy. But unlike stars, the first quasars to be discovered, in the early 1960s, were powerful sources of radio waves. Hence their name, a contraction of *quasi-stellar radio sources.* Many quasars were later found that do not give off radio waves but can be identified by their characteristic spectra.

Both the light and radio outputs of quasars fluctuate markedly, in some cases with a period of less than a day. Hundreds of quasars have been found, and there seem to be many more.

Quasar red shifts are usually large, corresponding to recession speeds of over three-quarters the speed of light. Such speeds mean that quasars are far away, so far away that the light reaching us from them was emitted early in the life of the universe. Because quasars are distant objects, the intensities of the light and radio waves we receive from them imply that they are giving off energy at colossal rates, billions or trillions of times more than ordinary stars. Where does the energy come from?

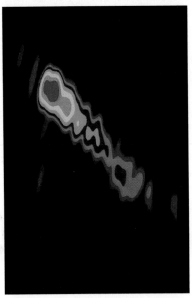

Quasars radiate many times more energy than ordinary galaxies like our own Milky Way, yet are far smaller in size. It is likely that quasars have black holes at their centers. The quasar shown here in a false-color radio image is 3 billion light-years away and emits energy at a rate of 10^{40} watts, equivalent to the output of 1000 ordinary galaxies. The jet of matter that extends from this quasar is 250,000 light-years long, although the quasar itself is only about 1 light-year across. The image of the quasar appears much larger here than it should because the exposure time of the photograph was increased to show the jet.

What Is a Quasar?

Quasars vary in brightness too rapidly for them to be large objects. If a quasar were 100,000 light-years across, as our galaxy is, even a change in its energy output that took only a week to occur (if such a thing were possible) would appear to us as spaced out over 100,000 years. The observed short-period fluctuations point to a diameter for most quasars that is smaller than that of the solar system—but each quasar's energy output may be thousands of times the output of our entire galaxy.

Thus quasars must be small even though more powerful than any other energy sources in the universe.

Most astronomers believe that at the heart of each quasar is a black hole whose mass is at least that of 100 million suns. As nearby stars are pulled toward the black hole by its strong gravitational field, their matter is compressed and heated to temperatures such that radiation is given off in abundance. (Once a star is inside the black hole, of course, it and its radiation vanish forever.) As it is being swallowed, a star may liberate 10 times as much energy as it would have given off had it lived out a normal life.

A diet of a few stars a year seems to be enough to keep a quasar blazing away as observed. Exactly how the various radiations emitted by a quasar are produced is not completely certain, but only a supermassive black hole seems able to account for them.

Many astronomers believe that quasars are the cores of newly formed galaxies. Did all galaxies, including ours, once undergo a quasar phase? Nobody can say as yet, but it seems possible.

EVOLUTION OF THE UNIVERSE

According to the aptly named **big bang theory,** the universe began as a tiny fireball of unimagineably dense matter that exploded violently billions of years ago. As the material spread out and cooled, local concentrations formed and became the galaxies. If we start from the assumption that the big bang did occur, no new physical principles seem to be needed to account for the evolution of the universe as it is today, and the observed expansion of the universe follows naturally.

18-8 DATING THE UNIVERSE

When Did the Big Bang Occur? It would seem that we could use the Hubble's law figure of approximately 17 km/s per million light-years for the rate of expansion of the universe to calculate backward to a time when the galaxies were together in one place. Let us consider a galaxy 1 million light-years away from us, which is 9.5×10^{21} m. The galaxy's speed is 17 km/s = 1.7×10^4 m/s, so it must have started its outward motion at a time

$$T = \frac{\text{distance}}{\text{speed}} = \frac{9.5 \times 10^{21} \text{ m}}{1.7 \times 10^4 \text{ m/s}} = 5.6 \times 10^{17} \text{ s}$$

ago. Since there are 3.2×10^7 s in a year, this would make the age of the universe 1.7×10^{10} years—17 billion years.

There are two problems with this calculation. The first is that the figure 17 km/s per million light-years is not exact because it is hard to establish the distance of remote galaxies independently of their red shifts. Second, the calculation ignores the effect of gravity on the expanding matter. If the universe were empty, its age would be roughly 17 billion years. Because it contains matter, the rate of expansion must be getting slower and slower, just as a ball thrown up into the air slows down. Since the expansion was faster in the past than it is now, the universe must be younger than the 17 billion years obtained from Hubble's law. An age of roughly 15 billion years seems about right when the effect of gravity is considered.

18-9 THE CYCLIC UNIVERSE

No Beginning and No End? The laws of physics as we know them today can take us from the present universe all the way back to about 10^{-35} s after the big bang. Plausible guesses can get us even closer to that moment of creation. But how did such a concentration of matter and energy arise? What was there before the big bang?

One suggestion is that the big bang was only the latest of a series of big bangs that extends infinitely far into the past, a series that will continue infinitely far into the future.

If the cosmic explosion that led to the present expansion of the universe was not violent enough, the expansion will not continue forever—the universe will be **closed** rather than **open.** In this case gravity will eventually slow down the expansion until it stops. The universe will then begin to collapse, and finally all the matter and energy in the universe will come together in a **big crunch.** Then another big bang will occur, and another cycle of expansion and contraction will start (Fig. 18-4).

AN ASTRONOMER AT WORK

Wendy Freedman (Carnegie Observatories, Pasadena). I can still recall vividly a warm summer night in the country when I was seven years old. Under a dark, star-filled sky, my father explained to me that the light coming to our eyes then had left those stars years or even thousands of years ago. We were seeing those stars as they were at some time in the distant past; perhaps they were not even there any more. It piqued my curiosity; I wanted to know more about this universe we live in, how it formed and changed over time.

I now find my days as an astronomer to be extremely stimulating, interesting, and challenging. I use large telescopes to make observations of distant stars and galaxies. These telescopes are located in places like the Andes mountains in Chile, Mount Palomar in southern California, and 14,000 foot Mauna Kea in Hawaii. At present, my research interest centers on learning some of the basic properties of the universe. I am a leader of an international team of 27 astronomers located in the U. S., Canada, Australia, and Great Britain who use the Hubble Space Telescope to measure how far away galaxies are and ultimately to determine the age and size of the universe.

Images of our chosen galaxies are sent to us by the Space Telescope Science Institute in Baltimore, Maryland. On a given day, I may analyze these images using computer software designed to extract the information needed for measuring distances to remote galaxies. As the project has evolved, more of my day now goes into interacting with various members of our team, helping to solve problems in analysis, helping to coordinate interaction amongst team members, even helping to solve disputes. Much of this interaction is done via E-mail, or by telephone, or during quarterly meetings of the team.

Also during the course of a day I will read about recent work done by other astronomers to keep abreast of new results and discoveries. For me, interpreting the results from both the Hubble Space telescope and ground-based telescopes is one of the most satisfying and challenging aspects of my work, an opportunity to learn completely new and fascinating things about our universe.

Often, after putting the kids to bed, I will stay up late into the night working on a problem or writing a paper. I do this because I love it and I want to do it. There are very few things in life more rewarding than finding a career that you enjoy.

The cyclic model of the universe is attractive in many ways, but is it correct? The key to the answer is how the present average density of the universe compares with a certain critical value that depends on how fast the universe is currently expanding. If the average density is less than the critical value, which is equivalent to several hydrogen atoms per cubic meter, the universe is open and will continue to expand forever. If the density is greater than the critical value, the universe is closed and some day a big crunch will come, presumably followed by another big bang.

FIG. 18-4 If the universe is "open," its expansion will continue forever. If it is "closed," the universe will eventually stop expanding and then collapse. In that case, a new big bang may conceivably occur that would start a new expansion.

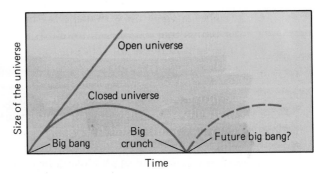

Dark Matter   The actual density of the luminous matter in the universe is just a percent or so of the critical density. But does luminous matter—the stars and galaxies we see in the sky—make up most of the matter in the universe?

Apparently not. Very strong evidence of several kinds indicates that a large amount of **dark matter** is also present. So much, in fact, that at least 90 percent of all matter in the universe gives off no light. For instance, the rotation speeds of the outer stars in spiral galaxies are unexpectedly high, which suggests that a lot of invisible matter must be inside each galaxy to keep them from flying off into space. Similarly, the motion of individual galaxies within clusters of galaxies implies gravitational fields about 10 times more powerful than the visible matter of the galaxies provides.

What can the dark matter be? The most obvious candidate is ordinary matter in various known forms. Examples are planetlike lumps ("brown dwarfs") too small to support the fusion reactions that would make them stars, burned-out dwarf stars, and black holes. The snag here is that, in the required numbers, such objects would certainly have been found already.

There is no shortage of other possibilities for the dark matter. However, all of them involve hypothetical elementary particles with unusual properties. Until any of these particles is actually discovered, or some other explanation is found, the nature of the dark matter in the universe will remain one of the most fundamental of all outstanding scientific mysteries. And we will probably have to wait for this mystery to be solved before we know whether we live in an open or a closed universe.

18-10   THE PRIMEVAL FIREBALL

A Glimmer of It Still Exists   Figure 18-5 shows some stages in the history of the universe. Just after the big bang, the universe was a compact, intensely hot mixture of matter and energy. Particles and antiparticles were constantly annihilating each other to form photons of radiation, and just as often photons were materializing into particle-antiparticle pairs. The particles and antiparticles at this time were the quarks and leptons described in Sec. 7-16.

As the fireball expanded and cooled, the energies of the photons decreased. (We recall from Fig. 17-1 that the average wavelength of the radiation from a hot object increases as its temperature drops, which corresponds to a decrease in frequency and hence to a decrease in quantum energy hf.) Finally, a few seconds after the big bang, the photons had too little energy to create any more particle-antiparticle pairs. The annihilation of the existing particles and antiparticles continued, however. Current theories of elementary particles suggest that particles outnumbered antiparticles, perhaps a billion and one particles for every billion antiparticles of the same kind. When the annihilation was finished, the surplus of particles—protons, neutrons, and electrons—remained.

When the universe was about a minute old, nuclear reactions began to form helium nuclei. Calculations show that the ratio of hydrogen

Age 0

The big bang. The universe begins to expand, at first very rapidly.

Age 10⁻³⁵ second

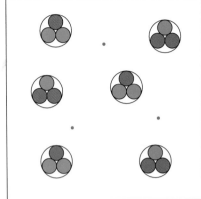

Quarks and leptons come into being. Particles outnumber antiparticles. All the antiparticles are eventually annihilated to leave a matter universe with no antimatter.

Age 1 second

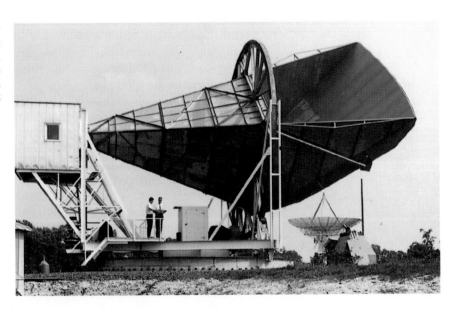

Quarks have joined to form neutrons and protons, some of which then react to form nuclei of light elements, mainly helium. There are as many electrons as protons, so the universe remains electrically neutral.

FIG. 18-5 Snapshots from the youth of the universe.

nuclei (protons) to helium nuclei ought to have ended up as 3:1, the same ratio found in most of the universe today. Although nearly all the helium in the universe was formed in the first few minutes after the big bang, with some created later in the nuclear reactions that power stars, the helium on the earth mainly came from the alpha decay of radioactive elements in its interior.

After 300,000 years the temperature of the universe was down to about 3000 K, cool enough for electrons and nuclei to combine into atoms. Since photons interact strongly with charged particles but only weakly with neutral atoms, at this time matter and radiation were "decoupled" and the universe became transparent. The radiation that

Radio waves that originated early in the history of the universe were first detected by Arno Pentias and Robert Wilson as a persistent hiss in a sensitive receiver attached to this 15-m-long antenna at Holmdel, New Jersey.

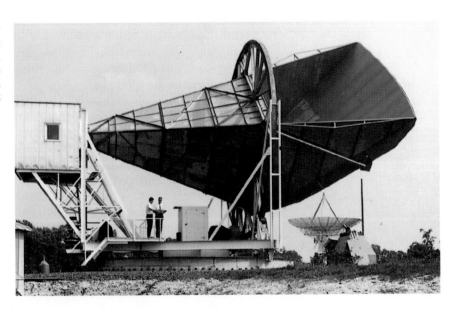

Age 3 minutes

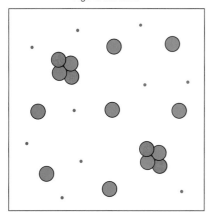

Nuclear reactions stop. The hydrogen: helium ratio is 3:1 as in most of today's universe. The universe is still too hot (10^9 K) for atoms to exist.

Age 300,000 years

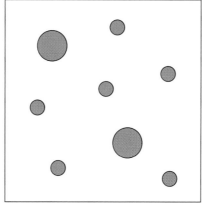

The universe has now cooled enough for nuclei and electrons to have joined to form neutral atoms. With few ions left to absorb it, radiation from this time was Doppler-shifted by the expansion of the universe to become radio waves that fill the universe today.

Age 1 billion years

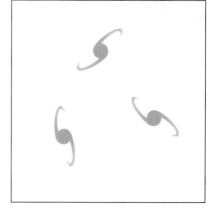

Under the influence of gravity, local concentrations of gas are growing and condensing into stars and galaxies. The universe continues to expand.

RIPPLES IN SPACETIME As well as can be determined by instruments on the earth, the cosmic background radiation appears the same in all directions in the sky. This finding troubled astronomers, because only a completely smooth distribution of matter in the early universe could lead to a perfectly uniform afterglow—and a smooth distribution would leave no way to explain how such irregularities as galaxies could have developed. To study the question further, NASA launched the Cosmic Background Explorer satellite in 1989. Three years later the results were announced: there are indeed very small variations in the radiation that are the result of "tiny ripples in the fabric of spacetime put there by the primeval explosion process," in the words of George Smoot, one of the astronomers involved in the project. These ripples, which correspond to clumps of gas denser than the rest in the early universe, were the seeds from which galaxies later grew.

was left then continued to spread out with the rest of the universe, so that even today remnants of it must be everywhere.

Because the universe is expanding, an observer today would expect to find these remnants to have undergone a red shift to long wavelengths, in the range of radio waves. The radiation would not be easy to find, since it would be very weak. However, it ought to have two distinctive characteristics that would permit it to be identified: it should come equally strongly from all directions, and its spectrum should be the same as that which an object at about 2.7 K would radiate.

Remarkably enough, radiation of exactly this kind has been discovered. Some of the "snow" that appears on television screens when no picture is being transmitted is due to this remnant radiation from the big bang. The universe is bathed in a sea of radio waves whose ultimate source seems to have been the primeval fireball. Thus we have three observations that strongly support the big bang theory of the origin of the universe:

1. The uniform expansion of the universe

2. The relative abundances of hydrogen and helium in the universe

3. The cosmic background radiation

18-11 ORIGIN OF THE EARTH

It Began with a Whisper, Not a Bang A billion or two years after the big bang much of the hydrogen and helium of the universe had accumulated in separate clouds under the influence of gravity. These clouds were the ancestors of today's galaxies. As the young galaxies contracted, local blobs of gas formed that became the first stars. Some of the early stars were very massive and went through their life cycles rapidly to end as supernovas. The explosions of these supernovas dumped elements heavier than hydrogen and helium into the remaining galactic gas. As a result the matter from which later stars condensed was a mixture of all the elements, not just hydrogen and helium. By the time the sun came into being, billions of years after the first generation of stars, the loose material of our galaxy contained between 1 and 2 percent of the heavier elements. These were in the form of small solid "dust" grains, some of ice, some of rock.

FIG. 18-6 (a) The solar system began as a spinning cloud of gas and grains of ice (frozen gases) and rock. (b) When the protosun began to shine, the ice grains near it were heated and vaporized. (c) The rock and ice grains collided and stuck together to form planetesimals, which were rocky near the sun and a mixture of rock and ice farther away. (d) The planetesimals themselves collided and stuck together to form the planets. The solar wind blew away the gas and other material that had not become incorporated in the planets.

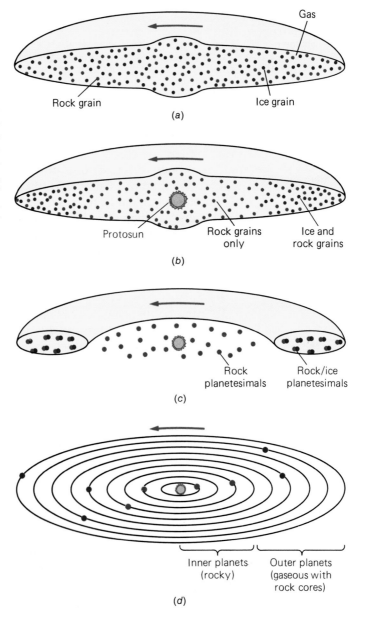

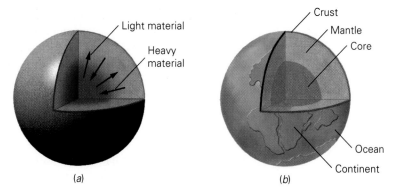

FIG. 18-7 (a) When the earth was young, its interior melted. Under the influence of gravity, heavy materials then migrated inward to form the core and lighter ones were forced outward to form the mantle and crust. (b) Eventually the separation into core, mantle, and crust was complete, the outer parts cooled and hardened, and the early oceans and continents came into being.

The swirling cloud that became the sun was originally much larger than the present solar system. As it shrank in the process of becoming a star, this **protosun** left behind a spinning disk of gas and dust (Fig. 18-6). Bits of matter in the disk collided and stuck together to form larger and larger grains, perhaps the size of pebbles. In time these grains formed larger bodies called **planetesimals** a few kilometers across that ultimately joined to become the planets. Near the protosun, which was heated by gravitational compression, it was too warm for gaseous elements to collect in any great amounts on the planetesimals, which is why the inner planets are rocky bodies. Farther out temperatures were lower, so the planetesimals there mirrored the composition of the original cloud by being mainly frozen gases.

As the planets were developing with their satellites around them, the protosun was gathering up more and more material. Eventually its internal temperature and pressure became sufficient for fusion reactions to occur in its hydrogen content, and the protosun became a star. Fast ions and electrons then began to stream out of the sun, much like today's solar wind but far more intense. This wind swept the solar system free of the gas and dust that had not yet been incorporated in the sun, the planets, and the satellites.

The infant planets were all heated by gravitational compression, which was naturally the most effective in giant Jupiter and Saturn. Radioactivity was important as well in heating the inner planets. The earth became hot enough at this time to melt and separate into a dense iron core and a lighter rocky mantle (Fig. 18-7). The remaining planetesimals in the solar system bombarded the planets and satellites heavily, leaving craters on the solid ones that, in the case of the earth, erosion would later erase.

EXTRATERRESTRIAL LIFE

The recipe for creating life begins with a warm rocky planet whose gravity is able to hold an atmosphere and whose distance from its parent star is such that it receives enough but not too much radiant energy. Then provide an atmosphere of certain quite common gases, allow the planet to cool down a bit so that water condenses out, and wait perhaps a billion years. This recipe was followed successfully on the earth. Let us now see how likely it is to have been followed elsewhere in the universe.

18-12   OTHER PLANETARY SYSTEMS

We Are Not Alone   The modern view of the solar system is that it is a natural product of the evolution of the sun. Since the sun is by no means an exceptional star in any other respects, it is reasonable to suppose that other stars are also attended by planetary systems. Direct evidence in support of this idea was the discovery by the Hubble Space Telescope of disks of dust and gas around a number of newly formed stars, disks like the one pictured in Fig. 18-6a. In the region of the Orion Nebula that was studied, about 40 percent of the young stars (less than a million years old) there had such disks.

Many of the 100 billion stars in the Milky Way are similar to the sun. Some fraction of these stars, even if not 40 percent, almost certainly have planetary systems. If the fraction is large, the total number of planets in our galaxy equals or exceeds the number of stars. If the fraction is small—shall we say one in a thousand?—then the number of planets still comes to a billion or so. And there are billions of other galaxies in the known universe. Even a conservative guess at the lower limit to the number of planets in the universe thus yields more than a billion billion.

Why Other Planetary Systems Are Hard to Find   If planets are such common objects, why are we certain only of the nine in our own solar system? Three interrelated factors stand in the way of identifying planets that belong to other stars. The first is distance: the closest star to us is thousands of times farther away than Pluto, and Pluto is hard enough to detect. The second is that planets shine by reflected light and not by themselves. This makes them very dim compared with their parent stars, far too dim to be seen directly.

The third difficulty is size. Planets are necessarily small relative to stars, because if they were not, they would have become stars. The small size of planets helps make them impossible to see at stellar distances, and their small mass makes the wobbles they cause in the paths of their parent stars very hard to find.

In spite of these problems several planetary systems have been unambiguously detected elsewhere in our galaxy. The first such system to be found consists of three planets in orbit around a pulsar 1400 light-years away. Two of the planets have more mass than the earth while the other has less; their periods are 25, 67, and 98 earth days long. The other planetary systems circle more normal stars.

From the point of view of life in the universe, it would be nice to actually observe planetary systems around other stars, but not crucial. There are so very many stars in the universe that it is unlikely that only our own sun has a planetary system. Whether there are thousands of billions of billions of other planets or merely billions is not really important. What is important is that a certain proportion of these planets surely meet, in the words of Harlow Shapley, "the happy requirement of suitable distance from the star, near-circular orbit, proper mass, salubrious atmosphere, and reasonable rotation period—all of which are necessary for life as we know it on earth." Although the hypothesis of life on other worlds may not be directly verified in the near future, no serious arguments that dispute this hypothesis have been advanced. It seems certain that we are not alone in the universe.

18-13   INTERSTELLAR TRAVEL

Not Now, Probably Not Ever   Astronauts have already visited the moon and spacecraft have traveled to other planets. If the money for such a project were available, there is no reason why people should not be able to set foot on Mars a few years after a decision to do so is made and see for themselves whether life existed there in the past. But travel to the planetary systems of other stars is much more of an undertaking.

Suppose we wish to go to Proxima Centauri, the star nearest to the sun, to check on whether it has a planet that contains living things. This star is 4×10^{13} km away, a little over 4 light-years. The various Apollo spacecraft took 3 days to reach the moon, which is 100 million times closer than Proxima Centauri. To reach this star at the same rate, a similar spacecraft would therefore need 300 million days, nearly a million years. So existing technology is out of the question.

What about the future? According to Einstein's special theory of relativity, which has been found to be completely accurate in all its predictions, the ultimate limit to the speed that anything can have is the speed of light. This speed is about 300,000 km/s. At a speed 1 percent short of this, a spacecraft could reach Proxima Centauri in a few years, a reasonable enough period of time. However, the energy required would be fantastic, more than all the energy currently used in the entire world per year for each ton of spacecraft weight—and the spacecraft would have to weigh many tons. How can all this energy be produced? How can it be concentrated at one time and place? How can it be given to the spacecraft? And how could a like amount of energy be provided in the vicinity of Proxima Centauri to return the spacecraft to the earth?

These questions are not quibbles. They are fundamental to such an enterprise—and they have no answers at present. As for a space traveler visiting the stars that lie millions and billions of times farther away than Proxima Centauri, there does not seem to be any hope whatever.

The laws of physics hold everywhere in the universe, which means that the extreme unlikelihood of travel from the earth to another planetary system is also true for travel from such a system to the earth. It is easy to attribute ancient legends and archeological findings that are hard to explain in terms of known events to visitors from another world. However, all such attributions have turned out to lack any real evidence to support them. A good story does not constitute proof of anything except the imagination of its author.

18-14   INTERSTELLAR COMMUNICATION

Why Not?   Although travel to other worlds is almost surely impossible, by contrast communication with them seems quite feasible. The first serious proposal was made a century and

This extremely sensitive 300-m radiotelescope at Arecibo in Puerto Rico was used in the SETI (Search for Extraterrestrial Intelligence) program.

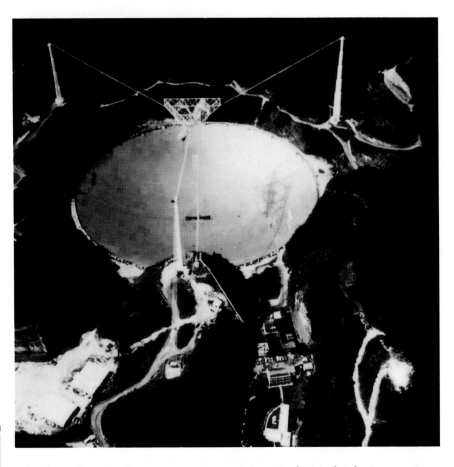

THE FUTURE OF THE EARTH According to an old Norse legend, the world will come to an end when "an icy winter will seize the earth in its grip, and a great wolf will devour the sun." A somewhat different fate seems more likely. In 5 billion years or so the sun will swell into a red giant (see Fig. 17-9), and its radiation will increase to a hundred times its present rate. The earth will grow warmer and the oceans will begin to boil. Steam will fill the atmosphere and all life will perish. Eventually the sun will become larger around than the earth's orbit, but this far out its substance will be a thin gas so the earth will survive as a cinder. Later the sun will shrink to a feeble white dwarf, which will ultimately fizzle out to a black dwarf if the universe has not undergone a big crunch before then.

a half ago by the German mathematician Karl Friedrich Gauss. Gauss suggested that trees be planted on the bare landscape of Siberia to form a right triangle large enough to be seen from the moon, which he thought might be inhabited. A better method for interstellar distances is to use radio waves.

Directed by NASA, a search for extraterrestrial intelligence (SETI) began in 1992 in which radio telescopes around the world sought messages from far away. Microwave signals were listened for at frequencies between 1 and 10 GHz, a relatively quiet part of the radio spectrum. Part of the program was targeted on the approximately 800 nearest stars that resemble the sun. The entire sky was also scanned, though necessarily with less sensitivity. The telescopes automatically looked for patterns of microwave emissions that are not the result of natural processes, such as pulsar bursts. Of course, intelligent living things elsewhere that do not send out microwave messages cannot be found in this way. Still, with so many possible homes for intelligent life in the universe, the SETI project is certainly a reasonable approach. The U.S. Congress canceled the program in 1993, but a similar effort has begun with private funding.

The next step would be to send out messages ourselves for other worlds to receive. By virtue of the long distances involved, years at best, more likely centuries would be needed for an interchange of information to take place. But what an extraordinary, exhilarating prospect!

IMPORTANT TERMS AND IDEAS

Our **galaxy** is a huge, rotating disk-shaped group of stars that we see in the sky as the **Milky Way** from our location about two-thirds of the way out from the center. Most of the stars of the galaxy are found in two **spiral arms** that extend outward from the center. These stars are **Population I** stars and are of all ages. **Globular clusters** of very old stars occur outside the central disk of the galaxy. These stars are **Population II** stars. **Spiral galaxies** are other collections of stars that resemble our galaxy. The universe apparently consists of widely separated galaxies of stars.

A **radio telescope** is a directional antenna connected to a sensitive radio receiver. Radio waves from space are produced by extremely hot gases, by fast electrons that move in magnetic fields, and by atoms and molecules excited to radiate. Especially notable sources are **quasars,** distant objects that emit both light and radio waves strongly and that may be powered by supermassive black holes at their centers.

Cosmic rays are atomic nuclei, mostly protons, that travel through the galaxy at speeds close to that of light. They probably were ejected during supernova explosions and are trapped in the galaxy by magnetic fields.

The spectral lines of distant galaxies show a doppler shift to the red arising from motion away from the earth. Since the speed of recession is observed to be proportional to distance, the **red shift** means that all the galaxies in the universe are moving away from one another. This **expansion of the universe** began about 15 billion years ago.

The **big bang** theory holds that the universe originated in a great explosion about 15 billion years ago. Radiation left over from the big bang and doppler-shifted to radio frequencies has been detected. If the explosion was violent enough, the expansion of the universe will continue forever; if not, the universe will eventually begin to contract and will end up in a **big crunch** after which another cycle of expansion and contraction may occur. It is likely that the solar system originated as part of an evolutionary process, and that planetary systems are quite common elsewhere in the universe.

There is apparently much more **dark matter** in the universe than the luminous matter, such as stars, that we can see. The nature of the dark matter is unknown.

EXERCISES: MULTIPLE CHOICE

1. The stars in space are
   a. uniformly spread out
   b. distributed completely at random
   c. chiefly in the Milky Way
   d. mostly contained within widely separated galaxies

2. The Milky Way is
   a. a gas cloud in the solar system
   b. a gas cloud in the galaxy of which the sun is a member
   c. the galaxy of which the sun is a member
   d. a nearby galaxy

3. Relative to the center of our galaxy,
   a. its stars are stationary
   b. its stars move entirely at random
   c. its stars revolve
   d. Population I stars are stationary and Population II stars revolve

4. Our galaxy is approximately
   a. 4 light-years across
   b. 15,000 light-years across
   c. 100,000 light-years across
   d. 4 million light-years across

5. The number of stars in our galaxy is roughly
   a. 1000   c. 1 billion
   b. 1 million   d. 100 billion

6. Our galaxy has
   a. 1 main spiral arm   c. 3 main spiral arms
   b. 2 main spiral arms   d. 6 main spiral arms

7. Our galaxy is shaped roughly like a fried egg whose diameter is about
   a. twice its thickness
   b. 10 times its thickness
   c. 100 times its thickness
   d. 1 million times its thickness

8. Each cubic centimeter of space between the stars in our galaxy contains on the average about
   a. 1 atom or molecule
   b. 1 million atoms or molecules
   c. 1 mg of matter
   d. 1 g of matter

9. Evidence of various kinds suggests that at the center of our galaxy is a
   a. quasar   c. neutron star
   b. pulsar   d. black hole

10. Clouds of luminous gas and dust in our galaxy are called
    a. galactic clusters   c. quasars
    b. galactic nebulas   d. cosmic rays

11. The globular clusters of our galaxy
    a. contain around a hundred stars each
    b. contain around a million stars each
    c. contain around a billion stars each
    d. lie only in its central disk

12. The Population II stars in globular clusters are
    a. all very young
    b. all very old
    c. all about as old as the sun
    d. of all ages

13. Compared with the Population II stars in globular clusters, the Population I stars in the galactic disk are
    a. richer in hydrogen and helium
    b. richer in the heavier elements
    c. older
    d. closer together

14. A radio telescope is basically a(an)
    a. device for magnifying radio waves
    b. telescope remotely controlled by radio
    c. directional antenna connected to a sensitive radio receiver
    d. optical telescope that uses electronic techniques to produce an image

15. Radio waves from space never originate in
    a. extremely hot gases
    b. fast electrons moving through magnetic fields
    c. molecules
    d. cosmic rays

16. Spiral galaxies
    a. are readily visible to the naked eye
    b. may be dark or bright
    c. are usually similar to our galaxy
    d. originate in supernova explosions

17. The stars in a galaxy are
    a. moving outward from its center
    b. moving inward toward its center
    c. revolving around its center
    d. stationary relative to its center

18. Cosmic rays carry energy to the earth at about the same rate as
    a. sunlight   c. starlight
    b. moonlight   d. meteoroids

19. Cosmic rays
    a. circulate freely through space
    b. are trapped in our galaxy by electric fields
    c. are trapped in our galaxy by magnetic fields
    d. are trapped in our galaxy by gravitational fields

20. Primary cosmic rays are composed largely of very fast
    a. protons   c. electrons
    b. neutrons   d. gamma rays

21. The red shift in the spectral lines of light reaching us from other galaxies implies that these galaxies
    a. are moving closer to one another
    b. are moving farther apart from one another
    c. are in rapid rotation
    d. consist predominantly of red giant stars

22. According to Einstein's general theory of relativity, the universe
    a. must be expanding
    b. must be contracting
    c. must be either expanding or contracting
    d. may be neither expanding nor contracting

23. Supernova explosions have no connection with
    a. the formation of heavy elements
    b. cosmic rays
    c. pulsars
    d. quasars

24. Current ideas suggest that what is responsible for the observed properties of a quasar is a massive
    a. neutron star   c. spiral galaxy
    b. black hole   d. star cluster

25. Quasars do not
    a. emit radio waves
    b. exhibit red shifts in their spectra
    c. vary in their output of radiation
    d. occur in the solar system

26. The age of the universe is probably in the neighborhood of
    a. 15 million years   c. 15 billion years
    b. $4\frac{1}{2}$ billion years   d. 30 billion years

27. The term big bang refers to
    a. the origin of the universe
    b. the ultimate fate of the universe
    c. a supernova explosion
    d. the formation of a quasar

28. Since the big bang the rate of expansion of the universe has
    a. been constant
    b. increased
    c. decreased
    d. alternately increased and decreased

29. The matter in the early universe that eventually condensed into galaxies and then into stars consisted of

a. only hydrogen

b. hydrogen and helium in an approximately 3:1 ratio

c. hydrogen and helium in equal amounts

d. hydrogen and helium in an approximately 1:3 ratio

30. The elements heavier than hydrogen and helium of which the planets are composed probably came from the

a. sun

b. debris of supernova explosions that occurred before the solar system came into being

c. big bang

d. big crunch

31. Soon after it came into being, the universe contained

a. only matter

b. only antimatter

c. equal amounts of matter and antimatter

d. slightly more matter than antimatter

32. Today the universe apparently contains

a. only matter

b. only antimatter

c. equal amounts of matter and antimatter

d. slightly more matter than antimatter

33. Radiation from the early history of the universe was doppler-shifted by the expansion of the universe until today it is in the form of

a. x-rays   c. infrared waves

b. ultraviolet waves   d. radio waves

34. Present evidence suggests that most of the mass of the universe is in the form of

a. dark matter   c. cosmic rays

b. luminous matter   d. black holes

35. It is likely that the planets, satellites, and other members of the solar system were formed

a. together with the sun

b. later than the sun from material it ejected

c. later than the sun from material it captured from space

d. elsewhere and were captured by the sun

36. Planets are always small compared with stars because otherwise

a. the rotation of the planets would cause them to disintegrate

b. the great mass of the planets would cause them to be pulled into their parent star

c. the great mass of the planets would prevent them from being held in orbit and they would escape

d. the planets would be stars themselves

37. In the next phase of its evolution, the sun will become a

a. red giant   c. neutron star

b. white dwarf   d. quasar

38. The least likely reason why planetary systems have not been directly observed around stars other than the sun is that

a. planets are small

b. planets shine by reflected light

c. planetary systems are rare

d. other stars are far away

QUESTIONS

1. Are the stars uniformly distributed in space?

2. What is the Milky Way?

3. Why is the sun considered to be located in the central disk of our galaxy?

4. The earth undergoes three major motions through space. What are they?

5. What are the properties of globular clusters?

6. Both galactic nebulas and globular clusters occur in our galaxy. Distinguish between them.

7. Distinguish between Population I and Population II stars.

8. A number of elliptically shaped galaxies are known that seem to contain only Population II stars. Would you expect such galaxies to contain abundant gas and dust?

9. Where is most of the interstellar gas in our galaxy located? What is its chief constituent?

10. How can the rotation of a spiral galaxy be determined?

11. Are most galaxies smaller, larger, or about the same size as our galaxy?

12. Do radio telescopes magnify anything? If not, why are larger and larger ones being built?

13. List the three ways in which radio waves from space originate.

14. Experiments have been carried out to monitor the 21-cm radio waves emitted by neutral hydrogen in the universe to seek signals produced by intelligent beings on planets outside our solar system. Why do you suppose this particular wavelength was chosen for monitoring?

15. Why are radio-astronomical studies of the distribution of hydrogen in the universe of greater

interest than studies of the distribution of other elements?

16. What kind of evidence supports the belief that molecules of various kinds, including some fairly complex ones, exist in space?

17. Cosmic-ray intensity varies around the world in a manner that is correlated with the earth's magnetic field. Why is such a correlation plausible? Account for the greater number of cosmic rays that reach the earth near the polar regions than near the equator.

18. There is no day-night difference in cosmic-ray intensity. How does this observation bear on the possibility of a solar origin for cosmic rays?

19. What do you think ultimately becomes of the protons and neutrons knocked out of atmospheric atoms by cosmic rays?

20. Cosmic-ray primaries are mostly protons, but few protons are in the cosmic rays that reach the earth's surface. Why?

21. What traps cosmic rays in our galaxy?

22. What is Hubble's law?

23. Why is the universe believed to be expanding?

24. Why must the rate of expansion of the universe be slowing down?

25. What is a quasar?

26. The spectra of quasars exhibit red shifts, never blue shifts. Why does this suggest that quasars are not members of our galaxy?

27. Why are quasars thought to be relatively small in size?

28. What is the observational evidence in favor of the big bang?

29. What are the origins of the helium found in the universe?

30. What are the origins of elements heavier than helium?

31. What is the observational evidence in favor of there being a great deal of dark matter in the universe?

32. How do measurements of the density of matter in space bear on the question of whether the universe will expand forever or will eventually begin to contract?

33. Did the sun begin as a small body that grew to its present size, or as a large one that subsequently shrank?

34. The sun and the giant outer planets contain hydrogen and helium in abundance; the inner planets, very little. Why?

35. Would you summarize the most likely destiny for the earth as fire or ice?

36. Why is it reasonable to suppose that many other stars have planetary systems circling them?

37. Why is it hard to detect the planetary systems of other stars?

38. Is there an ultimate limit to spacecraft speeds? If so, what is it?

ANSWERS TO MULTIPLE CHOICE			
1. d	**11.** b	**21.** b	**31.** d
2. c	**12.** b	**22.** c	**32.** a
3. c	**13.** b	**23.** d	**33.** d
4. c	**14.** c	**24.** b	**34.** a
5. d	**15.** d	**25.** d	**35.** a
6. b	**16.** c	**26.** c	**36.** d
7. b	**17.** c	**27.** a	**37.** a
8. a	**18.** c	**28.** c	**38.** c
9. d	**19.** c	**29.** b	
10. b	**20.** a	**30.** b	

MATH

REFRESHER

A little basic mathematics is needed to appreciate much of physical science. This review is included mainly to help those readers whose mathematical skills have become rusty, but it is sufficiently self-contained to introduce such useful ideas as powers-of-10 notation to those who have not been exposed to them elsewhere.

ALGEBRA

Algebra is the arithmetic of symbols that represent numbers. Instead of being limited to relationships among specific numbers, algebra can show more general relationships among quantities whose numerical values need not be known.

To give an example, in the theory of relativity it is shown that the "rest energy" of any object—that is, the energy it has due to its mass alone—is

$$E = mc^2$$

What this formula does is give a way to calculate the rest energy E in terms of mass m and speed of light c. The formula is not restricted to a particular object, but can be applied to any object whose mass is known. What is being shown is the way in which rest energy E varies with mass *m in general.*

If we are told only that the rest energy E of some object is 5 joules, we do not know upon what factors E depends or precisely how the value of E varies with those factors. (The joule is a unit of energy widely used in physics; it is equal to 1 kg $\cdot$m^2/s^2.) The quantities E and m are **variables,** since they have no fixed values. On the other hand, c^2 is a **constant,** since it is the square of the speed of light c and the value of c—almost 300 million m/s, or about 186,000 mi/s—is the same everywhere in the universe. Thus the formula $E = mc^2$ tells us, in a simple and straightforward way, that the rest energy of something varies only with its mass and also how to find the numerical value of E if we are given the mass m of a particular object.

The convenience of algebra in science is increased by the use of standard symbols for constants of nature. Thus c always represents the speed of light, π always represents the ratio between the circumference and diameter of a circle, e always represents the electric charge of the electron, and so on.

Before we go further, it is worth reviewing how the arithmetical operations of addition, subtraction, multiplication, and division are expressed in algebra. Addition and subtraction are straightforward:

$$x + y = a$$

means that we obtain the sum a by adding the two quantities x and y together, while

$$x - y = b$$

means that we obtain the difference b when quantity y is subtracted from quantity x.

In algebraic multiplication, no special sign is ordinarily used, and the symbols of the quantities to be multiplied are merely written together. Thus these four expressions have the same meaning:

$$xy = c \qquad x(y) = c \qquad (x)(y) = c \qquad x \times y = c$$

When the quantity x is to be divided by y to yield the quotient e, we write

$$\frac{x}{y} = e$$

which can also be expressed as

$$x/y = e$$

whose meaning is the same.

If several operations are to be performed in a certain order, parentheses ( ), brackets [ ], and braces { } are used to indicate this order. For instance $a(x + y)$ means that we are first to add x and y together and then to multiply their sum $(x + y)$ by a. In essence $a(x + y)$ is an abbreviation for the same quantity written out in full:

$$a(x + y) = ax + ay$$

Let us find the value of

$$v = 5 \left[\frac{(x - y)}{z} \right] + w$$

when $x = 15$, $y = 3$, $z = 4$, and $w = 10$. We proceed as follows:

1. Subtract y from x to give

$$x - y = 15 - 3 = 12$$

2. Divide $(x - y)$ by z to give

$$\frac{(x - y)}{z} = \frac{12}{4} = 3$$

3. Multiply $[(x - y)/z]$ by 5 to give

$$5 \left[\frac{(x - y)}{z} \right] = 5 \times 3 = 15$$

4. Add w to $5[(x - y)/z]$ to give

$$v = 5\left[\frac{(x-y)}{z}\right] + w = 15 + 10 = 25$$

POSITIVE AND NEGATIVE QUANTITIES

The rules for multiplying and dividing positive and negative quantities are simple. If the quantities are both positive or both negative, the result is positive; if one is positive and the other negative, the result is negative. In symbols,

$$(+a)(+b) = (-a)(-b) = +ab$$

$$\frac{+a}{+b} = \frac{-a}{-b} = +\frac{a}{b}$$

$$(-a)(+b) = (+a)(-b) = -ab$$

$$\frac{-a}{+b} = \frac{+a}{-b} = -\frac{a}{b}$$

Here are some examples:

$$(-3)(-5) = 15 \qquad \frac{-16}{-4} = 4$$

$$2(-4) = -8 \qquad \frac{10}{-5} = -2$$

$$(-12)6 = -72 \qquad \frac{-24}{4} = -6$$

To find the value of

$$w = \frac{xy}{x+y}$$

when $x = 5$ and $y = -6$, we begin by finding xy and $x + y$. These are

$$xy = (5)(-6) = -30$$

$$x + y = 5 + (-6) = 5 - 6 = -1$$

Hence

$$w = \frac{xy}{x+y} = \frac{-30}{-1} = 30$$

An example of the use of positive and negative quantities occurs in physics, where there are two kinds of electric charge. One kind is designated positive and the other negative. The force F that a charge Q_1 exerts on another charge Q_2 a distance r away is given by Coulomb's law as

$$F = K\frac{Q_1 Q_2}{r^2}$$

where K is a universal constant. By convention, a positive value of F means a repulsion between the charges—the force tends to push Q_1 and

Q_2 apart. A negative value of F means an attraction between the charges—the force tends to pull Q_1 and Q_2 together. A positive (=repulsive) force acts when *either* both charges are + *or* both are –: "like charges repel." When one charge is + and the other one –, the force is negative (=attractive): "opposite charges attract." Both the above observations about the types of force that occur together with the way in which the strength of F varies with the magnitudes of Q_1 and Q_2 and with their separation r are included in the simple formula $F = KQ_1Q_2/r^2$.

EXERCISES

A. Evaluate the following. The answers are given at the end of the Math Refresher.

1. $\dfrac{3(x+y)}{2}$ when $x = 5$ and $y = -2$

2. $\dfrac{1}{x-y} - \dfrac{1}{x+y}$ when $x = 3$ and $y = 2$

3. $\dfrac{4xy}{y+3x} + 5$ when $x = 1$ and $y = -2$

4. $\dfrac{x+y}{2z} + \dfrac{z}{x-y}$ when $x = -2$, $y = 2$, and $z = 4$

5. $\dfrac{x+z}{y} + \dfrac{xy}{2}$ when $x = 2$, $y = 8$, and $z = 10$

6. $\dfrac{3(x+7)}{y+2}$ when $x = 3$ and $y = -6$

7. $\dfrac{5(3-x)}{2(x+y)}$ when $x = -5$ and $y = 7$

EQUATIONS

An equation is a statement of equality: whatever is on the left-hand side of an equation is equal to whatever is on the right-hand side. An example of an arithmetical equation is

$$3 \times 9 + 8 = 35$$

since it contains only numbers. An example of an algebraic equation is

$$5x - 10 = 20$$

since it contains a symbol as well as numbers.

The symbols in an algebraic equation usually must have only certain values if the equality is to hold. To *solve* an equation is to find the possible values of these symbols. The solution of the equation $5x - 10 = 20$ is $x = 6$ since only when $x = 6$ is this equation a true statement:

$$5x - 10 = 20$$

$$5 \times 6 - 10 = 20$$

$$30 - 10 = 20$$

$$20 = 20$$

The methods that can be used to solve an equation are based on this principle:

> **Any operation carried out on one side of an equation must be carried out on the other side as well.**

Thus an equation remains valid when the same quantity is added to or subtracted from both sides or when the same quantity is used to multiply or divide both sides.

Two helpful rules follow from the above principle. The first is:

> **Any term on one side of an equation may be shifted to the other side by changing its sign.**

To check this rule, let us consider the equation

$$a + b = c$$

If we subtract b from each side of the equation, we obtain

$$a + b - b = c - b$$

$$a = c - b$$

Thus b has disappeared from the left-hand side and $-b$ is now on the right-hand side. Similarly, if

$$a - d = e$$

then it is true that

$$a = e + d$$

The second rule is:

> **A quantity that multiplies one side of an equation may be shifted so as to divide the other side, and vice versa.**

To check this rule, let us consider the equation

$$ab = c$$

If we divide both sides of the equation by b, we obtain

$$\frac{ab}{b} = \frac{c}{b}$$

$$a = \frac{c}{b}$$

Thus b, which was a multiplier on the left-hand side, is now a divisor on the right-hand side. Similarly, if

$$\frac{a}{d} = e$$

then it is true that

$$a = ed$$

Let us use the above rules to solve $5x - 10 = 20$ for the value of x. What we want is to have just x on the left-hand side of the equation. The first step is to shift the -10 to the right-hand side, where it becomes $+10$:

$$5x - 10 = 20$$

$$5x = 20 + 10 = 30$$

Now we shift the 5 so that it divides the right-hand side:

$$5x = 30$$

$$x = \frac{30}{5} = 6$$

The solution is $x = 6$.

When each side of an equation consists of a fraction, all we need do is **cross multiply** to remove the fractions:

$$\frac{a}{b} = \frac{c}{d} \qquad \frac{a}{b} \diagup\!\!\!\!\diagdown \frac{c}{d} \qquad ad = bc$$

For practice, let us solve the equation

$$\frac{5}{a + 2} = \frac{3}{a - 2}$$

for the value of a. We proceed as follows:

Cross multiply to give $\qquad\qquad 5\,(a - 2) = 3(a + 2)$

Multiply out both sides to give $\qquad 5a - 10 = 3a + 6$

Shift the -10 and the $3a$ to give $\qquad 5a - 3a = 6 + 10$

Carry out the indicated addition
and subtraction to give $\qquad\qquad\qquad 2a = 16$

Divide both sides by 2 to give $\qquad\qquad a = 8$

EXERCISES

B. Solve each of the following equations for the value of x:

1. $3x + 7 = 13$

2. $5x - 8 = 17$

3. $2(x + 5) = 6$

4. $7x - 10 = 0.5$

5. $\dfrac{x + 7}{6} = x + 2$

6. $\dfrac{4x - 35}{3} = 9(1 - x)$

7. $\dfrac{3x - 42}{9} = 2(7 - x)$

8. $\dfrac{1}{x + 1} = \dfrac{1}{2x - 1}$

9. $\dfrac{3}{x - 1} = \dfrac{5}{x + 1}$

10. $\dfrac{1}{3x + 4} = \dfrac{2}{x + 8}$

EXPONENTS

There is a convenient shorthand way to express a quantity that is to be multiplied by itself one or more times. In this scheme a superscript number called an **exponent** is used to show how many times the multiplication is to be carried out, as follows:

$$a = a^1$$
$$a \times a = a^2$$
$$a \times a \times a = a^3$$
$$a \times a \times a \times a = a^4$$

and so on. The quantity a^2 is read "a squared" because it is equal to the area of a square whose sides are a long. The quantity a^3 is read as "a cubed" because it is equal to the volume of a cube whose edges are a long. Past an exponent of 3 we read a^n as "a to the nth power," so that a^5 is "a to the fifth power."

Suppose we have a quantity raised to some power, say a^n, that is to be multiplied by the same quantity raised to another power, say a^m. In this event the result is that quantity raised to a power equal to the sum of the original exponents:

$$a^n \times a^m = a^n a^m = a^{n+m}$$

To convince ourselves that this is true, we can work out $a^3 \times a^4$:

$$(a \times a \times a)(a \times a \times a \times a) = a \times a \times a \times a \times a \times a \times a$$
$$a^3 a^4 = a^7$$

Because the process of multiplication is basically one of repeated addition,

$$(a^n)^m = a^{nm}$$

where $(a^n)^m$ means that a^n is to be multiplied by itself the number of times indicated by the exponent m. Thus

$$(a^2)^4 = a^{2 \times 4} = a^8$$

because

$$(a^2)^4 = a^2 \times a^2 \times a^2 \times a^2 = a^{2+2+2+2} = a^8$$

Reciprocal quantities are expressed according to the above scheme but with negative exponents:

$$\frac{1}{a} = a^{-1} \qquad \frac{1}{a^2} = a^{-2} \qquad \frac{1}{a^3} = a^{-3} \qquad \frac{1}{a^4} = a^{-4}$$

ROOTS

When the **square root** of a quantity is multiplied by itself, the product is equal to the quantity. The usual symbol for the square root of a quantity a is $\sqrt{a}$. Thus

$$\sqrt{a} \times \sqrt{a} = a$$

Here are some examples of square roots:

$$\sqrt{1} = 1 \qquad \text{because} \qquad 1 \times 1 = 1$$
$$\sqrt{4} = 2 \qquad \text{because} \qquad 2 \times 2 = 4$$
$$\sqrt{9} = 3 \qquad \text{because} \qquad 3 \times 3 = 9$$
$$\sqrt{100} = 10 \qquad \text{because} \qquad 10 \times 10 = 100$$
$$\sqrt{30.25} = 5.5 \qquad \text{because} \qquad 5.5 \times 5.5 = 30.25$$
$$\sqrt{16B^2} = 4B \qquad \text{because} \qquad 4B \times 4B = 16B^2$$

In the case of a number smaller than 1, the square root is larger than the number itself:

$$\sqrt{0.01} = 0.1 \qquad \text{because} \qquad 0.1 \times 0.1 = 0.01$$
$$\sqrt{0.25} = 0.5 \qquad \text{because} \qquad 0.5 \times 0.5 = 0.25$$

Similarly, multiplying the **cube root** $\sqrt[3]{a}$ of a quantity a by itself twice yields the quantity:

$$\sqrt[3]{a} \times \sqrt[3]{a} \times \sqrt[3]{a} = a$$

An expression of the form $\sqrt[n]{a}$ is read as "the nth root of a"; for instance, $\sqrt[4]{16}$ is "the fourth root of 16" and is equal to 2 because $2 \times 2 \times 2 \times 2 = 16$.

Although procedures exist for finding square and cube roots arithmetically, in practice electronic calculators or printed tables are normally used nowadays.

Here is an example of how a square root arises naturally in physics. Let us solve for r the equation

$$F = K\frac{Q_1Q_2}{r^2}$$

which expresses Coulomb's law of electric force. What we do is this:

Multiply both sides by r^2 to give $\qquad F r^2 = KQ_1Q_2$

Divide both sides by F to give $\qquad r^2 = \dfrac{KQ_1Q_2}{F}$

Take square root of both sides to give $\qquad r = \sqrt{\dfrac{KQ_1Q_2}{F}}$

In algebra, a fractional exponent is used to indicate a root of a quantity. In terms of exponents we would write the square root of a as

$$\sqrt{a} = a^{1/2}$$

because

$$a^{1/2} \times a^{1/2} = (a^{1/2})^2 = a^{2 \times 1/2} = a^1 = a$$

In a similar way the "cube root" of a, which is $\sqrt[3]{a}$, is indicated by the exponent $\frac{1}{3}$ because

$$a^{1/3} \times a^{1/3} \times a^{1/3} = (a^{1/3})^3 = a^1 = a$$

In general, the nth root of any quantity is indicated by the exponent $1/n$:

$$\sqrt[n]{a} = a^{1/n}$$

A few examples will indicate how fractional exponents fit into the general pattern of exponential notation:

$$(a^6)^{1/2} = a^{(1/2)\times 6} = a^3$$

$$(a^{1/2})^6 = a^{6\times 1/2} = a^3$$

$$(a^3)^{-1/3} = a^{(-1/3)\times 3} = a^{-1}$$

$$a^6 a^{1/2} = a^{6+1/2} = a^{6\ 1/2}$$

POWERS OF 10

There is a convenient and widely used method for expressing very large and very small numbers that makes use of powers of 10. Any number in decimal form can be written as a number between 1 and 10 multiplied by some power of 10. The power of 10 is positive for numbers larger than 10 and negative for numbers smaller than 1. Positive powers of 10 follow this pattern:

$10^0 = 1$	= 1 with decimal point moved 0 places
$10^1 = 10$	= 1 with decimal point moved 1 place to the right
$10^2 = 100$	= 1 with decimal point moved 2 places to the right
$10^3 = 1000$	= 1 with decimal point moved 3 places to the right
$10^4 = 10,000$	= 1 with decimal point moved 4 places to the right
$10^5 = 100,000$	= 1 with decimal point moved 5 places to the right
$10^6 = 1,000,000$	= 1 with decimal point moved 6 places to the right

and so on. The exponent of 10 in each case indicates the number of places through which the decimal point is moved to the right from 1.00000 . . . . Equivalently, the exponent gives the number of zeroes that follow the 1.

Negative powers of 10 follows a similar pattern:

$10^0 =$	1 = 1 with decimal point moved 0 places
$10^{-1} =$	0.1 = 1 with decimal point moved 1 place to the left
$10^{-2} =$	0.01 = 1 with decimal point moved 2 places to the left
$10^{-3} =$	0.001 = 1 with decimal point moved 3 places to the left
$10^{-4} =$	$0.000,1$ = 1 with decimal point moved 4 places to the left
$10^{-5} =$	$0.000,01$ = 1 with decimal point moved 5 places to the left
$10^{-6} =$	$0.000,001$ = 1 with decimal point moved 6 places to the left

and so on. Here the exponent of 10 in each case shows the number of places through which the decimal point is moved to the left from 1. The number of zeroes between the decimal point and the 1 is one less than the exponent, that is, $n - 1$.

Here are some examples of powers-of-10 notation:

$$8000 = 8 \times 1000 = 8 \times 10^3$$

$$347 = 3.47 \times 100 = 3.47 \times 10^2$$

$$8{,}700{,}000 = 8.7 \times 1{,}000{,}000 = 8.7 \times 10^6$$

$$0.22 = 2.2 \times 0.1 = 2.2 \times 10^{-1}$$

$$0.000{,}035 = 3.5 \times 0.000{,}01 = 3.5 \times 10^{-5}$$

An advantage of powers-of-10 notation is that it makes calculations involving large and small numbers easier to carry out. The rules for working with exponents that were reviewed in the previous section hold for exponents of 10. We have here

Multiplication: $\quad 10^n \times 10^m = 10^{n+m}$

Division: $\quad \dfrac{10^n}{10^m} = 10^{n-m}$

Raising to power: $\quad (10^n)^m = 10^{nm}$

Taking a root: $\quad (10^n)^{1/m} = 10^{n/m}$

An example will show how a calculation involving powers of 10 is worked out:

$$\frac{460 \times 0.000{,}03 \times 100{,}000}{9000 \times 0.006{,}2} = \frac{(4.6 \times 10^2) \times (3 \times 10^{-5}) \times (10^5)}{(9 \times 10^3) \times (6.2 \times 10^{-3})}$$

$$= \frac{4.6 \times 3}{9 \times 6.2} \times \frac{10^2 \times 10^{-5} \times 10^5}{10^3 \times 10^{-3}}$$

$$= 0.25 \times \frac{10^{2-5+5}}{10^{3-3}} = 0.25 \times \frac{10^2}{10^0}$$

$$= 25$$

Another virtue of this notation is that it permits us to express the accuracy with which a quantity is known in a clear way. The speed of light in free space c is often given as simply 3×10^8 m/s. If c were written out as 300,000,000 m/s we might be tempted to think the speed is precisely equal to this number, right down to the last zero. Actually, the speed of light is 299,792,458 m/s. For our purposes we do not need this much detail. By writing just $c = 3 \times 10^8$ we automatically indicate both how large the number is (the 10^8 tells how many decimal places are present) and how precise the quoted figure is (the single digit 3 means that c is closer to 3×10^8 than it is to either 2×10^8 or 4×10^8 m/s). If we wanted more precision, we could write $c = 2.998 \times 10^8$ m/s. Again how large c is and how precise the quoted figure is are both obvious at a glance.

To be sure, sometimes one or more zeroes in a number are meaningful in their own right and not solely decimal-point indicators. In the case of the speed of light, we can legitimately state that, to three-digit accuracy

$$c = 3.00 \times 10^8 \text{ m/s}$$

since c is closer to this figure than to 2.99×10^8 or 3.01×10^8 m/s. In the sample calculation of the preceding paragraph, the quantity $(4.6 \times 3)/(9 \times 6.2)$ actually equals 0.2473118. . . . It is rounded off to 0.25 because the result of a calculation may have no more significant digits than those in the least precise of the numbers that went into it.

EXERCISES

C. Express the following numbers in powers-of-10 notation:

1. 720 =

2. 890,000 =

3. 0.02 =

4. 0.000,062 =

5. 3.6 =

6. 0.4 =

7. 49,527 =

8. 0.002,943 =

9. 0.0014 =

10. 49,000,000,000 =

11. 0.000,000,011 =

12. 1.4763 =

D. Express the following numbers in decimal notation:

1. 3×10^{-4} =

2. 7.5×10^3 =

3. 8.126×10^{-5} =

4. 1.01×10^8 =

5. 5×10^2 =

6. 3.2×10^{-2} =

7. 4.32145×10^3 =

8. 6×10^6 =

9. 5.7×10^0 =

10. 6.9×10^{-5} =

E. Evaluate the following in powers-of-10 notation:

1. $\dfrac{30 \times 80,000,000,000}{0.0004} =$

2. $\dfrac{30,000 \times 0.000,000,6}{1000 \times 0.02} =$

3. $\dfrac{0.0001}{60,000 \times 200} =$

4. $5000 \times 0.005 =$

5. $\dfrac{5000}{0.005} =$

6. $\dfrac{200 \times 0.000,04}{400,000} =$

7. $\dfrac{0.002 \times 0.000,000,05}{0.000,004} =$

8. $\dfrac{500,000 \times 18,000}{9,000,000} =$

ANSWERS

A.
1. 4.5
2. 0.8
3. −3
4. −1
5. −6.5
6. −7.5
7. 10

B.
1. 2
2. 5
3. −2
4. 1.5
5. −1
6. 2
7. 8
8. 2
9. 4
10. 0

C.
1. 7.2×10^2
2. 8.9×10^5
3. 2×10^{-2}
4. 6.2×10^{-5}
5. 3.6×10^0
6. 4×10^{-1}
7. 4.9527×10^4
8. 2.943×10^{-3}
9. 1.4×10^{-3}
10. 4.9×10^{10}
11. 1.1×10^{-8}
12. 1.4763×10^0

D.
1. 0.0003
2. 7500
3. 0.000,081,26
4. 101,000,000
5. 500
6. 0.032
7. 4321.45
8. 6,000,000
9. 5.7
10. 0.000,069

E.
1. 6×10^{15}
2. 9×10^{-4}
3. 8.3×10^{-11}
4. 2.5×10^1
5. 10^6
6. 2×10^{-8}
7. 2.5×10^{-5}
8. 10^3

THE ELEMENTS

Atomic Number	Element	Symbol	Atomic Mass*	Atomic Number	Element	Symbol	Atomic Mass*	Atomic Number	Element	Symbol	Atomic Mass*
1	Hydrogen	H	1.008	38	Strontium	Sr	87.62	75	Rhenium	Re	186.2
2	Helium	He	4.003	39	Yttrium	Y	88.91	76	Osmium	Os	190.2
3	Lithium	Li	6.941	40	Zirconium	Zr	91.22	77	Iridium	Ir	192.2
4	Beryllium	Be	9.012	41	Niobium	Nb	92.91	78	Platinum	Pt	195.1
5	Boron	B	10.81	42	Molybdenum	Mo	95.94	79	Gold	Au	197.0
6	Carbon	C	12.01	43	Technetium	Tc	(97)	80	Mercury	Hg	200.6
7	Nitrogen	N	14.01	44	Ruthenium	Ru	101.1	81	Thallium	Ti	204.4
8	Oxygen	O	16.00	45	Rhodium	Rh	102.9	82	Lead	Pb	207.2
9	Fluorine	F	19.00	46	Palladium	Pd	106.4	83	Bismuth	Bi	209.0
10	Neon	Ne	20.18	47	Silver	Ag	107.9	84	Polonium	Po	(209)
11	Sodium	Na	22.99	48	Cadmium	Cd	112.4	85	Astatine	At	(210)
12	Magnesium	Mg	24.31	49	Indium	In	114.8	86	Radon	Rn	(222)
13	Aluminum	Al	26.98	50	Tin	Sn	118.7	87	Francium	Fr	(223)
14	Silicon	Si	28.09	51	Antimony	Sb	121.8	88	Radium	Ra	226.0
15	Phosphorus	P	30.97	52	Tellurium	Te	127.6	89	Actinium	Ac	(227)
16	Sulfur	S	32.06	53	Iodine	I	126.9	90	Thorium	Th	232.0
17	Chlorine	Cl	35.45	54	Xenon	Xe	131.3	91	Protactinium	Pa	231.0
18	Argon	Ar	39.95	55	Cesium	Cs	132.9	92	Uranium	U	238.0
19	Potassium	K	39.10	56	Barium	Ba	137.3	93	Neptunium	Np	(237)
20	Calcium	Ca	40.08	57	Lanthanum	La	138.9	94	Plutonium	Pu	(244)
21	Scandium	Sc	44.96	58	Cerium	Ce	140.1	95	Americium	Am	(243)
22	Titanium	Ti	47.90	59	Praseodymium	Pr	140.9	96	Curium	Cm	(247)
23	Vanadium	V	50.94	60	Neodymium	Nd	144.2	97	Berkelium	Bk	(247)
24	Chromium	Cr	52.00	61	Promethium	Pm	(145)	98	Californium	Cf	(251)
25	Manganese	Mn	54.94	62	Samarium	Sm	150.4	99	Einsteinium	Es	(254)
26	Iron	Fe	55.85	63	Europium	Eu	152.0	100	Fermium	Fm	(257)
27	Cobalt	Co	58.93	64	Gadolinium	Gd	157.3	101	Mendelevium	Md	(258)
28	Nickel	Ni	58.70	65	Terbium	Tb	158.9	102	Nobelium	No	(255)
29	Copper	Cu	63.54	66	Dysprosium	Dy	162.5	103	Lawrencium	Lr	(260)
30	Zinc	Zn	65.38	67	Holmium	Ho	164.9	104	Unnilquadium	Unq	(261)
31	Gallium	Ga	69.72	68	Erbium	Er	167.3	105	Unnilpentium	Unp	(262)
32	Germanium	Ge	72.59	69	Thulium	Tm	168.9	106	Unnilhexium	Unh	(263)
33	Arsenic	As	74.92	70	Ytterbium	Yb	173.0	107	Unnilseptium	Uns	(262)
34	Selenium	Se	78.96	71	Lutetium	Lu	175.0	108	Unniloctium	Uno	(265)
35	Bromine	Br	79.90	72	Hafnium	Hf	178.5	109	Unnilennium	Une	(266)
36	Krypton	Kr	83.80	73	Tantalum	Ta	180.9	110	Unununnilium	Uun	(269)
37	Rubidium	Rb	85.47	74	Tungsten	W	183.9				

*Masses in parentheses are those of the most stable isotopes of the elements.

ANSWERS TO ODD-NUMBERED QUESTIONS AND PROBLEMS

CHAPTER 1 QUESTIONS

1. None. The scientific method is based on observation and experiment together with reasoning from their results.
3. When first proposed, a scientific interpretation is called a hypothesis. If the hypothesis states a regularity or relationship, it is called a law after it has been verified by further observation and experiment. If the hypothesis uses general considerations to account for specific findings, it is called a theory after it has been verified.
5. At the north or south pole.
7. The moon's apparent diameter remains constant, and its eastward motion through the sky is uniform.
9. In the ptolemaic model, the sun, moon, and other planets all revolve around the earth. In the copernican model, the moon revolves around the earth, and the earth and the other planets revolve around the sun with the earth rotating daily on its axis. There is direct observational evidence that supports the copernican model.
11. Ptolemaic system: (a) the sun circles the earth every day; (b) the stars circle the earth in a little less than a day, so that the difference between the speeds of the sun and stars causes the sun to move eastward relative to the stars; (c) the moon circles the earth.

   Copernican system: (a) the earth turns on its axis once a day; (b) the earth moves around the sun once a year; (c) the moon circles the earth as the earth moves around the sun.
13. Yes, because $T_e^2/R_e^3 = T_a^2/R_a^3$.
15. The earth would then be more flattened at the poles and bulge to a greater extent at the equator.
17. The moon revolves around the earth in the same direction as the earth's rotation. Hence the moon is in the same place relative to a point on the earth's surface about 50 minutes later every day.

CHAPTER 2 QUESTIONS

1. Centimeter; yard; kilometer.
3. No. The distinction between vector and scalar quantities is simply that vector quantities have directions associated with them, and both kinds of quantity are found in the physical world.

5. Its speed increases; its acceleration remains constant.
7. The ball will remain stationary with respect to the barrel, since both are falling with the same acceleration.
9. (a) The same time. When dropped, the coin has the same initial speed as the elevator, so this constant speed does not affect the motion of the coin relative to the elevator. (b) More time. The downward acceleration of the elevator is not shared by the coin; hence the coin's acceleration relative to the elevator is less than g by the amount of the elevator's acceleration. (c) The same time for the reason given in (a). (d) Less time. The upward acceleration of the elevator is not shared by the coin, hence the coin's acceleration relative to the elevator is more than g by the amount of the elevator's acceleration.
11. If an equal and opposite force acts on the same object as another force does, the object will not be accelerated. Only a net, or unbalanced, force can produce an acceleration.
13. The mass of an object is a measure of its inertia, which is the resistance it offers to any change in its state of rest or of motion. The object's weight is the gravitational force with which the earth attracts it.
15. Action and reaction forces always act on different bodies.
17. (a) The reaction force is the upward force exerted by the table on the book; without this force the book would fall to the floor. (b) The reaction force is the gravitational pull the book exerts on the earth.
19. A propeller works by pushing backward on the air, whose reaction force in turn pushes the propeller itself and the airplane it is attached to forward. No air, no reaction force, so the idea is no good.
21. Under no circumstances.
23. Sprinters could not improve their time in the 100-m dash on the moon because their masses are the same there as on the earth. With the force that their legs can exert also unchanged, their acceleration will be the same, and hence their motion will not differ from that on the earth.
25. At an altitude of two earth's radii, the mass would be the same, but the weight would be $\frac{1}{9}$ as great because the distance from the earth's center is three times as great and gravitational force varies as $1/r^2$.
27. The sun's gravitational pull on the earth varies during the year since the distance from the earth to the sun varies.

29. The earth must travel faster when it is nearest the sun in order to counteract the greater gravitational force of the sun.

31. The time of revolution will be shorter, since the satellite must revolve faster about the earth than the moon does owing to the greater gravitational pull of the earth at the smaller distance.

33. No, because they and the airplane are "falling" at the same rate.

CHAPTER 2 PROBLEMS

1. (291 km)(1000 m/km) = 291,000 m; (291 km)(0.621 mi/km) = 181 mi.

3. (80 km)(0.621 mi/km) = 50 mi, hence 80 km/h = 50 mi/h.

5. $t = d/v = 4.37$ s.

7. $v = d_1/t_1 = 3.14$ m/s; $t_2 = d_2/v = 1.6$ s.

9. 81 m

11. $a = v/t = 2.23$ m/s^2.

13. $a = (80$ km/h$)/20$ s $= 4($km/h$)/$s; hence with $v_0 = 80$ km/h and $v = 130$ km/h, $t = (v - v_0)/a = 12.5$ s.

15. (a) $a = (v_f - v_0)/t = -3.5$ m/s^2. (b) $t = (v_f - v_0)/a = 5.71$ s. (c) $t = (v_f - v_0)/a = 2.86$ s.

17. $v = v_0 + gt = 21.8$ m/s.

19. (a) $v_0 = 30$ m/s and at the highest point $v = 0$. Since $v = v_0 - gt$, $t = (v_0 - v)/g = 3.06$ s. (b) The time of fall equals the time of rise, so total time is (2)(3.06 s) = 6.12 s. (c) 30 m/s.

21. $a = v/t = 1.5$ m/s^2; $F = ma = 120$ N.

23. $m = F_1/a_1 = 4$ kg; $F_2 = ma_2 = 4$ N; $F_3 = ma_3 = 40$ N.

25. $a = F/m = 3.33$ m/s^2; $t = v/a = 7.2$ s.

27. $m = w/g = 71.4$ kg; $F = w = 700$ N; $a = g = 9.8$ m/s^2.

29. $v_0 = -15$ m/s and $v = 20$ m/s, so $a = (v - v_0)/t = (20$ m/s$) - (-15$ m/s$)/0.005$ s $= (35$ m/s$)/0.005$ s $= 7000$ m/s^2 and $F = ma = 420$ N.

31. The force is her weight of mg plus her mass times her upward acceleration; hence $F = mg + ma = 708$ N.

33. $F = mv^2/r = 12.5$ N.

35. $v = \sqrt{Fr/m} = 6.3$ m/s.

37. $v_0 = (80$ km/h$)(1000$ m/km$)/3600$ s/h $= 22.2$ m/s and $v = 0$. Hence $a = (v - v_0)/t = (0 - 22.2$ m/s$)/3$ s $= -7.4$ m/s^2.

39. The car will leave the hump when the required centripetal force mv^2/r is more than the car's weight of mg; hence $mv^2/r = mg$, $v = \sqrt{rg} = 10.8$ m/s.

41. $F = Gm_1 m_2/r^2 = 7.4 \times 10^{-8}$ N. Since 1 g = 10^{-3} kg, 1 g of anything weighs $w = mg = 9.8 \times 10^{-3}$ N, which is more than 10^5 (a hundred thousand) times as great as the gravitational force of the lead on the cheese.

43. A satellite must have the minimum speed $v = \sqrt{rg}$. For Jupiter, $r = 0.715 \times 10^8$ m and $g = (2.6)(9.8$ m/s$^2) = 1.6$ m/s^2, so $v = 4.27 \times 10^4$ m/s.

CHAPTER 3 QUESTIONS

1. Yes, because all changes require that work be done.

3. When it is farthest from the sun; when it is closest to the sun. The work needed to pull a planet away from the sun to a given distance increases with the distance, so the planet's PE is greatest the farthest it is from the sun. The gravitational force of the sun on a planet is greatest when it is closest to the sun; hence its speed is also greatest there in order that gravitational and centripetal forces be in balance.

5. Their speeds are the same. The golf ball, which has the greater mass, has the greater KE, PE, and momentum.

7. Ball A reaches the bottom first because all of its original PE becomes KE of downward motion. Part of B's original PE becomes KE of its rotation, so there is less PE available to become KE of downward motion. As a result B moves more slowly than A and reaches the bottom first.

9. Most of the work done in hammering the nail is dissipated as heat owing to friction between the nail and the wood.

11. Yes; yes.

13. Since $2 \times (\frac{1}{2}mv^2) = \frac{1}{2}m(\sqrt{2}\, v)^2$, the speed increases by $\sqrt{2}$ and so the momentum increases by $\sqrt{2}$ as well.

15. (a) The speed decreases as rainwater collects in the truck, since the total momentum must remain constant. (b) The reduced speed is unchanged because the water that leaked out carried with it the momentum it has gained.

17. The resulting water will add to the oceans, and more of the earth's mass will be distant from its axis than before. Conservation of angular momentum requires that the earth in that event spin more slowly to compensate, and the day will be longer.

19. (a) The laws of physics are the same in all frames of reference moving at constant velocity with respect to one another. (b) The speed of light in free space has the same value for all observers.

21. The rod appears longest to the stationary observer.

23. (a) c is the speed of light, (b) m_0 stands for rest mass, which is the mass of an object measured when it is at rest. m stands for the actual mass of an object regardless of its state of motion; m varies depending on the speed of the object, increasing with increasing speed.

25. Mass must be considered as a form of energy.

27. Nuclear energy; geothermal energy.

CHAPTER 3 PROBLEMS

1. Since the onions do not move, no work is done.

3. $Fd = mgh = 32$ kJ.

5. $m = W/gh = (490$ J$)/(9.8$ m/s$^2)(10$ m$) = 5$ kg.

7. $P = W/t = mgh/t = 65$ W since 10 h = 36,000 s.

9. $d = $ KE$/F = 5$m.

11. Less work is needed.

13. $Fd = \frac{1}{2}mv^2$, $F = mv^2/2d = 0.96$ kN.

15. $P = W/t = mgh/t = 169$ W.

17. KE = PE = $mgh = 1.5$ kJ.

19. The difference in height here is $h = 1$ m. Setting KE and PE equal yields $mgh = mv^2/2$, so $v = \sqrt{2gh} = 4.4$ m/s. This is the girl's maximum speed, which occurs at the lowest point of her path through the air.

21. (a) $W = Fd = 1760$ J. (b) PE = $mgh = 1568$ J. (c) 192 J was lost as heat due to friction in the pulleys.

23. $t = d/v = 1250$ s, $W = Pt = 1.75 \times 10^6$ J; hence 1.75×10^6 J$/(4 \times 10^4$ J/g$) = 44$ g of fat is metabolized.

25. $v_1 = m_2 v_2 / m_1 = 0.7$ m/s.
27. Let Δ stand for "change in." Then $m_1 \Delta v_1 = m_2 \Delta v_2$, Δv_2 $= \Delta v_1 (m_1/m_2) = 17$ m/s; $v_f = v_1 - \Delta v = 28$ m/s.
29. $m = E/c^2 = 3.35 \times 10^5$ J/$(3 \times 10^8$ m/s$)^2 = 3.7 \times 10^{-12}$ kg.
31. The total rest energy of 1 kJ of anything is $mc^2 = 9 \times 10^{16}$ J. Here $(5.4 \times 10^6)/(9 \times 10^{16}) = 6 \times 10^{-11}$, which is 6×10^{-9} percent.

CHAPTER 4 QUESTIONS

1. The metal expands more than the glass when heated, thereby loosening the lid.
3. No. The only temperature scale on which such a comparison might make sense is the absolute temperature scale.
5. The person reduces the pressure in the straw by sucking on it, and atmospheric pressure then forces the liquid upward.
7. As the steam inside the can condenses, the internal pressure falls below atmospheric pressure.
9. Since the heights of the water columns are the same, the pressures are the same.
11. The water level is unchanged.
13. The water level falls. The floating canoe displaced a volume of water V whose weight equaled its own weight. The sunken canoe, however, displaces a volume of water equal to the volume of its aluminum shell, which is smaller than V because a given volume of aluminum weighs more than the same volume of water.
15. The pressure in the water decreases toward the top; hence the steam bubbles expand.
17. The molecules themselves occupy volume.
19. Lower.
21. The thermal energy of a solid resides in oscillations of its particles about fixed positions.
23. A piece of ice at 0°C is more effective in cooling a drink than the same weight of water at 0°C because of the heat of fusion that must be added to the ice before it melts. Hence the ice will absorb more heat from the drink than the cold water.
25. As the water in the cloth evaporates, the canteen is cooled.
27. (a) Increase the liquid temperature, which increases the average molecular KE while leaving unchanged the intermolecular attractive forces. (b) Reduce the pressure above the liquid, which decreases the likelihood that vapor molecules will return to the liquid after colliding with air molecules. (c) Arrange for a current of air to blow over the liquid surface, which will remove vapor molecules before they can return to the liquid. (d) Increase the area of the liquid surface, which brings more liquid molecules to the surface, where it is easiest for them to leave.
29. Both reservoirs are needed to provide a flow of heat, part of which the engine turns into work or mechanical energy.
31. No. "Cold" is a relative deficiency of heat and not something in its own right, as heat (a form of energy) is. What a refrigerator does is to take heat from a region at low temperature (the inside of the refrigerator box) and transfer it to a region at a higher temperature (the outside world).
33. The kitchen will warm up because the refrigerator gives off more heat than it absorbs. Leaving its door open means that it will run continuously and hence add even more heat to the kitchen.

CHAPTER 4 PROBLEMS

1. $T_F = \frac{9}{5}(37°) + 32° = 98.6°$F.
3. $T_C = \frac{5}{9}(T_F - 32°) = -80°$C.
5. The temperature difference is 80°C and 4.19 kJ/kg is needed per 1°C change in temperature. Hence $E = (4.19$ kJ/kg $\cdot$ °C$)(80$°C$)(0.2$ kg$) = 67$ kJ.
7. The stone's energy is $mgh = 9800$ J $= 9.8$ kJ. Since 4.19 kJ increases the temperature of 1 kg of water by 1°C and here there is 10^4 kg of water, the temperature rise is $(9.8$ kJ$)/(4.19$ kJ/kg $\cdot$ °C$)(10^4$ kg$) = 2.3 \times 10^{-4}$°C.
9. $m = dV = dLWH = 78$ kg.
11. The density of the bracelet is $d = m/V = 12.5$ g/cm$^3 = 1.25 \times 10^4$ kg/m^3. Since the density of gold is 1.9×10^4 kg/m^3, the bracelet cannot be pure gold.
13. (a) $V = m/d = 0.055$ m$^3 = 55$ L. (b) $V = m/d = 140$ m^3.
15. The total area of the biting edges of the two teeth is $A = 2(0.01$ m$)(0.001$ m$) = 2 \times 10^{-5}$ m. Hence $p = F/A = 1.5 \times 10^6$ Pa $= 1.5$ MPa.
17. (a) $V_2 = p_1 V_1 / p_2 = 333$ cm^3. (b) $T_1 = 273$ K, $T_2 = 546$ K, $V_2 = V_1 T_2 / T_1 = 2000$ cm^3.
19. $T_1 = 293$ K, $T_2 = T_1 V_2 / V_1 = 586$ K $= 313$°C.
21. $T_1 = 293$ K, $T_2 = p_2 T_1 / p_1 = 410$ K $= 137$°C.
23. $T = 27$°C $= 300$ K. Since the average energy is proportional to the average temperature, doubling the energy doubles the temperature to 600 K $= 327$°C.
25. The heat of vaporization of water is 2260 kJ/kg and (80 kg)(4.19 kJ/kg) is lost by the water in going from 100°C to 20°C, for a total of 2595 kJ $= 2.6$ MJ.
27. To melt the ice requires $(335$ kJ/kg$)(0.05$ kg$) = 17$ kJ. To raise the temperature of the melted ice from 0°C to 20°C requires $(20$°C$)(4.19$ kJ/kg$)(0.05$ kg$) = 4$ kJ. The total is 21 kJ.
29. If the heat lost is $H = 335$ kJ $= 335,000$ J, then $\frac{1}{2}mv^2 = H$, $v^2 = 2H/m$, $v = \sqrt{2H/m} = 819$ m/s.
31. An engine operating between $T_1 = 473$ K and $T_2 = 323$ K has a maximum efficiency of $1 - T_2/T_1 = 0.317 = 31.7$ percent.
33. Maximum efficiency is $1 - T_2/T_1 = 0.65 = 65$ percent; hence the actual efficiency is 62 percent of the maximum possible.

CHAPTER 5 QUESTIONS

1. Experiments show that every charge is either attracted or repelled by a + or a − charge; a charge that is attracted by a + charge is always repelled by a − charge, and vice versa; every charge obeys Coulomb's law when brought near a known charge; and so forth. Since all electrical phenomena can be accounted for on the basis of two kinds of charge only, there is no reason to suppose any other kind of charge exists.

3. Electrons.

5. No.

7. (a) Electric fields may be attractive or repulsive, while gravitational fields are always attractive. Electric fields may be created or destroyed, while gravitational fields are present whenever mass is present. Electric fields affect only certain objects, while gravitational fields affect everything with mass. (b) Compare the forces each field exerts on a charged and on an uncharged object.

9. No; yes.

11. Metals are good electrical conductors, nonmetals good electrical insulators. Liquids may be either. Gases are good insulators if un-ionized. Solids that are good electrical conductors are good heat conductors, and vice versa.

13. With a single wire, charge would be permanently transferred from one end to the other, and soon so large a charge separation would occur that the electric field that produced the current would be canceled out. With two wires, charge can be sent on a round trip, so to speak, and a net flow of energy from one place to another can occur without a net transfer of charge.

15. Energy.

17. The circuit elements might get too hot and start a fire or even melt. The criterion is therefore how much heat the circuit can tolerate without hazard. In the case of house wiring, this depends upon the thickness of the wire used and the nature of its insulation.

19. Either pole of the magnet attracts opposite poles within the iron, leading to a net attractive force since these poles line up facing the external magnet.

21. At a given point a test body free to move travels along a field line, by definition, and it can travel in only one direction at that point.

23. Direction D.

25. 0° or 180°; 90°.

27. The magnetic field through the loop undergoes both an increase and a decrease in each half of a complete rotation.

29. Electric motor: A, B. Generator: A, C. Transformer: C. Tape recording head: A. Tape playback head: C.

31. The changing magnetic field produced by an alternating current.

33. When the connection is made, a momentary current will occur in the secondary winding as the current in the primary builds up to its final value. Afterward, since the primary current will be constant and hence its magnetic field will not change, there will be no current in the secondary.

CHAPTER 5 PROBLEMS

1. The number of protons is $N = m/m_p = 6.0 \times 10^{23}$ protons, so $Q = Ne = 1.6 \times 10^4$ C.

3. $F = Kq_1q_2/r^2 = 5.4 \times 10^{-7}$ N.

5. $F_2/F_1 = (R_1/R_2)^2$, $R_2 = R_1\sqrt{F_1/F_2} = 20$ mm.

7. (a) $F = Kq_1q_2/r^2 = 180$ N. (b) When the charges are brought into contact, the 1×10^{-5} C positive charge cancels out this amount of negative charge to leave -1×10^{-5} C. The latter charge is equally divided between

the spheres, so when they are separated, the force between them is $F = (9 \times 10^9)(5 \times 10^{-6})(5 \times 10^{-6})/(0.1)^2$ N = 23 N and is now repulsive.

9. When the forces balance, $Gm_em_m/d^2 = KQ^2/d^2$ and so $Q = \sqrt{Gm_em_m/K} = 5.7 \times 10^{13}$ C.

11. A current of 2 A means a flow of 2 C/s. The charge on an electron is 1.6×10^{-19} C, so (2 C/s)(1.6 × 10⁻¹⁹ C/electron) = 1.25×10^{19} electron/s flow past the point.

13. $V = IR = 75$ V.

15. $I = P/V = 1.7$ A.

17. $P = IV = 3600$ W; 36 bulbs.

19. (a) $I = P/V = 0.625$ A. (b) $R = V/I = 192$ Ω. (c) $P = 75$ W.

21. $W = Pt = IVt = 2.7$ kWh = 9.7 MJ.

23. (a) $I = P/V = 7.4 \times 10^{-5}$ A. (b) $Q = 1.5$ C/s × 3600 s = 5400 C; $W = QV = 7290$ J; $t = W/P = 7.29 \times 10^3$ J/10⁻⁴ W = 7.29×10^7 s = 844 days = 2.3 years since 1 day = 86,400 s and 1 year = 365 days.

25. In an hour, the resistance element gives off $W = Pt$ = (2000 W)(3600 s) = 7200 kJ. To raise the temperature of 1 kg of water through 60°C requires (4.19 kJ/kg · °C)(60°C) = 251.4 kJ; hence (7200 kJ)/(251.4 kJ/kg) = 28.6 kg of water will be so heated in an hour.

27. The secondary voltage is $V_2 = (N_2/N_1)V_1 = 30$ V and so the secondary current is $I_2 = V_2/R = 0.30$ A; hence the primary current is $I_1 = (N_2/N_1)\ I_2 = 0.075$ A.

29. (a) $N_1/N_2 = V_1/V_2 = (5000$ V$)/(240$ V$) = 20.8$. (b) $P = IV$ so $I_2 = P/V_2 = (10,000$ W$)/(240$ V$) = 41.7$ A.

CHAPTER 6 QUESTIONS

1. (a) In a longitudinal wave, the particles of the medium vibrate parallel to the direction in which the wave is moving. In a transverse wave, the particles vibrate perpendicularly to this direction. (b) No; water waves are a combination of both.

3. Sound travels fastest in solids because their constituent particles are more tightly bound together than those of liquids and gases. Sound travels slowest in gases because their molecules interact only during random collisions.

5. All types of waves can be refracted. Refraction occurs when a wave passes at an oblique angle from one region to another in which its speed is different.

7. The moon has no atmosphere to transmit sound waves.

9. The decibel (dB).

11. In all kinds of waves.

13. The pitch change is too small to be detected when the speed of relative motion is a small percentage of the speed of sound.

15. When light is absorbed, the absorbing material is heated.

17. The electric and magnetic fields of an electromagnetic wave are perpendicular to each other and to the direction of propagation.

19. The sun does not lose momentum due to the light it emits because momentum is a vector quantity and the sun radiates equally in all directions.

23. The mirror must be half your own height. It does not matter how far away you are.

25. More light.
27. Light that is perpendicularly incident is not deflected, so the component wavelengths remain together in the beam.
29. (a) The stars would be red on a black field; instead of having stripes, the rest of the flag would be solid red. (b) Instead of white stars in a blue field, the upper left-hand part of the flag would be solid blue; the rest of the flag would have blue and black stripes.
31. Black.
33. The wavelengths in visible light are very small relative to the size of a building, whereas those in radio waves are more nearly comparable.

CHAPTER 6 PROBLEMS

1. Sound takes $t = d/v = 30$ m/(343 m/s) $= 0.87$ s to travel 30 m, but radio waves take only 5×10^6 m/(3×10^8 m/s) $= 0.017$ s to travel 5000 km.
3. The distance of the man from the spike is $d = v_s t_1 = 686$ m, where $v_s = 343$ m/s is the speed of sound in air. Hence the speed of sound v in the rail is $v = d/t_2 = 686$ m/0.14 s $= 4900$ m/s.
5. $f = v/\lambda = (343$ m/s)/0.25 m $= 1372$ Hz.
7. $\lambda = v/f = 137$ m.
9. $f = 1/(1.2$ s) $= 0.83$ Hz, $v = f\lambda = 5$ m/s.
11. $\lambda = v/f = (343$ m/s)/1044 Hz $= 0.329$ m. Hence there are $d/\lambda = 10$ m/0.329 m $= 30$ waves in a wavetrain 10 m long, which means that the string vibrates 30 times while its sound travels 10 m.
13. $f = c/\lambda = 1.2 \times 10^{10}$ Hz; $N = d/\lambda = vt/\lambda = 600$ waves.

CHAPTER 7 QUESTIONS

1. Nearly all the mass of an atom is located in its small central nucleus. Its positive charge is also in the nucleus. The electrons that carry the atom's negative charge move about the nucleus a relatively large distance away.
3. (a) This statement is no longer exact, since it does not take into account the conversion of mass into energy and energy into mass. A correct statement would be that the total amount of energy plus mass energy in the universe is constant. (b) This statement is still correct. (c) The statement is no longer exact. Experiments involving the bombardment of atomic nuclei with other particles have shown that atoms may be changed into other atoms or even be completely decomposed into their constituent neutrons, protons, and electrons. A correct statement would be that atoms are indivisible and indestructible by ordinary chemical or physical means. (d) This statement is no longer exact, since experiments have shown that isotopes, which are atoms of an element having different atomic masses, exist. A correct statement would be that all atoms of any one element have the same number of protons in their nuclei and the same number of electrons in their electron shells.
5. 3p, 3n; 6p, 7n; 15p, 16n; 40p, 54n.

7. The limited range of the strong interaction.
9. (a) A $^4_2$He nucleus, which consists of two protons and two neutrons; an electron or positron; a high-frequency electromagnetic wave. (b) Gamma rays are the most penetrating in general, alpha particles, the least.
11. Z decreases by 2; A decreases by 4.
13. (a) A nucleus emits an electron when it contains too many neutrons to be stable, and it emits a positron when it contains too many protons. (b) $^{14}_8$O emits a positron and $^{19}_8$O emits an electron.
15. It remains the same.
17. This nucleus has the highest binding energy per nucleon.
19. The binding energy per nucleon is greatest for nuclei of intermediate size.
21. Collisions with the nuclei of the moderator slow the fast neutrons produced in fission. This is desirable because ^{235}U undergoes fission more readily when struck by slow neutrons than by fast ones; hence the presence of a moderator promotes a chain reaction when this nuclide is the fuel. In addition, ^{238}U absorbs fast neutrons without undergoing fission but has little tendency to absorb slow neutrons.
23. Plutonium.
25. Both are neutral electrically. The neutron is associated with both the strong and weak nuclear interactions, the neutrino with the weak interaction only. The neutron has mass, while the neutrino does not. Both have antiparticles. The neutrino is stable; the neutron beta-decays in free space into a proton, an electron, and a neutrino.
27. Leptons are not subject to the strong interaction and are point particles with no detectable size. Hadrons are subject to the strong interaction, have definite sizes, and apparently consist of various combinations of quarks.
29. No; it is possible that there is a reason why quarks cannot exist except in combination with each other as hadrons.

CHAPTER 7 PROBLEMS

1. $84 - 2 = 82$; $210 - 4 = 206$.
3. 92; 233; uranium.
5. $^{207}_{82}$Pb.
7. If $\frac{1}{4}$ of the original amount remains after 10 years, the half-life must be 5 years since $\frac{1}{4} = (\frac{1}{2})(\frac{1}{2})$.
9. KE $= \frac{1}{2}mv^2 = (\frac{1}{2})(9.1 \times 10^{-31}$ kg)(10^6 m/s)$^2 = 4.55 \times 10^{-19}$ J. Since 1 eV $= 1.6 \times 10^{-19}$ J, this energy is equal to 2.84 eV.
11. 1 eV $= 1.6 \times 10^{-19}$ J, so 26 eV $= 4.16 \times 10^{-18}$ J. Since KE $= \frac{1}{2}mv^2$, $v = \sqrt{2KE/m} = 3.02 \times 10^6$ m/s.
13. (a) Emitting an electron increases the atomic number of a nucleus by 1, so the resulting neutral atom has an additional electron. Therefore the atomic mass does not change. (b) Now the atomic number decreases by 1, so the resulting neutral atom has 1 less electron. Therefore the atomic mass decreases by 2 electron masses.

15. $2m_H + 2m_n = 4.0330$ u and so the mass difference is 0.0304 u = 28.3 MeV. There are four nucleons in $_2^4$He, hence the binding energy per nucleon is 7.1 MeV.

17. (a) $m = Pt/c^2 = 9.6 \times 10^{-4}$ kg = 0.96 g. (b) Fissions per second = (total power)/(energy per fission) = 3.1×10^{19} per second.

19. $m_p + m_e = 1.0078$ u. The difference between this and m_n is 0.009 u = 0.8 MeV, which is less than the observed binding energies per nucleon in stable nuclei. Hence neutrons do not decay inside nonradioactive nuclei.

CHAPTER 8 QUESTIONS

1. Electrons have mass, while photons do not. Electrons have charge, while photons do not. Electrons may be stationary or move with speeds of up to almost the speed of light, while photons always travel with the speed of light. Electrons are constituents of ordinary matter, while photons are not. The energy of a photon depends upon its frequency, while that of an electron depends upon its speed.

3. (a) Interference and diffraction phenomena and agreement with the electromagnetic theory of light argue for a wave nature, whereas the photoelectric effect and the nature of line spectra argue for a particle nature. (b) Diffraction and interference are much easier to demonstrate than such quantum phenomena as the photoelectric effect.

5. Even a faint light involves a great many photons.

7. The photon energies are greater; their speeds and number per second are unchanged.

9. It is smaller.

11. No.

13. The proton's KE may be less than, equal to, or more than the photon energy, depending upon what the wavelength is.

15. (a) A continuous emission spectrum. (b) An emission line spectrum. (c) An absorption line spectrum.

17. A negative total energy signifies that the electron is bound to the nucleus. The KE of the electron is, of course, a positive quantity; the PE of the electron is sufficiently negative to make the total energy negative.

19. The Bohr theory assumes that the position and velocity of each electron in an atom may be definitely known at the same time, which is prohibited by the uncertainty principle.

21. (a) The state of lowest energy. (b) $n = 1$.

23. It is an excited state whose lifetime is longer than that of a normal excited state.

25. The light waves from a laser are coherent; that is, they are exactly in step with one another.

27. The results of quantum mechanics are in better quantitative agreement with experiment than those of the Bohr theory and can be applied to a greater variety of situations.

29. Under all circumstances.

CHAPTER 8 PROBLEMS

1. $hf = 1.3 \times 10^{-17}$ J; $hf = 1.3 \times 10^{-28}$ J.

3. 3 photons.

5. The frequency corresponding to a wavelength of 5.5×10^{-7} m is $f = c/\lambda = (3 \times 10^8$ m/s)$/5.5 \times 10^{-7}$ m = 5.45×10^{14} s^{-1}, so $hf = (6.63 \times 10^{-34}$ J $\cdot$ s)$(5.45 \times 10^{14}$ s^{-1}) = 3.61×10^{-19} J. Hence $(1400$ J/m$^2 \cdot$ s$)/(3.61 \times 10^{-19}$ J/photon) = 3.9×10^{21} photons/m$^2 \cdot$ s reach the earth from the sun.

7. Energy per pulse = $Pt = 0.01$ J; energy per photon = $hf = hc/\lambda = 3.1 \times 10^{-19}$ J/photon; photons per pulse = $(0.01$ J)$/(3.1 \times 10^{-19}$ J/photon) = 3.2×10^{16} photons.

9. (a) $E = hf$, so $f = E/h = 4 \times 10^{-19}$ J$/6.63 \times 10^{-34}$ J $\cdot$ s = 6.03×10^{14} s^{-1} = 6.03×10^{14} Hz. (b) $f = c/\lambda = (3 \times 10^8$ m/s)$/2 \times 10^{-7}$ m = 1.5×10^{15} Hz. The energy of a photon of light of this frequency is $E = hf = (6.63 \times 10^{-34}$ J $\cdot$ s)$(1.5 \times 10^{15}$ s^{-1}) = 9.95×10^{-19} J, so the maximum energy of the photoelectrons is $(9.95 - 4) \times 10^{-19}$ J = 6.0×10^{-19} J.

11. $\lambda = h/mv = 3.6 \times 10^{-11}$ m. This wavelength is comparable with atomic dimensions; hence the wave character of the electron will affect any interactions it has with atoms in its path.

13. Since 1 eV = 1.6×10^{-19} J, the energy of each electron is 6.4×10^{-15} J and from KE = $\frac{1}{2}mv^2$ its speed is

$$v = \sqrt{\frac{2 \times 6.4 \times 10^{-15} \text{ J}}{9.1 \times 10^{-31} \text{ kg}}} = 1.2 \times 10^8 \text{ m/s}$$

The corresponding de Broglie wavelength is $\lambda = h/mv$ = 6.63×10^{-34} J $\cdot$ s$/(9.1 \times 10^{-31}$ kg $\times 1.19 \times 10^8$ m/s) = 6.1×10^{-10} m.

15. The centripetal force mv^2/r on the electron is provided by the electrical attraction Ke^2/r^2 of the hydrogen nucleus. Hence $mv^2/r = Ke^2/r^2$ and $v = \sqrt{Ke^2/mr}$. In the $n = 1$ orbit, $r = 5.3 \times 10^{-11}$ m, so

$$v = \sqrt{\frac{9 \times 10^9 (1.6 \times 10^{-19})^2}{9.1 \times 10^{-31} \times 5.3 \times 10^{-11}}} \text{ m/s} = 2.2 \times 10^6 \text{ m/s}$$

CHAPTER 9 QUESTIONS

1. (a) The change from water to ice is a physical change because chemically the substance remains the same; the only differences between ice and water are in their physical properties. (b) The change from iron to rust is a chemical change because the chemical compositions of the two substances are different.

3. Although heating is a physical process, it can produce chemical changes such as the decomposition of mercuric oxide into its constituent elements mercury and oxygen.

5. Elements: mercury, liquid hydrogen. Compounds: alcohol, pure water. Solutions: seawater, beer.

7. Hydrogen; oxygen.

9. Sodium is a very active metal, and so it combines

readily, whereas platinum is highly inactive and there-fore does not tend to combine at all.

11. (a) Solid. (b) 2. (c) Slightly soluble. (d) HAt. (e) KAt, $CaAt_2$. (f) Less stable.

13. (a) Each outer shell has a single electron outside filled inner shells. (b) Each outer shell lacks an electron of being filled. (c) Each outer shell is filled.

15. Both F and Cl atoms lack one electron of having closed outer shells.

17. Electrons are liberated from metals illuminated by light more easily than from nonmetals because the outer electrons of metal atoms are less tightly bound, which is also the reason they tend to form positive ions. Electrons are most readily liberated from metals in group 1 of the periodic table.

19. No. There are no gaps in the atomic-number sequence of the elements that a new group could fit into.

21. The two isotopes of chlorine are identical in atomic structure except for a difference in the number of neutrons in their respective nuclei. Since their electron structures are the same, the chemical behavior of the two isotopes is the same.

23. A chlorine ion has a closed outer shell, whereas a chlorine atom lacks an electron of having a closed outer shell.

25. The attractive force of the two protons in an H_2 molecule is greater than the attractive force of the single proton in an H atom. The mutual repulsion of the two electrons in H_2 means that they tend to be on opposite ends of the molecule, and so this repulsive force is smaller than the increased attractive force of the two protons on each electron.

27. Inert gas atoms contain only closed outer shells and so they cannot accommodate other electrons, as would occur in covalent bonding.

29. Energy is absorbed by the molecules as they break up.

31. +1; −1; −2.

33. These formulas represent the ratios in which the atoms of the various elements are present in the respective compounds. They do not provide information on the structures of the individual molecules or crystals, on the physical properties of the compounds, or on how to prepare the compounds.

35. 14; 8.

37. Barium hydride; lithium phosphate; lead(II) oxide; copper(II) bromide; potassium hydroxide.

1. $+2e$; relatively easy.

3. BaI_2; NH_4ClO_3; $SnCrO_4$; $LiPO_4$.

5. a, c, d.

7. (a) 2; (b) 3; (c) 8; (d) 7.

9. $SO_3 + H_2O \rightarrow H_2SO_4$

11. $2Na + 2H_2O \rightarrow 2NaOH + H_2$

13. $2Al + 3Cl_2 \rightarrow 2AlCl_3$

15. $2Al + 6HCl \rightarrow 3H_2 + 2AlCl_3$

17. $2C_2H_2 + 5O_2 \rightarrow 4CO_2 + 2H_2O$

1. (a) Ionic, NaCl; covalent, diamond; van der Waals, ice; metallic, copper. (b) In each case the bonding is due to electric forces. (c) ions; atoms; molecules; ions.

3. By heating it gradually; if it is glass, it will sag slowly, but this will not occur if it is a crystalline solid.

5. These forces are too weak to hold inert gas atoms together to form molecules against the forces exerted during collisions in the gaseous state.

7. Ca^{2+}, F^-; K^+, I^-.

9. (a) Add some additional sugar and see if it dissolves. (b) Cool the solution and see if any sugar crystallizes out.

11. The solubility of a gas in a liquid decreases with increasing temperature.

13. A molecular ion is a molecule that has gained or lost one or more electrons relative to its normal number and carries a net electric charge as a result. A polar molecule is a molecule that is electrically neutral but has its electrons distributed unevenly so that it behaves as though one end is negatively charged and the other end positively charged.

15. A solution of an electrolyte conducts electricity, a solution of a nonelectrolyte does not.

17. (a) When Ag^+ is added to a solution containing Cl^-, AgCl is precipitated, but nothing happens if Ag^+ is added to a solution of NO_3^- since $AgNO_3$ is soluble. (b) When Cl^- is added to a solution containing Ag^+, AgCl is precipitated, but nothing happens if Cl^- is added to a solution of Na^+ since NaCl is soluble. (c) Cu^{2+} is blue in color, while Ca^{2+} is colorless.

19. NaCl will precipitate out since its solubility is less than that of KCl whereas the solubilities of $NaNO_3$ and KNO_3 are greater than that of KCl. (KCl is the less soluble of the two initial compounds.)

21. Pure acids in the liquid state consist of neutral covalently bonded molecules. They form H^+ ions by reaction with water.

23. 3; 10.

25. Water dissociates to a very small extent into H^+ and OH^- ions; hence it is both a weak acid and a weak base.

27. HBr is a strong acid because, like HCl, it is completely dissociated into ions in solution.

1. The ionic equation is $K^+ + OH^- + H^+ + NO_3^- \rightarrow H_2O + K^+ + NO_3^-$. The actual chemical change is the combination of H^+ and OH^- to form H_2O.

3. (a) Calcium chloride, $CaCl_2$. (b) $2HCl + Ca(OH)_2 \rightarrow CaCl_2 + 2H_2O$.

5. When $AlCl_3$ is dissolved in water, it first dissociates into Al^{3+} and Cl^- ions: $AlCl_3 \rightarrow Al^{3+} + 3Cl^-$. Some of the Al^{3+} ions then react with water to form $AlOH^{2+}$ and H^+ ions: $Al^{3+} + H_2O \rightarrow AlOH^{2+} + H^+$. Hence the solution contains Al^{3+}, $AlOH^{2+}$, H^+, and Cl^- ions and is acidic.

CHAPTER 11 QUESTIONS

1. Air provides the oxygen needed for combustion.
3. Exothermic; *a, b, e, f.*
5. In an atomic-bomb explosion, the liberated energy comes from rearrangements of particles within atomic nuclei, whereas in a dynamite explosion the liberated energy comes from rearrangements within the electron clouds of atoms.
7. The higher the temperature, the greater the number of activated molecules.
9. Graphite is more stable under ordinary conditions because its combustion evolves less energy. Hence diamonds must have been formed under conditions different from those corresponding to room temperature and atmospheric pressure. In fact, they are formed in the earth's interior where the temperature and pressure are both very high.
11. (a) The explosion of dynamite; the precipitation of $AgCl$ when solutions containing Ag^+ and Cl^- are mixed. (b) The rusting of iron; the formation of ammonia gas from a solution of NH_4OH.
13. At room temperature few of the molecules will have energies as great as the activation energy, and since only these few molecules can react, the process is a slow one.
15. The time is virtually independent of temperature.
17. When one of the products of a reaction leaves the system, the reaction must go to completion since the reverse reaction cannot then occur. A reaction in a liquid will go to completion if one of the products is (a) a gas which escapes; (b) an insoluble precipitate; or (c) composed of molecules that do not dissociate when the reaction involves ions.
19. (a) Increased, because the greater the gas pressure, the more of it dissolves. (b) Decreased, because the solubility of gases decreases with increasing temperature. (c) Decreased, because KOH is a strong base. (d) Increased, because removing S^{2-} ions by the precipitation of Ag_2S reduces the rate at which H_2S leaves the solution without affecting the rate at which H_2S enters it.
21. Decrease the yield because the reaction is exothermic; increase the yield because in the reaction three molecules combine to form only two; use a catalyst.
23. Oxidation-reduction reactions.
25. An element that loses electrons in a reaction is oxidized, and one that gains electrons is reduced. Hence Zn, Fe, Br^-, and Fe^{2+} are oxidized, and Cu^{2+}, H^+, and Cl_2 are reduced.
27. Na; Al; I^-; Cl; Hg^{2+}.
29. To show that magnesium is a better reducing agent than hydrogen, it may be placed in an acid solution. Hydrogen gas is evolved, meaning that the magnesium has reduced hydrogen ions in the solution.
31. Primarily because of electrolysis of the water content of the battery electrolyte when the battery is being charged. Evaporation also contributes to water loss to some extent.
33. A dry cell or storage battery cannot operate when its initial supply of reactants is exhausted and must be replaced or recharged. A fuel cell is supplied continuously with reactants.

CHAPTER 11 PROBLEMS

1. A cube has six square faces. The area of a 1-cm square is 1 cm^2, so the surface area of a 1-cm cube is 6 cm^2. The area of a 0.5-cm square is (0.5 cm)(0.5 cm) = 0.25 cm^2, so the surface area of a 0.5-cm cube is (6)(0.25 cm^2) = 1.5 cm^2. There are eight 0.5-cm cubes in a 1-cm cube, hence the total surface area of the small cubes is (8)(1.5 cm^2) = 12 cm^2. This is twice the area of the large cube, whose mass is the same. Since the speed of a chemical reaction, all else being the same, is proportional to the surface area involved, the eight small cubes should dissolve twice as fast as the one large cube. The time therefore should be 0.5 h.
3. The negative electrode is the one through which the hydrogen enters the cell; the positive electrode is the one through which the oxygen enters the cell. At the negative electrode, the reaction is $H_2 + 2OH^- \rightarrow 2H_2O + 2e^-$; at the positive electrode, the reaction is $O_2 + 2H_2O + 4e^- \rightarrow 4OH^-$. The first of these reactions must occur twice as many times as the second.

CHAPTER 12 QUESTIONS

1. There are more carbon compounds than compounds of any other element because of the ability of carbon atoms to form covalent bonds with one another.
3. Covalent bonds that consist of shared electron pairs.
5. Fractional distillation.
7. Structural formulas are more useful in organic chemistry than in inorganic chemistry because more than one organic compound may have the same molecular composition and because many properties of organic compounds depend upon the precise arrangement of the atoms in their molecules.
9. Such molecules are extremely reactive.
11. Two bonds, so that the structure of CO_2 is O=C=O.
13. The molecules of aromatic compounds contain rings of six carbon atoms; the molecules of aliphatic compounds do not.
15. There is no way these atoms can be arranged so that each carbon atom participates in four covalent bonds and each hydrogen atom participates in one covalent bond.
17. Yes.
19. One carbon-carbon bond is double.
21. The compound is an aldehyde, namely proprionaldehyde.
23. Trichloroethylene.
25. Ethyl acetate and nitroglycerin are esters. Acetic acid and citric acid are organic acids. Ethyl alcohol and glycerin are alcohols. Glucose and sucrose are sugars. Acetic acid and methyl chloride are derivatives of methane.
27. Esters are nonelectrolytes, while salts in solution are electrolytes. Salts (such as sodium chloride) are crystals in their pure state, while the simpler esters (such as methyl acetate) are liquids or gases.

29. In plants, carbohydrates are obtained by photosynthesis from CO_2 and H_2O; in animals, they are obtained by eating plants or foods derived from plants. In both plants and animals, fats are synthesized from carbohydrates.

31. Photosynthesis.

33. Thermal insulation and protection from mechanical injury.

35. Amino acids; some are synthesized by the body, others must be present in food.

CHAPTER 12 PROBLEMS

1. The two isomers are

What might seem to be a third isomer,

is really the first of the above reversed, which is not a true difference.

3.

5.

normal pentane   isopentane

neopentane

7. There are two isomers of bromopropane:

CHAPTER 13 QUESTIONS

1. (a) In photosynthesis, plants manufacture carbohydrates from atmospheric carbon dioxide and water with oxygen as a by-product. (b) By absorbing infrared radiation emitted by the earth, carbon dioxide is an intermediary in the process by which solar energy is transferred to the lower atmosphere and carried around the world by winds.

3. At the tropopause the temperature has decreased to a minimum and is about to increase; at the stratopause it has increased to a maximum and is about to decrease; at the mesopause it has decreased to a minimum and is about to increase.

5. The CFC gases catalyze the breakdown of ozone in the upper atmosphere.

7. In order of decreasing volume of freshwater: ice caps and glaciers; groundwater; lakes and rivers; atmospheric moisture. The volume of seawater is many times greater than that of freshwater.

9. (a) 100 percent. (b) The air remains saturated and so the relative humidity remains 100 percent, while the excess water vapor condenses out. (c) The relative humidity decreases.

11. On such a night the earth's surface cools by radiation. The air in contact with the surface also cools and may become saturated with water vapor, which then condenses into droplets of liquid water.

13. Stratus clouds are characteristic of a warm front.

15. The dust will increase the amount of scattering of sunlight and so produce exceptionally colorful sunrises and sunsets. Worldwide rainfall will increase. There will be more clouds than usual, and the clouds and dust will reflect more sunlight back into space and so lead to a general cooling of the atmosphere.

17. The interior of a greenhouse is warmer than the outside air because sunlight can enter through its windows, but the infrared radiation that the warm interior gives off cannot escape through them. The carbon dioxide and water vapor contents of the atmosphere act as a trap of this kind for the earth as a whole. The atmosphere is transparent to visible light, which is absorbed by the earth's surface. The temperature of the surface is thereby increased, which in turn increases the rate at which it emits infrared radiation. The carbon dioxide and water vapor in the atmosphere absorb the infrared radiation, which leads to a warming of the lower atmosphere.

19. Infrared radiation from the earth absorbed by the atmosphere.

21. Winds and ocean currents carry energy around the earth in the forms of warm air and warm water, respectively. Winds are more effective in energy transport than ocean currents.

23. A large amount of heat must be absorbed or lost by a region of the earth's surface before it reaches its final temperature when the rate of arrival of solar energy changes. Since the difference between the rates of energy absorption and energy loss is always small compared with the heat content of the earth's surface, the temperature of the surface cannot change rapidly enough to keep pace with changes in the rate at which solar energy arrives; hence the time lags in seasonal weather conditions.

25. Twice a year.

27. The jet streams occur near the top of the troposphere and flow from west to east.

29. The Chicago-London flight takes less time because high-altitude winds at that latitude are westerly.

31. (a) Anticyclones. (b) The winds spiral clockwise outward from its center. (c) The winds spiral counterclockwise outward from its center.

33. Anticyclonic weather is generally steady with relatively constant temperature, clear skies, and light winds. Cyclonic weather is unsettled with rapid changes in temperature that accompany the passages of cold and warm fronts, cloudy skies, rain, and fairly strong, shifting winds.

35. Cyclonic winds in the northern hemisphere are counterclockwise; hence when you face the wind, the center of low pressure will be on your right. Cyclonic winds in the southern hemisphere are clockwise; hence when you face the wind, the center of low pressure will be on your left.

37. Northwest → west → southwest.

39. Because their waters must be thoroughly mixed in the course of time to obtain a uniform composition of ions, the seas and oceans of the world cannot be static bodies but must exhibit large-scale currents, both vertical and horizontal.

41. (a) The greater the wind velocity, the higher the waves. (b) The longer the period of time during which the wind blows, the higher the waves. (c) The greater the distance (fetch) over which the wind blows across the water, the higher the waves. Each of the above factors ceases to have a strong effect on wave height after a certain point; for example, after a day or two the waves will have reached very nearly the maximum height possible for the wind velocity and fetch of a given situation.

43. (a) At first northward with the Gulf Stream, then northeastward with the North Atlantic Drift. (b) At first southward with the Canary Current, then westward and finally northwestward with the North Equatorial Current and the Gulf Stream.

45. When warm moist air from the west blows over the colder California Current, its temperature drops and moisture from the now supersaturated air condenses into tiny droplets to form a fog.

47. The doldrums are at the equator, so it is quite warm there with considerable evaporation of water and thus high humidity. The air flow is largely upward, so surface winds are light and erratic. The rising currents of moist air lead to considerable rainfall.

49. Milankovitch pointed out that what is important is the amount of solar energy reaching the polar regions in summer, which varies by up to 20 percent. Too little summer sunshine would not melt all the snow that fell in the winter before, and this snow would accumulate and turn into great sheets of ice that would move equatorward to produce an ice age.

51. The presence in the atmosphere of sulfate particles from the burning of fossil fuels partly blocks sunlight in these regions.

53. It is driven by the northeasterly and southeasterly trade winds on either side of the equator.

CHAPTER 14 QUESTIONS

1. A rock is an aggregate of grains of one or more minerals.

3. Crystal form refers to the shape of a crystal, which is determined by the pattern in which its constituent particles are linked together. Cleavage refers to the tendency, if any, of a crystal to break apart in a regular way, which is determined by the presence of weak bonds in certain directions in its structure.

5. Igneous rocks.

7. To have the amorphous structure of a liquid, obsidian must have solidified so rapidly that crystals had no chance to develop. This can only have occurred by the cooling of a lava flow at the earth's surface.

9. Diorite is a coarse-grained rock and andesite is a fine-grained rock.

11. The density increases because the pressures under which metamorphism occurs lead to more compact rearrangements of the atoms in the various minerals.

13. (a) Foliation. (b) Foliation results from the growth of platy or needlelike crystals along planes of movement in a rock produced by directed pressure (stress).

15. Chert consists largely of microscopic quartz crystals and hence is hard and durable.

17. (a) Granite, which contains only a small proportion of ferromagnesian material, is light in color, whereas gabbro, with much ferromagnesian material, is dark. (b) Limestone reacts readily with an acid, unlike basalt. (c) Schist is foliated, whereas diorite is not.

19. Rhyolite; slate; shale.

21. P and S waves both can travel through a solid medium, but only P waves can travel through a liquid. P waves are longitudinal and, like sound waves, consist of pressure fluctuations; S waves are transverse and are analogous to waves in a stretched string. P waves are faster than S waves in the same medium.

23. The mantle is believed to be solid because earthquake S waves can pass through it.

25. The core's radius is about half the earth's radius.

27. (a) Measurements made in mines and wells indicate that temperature increases with depth. (b) Molten rock from the interior emerges from volcanoes. (c) The outer core is liquid, which means that it must be at a high temperature.

29. Ferromagnetic materials lose their magnetic properties at high temperatures, and sufficiently high temperatures exist throughout all of the earth's interior except near the surface of the crust to cause such a loss. Also, both the direction and strength of the field are observed to vary, and in fact the field has reversed its direction many times in the past, which cannot be reconciled with the notion of a permanent magnet in the interior.

31. The rock debris produced by weathering is the principal constituent of soil.

33. Marble exposed to the atmosphere weathers fairly readily because its calcite content is soluble in rainwater that contains dissolved carbon dioxide. Slate consists largely of clay minerals that have metamorphosed to white mica, which is nearly as resistant to weathering as quartz.

35. The ultimate source of the energy that goes into erosion is the sun. Solar energy evaporates surface water, some of which subsequently falls as rain and snow on high ground. The potential energy of the latter water turns into kinetic energy as it flows downhill, and some of the kinetic energy becomes work done in eroding landscapes along its path.

37. The maximum depth to which streams can erode a landscape is sea level since streams flow downhill into the seas. Glacial erosion is not limited in this way, and glaciers can wear away landscapes to depths well below sea level.

39. A glacier forms when the average annual snowfall in a region exceeds the annual loss by evaporation and melting.

41. Stones and boulders are embedded in glaciers, some of which are hard enough to erode the bedrock.

43. The ocean floors.

45. Quartz is resistant to chemical attack and so survives weathering and erosion. Feldspar, the most common mineral, is converted into clay minerals by the carbonic acid of surface waters.

47. (a) Precipitate from groundwater. (b) Stream deposits. (c) Sand dunes.

49. (a) Basalt, rhyolite, andesite, obsidian. (b) Basalt.

51. Such holes were produced by bubbles of gas trapped in lava as it solidified.

53. Volcanic mountains are conical, steepening toward the top, with craters at the summit. The presence of isolated, conical mountains, lava flows, hot springs, geysers, and steam vents, as well as volcanic rocks, indicates a region where volcanoes had once been active.

55. (a) The intruded magma that solidifies to form a pluton is very hot, and thus nearby rocks often undergo thermal metamorphism. (b) Near a batholith, because of the greater heat that had to be dissipated in its cooling.

57. Metamorphic rocks are foliated because, under heat and pressure, their mineral grains grow in long needles or thin flakes as the rocks are squeezed. The stratification of sedimentary rock, however, arises because the various layers were formed at different times, perhaps from somewhat different materials. In sedimentary rocks the layering is caused by slight variations in color or grain size; in metamorphic rocks the foliation is caused by the alignment of mineral grains.

CHAPTER 15 QUESTIONS

1. The earth's interior.

3. Both normal and thrust faults produce cliffs called fault scarps. A strike-slip fault is often marked by a rift, which is a trench or valley caused by erosion of the disintegrated rock produced during the faulting.

5. Erosion.

7. (a) The crust is distinguished from the mantle beneath it by a sharp difference in seismic-wave velocity, which suggests a difference in the composition of the minerals involved or in their crystal structures or in both. The lithosphere is distinguished from the asthenosphere beneath it by a difference in their behavior under stress: the lithosphere is rigid whereas the asthenosphere is capable of plastic flow.

(b) The asthenosphere is plastic because its material is close to its melting point under the conditions of temperature and pressure found in that region of the mantle. Above the asthenosphere the temperature is too low and below it the pressure is too high for the material of the mantle to be plastic.

(c) When a large stress is applied over a long period of time, the asthenosphere gradually flows in response to it. When brief, relatively small forces are applied, as is the case with seismic waves, the asthenosphere is rigid enough to transmit them as a solid does.

9. The ocean floors are relatively recent in origin; the oldest sediments date back only about 135 million years. Continental rocks, in contrast, date back as much as 4000 million years. The reason is that, owing to their low density and consequent buoyancy, the continental blocks are not forced down into the mantle in subduction zones but remain as permanent features of the lithosphere plates they are part of. The ocean floors, on the other hand, are continually being destroyed in such zones as new ocean floors are deposited at mid-ocean ridges.

11. If South America and Africa were once joined together, there should be similar geologic formations and fossils of the same kinds at corresponding locations along their respective east and west coasts. This is indeed found for material deposited up to about 100 million years ago, which is when these continents must have begun to separate.

13. Tethys Sea; yes; the Mediterranean, Black, and Caspian seas.

15. Iceland is less than 70 million years old—much younger than North America, Greenland, and Eurasia—and was formed after the breakup of Laurasia from magma rising through the rift in the Mid-Atlantic Ridge.

17. The Himalayas.

19. There is relative motion between the two plates along their boundary; the Pacific plate is moving northwestward relative to the American plate.

21. The boundary of the western Atlantic plate is along the western edge of South America, and the Andes are the

result of the collision between this plate, which is drifting westward, and the eastern Pacific plate, which is drifting eastward. The eastern edge is not near a plate boundary.

23. 1.9 cm/year.

25. (a) An unconformity is an eroded surface buried under rocks that were subsequently deposited. (b) An unconformity is at *J*.

27. The ratio between the radiocarbon and ordinary carbon contents of all living things is the same. When a plant or animal dies, its radiocarbon content decreases at a fixed rate. Hence the ratio between the radiocarbon and ordinary carbon contents of an ancient specimen of organic origin will reveal its age.

29. (a) Actual plant or animal tissues, usually of a hard nature such as teeth, bones, hair, and shells. Entire insects have been found preserved in amber. (b) Plant tissues that have become coal through partial decay but that retain their original forms. (c) Tissues that have been replaced by material (such as silica) deposited from groundwater; petrified wood is an example. Sometimes a porous tissue such as bone will have its pore spaces filled with a deposited mineral. (d) Impressions that remain in a rock of plant or animal structures that have themselves disappeared. (e) Footprints, wormholes, or other cavities produced by animals in soft ground that have later filled with a different material and so can be distinguished today.

31. Igneous rocks have hardened from a molten state, and no fossil could survive such temperatures. Metamorphic rocks have been altered under conditions of heat and pressure severe enough to distort or destroy most fossils.

33. Precambrian time; the Cenozoic Era.

35. Abundant fossils exist in rocks belonging to the Phanerozoic Eon, which permits tracing the evolution of living things during this span of time. Few fossils exist from the Cryptozoic Eon, making it difficult to determine the forms of life that were present then and how they developed.

37. Precambrian geologic activity must have been similar to that of today.

39. Precambrian sedimentary rocks contain few fossils, whereas later sedimentary rocks usually contain abundant fossils.

41. Much of the area of the continents must have been near or below sea level during at least part of the Paleozoic since shallow seas must have been widespread on their surfaces.

43. Mesozoic Era; Cenozoic Era.

45. Petroleum is thought to have originated in the remains of marine animals and plants that became buried under sedimentary deposits. After bacterial decay in the absence of oxygen, low-temperature chemical reactions produced further modifications. Then complex hydrocarbons were "cracked" under the influence of temperatures of 70 to 130°C to the straight-chain alkane hydrocarbons found in petroleum. When the temperatures were higher, the result was the smaller alkanes that make up natural gas.

47. Reptiles; all sizes.

49. Mesozoic Era; reptiles.

51. During the Mesozoic Era today's continents were joined together so the animal populations (which were largely reptiles) could move freely among them. During the Cenozoic Era the continents were split apart, and the evolution of some of the mammals that replaced the reptiles proceeded differently on the various landmasses.

53. The Ice Ages involved the formation of ice sheets that covered large areas of the earth's surface. In the most recent of the Ice Ages there were four major episodes during which ice advanced across the continents, separated by interglacial periods during which the ice retreated poleward. The glacial advances took place during the past 2 million years, that is, the Pleistocene Epoch of the Quaternary Period of the Cenozoic Era. In the latest of the glacial episodes, ice covered much of Canada and northeastern United States and began to recede only about 20,000 years ago.

55. When the thick sheet of ice that covered this region melted, the continental block became lighter and its buoyancy provided an upward force that has been raising it toward a level of equilibrium.

CHAPTER 16 QUESTIONS

1. Yes; no.

3. The density of a comet is extremely low when it is in the vicinity of the earth, and in a collision most or all of the comet material would simply be absorbed in the upper atmosphere.

5. Stony meteorites resemble ordinary rocks, whereas iron ones are conspicuously different; also, stony meteorites are more readily eroded than iron ones.

7. (a) When viewed over a period of time, a planet will be seen to change its position in the sky relative to the stars. (b) Seen through a sufficiently powerful telescope, the planets appear as disks whereas the stars, which are much farther away, appear as points of light.

9. The planets shine because of the sunlight they reflect.

11. Venus is closer to the sun than Mars and hence receives more sunlight to reflect. It is larger than Mars, so the reflecting surface is greater in area. Venus is surrounded by clouds whereas Mars has none, and these clouds constitute a better reflector of sunlight than the Martian surface; the white polar caps on Mars are too small to make much difference in this respect. As a result of all these factors, Venus not only is brighter than Mars but also is at times the brightest object in the sky after the sun and moon.

13. The seasons would be visible, due to the widespread ice and snow of winter and the green of summer vegetation. The seasonal changes in the weather systems of the middle latitudes would be obvious from the cloud patterns.

15. Jupiter; Mercury; Mercury; Pluto.

17. Less: Mercury, Venus, Mars, Pluto. More: Jupiter, Saturn, Uranus, Neptune.

19. No.

21. Mars, 12.3 h; Jupiter, 4.9 h.

23. The sunlit side of Mercury is too hot and its dark side is too cold for life to exist. Also, Mercury has only a trace of an atmosphere, and it contains only inert gases.

25. Because running water would fill craters with sediments and level their raised rims, the presence of many meteoroid craters means that there has been no running water for a long time on the surface of Mars.

27. (a) Jupiter, Uranus, and Neptune. (b) Fairly small particles, as in the case of Saturn's rings.

29. Hydrogen and helium.

31. Asteroids follow orbits between those of Mars and Jupiter whereas meteoroid orbits are very much larger in size. In general, asteroids are larger than meteoroids—many asteroids are tens or hundreds of km across, whereas few meteoroids are as much as a meter across.

33. Not only does the sun provide daylight, but the light of the moon is reflected sunlight.

35. Two weeks.

37. No; yes.

39. Two weeks; two weeks.

41. The period would be shorter since the greater gravitational attraction of Jupiter would require a higher orbital speed if the orbit were to be stable.

43. The maria are approximately circular depressions covered with pulverized rocks. They are apparently lava flows that were broken up by meteorite impacts.

45. The moon's interior cannot be very hot since it is heat from the earth's interior that is ultimately responsible for earthquakes.

47. Because the moon apparently has a more or less uniform density, it probably never was entirely molten as the earth was. If the moon had melted all the way through, its heavier constituents would have become concentrated by gravity in a core with the lighter constituents forming a mantle around it, as in the case of the earth.

CHAPTER 17 QUESTIONS

1. The larger the telescope, the more light it can gather (thereby revealing faint objects in the sky) and the sharper the images it can produce (thereby resolving—that is, separating—objects that are close together).

3. The earth's atmosphere absorbs light of various frequencies, notably in the ultraviolet. The entire spectrum can be received by a telescope in orbit outside the atmosphere. Such a telescope is also unaffected by clouds and by the scattering of light by atmospheric dust.

5. Blue, white, yellow, red.

7. The star whose spectral lines are displaced farther to the red is moving away from the earth faster than the other star.

9. Sunspots appear dark only by comparison with the rest of the photosphere, whose temperature is higher.

11. Their number increases and then decreases.

13. The 6000 K temperature of the sun's visible surface suggests that the sun is almost wholly gaseous.

15. The solar spectrum is an absorption one, indicating that the surface layers are cooler than those below. An even stronger argument is that only in the sun's interior could conditions of pressure and temperature occur suitable for energy production by nuclear fusion reactions.

17. The presence of the spectral lines of a particular element in the solar spectrum means that this element must be present in the sun.

19. A helium nucleus has less mass than the total mass of the four hydrogen nuclei (protons) that combine to form it, and the "missing" mass appears as energy.

21. Stellar sizes vary more than stellar masses.

23. Black holes, neutron stars, and white dwarfs have the highest densities and giant stars the lowest.

25. One would begin by measuring the apparent brightnesses and periods of the Cepheid variables. From the known relationship between the period of a Cepheid and its intrinsic brightness, the latter can be computed, and a comparison of the intrinsic and apparent brightnesses then yields the distance of the star and, hence, of the cluster.

27. The surface temperature of a star determines the radiation it emits per unit area, while its intrinsic brightness is a measure of its total radiation; knowing both quantities permits computing the star's surface area and hence its diameter.

29. The distance of the star from the earth.

31. The star must be a member of a binary system ("double star"), and the periodic doppler shifts occur as it revolves around the center of mass of the system.

33. Such an object must contract owing to gravity, which causes both a rise in temperature and an increase in density. As a result the hydrogen present begins to react to form helium with the release of considerable energy. Thus any object with the mass and composition of the sun must radiate energy like the sun.

35. A star on the main sequence is in an equilibrium condition with its tendency to expand owing to high temperature exactly balanced by its tendency to contract gravitationally. The condition lasts until the star's hydrogen content decreases beyond a certain proportion, which requires a relatively long time compared with its earlier and later phases. Therefore most stars are members of the main sequence simply because this is the longest stage in a star's evolution.

37. A star whose mass is less than about $\frac{1}{40}$ the sun's mass is not able by gravitational contraction to reach a high enough temperature for nuclear reactions in its interior to occur. A star whose mass is more than about 100 times the sun's mass would become so hot as a result of accelerated nuclear reactions that the outward pressure would exceed the inward force of its gravitation, and it would not be stable.

39. (a) It is large, heavy, hot, and bright, with prominent hydrogen and helium lines in its spectrum. (b) It is small, exceedingly dense, very hot, and dim. (c) It is huge, diffuse, cool, and bright. (d) It is moderately small with moderate temperature, density, and mass, with a spectrum in which lines of metallic elements are prominent.

41. (a) It is very hot. (b) Its average density is low. (c) Upper end of main sequence.

43. Heavy stars use up their hydrogen rapidly, so their lifetimes in the main sequence are shorter than those of smaller stars.

45. Main-sequence stars; supernovas.

47. Near the end of its life cycle.

49. Diagonally downward (since the star's luminosity will decrease) and to the right (since its temperature will decrease).

51. (a) A pulsar emits bursts of radio waves at regular intervals. (b) Pulsars are believed to be very small, dense stars that consist almost entirely of neutrons. (c) Pulsars are believed to originate in supernova explosions.

53. A few km in diameter. Only stars with more than several solar masses can become black holes.

CHAPTER 18 QUESTIONS

1. No. Stars are concentrated in galaxies that are relatively far apart from one another.

3. The Milky Way is composed of stars in the spiral arms of our galaxy and so defines its central disk. Since the earth is close to the plane of the Milky Way, the sun must be part of the central disk of the galaxy.

5. A typical globular cluster is an assembly of hundreds of thousands of Population II stars that are relatively close together. They are found in all galaxies; in spiral galaxies; they are mostly located in the corona outside the central disk and move at high speeds about the galactic center. Since globular clusters are much smaller than galaxies and are always found as members of them, they cannot be considered as being themselves galaxies.

7. Population I stars are found in the spiral arms of spiral galaxies and are of all ages, including very young stars. Population II stars are found outside the arms of spiral galaxies and are all very old.

9. Most of the interstellar gas is located in the spiral arms of the galaxy, and its chief constituent is hydrogen.

11. Most galaxies are smaller than our galaxy, which is one of the largest known.

13. Random thermal motion of ions and electrons in a very hot gas; fast electrons moving in a magnetic field; spectral lines of atoms and molecules.

15. Hydrogen is by far the most abundant element in the universe.

17. (a) Primary cosmic rays are atomic nuclei and so are electrically charged particles, and moving charged particles experience forces in a magnetic field. (b) A charged particle approaching the equatorial regions moves perpendicularly through the lines of the geomagnetic field, which means that the maximum deflecting force is exerted on them. A charged particle approaching near the poles, on the other hand, is not deflected or is little deflected, since it moves parallel or nearly parallel to the lines of the geomagnetic field there.

19. The protons pick up electrons and become hydrogen atoms, while most of the neutrons are absorbed by carbon nuclei to form radiocarbon. Some neutrons escape from the earth entirely and decay into protons and electrons in space.

21. Magnetic fields.

23. Red shifts in the spectra of galaxies indicate that they are all moving apart.

25. A quasar appears in a telescope as a point of light, as a star does, but is a far more powerful source of radio waves than any known star. Quasar spectra show large red shifts, which suggests that they are far away and emit energy at enormous rates. The outputs of quasars sometimes change in periods of a few weeks, so they must be small. The energy source of a quasar is probably a giant black hole.

27. The radiations from quasars vary too rapidly for them to be large in size.

29. Most helium was formed in nuclear reactions soon after the big bang, and some was formed later in nuclear reactions inside stars. A little came from the alpha decay of radioactive nuclei.

31. Two examples are the high speeds of outer stars in spiral galaxies and the motion of individual galaxies in clusters of galaxies.

33. The young sun was much larger than it is today, perhaps as far across as the entire present solar system.

35. Fire, in the sense that the earth will be strongly heated when the sun swells into a red giant.

37. Other stars are all very far away; planets are small in size; planets are dim objects because they shine by reflected light.

PHOTO CREDITS

Camp & Associates. **Page 502:** DPA/Image Works. **Page 503:** George Hall/Woodfin Camp & Associates. **Page 505:** Courtesy Department Library Services, American Museum of Natural History. **Page 506:** Courtesy Down House & The Royal College of Surgeons of England. **Page 507:** Joe Sohm/Image Works. **Page 508:** Martin Miller/Earthlens. **Page 509:** Janet McCoy, University of Arizona. **Page 511:** left: Courtesy of the Cleveland Museum of Natural History; right: E. R. Degginger. **Page 515:** Kenneth W. Fink/Bruce Coleman. **Page 517:** Smithsonian Institution. **Page 518:** left: National Coal Association; right: American Petroleum Institute. **Page 520:** British Museum of Natural History, London. **Page 522:** K. G. Cox/Oxford. **Page 524:** The Bettmann Archive.

CHAPTER 16

Page 533: NASA/Science Source/Photo Researchers. **Page 538:** left: Ronald E. Royer & Steve Padilla; right: Max Planck Institute. **Page 539:** Courtesy of Hale Observatories. **Page 540:** Jonathan Blair/Woodfin Camp & Associates. **Page 541:** Francois Gohier/Photo Researchers. **Page 542:** Courtesy NASA. **Page 544:** top, left and right: Jet Propulsion Laboratory; bottom: Courtesy NASA. **Page 545:** top: Courtesy NASA; bottom: Lick Observatory. **Page 546:** Courtesy NASA. **Page 548:** Courtesy NASA. **Page 549:** Courtesy JPL/NASA. **Page 550:** UPI/Bettmann. **Page 551:** top: NASA/Science Source/Photo Researchers; bottom: Courtesy Space Telescope Science Institute/NASA. **Page 553:** Courtesy NASA. **Page 556:** The Bettmann Archive. **Page 557:** Courtesy NASA. **Page 559:** Courtesy Hale Observatories. **Page 560:** National Optical Astronomy Observatory. **Page 561:** Courtesy Hale Observatories. **Page 562:** Courtesy NASA. **Page 563:** Courtesy NASA. **Page 564:** top: Courtesy NASA; bottom: Courtesy NASA.

CHAPTER 17

Page 571: Courtesy Lick Observatories. **Page 573:** Roger Ressmeyer/Starlight. **Page 574:** Courtesy Lowell Observatories. **Page 576:** Courtesy Hale Observatories. **Page 578:** Our Universe/JPL/NASA. **Page 579:** top: Michael Giannechini/Photo Researchers; bottom: NASA/Image Works. **Page 580:** Courtesy Hale Observatories. **Page 583:** Mark Marten/Photo Researchers. **Page 584:** Courtesy Department Library Services of the American Museum of Natural History. **Page 585:** top and bottom: Courtesy Yerkes Observatory. **Page 586:** Topham/Image Works. **Page 588:** Courtesy Hale Observatories. **Page 589:** left: Courtesy Yerkes Observatory; right: AIP Niels Bohr Library. **Page 591:** Photography by B.W. Hadley, FBIPP, Royal Observatory Edinburgh from original negatives by UK Schmidt Telescope. © 1985 Royal Observatory, Edinburgh. **Page 592:** California Institute of Technology & Carnegie Institute of Washington. **Page 594:** Courtesy Lick Observatories.

CHAPTER 18

Page 601: Julian Baum/Nigel Henbest/Julian Baum Astronomical Art, Chesire, England. **Page 602:** Courtesy Hale Observatories. **Page 603:** Julian Baum/Nigel Henbest/Julian Baum Astronomical Art. **Page 604:** top: Courtesy Hale Observatories; bottom: Hayden Planetarium of The American Museum of Natural History. **Page 606:** David Parker/Science Photo Library/Photo Researchers. **Page 607** left: Courtesy Hale Observatories; right: Science Source/Photo Researchers. **Page 608:** top: Courtesy NASA; bottom: The Bettmann Archive. **Page 609:** The Bettmann Archive. **Page 612:** Dr. Stephen Unwin/Science Photo Library/Photo Researchers. **Page 616:** Courtesy Bell Labs. **Page 622:** Courtesy NASA.

INDEX